Peptide-Based Drug Design

About the Cover

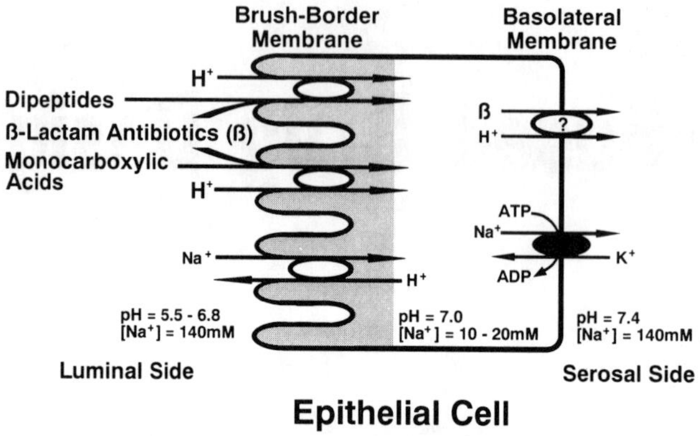

This figure illustrates a model for intestinal absorption of β-lactam antibiotics that utilizes the intestinal peptide transporter (*see* Chapter 5).

Peptide-Based Drug Design

Controlling Transport and Metabolism

Michael D. Taylor, EDITOR
TSRL, Inc.

Gordon L. Amidon, EDITOR
University of Michigan

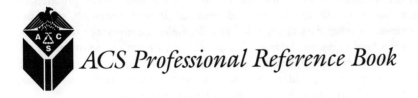

ACS Professional Reference Book

American Chemical Society, Washington, DC 1995

Library of Congress Cataloging-in-Publication Data

Peptide-based drug design : controlling transport and metabolism /
Michael D. Taylor, Gordon L. Amidon, editors.
p. cm. — (Professional reference book)
Includes bibliographical references and index.
ISBN 0-8412-3058-7
1. Peptide drugs—Physiological transport. 2. Peptide drugs—Metabolism.
I. Taylor, Michael D., 1954- . II. Amidon, Gordon L.
III. Series: ACS professional reference book.
[DNLM: 1. Peptides—pharmacokinetics. 2. Drug Design. 3. Drug Delivery Systems. QU 68 P4239 1994]
RS431.P38P454 1994
615′.19—dc20
DNLM/DLC
for Library of Congress 94-40969
CIP

PRINTED IN THE UNITED STATES OF AMERICA

1995 Advisory Board

About the Editors

MICHAEL D. TAYLOR received a B.S. in Chemistry (magna cum laude) from the University of South Florida in 1976 and a Ph.D. in Medicinal Chemistry from the State University of New York at Buffalo in 1981. He was a National Institutes of Health Postdoctoral Fellow at the University of Pennsylvania from 1981 to 1983. He then joined the Parke-Davis Pharmaceutical Research Division of Warner-Lambert Company as a research scientist. Taylor's career at Parke-Davis included research management positions in cardiovascular and neuroscience drug discovery, and, most recently, he was Senior Director of Neuroscience Medicinal Chemistry. In July 1994, Taylor became president of TSRL, Inc., a drug delivery research and technology firm located in Ann Arbor, Michigan. He has extensive drug delivery experience based on several drug discovery and development projects involving poorly available drugs. Taylor has published 30 papers and chapters in organic and medicinal chemistry, cardiovascular drug discovery, and drug delivery, and he has given several invited lectures on drug delivery. He also holds the position of Adjunct Associate Professor of Medicinal Chemistry and Pharmaceutics in the College of Pharmacy at the University of Michigan.

GORDON L. AMIDON received his B.S. in Pharmacy from the State University of New York, Buffalo (1967) and his M.A. in Mathematics (1970) and Ph.D. in Pharmaceutical Chemistry (1971) from the University of Michigan. From 1971 to 1981 Amidon was on the faculty at the University of Wisconsin, where he also served as Assistant Dean for Educational Planning and Policy. In 1981 he became Director of Pharmaceutical Research at Merck/INTERx in Lawrence, Kansas. Amidon was appointed Professor of Pharmaceutics at the University of Michigan in 1983 and is currently the Charles R. Walgreen, Jr., Professor of Pharmacy at the University of Michigan. He is internationally known for his research in the fields of solubility, transport phenomena, prodrugs, and drug absorption. He has published more than 300 papers and abstracts, coauthored one book, and contributed chapters to over 15 books and monographs. Amidon has presented numerous invited lectures and participated in symposiums around the world. His awards include the Ebert Prize of the American Pharmaceutical Association in 1974, 1981, and corecipient in 1984. He is a Fellow and member of several societies, and he served as President of the Controlled Release Society from 1993–1994. Amidon spent a sabbatical year at the U.S. Food and Drug Administration and at the University of California, San Francisco.

Contents

INTESTINAL PEPTIDE TRANSPORT

BLOOD–BRAIN BARRIER PEPTIDE TRANSPORT

PEPTIDE TRANSPORT IN MICROORGANISMS

PEPTIDE METABOLISM

METHODS AND SYSTEMS FOR EVALUATING TRANSPORT AND METABOLISM OF PEPTIDE–BASED DRUGS

Contributors

Gordon L. Amidon
College of Pharmacy
The University of Michigan
428 Church Street
Ann Arbor, MI 48109–1065

Jeffrey M. Becker
Microbiology Department
University of Tennessee
Knoxville, TN 37996

Nicholas Bodor
Center for Drug Discovery
College of Pharmacy
University of Florida
Gainesville, FL 32610–0497

Ronald T. Borchardt
Department of Pharmaceutical Chemistry
School of Pharmacy
University of Kansas
Lawrence, KS 66045

Elsbeth G. Chikhale
Department of Pharmaceutical Chemistry
School of Pharmacy
University of Kansas
Lawrence, KS 66045

Roger H. Erickson
Department of Veterans Affairs Medical Center
Gastrointestinal Research Laboratory, 151–M2
4150 Clement Street
San Francisco, CA 94121

David Fleisher
College of Pharmacy
The University of Michigan
428 Church Street
Ann Arbor, MI 48109–1065

Frank Girbig
SBU Metabolism
Hoechst Aktiengesellschaft
D–65926 Frankfurt am Main
Germany

Gregory J. Gores
Mayo Clinic
200 First Street, S.W.
Rochester, MN 55905

Ulrike Gutjahr
SBU Metabolism
Hoechst Aktiengesellschaft
D–65926 Frankfurt am Main
Germany

Akira Kato
Department of Pharmaceutical Chemistry
School of Pharmacy
University of Kansas
Lawrence, KS 66045

Yukio Kato
Faculty of Pharmaceutical Sciences
University of Tokyo
Hongo 7–3–1, Bunkyo-ku
Tokyo 113, Japan

Simone Kowalewski
SBU Metabolism
Hoechst Aktiengesellschaft
D–65926 Frankfurt am Main
Germany

Werner Kramer
SBU Metabolism
Hoechst Aktiengesellscaft
D–65926 Frankfurt am Main
Germany

Ramesh Krishnamoorthy
Department of Industrial and Physical Pharmacy
School of Pharmacy and Pharmacal Sciences
Purdue University
West Lafayette, IN 47907

Nicholas F. LaRusso
Mayo Clinic
200 First Street, S.W.
Rochester, MN 55905

Hye J. Lee
Alkermes, Inc.
64 Sidney Street
Cambridge, MA 02139–4136

David L. Marks
Mayo Clinic
200 First Street, S.W.
Rochester, MN 55905

Donald W. Miller
Department of Pharmaceutical Chemistry
School of Pharmacy
University of Kansas
Lawrence, KS 66045

Ashim K. Mitra
Department of Industrial and Physical Pharmacy
School of Pharmacy and Pharmacal Sciences
Purdue University
West Lafayette, IN 47907

Judi Møss
Royal Danish School of Pharmacy
Universitetsparken 2
DK-2100 Copenhagen, Denmark

Ka-yun Ng
Department of Pharmaceutical Chemistry
School of Pharmacy
University of Kansas
Lawrence, KS 66045

Fred R. Naider
Chemistry Department
College of Staten Island
City University of New York
Staten Island, NY 10314

John J. Nestor, Jr.
Institute of Bio-Organic Chemistry
Syntex Discovery Research
3401 Hillview Avenue, P.O. Box 10850
Palo Alto, CA 94304

Dale L. Oxender
Department of Biotechnology
Parke-Davis Pharmaceutical Research Division
Warner-Lambert Company
2800 Plymouth Road
Ann Arbor, MI 48106–1047

William M. Pardridge
Brain Research Institute
Department of Medicine
University of California at Los Angeles
Los Angeles, CA 90024

J. W. Payne
School of Biological Sciences
University of Wales
Bangor, Gwynedd LL57 2UW, United Kingdom

Laszlo Prokai
Center for Drug Discovery
College of Pharmacy
University of Florida
Gainesville, FL 32610–0497

Gregory J. Russell-Jones
Biotech Australia Pty. Limited
28 Barcoo Street, P.O. Box 20
Roseville, New South Wales 2069
Australia

Mary J. Ruwart
Drug Delivery Systems Research
The Upjohn Company
Kalamazoo, MI 49001

T. K. Sawyer
Department of Chemistry
Parke–Davis Pharmaceutical Research Division
Warner–Lambert Company
2800 Plymouth Road
Ann Arbor, MI 48106–1047

Barbra H. Stewart
Department of Pharmacokinetics and Drug Metabolism
Parke-Davis Pharmaceutical Research Division
Warner-Lambert Company
2800 Plymouth Road
Ann Arbor, MI 48106–1047

Yuichi Sugiyama
Faculty of Pharmaceutical Sciences
University of Tokyo
Hongo 7–3–1, Bunkyo-ku
Tokyo 113, Japan

Michael D. Taylor
TSRL, Incorporated
540 Avis Drive, Suite A
Ann Arbor, MI 48108

Tetsuya Terasaki
Department of Pharmaceutics
Faculty of Pharmaceutical Sciences
University of Tokyo
Hongo 7–3–1, Bunkyo-ku
Tokyo 113, Japan

Akira Tsuji
Department of Pharmaceutics
Faculty of Pharmaceutical Sciences
Kanazawa University
13–1 Takara-machi
Kanazawa 920, Japan

Rufus M. Williamson
Department of Biotechnology
Parke-Davis Pharmaceutical Research Division
Warner-Lambert Company
2800 Plymouth Road
Ann Arbor, MI 48106–1047

Shiyin Yee
Pfizer Inc.
Drug Metabolism
Eastern Point Road
Groton, CT 06340

Preface

This volume had as its stimulus a 1992 symposium on prodrugs of peptidomimetics at a National Meeting of the American Chemical Society. Although that initial focus has been incorporated into the book, the scope of the work expanded considerably to cover a broad range of topics on the factors involved in transport and metabolism of peptide-based drugs. This area arguably is the most critical aspect of design and development of peptide-based drugs. Drugs designed from peptides or proteins, in addition to interacting at receptors in a manner similar to the natural peptide, also interact with peptide-metabolizing enzymes and peptide transport systems responsible for processing peptides and proteins. Understanding the mechanisms of peptide transport and metabolism provides essential information for drug discovery research and the optimization of delivery properties of new therapeutic agents with peptide-like properties.

The book is divided into seven sections including an introductory section that covers some of the basic advances in molecular biology and more classically known information about drug metabolism and drug transport of amino acids and peptides. Subsequent sections focus on intestinal peptide transport, liver peptide transport, peptide delivery through the blood–brain barrier, and peptide transport in microorganisms. A section on approaches to limiting peptide metabolism is included because this subject is of such paramount importance in designing peptide-based therapeutic agents. Each of these sections includes chapters that either review the state of knowledge in the field to provide specific examples where knowledge of peptide transport or metabolism was used to improve delivery of specific classes of agents. Finally, because drug design is an interactive process in which incremental changes in structure are evaluated and the results are incorporated into the next generation of compounds, we included a section detailing experimental systems used to evaluate peptide transport and metabolism.

The main theme of this book is designing transport and metabolism properties into peptide-based therapeutic agents in much the same manner as intrinsic activity. Given the relative importance of transport and metabolism to the development of such agents, it is essential to include these considerations at the design stage because overcoming stability or transport problems frequently, but not always, becomes either an analogue- or prodrug-based approach. The alternative of employing formulation strategies or alternate routes of administration has limited scope. Including transport and metabolism considerations early in the discovery process would lead to an increased likelihood of producing a deliverable compound when entering phase I studies in humans, particularly if the goal is an orally active compound.

This field is advancing rapidly with progress in combinatorial peptide libraries and high-volume automated screening, on the one hand producing a diversity of potential therapeutic agents to screen, and on the other hand the advances in molecular and cellular biochemistry leading to an increasing understanding of cellular and subcellular processing of peptides and proteins. In the very near future, we will have cloned transporters and enzymes available for optimizing drug transport and metabolism just as we use cloned receptors to assist in the optimization of intrinsic activity. Thus, we can expect rapid advances in our ability to design effective therapeutic agents over the next five to ten years in this field.

We thank the authors who contributed to this work. Largely through their efforts, we are confident that a broad spectrum of readers will gain a better understanding of the importance of drug delivery considerations in peptide-based drug design.

Michael D. Taylor
TSRL, Inc.
Suite A
540 Avis Drive
Ann Arbor, MI 48108

Gordon L. Amidon
College of Pharmacy
University of Michigan
Ann Arbor, MI 48109

September 1994

Introduction

CHAPTER 1

Molecular Biology of Amino Acid, Peptide, and Oligopeptide Transport

Rufus M. Williamson
and Dale L. Oxender

The ability of the cell membrane to permit selective passage of solutes is mediated by several classes of transmembrane proteins. Before molecular biology approaches permitted the cloning and sequencing of specific transporters, researchers were restricted to the use of biochemical and kinetic approaches to study membrane transport. As the technology improved, the central questions for the membrane transport field became increasingly more sophisticated. Three decades ago, the major problem facing researchers in the field was proving that a solute was actively transported and not simply bound to intracellular components (*1*). Reflecting this earlier conceptual problem, several terms were introduced to describe transporters, such as carriers, permeases, pumps, and translocaters. At that time it was generally assumed, for example, that a common transporter served for all neutral amino acids. A decade later, kinetic analyses with a large number of inhibitors were required to establish that at least two distinct systems, referred to as systems A and L, served for the accumulation of neutral amino acids into most mammalian cells.

The development of molecular biology approaches has had a major impact on the transport field. It is now possible to clone transporter genes and to overexpress a specific transporter in permanent cell lines. Membrane vesicles can be prepared from the cell overexpressing a given transporter for determining the molecular characteristics of that transporter. More than 100 transporters have been cloned and characterized using these procedures. This application of molecular biology to the study of transport has established that large families of sequence and structurally related transporters exist. The use of molecular biological techniques has also helped to show that different tissues often display a unique spectrum of transporters. Through these approaches it has also been established that the relatedness or similarity of transporters cut across species lines, making it difficult a priori to distinguish a mammalian cell transporter from a bacterial transporter. Most transporters show characteristic structural features, including transmembrane spanning regions that are capable of creating a putative hydrophilic pore. Transporters can be oriented so as to produce net efflux of a ligand from a cell, as exemplified by the P-glycoproteins responsible for multidrug resistance, or to display net influx of a ligand, as demonstrated by a large number of nutritional transporters. It is now possible to use the database of cloned and sequenced transporters to develop strategies for identifying related new transporters through expression cloning or polymerase chain reaction techniques. With the increase in knowledge of the molecular nature of transport, the major questions and problems facing the field have shifted from those addressed in earlier years and now include studies of the molecular nature of substrate specificity, the basis for tissue specific expression of transport, and the molecular mechanisms governing the regulation of transport activity. In this chapter we will discuss and summarize the current status of amino acid, peptide, and oligopeptide transport.

Molecular Biology of Amino Acid Transport

Definitions and Terminology

Processes involving the uphill transport of a ligand against its concentration gradient requires an input of energy and defines active transport. When an energy-generating reaction, such as the hydrolysis of adenosine 5′-triphosphate (ATP), is directly coupled to the transport of a ligand, the process is referred to as primary active transport. Primary active transporters frequently generate an electrochemical potential of a transported ion across the cell membrane. The electrochemical potential generated is higher on the side of the membrane for which the primary active transporter is oriented. This creates a driving force for the downhill movement of an ion that can be subsequently coupled to the transport of a second ligand. The coupled transport of a second ligand with such a preexisting gradient is referred to as secondary or facilitated active transport. Many of the amino acid transporters described in this chapter belong to this class of transport, and our discussion of the molecular biology of amino acid transport will be restricted to facilitated active transport. Detailed descriptions of amino acid transport processes, their energetics, and historical developments have been discussed elsewhere by Guidotti and Gazzola (*2*), and Eddy (*3*).

Sequence and Structural Similarities among Amino Acid Transporters

Mitchell (*4*) proposed a model for facilitated transporters that assumes the existence of a central binding site and a mobile barrier, where the critical step in the translocation process is the movement of the barrier within the channel. A family of proteins that execute this mode of transport has been identified in recent years. These proteins display a common predicted folding pattern, which includes 10–12 hydrophobic transmembrane α-helices. Various prokaryotic and eukaryotic facilitated amino acid transporters for which sequence information is available are listed in Table I. A comparison of the group, shown in Figure 1, indicates that transport proteins that show close sequence relationships tend to transport structurally similar amino acids. Examples include transport proteins for basic, aromatic, and branched-chain amino acids. Such relationships among transport proteins may reflect amino acid sequence conservation for maintenance of protein topology and ligand specificity. The comparison also reveals that a common ligand specificity may, in some cases, be shared by proteins that display little overall sequence similarity. The transporters responsible for the uptake of proline, which may represent a likely example of convergence of proteins to a common ligand specificity, are a case in point.

Another salient feature that emerges from a comparison of amino acid transporter sequences as well as from results of sequence comparisons reported in the literature is the lack of a fundamental distinction between bacterial and eukaryotic transporters. For example, Storck et al. (*5*) reported significant similarities between a sodium-dependent glutamate–aspartate transporter from rat brain, and glutamate transporters from the bacterial species

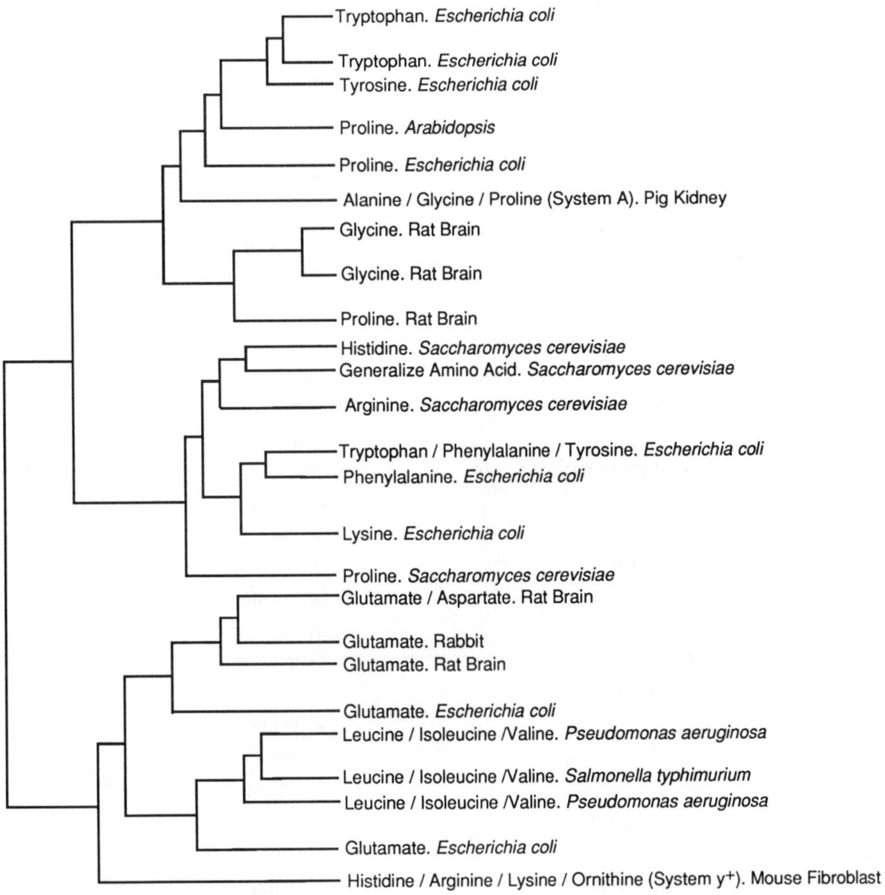

Figure 1. Relationship of amino acid transporters. Relationships among transporters were established by implementation of the unweighted pair group method (139).

Table I. Cloned Facilitated Amino Acid Transporter Genes in Different Organisms

Organism	*Amino Acid*	*Reference*
E. coli K-12	Tryptophan	130
E. coli K-12	Tryptophan	132
E. coli K-12	Tyrosine	133
Arabidopsis thaliana	Proline	137
E. coli K-12	Proline	131
Pig kidney	Alanine Glycine Proline	17
Rat brain	Glycine	20
Rat brain	Glycine	23
Rat brain	Proline	26
Sacch. cerevisiae	Histidine	12
Sacch. cerevisiae	General[a]	126
Sacch. cerevisiae	Arginine	11
E. coli K-12	Tyrosine Tryptophan Phenylalanine	134
E. coli K-12	Phenylalanine	129
E. coli K-12	Lysine	128
Sacch. cerevisiae	Proline	125
Rat brain	Glutamate Aspartate	5
Rabbit intestine	Glutamate	25
Rat brain	Glutamate	24
E. coli K-12	Glutamate Aspartate	127
Pseudomonas aeruginosa PAO	Leucine Isoleucine Valine	135
Pseudomonas aeruginosa PAO	Leucine Isoleucine Valine	136
S. typhimurium	Leucine Isoleucine Valine	138
E. coli B	Glutamate	18
Mouse fibroblast	Histidine Arginine Lysine Omithine	9
Human hippocampus	Serine Alanine Threonine	19

[a] Denotes generalized specificity for all amino acids.

Escherichia coli, *Bacillus stearothermophilus*, and *Bacillus caldotenax*. Additionally, Hediger et al. (*6*) brought attention to the sequence similarities observed between the *E. coli* proline transporter and a transporter from human intestine. Thus, information derived from a study of prokaryotic transporters can be expected to offer insights into the molecular basis for amino acid transport among eukaryotes. A comprehensive overview of prokaryotic amino acid transport is beyond the scope of this discussion. For a review of the molecular biology of prokaryotic transporters, the reader is directed to a concise discussion by Nikaido and Saier (*7*) and a more comprehensive review by Haney and Oxender (*8*).

Mammalian Amino Acid Transporters

Investigations into the nature of amino acid transport in mammals has yielded a large body of information on the nature of ligand specificities as well as on the kinetic parameters of individual amino acid transport systems. However, the determination of molecular details has been complicated by the relatively low abundance of transport proteins in membranes and the existence of multiple transport systems with overlapping specificities for individual amino acids. Nonetheless, several amino acid transporters and proteins that alter their activity have been identified from mammalian cells. In the following discussion, the progress made in the cloning and characterization of genes for amino acid transporters will be summarized. The genes identified thus far play roles in the transmembrane movement of nutrients and in the physiology of neurotransmission.

System y^+ Cationic Amino Acid Transporter from Mouse Fibroblast

Infection of mouse cells by the murine type C ectropic retrovirus is initiated by the binding of the viral envelope to a glycoprotein membrane receptor. The binding of the viral envelope to a membrane receptor is required for fusion of the virus to the target cell. Cunningham and co-workers (*9*) identified a cDNA clone that encodes a membrane receptor for the murine type C ecotropic retrovirus using a strategy that combined gene transfer of mouse NIH 3T3 DNA into nonpermissive human EJ cells and selection of EJ clones that acquired susceptibility to retroviral infection. An open reading frame containing the coding capacity for a 622 amino acid polypeptide was identified from the cDNA library. The deduced polypeptide was predicted to be extremely hydrophobic, with the potential to form 14 transmembrane helices and containing two potential extracellular *N*-linked glycosylation sites (i.e., Asn-X-Ser–Thr). Some of the molecular details of the retroviral receptor protein were understood following the cloning and determination of the nucleotide sequence of the receptor gene, but the normal cellular function of the membrane receptor was initially unclear.

Insight into a potential normal function of the receptor was provided by the observations of Cunningham and co-workers (*10*) when they noticed a coincidence in the positions of the first eight putative transmembrane helices predicted for the virus receptor and those transmembrane helices predicted to occur among the arginine (*11*) and histidine (*12*) transporters of *Saccharomyces cerevisiae*. Similarly, Wang et al. (*13*) proposed that the normal function of the virus receptor might be to transport an essential metabolite because the glycoprotein receptor occurred on many murine cells and was probably essential for viability of cultured fibroblasts. Both groups of investigators reported increased saturable uptake of L-arginine, L-lysine, and L-ornithine following injection of viral receptor-encoding mRNA into *Xenopus laevis* oocytes. Additionally, Wang et al. (*13*) observed an inward-directed current in oocytes in the presence of a mixture of the amino acids. These observations were collectively consistent with the cellular uptake of cationic L-amino acids by the previously known sodium-independent y^+ transport system and the demonstration of the use of the y^+ system transporter as a retroviral receptor.

Other cloned cell-surface receptors for viruses might serve normal cellular functions as transporters. For example, Macleod et al. (*14*) isolated a cDNA clone from T-lymphoma cells that encoded a potential polypeptide that has subsequently been shown to display a high degree of sequence similarity to the system y^+ transporter.

Sodium-Dependent Neutral Amino Acid Transporters

The major components for neutral amino acid uptake in mammalian cells are systems A, ASC, and L, which were first identified in Ehrlich ascites cells (*15, 16*). System L is sodium-independent and shows special reactivity toward branched-chain and aromatic amino acids. System ASC is sodium-dependent and shows a strong preference for alanine, serine, and cysteine. System A is sodium-dependent and serves predominantly for uptake of amino acids with short, polar, or linear side chains.

Kong et al. (*17*) reported the identification of a cDNA clone encoding a mammalian sodium-dependent neutral amino acid transporter from LLC-PK_1 cells derived from pig kidney. An open reading frame with the coding capacity for a 667 amino acid hydrophobic polypeptide was identified, and the amino acid sequence displayed 89% similarity to the amino acid sequence of the sodium-dependent glucose transporter SGLT1. Expression of the cDNA clone in COS-7 cells was associated with increased sodium-dependent uptake of 2-(methylamino)isobutyric acid, a specific substrate for the system A transporter. These investigators also observed that the expressed transport activity for 2-(methylamino)isobutyric acid was inhibited by a range of amino acids that are transported by system A. These observations suggest that the cDNA clone encodes a mRNA for system A transport activity. The level of similarity observed between the system A neutral amino acid and glucose transporters is striking owing to the structural difference in the transported ligand. Also noteworthy is the presence of a conserved set of amino acids that appear common among sodium-dependent transporters (*18*). Potentially, the conserved similarity could imply similarities in the mechanism of sodium-dependent transport of carbohydrates and amino acids.

Shafqat et al. (*19*) reported the identification of a cDNA clone encoding a sodium-dependent, chloride-independent, neutral amino acid transporter from human hippocampus. An open reading frame with the coding capacity for a 529 amino acid hydrophobic polypeptide that displayed significant similarities to mammalian L-glutamate transporters was identified. Expression of the cDNA clone in HeLa cells was associated with the stereospecific transport of L-serine, L-alanine, and L-threonine, but not L-cysteine, L-glutamate, or related dicarboxylates. Other studies (*19*) demonstrated that uptake of the transported amino acids was not inhibited in the presence of 2-(methylamino)isobutyric acid. Northern blot hybridization analysis revealed high levels of expression of the transporter mRNA in human skeletal muscle, pancreas, and brain, with lower levels observed in heart, liver, placenta, lung, and kidney tissues. Shafqat and co-workers (*19*) also noted a resemblance between the transport characteristics of the identified transporter and those reported for the sodium-dependent system ASC, leading to the postulate that multiple transporters with characteristics similar to system ASC may exist in a diversity of tissues and cell lines.

Excitatory Amino Acid Neurotransmitter Transporters

Nerve cells interact with other nerve cells at synaptic cleft junctions. Nerve impulses are transmitted across most synapses by small, diffusible neurotransmitter molecules. Termination of synaptic activity is postulated to result from the removal of neurotransmitters from the synaptic cleft by ion-dependent, high-affinity neurotransmitter transport proteins. These proteins are thought to be responsible for the reaccumulation of neurotransmitters into presynaptic terminals. Recently, several genes for the excitatory amino acid neurotransmitter transporters have been identified. The results of a comparison of the amino acid sequences of these transporters, shown in Figure 2, suggest that the neurotransmitter transporters constitute a subfamily of related proteins with similar ion dependence and protein topology.

Smith et al. (*20*) reported the identification of a cDNA clone encoding a glycine transporter from rat brain. An open reading frame with the coding capacity for a 638 amino acid polypeptide was identified. The deduced polypeptide displayed six potential *N*-linked glycosylation sites, and hydropathy analysis suggested the presence of 12 hydrophobic transmembrane helices. The amino acid sequence showed a high degree of similarity to the deduced amino acid sequences for the γ-amino-butyric acid (GABA) (*21*) and norepinephrine (*22*) transporters. A second cDNA clone encoding a different glycine transporter was identified from rat brain by Guastella et al. (*23*). Synthesis of mRNA from the cDNA clone was shown to direct the expression of sodium- and chloride-dependent transport of [^{3}H]glycine in *Xenopus* oocytes. The cDNA clone also contained the coding capacity for a 633 amino acid polypeptide with similarity to the GABA and norepinephrine transporters. It is noteworthy that the amino acid sequence and brain distribution of the two identified glycine transporters were similar, but not identical, suggesting the existence of multiple isoforms of the glycine transporter in brain tissue.

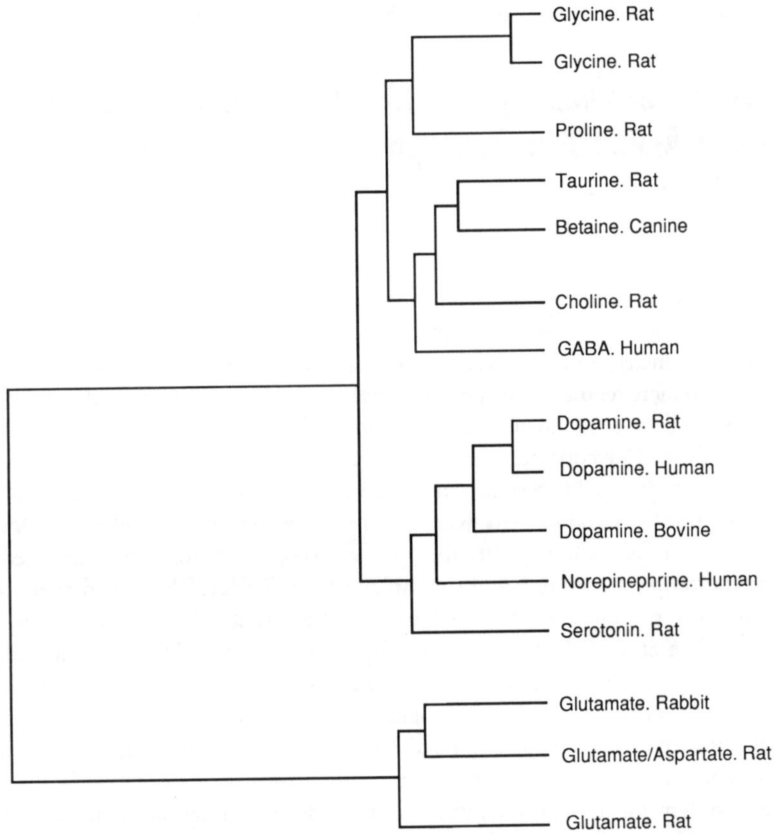

Figure 2. Relationship between excitatory amino acid and other neurotransmitter and brain transporters.

The nucleotide and deduced amino acid sequences for a family of glutamate transporters has also been recently identified. Storck et al. (*5*) reported the isolation of a cDNA clone from rat brain for which mRNA synthesis directed the expression of sodium-dependent uptake of L-glutamate and L-aspartate in *Xenopus* oocytes. The deduced amino acid sequence encoded a potential polypeptide containing 543 amino acids and revealed significant similarities to glutamate transporters found in bacteria. Pines et al. (*24*), using an antibody directed against a glial glutamate transporter from rat brain, isolated a cDNA clone for a glutamate transporter. The cDNA nucleotide sequence had an open reading frame with the coding capacity for a 573 amino acid polypeptide, with 8–9 potential transmembrane helices. Additionally, Kanai and Hediger (*25*) reported the isolation of a cDNA encoding a glutamate transporter from rabbit intestine using expression cloning in *Xenopus* oocytes. The cDNA contained an open reading frame for a potential polypeptide of 524 amino acids, and mRNA transcripts for the transporter were found distributed in specific neuronal structures as well as in the small intestine, kidney, liver, and heart. An interesting aspect of the three mammalian glutamate transporters is the variation in size and predicted configuration with either six, eight, or ten transmembrane helical regions postulated to occur.

Fremeau et al. (*26*) reported the molecular cloning and expression of a high affinity proline transporter from rat brain. As previously mentioned, this proline transporter displays little overall sequence similarity to the proline transporters from *E. coli* or *Sacch. cerevisiae.* It is interesting that the amino acid sequence for this proline transporter displays a high degree of sequence relationship to the GABA, norepinephrine (*27*), serotonin (*28*), and dopamine transporters, leading to the hypothesis of a role for proline in neurotransmission.

Relationship between Type II Membrane Glycoproteins and Mammalian Amino Acid Transport

The use of expression cloning as described in the previous discussions has proven to be a valuable approach in the identification of transport genes. The approach is based on detection of increased transport activity following the introduction of heterologous mRNA into *Xenopus* oocytes. Two classes of genes, in which expression of their corresponding mRNAs are associated with increased transport activity, could be identified by the expression cloning approach. One class of genes encodes transporters and the second class corresponds to genes whose products regulate transport activity. Recently, several genes that may be classified as members of the second class have been identified.

Type II glycoproteins represent a class of membrane proteins that typically are arranged with their NH_2 termini on the cytoplasmic face of the membrane and that contain a single transmembrane helix with an *N*-glycosylated extracellular COOH termini (*29*). cDNA clones encoding type II membrane glycoproteins have been isolated. Following in vitro transcription, capping of the complementary RNA (cRNA), and injection into *Xenopus laevis* oocytes, these cDNA clones induced the activity of several transport systems.

Tate et al. (*30*) first reported the cloning of a cDNA from rat kidney that when expressed in oocytes was associated with increased sodium-independent uptake of branched-chain and aromatic neutral amino acids. The cDNA sequence contained an open reading frame with the coding capacity for a 683 amino acid polypeptide, and the transport characteristics observed upon expression in oocytes were similar, but not exactly the same as that of the L-system for uptake of neutral amino acids. Examination of the amino acid sequence revealed that the potential polypeptide would not be predicted to form multiple transmembrane helices common to the known transporters. Subsequent to the report by Tate et al.

(*30*), Bertran et al. (*31*) reported the isolation of a cDNA clone (rBAT) from rabbit kidney that upon in vitro transcription and injection into oocytes also induced the expression of a single, sodium-independent transport system for dibasic and neutral amino acids, including cystine. The spectrum of the transported amino acids for the rBAT clone resembled the activity of system $b^{0,+}$, a system that had been observed in mouse blastocytes (*32, 33*). The mRNA from the rBAT clone displayed the coding potential for a 77.8 kDa protein, with only one putative transmembrane helix and seven potential *N*-glycosylation sites. Northern blot analysis indicated that tissue expression of the rBAT mRNA was mainly confined to kidney and intestinal mucosa. The amino acid sequence of the putative protein displayed a high degree of similarity to a family of carbohydrate-metabolizing enzymes, the α-glucosidases, as well as, to the 4F2 heavy-chain of the human and mouse surface antigens. Interestingly, addition of microsomes to an in vitro translation system containing rBAT cRNA resulted in a shift of the molecular mass of the protein product to a higher value, and treatment with endoglycosidase H resulted in a shift of the protein to the original molecular mass. These results provided experimental support for membrane association of the protein.

Other type II glycoproteins have been identified whose levels of expression are associated with increased amino acid transport activity. Wells and Hediger (*34*), using expression cloning in *Xenopus* oocytes and assaying for the uptake of [^{14}C]-L-cystine, isolated a cDNA, designated as D2, from a rat kidney library that induced the transport of neutral and cationic amino acids. The D2 cDNA contained an open reading frame with the coding potential for a 683 amino acid protein, having a molecular mass of 78 kDa. The deduced amino acid sequence displayed the potential for one transmembrane helix and seven potential *N*-linked glycosylation sites. Similar to the deduced amino acid sequence from the corresponding mRNA of the rBAT cDNA, the putative extracellular region of the D2 encoded protein also displayed a high degree of similarity to the 4F2 heavy-chain cell surface antigen and to a family of α-glucosidases. Interestingly, injection of in vitro-derived cRNA transcripts for the human 4F2 surface antigen into oocytes resulted in the stimulation of a system y^+-like amino acid transport activity.

A comparison of the type II glycoproteins associated with increased amino acid transport, shown in Figure 3, not only reveals a conservation of putative topological features

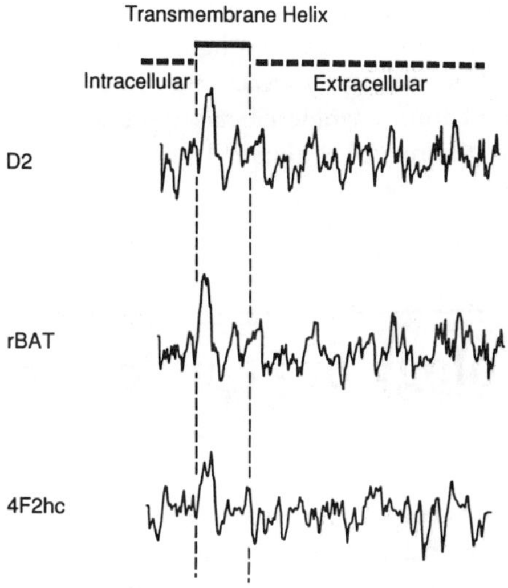

Figure 3. Comparison of the predicted topologies of Type II glycoproteins based on hydropathy profile analysis.

but also demonstrates the lack of similarity to the paradigm of multiple transmembrane helices common to the transporters that have thus far been identified. This lack of similarity has prompted some investigators to propose that the type II glycoproteins represent a new family of transport proteins. Others, however, have presented the alternative idea that the glycoproteins may be activators or regulatory subunits of transporters. Whereas the group of glycoproteins currently identified display sequence similarity to the α-glucosidases, none of the proteins has the conserved aspartic acid or glutamic acid residues that are postulated to play a crucial role in the catalytic site for α-glucosidases, amylases, or transglucanosylases (*35, 36*). Thus, the function of this group of proteins and the relationship of the proteins to transport remains unknown.

γ-Glutamyl Transpeptidase and Mammalian Amino Acid Transport

From the previous discussion it is relevant to mention that another protein, γ-glutamyl transpeptidase, which lacks topological similarity to facilitated transporters, has been implicated to play a role in amino acid transport. γ-Glutamyl transpeptidase catalyzes the hydrolysis and transfer of the γ-glutamyl moiety from glutathione to other amino acids and peptides. In addition to natural substrates, xenobiotic compounds that form conjugates with glutathione are also processed by γ-glutamyl transpeptidase and may contribute to the formation of toxic metabolites (*37*). The enzyme, composed of 569 amino acids, is a membrane-associated, heterodimeric protein consisting of two glycosylated subunits (*38, 39*). It has been postulated that γ-glutamyl transpeptidase plays a role in the transport of amino acids across the cell membrane via transpeptidation, although its precise role in amino acid transport has been the subject of considerable debate (*40–43*). The nucleotide and deduced amino acid sequences for genes encoding γ-glutamyl transpeptidase have been identified from rat kidney (*44*), human placenta (*45*), and isolated porcine brain capillaries (*46*). Additionally, unique mRNA transcripts for γ-glutamyl transpeptidase have been detected in human lung tissue (*47*). Comparison of the amino acid sequences of the enzymes derived from various tissues indicate a high degree of conservation (*46*). Additionally, hydropathy profiles suggest that the γ-glutamyl transpeptidase polypeptide has a potential transmembrane helical domain located at the NH_2-terminus, and are topologically analogous to the class of type II glycoproteins previously described. However, unlike type II glycoproteins, γ-glutamyl transpeptidase consists of a light-subunit that harbors the putative γ-glutamyl binding region and noncovalently associates with a larger heavy subunit (*48, 49*). The two subunits are derived from a single precursor protein (*50–53*). Although some molecular details are known, the role of γ-glutamyl transpeptidase in the transport process remains poorly understood.

Molecular Biology of Peptide and Oligopeptide Transport

The concept of transmembrane transport of peptides and oligopeptides has been implicit in cell physiology for nearly a century, and its history has been the subject of extensive reviews (*54, 55*). The transport of peptides and oligopeptides has been recognized not only among mammalian species, but also in fungi (*56*), higher plants (*57*), and bacteria (*58*). An understanding of the molecular basis of these transport processes has undergone a

Table II. Proteins Implicated in the Transport of Peptides and Oligopeptides

Protein	Putative Role	Organism	Reference
Sec61	Endoplasmic reticular peptide transport	*Sacch. cerevisiae*	68
HylB	Hemolysin peptide transport	*E. coli*	78
STE6	Pheromone peptide transport	*Sacch. cerevisiae*	79
MDR-1	Transport of N-acetyl-leucyl-leucyl-norleucinal	Chinese hamster ovary cell	91
PMP70	Peroxisomal peptide transport	Rat liver	97
SecY	Peptide export	*E. coli*	71
Peptide supply factor (Ring 4)	Transport of oligopeptides for major histocompatability complex	Human	109
Ring 11	Transport of oligopeptides for major histocompatability complex	Human	112
Oligopeptide permease	Oligopeptide transport	*E. coli*	116
Oligopeptide permease	Oligopeptide transport	*S. typhimurium*	117
Oligopeptide permease	Oligopeptide transport	*Bacillus subtilis*	118
Dipeptide permease	Dipeptide transport	*E. coli*	121, 122

revolution recently with the cloning and characterization of several transporters, listed in Table II, that are involved in key biological processes. In the following discussion, we will outline some of the progress made in understanding the molecular basis of peptide and oligopeptide transport.

Polypeptide Transport across the Endoplasmic Reticular Membrane

A large number of eukaryotic polypeptides that are destined for secretion or export are synthesized in the cytoplasm, and are concomitantly transported across the endoplasmic reticular (ER) membrane. These proteins are targeted for the ER membrane by amino terminal signal sequences, characterized as a continuous stretch of 6 to 20 nonpolar amino acids, and by a ribonucleoprotein complex known as the signal recognition particle. The signal recognition particle, composed of a molecule of 7S RNA and six polypeptide subunits, binds to the signal sequence of a growing polypeptide chain (*59*). Through interactions with a membrane-bound receptor for the signal recognition particle (*60, 61*), the growing polypeptide chain is eventually targeted to a translocation site consisting of several proteins, including a hypothetical protein-conducting channel. Even though the targeting process in protein transport has been studied in great detail (*62*), little was known, until recently, about the actual molecular mechanism of how proteins are transported across the ER membrane. The current understanding of the targeting cycle in protein transport has been the source of several recent reviews (*63, 64*). Discussions here will focus only on the evidence for a protein-conducting channel as a transfer mechanism in protein translocation.

Genetic studies in *Sacch. cerevisiae* identified several temperature-sensitive mutations that affect peptide translocation across the ER membrane (*65–69*). One of the identified mutations, SEC61, was associated with the accumulation of precursor molecules for both membrane and secretory proteins (*68*). The Sec61 polypeptide is a 52 kDa hydrophobic integral membrane protein (*68*) that can be cross-linked with various secretory proteins in the ER membrane (*70, 71*). Sec61 has sequence and topological similarities to SecY of *E. coli* (*72*), an integral membrane protein involved in protein export. Hydrophobicity analysis suggests that both proteins have the capacity to form 10 transmembrane helices and are topologically analogous to the facilitated amino acid transporters. Rapoport (*63*) considered several features of the Sec61 protein that suggest that the protein may be a constituent of a hypothetical protein-conducting channel. First, the Sec61 protein is found adjacent to translocating peptides in different organisms. Second, the protein has the potential to form a hydrophilic environment within the membrane. Third, its structure is highly conserved in evolution. Experimental support for the existence of a protein-conducting channel in the ER membrane has been provided by Simon and Blobel (*73*) with electrophysiological studies on pancreatic rough microsome vesicles. However, direct demonstration of the involvement of the Sec61 protein or related proteins in formation of a hypothetical protein-conducting channel awaits experimental verification.

Multidrug Resistance and P-Glycoproteins

The acquisition by mammalian cells of multidrug resistance (MDR) is a phenomenon associated with tumor cells and often is a contributing factor in the failure of some clinical cancer chemotherapy. The characteristics of MDR have been well established from studies of cell lines that were originally isolated for resistance to a single cytotoxic drug and that were subsequently shown to elicit cross-resistance to numerous other drugs (*74, 75*). It has been established that a major contributing mechanism for MDR is increased expression of P-glycoprotein, a 170 kDa plasma membrane protein. The P-glycoprotein is postulated to function as an ATP-hydrolyzing transmembrane efflux pump for many structurally diverse compounds.

Cloning of *mdr* genes has revealed the existence of a family of related genes that encode the P-glycoproteins associated with MDR (*75, 76, 77*). Analyses of the deduced translation products of the *mdr* genes indicate a high degree of similarity in the amino acid sequences among the P-glycoproteins, suggesting derivation from a common ancestral source. Predictions from hydropathy analyses and topological studies also suggest that the family of P-glycoproteins have the capacity to form 12 transmembrane helices and contain sites for *N*-linked glycosylation. All of the known P-glycoproteins are predicted to contain two ATP binding sites, presumably located on the cytoplasmic side of the membrane. These ATP binding sites are similar to sites found among a class of bacterial transport proteins involved in nutrient uptake (*78*). Similar MDR-like proteins containing ATP binding sites have also been observed among bacterial and yeast proteins involved in the efflux of substances, such as the hemolysin peptide by the HylB transporter of *E. coli* (*79*), and the efflux of a-factor, a peptide pheromone, by the STE-6 protein in *Sacch. cerevisiae* (*80*). The term "ABC" (ATP-binding cassette) protein has been introduced to describe this increasingly expanding family of proteins (*78*) that now consists of over 100 members.

The broad-spectrum species distribution of P-glycoprotein-like molecules became apparent with the identification of *mdr*-like genes among non-mammalian cells. Guilfoile and Hutchinson (*81*) and Colombo et al. (*82*) identified a bacterial analog of the *mdr* gene of mammalian tumor cells in the organism *Streptomyces peucetius.* A gene conferring efflux-mediated MDR has also been reported to exist in *Bacillus subtilis* (*83*). Saier and co-workers (*84*) called attention to a new subfamily of transporters that catalyzes the export of drugs and carbohydrates among the species *Rhizobium leguminosarum, Haemophilus influenza,* and

Neisseria meningitidis. These observations suggest that transport systems that catalyze export of drugs and other substances may be quite common among bacteria.

Among higher organisms, the first observation of a putative P-glycoprotein homologue in plants, providing new insights into the nature of cross-resistance to herbicides, was reported by Dudler and Hertig (*85*). Overexpression of P-glycoproteins in the trypanosomatid protozoa *Leishmania* has been shown to be associated with the development of resistance to heavy metals (*86*) and the acquisition of resistance to methotrexate (*87*). Additionally, among protozoans, amplification of a MDR phenotype has been associated with the development of chloroquine resistance among isolates of *Plasmodium falciparum* (*88*). Owing to the close similarities among the ATP-binding domains and the function of the various proteins in the process of transport, Ames (*89*) has advocated the use of bacterial transporters as representative models of this large and species-diverse family of proteins. The wide species distribution of this class of proteins would suggest that P-glycoproteins and their associated phenotypes arose early in evolution. Furthermore, the apparent maintenance of this class of glycoproteins across species would imply some potentially critical role for P-glycoproteins and similar proteins in normal cell function.

P-Glycoproteins and Relationship to Oligopeptide Transport

A persistent and difficult problem facing the elucidation of the normal cellular role(s) of the P-glycoproteins is the unusually broad spectrum of substances that can serve as ligands for efflux. Until recently, it was thought that P-glycoprotein-mediated efflux of compounds was limited to only small lipophilic compounds (*90*) and possibly certain classes of hormones (*91*). However, Sharma et al. (*92*) have presented evidence that suggests that small oligopeptides can also serve as ligands for ATP-dependent efflux catalyzed by P-glycoproteins. These investigators showed that a Chinese hamster ovary cell line, either selected for resistance to the synthetic cytotoxic tripeptide *N*-acetyl-leucyl-leucyl-norleucinal or transfected with the *mdr1* gene, displayed the classical MDR phenotype including amplification of an *mdr* gene and overexpression of its product. They also found that a mouse embryo cell line that overexpressed P-glycoprotein displayed similar resistance to the tripeptide.

Another line of evidence suggesting a role of P-glycoproteins in peptide transport came from studies reported by Raymond et al. (*93*). These investigators demonstrated that expression of a mammalian *mdr* gene complemented mutation of the *ste6* gene whose translation produce mediates export of the a-factor mating peptide in *Sacch. cerevisiae.* Additional support for the role of P-glycoprotein in peptide transport was provided by Naito et al. (*94*), who showed that overproduction of human P-glycoprotein confers resistance against gramicidin D, a peptide ionophore. Furthermore, Saeki et al. (*95*) demonstrated that cyclosporin A, a cyclic undecapeptide, displayed saturable transport kinetics in LLC-PK_1 cells that were transfected with the human *mdr1* gene for P-glycoprotein. A detailed overview of cyclosporin as a drug resistance modifier has been presented by Twentyman (*96*). These results collectively raise the possibility that *mdr* or *mdr*-like genes present in the genome of mammalian and non-mammalian cells may be involved in the secretion of oligopeptides or cellular proteins.

P-Glycoproteins and Intercompartmental Translocation of Proteins

Another line of investigation suggest a role of P-glycoprotein and related proteins in the intercompartmental translocation of proteins. The subcellular compartmentalization of proteins into organelles is a feature common to most eukaryotic cells. However, the mechanism for protein translocation has been poorly understood. Insights into a molecular basis for protein translocation into organelles has been provided from studies on peroxisomal organelles. Peroxisomes are small organelles that are bound to a single membrane and are postulated to form by division of preexisting peroxisomes. Gould et al. (*97*) demonstrated that transport of proteins into peroxisomes is conserved between yeast, plants, insects, and mammals, suggesting that at least one mechanism of protein translocation has been conserved throughout eukaryotic evolution.

Imanaki et al. (*98*) provided evidence for the requirement of ATP hydrolysis in the transport of acyl-CoA oxidase as well as several other enzymes into peroxisomes. Other studies (*98*) showed that a 70 kDa protein was also required for protein transport. This protein is one of the major integral peroxisomal membrane proteins. Peroxisomes depleted of intact 70 kDa protein by proteinase K digestion were found to be deficient in the transport of acyl-CoA oxidase in vitro (*99*). cDNA clones for the 70 kDa protein were isolated and sequenced by Kamijo et al. (*99*). An open reading frame with the coding capacity for a 659 amino acid hydrophobic polypeptide was identified. The deduced amino acid sequence displayed significant similarities to the ABC protein family, with a high degree of sequence similarity to P-glycoproteins. The results lead to the proposal that the major 70 kDa protein of peroxisomal membrane associated with protein transport into the organelle is a member of the P-glycoprotein superfamily. These observations are consistent with the previously mentioned roles of P-glycoproteins in the transport of peptides across biological membranes.

Oligopeptide Transporters from the Major Histocompatibility Complex

The human major histocompatability complex (MHC) consists of an array of genes that together are essential for regulation of the immune response. Two classes of cell surface antigens are known to be encoded by the MHC. One type of molecule (class I) is found on lymphocytes, macrophages, and the majority of nucleated cells. These antigens are involved in the recognition and rejection of tissue grafts. A second type of molecule (class II) has a more restricted distribution, and members are found only on B lymphocytes, some macrophages, and activated T lymphocytes. Class II molecules are essential for the interactions of T and B lymphocytes and probably for macrophage–lymphocyte interactions.

Class I MHC molecules present antigen to cytotoxic T lymphocytes. This process proceeds in several stages that are initiated by the cytosolic degradation of polypeptides to oligopeptides, followed by assembly of oligopeptides and class I MHC molecules into complexes. A number of studies (*100–109*) suggest that assembly of oligopeptide–class I MHC complexes occur in the endoplasmic reticulum or in an undefined subcompartment. Spies et al. (*110*) reported evidence indicating that oligopeptides derived from protein degradation are transported into these compartments by a peptide supply factor whose gene, designated *RING4,* is located in the MHC II region. A deduced amino sequence from partial sequencing of a cDNA clone harboring the gene for the factor revealed striking similarities to the MDR family of transporters. Spies and DeMars (*111*) demonstrated by gene transfer experiments that expression of the putative transporter gene in a human lymphoblastoid cell line mutant restored normal levels of surface HLA antigens. In subsequent studies, the *RING4* gene product was shown to be necessary for the efficient assembly of complexes between class I molecules and degraded oligopeptides. Furthermore, immunochemical studies and genetic data (*112–114*) demonstrated that the *RING4* gene product was associated with a second putative transport protein, designated *RING11,* and that presentation of MHC class I molecules was dependent on the peptide transporter heterodimer.

Homologous genes for oligopeptide transporters have also been identified in the class II MHC regions of rat and mouse (*115*).

Oligopeptide Transport in Prokaryotes

The genes and corresponding proteins for several oligopeptide transport systems have been genetically identified and cloned in prokaryotes. These active transport systems characteristically consist of multiple protein assemblies, are dependent on periplasmic binding proteins, and differ to some extent from the oligopeptide transporters of eukaroytes thus far discussed.

Oligopeptide transport in bacteria plays a role in both providing peptides as a carbon and nitrogen source for nutrition as well as in the recycling of cell wall peptides during growth. These transport systems are responsible for the uptake of oligopeptides that are no greater than five natural L-amino acids in length. The oligopeptide transporters do not serve for the transport of free amino acids (*116*). Andrews and Short (*117*) cloned and characterized five genes from *E. coli* that are essential for oligopeptide transport activity. A homologous set of genes has also been identified and cloned from the related bacterium *Salmonella typhimurium* (*118*). Additionally, Perego et al. (*119*) reported the nucleotide and deduced amino acid sequences for an oligopeptide transport system from *B. subtilis* and identified a role of oligopeptide transport in the initiation of sporulation.

Each set of genes from the various species encode a periplasmic binding protein and at most four membrane-associated proteins. Furthermore, each of the sets of proteins display a high degree of interspecies sequence similarity. How oligopeptides traverse the membrane by this class of transporter is unclear, but the process appears to involve ATP hydrolysis. Also unclear is the nature of the generalized ligand specificity for oligopeptides. Interestingly, the oligopeptide transporters may have specificities that extend beyond that of small peptides. Kashiwagi et al. (*120*) found that uptake of certain aminoglycoside antibiotics occur via the oligopeptide transporter in *E. coli*.

Genes responsible for the binding protein-dependent transport of dipeptides in enteric organisms have also been identified. The dipeptide transport systems of *E. coli* and *S. typhimurium* are involved in the uptake of dipeptides and are required for a peptide chemotactic response elicited by some enteric organisms (*121*). Abouhamad et al. (*122*) cloned and characterized genes for dipeptide periplasmic binding protein-dependent transport systems from these organisms. Olson et al. (*123*) independently characterized the periplasmic binding protein gene of the dipeptide transport system of *E. coli* and reported that amino acid similarities occur between the dipeptide binding protein of *E. coli* and the oligopeptide binding protein from *S. typhimurium*. The amino acid similarities suggest some commonalities may exist between binding specificities for di- and oligopeptide interactions with binding proteins.

Conclusions

The role of transport proteins in regulating the permeability barrier of biological membranes for amino acids, peptides, and oligopeptides has been well established. Decades of extensive biochemical and kinetic characterizations of transporters activities laid the foundations for the cloning of transporter genes. With new molecular biology approaches to isolate and study transporters, several basic concepts have emerged about the transport process. Most transporters involved in the facilitated transport of ligands are predicted to have similar protein topologies. Whereas this generalization may, in part, result from limited high-resolution structural information from crystallographic studies of membrane proteins, transporters that show close sequence relationships tend to transport structurally similar ligands. Similar predicted protein topologies may imply an underlying common mechanism for the translocation of various structurally diverse ligands. It is also true in some cases that proteins from different cell types that mediate the transport of the same ligand display little sequence similarity and appear to have converged to a common function. Therefore, a difficult challenge for future research will be the determination of how ligand specificity is acquired in this large superfamily of proteins. Such an understanding of ligand specificity will become increasingly important as reports on the diversity of transporters with potential overlapping ligand specifities continue to increase. The described connection between thyroid hormone transport and the transport of aromatic amino acids in erythrocytes (*124*) serves as an example. Equally challenging is the determination of the molecular mechanism for the transport of a diversity of amino acids, peptides, and oligopeptides by common transport proteins. Many xenobiotic compounds and certain viruses clearly have exploited transporters for uptake into particular target tissues. An understanding of the molecular nature of ligand specificities of transporters should provide a rational basis for understanding how transporters contribute to the bioavailability of xenobiotic compounds and facilitate the future design of such compounds.

References

1. Oxender, D. L.; Christensen, H. N. *J. Biol. Chem.* **1959,** *234,* 2321–2324.
2. Guidotti, G. G.; Gazzola, G. C. In *Mammalian Amino Acid Transport: Mechanisms and Control;* Kilberg, S.; Haussinger, D., Eds.; Plenum: New York, 1992; pp 3–30.
3. Eddy, A. A. In *Mammalian Amino Acid Transport: Mechanisms and Control;* Kilberg, S.; Haussinger, D., Eds.; Plenum: New York, 1992; pp 31–49.
4. Mitchell, P. In *Comprehensive Biochemistry;* Florkin, M.; Stotz, E. H., Eds.; Elsevier: Amsterdam, The Netherlands, 1967; Vol. 22, p 167.
5. Storck, T.; Schulte, S.; Hofmann, K.; Stoffel, W. *Proc. Natl. Acad. Sci. U.S.A.* **1992,** *89,* 10955–10959.
6. Hediger, M. A.; Turk, E.; Wright, E. M. *Proc. Natl. Acad. Sci. U.S.A.* **1989,** *86,* 5748–5752.
7. Nikaido, H.; Saier, M. H., Jr. *Science (Washington, DC)* **1992,** *258,* 936–942.
8. Haney, S. A.; Oxender, D. L. In *Molecular Biology of Receptors and Transporter: Bacterial and Glucose Transporters;* Friendlander, M.; Mueckler, M., Eds.; Academic: Orlando, FL, 1992; Vol. 137A, pp 37–45.
9. Albritton, L. M.; Tseng, L.; Scadden, D.; Cunningham, J. M. *Cell* **1989,** *57,* 659–666.
10. Kim, J. W.; Closs, E. I.; Albritton, L. M.; Cunningham, J. M. *Nature (London)* **1991,** *352,* 725–728.
11. Hoffmann, W. *J. Biol. Chem.* **1985,** *260,* 11831–11837.
12. Tanaka, J.; Fink, G. R. *Gene* **1985,** *38,* 205–214.
13. Wang, H.; Kavanaugh, M. P.; North, R. A.; Kabat, D. *Nature (London)* **1991,** *352,* 729–731.
14. Macleod, C. L.; Finley, K.; Kakuda, D.; Kozak, C. A.; Wilkinson, M. F. *Mol. Cell. Biol.* **1990,** *10,* 3663–3674.
15. Oxender, D. L.; Christensen, H. N. *J. Biol. Chem.* **1963,** *238,* 3686–3699.

16. Christensen, H. N.; Liang, M.; Archer, E. G. *J. Biol. Chem.* **1967,** *242,* 5237–5246.

17. Kong, C.-T.; Yet, S.-F.; Lever, J. E. *J. Biol. Chem.* **1993,** *268,* 1509–1512.

18. Deguchi, Y.; Yamato, I.; Anraku, Y. *J. Biol. Chem.* **1990,** *265,* 21704–21709.

19. Shafqat, S.; Tamarappoo, B. K.; Kilberg, M. S.; Puranam, R. S.; McNamara, J. O.; Guadano-Ferraz, A.; Fremeau, R. T. *J. Biol. Chem.* **1993,** *268,* 15351–15355.

20. Smith, K. E.; Borden, L. A.; Hartig, P. R.; Branchek, T.; Weinshank, R. L. *Neuron* **1992,** *8,* 927–935.

21. Pacholczyk, T.; Blakeley, R. D.; Amara, S. G. *Nature (London)* **1991,** *350,* 350–353.

22. Ritz, M. C.; Lamb, R. J.; Goldberg, S. R.; Kuhar, M. J. *Science (Washington, DC)* **1987,** *237,* 1219–1223.

23. Guastella, J.; Brecha, N.; Weigmann, C.; Lester, H. A.; Davidson, N. *Proc. Natl. Acad. Sci. U.S.A.* **1992,** *89,* 7189–7193.

24. Pines, G.; Danbolt, N. C., Bjoras, M.; Zhang, Y.; Bendahan, A.; Eide, L.; Koepsell, H.; Storm-Mathisen, J.; Seeberg, E.; Kanner, B. I. *Nature (London)* **1992,** *360,* 464–467.

25. Kanai, Y.; Hediger, M. A. *Nature (London)* **1992,** *360,* 467–471.

26. Fremeau, R. T.; Caron, M. G.; Blakeley, R. D. *Neuron* **1992,** *8,* 915–926.

27. Kilty, J. E.; Lorang, D.; Amara, S. G. *Science (Washington, DC)* **1991,** *254,* 578–579.

28. Blakely, R. D.; Berson, H. E.; Fremeau, R. T.; Caron, M. G.; Peek, M. M.; Prince, H. K.; Bradley, C. C. *Science (Washington, DC)* **1991,** *354,* 66–70.

29. Holland, E. C.; Drickamer, K. *J. Biol. Chem.* **1986,** *261,* 1286–1292.

30. Tate, S. S.; Yan, N.; Udenfriend, S. *Proc. Natl. Acad. Sci. U.S.A.* **1992,** *89,* 1–5.

31. Bertran, J.; Werner, A.; Moore, M. L.; Stange, G.; Markovich, D.; Biber, J.; Testar, X.; Zorzano, A.; Palacin, M.; Murer, H. *Proc. Natl. Acad. Sci. U.S.A.* **1992,** 89, 5601–5605.

32. Van Winkle, L. J.; Campione, A. L.; Gorman, J. M. *J. Biol. Chem.* **1988,** *263,* 3150–3163.

33. Bertran, J.; Werner, A.; Stange, G.; Markovich, D.; Biber, J.; Testar, X.; Zorzano, A.; Palacin, M.; Murer, H. *Biochem. J.* **1992,** *281,* 717–723.

34. Wells, R. G.; Hediger, M. A. *Proc. Natl. Acad. Sci. U.S.A.* **1992,** *89,* 5596–5600.

35. Bertran, J.; Magagnin, S.; Werner, A.; Markovich, D.; Biber, J.; Testar, X.; Zorzano, A.; Kuhn, L. C.; Palacin, M.; Murer, H. *Proc. Natl. Acad. Sci. U.S.A.* **1992,** *89,* 5606–5610.

36. Svensson, B. *FEBS Lett.* **1988,** *230,* 72–76.

37. Monks, T. J.; Anders, M. W.; Dekant, W.; Stevens, J. L.; Lau, S. S.; van Bladeren, P. J. *Toxicol. Appl. Pharmacol.* **1990,** *106,* 1–19.

38. Tate, S. S.; Meister, A. *Mol. Cell. Biochem.* **1982,** *39,* 357–368.

39. Tate, S. S.; Khadse, V. *Biochem. Biophys. Res. Commun.* **1986,** *141,* 1189–1194.

40. Meister, A.; Anderson, E. M. *Annu. Rev. Biochem.* **1983,** *52,* 711–760.

41. Meister, A. *Trends Biochem. Sci.* **1981,** *6,* 231–234.

42. Tate, S. S.; Meister, A. *Annu. Rev. Biochem.* **1976,** *45,* 559–604.

43. Smith, T. K.; Gibson, C. L.; Howlin, B.; Pratt, J. M. *Biochem. Biophys. Res. Commun.* **1991,** *178,* 1028–1035.

44. Laperche, Y.; Bulle, F.; Aissani, T.; Chobert, M.-N.; Aggerbeck, M.; Hanoune, J.; Guellaen, G. *Proc. Natl. Acad. Sci. U.S.A.* **1986,** *83,* 937–941.

45. Rajpert-DeMeyts, E.; Heisterkamp, N.; Groffen, J. *Proc. Natl. Acad. Sci. U.S.A.* **1988,** *85,* 8840–8844.

46. Papandrikopoulou, A.; Frey, A.; Gassen, H. G. *Eur. J. Biochem.* **1989,** *183,* 693–698.

47. Wetmore, L. A.; Gerard, C.; Drazen, J. M. *Proc. Natl. Acad. Sci. U.S.A.* **1993,** *90,* 7461–7465.

48. Hughey, R. P.; Coyle, P. J.; Curthoys, N. P. *J. Biol. Chem.* **1979,** *254,* 1124–1128.
49. Tsuji, A.; Matsuda, Y.; Katunuma, N. *J. Biochem. (Tokyo)* **1980,** *87,* 1567–1571.
50. Finidori, J.; Laperche, Y.; Haguenauer-Tsapis, R.; Barouki, R.; Guellaen, G.; Hanoune, J. *J. Biol. Chem.* **1984,** *259,* 4687–4690.
51. Barouki, R.; Finidori, J.; Chobert, M. N.; Aggerbeck, M.; Laperche, Y.; Hanoune, J. *J. Biol. Chem.* **1984,** *259,* 7970–7974.
52. Nash, B.; Tate, S. S. *J. Biol. Chem.* **1982,** *257,* 585–588.
53. Capraro, M. A.; Hughey, R. P. *FEBS Lett.* **1983,** *157,* 139–143.
54. Matthews, D. M. *Gastroenterology* **1977,** *73,* 1267–1279.
55. Matthews, D. M. *Ther. Clin. Nutr.* **1987,** *17,* 6–53.
56. Naider, F.; Becker, J. M. *Curr. Top. Med. Mycol.* **1988,** *2,* 170–198.
57. Walker-Smith, D. J.; Payne, J. W. *Planta* **1984,** *162,* 166–173.
58. Payne, J. W.; Gilvarg, C. In *Bacterial Transport;* Rosen, B. P., Ed.; Marcel Dekker: New York, 1978; pp. 325–383.
59. Walter,P.; Lingappa, V. R. *Annu. Rev. Cell Biol.* **1986,** *2,* 499–516.
60. Meyer, D. I.; Krause, E.; Dobberstein, B. *Nature (London)* **1982,** *297,* 503.
61. Gilmore, R.; Walter, P.; Blobel, G. *J. Cell Biol.* **1982,** *95,* 470.
62. Connolly, T.; Rapiejko, P. J.; Gilmore, R. *Science (Washington, DC)* **1991,** *250,* 1171–1173.
63. Rapoport, T. A. *Science (Washington, DC)* **1992,** *258,* 931–936.
64. Sanders, S. L.; Schekman, R. *J. Biol. Chem.* **1992,** *267,* 13791–13794.
65. Deshaies, R. J.; Schekman, R. *J. Cell Biol.* **1987,** *105,* 633–645.
66. Deshaies, R. J.; Schekman, R. *J. Cell Biol.* **1989,** *109,* 2653–2664.
67. Rothblatt, J. A.; Deshaies, R. J.; Sanders, S. L.; Daum, G.; Schekman, R. *J. Cell Biol.* **1989,** *109,* 2641–2652.
68. Stirling, C. J.; Rothblatt, J.; Hosobuchi, M.; Deshaies, R.; Schekman, R. *Mol. Cell. Biol.* **1992,** *3,* 129–142.
69. Toyn, J.; Hibbs, A. R.; Sanz, P.; Crowe, J.; Meyer, D. I. *EMBO J.* **1988,** *7,* 4347–4353.
70. Sanders, S. L.; Whitfield, K. M.; Vogel, J. P.; Rose, M. D.; Schekman, R. *Cell* **1992,** *69,* 353–365.
71. Musch, A.; Wiedmann, M.; Rapoport, T. A. *Cell* **1992,** *69,* 343–352.
72. Akiyama, Y.; Ito, K. *EMBO J.* **1987,** *6,* 3465–3470.
73. Simon, S. M.; Blobel, G. *Cell* **1991,** *65,* 371–380.
74. Ling, V.; Endicott, J. A. *Annu. Rev. Biochem.* **1989,** *58,* 137–171.
75. Gottesman, M. M.; Pastan, I. *Annu. Rev. Biochem.* **1993,** *62,* 385–427.
76. Gros, P.; Shustik, C. *Cancer Invest.* **1991,** *9,* 563–569.
77. Juranka, P. F.; Zastawny, R. L.; Ling, V. *FASEB J.* **1989,** *3,* 2583–2592.
78. Hyde, S. C.; Emsley, P.; Hartshorn, M. J.; Mimmack, M. M.; Gileadi, U.; Pearce, S. R.; Gallagher, M. P.; Gill, D. R.; Hubbard, R. E.; Higgins, C. F. *Nature (London)* **1990,** *346,* 362–365.
79. Kuchler, K.; Sterne, R. E.; Thorner, J. *EMBO J.* **1989,** *8,* 3973–3984.
80. McGrath, J. P.; Varshavsky, A. *Nature (London)* **1989,** *340,* 400–403.
81. Guilfoile, P. G.; Hutchinson, C. R. *Proc. Natl. Acad. Sci. U.S.A.* **1991,** *88,* 8553–8557.
82. Colombo, A. L.; Solinas, M. M.; Perini, G.; Biamonti, G.; Zanella, G.; Caruso, M.; Torti, F.; Fillippini, S.; Inventi-Solari, A.; Garofano, L. *J. Bacteriol.* **1992,** *174,* 1641–1646.
83. Neyfakh, A. A.; Bidnenko, V. E.; Chen, L. B. *Proc. Natl. Acad. Sci. U.S.A.* **1991,** *88,* 4781–4785.

84. Reizer, J.; Reizer, A.; Saier, M. H. *Protein Sci.* **1992,** *1,* 1326–1332.

85. Dudler, R.; Hertig, C. *J. Biol. Chem.* **1992,** *267,* 5882–5888.

86. Callahan, H. L.; Beverly, S. M. *J. Biol. Chem.* **1991,** *266,* 18427–18430.

87. Ouellette, M.; Fase-Fowler, F.; Borst, P. *EMBO J.* **1990,** *9,* 1027–1033.

88. Foote, S. J.; Thompson, J. K.; Cowman, A. F.; Kemp, D. J. *Cell* **1989,** *57,* 921–930.

89. Ames, G. F-L. In *Molecular Biology of Receptors and Transporter: Bacterial and Glucose Transporters.;* Friedlander, M.; Mueckler, M., Eds.; Academic: Orlando, FL, 1992; Vol. 137A, pp 1–35.

90. Hait, W. N.; Aftab, D. T. *Biochem. Pharm.* **1992,** *43,* 103–107.

91. Ueda, K.; Okamura, N.; Hirai, M.; Tanigawara, Y.; Saeki, T.; Kioka, N.; Komano, T.; Hori, R. *J. Biol. Chem.* **1992,** *267,* 24248–24252.

92. Sharma, R. C.; Inoue, S.; Roitelman, J.; Schimke, R. T.; Simoni, R. *J. Biol. Chem.* **1992,** *267,* 5731–5734.

93. Raymond, M.; Gros, P.; Whiteway, M.; Thomas, D. Y. *Science* **1992,** 232–234.

94. Naito, M.; Hoshino, T.; Matsushita, Y.; Hirai, R.; Tsuruo, T. J. *Cell. Pharmacol.* **1991,** *2,* 263–267.

95. Saeki, T.; Kazumitsu, U.; Tanigawara, Y.; Hori, T.; Komano, T. *J. Biol. Chem.* **1993,** *268,* 6077–6080.

96. Twentyman, P. R. *Biochem. Pharm.* **1992,** *43,* 109–117.

97. Gould, S. J.; Keller, G.-A.; Schneider, M.; Howell, S. H.; Garrard, L. L.; Goodman, J. M.; Distel, B.; Tabak, H.; Subramani, S. *EMBO J.* **1990,** *9,* 85–90.

98. Imanaka, T.; Small, G. M.; Lazarow, P. B. *J. Cell. Biol.* **1987,** *105,* 2915–2922.

99. Kamijo, K.; Taketani, S.; Yokota, S.; Osumi, T.; Hashimoto, T. *J. Biol. Chem.* **1990,** *265,* 4534–4540.

100. Towsend, A. R. M.; Gotch, F. M.; Davey, J. *Cell* **1985,** *42,* 457–467.

101. Morrison, L. A.; Lukacher, A. E.; Braciale, V. L.; Fan, D. P.; Draciale, T. J. *J. Exp. Med.* **1986,** *163,* 903–921.

102. Bjorkman, P. J.; Saper, M. A.; Samraouri, B.; Bennett, W. S.; Strominger, J. L.; Wiley, D. C. *Nature (London)* **1987,** *329,* 512–516.

103. Moore, M. W.; Carbone, F. R.; Bevan, M. J. *Cell* **1988,** *54,* 777–785.

104. Van Bleek, G. M.; Nathenson, S. G. *Nature (London)* **1990,** *348,* 213–216.

105. Townsend, A.; Ohlen, C.; Bastin, J.; Ljunggren, H. G.; Foster, L.; Karre, K. *Nature (London)* **1989,** *340,* 443–448.

106. Cerundolo, V.; Alexander, J.; Anderson, K.; Lamb, C.; Cresswell, P.; McMichael, A.; Gotch, F.; Townsed, A. *Nature (London)* **1990,** *345,* 449–452.

107. Salter, R. D.; Cresswell, P. *EMBO J.* **1986,** *5,* 943–949.

108. Towsend, A.; Elliott, T.; Cerundolo, V.; Foster, L.; Barber, B.; Tse, A. *Cell* **1990,** *62,* 285–295.

109. Schumacher, T. N. M.; Heemels, M. T.; Neefjes, J. J.; Kast, W. M.; Melief, C. J. M.; Ploegh, H. L. *Cell* **1990,** *62,* 563–567.

110. Spies, T.; Bresnahan, M.; Bahram, S.; Arnold, D.; Blanck, G.; Mellins, E.; Pious, D.; DeMars, R. *Nature (London)* **1990,** *348,* 744–747.

111. Spies, T.; DeMars, R. *Nature (London)* **1991,** *351,* 323–324.

112. Spies, T.; Cerundolo, V.; Colonna, M.; Cresswell, P.; Townsend, A.; DeMars, R. *Nature (London)* **1992,** *355,* 644–646.

113. Powis, S. H.; Mockridge, I.; Kelly, A.; Kerr, L.-A.; Glynne, R.; Gileadi, U.; Beck, S.; Trowsdale, J. *Proc. Natl. Acad. Sci. U.S.A.* **1992,** *89,* 1463–1467.

114. Kelly, A.; Powis, S. H.; Kerr, L.-A.; Mockridge, I.; Elliott, T.; Bastin, J.; Uchanska-Ziegler, B.; Ziegler, A.; Trowsdale, J.; Townsend, A. *Nature (London)* **1992,** *355,* 641–644.

115. Deverson, E. V.; Gow, I. R.; Coadwell, W. J.; Monaco, J. J.; Butcher, G. W.; Howard, J. C. *Nature (London)* **1990,** *348,* 738–741.

116. Guyer, C. A.; Morgan, D. G.; Staros, J. V. *J. Bacteriol.* **1986,** *168,* 775–779.

117. Andrews, J. C.; Short, S. A. *J. Bacteriol.* **1985,** *161,* 484–492.

118. Hiles, I. D.; Gallagher, M. P.; Jamieson, D. J.; Higgins, C. F. *J. Mol. Biol.* **1987,** *195,* 125–142.

119. Perego, M.; Higgins, C. F.; Pearce, S. R.; Gallagher, M. P.; Hoch, J. A. *Mol. Microbiol.* **1991,** *5,* 173–185.

120. Kashiwagi, K.; Miyaji, A.; Ikeda, S.; Tobe, T.; Sasakawa, C.; Igarashi, K. *J. Bacteriol.* **1992,** *174,* 4331–4337.

121. Manson, M. D.; Blank, V.; Brade, G.; Higgin, C. F. *Nature (London)* **1986,** *321,* 253–256.

122. Abouhamad, W. N.; Manson, M.; Gibson, M. M.; Higgins, C. F. *Mol. Microbiol.* **1991,** *5,* 1035–1047.

123. Olson, E. R.; Dunyak, D. S.; Jurss, L. M.; Poorman, R. A. *J. Bacteriol.* **1991,** *173,* 234–244.

124. Zhou, Y.; Samson, M.; Osty, J.; Francon, J.; Blondeau, J.-P. *J. Biol. Chem.* **1990,** *265,* 17000–17004.

125. Vandenbol, M.; Jauniaux, J.-C.; Grenson, M. *Gene* **1989,** *83,* 153–159.

126. Jauniaux, J.-C.; Grenson, M. *Eur. J. Biochem.* **1990,** *190,* 39–44.

127. Tolner, B.; Poolman, B.; Wallace, B.; Konings, W. N. *J. Bacteriol.* **1992,** *174,* 2391–2393.

128. Steffes, C.; Ellis, J.; Wu, J.; Rosen, B. P. *J. Bacteriol.* **1992,** *174,* 3242–3249.

129. Pi, J.; Wookey, P. J.; Pittard, A. J. *J. Bacteriol.* **1991,** *173,* 3622–3629.

130. Heatwole, V. M.; Somerville, R. L. *J. Bacteriol.* **1991,** *173,* 108–115.

131. Nakao, T.; Yamato, I.; Anraku, Y. *Mol. Gen. Genet.* **1987,** *208,* 70–75.

132. Sarsero, J. P.; Wookey, P. J.; Gollnick, P.; Yanofsky, C.; Pittard, A. J. *J. Bacteriol.* **1991,** *173,* 3231–3234.

133. Wookey, P. J.; Pittard, A. J. *J. Bacteriol.* **1988,** *170,* 4946–4949.

134. Honore, N.; Cole, S. T. *Nucleic Acids Res.* **1990,** *18,* 653.

135. Hoshino, T.; Kose-Terai, K.; Uratani, Y. *J. Bacteriol.* **1991,** *173,* 1855–1861.

136. Hoshino, T.; Kose, K. *J. Bacteriol.* **1990,** *172,* 5531–5539.

137. Frommer, W. B.; Hummel, S.; Riesmeier, J. W. *Proc. Natl. Acad. Sci. U.S.A.* **1993,** *90,* 5944–5948.

138. Ohnishi, K.; Hasegawa, A.; Matsubara, K.; Date, T.; Okada, T.; Kiritani, K. *Jpn. J. Genet.* **1988,** 63, 343–357.

139. Nei, M. In *Molecular Evolutionary Genetics;* Columbia University: New York, 1987; pp 293–298.

RECEIVED for review July 12, 1993. ACCEPTED revised manuscript October 1, 1993.

CHAPTER 2

Peptide Metabolism at Brush-Border Membranes

Roger H. Erickson

Brush-border (microvillus) membranes were among the earliest cell "structures" identified by microscopy 150 years ago at the luminal surface of intestinal epithelial cells (*1*). However, the first separation of brush-border membranes for biochemical investigation was not reported until 1961 (*2*). Since that time, interest in these specialized plasma membranes has expanded. Brush-border membranes, or microvillus structures, are associated with the epithelial cells of the intestine and kidney, the syncytiotrophoblast of the placenta, and the choroid plexus of the nervous system. The most intensively studied membranes are those of the intestine and kidney.

Many details regarding the general structure of the brush-border membranes of the intestine and kidney are known. Studies over the past 10–15 years have shown that a series of enzymes and carrier transport systems involved in the cell surface metabolism and absorption of various carbohydrates and peptides are associated with these membranes. Brush-border-membrane peptidases participate in the hydrolysis of a wide variety of peptides and can serve as useful in vitro models to examine this hydrolysis. Concomitantly, a heightened awareness about these enzymes has developed among drug-delivery scientists, pharmacologists, neuroscientists, immunologists, and others interested in peptide metabolism.

Also, the rate of discovery of proteins and peptides having potent and specific biological effects has increased exponentially. Peptides and proteins play important roles in many diverse biochemical processes such as neurotransmission, cell signaling, cellular growth and development, endocrine system function, and immune response. With the convergence of molecular biology and biotechnology, many of these bioactive molecules can be produced on a mass scale. Most of these peptides have obvious therapeutic potential, and considerable interest exists in designing systems to achieve effective delivery of those molecules to various in vivo targets. Major obstacles to this goal are the many proteases and peptidases that are present in most cells, tissues, organs, and circulating fluids and can rapidly degrade these types of bioactive molecules.

This problem is particularly acute when considering oral-delivery systems. Oral delivery of peptides and proteins, though very desirable, has two major obstacles that present major challenges to effective implementation and widespread use. The first obstacle is that absorption of intact, large peptides and proteins by the small-intestinal mucosa is relatively poor. Second obstacle is the presence in the intestine of a wide variety of enzymes that hydrolyze peptide bonds. These enzymes can be broadly categorized into two basic groups, those of gastric and pancreatic origin that are secreted into the intestinal lumen and the brush border membrane associated peptidases of small-intestinal epithelial cells. Even though there is widespread recognition of the enzymes in the first group and their role in protein degradation, less information has been available, until recently, regarding brush-border-membrane peptidases and their role in peptide metabolism.

The purpose of this volume is to address the different issues involved in the metabolic transport and stabilization of peptides and proteins of therapeutic value. Accordingly, I will provide a general overview of the types of peptidases associated with brush-border membranes and their roles in peptide metabolism. Much of this information comes from studies of kidney and intestinal peptidases, and the reader is encouraged to seek out details of interest in the many excellent reviews and articles that have appeared regarding this subject. Because the gastrointestinal tract is an obvious, important route for peptide–protein delivery, a primary focus of this article will be on the brush-border-membrane peptidases associated with small-intestinal epithelial cells of mammals.

Protein and Peptide Hydrolysis in the Intestine and Kidney

Each day, the mammalian gastrointestinal system hydrolyzes and assimilates a wide variety of protein from both exogenous and endogenous sources (*3*). In the human, the average daily North American diet provides approximately 70–100 g of exogenous protein alone. In normal individuals, 95–98% of the total protein is completely digested and absorbed. The efficiency of this protein assimilation is a testament to the numerous enzymes present in various areas of the gastrointestinal tract that are capable of hydrolyzing peptide bonds.

The process of protein digestion begins in the stomach, where at least two enzymes, pepsins I and II, initiate hydrolysis of the polypeptide backbone (*3–6*). Protein digestion continues in the lumen of the small intestine through the action of five potent proteolytic enzymes secreted by the pancreas, namely trypsin, chymotrypsin, elastase, and carboxypeptidases A (CP-A) and B (CP-B). The specificities of these pancreatic enzymes are complementary to one another and serve to convert proteins and large polypeptides to a mixture of free amino acids and small peptides of from 2 to 8 amino acids. Until recently, this process was thought to be very efficient and to result in the complete conversion of proteins to their individual amino acid constituents. However, we now know that the small-peptide fraction can make up 70% of the total hydrolysate and that the peptidases of the intestinal absorptive cell play a crucial role in completing the hydrolysis of these small peptides before intestinal absorption.

In the kidney, the polarized epithelial cells lining the proximal tubule possess microvilli that project into the lumen from the apical cell surface. Although the function of the peptidases expressed at this site is not entirely clear, they are thought to play a scavenging and protective role by removing peptides from the urine (*7, 8*). In this capacity, they may also be important in the termination of various hormonal signals (*9, 10*).

Histology and Structure of Small-Intestinal Epithelial Cells

A stylized diagram of the histologic organization of the single epithelial cell layer that lines the small intestine (*11*) shows that the cell layer is organized into a series of numerous, microscopic, finger-like projections called "villi" (Figure 1). In between adjacent villi are the areas called "crypts." The undifferentiated cells of the crypt rapidly proliferate, and the cells migrate toward the villus tip and undergo terminal differentiation to several cell types. Ultimately, the senescent cells are shed from the villus tip into the intestinal lumen. Thus, a complete turnover of the intestinal epithelium occurs approximately every 3–7 days.

One of the differentiated cell types composing the epithelium is the small-intestinal enterocyte or absorptive cell (*11*). These cells are highly polarized and have the general morphological characteristics shown in Figure 2. The brush-border or microvillus membrane faces the lumen of the small intestine. Associated with this membrane are a number of different hydrolases, such as the peptidases and carbohydrases, that participate in the terminal phases of the digestive process. Also present are various carrier transport systems for amino acids, dipeptides, tripeptides, and sugars that allow the cell to efficiently absorb the final products of digestion from the intestinal lumen. Separate and quite distinct is the basolateral membrane, which maintains contact with neighboring cells and contains a completely different complement of enzymes, proteins, and transport carriers when compared with the brush-border membrane. In addition, these cells are able to respond to various factors and to rapidly synthesize, process, transport, and segregate proteins destined for either of the two types of membranes.

The digestive peptidases that are expressed by intestinal epithelial cells are basically found in two different subcellular fractions, namely the brush-border membrane and the cytoplasm (*12*). The soluble enzymes of the cytoplasm (*13*) consist chiefly of dipeptidases, an aminotripeptidase, and proline dipeptidase, which serve to complete the intracellular hydrolysis of dipeptides and tripeptides absorbed by the enterocyte. The intestinal peptidases associated with the brush-border membrane consist of a number of distinct enzymes, each with a different substrate specificity.

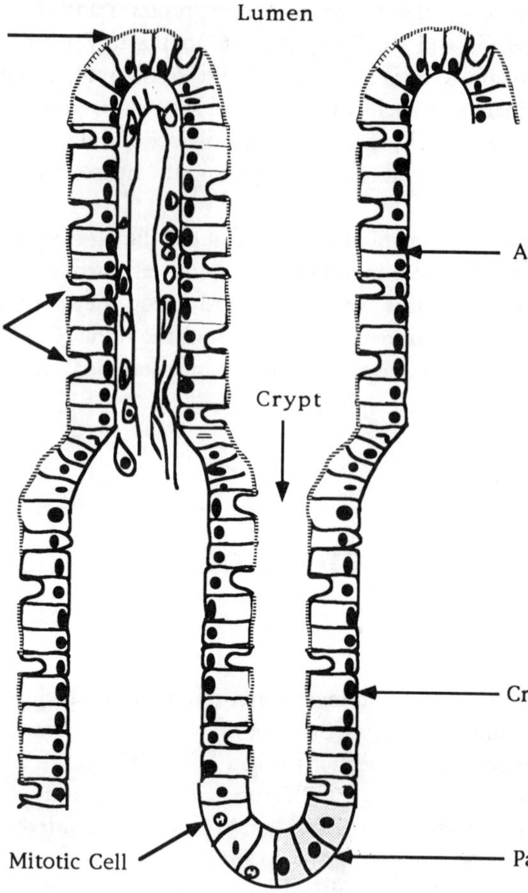

Figure 1. Histology of the small-intestinal mucosa.

Types of Brush-Border-Membrane Peptidases

The list of brush-border-membrane peptidases in Table I is a composite of those found in mammalian kidney (*7, 8*) and small intestine (*3–6, 12*). For simplicity, the peptidases have been divided into two basic categories, exopeptidases and endopeptidases. These two groups can be further subdivided into additional classes based on their substrate specificity and catalytic mechanism (*14*). Many of these enzymes were given different names over the years (*14*). For clarity and ease of recognition, a single name has been ascribed to most of the peptidases in Table I. To facilitate positive identification, the Enzyme Commission number for each peptidase is also presented when available (*7, 8, 14*).

Most of the data in Table I comes from studies of pig, rodent (mouse and rat), and human tissues. Although information is far from complete, species' differences probably exist and other peptidases probably have not been identified. Even though the brush-border membranes of kidney and intestine express the same complement of peptidases, there are some notable exceptions. For instance, enzymes such as enteropeptidase, γ-glutamyl carboxypeptidase, Gly–Leu dipeptidase, and zinc-stable dipeptidase are present in the intestine

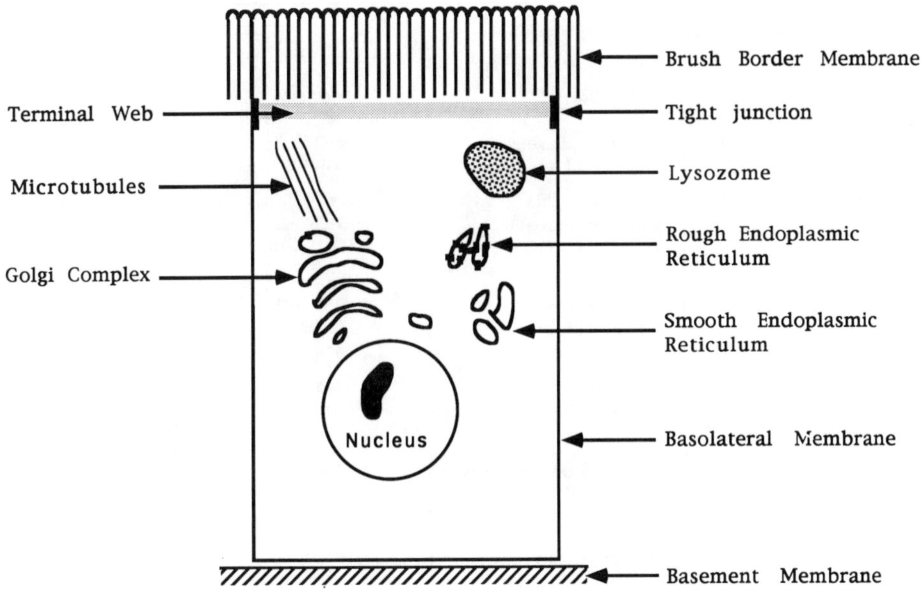

Figure 2. Structure of small-intestinal absorptive cells.

but have not been described in the kidney. In addition, differences in the levels of individual enzymes are generally observed between the two tissues.

Among the exopeptidases found associated with the brush-border membrane are those that hydrolyze peptides from the amino terminal end. All of these enzymes are present in both kidney and intestine. The first, aminopeptidase N (AP-N), is abundant in both tissues, contains zinc at the active site, and removes the amino terminal amino acid from dipeptides, tripeptides, and a variety of oligopeptides (*7, 14–17*). The enzyme is relatively nonspecific and is usually assayed with peptide derivatives of leucine or alanine. Although AP-N can act on dipeptides, higher activities are usually observed with longer chain-length oligopeptides (*15*).

The best substrates for aminopeptidase A (AP-A) are peptides with N-terminal glutamic and aspartic acid residues. AP-A has calcium at the active site and is activated by the calcium cation (*18*). Aminopeptidase P(AP-P) hydrolyzes prolyl peptide bonds when proline is in the penultimate position and is stimulated by manganese (*19–21*). AP-P removes the amino terminal amino acid from peptides such as Gly–Pro–Hyp shown in Table I. Aminopeptidase W(AP-W) a newly described peptidase, removes the amino terminal amino acid from peptides with tryptophan in the penultimate position and tends to favor dipeptides as substrates (*22, 23*). AP-A, -P, and -W are much more restricted in their substrate specificity when compared with AP-N. Included in this category of exopeptidases is a peptidase whose catalytic mechanism differs from the first four, dipeptidyl peptidase IV (DPPIV) (*24–26*). It removes the amino terminal dipeptide from peptides with either proline or alanine in the penultimate position. It is the only serine brush-border-membrane peptidase described to date (except for intestinal enteropeptidase). Similarly, γ-glutamyl transpeptidase is classified as a transferase because it transfers γ-glutamyl groups from various peptide donors to a wide variety of peptide acceptors (*14*).

Several enzymes that hydrolyze peptides from the carboxy terminus are known. The first, angiotensin-converting enzyme (ACE), is not a true carboxypeptidase. It cleaves dipeptides from the C-terminal end and is hence classified as a peptidyl dipeptidase (peptidyl

Table I. Brush-Border-Membrane Peptidases

Peptidase	*Enzyme*	*Specificity*	*Typical Assay Substrate*
Exopeptidase 1. NH_2 terminus	Aminopeptidase N (EC 3.4.11.2)	▼ ■-□-□-□- (many)	▼ Leu-- NNap
	Aminopeptidase A (EC 3.4.11.7)	▼ ■-□-□-□- (Asp, Glu)	▼ Asp--NNap
	Aminopeptidase P (EC 3.4.11.9)	▼ □-■-□-□- (Pro)	▼ Gly--Pro-Hyp
	Aminopeptidase W (EC 3.4.11.-)	▼ □-■-□-□- (Trp)	▼ Gly--Trp
	Dipeptidyl peptidase IV (EC 3.4.14.5)	▼ □-■-□-□- (Pro, Ala)	▼ Gly-Pro--NNap
	γ-Glutamyl trans-peptidase (EC 2.3.2.2)	▼ ■-□- □-□ (γ-Glu)	▼ γ-Glu--NNap
2. COOH terminus	Angiotensin-converting enzyme (EC 3.4.15.1)	▼ -□-□-■-■ (many)	▼ Bz-Gly--His-Leu
	Carboxypeptidase P (EC 3.4.17.-)	▼ -□-□-■-□ (Pro, Ala, Gly)	▼ Cbz-Gly-Pro--Leu
	Carboxypeptidase M (EC 3.4.17.12)	▼ -□-□-□-■ (Lys,Arg)	▼ Bz-Gly--Arg
	γ-Glutamyl carboxy-peptidase (EC 3.4.22.-)	▼ -■-■-■-■ (γ-Glu)n	▼ Pteroyl-(γ- Glu)n--γ -Glu
3. Dipeptidase	Microsomal dipeptidase (EC 3.4.13.19)	▼ ■-■ (many)	▼ Gly--D-Phe
	Gly-Leu peptidase	▼ ■-□ (neutral)	▼ Gly--Leu
	Zinc stable peptidase	▼ ■-□ (Asp, Met)	▼ Asp--Lys
Endopeptidase	Endopeptidase-24.11 (EC 3.4.24.11)	▼ -□-□-■-□- (hydrophobic)	Insulin β chain
	Endopeptidase-2 (EC 3.4.24.18, PABA-peptide hydrolase, Meprin)	▼▼ -□-■-□-□ (aromatic/hydrophobic)	▼ Bz-Tyr--pAB, Insulin β chain, Azocasein
	Endopeptidase-3 (?)	??	▼ Cbz-Ala-Arg--Arg-MNA
	Enteropeptidase (EC 3.4.21.9)	▼ -□-■-□-□- [pentapeptide, $(Asp)_4$-Lys]	Trypsinogen

NOTE: Bz = benzoyl, Cbz = benzyloxycarbonyl, MNA = 4-methoxy-β-naphthylamide, NNap = 2-naphthylamide, PAPA-peptide = Bz-tyr-p-aminobenzoate.

SOURCE: Data are adapted from references 7, 8, 14, 41, and 53.

dipeptidase A). It is the same enzyme that is well-known for its pivotal role in the regulation of blood pressure in other organs (*27*). In the intestine, ACE was shown to play a digestive function by hydrolyzing the C-terminal dipeptide from a number of different peptide substrates (*28, 29*). Its activity is somewhat analogous to that of DPP IV. Although the enzyme is relatively nonspecific for the C-terminal dipeptide, high levels of enzymatic activity are observed when the C-terminal amino acid is a proline residue (*28, 30*). For instance, hydrolytic rates are 4–5 times higher when Bz–Gly–Ala–Pro is used as substrate when compared with the corresponding His–Leu peptide. Substrates with proline in the penultimate position are not hydrolyzed by ACE because of the high degree of positional specificity ACE has for proline.

This enzyme requires chloride ions for full activity and has zinc at its active site. Carboxypeptidase P(CP-P) is a true carboxypeptidase and has a restricted substrate specificity. Similar to AP-P, it hydrolyzes peptides where the penultimate amino acid residue is either proline, alanine, or glycine and is present in both kidney and intestine (*31, 32*). Carboxypeptidase M is found in the kidney and was purified from human placental microvilli (*33, 34*). Its presence in intestinal brush-border membranes was also reported (*35*). CP-M hydrolyzes C-terminal lysine and arginine residues from peptides, and its properties are somewhat similar to CP-B (*34*). In the placenta, CP-M is thought to play a role in the modulation of peptide hormone activity by removing basic amino acids from peptides such as bradykinin (*34*). γ-Glutamyl carboxypeptidase catalyzes the sequential release of glutamyl groups from pteroylpolyglutamates and is important for dietary-folate absorption by the intestine (*14, 36*).

Dipeptidases are a subclass of the exopeptidases, and the name is simply based on the size of the substrate that is hydrolyzed. Microsomal dipeptidase (EC 3.4.13.11) is abundant in the kidney and contains zinc at the active site (*37*). This enzyme is thought to be responsible for the hydrolysis of glutathione, its S-derivatives, and leukotriene D_4 (*38*). It has also been implicated in the hydrolysis of β-lactam antibiotics (*39*). Interestingly, peptides such as Gly–D–Ala and Gly–D–Phe are excellent substrates for this dipeptidase (*38, 40*). Although most studies dealt with the renal form of the enzyme, it is also found in the intestine (*38*). Two other dipeptidases, Gly–Leu dipeptidase and zinc-stable Asp–Lys peptidase, were purified and characterized from the brush-border membrane of human small intestine (*41*).

Brush-border membranes contain several endopeptidases that hydrolyze the interior peptide bonds of peptides and, in some cases, larger proteins. Endopeptidase-24.11 (Endo-1) is improperly known as enkephalinase or is known by its Enzyme Commission number (Table I). Endo-1 was first identified in the kidney by its ability to hydrolyze insulin β-chain (*42, 43*). Its specificity is very similar to that of the microbial metalloendopeptidase thermolysin by hydrolyzing peptide bonds involving the amino group of hydrophobic amino acid residues. The active site of the enzyme contains zinc, is quite complex, and extends over a number of subsites (*7*). Like thermolysin, it is inhibited by the natural product inhibitor phosphoramidon (*44*).

A second type of endopeptidase, known as endopeptidase-24.18 (Endo-2) is also found in the brush-border membranes of kidney and intestine (*45, 46*). This enzyme is closely related to the PABA–peptide hydrolase described in humans (*47*) and meprin found in rodents (*48, 49*). Recent cloning and sequencing studies have shown that all three are related and belong to a separate group of endopeptidases referred to as the "astacin family" (*50, 51*). Astacin is a peptidase originally described in the digestive tract of the crayfish (*52*). This group of peptidases are quite different from Endo-1 and other thermolysin-like endopeptidases and are not inhibited by phosphoramidon. Hydrolysis of PABA-peptide indicates that these enzymes have a chymotrypsin-like activity. They can hydrolyze relatively large protein molecules such as azocasein, histone, and fibrinogen (*45, 48, 53*). Other endopeptidases include a third distinct intestinal endopeptidase, about which little is known (*53*), and enteropeptidase, a specialized serine peptidase found in the intestinal duodenum, where it activates trypsin from its inactive zymogen precursor (*4*).

Characteristics of Brush-Border-Membrane Peptidases

Although the list of brush-border-membrane peptidases includes a diverse group of enzymes, they share a number of basic features that are briefly described.

General Structure and Membrane Topology

All of the brush-border-membrane peptidases that have been extensively characterized can be called "ectoenzymes" (*54*). This term implies a certain set of characteristics relating to their structure and association with cell surface membranes. They are invariably membrane-bound or "integral" membrane proteins oriented with the active site at the external surface of the cell membrane. Numerous studies were performed to identify the structure and mode of association of these enzymes with the brush-border membrane (*55–57*). A number of peptidases such as Endo-1 (*58*), Endo-2 (*51*), DPP-IV (*59*), AP-N (*60, 61*), γ-glutamyl transpeptidase (*62*), CP-M (*63*), microsomal dipeptidase (*64, 65*), and ACE (*66*) were cloned and sequenced to afford a more detailed examination of their structure as deduced from their amino acid sequence. From these studies, a general picture of the structure of these peptidases has emerged (*see* Figure 3) as illustrated for AP-N and DPP-IV.

These proteins are large when compared with other proteolytic enzymes (i.e., pancreatic proteases) and have subunit molecular weights that range from 50 to 150 thousand. Many of these proteins exist in the brush-border membrane as homodimers or aggregates of two identical subunits. Therefore, purified preparations of enzymes such as AP-N and DPP-IV have native molecular weights ranging from 200 to 300 thousand (*15, 25, 55–57*). Another common feature of these peptidases is that they consist of several distinct domains. One domain is the large hydrophilic portion that consists of most of the molecule and protrudes from the membrane surface of the cell. This region contains the active site and attached oligosaccharide moieties (N-linked). These enzymes tend to be highly glycosylated and have both N- and O-linked carbohydrate, which can compose 10–35% of the weight of the molecule (*55, 57*). The small junction region of the polypeptide links the hydrophilic region to the hydrophobic membrane anchor. This linkage gives the enzymes a "stalked" appearance in electron micrographs, and the length of the stalk varies from 2 to 10 nm depending on the enzyme (*54, 57*). In some enzymes such as AP-N, this region contains serine and threonine amino acid residues that may act as potential sites for O-glycosylation (*60, 61*).

The transmembrane hydrophobic anchor peptide has a molecular weight of approximately 10 thousand and anchors the enzyme to the membrane through protein–lipid interaction. For AP-N and DPP-IV, the hydrophobic region (uncleaved signal peptide) occurs near the amino terminus of the molecule (Type II membrane glycoprotein). This region traverses the membrane once and ends in a small tail portion that is in contact with the interior (cytoplasm) of the cell. Many peptidases of this type can be readily solubilized from

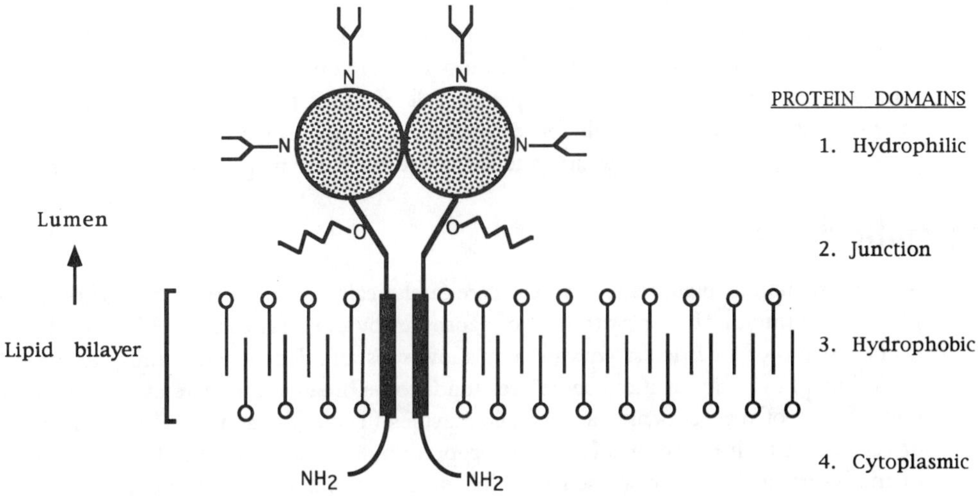

Figure 3. Membrane topology of brush-border-membrane aminopeptidase N and dipeptidyl peptidase IV.

the membrane with various detergents or by cleaving of the junction peptide region with a protease such as papain. Papain hydrolysis generally results in a fully soluble, stable, and active peptidase molecule.

Of course, variations are found in the general scheme depicted in Figure 3. Rabbit AP-N, AP-A, and Endo-1 exist in the membrane as enzymatically active monomeric units (*57*). Therefore, the significance of the dimeric structure found in other species is unclear. Human CP-M is also a monomeric enzyme (*34*). Similarly, ACE is a single polypeptide, but its hydrophobic membrane anchor is near the carboxy terminus (Type I membrane protein) (*29, 66–69*). AP-P, microsomal dipeptidase, and CP-M are associated with the brush-border membrane by a C-terminal glycosyl–phosphatidylinositol anchor (*21, 40, 70, 71*). These peptidases can be enzymatically released from membranes by phosphatidylinositol-specific phospholipase D. Endo-2, a Type I membrane protein, has a tetrameric structure containing two α- and two β-subunits linked together by disulfide bonds (*45, 51*). Other genetically influenced tetrameric combinations were described for the related enzyme meprin found in the mouse kidney (*72, 73*).

Biosynthesis of Brush-Border-Membrane Peptidases

Brush-border-membrane peptidases undergo similar modes of biosynthesis in kidney and intestinal epithelial cells (*56, 73–75*). Initial translation, translocation, and high-mannose glycosylation occur in the endoplasmic reticulum. Modification of the carbohydrate moieties to the complex type and O-glycosylation occurs in the Golgi membranes. The mature proteins are then transported directly to the brush-border membrane while some molecules go by an indirect route via the basolateral membrane. A portion of newly synthesized peptidases may also be directly targeted to lysozomes, and some recycling of peptidases associated with brush-border membranes occurs through endocytotic pathways (*76, 77*). No intracellular proteolytic processing of AP-N and DPP-IV occurs, although the amino terminal-signal peptide of peptidases such as ACE (Type I membrane protein) is removed during biosynthesis (*67*).

Most brush-border-membrane peptidases are glycoproteins containing approximately 10–35% carbohydrate. They are predominantly of the N-linked type, but O-linked carbohydrate side chains may also be present. The N-linked oligosaccharide side chains of AP-N and DPP-IV were sequenced and shown to contain many different types of complex carbohydrates (*78, 79*). Although the precise role of the carbohydrate moieties is not clear, the moieties may serve as signals for intracellular routing, provide resistance to proteolytic cleavage, stabilize protein structure, and influence the catalytic properties of the enzyme.

Enzymatic Activity

Most brush-border-membrane peptidases are catalytically active at neutral to slightly alkaline pH. The optimum pH for hydrolysis of azocasein by mouse meprin is 9.0 (*80*). The pH characteristics of a particular enzyme may show substrate dependence similar to the ACE (*30*). The pH profile for these peptidases tends to be broad and to extend over several pH units. Many of the peptidases are metalloenzymes (Table VI) and have zinc, manganese, or calcium at their active sites (*7*). This property makes them susceptible to certain types of inhibitors, which are discussed in a following section. Dipeptidyl peptidase IV is an exception because it is a serine peptidase. In many instances, the detailed kinetic parameters (Michaelis constant, K_m; limiting velocity, V_{max}) have been determined for those enzymes that have been purified. These constants are indicative of the binding of an enzyme to a particular peptide substrate and give a measure of the efficiency with which various peptides are hydrolyzed.

Peptidase Levels in Other Tissues

Levels of some of the peptidases in rat-intestinal brush-border membranes are shown in Table II. These results are highly dependent on the substrates used, the substrate concentration, and the assay conditions. Despite these limitations, Table II serves to give a general, comparative profile of the enzymes that are present in the small intestine. Enzymes such as AP-N, AP-A, DPP-IV, and ACE are major constituents. High levels of intestinal AP-P are also present. By comparison, low levels of CP-P are observed, and Endo-1 is present at much lower levels than reported for the kidney (*81*). Endo-2 is expressed at

Table II. Levels of Rat-intestinal Brush-Border-Membrane Peptidases

Enzyme	*Substrate*	*Specific Activity (nmole/min per mg protein)*
Aminopeptidase N	Leu–NNap	1100
Aminopeptidase P	Gly–Pro–Hyp	1000
Aminopeptidase A	Glu–NNap	600
Dipeptidyl peptidase IV	Gly–Pro–NNap	600
Angiotensin-converting enzyme	Bz–Gly–His–Leu	150
Carboxypeptidase P	Cbz–Gly–Pro–Leu	40
Endopeptidase-24.11	Glutaryl–Ala–Ala–Ala–MNA	20
Endopeptidase-24.18	Azocasein[a]	65
Endopeptidase-3	Cbz–Ala–Arg–Arg–MNA	35

NOTE: Bz = benzoyl, Cbz = benzyloxycarbonyl, MNA = 4-methoxy-β-naphthylamide, and NNap = 2-naphthylamide.
[a] Azocasein hydrolysis is expressed as μg/min per mg protein.
SOURCE: Data are from references 28, 32, 53, 83, and unpublished observations (R. H. Erickson).

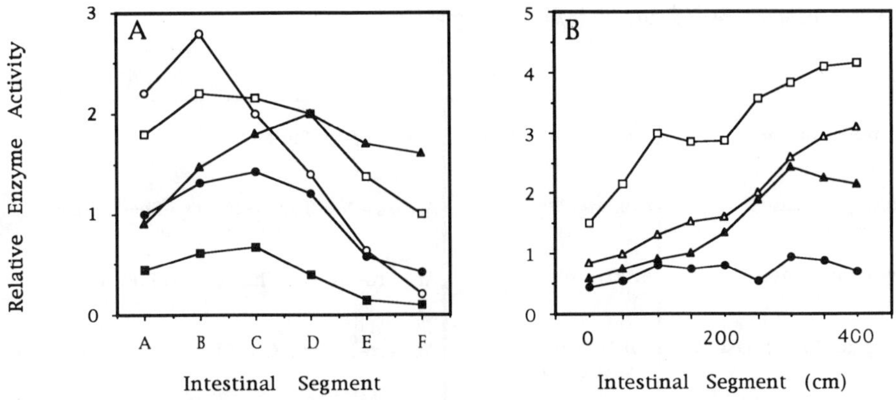

Figure 4. Longitudinal distribution of brush-border-membrane peptidases: (A) rat small intestine and (B) human small intestine. Key: ○—○, angiotensin-converting enzyme, □—□, aminopeptidase; ▲—▲, dipeptidyl peptidase IV; △—△, aminopeptidase A; ●—●, carboxypeptidase P; and ■—■, endopeptidase-24.11. (Data courtesy of references 28 and 82–84.)

higher levels than Endo-1 as judged by its ability to hydrolyze the substrate, azocasein. Levels of Endo-2 are comparable to those reported for meprin in the mouse kidney (*80*).

The expression of brush-border-membrane peptidases varies along the longitudinal (proximal-distal) axis of the small intestine (Figure 4). In the rat, higher levels are usually observed in the proximal-to-middle regions (*82, 83*). ACE displays a sharp proximal-distal gradient of activity (*28*), whereas DPP-IV activity is highest in the middle-to-distal region. Less information is available regarding the distribution of these peptidase in the human intestine. Data from one study (*84*) indicate that higher levels of these enzymes are found in the distal intestine. Similarly, higher levels of AP-W are found in the ileum of pig intestine when compared with the duodenum and jejunum (*22*).

Many of the peptidases are particularly abundant in renal microvillus membranes. This abundance is not immediately obvious from a functional standpoint (*8*). These enzymes are also found in many other organs and cell types (*85*). The microvilli of the placental syncytiotrophoblast (*86, 87*) and choroid plexus (*88, 89*) contains a number of these enzymes. These enzymes are also found in lymph nodes, lungs, liver, salivary glands, pancreas, cartilage, and different areas of the nervous system (*10*). Recent studies have shown that cell surface AP-N (CD13), DPP-IV (CD26), and Endo-1 (CD10) play a role in the activation of T-lymphocytes (*90–92*). As shown by this list, a great variety of cells express various peptidases. The widespread distribution of peptidases suggests that these enzymes participate in a number of physiological functions dependent on the metabolism of bioactive peptides.

Hydrolysis of Bioactive Peptides by Individual Peptidases

Table III shows a list of several bioactive peptides and their hydrolysis cites by individual brush-border-membrane peptidases. This list is meant to serve as a general illustration of how a few selected peptides are hydrolyzed by these enzymes and is not intended to be

Table III. Peptides Hydrolyzed by Individual Peptidases

Enzyme	*Peptide*	*Bond(s) Hydrolyzed*
Aminopeptidase N	Enkephalin (met)	Tyr↓Gly-Gly-Phe-Met
Aminopeptidase A	Angiotensin I	Asp↓Arg-Val-Tyr-Ile-His-Pro-Phe-His-Leu
Aminopeptidase P	Bradykinin	Arg↓Pro-Pro-Gly-Phe-Ser-Pro-Phe-Arg
Dipeptidyl peptidase IV	Substance P (deamidated)	Arg-Pro(1↓)Lys-Pro(2↓)Gln-Gln-Phe-Phe-Gly-Leu-Met
	β-Casomorphin	Tyr-Pro(1↓)Phe-Pro(2↓)Gly-Pro(3↓)Ile
Angiotensin-converting enzyme	Angiotensin I	Asp-Arg-Val-Tyr-Ile-His-Pro-Phe↓His-Leu
	Bradykinin	Arg-Pro-Pro-Gly-Phe(2↓)Ser-Pro(1↓)Phe-Arg
	Enkephalin (met)	Tyr-Gly-Gly↓Phe-Met
	Substance P (deamidated)	Arg-Pro-Lys-Pro-Gln(3↓)Gln-Phe(2↓)Phe-Gly(1↓)Leu-Met
Carboxypeptidase P	Angiotensin II	Asp-Arg-Val-Tyr-Ile-His-Pro↓Phe
Carboxypeptidase M	Bradykinin	Arg-Pro-Pro-Gly-Phe-Ser-Pro-Phe↓Arg
	Enkephalin-Met^{5}-Arg^{6}	Tyr-Gly-Gly-Phe-Met↓Arg
Endopeptidase-24.11	Enkephalin (met)	Tyr-Gly-Gly↓Phe-Met
	Bradykinin	Arg-Pro-Pro-Gly↓Phe-Ser-Pro↓Phe-Arg
	Substance P (deamidated)	Arg-Pro-Lys-Pro-Gln-Gln↓Phe↓Phe-Gly↓Leu-Met
	α-Atrial Natriuretic Factor (human)	Ser-Leu-Arg-Arg↓Ser-Ser-Cys↓Phe-Gly-Gly- ↓Arg↓Met-Asp-Arg↓Ile-Gly-Ala↓Gln-Ser-Gly↓ Leu-Gly-Cys-Asn-Ser↓Phe-Arg-Tyr (Cys–Cys bridge)
Endopeptidase-24.18	Angiotensin I	Asp-Arg-Val-Tyr↓Ile-His-Pro-Phe-His↓Leu
	Bradykinin	Arg-Pro-Pro-Gly↓Phe↓Ser-Pro-Phe-Arg
	Substance P (deamidated)	Arg-Pro-Lys-Pro-Gln-Gln↓Phe↓Phe↓Gly-Leu-Met

SOURCE: Data are from references 8–10 and 93–107.

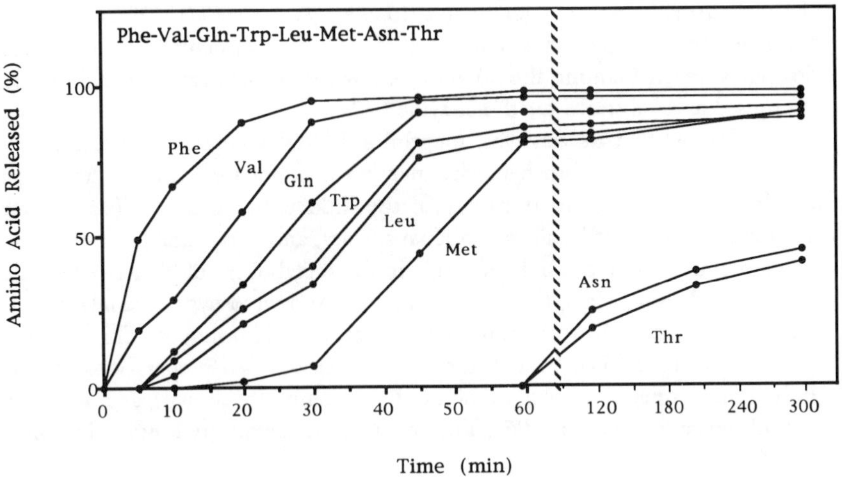

Figure 5. Sequential hydrolysis of an octapeptide by purified intestinal aminopeptidase N. (Reproduced with permission from reference 16. Copyright 1976 Americal Society for Biochemistry.)

comprehensive in its scope. AP-N was shown to be involved in the degradation of the opiate enkephalin (*93*) by removing the N-terminal tyrosine. Although AP-N has a wide substrate specificity, it does not readily hydrolyze the Gly^2–Gly^3 bond of enkephalin, which prevents subsequent hydrolysis by AP-N. With other peptides, hydrolysis may proceed in a stepwise fashion. This process is illustrated in Figure 5, which shows the action of purified AP-N on a model peptide. When the release of amino acids from this octapeptide was followed at various times during incubation with the enzyme, a sequential pattern of release was observed (*16*). Phenylalanine, the N-terminal amino acid, was released first, followed by the sequential release of valine, glutamine, tryptophan, and so on until threonine, the C-terminal amino acid, was released last. The end result was the complete hydrolysis of this particular peptide by AP-N. Similarly, the N-terminal aspartic acid residue of peptides such as angiotensins I and II are subject to hydrolysis by AP-A (*9*), and AP-P hydrolyzes bradykinin, substance P, casomorphin, and melanostatin (*94, 95*). Because of the restricted substrate specificity of these two enzymes, only the N-terminal amino acid is removed. DPP-IV was shown to sequentially remove dipeptides from substance P (*96*). The opiate peptide fragment β-casomorphin, found in the milk protein casein, is an ideal peptide substrate for DPP-IV with alternating proline residues in its sequence.

ACE is best known for its role in blood-pressure regulation in the lung and vascular endothelium by hydrolyzing angiotensin I to the potent vasoconstrictor angiotensin II. This enzyme also hydrolyzes substance P in the central nervous system but not the related tachykinins neurokinin A and B (*97*). It also cleaves the Phe^8–Gly^9 bond of amidated substance P to release the C-terminal tripeptide and then successive dipeptides. Leu^5-enkephalinamide is also hydrolyzed by ACE, a phenomenon that demonstrates the endopeptidase-like activity of the enzyme (*97*). Amidated peptides, however, tend to be less efficiently hydrolyzed than those with a free-carboxy terminus. ACE also hydrolyzes peptides such as the enkephalins, luliberin, bradykinin, endorphin, and neurotensin (*10, 97–100*).

Endo-1 plays an important role in the metabolism of neuropeptides in a number of tissues. Initially identified in the kidney, it was subsequently shown to be identical to the "enkephalinase" responsible for the hydrolysis of these opiate neuropeptides in the brain (*101*). Subsequent studies have shown that it can hydrolyze many different neuropeptides of short-to-medium chain length (*8, 10, 89, 99, 102–104*). Often, hydrolysis occurs at multiple sites and generates a number of peptide fragments. For instance, 10 potential cleavage sites exist among the 30 amino acids that constitute the insulin β-chain, which is often used as a substrate for the enzyme (*42*).

Endo-2 and its relatives, meprin and PABA-peptide hydrolase, are more recent additions to the list of brush-border-membrane peptidases. Endo-2 has a restricted tissue distribution and is found primarily in the kidney and intestine (*46*). Furthermore, the enzyme is not present in all mammalian species, and is noticeably absent from pig kidney. Therefore, its precise physiological role in the metabolism of bioactive peptides remains to be clarified. Nevertheless, this enzyme has been shown to hydrolyze relatively large peptides and proteins, such as azocasein, α-casein, histone, and fibrinogen. This ability is characteristic of a proteinase (*53*). In addition (Table III), purified preparations of Endo-2 can hydrolyze smaller peptides, such as the insulin β-chain (not shown), angiotensin I, bradykinin, and substance P (*45, 105, 106*). The enkephalins are not hydrolyzed by this enzyme (*45*).

Concerted Hydrolysis of Peptides by Brush-Border-Membrane Peptidases

In the preceding sections, the discussion focused on the substrate specificities of individual peptidases and the peptide bonds that are hydrolyzed by purified preparations of these enzymes. This information however does not promote an adequate understanding of the way peptides are degraded when all of the peptidases act together in concert. To illustrate this point, Figures 6 and 7 show what can happen during the concerted hydrolysis of bradykinin and substance P, two peptides used in Table III to show individual enzyme specificities.

In the case of bradykinin, the solid arrows in Figure 6 show the potential initial sites of hydrolysis by some of the individual peptidases listed in Table III. Cleavage by one or more of these peptidases exposes other bonds of the peptide molecule to secondary attack by peptidases at sites shown by the striped arrows. Therefore, peptidases that normally would not be able to hydrolyze intact bradykinin, such as DPP-IV, AP-N, or CP-P, are able to hydrolyze the molecule after the initial hydrolytic cleavage. The end result is the complete hydrolysis of bradykinin, except perhaps for the Pro–Pro dipeptide.

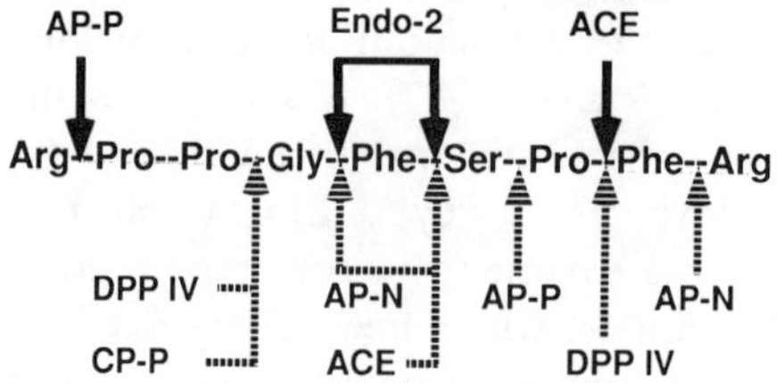

Figure 6. Concerted hydrolysis of bradykinin by brush-border-membrane peptidases: sold arrows, potential initial sites of hydrolysis; broken arrows, secondary sites of hydrolysis. Key: ACE = angiotensin-converting enzyme, AP-N = aminopeptidase N, AP-P = aminopeptidase P, CP-P = carboxypeptidase P, DPP IV = dipeptidyl peptidase IV, Endo-1 = endopeptidase-24.11, and Endo-2 = endopeptidase-24.18.

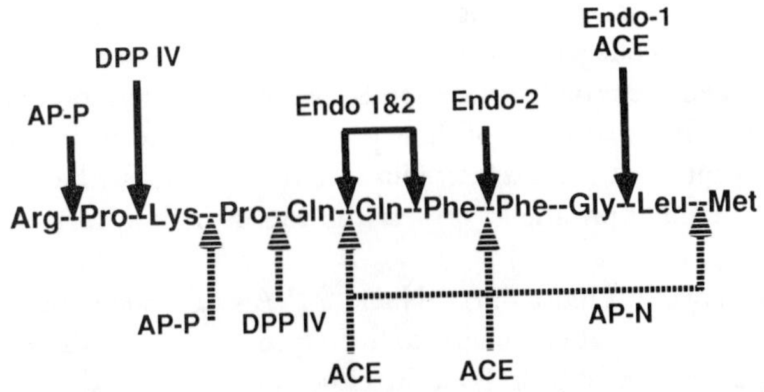

Figure 7. Concerted hydrolysis of substance P by brush-border-membrane peptidases: solid arrows, potential initial sites of hydrolysis; broken arrows, secondary sites of hydrolysis. Key: ACE = angiotensin-converting enzyme, AP-N = aminopeptidase N, AP-P = aminopeptidase P, DPP IV = dipeptidyl peptidase IV, Endo-1 = endopeptidase-24.11, and Endo-2 = endopeptidase-24.18.

A similar situation can be shown for substance P; the initial points of hydrolysis are shown in Figure 7 above the peptide, and the potential secondary points of cleavage are shown below. Again, the end result is the complete hydrolysis of substance P. The enzymes that play the dominant, crucial role in this process are the endopeptidases, because they expose the interior regions of the peptide molecule to subsequent attack by the various exopeptidases (*107*).

Not all peptides are equally susceptible to hydrolysis by the potent mixture of brush-border-membrane peptidases (*107*). The half-lives of several neuropeptides were determined (*107*) during incubation with kidney brush-border membranes and are shown in Table IV. Substance P and bradykinin tend to be rapidly hydrolyzed, whereas oxytocin, vasopressin, and insulin have comparatively long half-lives. The β-chain of insulin is rapidly hydrolyzed, even though insulin itself is somewhat resistant. Stephenson and Kenny (*107*) suggested that the secondary and tertiary structure of peptides and proteins may be important in protecting an otherwise susceptible molecule.

In the small intestine, the process of concerted hydrolysis is particularly important

Table IV. Hydrolysis of Neuropeptides by Kidney Brush-Border Membranes

Peptide	*Half-life (min)*
Substance P	5
Bradykinin	8
Angiotensin III	19
Angiotensin II	21
Insulin β-chain	26
Angiotensin I	34
Oxytocin	510
Vasopressin	24 h
Insulin	NH[a]

[a] NH = no hydrolysis after 24 h.

SOURCE: Reproduced with permission from reference 107. Copyright 1987 Portland Press.

in the case of peptides containing high amounts of the amino acid proline. Important dietary proteins such as collagen, gliadin, and α-casein contain relatively high amounts of this amino acid. Because prolyl peptide bonds are not readily hydrolyzed by gastric and pancreatic proteases, the final hydrolysis of these peptides occurs in the intestine by the brush-border-membrane peptidases (*3*). At least four intestinal peptidases have high hydrolytic rates with peptides containing proline at or near the site of hydrolysis (i.e., AP-P, DPP-IV, ACE, and CP-P).

As shown in Figure 8, AP-N and AP-A can sequentially remove amino acids from the amino terminus but are unable to hydrolyze peptide bonds involving proline. However, these bonds are readily hydrolyzed by either DPP-IV or AP-P (*108, 109*). Similarly, the two enzymes that degrade peptides from the carboxy terminus, ACE and CP-P, have high hydrolytic rates with prolyl peptides (*28, 32*). In this simplified scheme, ACE readily removes the dipeptide shown in Figure 8 with proline at the carboxy terminus (*28*). However, ACE is unable to hydrolyze the next dipeptide in the series with proline in the second position. The Pro–Y bond, on the other hand, is an ideal substrate for CP-P. Removal of the Y amino acid thus exposes another C-terminal proline residue, and the X–proline dipeptide is subsequently released by ACE (*110*). Thus, these four proline-specific enzymes can act in concert to effectively degrade proline containing peptides from both the amino and carboxy terminal ends.

Certain large, intact protein molecules can be readily hydrolyzed by rat and human intestinal brush-border-membrane preparations (*53*). Intact α-casein, histone, and fibrinogen are particularly susceptible to hydrolysis, whereas other proteins such as hemoglobin, bovine serum albumin, and ovalbumin are comparatively resistant to hydrolysis (Table V). Figure 9 shows the electrophoretic profile of α-casein after incubation with purified rat-intestinal brush-border membranes for increasing periods of time. Intact α-casein has a molecular weight of approximately 29 thousand. The intensity of the α-casein component gradually decreases and the intensity of the lower molecular weight species gradually increases. After 4 h of incubation, most of the α-casein hydrolysate consists of small peptides

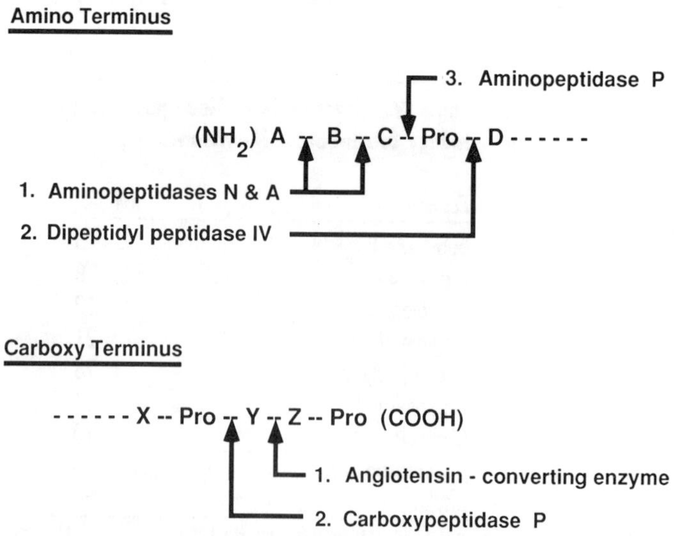

Figure 8. Concerted hydrolysis of prolyl peptides by intestinal brush-border-membrane peptidases.

Table V. Hydrolysis of Protein Substrates by Rat- and Human-Intestinal Brush-Border Membranes

Substrate	*Rat*	*Human*
α-Casein	1740 ± 177	1761 ± 117
Histone	1184 ± 83	1303 ± 128
Fibrinogen	496 ± 63	735 ± 182
Hemoglobin	160 ± 23	255 ± 203
Serum albumin	96 ± 83	0
Ovalbumin	130 ± 15	0

NOTE: Values are specific enzyme activities (units/mg protein).
SOURCE: Reproduced with permission from reference 53. Copyright 1988 American Physiological Society.

and amino acids. Figure 10 is an interpretation of what happens during brush-border-membrane hydrolysis of these large protein molecules. Hydrolysis is initiated by the endopeptidases, and studies (*53*) suggest that in the intestine, Endo-2 or related phosphoramidon-insensitive peptidases play the primary role in this hydrolysis. Hydrolysis of the polypeptide backbone is followed with a concerted attack on smaller oligopeptides by the N- and C-terminal exopeptidases that ultimately results in the complete degradation of the oligopeptides to a mixture of di- and tripeptides and free amino acids. These hydrolytic products are eventually absorbed by the intestinal enterocyte. Thus, the small intestine has the capacity to hydrolyze and assimilate relative large, intact protein molecules in the absence of gastric and pancreatic proteases.

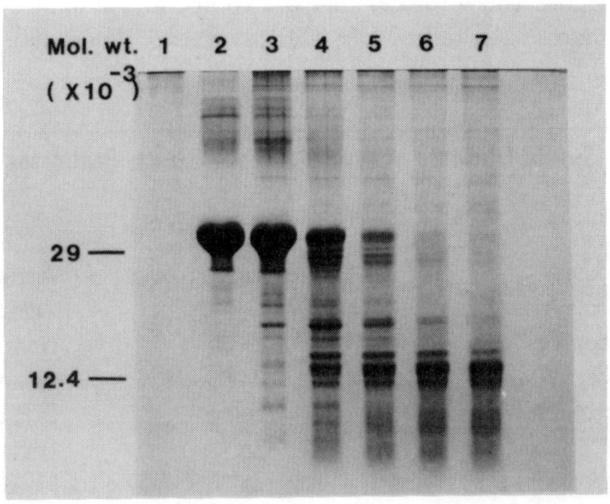

Figure 9. Hydrolysis of α-casein by human-intestinal brush-border-membrane peptidases. α-Casein was incubated with purified human-intestinal brush-border membranes for up to 4 h. After incubation, samples were removed and subjected to sodium dodecyl sulfate–polyacrylamide gel electrophoresis: Lane 1, brush-border membrane control; Lane 2, α-casein control; Lanes 3–7, digests of α-casein (100 ug) for 0, 1, 2, 3, and 4 h, respectively. (Reproduced with permission from reference 53. Copyright 1988 American Physiological Society.)

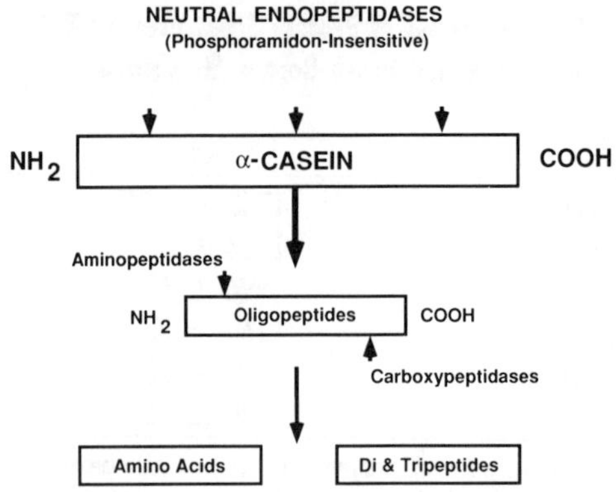

Figure 10. Mechanism of hydrolysis of α-casein by intestinal brush-border-membrane peptidases.

Inhibitors of Brush-Border-Membrane Peptidases

A discussion of the activities and specificities of the intestinal peptidases would not be complete without considering the sensitivity of these enzymes to a range of different types of inhibitors (*7, 111–117*). Although highly concerted efforts have been directed toward designing specific inhibitors for ACE (*111, 113*) and Endo-2 (*111, 116, 117*), the same is not generally true for many of the other peptidases. Consequently, the list of known specific inhibitors for some of these enzymes is limited. The aminopeptidases can be inhibited specifically by a series of small molecular weight compounds that were isolated from microbial sources (*111*), namely actinonin, amastatin, and bestatin (Table VI). DPP-IV is inhibited by the serine protease inhibitor diisopropyl fluorophosphate and diprotin A, a prolyl tripeptide (Ile–Pro–Ile) that is a competitive inhibitor (*112*). ACE is inhibited by the well-known series of highly specific, synthetic, active-site inhibitors shown in Table VI

Table VI. Specific Inhibitors of Brush-Border-Membrane Peptidases

Enzyme	*Active Site*	*Specific Inhibitor*
Aminopeptidase N	Zn^{++}	Amastatin, bestatin, actinonin
Aminopeptidase A	Ca^{++}	Amastatin, bestatin
Aminopeptidase P	Zn^{++}	None (chelating agents)
Aminopeptidase W	Zn^{++}	Amastatin, bestatin
Dipeptidyl peptidase IV	Serine	DFP, diprotin A
Angiotensin-converting enzyme	Zn^{++}	Captopril, lisinopril, enalapril, etc.
Carboxypeptidase P	Zn^{++}	None (chelating agents)
Carboxypeptidase M	Zn^{++}	GEMSA, MGTA
Microsomal dipeptidase	Zn^{++}	Cilastatin
Endopeptidase-24.11	Zn^{++}	Phosphoramidon, CPAB, thiorphan, etc.
Endopeptidase-24.18	Zn^{++}	Actinonin

NOTE: DFP = diisopropyl fluorophosphate, GEMSA = guanidinoethylmercaptosuccinic acid, MGTA = DL-2-mercaptomethyl-3-guanidinoethylthiopropanoic acid, CPAB = *N*-[1(*R,S*)-carboxy-2-phenylethyl]–Phe–*p*-aminobenzoate.
SOURCE: Data adapted from references 7, 106, 111–116.

(*111, 113*). These ACE inhibitors have been used as effective antihypertensive agents for a number of years.

Endo-1 is inhibited specifically by phosphoramidon, a microbial-derived inhibitor, and thiorphan, a synthetically designed mercapto compound (*111, 116*). No specific inhibitors of Endo-2 have been identified, although actinonin was reported to be very effective (*106*). Because many of these peptidases are metalloenzymes, they can be inhibited readily by metal-chelating agents such as ethylenediaminetetraacetic acid (EDTA) and 1,10-phenanthroline. Sulfhydryl agents such as 2-mercaptoethanol and dithiothreitol are also effective, nonspecific inhibitors because they form a coordination complex with the active-site metal ion.

None of the brush-border-membrane peptidases has been crystallized; therefore, detailed X-ray analyses of the active site and enzyme–substrate interactions have not been possible. This deficiency should not be a hindrance to the development of very specific active-site inhibitors for these enzymes, however. Spectacular success was achieved in inhibitor design and development for ACE. These developments for ACE drew heavily on detailed knowledge of the active site of other zinc peptidases such as CP-A and thermolysin. Because many of the peptidases are zinc-containing metalloenzymes, they probably share similar basic structural features. A number of brush-border-membrane peptidases have recently been cloned and sequenced (*51, 58–66*). The additional details regarding the active sites of many of these enzymes should foster growth in the development of a variety of peptidase-directed inhibitors. The therapeutic potential of such inhibitors is enormous. One only has to consider the success achieved by the ACE inhibitors. Additional inhibitors will not only pave the way to a better understanding of the role peptidases plays in various biochemical and physiological processes, they will also lead to the creation of new and interesting drugs.

Summary and Conclusions

The following four points were emphasized in this chapter:

1. Cell-surface peptidases are particularly abundant on the brush-border membrane of kidney and intestinal epithelial cells and are widely distributed among other cells and tissues.
2. This group of enzymes constitutes a wide variety of distinct exo- and endopeptidases that each have a different substrate specificity.
3. These peptidases can act in concert to effectively degrade many types of peptides and proteins.
4. Brush-border-membrane peptidases are potentially a formidable barrier to effective delivery of therapeutic peptides and proteins, particularly through the small intestine.

Currently, an accurate prediction of the peptidases that will be important in the hydrolysis of a particular peptide or protein or the stability of bioactive peptides in vivo is not possible. Fortunately, purified preparations of brush-border membranes are relatively easy to prepare, and they can be a valuable in vitro model system for examining some of these parameters during the early phases of peptide drug development.

Acknowledgments

This work was supported by Grant DK 17938 from the National Institutes of Health and by the Department of Veterans Affairs Medical Research Service. The assistance of Rita Burns in the preparation of this manuscript is gratefully acknowledged.

References

1. Parsons, D. S. In *Brush Border Membranes;* Ciba Foundation Symposium 95; Porter, R. G.; Collins, G. M., Eds.; Pitman: London, 1983; pp 3–11.
2. Miller, D.; Crane, E. K. *Biochim. Biophys, Acta* **1961,** *52,* 293–298.
3. Erickson, R. H.; Kim, Y. S. *Annu. Rev. Med.* **1990,** *41,* 133–139.
4. Van Dyke, R. In *Gastrointestinal Disease, Pathophysiology, Diagnosis, Management,* 4th ed.; Sleisenger, M.; Fordtran, J., Eds.; Saunders: Philadelphia, PA, 1989; pp 1062–1088.
5. Alpers, D. In *Physiology of the Gastrointestinal Tract,* 2nd ed.; Johnson, L., Ed.; Raven: New York, **1987;** *Vol. 2,* pp 1469–1487.
6. Gray, G. M. In *The Exocrine Pancreas: Biology, Pathobiology and Diseases;* Go. V. L. W., Ed.; Raven: New York, 1986; pp 375–386.
7. Kenny, A. J.; Stephenson, S. L.; Turner, A. J. In *Mammalian Ectoenzymes;* Kenny, A. J., and Turner, A. J., Eds.; Elsevier Science Publishers B. V.: Amsterdam, The Netherlands, 1987; pp 169–210.
8. Kenny, A. J.; Stephenson, S. L. *FEBS Lett.* **1988,** *232,* 1–8.
9. Erdös, E. G.; Skidgel, R. A. *Kidney Int.* **1990,** *38,* S24–S27.
10. Turner, A. J.; Hooper, N. M.; Kenny, A. J. In *Mammalian Ectoenzymes;* Kenny, A. J.; Turner, A. J., Eds.; Elsevier Science Publishers B. V.: Amsterdam, The Netherlands, 1987; pp 211–248.
11. Madara, J. L.; Trier, J. S. In *Physiology of the Gastrointestinal Tract;* 2nd ed.; Johnson, L. R., Ed.; Raven: New York, 1987; pp 1209–1249.
12. Adibi, S. A.; Kim, Y. S. In *Physiology of the Gastrointestinal Tract;* Johnson, L. R., Ed.; Raven: New York, 1981; Vol. 2, pp 1073–1095.
13. Sjöström, H.; Noren, O. In *Molecular and Cellular Basis of Digestion;* Desnuelle, P.; Sjöström, H.; Noren, O., Eds.; Elsevier Science Publishers B. V.: Amsterdam, The Netherlands, 1986; pp 367–379.
14. McDonald, J. K.; Barrett, A. J. *Mammalian Proteases. A Glossary and Bibliography; Vol. 2. Exopeptidases;* Academic: London, 1986.
15. Kim, Y. S.; Brophy, E. J. *J. Biol. Chem.* **1976,** *251,* 3199–3205.
16. Kim, Y. S.; Brophy, E. J.; Nicholson, J. A. *J. Biol. Chem.* **1976,** *251,* 3206–3212.
17. Gray, G. M.; Santiago, N. A. *J. Biol. Chem.* **1977,** *252,* 4922–4928.
18. Benajiba, A.; Maroux, S. *Eur. J. Biochem.* **1980,** *107,* 381–388.
19. Dehm, P.; Nordwig, A. *Eur. J. Biochem.* **1970,** *17,* 304–371.
20. Lasch, J.; Koelsch, R.; Ladhoff, A.-M.; Hartrodt, B. *Biomed. Biochim. Acta* **1986,** *45,* 833–843.
21. Hooper, N. M.; Hryszko, J.; Turner, A. J. *Biochem. J.* **1990,** *267,* 509–515.
22. Gee, N. S.; Kenny, A. J. *Biochem. J.* **1985,** *230,* 753–764.
23. Gee, N. S.; Kenny, A. J. *Biochem. J.* **1987,** *246,* 97–102.

24. Macnair, R. D. C.: Kenny, A. J. *Biochem. J.* **1979,** *179,* 379–395.

25. Erickson, R. H.; Bella, A. M.; Brophy, E. J.; Kobata, A.; Kim, Y. S. *Biochim. Biophys. Acta* **1983,** *756,* 258–265.

26. Bella, A. M., Jr.; Erickson, R. H.; Kim, Y. S. *Arch. Biochem. Biophys.* **1982,** *218,* 156–162.

27. Ehlers, M. R.; Riordan, J. F. *Biochemistry* **1989,** *28,* 5311–5318.

28. Yoshioka, M.; Erickson, R. H.; Woodley, J. F.; Gulli, R.; Guan, D.; Kim, Y. S. *Am. J. Physiol.* **1987,** *253,* G781–G786.

29. Erickson, R. H.; Suzuki, Y.; Sedlmayer, A.; Song, I. S.; Kim, Y. S. *Am. J. Physiol.* **1992,** *263,* G466–G473.

30. Cheung, H.-S.; Wang, F.-L.; Ondetti, A.; Sabo, E. F.; Cushman, D. W. *J. Biol. Chem.* **1980,** *255,* 401–407.

31. Hedeager-Sorensen, S.; Kenny, A. J. *Biochem. J.* **1985,** *229,* 251–257.

32. Erickson, R. H.; Song, I. S.; Yoshioka, M.; Gulli, R.; Miura, S.; Kim, Y. S. *Dig. Dis. Sci.* **1989,** *34,* 400–406.

33. Skidgel, R. A.; Johnson, A. R.; Erdös, E. G. *Biochem. Pharmacol.* **1984,** *33,* 3471–3478.

34. Skidgel, R. A.; Davis, R. M.; Tan, F. *J. Biol. Chem.* **1989,** *264,* 2236–2241.

35. Skidgel, R. A.: Anders, R. A.; Deddish, P. A.; Erdös, E. G. *FASEB J. (Abstr.)* **1991,** *5,* A1578.

36. Reisenauer, A. M.; Krumdieck, C. L.; Halsted, C. H. *Science (Washington, DC)* **1977,** *198,* 196–197.

37. Campbell, B. J. *Methods Enzymol.* **1970,** *19,* 722–729.

38. Kozak, E. M.; Tate, S. S. *J. Biol. Chem.* **1982,** *257,* 6322–6327.

39. Campbell, B. J.; Forrester, L. J.; Zahler, W. L.; Burks, M. *J. Biol. Chem.* **1984,** *259,* 14586–14590.

40. Littlewood, G. M.; Hooper, N. M.; Turner, A. J. *Biochem. J.* **1989,** *257,* 361–367.

41. Tobey, N.; Heizer, W.; Yeh, R.; Huang, T.-I.; Hoffner, C. *Gastroenterology* **1985,** *88,* 913–926.

42. Kerr, M. A.; Kenny, A. J. *Biochem. J.* **1974,** *137,* 477–488.

43. Kerr, M. A.; Kenny, A. J. *Biochem. J.* **1974,** *137,* 489–495.

44. Kenny, A. J. In *Proteinases in Mammalian Cells and Tissues;* Barrett, A. J., Ed.; Elsevier–North Holland Biomedical: Amsterdam, The Netherlands, 1977; pp 393–444.

45. Kenny, A. J.; Ingram, J. *Biochem. J.* **1987,** *245,* 515–524.

46. Barnes, K.; Ingram, J.; Kenny, A. J. *Biochem. J.* **1989,** *264,* 335–346.

47. Sterchi, E. E.; Niam, H. Y.; Lentze, M. J.; Hauri, H.-P.; Fransen, J. A. M. *Arch. Biochem. Biophys.* **1988,** *265,* 105–118.

48. Butler, P. E.; McKay, M. J.; Bond, J. S. *Biochem. J.* **1987,** *241,* 229–235.

49. Bond, J. S.; Beynon, R. J. *Curr. Top. Cell. Regul.* **1986,** *28,* 263–290.

50. Dumermuth, E.; Sterchi, E. E.; Jiang, W.; Wolz, R. L.; Bond, J. S.; Flannery, A. V.; Beynon, R. J. *J. Biol. Chem.* **1991,** *266,* 21381–21385.

51. Corbeil, D.; Gaudoux, F.; Wainwright, S.; Ingram, J.; Kenny, A. J.; Bioleau, G.; Crine, P. *FEBS Lett.* **1992,** *309,* 203–208.

52. Titani, K.; Torff, H.-J.; Hormel, S.; Kumar, S.; Walsh, K. A.; Rödl, J.; Neurath, H.; Zwilling, R. *Biochemistry* **1987,** *26,* 222–226.

53. Guan, D.; Yoshioka, M.; Erickson, R.; Heizer, W.; Kim, Y. S. *Am. J. Physiol.* **1988,** *255,* G212–G220.

54. Kenny, A. J.; Turner, A. J. In *Mammalian Ectoenzymes;* Kenny, A. J.; Turner, A. J., Eds.; Elsevier Science Publishers B.V.: Amsterdam, The Netherlands, 1987; pp 1–13.

55. Kenny, A. J.; Maroux, S. *Physiol. Rev.* **1982,** *62,* 91–128.

56. Semenza, G. *Annu. Rev. Cell Biol.* **1986,** *2,* 255–313.

57. Maroux, S. In *Mammalian Ectoenzymes;* Kenny, A. J.; Turner, A. J., Eds.; Elsevier Science Publishers B.V.: Amsterdam, The Netherlands, 1987; pp 15–45.

58. Malfroy, B.; Schofield, P. R.; Kuang, W.-J.; Seeburg, P. H.; Mason, A. J.; Henzel, W. J. *Biochem. Biophys. Res. Commun.* **1987,** *144,* 59–66.

59. Ogata, S.; Misumi, Y.; Ikehara, Y. *J. Biol. Chem.* **1989,** *264,* 3596–3601.

60. Olsen, J.; Cowell, G. M.; Konigshofer, E.; Danielsen, E. M.; Moller, J.; Laustsen, L.; Hansen, O. C.; Welinder, K. G.; Engberg, J.; Hunziker, W.; Spiess, M.; Sjöström, H.; Noren, O. *FEBS Lett.* **1988,** *238,* 307–314.

61. Watt, V. M.; Yip, C. C. *J. Biol. Chem.* **1989,** *264,* 5480–5487.

62. Laperche, Y.; Bulle, F.; Aissani, T.; Chobert, M.-N.; Aggerbeck, M.; Hanoune, J.; Guellaën, G. *Proc. Natl. Acad. Sci. U.S.A.* **1986,** *83,* 937–941.

63. Tan, F.; Chan, S. J.; Steiner, D. F.; Schilling, J. W.; Skidgel, R. A. *J. Biol. Chem.* **1989,** *264,* 13165–13170.

64. Rached, E.; Hooper, N. M.; James, P.; Semenza, G.; Turner, A. J.; Mantei, N. *Biochem. J.* **1990,** *271,* 755–760.

65. Adachi, H.; Tawaragi, Y.; Inuzuka, C.; Kubota, I.; Tsujimoto, M.; Nishihara, T.; Nakazato, H. *J. Biol. Chem.* **1990,** *265,* 3992–3995.

66. Soubrier, F.; Alhenc-Gelas, F.; Hubert, C.; Allegrini, J.; John, M.; Tregear, G.; Corvol, P. *Proc. Natl. Acad. Sci. U.S.A.* **1988,** *85,* 9386–9390.

67. Wei, L.; Alhenc-Gelas, F.; Soubrier, F.; Michaud, A.; Corvol, P.; Clauser, E. *J. Biol. Chem.* **1991,** *266,* 5540–5546.

68. Naim, H. Y. *Biochem. J.* **1992,** *286,* 451–457.

69. Hooper, N. M.; Keen, J.; Pappin, D. J. C.; Turner, A. J. *Biochem. J.* **1987,** *247,* 85–93.

70. Low, M. G.; Saltiel, A. R. *Science (Washington, DC)* **1988,** *239,* 268–275.

71. Skidgel, R. A.; Tan, F.; Deddish, P. A.; Li, X.-Y. *Biomed. Biochim. Acta* **1991,** *50,* 815–820.

72. Kounnas, M. Z.; Wolz, R. L.; Gorbea, C. M.; Bond, J. S. *J. Biol. Chem.* **1991,** *266,* 17350–17357.

73. Jiang, W.; Gorbea, C. M.; Flannery, A. V.; Beynon, R. J.; Grant, G. A.; Bond, J. S. *J. Biol. Chem.* **1992,** *267,* 9185–9193.

74. Danielsen, E. M.; Cowell, G. M.; Noren, O.; Sjöström, H. *Biochem. J.* **1984,** *221,* 1–14.

75. Danielsen, E. M.; Cowell, G. M.; Noren, O.; Sjöoström, H. In *Mammalian Ectoenzymes;* Kenny, A. J.; Turner, A. J., Eds.; Elsevier Science Publishers B.V.: Amsterdam, The Netherlands, 1987; Chapter 3.

76. Matter, K.; Brauchbar, M.; Bucher, K.; Hauri, H.-P. *Cell* **1990,** *60,* 429–437.

77. Matter, K.; Stieger, B.; Klumperman, J.; Ginsel, L.; Hauri, H.-P. *J. Biol. Chem.* **1990,** *265,* 3503–3512.

78. Takasaki, S.; Erickson, R. H.; Kim, Y. S.; Kochibe, N.; Kobata, A. *Biochemistry* **1991,** *30,* 9102–9110.

79. Yamashita, K.; Tachibana, Y.; Matsuda, Y.; Katunama, N.; Kochibe, N.; Kobata, A. *Biochemistry* **1988,** *27,* 5565–5573.

80. Beynon, R. T.; Shannon, J. D.; Bond, J. S. *Biochem. J.* **1981,** *199,* 591–598.

81. Gee, N. S.; Bowes, M. A.; Buck, P.; Kenny, A. J. *Biochem. J.* **1985,** *228,* 119–126.

82. Miura, S.; Song, I.; Morita, A.; Erickson, R. H.; Kim, Y. S. *Biochim. Biophys. Acta.* **1983,** *761,* 66–75.

83. Song, I.-S.; Yoshioka, M.; Erickson, R. H.; Miura, S.; Guan, D.; Kim, Y. S. *Gastroenterology* **1986,** *91,* 1234–1242.

84. Skovbjerg, H. *Clin. Chim. Acta* **1981,** *112,* 205–212.
85. Kenny, A. J. *Trends Biochem. Sci.* **1986,** *11,* 40–42.
86. Johnson, A. R.; Skidgel, R. A.; Gafford, J. T.; Erdös, E. G. *Peptides* **1984,** *5,* 789–796.
87. Püschel, G.; Mentlein, R.; Heymann, E. *Eur. J. Biochem.* **1982,** *126,* 359–365.
88. Bourne, A.; Kenny, A. J. *Biochem. J.* **1990,** *271,* 381–385.
89. Bourne, A.; Barnes, K.; Taylor, B. A.; Turner, A. J.; Kenny, A. J. *Biochem. J.* **1989,** *259,* 69–80.
90. Kenny, A. J.; O'Hare, M. J.; Gusterson, B. A. *Lancet* **1989,** *2,* 785–787.
91. Gorvel, J.-P.; Vivier, I.; Naguet, P.; Brekelmans, P.; Rigal, A.; Pierres, M. *J. Immunol.* **1990,** *144,* 2899–2907.
92. Marguet, D.; Bernard, A.-M.; Vivier, I.; Darmoul, D.; Naquet, P.; Pierres, M. *J. Biol. Chem.* **1992,** *267,* 2200–2208.
93. Gros, C.; Giros, B.; Schwartz, J.-C. *Biochemistry* **1985,** *24,* 2179–2185.
94. Harbeck, H.-T.; Mentlein, R. *Eur. J. Biochem.* **1991,** *198,* 451–458.
95. Simmons, W. H.; Orawski, A. T. *J. Biol. Chem.* **1992,** *267,* 4897–4903.
96. Heymann, E.; Mentlein, R. *FEBS Lett.* **1978,** *91,* 360–364.
97. Hooper, N. M.; Turner, A. J. *Biochem. J.* **1987,** *241,* 625–633.
98. Skidgel, R. A.; Defendini, R.; Erdös, E. G. In *Neuropeptides and Their Peptidases;* Turner, A. J., Ed.; Ellis Horwood: Chichester, England, 1987; pp 165–182.
99. Skidgel, R. A.; Engelbrecht, S.; Johnson, A. R.; Erdös, E. G. *Peptides* **1984,** *5,* 769–776.
100. Drapeau, G.; Chow, A.; Ward, P. E. *Peptides* **1991,** *12,* 631–638.
101. Matsas, R.; Fulcher, I. S.; Kenny, A. J.; Turner, A. J. *Proc. Natl. Acad. Sci. U.S.A.* **1983,** *80,* 3111–3115.
102. Turner, A. J. In *Neuropeptides and Their Peptidases;* Turner, A. J., Ed.; Ellis Horwood: Chichester, England, 1987; pp 183–201.
103. Matsas, R.; Kenny, A. J.; Turner, A. J. *Biochem. J.* **1984,** *223,* 433–440.
104. Katayama, M.; Nadel, J. A.; Bunnett, N. W.; Di Maria, G. V.; Haxhiu, M.; Borson, D. B. *Peptides* **1991,** *12,* 563–567.
105. Stephenson, S. L.; Kenny, A. J. *Biochem. J.* **1988,** *255,* 45–51.
106. Choudry, Y.; Kenny, A. J. *Biochem. J.* **1991,** *280,* 57–60.
107. Stephenson, S. L.; Kenny, A. J. *Biochem. J.* **1987,** *241,* 237–247.
108. Morita, A.; Chung, Y. C.; Freeman, H. J.; Erickson, R. H.; Sleisenger, M. H.; Kim, Y. S. *J. Clin. Invest.* **1983,** *72,* 610–616.
109. Erickson, R. H.; Suzuki, Y.; Sedlmayer, A.; Kim, Y. S. *J. Biol. Chem.* **1992,** *267,* 21623–21629.
110. Yoshioka, M.; Erickson, R. H.; Kim, Y. S. *J. Clin. Invest.* **1988,** *81,* 1090–1095.
111. Thorsett, E. D.; Wyvratt, M. J. In *Neuropeptides and Their Peptidases;* Turner, A. J., Ed.; Ellis Horwood: Chichester, England, 1987; pp 229–292.
112. Rahfeld, J.: Schierhorn, M.; Hartrodt, B.; Neubert, K.; Heins, J. *Biochim Biophys. Acta* **1991,** *1076,* 314–316.
113. Cushman, D. W.; Ondetti, M. A. In *Progress in Medicinal Chemistry;* Ellis, G. P.; West, G. B., Eds.; Elsevier–North Holland Biomedical: Amsterdam, The Netherlands, 1980; pp 41–104.
114. Nagae, A.; Deddish, P. A.; Becker, R. P.; Anderson, C. H.; Abe, M.; Tan, F.; Skidgel, A.; Erdös, E. G. *J. Neurochem.* **1992,** *59,* 2201–2212.
115. Deddish, P. A.; Skidgel, R. A.; Erdös, E. G. *Biochem. J.* **1989,** *261,* 289–291.
116. Chipkin, R. E. *Drugs Future* **1986,** *11,* 593–606.
117. Fournie-Zaluski, M. C.; Soleilhac, J. M.; Turcaud, S.; Lai-Kuen, R.; Crine, P.; Beaumont, A.; Roques, B. P. *Proc. Natl. Acad. Sci. U.S.A.* **1992,** *89,* 6388–6392.

RECEIVED for review July 12, 1993. ACCEPTED revised manuscript October 14, 1993.

CHAPTER 3

Peptide Metabolism by Gastric, Pancreatic, and Lysosomal Proteinases

Ramesh Krishnamoorthy and Ashim K. Mitra

The oral route of administration poses a challenge to peptide-drug absorption due primarily to the enzymatic and cellular transport barriers of the intestinal mucosa. The delivery of macromolecules such as peptide-based drugs across the gastrointestinal tract (GIT) is a difficult task. The drugs must be delivered in a reliable manner to provide therapeutic concentrations in the systemic circulation or at the site of action.

An understanding of the mechanisms of peptide hydrolysis and absorption in the GIT is a prerequisite for the development of a commercially feasible delivery system for peptide drugs. During process development of such a system for the oral delivery of peptide and protein drugs, the following three parameters must be considered:

1. The peptide must be protected from the hydrolytic enzymes in the GIT.
2. The mucosal transport of the peptide must be enhanced simultaneously in many cases.
3. A controlled-release form of the drug in some therapeutic indications may be desirable.

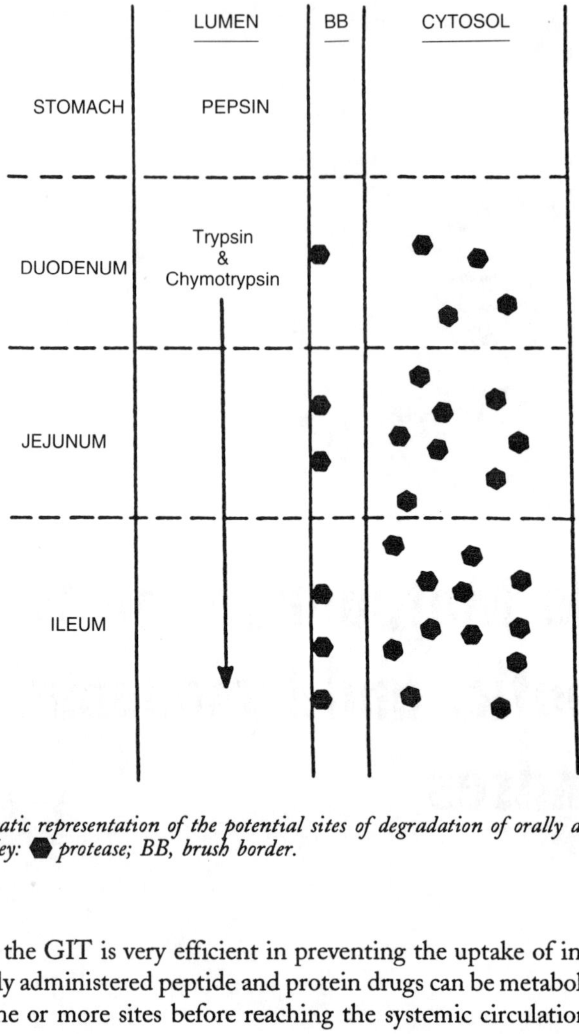

Figure 1. Diagramatic representation of the potential sites of degradation of orally administered protein drugs by proteases. Key: ⬣ *protease; BB, brush border.*

By design, the GIT is very efficient in preventing the uptake of intact polypeptides and proteins. Orally administered peptide and protein drugs can be metabolized by a number of peptidases at one or more sites before reaching the systemic circulation (Figure 1).

- Intraluminal metabolism can occur by pepsin secreted by the gastric cells and by trypsin, chymotrypsin, and elastase secreted by pancreatic cells.
- Brush-border enzymes like carboxypeptidases can also act on oligopeptides having 2–6 amino acid residues.
- The cytosolic fraction of enterocytes can also convert peptides to their constituent amino acids primarily on account of the peptidases present within the fraction.

The peptide and protein molecules that escape intraluminal and brush-border hydrolysis can enter mucosal cells by a carrier-mediated process, and this process can be followed by intracellular hydrolysis to constituent amino acids.

The large size and polar nature of the peptide and protein molecules coupled with the high diffusional resistance of the gastrointestinal mucosa to these molecules make transport difficult. Even if these compounds are taken up intact, these macromolecules will probably undergo storage or metabolism within the enterocytic cell without being transferred to the systemic circulation. However, if this barrier is overcome, the compound can still be metabolized by the liver or excreted into the bile.

Table I. Representative Examples of Enzymes Involved in Presystemic Metabolism of Peptide and Protein Drugs

Protein or Peptide Drug	*Metabolizing Enzyme*
Cyclosporin	Intestinal peptidases
Cholecystokinin	Trypsin-like, kallakrein-like enzymes
Desmopressin[a]	Intestinal peptidases, hepatic enzymes
Insulin	Trypsin, chymotrypsin, cytosolic enzymes
Leucine enkephalin	Aminopeptidases, endopeptidase 24.11
Substance P	Endopeptidase 24.11, angiotensin-converting enzyme, metalloendoproteinase
L-Arg^8 vasopressin	Trypsin, carboxypeptidase
Thyrotropin-releasing hormone	Pyroglutamyl aminopeptidase, prolyl endopeptidase, aminopeptidase
Morphiceptin	Dipeptidylpeptidase

[a] Desmopressin is a vasopressin analogue.

Because one of the physiological functions of the GIT is to digest ingested proteins until absorbable tri- and dipeptides and amino acids remain, the problem of metabolism of peptides and proteins by digestive enzymes is a difficult one to solve. The potential sites of presystemic metabolism are the intestinal lumen, brush border, cytosol of the epithelial cells (enterocytes), and the liver (*1*). Various luminal, brush-border, intestinal-cytosolic, and liver enzymes could each be involved, as shown by the examples in Table I. The liver is the primary site of xenobiotic metabolism, yet the intestinal enterocytes are qualitatively similar in enzymic activity and are capable of performing many of the same enzymatic reactions (*2*). The metabolism by cytosolic enzymes can be studied with intestinal homogenates. These activities vary along the length of the intestinal tract. The relative contributions of the liver and intestinal mucosa to presystemic drug elimination can be evaluated by comparing portal and oral bioavailabilities.

This chapter will address the various enzymatic barriers to peptide- and protein-drug delivery excluding the brush-border region, which is covered in detail in another chapter. We will attempt to focus on the metabolic fate of this class of compounds as they traverse through the GIT after oral delivery. We will also focus on the enzymes they will encounter along that path. Conclusions and suggestions for circumventing this barrier and enhancing oral bioavailability of peptide and protein drugs will also be presented.

Barriers to Oral Delivery of Protein and Peptide Drugs

Enzymatic Barriers to Protein Delivery

A major step toward enhanced oral peptide and protein delivery can be achieved by circumventing the enzymatic barriers that limit the amount of these compounds from reaching the target site. Proteases are essentially hydrolases and rarely show absolute specificity in their action. The kinetics and specificity of proteases in vivo are likely to be different from those in vitro (*3*). The aspartic proteases, i.e., renin, cathepsin D, and pepsin, are all restricted in their action either by pH or intolerance to the medium, which sometimes behaves in a protective manner. Inactivation of the protease may also occur in the neutral–alkaline environment of the duodenum, as seen in the denaturation of pepsin. Also, peptidases in vivo may be kept in check by an excess of proteinase inhibitors such as α_1-antitrypsin, α_2-antiplasmin, and α_2-macroglobulin (*4*). The relative distribution, and hence the proteinase–inhibitor concentration ratio, may vary considerably in body fluids, particularly in the splenic microenvironment where the concentration of macrophages is high. Thus, results from an in vitro model to assess the specificity of proteases need to be analyzed with caution. Such an in vitro method does not allow direct in vivo extrapolation, because the peptides and proteins may be degraded by peptidases acting in concert. One also needs to exercise caution in using appropriate experimental conditions when performing in vitro studies so as not to denature, inactivate, or potentiate the proteases.

Characteristics of the Enzymatic Barrier

The enzymic barrier comprising the exopeptidases that cleave peptides and proteins at the N- and C-termini and the endopeptidases that cleave the macromolecules at an internal peptide bond is well-characterized for some enzymes. Examples of exopeptidases are aminopeptidase (EC 3.4.11), dipeptidase (EC 3.4.13), diaminopeptidase (EC 3.4.14), serine carboxypeptidase (EC 3.4.16), metallocarboxypeptidase (EC 3.4.17), and cysteine carboxypeptidase (EC 3.4.18). Examples of endopeptidases include serine proteinase (EC 3.4.21), cysteine proteinase (EC 3.4.22), aspartic proteinase (EC 3.4.23), and metalloproteinase (EC 3.4.24). To better understand the role of the enzymatic barrier in peptide and protein absorption, the types and distribution of proteases, their substrate specificities, and their probability of coming in contact with a peptide or protein must be known.

Because of the ubiquitous nature of proteases, such degradations may occur at a multiplicity of sites. A typical example is that of the metabolism of insulin by glutathione transhydrogenase in various tissues (Table II) (*5*). Consequently, peptides and proteins need to be protected against degradation at more than one site for them to reach their target site intact. For a given protein or peptide, more than one degradation site may exist along the structure, and each locus may be mediated by a certain peptidase, as seen in the case of

Table II. Glutathione–Insulin Transhydrogenase Activity in Various Tissues of Rat Relative to Liver

Tissue	*Activity per mg Protein*
Liver	100
Pancreas	422.6 ± 38.9
Intestine	77.0 ± 6.6
Kidney	28.2 ± 2.9
Lung	22.1 ± 3.5
Brain	17.1 ± 2.8

SOURCE: Reproduced with permission from reference 5. Copyright 1988 CRC.

Table III. Proteases Involved in Hydrolysis of Substance P

Protease	*Bonds Cleaved*
Endopeptidase 24.11	6-7, 7-8, 9-10
Angiotensin-converting enzyme	8-9, 9-10
Dipeptidyl aminopeptidase IV	2-3, 4-5
Prolyl endopeptidase	4-5
Substance pase	6-7, 7-8, 8-9

SOURCE: Data are from reference 5.

hydrolysis of substance P (Table III) (*6*). A subtle modification of one linkage to circumvent one protease may make the rest of the molecule vulnerable to other proteases. Moreover, all the proteases capable of degrading a peptide or protein are usually present at an anatomical site. Because transport is the rate-limiting barrier to a number of peptide drugs, complete absorption of enzymatically stable peptides and proteins will also be an exception rather than a rule.

Physicochemical Properties of Peptides and Proteins

Peptides and proteins present a unique problem for delivery across the intestinal mucosa. High molecular weights coupled with unique molecular properties (e.g., hydrogen bonding leading to low thermodynamic activity and thus low escaping tendency from aqueous phase) make the task more difficult. A peptide or protein may demonstrate aggregation and exhibit conformational changes (unfolding and denaturation) that in turn may have a direct bearing on the biological response. A typical example is that of immunoglobulin G (IgG), which has two flexible domains (Fab_2 regions) that are believed to be responsible for antigen binding and that lead to immunogenic responses because of conformational changes. Self association involving hydrogen bonds and hydrophobic interactions can also lead to decreased solubility of the peptides and reduced availability. Finally, aggregation and polymerization of the molecule may also lower their potential for oral delivery.

There appears to be a strong interplay between penetration and proteolytic barriers in the absorption of macromolecules across the gut. Macromolecules, such as polyethylene glycols (PEGs), were used to study permeability across the intestine (*7*). Because these PEGs traverse the intestinal mucosa by a mechanism different from that of proteins, they may not be good markers for assessing membrane permeability of proteins.

The transient increase in permeability to macromolecules may be attributed to an interaction of several factors:

- enhanced endocytosis of macromolecules,
- increase membrane fluidity, and
- decreased intestinal proteolysis.

Proteolysis is not a dominant factor limiting macromolecular antibody absorption at a young age (*8*). The more extensive absorption of IgG, despite its molecular size, is attributed to the involvement of its specific receptor in its uptake.

Protein Absorption and Metabolism

Physiological Importance of Peptide Absorption

The intestinal assimilation of dietary protein occurs primarily through breakdown, followed by di-, and tripeptide and amino acid absorption. Several studies (*9–11*) indicated that the products of protein digestion accumulate in the gut lumen. Further clinical studies (*12, 13*) in patients with hereditary deficiencies in amino acid transport also tend to support this hypothesis.

Patients with cystinuria and Hartnup syndrome have a greatly reduced capacity for absorption of basic and neutral amino acids, respectively. However, the dipeptide absorption was normal when studied (*14, 15*) in these patients, a finding that indicates that protein nutrition can be maintained with an adequate number of amino acid dipeptide carrier systems. Therefore, the peptide absorption can become largely responsible for dietary protein assimilation.

Among the parameters that influence the peptide absorption, the maximal number of amino acid residues, the length of the amino acid side chain, and the stereoisomerism of the side chain are among the most important. Di- and triglycines are not hydrolyzed in human jejunal fluid, whereas tetra- to hexaglycines are hydrolyzed; however, the differences in the hydrolytic rates of the last three peptides do not seem to be statistically significant. Similarly, di- and triglycines are stable in the luminal fluid, whereas their leucine counterparts are hydrolyzed in the lumen. Humans and monkeys have one dipeptide uptake system with an extremely broad specificity (*16*). Intestinal mucosal uptake of oligopeptides is limited to di- and tripeptides only. This uptake is mediated by a specific carrier system that shows preference for peptides with bulky side chains and L-stereoisomer amino acid residues at both the N and C terminals.

The peptide that does not possess the molecular features for interaction with the carrier system probably gets absorbed across the intestine by simple diffusion. Those peptides that exceed a molecular weight of 1 kDa cannot use this aqueous pore pathway and are probably absorbed across the intestinal wall by endocytosis. However, the fate of such peptides following receptor-mediated endocytosis is postinternalization degradation by the lysosomes (*17*). This type of degradation will be discussed at length later in the chapter.

Peptide Metabolism by Luminal, Enterocytic, and Lysosomal Proteinases

The metabolism of peptide and protein drugs by the intraluminal enzymes, i.e., gastric, pancreatic, and intestinal proteases, followed by enterocytic and hepatic lysosomal enzymes will be addressed in the following segments.

Gastric Enzymes

After oral administration of peptides and proteins, digestion is initiated in the gastric juice by a family of aspartic proteinases called pepsins, which are most active at pH 2–3 but become inactive at a pH above 5. However, pepsins rarely degrade peptides or proteins to their constituent amino acids.

Hydrolysis of β-casein by gastric proteases was studied (*18*) by comparing the activity of bovine chymosin and pepsin A. The cleavage patterns revealed that only six of the peptide bonds were hydrolyzed by chymosin, and seven others were hydrolyzed by pepsin. The results indicated that the preferential splitting occurred at the Leu–X, Ser–X, and Trp–X bonds by chymosin and the Leu–X, Met–X, and Thr–X bonds by pepsin A.

Pepsin has been known to exist in a variety of molecular forms in vivo that were studied (*19*) in preterm and post-term Nigerian infants. The levels of total enzyme activity and gastric acidity were lower in the preterm infants when compared to term and postterm infants. Differential development of the enzyme and the isoenzyme may have an important bearing on the subsequent pancreatic hydrolysis and may provide a basis for evaluating the gastric capacity for peptide hydrolysis and nutritional management.

Pancreatic Enzymes

The pancreas is a storage site for proteolytic enzymes, notably elastase, aminopeptidase (dipeptidyl aminopeptidase (IV), chymotrypsin, trypsin, and carboxypeptidase A. The first four enzymes are all endopeptidases, whereas the last one is an exopeptidase. The combined activity of pancreatic proteases against dietary proteins is quite remarkable. However, their activity toward small peptides is very much restricted, and much of the luminal fluid activity against small peptides is derived from either the brush border or the cytoplasm of the enterocyte. Functional alterations have also been observed in cases of uremia. In case of acute uremia, the levels of intestinal peptidases were elevated, especially amino- and jejunal peptidases (*20*).

Carboxypeptidase A is a well-characterized exopeptidase. The two requirements of the substrates for this enzyme are as follows:

- a free terminal carboxyl group.
- a C-terminal amino acid bearing a branched aliphatic or aromatic group.

The three endopeptidases work in concert to cleave almost all of the internal peptide linkages in most peptides and proteins. α-Chymotrypsin prefers to cleave peptide bonds near hydrophobic amino acids (e.g., leucine, methionine, tryptophan, and tyrosine). Trypsin, on the other hand, preferentially cleaves peptide bonds near basic amino acids such as arginine and lysine. Elastase complements the other two proteases by cleaving peptide bonds of amino acids bearing smaller, unbranched, nonaromatic side chains. All three proteases have an optimum pH of about 8.

Unlike the porcine and rat pancreatic elastase, human pancreatic elastase can be inhibited by an oxidized alpha 1-proteinase inhibitor. Such an inhibitor was effective in diminishing the elastolytic activity of elastase and preventing the degradation of blood vessels during acute hemorrhagic pancreatitis (*21*). Pancreatic proteases may also play an important role in the control of brush-border disaccharidase activities (*22*). The possible role of pancreatic proteases in the turnover of intestinal brush-border proteins was also suggested (*23*). The surface of intestinal absorptive cells is constantly renewed, and certain surface enzymes are removed in part from the membrane by the action of pancreatic proteases. This action denotes another important physiological role for elastase.

The role of bile in the regulation of intestinal proteolytic activity in rats was investigated (*24*). The diversion of bile and pancreatic juice from the intestine causes an increase in pancreatic enzyme secretion. Bile-duct obstruction results in a threefold increase in pancreatic juice chymotrypsin activities but causes a large decrease in intestinal mucosal trypsin, chymotrypsin, and total proteolytic activities. These results indicate that an interruption in bile flow causes an accelerated rate of degradation of pancreatic proteolytic enzymes. However, bile-duct obstruction also induces pancreatic proteolytic secretion in response to decreased intestinal proteolytic activity (*25*). Chronic diversion of bile pancreatic juice exerts hypergrowth of pancreas and hypersecretion of proteases in the fasting state. However, such diversion also imparts less sensitivity of pancreatic enzyme secretion to dietary feeding (*24*).

The role of cholecystokinin (CCK) was implicated (*26, 27*) in the putative feedback mechanism between intraduodenal pancreatic proteases and pancreatic enzyme secretion. Logsdon et al. (*28*) studied the effects of CCK on gene expression of endocrine pancreatic hormones. CCK at physiological postprandial plasma concentrations stimulates pancreatic protease gene expression but has no effect on gene expression of endocrine pancreatic hormones. The effects of pH and fasting on the pancreatic serine protease levels and on the extent of oral absorption in dogs and humans were examined (*29*). Chymotrypsin levels in fasted dogs are 10 times higher than in humans. Because pharmacokinetic studies and preclinical screening of peptidic candidates in animals and humans are routinely performed during the fasted state, this finding has significance in the correlation of bioavailability results between dogs and humans (*29*).

The pancreatic proteases were extensively investigated because they tend to inactivate one of the most widely studied model peptide drugs, insulin. We will summarize the relevant literature in this area and report some recent findings from our laboratories.

The proper management of Type I diabetes (and some Type II) requires daily subcutaneous insulin injections. Inconvenience and poor patient acceptance have prompted extensive investigation of alternative and nonparenteral insulin delivery pathways, such as rectal, enteral, nasal, and even transdermal routes of administration. Although other delivery modes may improve diabetic therapy, the oral route is still regarded as the most acceptable one. For this reason, intense investigation of oral insulin delivery have been pursued, such as the incorporation of enhancers, use of enteric coating, targeting insulin release to certain regions of the intestine, and the use of protease inhibitors.

To better design an effective insulin oral delivery system, several factors limiting the intestinal insulin uptake need to be fully explored and variables need to be defined. Such variables include the intrinsic permeability difference along the intestinal lumen, the enzymatic barrier difference in the GIT, and the effects of additives on intestinal permeability and insulin-transport characteristics. Early in vitro everted gut sac experiments using different segments of the rat intestine revealed (*30*) that a significant permeability difference exists for insulin. The duodenum showed little or no absorption, whereas the jejunum and the ileum absorbed low but significantly greater amounts of insulin. No significant insulin metabolism was observed during flux measurements, but significant metabolism was observed in tissue homogenates. Insulin uptake across the intestinal mucosa appears to be by passive transport only. This evidence tends to suggest that the oral delivery of insulin may be viable by selective release of insulin in the mid-jejunum to ileum segments, thereby exposing insulin to the optimal region of absorption. In situ closed-intestinal-loop studies (*31*) indicated bioavailability differences for insulin solution delivered to different regions of the rat intestine (*31*). With addition of sodium glycocholate and linoleic acid, absorption in the duodenum and medial and distal jejunum was increased eight-, three-, and twofold, respectively.

The enzymatic inactivation of insulin by luminal and brush-border enzymes is well known. Schilling and Mitra (*32*) conducted in vitro degradation studies of insulin by trypsin and α-chymotrypsin (Figure 2). Data (Table IV) suggest that the apparent K_m values (Michaelis constant) for the two enzymes were nearly the same, whereas the apparent V_{max} (limiting velocity) of chymotrysin was found to be 8.6 times higher than that for trypsin. These results indicate that α-chymotrypsin possesses much stronger proteolytic activity against insulin than trypsin. Initially, α-chymotrypsin attacks insulin at the carboxyl side of $B^{26\text{-Tyr}}$ and $A^{19\text{-Tyr}}$ residues and then at the $B^{16\text{-Tyr}}$, $B^{25\text{-Phe}}$, and $A^{14\text{-Tyr}}$ sites. The

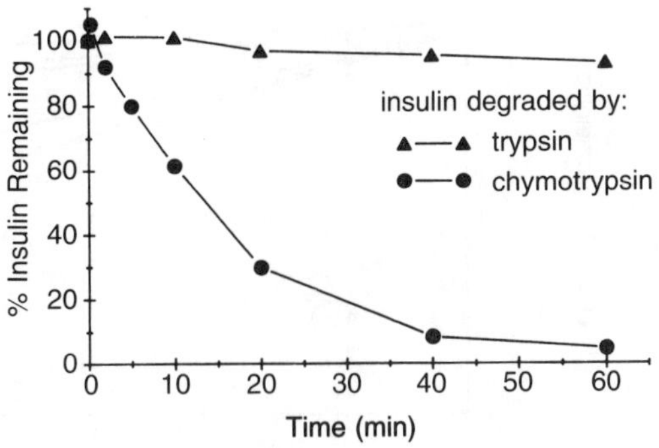

Figure 2. Percentage insulin remaining versus time when incubated with trypsin or chymotrypsin at 37 °C, pH 8; 17.24 μM insulin was incubated with 0.1 μM enzyme. (Reproduced with permission from reference 32. Copyright 1991 Plenum.)

extensive deactivation of insulin by α-chymotrypsin makes any structural modification of insulin for improved stability in the intestinal tract unattractive.

Because insulin molecules aggregate in aqueous solution to form dimers, hexamers, and large aggregates, the influence of insulin dissociation on enzymatic cleavage was studied. Liu et al. (*33*) reported that the rate constant for α-chymotrypsin-mediated insulin degradation was accelerated in the presence of ethylenediaminetetraacetic acid disodium salt (EDTA), a zinc-chelating agent. Complexation of zinc by EDTA essentially dissociates insulin hexamers to dimers in a concentration-dependent manner and thereby exposes more insulin cleaving sites to the enzyme. A study (*34*) using a naturally occurring bile salt, sodium glycocholate, indicates that this bile salt is also capable of completely dissociating insulin oligomers. This adjuvant, however, does not appear to inhibit the activity of α-chymotrypsin to any significant extent (Figure 3). An inhibiting effect of bile salts on nasal aminopeptidases was reported (*35*). Recently, Shao et al. (*36*) extended the initial study to evaluate the differential effects of anionic, cationic, and nonionic surfactants on α-chymotryptic degradation and enteral absorption of insulin oligomers. Results suggest that these surfactants exert their action on insulin dissociation and enzymatic degradation in a very different manner. Anionic surfactants such as sodium lauryl sulfate efficiently dissociate insulin oligomers and denature α-chymotrypsin at concentrations above sodium lauryl sulfate's critical micellar concentration. Cationic surfactants such as cetrimonium bromide (CTAB) interact with negatively charged insulin molecules at lower concentration, whereas then completely solubilize insulin at higher concentrations. CTAB solubilization of insulin

Table IV. Michaelis–Menten Kinetic Parameters for Initial Insulin Degradation

Enzyme	*Apparent K_m (mM)*	*Apparent V_{max} (M/min)*
α-Chymotrypsin	0.10	6.48×10^{-6}
Trypsin	0.099	0.75×10^{-6}

NOTE: k_m is the Michaelis constant. V_{max} is limiting velocity.
SOURCE: Reproduced with permission from reference 32. Copyright 1991 Plenum.

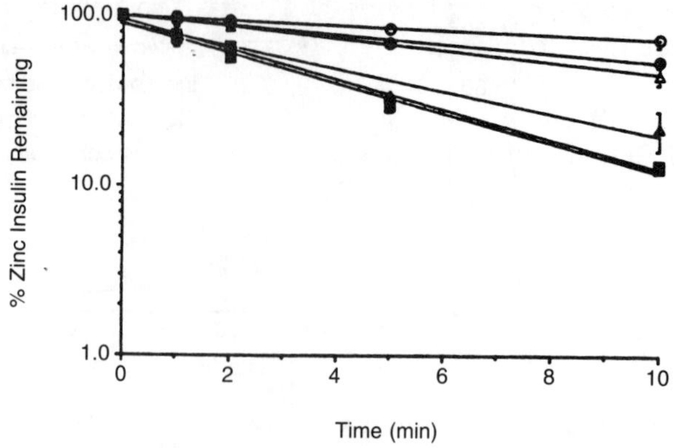

Figure 3. Semilogarithmic plots of zinc insulin biodegradation by α-chymotrypsin as a function of sodium glycocholate (NaGC) concentration. Values are mean ± standard deviation (n = 3). Key: ○, 0 mM NaGC; ●, 5 mM NaGC; △, 10 mM NaGC; ▲, 20 mM NaGC; □, 30 mM NaGC; and ■, 40 mM NaGC. (Reproduced with permission from reference 34. Copyright 1992 Plenum.)

tends to retard α-chymotryptic degradation, probably by repelling the positively charged enzyme molecule away from the substrate. Nonionic surfactants such as Tween 80 failed to exert any effect on insulin degradation (Figure 4). Anionic surfactants may serve as better insulin-absorption enhancers than other types of surfactants due to their ability to dissociate insulin oligomers and produce smaller diffusing species and to retard the presystemic degradation of insulin by luminal proteases.

Intestinal Proteases

Most of the intestinal mucosal proteases are of the brush-border variety and have been discussed in depth in preceding chapters. Regional protease distribution causes wide variability in peptide absorption from the intestinc. In addition to the anatomical differences in

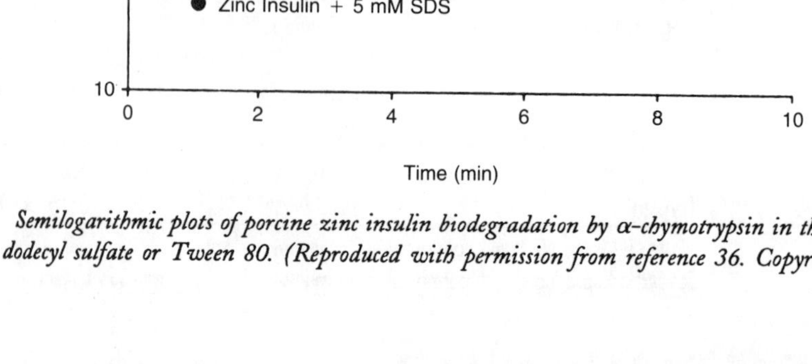

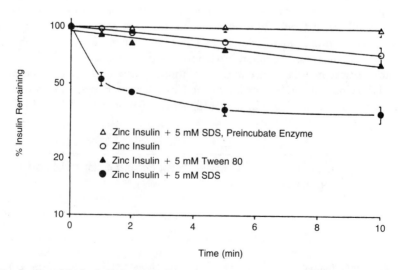

Figure 4. Semilogarithmic plots of porcine zinc insulin biodegradation by α-chymotrypsin in the presence of sodium dodecyl sulfate or Tween 80. (Reproduced with permission from reference 36. Copyright 1993 Plenum.)

the various segments, a pH differential also exists. Also, uremia and similar disorders lead to an increase in activities of intestinal proteases such as aminopeptidases. A significant increase in aminopeptidase N activity and a positive correlation between aminopeptidase N activity and serum urea was observed in uremic rats. However, the disaccharidase activity remains unaltered. These observations are also compatible with different regulation mechanisms for the brush-border peptidases and disaccharidases.

Other Metabolic Factors

Peptide Hydrolases along Small Intestinal Mucosa

The distribution of peptide hydrolase activity between brush-border membrane and cytosol varies considerably with species. Peptide hydrolases show dual location in the cells along the intestinal tract, i.e., brush-border membrane and cytosol; the exact proportion of activity in each location is often difficult to evaluate. The intestinal mucosa is the most likely source of luminal oligopeptide hydrolase activity. The enzymes associated with the cytosol and brush-border fractions of the intestinal mucosa are capable of hydrolyzing oligopeptides (Table V) (*37–41*). The peptide hydrolases bound to the intestinal brush borders are distinct from those in the cytosol (*42*). The jejunal–luminal peptide hydrolase activity originates from the cytosol fraction of the mucosal cells (*43, 44*). A major biological role for intestinal cytosolic peptide hydrolases may be the catabolism of intracellular proteins (*45*). Thus, the quantitative significance of the luminal peptidases in the digestion of exogenous proteins must be assessed by comparing luminal and total cellular enzymic activity.

Distribution of Peptide Hydrolases

Studies (*40*) on the subcellular distribution of peptide hydrolases have shown peptidase activities in both the soluble and brush-border fractions of the small-intestinal mucosa. The soluble fractions constitute about 85%, and the brush-border fractions constitute 10–15%

Table V. Peptide Hydrolases Distribution in Brush-Border Membrane and Cytosol in Homogenates of Human Intestine

Peptide	*Brush Border (%)*	*Cytosol (%)*
Tetrapeptide		
Phe–Gly–Gly–Phe	95	5
Tripeptides		
Tyr–Tyr–Tyr	87	13
Leu–Leu–Leu	81	19
Tyr–Gly–Gly	64	35
Phe–Gly–Gly	50	50
Leu–Gly–Gly	33	67
Dipeptides		
Tyr–Gly	55	45
Phe–Gly	42	58
Tyr–Tyr	36	64
Leu–Leu	25	75
Leu–Gly	14	86

Table VI. Dipeptidase Activity in Homogenates of Biopsied Specimens from Various Regions of Human Gastrointestinal Track

Region	*Infant*	*Adult*
Stomach	13.6 ± 8.3	57.1 ± 36.0
Proximal duodenum	11.5 ± 7.4	39.8 ± 18.7
Jejunum	26.3 ± 5.4	111.4 ± 13.7
Ileum	41.8 ± 9.2	211.6 ± 11.6
Colon	9.5 ± 3.7	44.5 ± 10.6

NOTE: Infant values are for a 1-week-old child. All values are units of activity per milligrams protein.
SOURCE: Data are from reference 5.

of the total activity. The subcellular distribution studies (*44*) of peptidase activities in the normal human jejunum against glycine and leucine homopeptides indicated a variability in the activities of these enzymes along the intestinal mucosa. Leucine and glycine aminopeptidases exhibit variation in the cytosolic distribution and the bioactivity varies markedly with the chain length of the substrate.

Zymogram studies of peptide hydrolases of cytosolic factions have yielded multiple bands indicating multiple zones of enzymic activity. Such zymogram patterns of the brush-border-membrane fractions, however, are quite different from those of the soluble fraction. This finding indicates that the enzymes from the two sources may be different (*40*).

Attempts have been made to determine indirectly the functional localization of peptidase activity in intact intestinal mucosa. The results and the interpretation of those studies have often been conflicting (*45–47*). In vitro studies that compared the rate of uptake of peptides and their constituent amino acids on intestinal segments suggested that hydrolysis precedes transport (*48, 49*). The relevance of these studies to the in vivo situation is not clear because the peptidases are rapidly released from the mucosa into the incubation medium (*50*).

The regional, cellular, and subcellular distribution patterns of aminopeptidases are variable (Table VI). The cytosolic fraction of the intestinal enterocyte contains aminopeptidases that are distinct from the membrane-bound varieties (*51*). Moreover, cytosolic aminopeptidases appear to be closely related to the brush-border-membrane peptidases, a characteristic that tends to suggest that cytosolic aminopeptidases are involved in the assembly of the final surface-membrane aminopeptidase (*52*).

Lysosomal Proteolytic Pathways

Lysosomes are responsible for degrading exogenous and endogenous proteins. Exogenous proteins enter the cells by endocytosis and are degraded within endosomes or rerouted to the lysosome for degradation (*53, 54*). This process is referred to as heterophagy. The role played by lysosomes in the breakdown of intracellular proteins in certain situations (nutritional depravation and pathological states) is well-known. However, aside from cathepsins, little is known about other lysosomal proteases and their mechanisms (*55–57*). Degradation of endogenous proteins by lysosomal pathways is referred to as autophagy, and these proteins enter the lysosomes by four distinct processes: macroautophagy, microautophagy, crinophagy, and transport mediated by the 73-kDa heat shock cognate protein (hsc73).

Macroautophagy involves sequestration of a portion of cytoplasm by a preexisting intracellular membrane to form an autophagic vacuole. Nascent autophagic vacuoles mature in a stepwise manner by acquiring lysosomal membranes and becoming acidified and then undergoing acquisition by lysosomal hydrolases and degradation of the vacuoular contents. Macroautophagy appears to modulate overall intracellular protein levels rather than specifically regulate the levels of certain proteins, as seen in states of amino acid deprivation. Microautophagy is a pathway for basal protein degradation defined biochemically under conditions where macroautophagy is suppressed. Microautophagic structures arise during in vitro incubation of lysosomes isolated from livers of starved rats. Crinophagy is the fusion of secretory granules with lysosomes resulting in digestion of the granule contents. Crinophagy can be induced in the liver by drugs that block secretion and, under normal conditions, occurs mainly in the endocrine glands. Crinophagy appears to be an efficient mechanism for preventing an overabundance of intracellular hormone when extracellular secretion is low.

Proteins sequestered by a nonselective bulk process within the lysosomes turn over with an apparent half-life of about 8 minutes. This rapid lysosomal proteolysis is initiated by endopeptidases, in particular by the cathepsins D and L. Even though information is still inadequate about catheptic enzymes, cathepsins D, E, and L appear to be the major endopeptidases, especially in the reticuloendothelial system. Cathespins that show mainly exopeptidase activity include cathepsins A, B, C, and H, and they may in turn attack the carboxy or amino terminus of proteins. Cathepsin H is most probably the only lysosomal aminopeptidase in many cell types (*58*). Cathepsin B was reported (*59*) to have some endopeptidase activity because it activates trypsinogen to trypsin under acid pH.

The role of lipoproteins and cholesterol levels in circulating blood have been implicated in heart disorders. High levels of high-density lipoproteins (HDL) are associated with decreased risk of heart disease, whereas increased levels of low-density lipoproteins (LDL) aid in atherosclerosis. Native LDLs are degraded by the proteases of the lysosomal extract but are not sensitive to cathepsin D. This degradation reaction was most rapid at pH values of 4 and 4.5 (*60*). Administration of compounds such as thyroxine and chloroquine tends to increase proteolysis. Chloroquine increased the number of autophagic vacuoles in the rat pancreas and led to increased proteolytic lysosomal enzyme activities, especially acid phosphatase and cathepsin B activities (*61*).

The degradation of metallothionein is greater in the rat-liver homogenates than in cytosol and predominates under acidic pH. The role of lysosomal proteases in degradation of metallothionein is now well-established (*62*). This degradation could be inhibited by leupeptin, which is a known blocker of the lysosomal proteases cathepsin B and L. However, another study (*63*) revealed that injections of leupeptin not only increased the activity of cathepsin A and D but also produced moderate increases in the activities of cathepsins B and L. Therefore, the lysosomal proteinase may be involved in the activation of some aldolases. The role of lysosomal proteinases in the modification of other cytosolic enzymes was also revealed (*63, 63a*).

A new cathepsin was isolated from rat liver and was shown to be a lysosomal protease (*64*). This particular cathepsin inactivates glucose-6-phosphate dehydrogenase and differs from cathepsin B because it scarcely hydrolyzes N-substituted derivatives of arginine.

Usually, all cathepsins are stable and are optimally active under acidic pH conditions. Cathepsins have long been thought to be virtually inactive under neutral and basic pH conditions. Nevertheless, cathepsin E has a proteolytic activity and cleavage specificity toward the B-chain of oxidized insulin at physiological pH (*65*). In an attempt to correlate alcohol consumption to the activity of cathepsins and thus to hepatic lysosomal protease activity, metabolism experiments (*66*) were conducted after acute ethanol administration. The study concluded that alterations in hepatic protein catabolism following ethanol administration was not related to the changes in the activities of cathepsins. This finding contradicted previous reports (*66*).

Thus, the initial events in the breakdown of proteins may occur outside the lysosome, but the final stages are inherently intralysosomal. Even though the route taken by a given protein may be different and may vary with hormonal status, environmental conditions, and nutritional status, the underlying pathway for final degradation is purely lysosomal.

Cytosolic Proteolytic Pathways

Proteolytic pathways in cells are highly regulated. In contrast to the lysosomal enzymes, any protease activity free in the normal cytosol must be under rigid control. Because the protein hydrolysis is a thermodynamically favored reaction, the enzymes participating in such cytosolic degradation must be more than simple catalysts for the hydrolytic process. Although the degradation may be modified by changing the susceptibility of individual substrates, the activity of other pathways such as the ubiquitin- or the calcium-dependent pathways can be modulated by the cell.

Ubiquitin, a highly conserved polypeptide of 76 amino acids, serves as a marker for vulnerable proteins. It is found free or covalently conjugated to a variety of protein targets within all eukaryotic cells (*67, 68*). Ubiquitin is covalently attached to certain proteins in a stable form that may alter a protein's conformation or aid in its assembly into macromolecular structures (*68*). Covalent links between substrate proteins and ubiquitin are thought to commit proteins to degradation. Ubiquitin-mediated protein degradation involves a series of intermediary reactions starting with an ATP-requiring ubiquitin-activation step. Ubiquitin moieties are attached to ϵ-amino groups of the lysine residues on the fated protein. Proteins so marked are then digested by specific cytosolic enzymes. Remarkably, ATP is needed for both the activation and the protein-cleavage steps.

Molecular determinants within proteins determine their susceptibility to ubiquitin conjugation (*69*). Hershko and Ciechanover (*70*) proposed that ubiquitin binding acts as a signal for attack by proteinases that are specific for ubiquitin–protein conjugates. Protein and peptide pharmaceuticals can be engineered so that they are not susceptible to degradation during synthesis. In addition, they can be modified to be substrates for a specific proteolytic pathway or not to be a substrate if they naturally are, so as to control their half-lives in the cell. Caution needs to be exercised in the fabrication of peptide- and protein-delivery systems so as not to potentially "mark" the protein pharmaceutical for rapid hydrolytic degradation.

In addition, two calcium-dependent, neutral proteases were isolated from eukaryotic cells (*71, 72*). Calpain I requires micromolar concentration of calcium, whereas calpain II requires millimolar concentrations of calcium. Calpains degrade mainly cytoskeletal or membrane proteins in vitro (*73*). The degree to which calpains contribute to overall proteolysis is not yet clear.

First-Pass Metabolism

The liver is a potential site for removal of macromolecules such as peptides and proteins following oral delivery. Because the liver is well-perfused and composed of several cell types, including hepatocytes, Kupffer cells, and endothelial cells, it is an important organ for protein metabolism. These cell types have receptors for different proteins that enable them to recognize and internalize proteins and peptide drugs.

The transendothelial passage of proteins and peptide drugs depends upon the physicochemical properties of the drug and also the capillaries. Uptake of peptides and proteins from plasma by hepatocytes occurs by two distinct, yet not entirely separable processes: receptor-mediated endocytosis and nonselective pinocytosis.

In receptor-mediated endocytosis, plasma-derived proteins become internalized postbinding by hepatocyte receptor proteins located within the plasma membrane. Receptor-mediated endocytosis is operative in hepatocytes for several proteins, including insulin, glucagon, growth hormone, intestinal and pancreatic peptides, and metallo- and hemoproteins (*74*).

Receptor-mediated endocytosis starts with plasma-derived proteins becoming internalized following specific recognition and binding by hepatocyte-receptor proteins. These receptors are typically integral membrane glycoproteins (*75*). The receptors can be of "functional" and "clearance" types, and therefore, clearance of a protein can be accentuated if a protein binds to both. The liver contains both of these receptors, so turnover of proteins is fast (*76*). In the indirect shuttle pathway of polypeptide internalization by hepatocytes, the ligand–receptor complexes proceed through coated vesicles to the endocytic compartment (*77*). This process regulates the receptor–ligand binding affinity and also enables the receptor protein to recycle efficiently. The receptor-mediated endocytosis process has a high capacity because of receptor recycling and a high affinity due to the highly specific recognition.

Proteins and peptide drugs may gain access to the cytoplasm of hepatocytes by another process known as nonselective pinocytosis (*78*). The amount of proteins internalized by this process is only a small fraction of the total proteins. Albumin, some pancreatic proteins, and glycoproteins are examples of proteins removed from plasma by hepatocytes through a nonreceptor-mediated process.

Therefore, the peripheral sites of degradation and the cellular mechanisms involved in regulating catabolism or processing of an administered protein must be determined, because they may contribute to the design of molecules possessing desirable kinetic or pharmacological properties.

Metabolism of Selected Peptide Drugs

Insulin

Insulin, one of the most widely studied peptide drugs, is variably absorbed in diabetics. Figure 5 illustrates the metabolism of insulin by the different segments along the intestinal tract. Degradation of insulin occurs following uptake into cells by receptor-mediated endocytosis. Two principal enzymes have been implicated in the degradation of insulin, namely, glutathione insulin transhydrogenase and insulin-degrading enzyme. Although the relative roles of these two enzymes are still controversial, glutathione insulin transhydrogenase is thought to cleave insulin at the disulfide bridges (*79*), whereas insulin-degrading enzyme cleaves the Tyr^{16}–Leu^{17} bond in the B-chain (*80*). Another widely held belief is that insulin-degrading enzyme initiates degradation that results in three peptide chains held together by disulfide bonds. This molecule is further degraded by nonspecific proteases (*81*). Of the many reports on the binding and subsequent degradation of insulin by skeletal-muscle preparation and liver homogenates, all have identified the involvement of the two previously mentioned enzyme systems.

Substance P

This peptide comprising 11 amino acids is found mainly in the central nervous system (CNS) as well as in the GIT. The action of this peptide is terminated by proteases at the end of a synapse, though the enzymes have not been identified. Membrane-bound proteases,

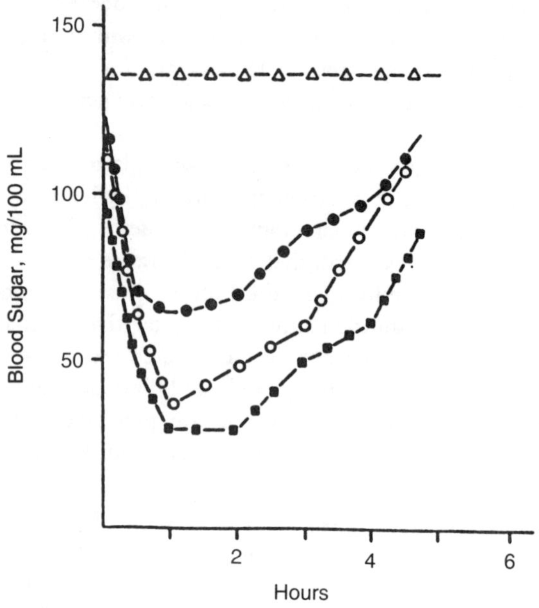

Figure 5. Time course of blood sugar concentration in rabbit following administration of 3 units of insulin to the stomach (△), jejunum (●), and ileum (○) of the GIT compared with an intramuscularly administered dose (■). (Reproduced with permission from reference 5. Copyright 1988 CRC.)

mainly metalloendopeptidases, are thought to be involved in the degradation (*82*). The role of angiotensin-converting enzyme has been implicated only in the CNS (*83*). This particular enzyme also cleaves the Phe^{8}–Gly^{9} bond of amidated substance P, an action that results in the release of the C-terminal tripeptide followed by removal of the successive dipeptides.

Thyrotropin-Releasing Hormone

This hypothalamic regulatory hormone stimulates release of thyrotropin, prolactin, and growth hormone from the pituitary. It has a pyroglutamyl residue at its amino terminus and an amidated carboxyl terminus, and thus, it is stable toward classical exopeptidases. However, the hormone is potentially susceptible to attack by aminopeptidases, such as pyroglutamyl aminopeptidase (*84–88*). This enzyme was found in the rat colon and other intestinal tissues please provide reference. Also, further degradation of the peptide occurs by deamidation of the terminal amide, and this reaction is assisted by prolyl endopeptidase, an enzyme found extensively in the liver, pancreas, ileum, lung, and skeletal muscle of rats. Studies (*89*) with some analogues have shown that they are resistant to degradation by gastrointestinal and liver enzymes.

Vasopressin

Arginine vasopressin is a neurohypophysial nonapeptide hormone with antidiuretic effect on the kidney (*90*). Vasopressin is stable in normal plasma in vitro, although it was shown to be degraded extensively in the proximal renal tubule (*91*). Rapid endocytotic uptake of the intact peptide followed by lysosomal degradation in the proximal tubule occurs in vivo. Several analogues with significant antidiuretic activity were synthesized and tested (*92–94*).

The enhanced activity of these analogues on oral administration was attributed to their resistance to proteolytic degradation before and after absorption and to improved membrane permeation.

Approaches to Circumventing the Metabolic Barrier

In the gastrointestinal tract, the enzymatic barrier efficiently digests proteins to a mixture of amino acids and small quantities of peptides consisting of 2–6 amino acid residues. The poor oral availability of peptides and proteins has prompted the examination of various other noninvasive pathways, including the nasal (*95, 96*), pulmonary (*97*), rectal (*98, 99*), ocular (*100, 101*), buccal (*102*), and vaginal routes (*103*). However, even with these routes, the enzymatic barrier is substantial.

Approaches to circumventing the protease action should be based entirely on the principle site of degradation of the peptide drug: intracellular, luminal, or brush-border. The approaches may include the following:

1. chemical modification of the peptide or protein structure,
2. coadministration of protease inhibitors,
3. formulation approaches to minimize the contact of the peptide or protein with the proteases.

Chemical Modifications of Peptide and Protein Drugs

Proteins are quite labile because of the susceptibility of the peptide backbone to proteolytic degradation; their large molecular size; and their complex secondary, tertiary, and sometimes even quaternary structures. Prodrugs of peptides have been prepared and tested in vitro (*104, 105*). Approaches to reducing proteolysis have included chemical modification of the peptide backbone, and N- and C-terminal modification or blocking (*106*). Bioreversible *N*-α-hydroxyalkylation of the peptide bond to effect protection against carboxypeptidases or other proteolytic enzymes has been one of these approaches. Additional structural modifications include substitution of nondisulfide-bonded cysteine residues with nonsulfur-containing amino acids (*107*) and chemical replacement of labile methionine residues with nonoxidizable analogues such as norleucine.

A classical approach to circumventing enzymatic degradation and increasing plasma half-life of peptides has been the covalent attachment of polyethylene glycol (PEG) to amino groups (*108–110*). PEG modifications of other nonenzyme proteins were also reported (*111, 112*). Some of the other modifications used to reduce peptide degradation include the following (*110*):

- olefin substitution
- carbonyl reduction
- D-amino acid substitution
- N- to C-terminal cyclization
- dehydro amino acid substitution

- retro-inversion modification
- thiomethylene modification.

The modification of the amino acid composition to afford stability against proteases led to enhanced peptide absorption, as shown in the case of Tyr–D-Ala–Gly–L-Phe–D-Leu, a leucine enkephalin analogue and metenkephamid (*113*). Certain nonapeptides (e.g., 1-deamino-8-D-arginine vasopressin and 1-deamino-2-tyrosin(*O*-ethyl)oxytocin) were shown to be orally effective in conscious dogs and human volunteers (*114*). This effectiveness was probably the result of higher permeability and thus reduced contact time with proteases (*114*).

Coadministration of Protease Inhibitors

Various protease inhibitors have been examined for their ability to suppress proteolytic activity. This approach has met with mixed results. Positive results were observed in the oral absorption of tetragastrin (*115*), insulin (*116*), arginine vasopressin (*117*), and a nonapeptide renin inhibitor (*118*). The inhibitors used included bacitracin (a nonspecific protease inhibitor), phosphoramidon (a metalloprotease inhibitor), *p*-hydroxymercuribenzoate (a cysteine protease inhibitor), aprotinin and diisopropyl fluorophosphate (serine proteinase inhibitors), and bestatin, puromycin, and α-aminoboronic acid derivatives (aminopeptidase inhibitors). Sodium glycocholate, a penetration enhancer, was shown to inhibit leucine aminopeptidase activity and protect insulin from proteolysis (*35*).

Also, ligating the pancreatic duct to exclude the pancreatic juices from the small intestine can cause a 12-fold increase in absorption of pentagastrin from the duodenum of rats (*115*).

Formulation Approach

A third strategy to circumvent the enzymatic barrier is the formulation approach, and studies (*119*) were performed using insulin as a model peptide. The delivery systems were designed to protect insulin from coming in contact with proteases, primarily in the lumen, and to release the protein only upon reaching a favorable area for absorption.

Several formulations tested include emulsions (*116, 117*), liposomes (*35, 118*), nanoparticles (*120*), and soft gelatin-coated capsules (*121*). An azo polymer capable of remaining stable along the gastrointestinal tract but that decomposes at the ileocecal junction to release insulin was attempted as a delivery system for insulin (*122*).

These and other approaches have been detailed elsewhere in this book.

Summary and Conclusions

Systemic availability of many peptide and proteins after oral administration has been extremely low (0.1–2%), a problem that renders this mode of drug administration unsuitable for such macromolecules. As described here, several luminal, pancreatic, cytosolic, and lysosomal proteases can severely limit the systemic absorption of these compounds. This enzymatic barrier sets an upper limit to the percent dose of a peptide or protein drug that can be orally absorbed. The design of better peptide drugs and protease inhibitors will become possible once we have a better understanding of the type, properties, and distribution of proteases at a given mucosal site. A knowledge of the rate and extent of various metabolite formations and an examination of the sequential steps in a metabolic scheme might aid in the design of peptidase-resistant prodrugs, analogues, and peptidomimetics. Formulation approaches can also be used to protect the compound from luminal proteases by coating, site-directed release, or inclusion of specific protease inhibitors. The synthetic and site-directed mutagenesis approach to peptide structure alteration coupled with the rapid development of the newer generation of nontoxic permeation enhancers and site-specific bioadhesive polymers may make the oral delivery of peptide and protein drugs commercially feasible in the near future.

References

1. Davis, S. S. In *Delivery Systems for Peptide Drugs;* Plenum; New York, 1986; p 1.
2. Chadler, M. L.; Varandani, P. T. *Biochim. Biophys. Acta* **1972,** *286,* 136–140.
3. Hanley, M. R. In *Degradation of Endogenous Opiods: Its Relevance in Human Pathology and Therapy;* Raven: New York, 1983; p. 129.
4. Travis, J.; Savesen, G. S. *Annu. Rev. Biochem.* **1983,** *52,* 655–709.
5. Lee, V. H. L. *CRC Crit Rev. Ther. Drug Carrier Syst.* **1988,** *5,* 69–97.
6. Ilett, K. F.; Tee, L. B. G.; Reeves, P. T.; Minchin, R. F. *Pharmacol. Ther.* **1990,** *46,* 67–93.
7. Donovan, M. D.; Flynn, G. L.; Amidon, G. L. *Pharm. Res.* **1990,** *7,* 863–868.
8. Telemo, E.; Weström, B. R.; Ekström, G.; Karlsson, B. W. *Biol. Neonate.* **1987,** *52,* 141–148.
9. Hare, J. F. *Biochim. Biophys. Acta* **1990,** *1031,* 71–90.
10. Bond, J. S.; Beynon, R. J. In *Proteolysis and Physiological Regulation;* Baum, H., Ed.; Pergamon; Oxford, England, 1987; pp 173–285.
11. Adibi, S. A. *J. Clin. Invest.* **1971,** *50,* 2266–2275.
12. Adibi, S. A.; Mercer, D. W. *J. Clin. Invest.* **1973,** *52,* 1586–1594.
13. Chung, Y. C.; Kim, Y. S.; Shadchehr, A.; Garrido, A.; MacGregor, I. L.; Sleisenger, M. H. *Gastroenterology* **1979,** *76,* 1415–1421.
14. Asatoor, A. M.; Crouchman, M. R.; Harrison, A. R.; Light, F. W.; Loughridge, L. W.; Milne, M. D.; Richards, A. *J. Clin. Sci.* **1971,** *41,* 23–33.
15. Asatoor, A. M.; Cheng, B.; Edwards, K. D. G.; Lant, A. F.; Matthews, D. M.; Milne, M. D.; Navab, F.; Richards, A. J. *Gut.* **1970,** *11,* 380–387.
16. Silk, D. B. A.; Perret, D.; Clark, M. L. *Gastroenterology* **1975,** *68,* 1426–1432.
17. Blay, J.; Brown, K. D. *Biochem. J.* **1985,** *225,* 85–94.
18. Guillou, H.; Miranda, G.; Pelissier, J. P. *Int. J. Pept. Protein. Res.* **1991,** *37,* 494–501.
19. Adamson, I.; Esangbedo, A.; Okolo, A. A.; Omene, J. A. *Biol. Neonate.* **1988,** *53,* 267–73.
20. Magnusson, M.; Sjostrom, H.; Noren, O.; Asp, N. G.; Denneberg, T. *Nephron* **1991,** *58,* 456–460.
21. Donohue, T. M., Jr.; Drey, M. L.; Zetterman, R. K. *Alcohol Alcohol.* **1988,** *23,* 265–270.
22. Padres, M.; Rabaud, M.; Bieth, J. G. *Biochim. Biophys. Acta* **1992,** *1118,* 174–178.
23. McNamara, D.; Teitelbaum, J.; Potier, M. *Biomedicine* **1981,** *35,* 122–124.
24. Green, G.; Nasset, E. S. *Am. J. Dig. Dis.* **1977,** *22,* 437–444.
25. Alpers, D. H.; Tedesco, F. J. *Biochim. Biophys. Acta* **1975,** *401,* 28–40.

26. Hara, H.; Kiriyama, S. *Proc. Soc. Exp. Biol. Med.* **1991,** *198,* 732–736.
27. Guicherit, O. R.; Gooszen, H. G.; Jansen, J. B.; van-der-Burg, M. P.; Lamers, C. B. *Digestion* **1990,** *47,* 226–231.
28. Rosewicz, S.; Riecken, E. O.; Logsdon, C. D. *Digestion* **1990,** *46,* 390–395.
29. Sinko, P. J. *Pharm. Res.* **1992,** *9,* 320–325.
30. Schilling, R. J.; Mitra, A. K. *Int. J. Pharm.* **1990,** *62,* 53–64.
31. Schilling, R. J.; Mitra, A. K. *Pharm. Res.* **1992,** *9,* 1003–1009.
32. Schilling, R. J.; Mitra, A. K. *Pharm. Res.* **1991,** *8,* 721–727.
33. Liu, F.-Y.; Kildsig, D. O.; Mitra, A. K. *Pharm. Res.* **1991,** *8,* 925–929.
34. Li, Y.; Shao, Z.; Mitra, A. K. *Pharm. Res.* **1992,** *9,* 864–869.
35. Hirai, S.; Yashiki, T.; Mima, H. *Int. J. Pharm.* **1981,** *9,* 173–184.
36. Shao, Z.; Krishnamoorthy, R.; Chermak, T.; Mitra, A. K. *Pharm. Res.* **1993,** *10,* 243–251.
37. Leonard, J. V.; Marrs, T. C.; Addison, J. M.; Burston, D.; Clegg, K. M.; Lloyd, J. K.; Matthews, D. M.; Seakins, J. W. *Pediatr. Res.* **1976,** *10,* 246–249.
38. Fujita, M.; Parsons, D. S.; Wojnarowska, F. *J. Physiol. (London)* **1972,** *227,* 377–394.
39. Peters, T. J. *Clin. Sci. Mol. Med.* **1973,** *45,* 803–816.
40. Silk, D. B. A.; Perret, D.; Clark M. L. *Clin Sci. Mol. Med.* **1975,** *49,* 523–526.
41. Kim, Y. S.; Birtwhistle, W.; Kim, Y. W. *J. Clin. Invest.* **1972,** *51,* 1419–1430.
42. Kim, Y. S.; Sleisenger, M. H.; Kim, Y. W. *Biochim. Biophys. Acta* **1974,** *380,* 283–296.
43. Heizer, W. D.; Kerley, R. L.; Isselbacher, K. J. *Biochim. Biophys, Acta* **1972,** *264,* 450–461.
44. Parson, D. S. In *Transport across the Intestine;* Churchill: London, 1972; pp 253–278.
45. Nicholson, J. A.; Peters, T. J. *Clin. Sci. Mol. Med.* **1978,** *54,* 205–207.
46. Newy, H.; Smyth, D. H. *J. Physiol (London)* **1960,** *152,* 367–371.
47. Peters, T. J.; Modha, K.; MacMahon, M. T. *Gut* **1969,** *10,* 1055–1058.
48. Matthews, D. M.; Lis, M. T.; Cheng, B.; Crampton, R. F. *Clin. Sci.* **1969,** *37,* 751–754.
49. Fern, E. B.; Hider, R. C.; London, D. R. *Biochem. J.* **1969,** *111,* 30–33.
50. Kushak, R. I.; Ugolev, A. M. *Dokl. Biol. Sci.* **1966,** *168,* 411–413.
51. Josefsson, L.; Sjöstrom, H. *Acta Physiol. Scand.* **1966,** *67,* 27–31.
52. Miura, S.; Song, I-S.; Morita, A.; Erickson, R.; Kim, Y. S. *Biochim. Biophys. Acta* **1983,** *761,* 66–75.
53. Maze, M.; Gray, G. M. *Biochemistry* **1980,** *19,* 2351–2358.
54. Das, M.; Radhakrishnan, A. N. *Biochem. J.* **1975,** *146,* 133–139.
55. Diment, S.; Leech, M. S.; Stahl, P. D. *J. Biol. Chem.* **1988,** *263,* 6901–6907.
56. Glaumann, H.; Ballard, F. J. In *Lysosomes: Their Role in Protein Breakdown;* Academic: Orlando, FL, 1986.
57. Mayer, R. J.; Doherty, F. *FEBS Lett.* **1986,** *198,* 181–193.
58. Mortimore, G. E.; Poso, A. R. *Fed. Proc.* **1984,** *34,* 1289–1294.
59. Bohley, P.; Seglen, P. O. *Experentia* **1992,** *48,* 151–157.
60. Greenbaum, L. M.; Hirshkowitz, A.; Shoichet, I. *J. Biol. Chem.* **1959,** *234,* 2885–2887.
61. Skrzydlewski, Z.; Worowski, K. *Acta. Biol. Acad. Sci. Hung.* **1978,** *29,* 19–22.
62. Yucel, L. T.; Jansson, H.; Glaumann, H. *Virchows Arch. B* **1991,** *61,* 141–145.
63. Choudhuri, S.; McKim, J. M.; Klassen, C. D. *Toxicol. Appl. Pharmacol.* **1992,** *115,* 64–71. (a) Kominami, E.; Hashida, S.; Katunuma, N. *Biochim. Biophys. Acta* **1981,** *659,* 378–389.

64. Katunuma, N.; Kominami, E.; Hashida, S.; Wakamatsu, N. *Adv. Enzyme Regul.* **1982,** *20,* 337–350.

65. Towatari, T.; Tanaka, K.; Yoshikawa, D.; Katunuma, N. *J. Biochem. Tokyo* **1978,** *84,* 659–672.

66. Athauda, S. B.; Takahashi, T.; Inoue, H.; Ichinose, M.; Takahashi, K. *FEBS Lett.* **1991,** *292,* 53–56.

67. Hershko, A. *J. Biol. Chem.* **1988,** *263,* 15237–15240.

68. *Ubiquitin,* Rechsteiner, M., Ed.; Plenum: New York, 1988.

69. Ciechanover, A.; Schwartz, A. L. *Trends Biochem. Sci.* **1989,** *14,* 483–488.

70. Hershko, A.; Ciechanover, A. *Annu. Rev. Biochem.* **1982,** *51,* 335–364.

71. Murachi, T.; Takano, E.; Maki, M.; Adachi, Y.; Hatanaka, M. *Biochem. Soc. Symp.* **1989,** *55,* 13–28.

72. Mellgren, R. L.; Renno, W. M.; Lane, R. D. *Cell Biol. Rev.* **1989,** *20,* 139–159.

73. Wang, K. K. W.; Villalobo, A.; Roufogalis, B. D. *Biochemistry* **1989,** *262,* 693–706.

74. Jones, A. L.; Renston, R. H.; Burwen, S. T. In *Progress in Liver Diseases;* Grune and Stratton: New York, 1982; Vol. 7, pp 51–69.

75. Jones, A. L.; Vierling, J. M.; Steer, C. J.; Recihen, J. In *Progress in Liver Diseases;* Popper, H.; Schaffner, F., Eds.; Grune and Stratton: New York, 1979; Vol. 6, pp 43–80.

76. Sugiyama, Y.; Hanano, M. *Pharm. Res.* **1989,** *6,* 192–202.

77. Stahl, P.; Schwartz, A. L. *J. Clin. Invest.* **1986,** *77,* 657–662.

78. Kaplan, J. *Science (Washington, DC)* **1981,** *212,* 14–20.

79. Varandani, P. T. *Biochim. Biophys. Acta* **1972,** *286,* 126–135.

80. Duckworth, W. C.; Stentz, F. B.; Heinemann, M.; Kitabachi, A. E. *Proc. Natl. Acad. Sci. U.S.A.* **1979,** *76,* 635–639.

81. Varandani, P. T.; Shroyer, L. A.; Naf, M. A. *Proc. Natl. Acad. Sci. U.S.A.* **1972,** *69,* 1681–1684.

82. Endo, S.; Yokosawa, H.; Ishii, S. *Biochim. Biophys. Res. Commun.* **1985,** *129,* 684–700.

83. Hooper, N. M.; Turner, A. J. *Biochem. J.* **1987,** *241,* 625–633.

84. Griffiths, E. C.; Kelly, J. A. *Mol. Cell. Endocrinol.* **1979,** *14,* 3–17.

85. Hersh, L. B.; McKelvy, J. F. *Brain Res.* **1979,** *168,* 553–564.

86. Prasad, C.; Peterkosfky, A. *J. Biol. Chem.* **1976,** *251,* 3229–3234.

87. Kreider, M. S.; Winodur, A.; Krieger, N. R. *Neur. Endocrin. Lett.* **1981,** *3,* 115–118.

88. Griffiths, E. C.; Kelly, J. A.; White, N.; Jeffcoate, S. L. *Acta Endocrinol.* **1980,** *93,* 385–391.

89. Morier, E. H.; Han, K. K.; Patsouris, L.; Mareau, O.; Rips, R. *Int. J. Peptide Protein Res.* **1981,** *18,* 513–515.

90. Gibbs, D. M. *Psychoneuroendocrinology* **1986,** *11,* 131–140.

91. Carone, F. A.; Christensen, E. J.; Flouret, G. *Am. J. Physiol.* **1987,** *253* F1120–F1127.

92. Vavra, I.; Machova, A.; Krejci, I. *J. Pharmacol. Exp. Ther.* **1974,** *188,* 241–247.

93. Sawyer, W. H.; Acosta, M.; Manning, M. *Endocrinology* **1974,** *95,* 140–145.

94. Rado, J. P.; Marosi, J.; Szende, L.; Borbely, L.; Tako, J.; Fischer, J. *Int. J. Clin. Pharmacol.* **1976,** *13,* 199–209.

95. Moses, A. C.; Gordon, G. S.; Carey, M. C.; Flier, J. S. *Diabetes* **1983,** *32,* 1040–1047.

96. Nagai, T.; Nishimoto, Y.; Nambu, N.; Suzuki, Y.; Sekine, K. *J. Controlled Release* **1984,** *1,* 15–22.

97. Yoshida, H.; Okumura, K.; Hori, R.; Anmo, I.; Yamaguchi, H. *J. Pharm. Sci.* **1979,** *68,* 670–671.

98. Ichikawa, K.; Ohata, I.; Mitomi, M.; Kawamura, S.; Maeno, H.; Kawata, H. *J. Pharm. Pharmacol.* **1980,** *32,* 314–318.

99. Nishita, T.; Okamura, Y.; Kamada, A.; Higuchi, T.; Yagi, T.; Shichiri, M. *J. Pharm. Pharmacol.* **1985,** *37,* 22–26.
100. Stratford, R. E.; Carson, L. W.; Dodda-Kashi, S.; Lee, V. H. L. *J. Pharm. Sci.* **1988,** 838–842.
101. Yamamoto, A.; Luo, A. M.; Dodda-Kashi, S.; Lee, V. H. L. *J. Pharmacol. Exp. Ther.* **1989,** *249,* 249–255.
102. Ishida, M.; Machida, Y.; Nambu, N.; Nagai, T. *Chem. Pharm. Bull.* **1981,** *29,* 810–816.
103. Okada, H.; Yamazaki, I.; Yashiki, T.; Mima, H. *J. Pharm. Sci.* **1983,** *72,* 75–78.
104. Bundgaard, H. In *Delivery Systems for Peptide Drugs;* Davis, S. S.; Illum, L.; Tomlinson, E., Eds.; Plenum: New York, 1986; pp 49–68.
105. Bundgaard, H.; Moss, J. *Pharm. Res.* **1990,** *70,* 885–892.
106. Samamen, J. M. In *Polymeric Material in Medication;* Plenum: New York, 1985, p 227.
107. Mark, D.; Lin, L.; Su, S. U.S. Patents 4 518 584, 1985, and 4 588 585, Human Recombinant Cysteine Depleted Interferon-b Muteins, 1986.
108. Chen, R. H-L.; Abuchowski, A.; van Es, T.; Palczuk, N. C.; Davis, F. F. *Biochim. Biophys. Acta* **1982,** *660,* 293–296.
109. Abuchowski, A.; McCoy, J. R.; van Es, T.; Palczuk, N. C.; Davis, F. F. *J. Biol. Chem.* **1977,** *252,* 3582–3586.
110. Davis, S.; Abuchowski, A.; Park, Y. K.; Davis, F. F. *Clin. Exp. Immunol.* **1981,** *46,* 649–653.
111. Katre, N. V.; Knauf, M. J.; Laird, W. F. *Proc. Natl. Acad. Sci. U.S.A.* **1987,** *84,* 1487–1492.
112. Ajisaka, K.; Iwashita, Y. *Biochim. Biophys. Res. Commun.* **1980,** *97,* 1076–1080.
113. Su, K. S. E.; Campanale, K. M.; Mendelsohn, L. G.; Kerchner, G. A.; Gries, C. L. *J. Pharm. Sci.* **1985,** *74,* 394–398.
114. Vilhardt, H.; Bie, P. *Eur. J. Pharmacol.* **1983,** *93,* 201–205.
115. Jennewein, H. M.; Waldeck, F.; Konz, W. *Arzneim. Forsch.* **1974,** *24,* 1225–1228.
116. Kidron, M.; Bar, O. J.; Berry, E. M.; Ziv, E. *Life Sci.* **1982,** *31,* 2837–2841.
117. Saffran, M.; Bedra, C.; Kumar, G. S.; Neckers, D. C. *J. Pharm. Sci.* **1988,** *77,* 33–38.
118. Takaori, K.; Burton, J.; Donowitz, M. *Biochim. Biophys. Res. Commun.* **1986,** *137,* 682–687.
119. Damg, C.; Michel, C.; Aprahamian, M.; Couvreur, P.; Devissaguet, J. P. *J. Controlled Release* **1990,** *130,* 233–237.
120. Oppenheim, R. C.; Stewart, N. F.; Gordon, L.; Patel, H. M. *Drug Dev. Ind. Pharm.* **1982,** *8,* 531–546.
121. Touitou, E.; Rubinstein, A. *Int. J. Pharm.* **1986,** *30,* 95–99.
122. Saffran, M.; Kumar, G. S.; Savariar, C.; Burnham, J. C.; Williams, F.; Neckers, D. C. *Science (Washington, DC)* **1986,** *233,* 1081–1084.

RECEIVED for review July 12, 1993. ACCEPTED revised manuscript October 5, 1993.

CHAPTER 4

Biopharmaceutical Properties and Pharmacokinetics of Peptide and Protein Drugs

Hye J. Lee

The discovery of diverse neuronal and extraneuronal peptides has increased interest in exploiting peptides and proteins as therapeutic drugs (*1–4*). Peptide or protein drugs are complex polymers that are derived from amino acids by peptide-bond linkages. A peptide containing less than eight amino acid residues is called a small peptide (*5*). Peptide drugs in this group include enalapril, lisinopril, and thyroid-releasing hormone analogues. The term polypeptide drugs refers to peptide drugs with eight to 50 amino acid residues, such as the drugs calcitonin, cyclosporin, leuproline, and luliberin. Proteins are large peptides. Polypeptide drugs containing from 50 to 2500 amino acid residues are protein drugs. These include insulin, growth hormone, and interferon. Some protein drugs such as insulin or IgG, which contain two or more polypeptide chains, are called oligomeric proteins. Their component chains are termed subunits or protomers (*5*). Peptide and protein drug molecules possess a wide spectrum of biological activities in such systems as the nervous system, the immunological system, and the gastrointestinal tract (GIT) as shown in Table I (*6*). The modes of action are wide-ranging and include receptor agonists or antagonists, enzyme inhibitors, and carrier functions.

Peptide drugs are generally difficult and expensive to synthesize. However, with the

Table I. Pharmacologically Active Peptide and Protein Drugs

Application	*Peptide and Protein Drugs*
Hormones	Insulin, relaxin, proinsulin, growth hormone, proinsulin, prolactin, parathyroid hormone, atrial natriuretic factor, LHRH, vasoactive intestinal peptides, calcitonins, vasopressin thyrotropin-releasing hormone
Immunoactive and antitumor	Cyclosporin, sandostatin, interferon, antibodies, gonadorelin analogues, cytokins, tumor necrosis factor, interleukins
Cardiovascular, thrombolytic	Factor VIII, factor IX, factor VII/VIIa, factor XIII, tissue plasminogen activator, streptokinase, urokinase
Hematopoietic	Erythropoietin, granulocyte stimulating factors
Blood volume replacement	Albumin, plasma proteins
Immunization	Vaccines, immunoglobulins, monoclonal antibodies, antitoxins
Antibiotics	

SOURCE: Data were taken from references 6 and 7.

rapid development of recombinant DNA technology, the list of available therapeutic agents is rapidly expanding to include interferon, macrophage activation factors, tissue plasminogen activator, neuropeptides, and experimental agents that may have applications for cardiovascular disease, inflammation, and contraception (7). Therefore, the design of successful delivery systems for these complex peptide and protein drugs represents a major challenge to scientists, because this development is generally limited by their rapid metabolism, disposition, and low bioavailability. Achievement of metabolic stability with a peptide structure does not, however, guarantee a long biological half-life. Nevertheless, chemical approaches to these problems are becoming more successful, and are increasing the understanding of peptide structures and the ability to design metabolically stable analogues that are excreted unchanged and have high potency. Moreover, research indicates that alternative routes and formulations for peptides are likely to show promise in two areas: overcoming the problems of low bioavailability and short duration of action, and providing rate-controlled delivery of peptides to their site of action. Recent progress in the development of specific and sensitive analytical technologies for the assay of peptide and protein drugs and biological samples is also accelerating drug development of peptide and proteins.

For most peptide and protein drugs, a small volume of distribution, metabolism, and hepatic, biliary, and renal clearance by glomerular filtration and tubular secretion result in rapid elimination from the body. To aid in the design of peptide drugs, determination of bioavailability and activity in vivo is essential. This chapter will review the biopharmaceutical and pharmacokinetic properties of peptide and protein drugs and refer specifically to route of administration, intestinal absorption, target-tissue distribution, metabolic stability, and renal versus biliary excretion.

Absorption of Peptide and Protein Drugs

The concerns that need to be addressed in pharmacokinetic studies of peptide and protein drugs are basically the same as those that are addressed for conventional therapeutic properties, including molecular size, susceptibility to proteolytic breakdown, rapid plasma clearance, immunogenicity, and denaturation, that make them difficult to systematically deliver using the normal absorption routes, such as the oral route. For the absorption of di- or tripeptides following oral administration, carrier-mediated transport is the suggested mechanism. The characterization of a carrier-mediated transport system for peptides is a relatively recent development, and differences between species and the effects of age, disease, and nutrition need to be investigated further.

However, with a knowledge of carrier-mediated specificity, the production of di- or tripeptide mimetic drugs that are well-absorbed and exhibit good oral activity should be possible. For larger peptides, the absorption mechanism is different from di- or tripeptides. Suggested mechanisms of intact absorption for these compounds are as follows (*8, 9*):

- diffusion through aqueous pores or intercellular spaces for hydrophilic peptides
- diffusion through membrane lipid for hydrophobic peptides
- uptake into epithelial cells by endocytosis or pinocytosis

Some of these mechanisms can be exploited by chemical modification and prodrug design. For metabolically stable somatostatin analogues, which are cyclic hexa- and octapeptides, low oral absorption (about 2%) has prompted research (*10, 11*) into an alternative route of administration.

Cyclosporin with the olive oil-based formulation results in 20 to 50% of absorption (*12*). The mechanism of absorption of this large molecular weight and hydrophobic peptide is not clearly understood, but the observation of a zero-order rate for absorption is perhaps indicative of a carrier-mediated or saturable process (*13*). Overall, the extent of absorption exhibited by intact peptides varies over a wide range and more research is required to identify key structural properties. To optimize the absorption of peptide and protein drugs across absorption barriers, several approaches are under investigation, for example, inhibition of their enzymatic degradation, increase in their permeability across the relevant membrane, and improvement in their resistance to breakdown by structural modification (*5*). Clearly, if the peptide is vulnerable to peptidase attack, oral bioavailability will be compromised severely.

Factors Affecting Absorption of Peptide and Protein Drugs

Permeability and stability of peptide and protein drugs are known to be major absorption barriers at absorptive areas. Molecular size, solubility, lipophilicity, charge, intestinal pH, absorption site, diet, age, sex, and disease state are factors that eventually affect either stability or permeability. Difference in species could also play a role in differences in absorption, possibly because of species differences in gut function and morphology in the stability of peptides (*14*). The dog is a poor model for thyrotropin-releasing factor (TRH) analogues (pGlu–His–Pro–NH_2) and amoxycillin, as indicated by bioavailability (F) values of 13 and 18%, respectively. Values for F in man are 2 and 72% (*85*). Systematic study of factors influencing peptide and protein drug absorption has not been established. In this section, a few factors affecting oral absorption will be discussed.

Stability

Both enzymatic and nonenzymatic breakdown result in the instability of peptide and protein drugs. Metabolism by various enzymes or degradation by an acidic pH environment results in the digestion of peptides and proteins into smaller fractions of peptides or amino acids. These processes are natural for the absorption of peptides and proteins into the body system.

Enzymatic breakdowns are discussed in greater detail in a later section. Because the specificity and distribution of proteolytic enzymes in the GIT are highly variable, generalizations regarding metabolism are more difficult to make. However, intestinal pH, absorption site, diet, fast and fed state, age, sex, and disease state partially or entirely contribute to the specificity and distribution of enzymes for peptide and protein drugs. For instance, the maximal absorptive ability for peptide and peptide drugs appears to be different along the length of the small intestine. This difference could be due to differences in permeability or proteolytic enzyme activity along the intestine.

Nonenzymatic breakdown results from chemical or physical changes. Physical changes include aggregation and precipitation and are usually induced by high concentrations of cosolvents that may be used in some formulations or by injudicious choice of ionic strengths. Change of conformation may not only lead to poor absorption but also to a loss of activity (*5*). Chemical changes include beta-elimination, deamidation, disulfide exchange, racemization, and oxidation in the vigorous GIT environment (*5*). Irrespective of which dosage form is used, peptide or protein decomposition may be a problem for absorption.

Permeability

Mucosal-cell permeability determines the rate of entry into the mucosa across the brush-border membrane. Molecular weight, lipophilicity, solubility, concentration, absorption site, pH, age, and disease state could affect the permeability of compounds.

Investigations (*15*) with inert polar substances show that the capillary membranes of different tissues vary appreciably in permeability. An understanding of the characteristic profiles of different membranes could be helpful in designing delivery systems for peptide and protein drugs via various routes and sites. The nasal membranes were shown (*16*) to be highly permeable to peptides of molecular weight up to 1000 kDa. The small intestine is much less permeable to peptides of this molecular weight, whereas the rectum is permeable to these peptides and may be similar to the nasal membranes (*17*). For higher molecular weight peptides, adjuvants can dramatically improve nasal bioavailability (*18*).

Molecular size and lipophilicity may affect the permeabilities of peptide and protein drugs. The absorption of high molecular weight compounds is believed to be low. Earlier studies (*19*) with two series of digestion-resistant radiolabeled peptides that varied in physical properties of molecular weight, lipophilicity, and hydrogen bonding sites suggested that intestinal absorption was most affected by hydrogen bonding sites, which are the major determinant of desolvation energy. Comparison of the biliary and urinary recovery of these radiolabeled peptides in rats was made following intravenous or intraduodenal doses to investigate whether this correlation could be confirmed in vivo (*19*). As a result, absorption was inversely correlated to the number of calculated hydrogen bonding sites for the model peptides. Clearance by liver and kidneys appeared to be unaffected by desolvation energy but was well-correlated with lipophilicity.

Peptide molecules of molecular weight up to 3000 kDa usually have a flexible structure in the aqueous phase and are water soluble because of polar peptide bonds and polar side-chains. Larger molecules have a folded structure that internalizes most of the nonpolar residues and results once again in a water-soluble molecule. Nonpolar groups can occur on the surface of folded peptides but are usually dispersed so that patches that could lead to aggression are not present. The net charge and total hydrophobic contents of peptides can vary appreciably (*20*).

Effect of Route of Administration

In addition to metabolism, the effect of route of administration on the bioavailability of peptide and protein drugs is a major issue in the designing of delivery systems. The route of administration of a given drug can have a significant influence upon whole-body distribution, bioavailability, and pharmacology (*21–23*). Table II compares the absolute bioavailability of a few selected peptides and proteins among various routes of administration. Although direct comparison of bioavailabilities is difficult because of differences in compounds, species, and assay methods, the extent of absorption appears intramuscular (im) = subcutaneous (sc) = intraperitoneal (ip) > nasal > pulmonary > rectal > ileal > vaginal > buccal > oral (po) (Table II).

The cyclic enkephalin derivative, [D-Pen^2, D-Pen^5]-enkephalin (DPDPE) acts through delta receptors and has shown (*23*) to inhibit GIT transit when given intrathecally (it) but not after intracerebralventricular (icv) or sc administration. DPDPE was shown to inhibit diarrhea through an antisecretory mechanism after icv or sc administration (*24*) and to produce an analgesic effect after icv, it, or intravenous (iv) administration (*25–27*). IV administration resulted in a significantly large amount of DPDPE in the small intestine, and the flush at 15 and 30 min postadministration suggested rapid biliary excretion. The highest level in the brain after iv administration occurred at 60 min (0.08%). After ip and sc administration, large amounts of DPDPE were found in the small intestine and the flush at 60 min postadministration, a finding that suggested a slower rate of absorption from the site of administration. Brain levels of DPDPE peaked 120 min after ip and sc administration (0.07 and 0.09%, respectively). The highest levels in the brain after po administration were

Table II. Comparison of Bioavailabilities of Selected Peptide and Protein Drugs among Various Routes of Administration

	Route of Administration						
Compounds	*im*	*sc*	*ip*	*Nasal*	*Oral*	*Miscellaneous*	*Subjects*
Calcitonin	70	70		2–7	<0.5	17 (lung)	Rat
Cyclosporin	Poor				20–50		Human
Cyclosporin		58.5	57.6				Rat
D-Ala–D-Leu–enkephalin		70	65	20–30	<0.5	2.5 (ileum)	Rat
Glucagon				70–90			Human
Growth hormone, releasin hormone				2–20			Rat, dog, human
Insulin		80		10–40	0.05	0.5 (buccal)	Human
α-Interferon	>80	>80					Human
β-Interferon	33			2.2			Rabbit
Leuprolide		65		2–3	0.05	8 (rectal)	Rat
Relaxin, human		76					
α-Tumor necrosis factor	20–75	20–83	Significant		Some		Human
Vasopressin				4–12		1.2 (vaginal)	Human

NOTE: im is intramuscular, sc is subcutaneous, and ip is intraperitoneal.
SOURCE: Data were taken from references 33, 34, 36, 186, and 187.

at 240 min (0.03%). Examination of regional brain-distribution data showed no significant differences in the levels of DPDPE among brain regions at any time point studied (*21*).

Other mucous membranes can be used as potential sites of drug delivery and alternative routes for parenteral administration. The various membrane sites have different permeabilities to water and peptide molecules because of the different junctions and pores between the cells. Furthermore, all the alternative routes of administration to the oral pathway except ip administration have the advantage of avoiding the effects of stomach acid, gut proteases, and "first-pass" metabolism by the liver. Once the desired biological activity of a peptide or protein molecule is achieved, the route of administration and formulation may need to be considered in conjunction with a knowledge of the target tissue and the required kinetic profile (*28*).

Parenteral Route

For systemic delivery of peptide and protein drugs, parenteral administration is the most widespread method used to achieve consistent therapeutic activity. This type of administration avoids the susceptibility of a drug to breakdown by gastric acid and the proteolytic enzymes in the GIT. In addition, peptides and proteins are high molecular weight substances and do not easily cross the intestinal mucosa. Of the parenteral routes, only iv administration is considered to be efficient in delivering protein and peptide drugs to systemic circulation. For example, optimal blood levels of protein or peptide drugs, such as γ-globulin, can be achieved by the iv route. Generally, im or sc injections are less efficient due to absorption and diffusion barriers presented by muscle mass and connective tissues under the skin. Even though most peptide and protein drugs can be delivered efficiently to systemic circulation by parenteral injections, the rapid elimination profile still leads to low therapeutic concentrations resulting in frequent dosing intervals. Such frequent injections are unpleasant to the patients and lead to usual complications such as thrombophlebitis and tissue necrosis. For parenteral administration of high potency hormones, consideration needs to be given to the rate of drug delivery and to the possible benefit of sustained release of peptide (*29*).

Oral Route

The development of oral delivery of peptide and protein drugs has been limited, in part, by rapid drug metabolism in the GIT and the inability to readily cross the membrane barrier. Stability of peptides in the hostile environment of the gut is an essential requirement for oral activity. Oral delivery can be successful if the two major obstacles, enzymatic degradation and absorption in the GIT, are overcome (*30*). To avoid proteolysis of the peptides in the gut, the following approaches have been attempted:

- coadministration with protease inhibitors or permeability promoters (*31–33*)
- delivery of the drug to a site in the GIT where the enzyme activity is minimal (*33, 34*)
- enhancement of absorption of the peptides with surfactants or mucolytic agents before proteolysis occurs
- encapsulation of the drug within liposomes (*35, 36*)
- structural modification of the peptide such as a prodrug approach to prevent its proteolysis (*37*)
- formulation of the drug with an oil–water emulsion (*38*)
- entrapment of the peptide in nanometer-sized polymer particles (*39*)

These approaches have increased oral activity of peptide molecules. However, the extent of absorption is still low. Cyclosporin, which is stable to the action of proteases, exhibits the greatest absorption (20–50% of dose) after administration in an olive oil based formulation (*12*), especially when given with food (*40*). The use of GIT absorption enhancers for therapeutic peptides and proteins would increase bioavailability (*41–43*), although questions remain concerning the safety of many of these systems. Examples of peptides that are biologically active upon oral or intraluminal administration include the following:

- lysine vasopressin, especially in the presence of trasylol protease inhibitor (*44*)
- *l*-Desamino-8-D-arginine vasopressin [0.7–1% of an oral dose is absorbed by humans compared with 11% of a nasal dose (*45, 46*)]
- a selective mini-somatostatin, which is a blocked octapeptide having three times more potency than somatostatin (*47, 48*)
- luliberin, a decapeptide with blocked termini (*49–51*)
- thyroliberin, a tripeptide with blocked termini (*52*)
- insulin with hypertonic solutions or trypsin inhibitors, or encapsulated in liposomes (*53–55*)

Transdermal Route

The transdermal route is an attractive alternative, because it offers a number of advantages over oral and nasal routes. It has fewer problems of local proteolytic degradation (*56*) and it has no hepatic first-pass metabolism. This route may then provide better control of delivery and maintenance of the therapeutic level of drug over a prolonged period of time. This route may prove the most popular with the recipient, because a transdermal formulation of peptide drugs can be applied easily to and removed from different parts of the body. However, because of the general impermeability of the skin and the large molecular size and ionic character of peptides, a means of increasing permeation must be found. One approach employs iontophoresis as an active driving force for the charged molecules (*57, 58*). Another approach involves the use of chemical enhancers (*59–61*). However, little is known about the proteolytic enzyme activity in the skin and the possible influences on mass transfer, even though other enzymes of the skin have been relatively well-characterized (*62, 63*). For example, the subcellular distribution of aminopeptidases was determined in homogenates of cultured mouse keratinocytes and mouse epidermis (*64*).

In an investigation (*65*) of the effects of the nonionic surfactant *n*-decylmethyl sulfoxide (NDMS) on the pH, the inhibitors of metabolism, and the permeation of amino acids, dipeptides, and pentapeptide enkephalin through hairless mouse skin, NDMS increased the permeability of all amino acids and peptides tested. At neutral pH, where the enzyme activity within the skin was such that no flux of leucine–enkephalin (YGGFL) was observed, the donor cell concentration of YGGFL decreased rapidly. The major cleavage occurred at the Tyr–Gly bond. At pH 5.0, the metabolic activity was reduced significantly and a substantial flux of YGGFL was observed. These results show that the complex proteolytic enzyme activities occurring during skin permeation are different from those in skin homogenates and that a combination of enhancer, pH adjustment, and inhibitors can increase the transdermal delivery of peptides (*65*).

Therefore, the large molecular size and charged character of protein and peptide drugs make them poor candidates for passive transdermal delivery (*66*). Luteinizing hormone-releasing factor (LHRH), a decapeptide of molecular weight 1182 and about +2 charge at pH 6.0, was delivered through pig skin by anodal iontophoresis. Although usually potent at very low concentrations, peptides generally have very short half-lives that necessitate frequent dosing. The achievement of a reliable transdermal delivery of peptides would offer several clinical advantages over other more conventional routes of delivery.

Nasal Route

The nasal mucosa has a high surface area due to numerous microvilli on the epithelial cells and an extensive vascular network. Therefore, the nasal cavity provides a good site for extensive absorption of lipophilic and hydrophilic drugs that are poorly absorbed by the oral route. Peptides and peptide-type drugs that were shown (*67, 68*) to be absorbed extensively by the nasal route include LHRH, Ala^1–Lys^{17}–adrenocorticotropic hormone (ACTH), cholecystokinin octapeptide, sulbenicilin, and cephalosporin (*67, 68*). Even larger peptides or proteins seem to be well-absorbed. For example, insulin absorption up to 30% in rats was reported (*67*). The rapidity of absorption has also created great interest in intranasal insulin administration as a means of achieving the desired insulin bioavailability profile (*69*). Although nasal delivery avoids the hepatic first-pass effect, the enzymatic barrier of the nasal mucosa creates a pseudo-first-pass effect.

The xenobiotic metabolic activity in the nasal epithelium was investigated (*70–72*) in humans and several other species. The cytochrome P-450 activity in the olfactory region of the nasal epithelium is even higher than in the liver, mainly because of a three- to fourfold higher reduced nicotinamide adenine dinucleotide phosphate (NADPH)-cytochrome P-450 reductase content (*70*). Phase II activity was also found in the nasal epithelium (*70*). The delivery of peptides and proteins has been hindered by the peptidase and protease activity in the nasal mucosa. The predominant enzyme among other exopeptidases and endopeptidases appears to be aminopeptidase. Therefore, the absorption of peptide drugs can be improved by using aminoboronic acid derivatives, amastatin, and other enzyme inhibitors as absorption enhancers. Some of the surfactants (e.g., bile salts) may increase absorption by inhibiting the proteolytic enzymes. Mean plasma levels of human growth hormone following intranasal administration of 1 mg/kg human growth hormone were significantly lower than levels when the hormone was combined with 0.015% each of the aminopeptidase inhibitors amastatin and bestatin (*71*). The surfactant sodium glycolate provided a bioavailability of 100% for corticotrophin-releasing hormone but only 7.1% for growth hormone releasing hormone. Bile salts appear to inhibit proteolytic activity by denaturing the enzyme and preventing the enzyme–substrate complex from undergoing the necessary conformational change that aligns the catalytic site on the protease with the susceptible bond of the substrate (*72*).

Miscellaneous Routes

Regions such as the colon possess less degradative enzymes and may offer a less harsh milieu to peptide and protein drugs before absorption (*73, 74*). Dosage forms that deliver peptide drugs to specific regions of the GIT, such as the colon and the ileum, have been investigated (*33, 34*). Alternative routes for systemic administration of peptide drugs have also been investigated, for example, buccal (*75*), vaginal (*76*), and ocular routes (*77*). A recent report (*78*) on intrauterine delivery of insulin and calcitonin shows that the intrauterine route could benefit not only bioavailability (equivalent to subcutaneous administration) but also controlled or pulsatile release fashion. The intravaginal route has shown good absorption (up to 40%) for peptides such as LHRH analogues (*79*) and insulin (*80*). Rectal administration provided good delivery for tetragastrin (16% of dose) and insulin (28% of dose) in the dog (*81, 82*). Although absorption of peptides by rectal administration generally does not compare favorably with intravaginal or intranasal routes (*83, 84*), the rectal route offers a number of advantages over the oral route.

Choices are available in the development of alternative routes, but the formulation of proteins and peptides into effective dosage forms for these routes has become a formidable task. Various approaches have been investigated for the development of safe and effective peptide delivery. Although alternative routes for peptide administration do not give good bioavailability, as do parenteral routes, they do offer the advantage of self-medication and the potential for much greater and more rapid absorption than from the oral route.

Distribution

The distribution profiles of peptide and protein drugs after iv administration, except for a few peptides such as relaxin, often follow two exponential decays. During the first phase of decay, the drug is rapidly distributed into a pool that has an apparent volume greater than plasma. Absence of the early phase for many peptides and proteins can often be attributed to insufficient number of samples collected. Following the first phase, drug levels usually decline exponentially. The half-lives of the first phase appear to be faster with smaller peptides and smaller animals (*33, 34, 36*) and range from less than one minute to a few minutes.

Because of their hydrophilicity and relatively large molecular weight, most peptides and proteins remain in the extracellular space unless active mechanisms of transport or elimination come into play. The extracellular space can be divided into plasma, the interstitial space of tissue, lymph, cerebrospinal fluid, fluid-bathing mucosal surfaces, and other minor components. The compartments are separated from each other by multicellular membranes. Transport across these endothelial and epithelial membranes can occur by a number of mechanisms. Passive mechanisms apply to all peptides and often sufficiently account for major distribution characteristics. The kinetic principles that apply to the distribution of drugs into cells and organs by passive diffusion are also generally applicable to peptides and proteins (*85*). Thus, the extent of drug in the tissue is dependent on the nature of drug binding in the tissue and the amount of drug transferred to the tissue compartment. The amount of drug transferred is dependent on free concentration in the blood, rate of transfer across membranes, and rate of drug elimination from blood and the body.

The diffusion of peptides through membranes is also influenced by the physicochemical properties of the molecules, as discussed for gastrointestinal absorption. For peptides, exceptions to these rules would occur if the peptides were involved in some active transport process into or out of tissues, which may be saturable, or in uptake into cells by some absorptive endocytotic mechanism (*86*). In fact, the facility of peptides to cross cell membranes is suggested by the rapidity with which some peptides undergo metabolism and elimination (*87, 88*). For example, the most prominent route of entry for leupeptin is by direct permeation of the plasma membrane, possibly through pores, and is directly dependent on external concentration (*87, 88*). However, some leupetin may be taken up by a specific process that is saturable and can be competitively inhibited by reduced leupeptin. Initially, leupeptin can be located in the soluble fraction of liver homogenates, but leupeptin is rapidly concentrated in the lysosome-rich fraction.

Tissue distribution studies (*89*) using radiolabeled relaxins in mice, guinea pigs, and rats showed that relaxin was accumulated extensively in the kidney, liver, lung, uterus, pituitary, and spleen. Generally, the tissue distribution of neuropeptides and their analogues reflects sites of metabolism and the specificity of binding at the site of action (*90*). The LHRH analogue buserelin has a plasma half-life in rats of 30 min and was observed (*90*) to accumulate preferentially in the anterior pituitary, liver, and kidneys. The liver and kidneys are high in peptidase activity.

Peptides have been used in nature as masking and carrier functions to deliver potent neuropeptides to their target receptors and to smuggle cytotoxic molecules into microorganisms (*91*). Tissue-specific delivery of natural neuropeptides often involves the use of tissue-selective peptidases to release the active moiety where it is needed and recognition of the moiety by specific cell-surface receptors. In the case of certain polypeptide hormones, these receptors are internalized via a coated-pit endocytotic process (*86*). The application of these targeting principles has been attempted in three main areas of drug design: antimicrobial "warhead" delivery systems, peptides capable of releasing cytotoxic agents in tumors, and renally selective dopamine peptides. Specific delivery of peptides to normal tissues is also possible if selective membrane-transport systems, specific localizations of tissue binding sites, or local metabolizing activity can be identified. L-Dopa and oxfenicine, which provide selective delivery to the central nervous system (CNS) and the heart mitochondria, respectively, are examples. The kidney can be targeted with certain peptides because of the high concentrations of renal γ-glutamyl transpeptidase, which can lead to accumulation and metabolism of γ-glutamyl derivatives of amino acids and peptides (*92*). For the sulfur-containing amino acids and peptides, thiol groups can form reversible disulfide bonds to create altered distribution and kinetic properties (*85*).

Central Nervous System Distribution

Most peptides do not enter the CNS for the following reasons:

- the hydrophilic character of the peptide
- the absence of specific transport system in the membrane
- the presence of proteolytic enzymes in the lipoidal blood–brain barrier

The capillaries in the brain parenchyma possess high-resistance, tight junctions between the endothelial cells (*76*). The cells also lack pores; thus, the brain capillary endothelium behaves like a continuous lipid bilayer. Diffusion through this layer, the physical blood–brain barrier, is largely dependent on the lipid solubility of the solute (*93*). Water-soluble molecules enter the brain almost exclusively by carrier-mediated transport. The blood–brain barrier is the major obstacle for the development of centrally active peptides.

One recent approach (*93*) to achieve brain delivery of a peptide conjugate is the placement of an opioid peptide (enkephalin) in a molecular environment that costumes the peptide nature and provides biolabile and lipophilic functions to penetrate the passive transport of the blood–brain barrier. Tissue distribution studies using whole-body autoradiography for the tripeptide RX-77,368 (*94*), the hexapeptide pepstantinyl glycine (*95*), and an octapeptide somatostatin analogue (*10*) confirm low penetration into the CNS. For Leu- and Met–enkephalin, penetration of the blood–brain barrier is low and does not involve an active transport process (*96*). One approach to increasing the availability of peptides to the CNS is to increase the rate of passive diffusion by using a more lipophilic prodrug as the carrier molecule.

Protein Binding

High degree of protein binding appears to significantly impair the distribution character of peptide and protein drugs and the pharmacological activity of the peptide (*97*). A peptide that is extensively and strongly bound to plasma proteins has limited access to cellular sites of elimination and undergoes a decrease in clearance. For instance, clearance of unbound [^{125}I]-rhuIGF (insulin-like growth factor; molecular weight, 50 kDa) is very rapid (168–204 mL min^{-1} kg^{-1}); clearance of radioactivity associated with the low molecular weight complex is intermediate (15.5–48 mL min^{-1} kg^{-1}); and clearance of radioactivity associated with the high molecular weight complex is slow (0.5–10 mL min^{-1} kg^{-1}). About 40–50% of growth hormone (GH; 22 kDa) binds to GH-binding protein (*89*). At higher GH concentrations, the fraction of bound GH gradually declines in vivo and in vitro because of saturation of the binding protein. Greater than 93% of lipopeptide daptomycin binds to protein, of which 30 to 40% binds to alpha-1 glycoprotein (*98*). As analytical difficulties have become resolved, the effects of protein binding on peptide and protein drugs pharmacokinetics and metabolism are now more actively investigated. For instance, insulin-like growth factor-I (IGF-I) was shown (*99, 100*) to be bound extensively in plasma to multiple plasma-carrier proteins. The clearances of free IGF-I and bound IGF-I differ; free IGF-I is cleared very rapidly compared with the bound forms (*101*). These binding proteins may modulate IGF-I actions by either inhibiting or potentiating its effect at the cellular level.

Metabolism

In addition to transport, another key factor playing an important role in the pharmacokinetic and biopharmaceutic properties of peptide and protein drugs is metabolism. Metabolisms of peptides and proteins usually result in loss of efficacy and require careful assessment of the true site of the loss. The sites for the presystemic hydrolysis of peptide and protein drugs following oral administration are the gastrointestinal lumen, mucosal cells including brush-border membranes and cytosol, portal blood, and the liver. Extensive studies on identification of participating enzymes in the metabolism of peptides and protein drugs have been in progress for several years. A major design goal in the synthesis of biologically active peptides and related drugs, whether derived from naturally occurring compounds or not, is to achieve metabolic stability and enhanced oral bioavailability and duration of action.

Peptide drugs that are stable to peptidase activity may still undergo metabolism by other detoxification systems in the body. For example, the cyclic undecapeptide cyclosporin is metabolized to at least nine metabolites (*102*). Biotransformation includes mono- and dihydroxylation and N-demethylation at various sites on the cyclosporin molecule. Biotransformation is thought to involve the cytochrome P-450 monoxygenase system because compounds known to inhibit these enzymes increase cyclosporin blood levels (*101*). Nevertheless, enzymatic hydrolysis is the dominant cause of metabolism of peptide and protein drugs. Therefore, metabolic sites and participating enzymes in the GIT, liver, and kidney will be discussed.

Gastrointestinal Tract

Luminal Enzymes

Typical luminal enzymes that can form, convert or inactivate peptide and protein drugs and their specificities are as follows: pepsin, hydrophobic; trypsin, basic; α-chymotrypsin, large hydrophobic; elastase, small aliphatic and aromatic; and carboxypeptidases A and B, basic C-terminal amino acids (*103*). Table III gives typical substrates for these enzymes. Because these enzymes autodigest as they traverse the intestine, oral peptide delivery strategies that focus on ileal or colonic delivery are intended, in part, to deliver the drug past the region of highest luminal-enzyme activity. This strategy can reduce the luminal metabolic component limiting oral peptide delivery. For this strategy to be successful, the drug must exhibit good

Table III. Luminal Proteolytic Enzymes and Example Substrates

Enzyme	*Substrate*
Pepsin	z-His–Phe–Phe–OMe
Trypsin	Benzoyl arginine methyl ester
α-Chymotrypsin	Benzoyl tryosine ethyl ester
Elastase	Ala–Ala–Ala methyl ester
Carboxypeptidase A	Hippuryl phenylalanine
Carboxypeptidase B	Hippuryl arginine

membrane permeability in the ileum and colon. This ability probably requires good passive permeability and solubility properties. If the peptide absorption mechanism is carrier-mediated and the transporter is expressed only in the small intestine, this strategy likely will not be successful. Some success with this strategy has been reported (*33, 104*).

Mucosal Cell Enzymes Bound to Brush-Border Membranes

Brush-border membranes (BBM) from the luminal surface of the epithelial cells of the intestine, liver, and kidney are rich in enzymes and transporters that are involved in the cell-surface metabolism and uptake of various peptides and peptide-type drugs. Typical gastrointestinal BBM enzymes are listed in Table IV (*105, 106*). BBM enzymes can be divided into four types based on their mode of action (Table IV): aminopeptidases, carboxypeptidases, dipeptidases, and endopeptidases. Aminopeptidase, carboxypeptidase, and dipeptidase belong to the exopeptidase class because these enzymes hydrolyze peptides sequentially from either the N- or C-terminal end of the molecule. The most abundant enzymes in the intestinal BBM are aminopeptidases. Dipeptides except glycyl-dipeptides are hydrolyzed in the brush border, particularly in the human jejunum (*107*). Aminopeptidases capable of hydrolyzing dipeptides containing proline are absent in the BBM. However, dipeptidylpepdidase-IV hydrolyzes Pro-containing oligopeptides from the C-terminus (*108*). Peptides of at least eight amino acid residues may be hydrolyzed by the indigenous peptidases of the BBM (*105*). The oligoaminopeptidases of the BBM do not hydrolyze very large peptides and have no action on the B-chain of insulin or on bovine serum albumin

Table IV. Typical Intestinal Brush-Border-Membrane Enzymes

Type	*Specificity*	*Enzyme*
Exopeptidase, N-terminus	Many amino acids	Aminopeptidase N
	Asp or Glu	Aminopeptidase A
	Amino acid–Pro	Aminopeptidase P
	Amino acid–Pro, –Ala	Dipeptidylpeptidase IV
	γ-Glu	γ-Glutamyltransferase
Exopeptidase, C-terminus	Many amino acids	Angiotensin converting enzyme
	Pro, Ala, Gly	Carboxypeptidase P
Exopeptidase dipeptidase	Many amino acids	Microsomal dipeptidase
Endopeptidase	Hydrophobic	Endopeptidase-24.11
	Aromatic	Endopeptidase-24.18

Table V. Intestinal Brush-Border Cytosol Enzymes with Typical Substrates

Enzymes	*Typical Substrate*
Dipeptidase	Neutral dipeptides
Aminopeptidase	Tripeptides with N-terminal Pro
Prolidase	Imidodipeptides with C-terminal Pro or Hyp
Prolinase	Imidodipeptides with N-terminal Pro or Hyp
Carnosinase	Carnosine (β-Ala–His)

(*105*). However, some endopeptidases in the BBM are still capable of hydrolyzing proteins and large polypeptides. Although BBMs, in general, are species- and substrate-dependent, the rate of hydrolysis for peptide-type drugs in the brush border increases in the order of dipeptide > tripeptide > oligopeptide (*109, 110*).

The typical BBM enzyme is aminopeptidase M (AmM, EC 3.4.11.2), which hydrolyzes basic and neutral amino acids (but not Glu or Asp) from the N-terminus of peptides, except where the penurious amino acid is Pro (*111*). Consistent with this substrate specificity, vascular AmM can convert kallidin to bradykinin and inactivate des(Asp[1])angiotensin I, angiotensin III, hepta (5–11) substance P, and Met–enkephalin (*112*). AmM does not hydrolyze bradykinin, angiotensin I, angiotensin II, saralasin, vasopressin, oxytocin, or any form of substance P containing the Arg–Pro–Lys–Pro sequence. Lowering the jejunal pH below 5.0 significantly reduces aminopeptidase activity and, to a lesser extent, endopeptidase activity (*113*). Metabolism of opioid peptides by cerebral microvascular AmM was reported (*114*). Degradation is inhibited by amastatin (50% ineffective dose [ID_{50}], 0.2 μM) and bestatin (ID_{50}, 10 μM). However, a number of other peptidase inhibitors, including captopril and phosphoramidon do not inhibit degradation. Rates of degradation are highest for the shorter peptides, whereas β-endorphin is almost completely resistant to N-terminal hydrolysis. Michaelis values (K_m) decreased significantly with increasing peptide length. Peptides known to be present within or in close proximity to cerebral vessels competitively inhibited enkephalin degradation [inhibitor constant (K_i) = 20.4 and 7.9 μM, respectively, for neurotensin and substance P]. These data suggest that cerebral microvascular AmM may play a role in vivo in modulating peptide-mediated, local, cerebral blood flow and in preventing circulating enkephalins from crossing the blood–brain barrier (*114*).

Cytosolic Enzymes

The cytosolic peptidases aminotripeptidase, prolidase, prolinase, dipeptidase, and carnosinase are listed in Table V. Substrate specificity of these enzymes is tabulated in Table V (*106*). Neutral dipeptides are hydrolyzed more readily than acidic or basic dipeptides. However, the detection of these peptides in the portal blood stream at low levels suggests that the enzyme activity is not completely effective. The hydrolysis of peptides of more than 3 amino acid residues in the cytosol is uncertain (*105*). Overall, the cytosolic enzymes prefer to hydrolyze dipeptides instead of tripeptides or larger peptides.

Liver

In addition to the GIT sites, the liver contributes to the presystemic loss of peptide and peptide-type drugs and is responsible for the metabolism of peptide drugs administered peripherally. The hepatic metabolism of peptide and peptide-type drugs is influenced by the route of administration (cf. po vs. iv) and physicochemical properties of the peptide molecules (*115–117*). Data from which to draw general rules regarding hepatic metabolism of peptides and proteins are lacking. However, one rough empirical conclusion based on data currently available is that smaller peptides with chain length of less than 8 amino acid residues result in higher hepatic extraction. For instance, radiolabeled cholecystokinin (CCK-33), CCK octapeptide (CCK-8), CCK-8 desulfate, and CCK-4 are extracted by the liver in a structurally specific manner (*118*). In a study (*119*) of the fate of the extracted, radiolabeled peptides by quantitating biliary excretion and determining the nature of the metabolites in the bile, CCK-8 desulfate and CCK-4 appeared in the bile in completely metabolized forms. In contrast, about 3 and 20% of the major forms of the label in bile were intact, labeled CCK-33 and CCK-8, respectively (*118*). Transforming growth factor beta (TGFβ with 3 amino acid), a recently discovered polypeptide, modulates growth of normal and neoplastic cells (*119*). After iv injection of [^{125}I]TGFβ in the rat, most of the label removed by the liver (83%) was excreted into bile.

The mechanism of hepatic uptake of peptides involves two general processes (*120, 121*): passive for hydrophobic peptides and active for less hydrophobic peptides. Active processes can be divided into two types of mechanisms: endocytosis and carrier-mediated transport (*120, 121*). Most water-soluble peptides that are not cleared by specific mechanisms are believed to enter liver cells by endocytosis. The representative liver enzymes for peptide and peptide-type drugs are cathepsin (D, B, L, and H), intracellular, and lysosomal proteinases (*122*). Some membrane-bound aminopeptidases are also found in the biliary canalicular membranes (*123*). Although limited information is available regarding participating liver enzymes, the characterization of the enzymes and transport mechanisms will aid greatly in the future design of orally active peptides and peptide-type drugs.

Kidney

The kidney plays a specific role in the clearance of many proteins and peptide hormones from the circulation (*124–127*). Larger proteins are cleared from the circulation after filtration by endocytosis and lysosomal degradation (*128*). Small proteins and peptide hormones are cleared by a different mechanism. After filtration at the glomerulus, they are hydrolyzed in the proximal tubule by brush-border enzymes (*126, 127*). The small peptides produced may then be hydrolyzed to their constituent amino acids and reabsorbed by specific amino acid transport systems (*129*). An alternative route, whereby proteins are first cleaved to dipeptides followed by transport and intracellular hydrolysis, was shown to exist in the intestinal mucosa (*130*) and may also be present in the epithelial cells of the proximal tubule (*131, 132*). Peritubular uptake of peptides may also be involved in the clearance of peptide hormones from the circulation (*133*). Proteins are generally agreed to be metabolized by a route involving pinocytosis by the proximal-tubule cells and digestion within the lysosomal system (*134*). Moreover, the microvillar peptidases seem to have little or no proteinase activity. In pig kidney microvilli, only one endopeptidase exists, endopeptidase-24.11 (EC 3.4.24.11) whose action is limited to peptides (*135*). Endopeptidase-24.11 is also the only endopeptidase in human and rabbit kidney brush borders (*136*). However, the endopeptidases "neutral endopeptidase-2" and meprin have been found in rats and mice, respectively. A range of neuropeptides are degraded efficiently by microvillar membranes. Endopeptidase-24.11 plays a key role in this process (*137*).

Endopeptidase-24.11

Endopeptidase-24.11 (*138*) found in renal brush borders is usually the only endopeptidase present among a host of exopeptidases and appears to control the rate-limiting step in the hydrolysis of a number of peptides, including bradykinin and angiotensins. Endopeptidase-

24.11 is widely, but not ubiquitously, distributed and present not only in renal and intestinal brush borders, but also in lymph nodes, glandular tissues, and the nervous system. The lymph nodes, glandular tissues, and nervous system are rich in neuropeptides. Endopeptidase-24.11 exhibits high specificity constants for a number of these potential substrates, including tachykinins, enkephalins, and bradykinin. In the brain, immunocytochemical studies (*138*) have shown colocalization of the enzymes and substance P. Thus, endopeptidase-24.11 has the appropriate topology, specificity, kinetic properties, and localization to play a role in the metabolism of regulatory peptides (*138*).

Approximately 50 known neuropeptides are located in peripheral tissues as well as in the nervous system. In many cases, primary structural features limit the possibility of attack by exopeptidases and require an endopeptidase for initiating degradation. The examples of bradykinin with pig kidney microvillar membranes (*139*) and substance P with striatal synaptic membranes (*139*) demonstrate that endopeptidase-24.11 is the crucial enzyme for the initiation of neuropeptide degradation. Pig kidney microvilli contain only one endopeptidase activity, and the substitution of the rat or mouse kidney microvilli (where a second endopeptidase is present) in such experiments might modify the conclusions. Interpretation of such data should be cautious in relation to the physiological condition where peptide concentrations are several orders of magnitude lower (*138*).

Endopeptidase-24.18 (Endopeptidase-2)

Endopeptidase-24.18, the second endopeptidase in rat kidney brush border, was characterized in regard to its specificity and its contribution to the hydrolysis of peptides by microvillar membrane preparations (*111*). Luliberin was hydrolyzed faster than other peptides tested, followed by substance P and bradykinin. Human alpha-atrial natriuretic peptide and the angiotensins were only slowly attacked (*140*). Oxytocin and [Arg^{8}]vasopressin were not hydrolyzed. No peptide fragments were detected on prolonged incubation with insulin, cytochrome c, ovalbumin, and serum albumin. In comparison with pig endopeptidase-24.11, the rates for the susceptible peptides were, with the exception of lulibrin, much lower for endopeptidase-24.18. Indeed, for bradykinin and substance P, the ratio of the catalysis constant (K_{cat}) to K_m was two orders of magnitude lower. Because both endopeptidases are present in rat kidney microvilli, an assessment was made of the relative contributions to the hydrolysis of lulibrin, bradykinin, and substance P. For lulibrin endopeptidase-24.18 was the dominant enzyme; for bradykinin it made an equal contribution; for substance P it made a minor contribution (*111*). Microvillar membranes derived from the brush border of the renal proximal tubule are very rich in peptidases. Pig kidney microvilli contain endopeptidase-24.11 that is associated with a battery of exopeptidases (*137*). Substance P, bradykinin, angiotensins I, II, and III, and insulin B-chain were rapidly hydrolyzed by kidney microvilli. Oxytocin was hydrolyzed much more slowly, but no products were detected from [Arg^{8}]vasopressin or insulin under the conditions used for other peptides (*137*).

Blood

Peptides and proteins are known to be unstable in blood. Su et al. (*34*) showed that the order of greatest blood metabolism of cholecystokinin is in vivo > ex vivo fresh blood > ex vivo blood. This order suggests that enzymes responsible for blood metabolism are either membrane bound, endothelial enzymes, or unstable once they are out of the body (*33, 34*). The half-lives of [d-Ala, d-Leu]-enkephalin, cholecystokinin octapeptide sulfate, and calcitonin in fresh rat blood are 4 h, 2.5 h, and 2–3 h, respectively (*33, 34, 36*). Enzymes that are known to be distributed in blood are dipeptidyl(amino)peptidase IV (DAP IV; EC

3.4.14.5) (*141, 142*), post-proline cleaving enzymes (PPCE; EC 3.4.21.26) (*143*), and aminopeptidase (*112, 144*). These enzymes are all either membrane bound, endothelial, or smooth muscle enzymes.

Vascular DAP IV specifically hydrolyzes X–Pro dipeptides from the N-terminus of polypeptides (*141, 142*). Substance P, which has sequential Arg–Pro and Lys–Pro dipeptides at its N-terminus, is the only vasoactive peptide that fits DAP IV specificity. Although bradykinin also has an Arg–Pro N-terminus, it is not hydrolyzed by DAP IV because of the presence of a second Pro in position three (*145*). Vascular plasma membrane DAP IV can specifically convert substance P to hepta(5–11)-substance P, and AmM can differentially metabolize a number of vasoactive peptides (*112, 144*). Thus, both enzymes have the correct subcellular localization and substrate specificity to modulate both the form and concentration of vasoactive peptides in the microenvironment of the vascular cell surface receptors.

Vasoactive peptides contain high proportions of proline residues, which make the peptides resistant to hydrolysis by many peptidases. However, PPCEs, which are proline-specific endopeptidases that specifically hydrolyze internal peptide bonds on the carboxyl side of proline residues, inactivate numerous vasoactive peptides such as angiotensins, kinins, substance P, vasopressin, and oxytocin (*143*). In the localization and characterization studies of PPCE-like activity in dog aorta and mesenteric artery, the subcellular distribution of vascular PPCE was essentially the same as that of the cytosolic marker enzyme lactic dehydrogenase (LDH). PPCE was enriched sixfold in the cytosolic fraction and, unlike the plasma membrane-bound proline-specific exopeptidase DAP IV; *EC 3.4.14.5*), little or no activity could be detected for PPCE in the microsomal or plasma membrane fractions. Similar to PPCE characterized from other sites, vascular PPCE was stabilized and activated by dithiothreitol and ethylenediaminetetraacetic acid and inhibited by diisopropyl fluorophosphate, *p*-chloromercuriphenyl sulfonic acid, L-1-tosylamido-2-phenylethylchloromethyl ketone, Cu^{2+}, Ca^{2+}, and Zn^{2+}. Vascular PPCE was unaffected by inhibitors of trypsin and kallikrein (aprotinin, ABTI), AmM (bestatin, amastatin), neutral endopeptidase (phosphoramidon), angiotensin I converting enzyme (captopril), or carboxypeptidase N (MERGETPA). These characteristics demonstrate that PPCE is present in vascular endothelium and smooth muscle. Although PPCE may participate in vascular metabolism of vasoactive peptides, its cytosolic localization and endopeptidase specificity suggests that its role is limited to peptide degradation after transport into the cell (*143*).

Approaches To Reduce Metabolism

Approaches that have been used to stabilize the parent peptide molecule to hydrolytic attack are as follows:

- chemical modification of peptide and protein structures
- coadministration with enzyme inhibitors and surfactants
- selection of a delivery site with less enzyme distribution

The methods used for chemical modification are as follows:

- Conformation constraints by steric or stereochemical means; i.e., the use of unnatural or D-amino acids or *N*-methyl amino acids (*85, 146*).
- Covalent constraints; i.e., the use of cyclic amino acids, bridged dipeptides, or cyclic peptides (*14, 147*).
- Retroenantiomer analogues; i.e., the direction of the peptide backbone is reversed and

the chirality of each amino acid is inverted (*147*) This approach is limited if the amide linkages are important in the interactions with a target receptor).

- Analogues with substituted C- and N-termini; i.e., acylation or alkylation of the N-terminus or alteration of the C-terminus by reduction or amide formation (*146*).
- Substitution or reduction of the peptide bond; for the angiotensin-converting-enzyme (ACE) inhibitor benzoyl–Phe–Gly–Pro as well as for the human immunodeficiency virus protease inhibitors, replacement of the CONH– group by $-COCH_2-$ enhanced the inhibitor activity (*147*).

Chemical modification of peptides to achieve stability against hydrolysis is likely to reduce absorption by selective carrier-mediated mechanisms; however, potential for the design of orally active molecules should increase. Nevertheless, the contribution of oral bioavailability to the overall activity of a synthetic peptide derivative can be difficult to determine because the metabolic stability or intrinsic potency at the site of action may be altered. Examples of the successful design of orally active derivatives of natural peptides include analogues of LHRH (*148, 149*), metenkephalin (*146*), pentagastrin (*150*), and adrenocorticotrophic hormone (*151*).

Another approach that uses enzyme inhibitors or surfactants is as follows: Metabolism of carbapenems by a renal dehydropeptidase can be blocked by coadministration of cilastatin, which was designed as a specific suicide inhibitor (*152*). Similarly, the deactivation of β-lactam antibiotics by β-lactamase-producing bacteria has led to the design of inhibitors, such as clavulanic acid or sulbactam, that can be used in combination with antibiotics (*2*).

Excretion

Regulatory peptides and proteins usually carry specific messages between cells and transmission channels in the extracellular space (*153*). The specificity of these molecules in vitro is primarily a property of the fit between the peptide and its receptor. However, the biological action of a peptide in a physiological environment is strongly affected by transport and elimination processes. Transport processes determine which cells are potentially accessible; elimination processes determine the duration of action of the molecule in certain environments. Transport and elimination processes depend qualitatively and quantitatively on tissue type and precise cellular environment and peptide structure. Thus, the peptide carries information that determines not only the receptor type that can be activated, but also a specific pattern of distribution and elimination.

For most peptide and protein drugs, a small volume of distribution, hepatic and biliary clearances, and renal clearance by glomerular filtration and tubular secretion mechanisms result in rapid elimination from the body. The excretion data on the somatostatin analogues and renin inhibitors, except the acidic prototype pepstatin, suggest that the features favoring rapid hepatic clearance in bile are high molecular weight and hydrophobicity combined with some degree of residual basicity and polarity. The kidney plays an important role in the disposition of protein drugs because proteins with molecular weights of less than 30 kDa are filtrated through glomeruli (*124, 154*) and proteins are also metabolically degraded in this organ (*155, 156*). Interferons, interleukin 2, granulocyte colony-stimulating factor, and erythropoietin have molecular weights within this range, and the kidney plays a major role in their clearance from the body (*157*). Renin tubular epithelial cells, particularly those in the proximal tubule, have the ability to reabsorb proteins from the tubular lumen.

In this process, proteins are bound to the luminal cell surface, undergo endocytosis, and then are digested in lysosomes of the cells. The reabsorption process was shown to be a saturable process because the ratio decreased with an increase in the dose. The nonlinear reabsorption could be explained by the saturation of intact proteins binding to the luminal surface as an initial step of endocytosis. Net charge may be an important determinant of renal cell uptake of proteins because cationic proteins were reported to be more susceptible to reabsorption than anionic proteins (*158, 159*). The difference in numbers of free amino groups or isoelectric point may account for the differences in uptake. These findings suggest that the renal uptake of proteins mainly occurred at the proximal tubule (*160*). The critical point might have existed at a molecular size of 20,000–30,000. Nevertheless, the present experimental system using the perfused rat kidney is useful for quantitative analysis of renal disposition of proteins and their various derivatives.

The hepatic and biliary clearance of cefoperazone reflects the hydrophobicity of the molecule or indicates that cefoperazone is recognized by some selective anion uptake system. For the "third generation" cephalosporins (*161*), the property of metabolic stability resulted in half-life values greater than 2 h for ceftriaxone only. Of these compounds, cefoperazone is unusual in that biliary excretion—and not renal clearance, as with moxalactam—is the predominant route of elimination. The dipeptide mimetics (e.g., captopril and enalapril) do not appear to be extensively excreted in bile, although the more recent lipophilic ACE inhibitors (e.g., ramipril) show increasing preference for the biliary route (*141*). An example of a larger acidic peptide excreted in bile is the renin inhibitor pepstatinyl–glycine, a hexapeptide. This compound is also excreted in urine (*95*). Thus, for the acidic peptides, the profile of biliary excretion seems similar to that described for many organic acids (*142*) where elimination in bile is favored for high molecular weight compounds (*118*).

However, the immunostimulant muramyl dipeptide, a glycopeptide with a molecular weight of nearly 500 that is predominantly excreted in urine (*162*), is an exception to the rule, presumably because the molecule is highly polar. TRH analogues are tripeptides containing a basic histidine residue and seem to be cleared predominantly in the urine, whereas the tripeptide leupeptins possessing two hydrophobic leucine residues and an amphoteric arginine residue are excreted in both urine (*163*) and bile (*88*). Even larger basic oligopeptides, such as some of the peptide inhibitors of renin, lack oral activity due to first-pass clearance by hepatic biliary excretion (*164*).

For somatostatin analogues, rapid clearance by biliary excretion is a problem in the design of effective therapeutic agents. Interestingly, the parent peptide somatostatin, a cyclic tetradecapeptide, is not excreted in bile (*10*), perhaps because of rapid hydrolysis (*87*). However, the cyclic octapeptide CGP-15425 and its α-naphthyl γ-amino butyric acid analogue are metabolically stable analogues of somatostatin (*10*) and undergo rapid elimination as intact peptides in bile. A slightly smaller hexapeptide analogue of somatostatin is also rapidly cleared in bile (*11*). The fact that these compounds are secreted in bile against a high concentration gradient points to a specific carrier-mediated active transport process (*142*) such as that known to exist for organic cations.

Excretion of Selected Peptide and Protein Drugs

Radiolabeled cholecystokinin (CCK-33), CCK octapeptide (CCK-8) (*118*), CCK-8 desulfate, and CCK-4 are extracted by the liver in a structurally specific manner. Rapid biliary excretion of labeled CCK-8, CCK-8-desulfate, and CCK-4 occurred in the isolated, perfused rat-liver system. By means of high-performance liquid chromatography and immunoprecipitation, CCK-8 desulfate and CCK-4 were shown to appear in bile in completely metabolized forms. In contrast about 20% of the major forms of the label for CCK-8 in bile was intact octapeptide. To gain insight into the subcellular sites of metabolism and transhepatic transport of CCK-8, Gores et al. (*165*) determined the effects of taurocholate, lysomotropic agents, and microtubule binding agents on biliary excretion. Taurocholate had no effect on the percent of extracted label or the nature of the metabolites appearing in bile. Two microtubule binding agents, vinblastine and colchicine, also did not affect the percent of extracted label appearing in bile. These results suggest that, after efficient first-pass hepatic extraction, cholecystokinin undergoes extensive metabolism, possibly at a subcellular site other than lysosomes. Also, cholecystokinin is rapidly and efficiently transported across the hepatocyte into bile, possibly by a nonvesicular transport process.

TGFβ, the aforementioned polypeptide, modulates growth of normal and neoplastic cells (*119*). First-pass hepatic extraction of [^{125}I]TGFβ in the isolated, perfused rat liver was efficient (36%).

Administration of leukotrienes ([^{3}H]leukotriene C) to rats resulted in a rapid (5 min) systemic clearance of radioactivity from the blood. Intravenous administration of [^{3}H]LTC4 also resulted in a significant time-related biliary excretion of 69 ± 4.1% (n = 6 over 60 min) of administered radioactivity and a urinary recovery of only 0.9 ± 0.3% (n = 3 over 60 min). These data suggest that the leukotrienes are eliminated from systemic circulation by an efficient mechanism in the liver resulting in biliary excretion as the major route of elimination. Characterization of these metabolites has allowed us to clearly demonstrate that peptide leukotrienes undergo N-acetylation followed by omega- and subsequent beta-oxidation in the rat.

Simultaneously measured hepatic extractions of endogenous and exogenously infused insulin were 43.8 ± 7.6 and 47.5 ± 4.4%, respectively, in dogs. The metabolic clearance rate of infused connecting peptide (C-peptide) was 11.5 ± 0.8 mL kg^{-1} min^{-1} and was constant over the concentration range usually encountered under physiological conditions (*166, 167*). After infusion of somatostatin 28 (SS28) and somatostatin 14 (SS14) in anesthetized dogs, the hepatic and renal extractions for SS28 were 11.0 ± 1.5 and 50 ± 4.8%, respectively, whereas those for SS14 were 43.1 ± 7.4 and 82.2 ± 6.6%, respectively (*168*).

Effect of Disease State on Absorption, Metabolism, Distribution, and Elimination

Some clinical conditions (e.g., essential hypertension, pulmonary hypertension, cardiomyopathy, acute myocardiac infarction, uremia, diabetes, pregnancy, cold, and exercise) may alter the pharmacokinetics of peptide and protein drugs.

In patients with celiac disease, tropical sprue, and short-bowel syndrome, jejunal absorption of dipeptides is less severely affected than is absorption of free amino acids. Protein–calorie malnutrition reduced absorption of leucine without affecting the dipeptide, glycine–leucine (*169*). With the induction of hypoproteinemia, the diet appeared to simulate water absorption when compared to control animals ($p < 0.01$). Luminal perfusion with high protein diet failed to attenuate net water secretion induced by hypoproteinemia. Capillary and mucosal albumin clearance was similar for all groups studied. These findings suggest that small molecular weight peptides may affect the rate of intestinal absorption in patients with acute kwashiorkor-like hypoalbuminemia (*169*).

The circulating blood levels of endothelins in patients are increased, usually by two- to threefold, in a variety of disease states (*170*). These disease states all represent increases in stress, as does ischemia. Endothelin release might be part of a stress response, either as an initiator or as a potentiator of the response manifestations. The clinical conditions in which plasma endothelin levels are elevated can be mimicked in animal models, except perhaps for pre-eclampsia. However, we do not know if these increased levels are a primary or secondary symptom or if they are even sufficient to produce significant systemic effects (*170*).

The protease activity of strained rumen fluid (SRF), expressed in milliliters, was slightly lower in faunated sheep than in celiate-free animals (*171*). However, this difference was apparently the result of a fall in protein content of SRF, because the activities per milligram of protein were not significantly different. Introduction of large protozoa to sheep with an existing population of small entodinia had no significant effect. The protozoa may have a much greater ecological role in the breakdown of bacterial protein, and possibly other proteins in small particulate form, rather than in the breakdown of soluble proteins (*171*).

Patients whose plasma most rapidly degraded human calcitonin (HCT) had milk–alkali syndrome, metastatic carcinoma of the colon, metastatic oat-cell carcinoma of the lung, or metastatic breast carcinoma (*172*). The human plasma contains one or more enzymes that degrade HCT. This hormone is degraded more rapidly in plasma from some patients with hypercalcemia (*172*). The relationship between blood calcium levels and HCT degradation rate in plasma was explored. At first, the effect of increasing and decreasing the calcium concentration in vitro was investigated; neither change altered the rate of HCT loss. In the second study, loss of HCT during incubation in plasma obtained from four healthy volunteers following a 4-h iv infusion of 15 mg Ca^{2+}/kg was compared with HCT loss in plasma obtained before infusion; no difference was observed. For rabbit base-pair peptides (fraction 43–88) administered to patients who had undergone bilateral nephrectomies, the elimination half-life increased from 50 to 500 min and the distribution half-life increased from 5 to 15 min, (*173*).

Infusion in normal subjects of exogenous insulin results in a decrease of glucagon and free fatty acids levels (*174*). In obese and diabetic patients, the presence of higher than normal concentrations of glucagon or free fatty acids, or the lack of an appropriate decrease of glucose in the presence of high insulin levels, could contribute to insulin resistance (*174*).

Plasma concentrations of peptide YY (PYY), which is produced by ileal and colonic endocrine cells, were measured in several groups of patients with digestive disorders after a standardized normal breakfast (*175*). PYY circulates in plasma, and its concentration rises in response to the physiologic stimulus of food (*176*). Levels of PYY were grossly elevated in patients with steatorrhea due to small intestinal mucosal atrophy (tropical sprue). Basal levels in patients were 79 ± 18 pmol, a level that was nearly 10-fold higher than levels seen in healthy controls (8.5 ± 0.8 pmol). Moderately elevated plasma PYY concentrations were seen in patients with inflammatory bowel disease and patients recovering from acute infective diarrhea. In contrast, patients with diverticular disease, duodenal ulcer, and functional bowel disease had normal PYY responses. These changes in the secretion of PYY appear to result from malabsorption and may shed light on the physiologic role of this newly discovered peptide and on intestinal adaptation to common digestive disorders (*175*). These digestive disorders are associated with changes in secretion and motility that may be influenced by PYY.

Pharmacokinetics and Biopharmaceutical Properties of Selected Peptide and Protein Drugs

Pharmacokinetic parameters of peptide and protein drugs reported in the literature may differ among investigators. This difference could be due to the differences in the experimental protocols, such as the use of animals, [^{125}I]labeled peptides, and dosing levels influencing the kinetics. Moreover, interpretation of the many studies cited is difficult due to the differences in experimental paradigms, species differences, mode of administration, lack of dose–response relationships, and state of consciousness of the animal studied (*177*). Although a limited database is available in the literature, Table VI is an attempt at the construction of a master table that summarizes pharmacokinetic parameters. Individual peptide and protein drugs with noticeable pharmacokinetic characteristics are discussed in the following sections.

Atrial Natriuretic Factors

The pharmacokinetics of the atrial natriuretic factor (ANF) were studied in various species (*89*). In rats, ANF is cleared from the body rapidly (about 100 mL min^{-1} kg^{-1}) with a large volume of distribution (2.35 L/kg). Similar clearance was reported in rabbits and dogs. This fast clearance and large volume of distribution of ANF are probably related to receptor-binding events (*89*).

Calcitonin

Arteriovenous differences of calcitonin were found in dogs (*178*) across the lung, kidney, liver, muscle, and bone, and these differences indicate slow distribution or tissue metabolism. The plasma distribution profile following iv bolus injection of 4 μg of salmon calcitonin (*36*) in rats peaked at 1–3 min. The kidney was the major organ for inactivation of salmon, porcine, and human calcitonins in dogs. The fivefold increase in plasma calcitonin concentrations with renal-failure patients supports these findings (*178*).

Cyclosporin

After iv administration, cyclosporin exhibits multiexponential decay (*179, 180*) in humans. Pharmacokinetic parameters of cyclosporin showed wide variations that were dependent on patient disease states, type of organ transplant, age, and drug interaction. More than 90% of the administered iv dose was excreted as metabolites in bile in humans, dogs, and rats (*181*). Rapid initial distribution (0.1 h half-life) in humans can be attributed to the high lipid solubility of cyclosporin and its ability to cross some biological membranes. Consistent with its lipophilicity, cyclosporin accumulates in body fat (*182*). In blood, 58% of cyclosporin is bound to erythrocytes.

Interferons

The distribution and elimination of interferons (IFNs) are similar across most species. This similarity is probably because IFNs are endogenous substances and protein processing is a natural function of all species (*89*). Volume distributions of IFN-α and IFN-β following

Table VI. Comparison of Pharmacokinetic Parameters of Selected Peptide and Protein Drugs

Drug	M_r *(kDa)*[a]	*Dose*	*α-Half-Life (min)*	*β-Half-Life (min)*	*Vdss (L/kg)*[b]	*Cl (mL min⁻¹ kg⁻¹)*[b]	Subject
ANF (99–126)	3.1 (28)	100 μg	1.7	13.3	0.062	21.7	Human
ANF (103–126)	(24)	0.25 μg	0.44		2.35	100	Rat
Calcitonin, salmon	3.4 (32)	0.4 μg	5–10	50–80	0.04–80	2.85–7.14	Human
Cilofungin			3.7 ± 0.2	12.9 ± 0.7	0.85 ± 0.23	30 ± 10	Rabbit
Cyclosporin	(11)		6	66	3–9		Human
Daptomycin					0.10–0.15	0.13	Human
Delta sleep-inducing peptide	0.84 (9)	0.1 mg/kg	6–8		0.67		Dog
Endothelins	2.6 (21)	30 pmol/kg	7[c]	35–46[c]			Rat
Erythropoietin	30.4 (26)		23.7 ± 5	540 ± 36	0.102 ± 0.016		Dog
Factor VIII	80–210			540–660	0.049–0.058	0.055–0.083	Human
Gastrin-releasing peptide	2.9 (27)		2.8 ± 0.4		133 ± 31		Human
Gastrin-releasing peptide			1.4	6.6			Dog
Human GH	20		4.1		0.19	16	Mice
Human GH					0.10	13.5	Rat
Human GH					0.083	3.87	Monkey
Human GH					0.059	2.18	Human
Interferon, α	(17–23)			240–960	0.17	5.71–24.5	Human
Interferon, β				60–120		22.1–44.3	
Interferon, γ				25–35	0.57	14.8–37.8	
Leukotrienes			5[c]				
LHRH	1.2 (10)			6.6	0.56 (Vdβ)		Rat
LHRH				33 ± 4.8	1.53		Monkey
Myelin basic protein	1.7 (15)		5	55			Rabbit
PYY	4.3 (36)			11.7 ± 0.6		13.8 ± 1.6	Human
Relaxin, human	(57)	88 mg/kg	5.3	36 (β)	0.22	15.7	Mice
				513 (γ)	0.67		
Relaxin, human		88–100 mg/kg	2.0	25 (β)	0.066–0.072	5.8	Rat
				76–293 (γ)	0.29–0.62		
Relaxin (rhesus monkey)	(57)	100 mg/kg	2.0 ± 0.5	24 ± 7 (β)	0.078 ± 0.025	4.1 ± 0.6	Rhesus monkey
				249 ± 49 (γ)	0.69 ± 0.22		
Somatostatin 14	1.5 (14)			1.7 ± 0.2	21.9 ± 6.5		Dog
Somatostatin 28	3.1 (28)			2.8 ± 0.3	9.9 ± 1.4		Dog
TPA	69 (530)	0.25 mg/kg	3.4 ± 0.6	34 ± 12	7.6 ± 1.9	17.6 ± 2.43	Human
TGFβ	(3)		2.2				
TNF, α	(17)	25–100 μg/m²		14–18	0.94–1.73		Human
Vasopressin	1.08 (9)	0.9 μg/kg	5.4				Rabbit
Vasopressin		1.4 μg/kg	4.1				Dog
Vasopressin			26.1				Human

NOTE: Vdβ is volume of distribution in terminal phase; β is terminal phase in two-exponential decay; γ is terminal phase in three-exponential decay; Vdss is volume of distribution at steady state; Cl is total body clearance; ANF is atrial natriuretic factor; GH is growth hormone; TPA is tissue plasminogen activator; TNF is tumor necrosis factor.

[a] Value in parentheses is number of amino acids.

[b] Values were normalized by body weights of the following: human, 70 kg; rat, 0.15 kg; mouse, 0.02 kg; and monkey, 3.8 kg.

[c] Values were determined by total radioactivity.

SOURCE: Data were taken from references 186 and 187.

iv administration range from 20 to 100% of body weight in mice, rats, rabbits, dogs, and monkeys (*183*). Although oral absorption is not successful, other routes of administration have shown good bioavailability and provided adequate concentrations of IFN in cerebrospinal fluid, lymph, nasal mucosa, and peritoneal fluid (*184*). However, clinical application of alternative routes is not successful and indicates the presence of more complex pharmacology.

Thyrotropin-Releasing Hormone

TRH is rapidly degraded in various tissues. The degradation rates were highly species-dependent (*14, 184*). The half-lives in rabbit and human were 2 and 0.67 h, respectively. The half-life in pig was 40 times faster than in guinea pig plasma. In dog, almost no degradation was observed. The metabolic sites for TRH were also species-dependent. The liver was an active tissue in rats and mice, but it was less active in humans and was inactive in dogs (<5% change in 6 h). TRH-degrading activity was low in all gut tissues of rats, humans, and mice, and again activity was absent in dogs. The brain was rich in TRH-metabolizing enzymes in mouse, rat, dog, and human. Of the dog tissues examined, (e.g., plasma, liver, gut, and brain) the brain was the only tissue containing TRH-degrading enzymes.

Tumor Necrosis Factor

The pharmacokinetics of tumor necrosis factor-α appear to be dose-dependent. At low dose-infusion in rhesus monkey, the elimination rate increased steadily and indicated time- and concentration-dependent kinetics. Plasma disposition profiles were biphasic and alpha and beta half-lives were 0.89 and 1.99 h in rabbits and of 1.29 and 4.52 h in monkeys, respectively (*185*).

References

1. Ringrose, P. S. *Biochem. Soc. Trans.* **1983,** *2,* 804.
2. Ringrose, P. S. *Soc. G. Microbiol.* **1985,** *35.*
3. Krieger, T. T. *Science (Washington, DC)* **1983,** *222,* 975.
4. Hruby, V. J.; Krstenansky, J. L.; Cody, W. L. *Annu. Rep. Med. Chem.* **1984,** *19,* Chapter 30.
5. Zhou, X. H.; Po, A. L. W. *Int. J. Pharm.* **1991,** *75,* 97–115.
6. Smith, L. C.; Pownall, H. J.; Gotto, A. J., Jr. *Annu. Rev. Biochem.* **1978,** *47,* 751.
7. Bristow, A. F. In *Polypeptide and Protein Drugs*; Hider, R. C.; Baelow, D., Eds.; Ellis Horwood Limited: London, 1991; p 54.
8. Rowland, R. N.; Woodley, J. F. *Biosci. Rep.* **1981,** *1,* 399.
9. Houston, J. B.; Wood, S. G. In *Progress in Drug Metabolism*; Bridges, J. W.; Chasseaud, L. F., Eds.; Wiley: New York, 1980; Vol. 4, p 57.
10. Baker, J. P.; Kemmense, B. H.; McMartin, C.; Peters, G. E. *Regul. Pept.* **1984,** *9,* 213.
11. Bell, J.; Peters, G. E.; McMartin, C.; Thomas, N. W.; Wilson, C. G. *J. Pharm. Pharmacol.* **1984,** *36,* 88.
12. Wood, A. J.; Maurer, G.; Niederberger, W.; Reveridge, T. *Transplant. Proc.* **1983,** *15,* 2409.

13. Voda, C. T.; Lemaire, M.; Sell, G. G.; Nussbaumer, K. *Biopharm. Drug Dispos.* **1983,** *4,* 113.

14. Brewster, D.; Waltham, K. *Biochem. Pharmacol.* **1981,** *30,* 619.

15. Taylor, A. E.; Granger, D. N. In *Handbook of Physiology*; Hamilton, W. F., Ed.; American Physiology Society: Washington, 1984; Vol. 6, pp 467–520.

16. McMartin, C.; Hutchinson, L. E. F.; Hude, R.; Peters, G. E. *J. Pharm. Sci.* **1987,** *76,* 535–540.

17. Peters, G. E.; Hutchinson, L. E. F.; Hyde, R.; McMartin, C.; Metcalf, S. B. *J. Pharm. Sci.* **1987,** *76,* 857–861.

18. Lee, W. A.; Longenecker, J. P. *J. Biopharm.* **1988,** *1,* 30–37.

19. Karls, M. S.; Rush, B. D.; Wilkinson, K. F.; Vidmar, T. J.; Burton, P. S.; Ruwart, M. J. *Pharm. Res.* **1991,** *8(12),* 1477–1481.

20. Samanen, J. M. In *Peptide and Protein Drug Delivery*; Lee, V. H. L., Ed.; Marcel Dekker: New York, 1991; Vol. 4, pp 137–202.

21. Weber, S. J.; Greene, D. L.; Hruby, N. J.; Tomlinson, E.; Inby, V. J.; Yamamura, H. I.; Porreca, F.; Davis, P. D. *J. Pharmacol. Exp. Therap.* **1992,** *263(3),* 1308–1316.

22. Benet, L. A. In *Drugs and the Pharmaceutical Sciences,* 2nd ed.; Banker, G. S.; Rhodes, C. T., Eds.; Marcel Dekker: New York, 1990; Vol. 40, pp 181–207.

23. Burkes, T. F.; Fox, D. A.; Hirining, L. D.; Shook, J. E.; Porreca, F. *Life Sci.* **1988,** *43,* 2177–2181.

24. Shook, J. E.; Lemcke, P. K.; Gehrig, C. A.; Hruby, V. J.; Burks, T. F. *J. Pharmacol. Exp. Ther.* **1989,** *249,* 83–90.

25. Porreca, F.; Mosberg, H. I.; Hurst, R.; Hruby, V. J.; Burks, R. F. *J. Pharmacol. Exp. Ther.* **1984,** *230,* 341–348.

26. Shook, J. E.; Pelton, J. T.; Hruby, V. J.; Burks, T. F. *J. Pharmacol. Exp. Ther.* **1987,** *243,* 492–500.

27. Weber, S. J.; Greene, D. L.; Sharma, S. D.; Yamamura, H. I.; Kramer, T. K.; Burks, T. F.; Hruby, V. J.; Hersh, L. B.; Davis, T. P. *J. Pharmacol. Exp. Ther.* **1991,** *259,* 1109–1117.

28. Ponsin, G. *Adv. Exp. Med. Biol.* **1988,** *243,* 139–147.

29. Hutchinson, F. G.; Furr, B. J. *Biochem. Soc. Trans.* **1985,** *13,* 520.

30. Patel, H. M. *Biochem. Soc. Trans.* **1989,** *17(5),* 931.

31. Ziv, E.; Kidron, M.; Berry, E.; Baron, H. *Life Sci.* **1981,** *29,* 803.

32. Fujii, S.; Yokoyama, T.; Ikegaya, K.; Sato, F.; Yokoo, N. *J. Pharm. Pharmacol.* **1985,** *37,* 545.

33. Lee, H. J.; Amidon, G. L. *Abstract of Papers,* International Symposium, Delivery of Protein Drugs—The Next 10 Years, Kyoto, Japan; Controlled Release Society, Inc.: Dearfield, IL, 1993.

34. Su, S.-F.; Lee, H. J.; Amidon, G. L. *Pharm. Res.* **1993,** *10,* 291S.

35. Gregoriadis, G. *N. Engl. J. Med.* **1976,** *295,* 704.

36. Lee, H. J.; Kim, J. S.; Amidon, G. L.; Chandrasekharan, R.; Weiner, N. D. *Pharm. Res.* **1993,** *10,* 291S.

37. Bai, J. P.; Hu, M.; Subramanian, P.; Mosberg, H. I.; Amidon, G. L. *J. Pharm. Sci.* **1992,** *81,* 113–116.

38. Shichiri, M.; Kamon, R.; Yoshida, M.; Etani, N.; Hoshi, M.; Izumi, K.; Shigeta, Y.; Abe, H. *Diabetes* **1975,** *24,* 971.

39. Couvreur, P.; Lenaerts, V.; Kante, B.; Roland, M.; Speiser, P. P. *Acta Pharm. Technol.* **1980,** *26,* 220.

40. Ptachcinski, R.; Venkataramanan, R.; Rosenthal, J.; Burckart, G.; Taylor, R.; Hakala, T. *Transplantation* **1985,** *40,* 174.

41. Lee, V. H. L.; Yamamoto, A.; Kompella, U. B. *CRC Crit. Rev. Ther. Drug Carrier Syst.* **1991,** *8,* 91–192.

42. Muranashi, S. *CRC Crit. Rev. Ther. Drug Carrier Syst.* **1990,** *7,* 1–34.

43. Ritschel, W. A. *Methods Find. Exp. Clin. Pharmacol.* **1991,** *13,* 202–220.

44. Saffran, M.; Franco-Saenz, R.; Kongm, A.; Papahadjopoulos, D.; Szoka, F. *Can. J. Biochem.* **1979,** *57,* 548–553.

45. Lundin, S.; Vilhardt, H. *Life Sci.* **1986,** *38,* 703–709.

46. Vilhardt, H.; Lundin, S. *Gen. Pharmacol.* **1986,** *17,* 481–483.

47. Bauer, W.; Briner, U.; Doepfner, W.; Haller, R.; Huguenin, E.; Marbach, P.; Petcher, T. J.; Pless, J. *Life Sci.* **1982,** *31,* 1133–1140.

48. Fuessl, H. S.; Domin, J.; Bloom, S. R. *Clin. Sci.* **1987,** *72,* 255–257.

49. Amoss, M.; Rivier, J.; Guillemin, R. *J. Clin. Endocrinol. Metab.* **1972,** *35,* 175–177.

50. Humphrey, R. R.; Dermody, W. C.; Brink, H. O.; Bousley, F. G.; Schottin, N. H.; Sakowski, R.; Vaitkus, J. W.; Veloso, H. T.; Reel, J. R. *Endocrinology* **1973,** *92,* 1515–1526.

51. Nishi, N.; Arimura, A.; Coy, D. H.; Vilchez-Martinez, J. A.; Schally, A. V. *Proc. Soc. Exp. Biol. Med.* **1975,** *148,* 1009–1012.

52. Ormiston, B. J. In *Thyrotropin Releasing Hormone*; Hall, R.; Werner, I.; Holgate, H., Eds.; Karger: Basel, Switzerland, 1972; pp 45–52.

53. Laskowski, M; Haessler, H. A.; Miech, R. P.; Peanasky, R. J.; Laskowski, M. *Science (Washington, DC)* **1958,** *127,* 1115–1116.

54. Fitzgerald, J. F.; Kottmeier, P. K.; Adamsons, R. J.; Butt, K. M.; Hochman, R. A.; Dennis, C. *Surg. Forum* **1968,** *19,* 297–299.

55. Patel, H. M.; Ryman, B. E. *FEBS Lett.* **1976,** *62,* 60–63.

56. Banga, A. K.; Chien, Y. W., *Int. J. Pharm.* **1988,** *48,* 15–50.

57. Chien, Y. W.; Liu, J.; Sun, Y.: Siddiqui, O. *J. Pharm. Sci.* **1987,** *76,* S60.

58. Burnette, R. R.; Marrero, D. *J. Pharm. Sci.* **1986,** *75(8),* 738–743.

59. Banerjee, P. S.; Ritschel, W. A. *Int. J. Pharm.* **1989,** *49,* 189–197.

60. Kazim, M.; Weber, C.; Strausberg, L.; LaForet, G.; Nicolson, J.; Reemtsma, K. *Diabetes* **1986,** *33(Suppl),* 181A.

61. Banerjee, P. S.; Ritschel, W. A. *Int. J. Pharm.* **1989,** *49,* 199–204.

62. Hadgraft, J In *Design of Prodrugs*; Bundgaard, H., Ed.; Elsevier: New York, 1985; pp 271–289.

63. Potts, R. O.; McNeil, S. C.; Desbonnet, C. R.; Wakshull, E. *Pharm. Res.* **1989,** *6,* 119–124.

64. Shah, P. K.; Borchardt, R. T. *Pharm. Res.* **1991,** *8(1),* 70–75.

65. Choi, H.; Flynn, G. L.; Amidon, G. L. *Pharm. Res.* **1990,** *7(11),* 1099–1106.

66. Heit, M. C.; Williams, P. L.; Jayes, F. L.; Chang, S. K.; Riviere, J. E. *J. Pharm. Sci.* **1993,** *82(3),* 240–243.

67. Hirai, S.; Yashiki, T.; Matsuzawa, T.; Mima, H. *Int. J. Pharm.* **1981,** *7,* 317.

68. Adelmann, H.; Graef, V.; Schatz, H. *Horm. Metab. Res.* **1984,** *16,* 55.

69. Schade, D. S.; Eaton, R. P. *N. Engl. J. Med.* **1985,** *312,* 1120.

70. Sarkar, M. A. *Pharm Res.* **1992,** *9(1),* 1–9.

71. O'Hagan, D. T.; Critchley, H.; Faraj, N. F.; Fisher, A. N.; Johansen, B. R.; Davis, S. S.; Illum. L. *Pharm. Res.* **1990,** *7,* 772–776.

72. Gallardo, D.; Longenecker, J. P.; Lee, V. H. L. *Abstracts of Papers,* Proceedings of the 14th International Symposium on Controlled Release of Bioactive Materials; Controlled Release Society, Inc.: Dearfield, IL, 1987; Abstract 30.

73. Mackay, M.; Tomlinson, E. In *Colonic Drug Absorption and Metabolism*; Bieck, P. R., Ed.; Marcel Dekker: New York, 1992.

74. Ritschel, W. A. *Exp. Clin. Pharmacol.* **1991,** *13,* 313–336.

75. Anders, R.; Merkle, H. P.; Schurr, W.; Ziegler, R. *J. Pharm. Sci.* **1983,** *72,* 1482–1483.

76. Ortega-Corona, B. G.; Garcia-Pineda, J.; Parra, A.; Gallegos, A. J. *Arch. Invest. Med.* **1989,** *20,* 239–242.

77. Straford, R. E., Jr.; Carson, L. W.; Dodda-Kashi, S.; Lee, V. H. L. *J. Pharm. Sci.* **1988,** *77,* 838–842.

78. Golomb, G.; Avramoff, A.; Hoffman, A. *Pharm. Res.* **1993,** *10,* 828.

79. Okada, H.; Yamazaki, I.; Ogawa, Y.; Harai, S.; Yaskiki, T.; Mima, H. *J. Pharm. Sci.* **1982,** *71,* 1367.

80. Yoshida, H.; Kumura, K. O.; Hori, R.; Anmo, T.; Yamaguchi, H. *J. Pharm. Sci.* **1979,** *68,* 670.

81. Jennewein, H. M.; Waldock, F.; Konz, W. *Arzneim. Forsch.* **1974,** *24,* 1225.

82. Yagi, T.; Hakui, N.; Yamasaki, Y.; Kawamori, R.; Schichiri, M.; Abe, H.; Kim, S.; Miyake, M.; Kamikawa, K.; Nishihata, T.; Kamada, A. *J. Pharm. Pharmacol.* **1983,** *35,* 177.

83. Mitsuma, T.; Nogimori, T. *Acta Endocrinol.* **1984,** *107,* 207.

84. Lonovics, J.; Narai, G.; Varro, V. *Mater. Med. Pol.* **1980,** *12,* 229.

85. Humphrey, M. J.; Ringrose, P. S. *Drug. Metab. Rev.* **1986,** *17(3,4),* 283–310.

86. Dickson, R. B. *Trends Pharm. Sci.* **1989,** *10,* 125–127.

87. Peters, G. E. *Regul. Pept.* **1982,** *31,* 361.

88. Demis, P. A.; Aronson, N. N. *Arch. Biochem. Biophys.* **1985,** *240,* 768.

89. Moore, J. A.; Wroblewski, V. J. In *Protein Pharmacokinetics and Metabolism*; Ferraiolo, B. L.; Mohler, M. A.; Gloff, C. A., Eds.; Plenum: New York, 1992; pp 93–117.

90. Sandow, J.; Eckert, H.; Stoll, W.; von Rechenberg, W. *J. Endocrinol.* **1977,** *73,* 33.

91. Ringrose, P. S. *Med. Microbiol.* **1983,** *3,* 179.

92. Magnan, S. D.; Shirota, F. N.; Nagasawa, H. T. *J. Med. Chem.* **1982,** *25,* 1018.

93. Bodor, N.; Prokai, L.; Wu, W.; Farag, H.; Jonalagadda, S.; Kawamura, M.; Simpkins, J. *Science (Washington, DC)* **1992,** *257,* 1698–1700.

94. Metcalf, G.; Dettmar, P. W.; Lynn, A.; Brewster, D.; Hauler, M. E. *Regul. Pept.* **1981,** *2,* 277.

95. Grant, D. A.; Ford, T. F.; McCulloch, R. J. *Biochem. Pharmacol.* **1982,** *31,* 2302.

96. Cornford, E. M.; Braun, L. D.; Crane, P. D.; Oldendorf, W. H. *Endocrinology* **1978,** *103,* 1297.

97. Rybak, M. J.; Bailey, E. M.; Lamp, K. C.; Kaantz, G. W. *Antimicrob. Agents Chemother.* **1992,** *36(5),* 1109–1114.

98. Woodworth, J. R.; Nyhart, E. H., Jr.; Brier, G. L.; Wolny, J. D.; Black, H. R. *Antimicrob. Agents Chemother.* **1992,** *36(5),* 318–325.

99. Ooi, G. T. *Mol. Cell. Endocrinol.* **1990,** *71,* C39–C43.

100. Zapf, J.; Kiefer, M.; Merryweather, J.; Masiarz, F.; Bauer, D.; Born, W.; Fischer, J. A.; Froesch, E. R. *J. Biol. Chem.* **1990,** *265,* 14892–14898.

101. Cook, J. E.; Ferraiolo, B. L.; Mohler, M. A. *Pharm Res.* **1989,** *6,* S30, Abstract BT219.

102. Maurer, G. *Transplant. Proc.* **1985,** *17 (Suppl. 1),* 19.

103. Banerjee, P. K.; Amidon, G. L. In *Design of Prodrugs*; H. Bundgaard, Ed.; Elsevier: Amsterdam, The Netherlands, 1985; pp 93–133.

104. Takaori, F.; Burton, J.; Donowitz, M. *Biochem. Biophys. Res. Commun.* **1986,** *137,* 682–687.

105. Matthews, D. M. *Protein Absorption*; John Wiley & Sons: New York, 1991.

106. Bai, J. P. F.; Amidon, G. L. *Pharm. Res.* **1992,** *9(8),* 969–978.

107. Steinhardt, H. J.; Adibi, S. A. *Gastroenterology* **1986,** *90,* 577.

108. Morita, A.; Chung, Y. C.; Freeman, H. J.; Erickson, R. H.; Sleisenger, M. M.; Kim, Y. S. *J. Clin. Invest.* **1983,** *72,* 610.

109. Wojnarowska, F.; Gray, G. M. *Biochim. Biophys. Acta* **1975,** *403,* 147.

110. Kania, R. K.; Santiago, N. A.; Gray, G. M. *J. Biol. Chem.* **1977,** *252,* 4928.

111. Kenny, A. J.; Booth, A. G. *Essays Biochem.* **1978,** *14,* 1–44.

112. Palmieri, F. E.; Petrelli, J. J.; Ward, P. E. *Biochem. Pharmacol.* **1985,** *34(13),* 2309–2317.

113. Friedman, D. I.; Amidon, G. L. *Pharm. Res.* **1991,** *8(1),* 93–96.

114. Churchill, L.; Bausback, H. H.; Gerritsen, M. E.; Ward, P. E. *Biochim. Biophys. Acta* **1987,** *923,* 35–41.

115. Coffey, R. J.; Kost, L. J.; Lyons, R. M.; Moses, H. L.; LaRusso, N. F. *J. Clin. Invest.* **1987,** *80,* 750–757.

116. Hagenbuch, B.; Stieger, B.; Foguet, M.; Lübbert, H.; Meier, P. *Proc. Natl. Acad. Sci. U.S.A.* **1991,** *88,* 10629–10633.

117. Premont, R. T. In *Peptide Hormone Receptors*; Kalimi, M. Y.; Hubbard, J. R., Eds.; Walter de Gruyter: New York, 1987; pp 129–177.

118. Millburn, P. In *The Hepatobiliary System: Fundamental and Pathological Mechanisms*; Taylor, W., Ed.; Plenum: New York, 1976.

119. Pradayrol, L.; Jornvall, H.; Mutt, V.; Ribet, A. *FEBS Lett.* **1980,** *109,* 55.

120. Meijer, D. K. F. J. *Hepatology* **1987,** *4,* 259–268.

121. Silverstein, S. C.; Steinman, R. M.; Cohn, Z. A. *Annu. Rev. Biochem.* **1977,** *46,* 669–722.

122. Neurath, H. In *Proteolytic Enzymes: A Practical Approach*; Beynon, R. J.; Bond, J. S., Eds.; IRL: New York, 1989; pp 1–13.

123. Boyer, J. L.; Meier, P. J. *Methods Enzymol.* **1990,** *192,* 517–545.

124. Maack, T.; Johnson, V.; Kau, S. T.; Figueriredo, J; Sigulem, D. *Kidney Int.* **1979,** *16,* 251–270.

125. Carone, F. A.; Peterson, D. R.; Oparil, S.; Pullman, T. N. *Kidney Int.* **1979,** *16,* 271–278.

126. Carone, F. A.; Peterson, D. R. *Am. J. Physiol.* **1980,** *288,* G151–G158.

127. Rabkin, R.; Kitaji, J. *Miner. Electrolyte Metab.* **1983,** *9,* 212–226.

128. Bourdeau, J. E.; Corone, F. A.; Ganote, C. E. *J. Cell. Biol.* **1972,** *54,* 382–398.

129. Silbernagl, S.; Volkl, H. *Curr. Probl. Clin. Biochem.* **1977,** *8,* 59–65.

130. Matthews, D. *Physiol. Rev.* **1975,** *55,* 537–608.

131. Nutzenadel, W.; Scriver, C. R. *Am. J. Physiol.* **1976,** *230,* 642–651.

132. Abidi, S. A.; Kryzysik, B. A. *Clin. Sci.* **1977,** *52,* 205–213.

133. Katz, A. I.; Rubenstein, A. HJ. *J. Clin. Invest.* **1973,** *52,* 1113–1121.

134. Emmanouel, D. S.; Katz, A. I.; Lindheimer, M. D. In *Renal Endocrinology*; Dunn, M. J., Ed.; Williams & Wilkins: Baltimore, MD, 1983; pp 474–504.

135. Kerr, M. A.; Kenny, A. J. *Biochem. J.* **1974,** *137,* 477–488.

136. Abbs, M. T.; Kenny, A. J. *Clin. Sci.* **1983,** *65,* 551–559.

137. Stephenson, S. L.; Kenny, A. J. *Biochem. J.* **1987,** *241,* 237–247.

138. Stephenson, S. L.; Kenny, A. J. *Biochem. J.* **1988,** *255,* 45–51.

139. Stephenson, S. L.; Kenny, A. J. *Biochem. J.* **1987,** *243,* 183–187.

140. Matsas, R.; Fulcher, I. S.; Kenny, A. J.; Turner, A. J. *Proc. Natl. Acad. Sci. U.S.A.* **1985,** *80,* 3111–3115.

141. Eckert, H. G.; Badian, M. J.; Gantz, D.; Kellner, H. M.; Totz, M. *Arzneim. Forsch.* **1984,** *34,* 1433.

142. Klaassen, C. D.; Watkins, J. B. *Pharmacol. Rev.* **1984,** *36,* 1.

143. Bausback, H. H.; Ward, P. E. *Adv. Exp. Med. Biol.* **1986,** *198A,* 397–404.

144. Palmieri, F. E.; Ward, P. E. *Biochim. Biophys. Acta* **1983,** *755,* 522–525.

145. Kato, T.; Nagatsu, T.; Fukasawa, K.; Harada, M.; Nagatsu, I.; Sakakibara, S. *Biochim. Biophys. Acta* **1978,** *525,* 417–422.

146. Roemer, D.; Pless, J. *Life Sci.* **1979,** *24,* 621.

147. Wyvratt, M. J.; Patchett, A. A. *Med. Res. Rev.* **1985,** *5,* 483.

148. Freidinger, R. M.; Veber, D. F. In *Conformationally Directed Drug Design: Peptides and Nucleic Acids as Templates or Targets*; Vida, J. A.; Gordon, M., Eds.; ACS Symposium Series 251; American Chemical Society, Washington, DC, 1984; pp 169–187.

149. Rippel, R. H.; Johnson, E. S.; White, W. F.; Fujino, M.; Fukuda, T.; Kobayashi, S. *Proc. Soc. Exp. Biol. Med.* **1975,** *148,* 1193.

150. Burchaladze, R. A.; Liepkaala, I. K.; Romanovskii, P. I.; Chipers, G. I. *Biul. Eksp. Biol. Med.* **1989,** *108,* 214–216.

151. Shcen, J.; Kisfaludy, L.; Nafradi, J.; Varga, L.; Varro, V. *Hoppe-Seylesis Z. Physiol. Chem.* **1978,** *359,* 917.

152. Kropp, H.; Sundelof, J. G.; Hajdu, R.; Kahan, F. M. *Antimicrob. Agents Chemother.* **1982,** *22,* 62.

153. McMartin, C. *Biochem. Soc. Trans.* **1989,** *17(5),* 931–934.

154. Mihara, K.; Mori, M; Hojo, T.; Takakura, Y.; Sezaki, H.; Hashida, M. *Biol. Pharm. Bull.* **1993,** *16(2),* 158–162.

155. Rabkin, R.; Ryan, M. P.; Duckworth, W. C. *Diabetologia* **1984,** *27,* 351–357.

156. Sato, H.; Yoshioka, K.; Terasaki, T.; Tsuji, A. *Biochim. Biophys.* **1991,** *1073,* 442–450.

157. Mihara, K.; Hojo, T.; Fujikawa, M.; Takakura, Y.; Sezaki, H.; Hashida, M. *Pharm. Res.* **1993,** *10(6),* 823–827.

158. Sumpio, B. E.; Maack, T. *Am. J. Physiol.* **1982,** *423,* F379–F392.

159. Christensen, E. I.; Rennke, H. G.; Carone, F. A. *Am. J. Physiol.* **1983,** *244,* F436–F441.

160. Cortney, M. A.; Sawin, L. L.; Weiss, D. D. *J. Clin. Invest.* **1970,** *49,* 1–4.

161. Barriere, S. L.; Flaherty, J. F. *Clin. Pharm.* **1984,** *3,* 352.

162. Parant, M; Parant, F.; Chedid, L.; Yapo, A.; Petit, J. F.; Loderer, E. *Int. J. Immunopharmacol.* **1979,** *1,* 35.

163. Aoyagi, T.; Miyata, S.; Nanbo, M.; Kojima, F.; Matsuzaki, M.; Ishizuka, M.; Takeuchi, T.; Umezawa, H. *J. Antibiot.* **1969,** *22,* 558.

164. Poe, M.; Perlav, D. S.; Boger, J. *J. Enzyme Inhib.* **1985,** *1,* 13–23.

165. Gores, G. J.; Larusso, N. F.; Miller, L. J. *Am. J. Physiol.* **1986,** *250,* G344–G349.

166. Madison, L. L.; Kaplan, L. *J. Lab. Clin. Med.* **1958,** *52,* 927–932.

167. Harding, P. E.; Bloom, G.; Field, J. B. *Am. J. Physiol.* **1975,** *228,* 1580–1588.

168. Chap, Z.; Ishida, T.; Jones, R. H.; Pena, J. R.; Vinik, A.; Field, J. B. *Am. J. Physiol.* **1988,** *254,* E214–E221.

169. Brinson, R. R.; Pitts, V. L.; Taylor, A. E. *Crit. Care Med.* **1989,** *17(7),* 657–660.

170. Miller, R. C.; Pelton, J. T.; Huggins, J. P. *Trends Pharmacol. Sci.* **1993,** *14,* 54–60.

171. Wallace, R. J.; Broderick, G. A.; BrAmMall, M. L. *Br. J. Nutr.* **1987,** *58,* 87–93.

172. Baylin, S. B.; Bailey, A. L., Hsu, T.; Giraud, V. F. *Metabolism* **1977,** *26(12),* 1345–1354.

173. Bashir, R. M.; Whitaker, J. N. *Neurology* **1980,** *30,* 1184–1192.

174. Chambrier, C.; Picard, S.; Vidal, H.; Cohen, R.; Riou, J. P.; Beylot, M. *Metabolism* **1990,** *39(9),* 976–984.

175. Adrian, T. E.; Savage, A. P.; Bacarese-Hamilton, A. J.; Wolfe, K.; Besterman, H. S.; Bloom, S. R. *Gastroenterology* **1986,** *90,* 379–384.

176. Adrian, T. E.; Ferri, G. L.; Bacarese-Hamilton, A. J.; Fuessl, H. S.; Polak, J. M.; Bloom, S. R. *Gastroenterology* **1985,** *89,* 1070–1077.

177. Lala, A.; Bouloux, P.; Tamburrano, G.; Gale, E. *J. Endocrinol. Invest.* **1987,** *10(95),* 95–104.

178. Singer, F. R.; Habener, J. F.; Green, E.; Godin, P.; Potts, J. T., Jr. *Nature (London) New Biol.* **1972,** *237,* 269–270.

179, Azria, M. In *The Calcitonins*; Karger: Basel, Switzerland, 1989.

180. Follath, F.; Wenk, M.; Vozeh, S.; Thiel, G.; Brunner, F.; Loertscher, R.; LeMaire, M.; Nussabaumer, K.; Niederbager, W.; Wood, A. *Clin. Pharmacol. Ther.* **1983,** *34,* 627–643.

181. Waters, M. R.; Albano, J. D. M.; Sharman, V. L.; Venkat Raman, G. *Clin. Sci.* **1986,** *71(Suppl. 15),* 2P.

182. Beveridge, T. In *Cyclosporin A*; White, D. J. G., Ed.; Elsevier Biomedical: New York, 1982; pp 5–17.

183. Gloff, C. A.; Wills, R. J. In *Protein Pharmacokinetics and Metabolism*; Ferraiolo, B. L.; Mohler, M. A.; Gloff, C. A., Eds.; Plenum: New York, 1992; pp 127–141.

184. Majno, G. In *Handbook of Physiology*; Hamilton, W. F.; Dow, P., Eds.; American Physiology Society: Washington, DC, 1965; pp 2293–2375.

185. Allison, A. C. *Nature (London)* **1960,** *188,* 37–40.

186. Kompella, U. B.; Lee, V. H. L. In *Peptide and Protein Drug Delivery*; Vincent H. L. L., Ed.; Marcel Dekker: New York, 1991; pp 391–483.

187. Ferraiolo, B. L.; Mohler, M. A.; Gloff, C. A. In *Protein Pharmacokinetics and Metabolism*; Plenum: New York, 1992.

RECEIVED for review September 9, 1993. ACCEPTED revised manuscript March 21, 1994.

Intestinal Peptide Transport

CHAPTER 5

Intestinal Absorption of β-Lactam Antibiotics

Akira Tsuji

β-Lactam antibiotics (i.e., penicillins and cephalosporin) have been extensively developed, and many researchers are still working on the synthesis of new derivatives with potent antibacterial activities against both gram-positive and gram-negative bacteria. The β-lactam antibiotics play a key role in combating bacterial diseases and infections because of their highly selective toxicity toward bacteria. However, surprisingly few of the derivatives are significantly active after oral administration (Chart I). This inactivity could be because of inadequate strategies in screening of orally active antibiotics and/or because the structural requirements for good intestinal absorption are different from those for potent bactericidal effect. In general, scientists in the pharmaceutical industries favor strategies to select candidate drugs with the highest potency against both gram-positive and gram-negative bacteria. This tendency may have led to the rejection of derivatives with good intestinal absorptive properties but with restricted antibacterial activity.

For the rational design of orally active drugs, an understanding of the mechanism of absorption from the gastrointestinal tract is necessary. Extensive studies have been performed to elucidate the intestinal transport mechanism of β-lactam antibiotics by various techniques, including in situ perfusion, in situ loop, in vitro everted sac, brush-border-

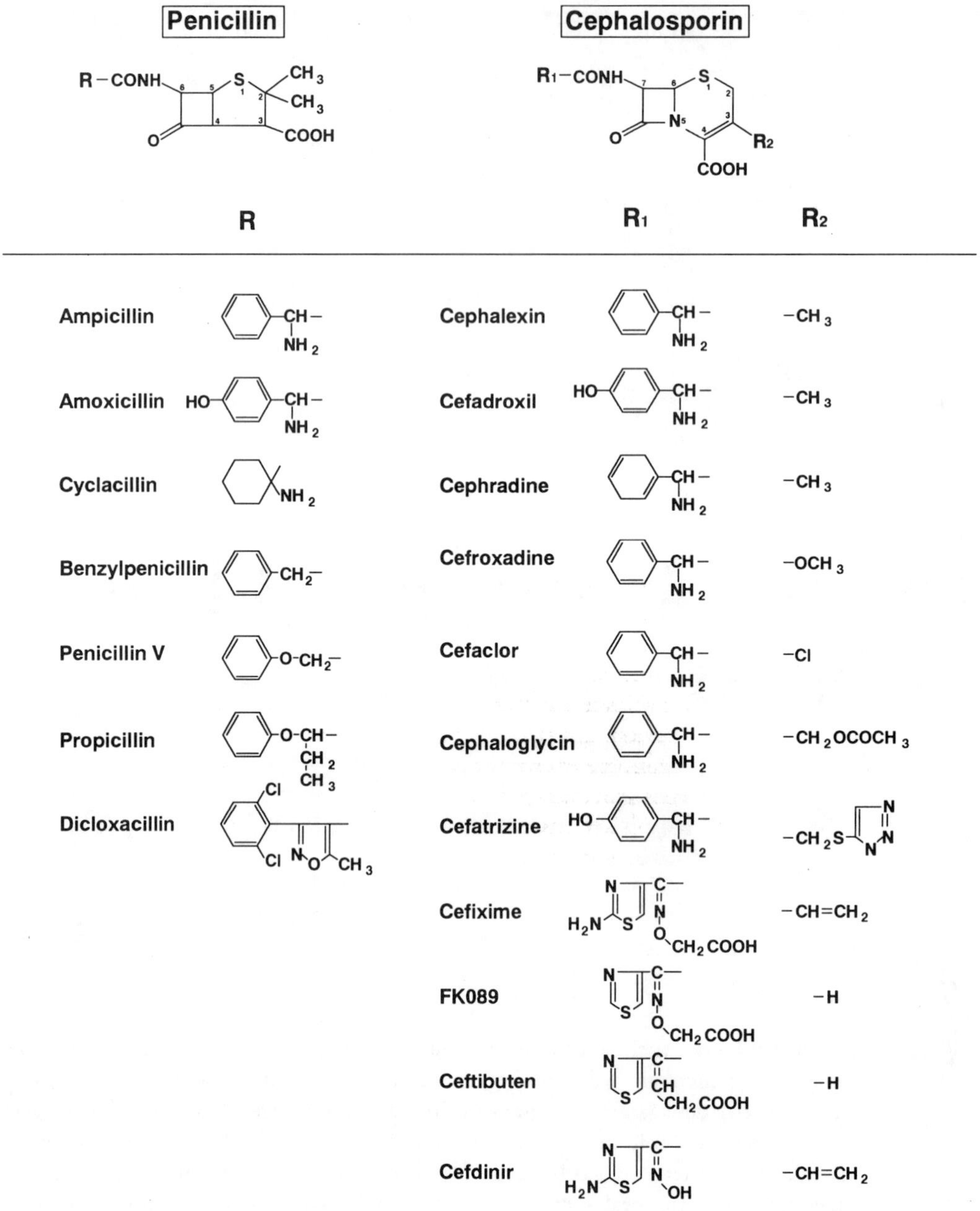

Chart I. Chemical structures of orally active penicillins and cephalosporins.

membrane vesicles (BBMVs), monolayers of cultured epithelial cell line, photoaffinity labeling with some photoactive probes, and expression of the transport protein in *Xenopus laevis* oocytes. Several groups have reported that aminopenicillins, aminocephalosporins, and dicarboxylic acid-type cephalosporins, such as cefixime and ceftibuten, are absorbed by a carrier-mediated process in rat and rabbit small intestine, and some of these antibiotics may share a common transport system with oligopeptides (di- or tripeptides).

In this chapter, the current state of knowledge of (1) the intestinal transport mechanism of β-lactam antibiotics, including the molecular mechanism of interaction with the transport system, and (2) the absorption–structure relationships of antibiotics are reviewed.

Transport Mechanism of Amino-β-Lactam Antibiotic Derivatives

Historical Background

Among orally active β-lactam antibiotics, aminopenicillins (e.g., ampicillin, amoxicillin, or cyclacillin) and aminocephalosporins (e.g., cephalexin, cefadroxil, or cephradine) have zwitterionic structures with an α-amino group in the substituent at position 6 or 7 of the penam or cephem nucleus and a free carboxyl group at position 3 or 4 (*see* structure in Chart I). At physiological pH, these amino-β-lactam antibiotics exist in zwitterionic forms with a positively charged NH_3^+ group and a negatively charged COO^- group. In spite of having very low lipid solubilities, these amino-β-lactam antibiotics are absorbed efficiently from the small intestine after oral administration.

The intestinal absorption mechanism of amino-β-lactam antibiotics has been studied for over 20 years. However, experiments with animal small intestine, using in vitro and in vivo techniques, have sometimes yielded conflicting results and conclusions. In 1972, Quay (*1*) first demonstrated the active transport of cephalexin across the isolated rat jejunum. Addison et al. (*2*) found in 1975 that cephalexin significantly inhibited the uptake of glycylsarcosylsarcosine, a tripeptide, and suggested that this peptide-like antibiotic is probably transported by the di- or tripeptide carrier system. On the other hand, Penzotti and Poole (*3*) reported in 1974 that β-lactam antibiotics (either penicillins or cephalosporins, including ampicillin, cyclacillin, cephalexin, and cephaloglycin) are transported across the everted intestine by passive diffusion.

Since 1977, the intestinal transport of β-lactam antibiotics has been extensively studied to clarify if specialized mechanisms participate (*4–76*). Dixon and Mizen (*6*) found that only cyclacillin was actively transported, whereas the other aminopenicillins, including ampicillin and amoxicillin, diffused passively across the everted intestine. On the contrary, Shindo et al. (*11*) obtained evidence of saturable absorption of ampicillin from the rat intestinal loop. Although we have shown (*4, 9, 13, 18, 19, 25, 26, 28, 29*) that the absorption rates of amoxicillin, cyclacillin, cephalexin, cephradine, cefadroxil, cefroxadine, cefaclor, and cefatrizine from an in situ intestinal loop or in situ intestinal perfusion solution followed saturable kinetics, similar experimental techniques in other laboratories resulted in the conflicting observations that ampicillin, amoxicillin, and cephalexin were absorbed in a nonsaturable manner (*5, 7, 8, 12, 15*).

Such apparently conflicting observations by different investigators are not surprising because it is now known that the relative contributions of saturable and nonsaturable processes to the net transport rate are dependent on the experimental systems and hydrodynamic conditions employed. For example, with the intestinal closed-loop method, the aqueous resistance rather than the membrane resistance limits absorption due to segment flow conditions. Sinko and Amidon (*43*) pointed out that the aforementioned inconsistencies among the previous β-lactam studies using the non-steady-state closed loop and intestinal perfusion methods can be explained by the considerable effect of the diffusion boundary layer. On the other hand, in the steady-state single-pass perfusion technique employed by Amidon's group (*42, 43, 52, 70*), the membrane represents the controlling resistance.

In Situ Single-Pass Intestinal Perfusion

Amidon and his co-workers (*42, 43, 52, 70*) employed the technique of in situ single-pass intestinal perfusion in the rat to determine the intrinsic membrane absorption parameters with a modified boundary layer model. Intrinsic wall permeabilities, P_w^*, of aminopenicillins (ampicillin, amoxicillin, and cyclacillin; Figure 1A) and aminocephalosporins (cefaclor, cefatrizine, cefadroxil, cephalexin, and cephradine; Figure 1B) show a dependence on wall concentration (C_w), suggesting the existence of a saturable absorption mechanism for amino-β-lactam antibiotics in rat jejunum. The experimental data were fitted to eq 1 to obtain the absorption parameters [i.e., the maximal flux (J_{max}^*), the intrinsic Michaelis constant (K_m) for the carrier transport, and the intrinsic nonsaturable membrane permeability (P_m^*)]. A summary of the intrinsic absorption parameters is given in Table I.

$$P_w^* = \frac{J_{max}^*}{K_m + C_w} + P_m^* \tag{1}$$

where C_w and P_m^* were obtained from eqs 2 and 3, respectively:

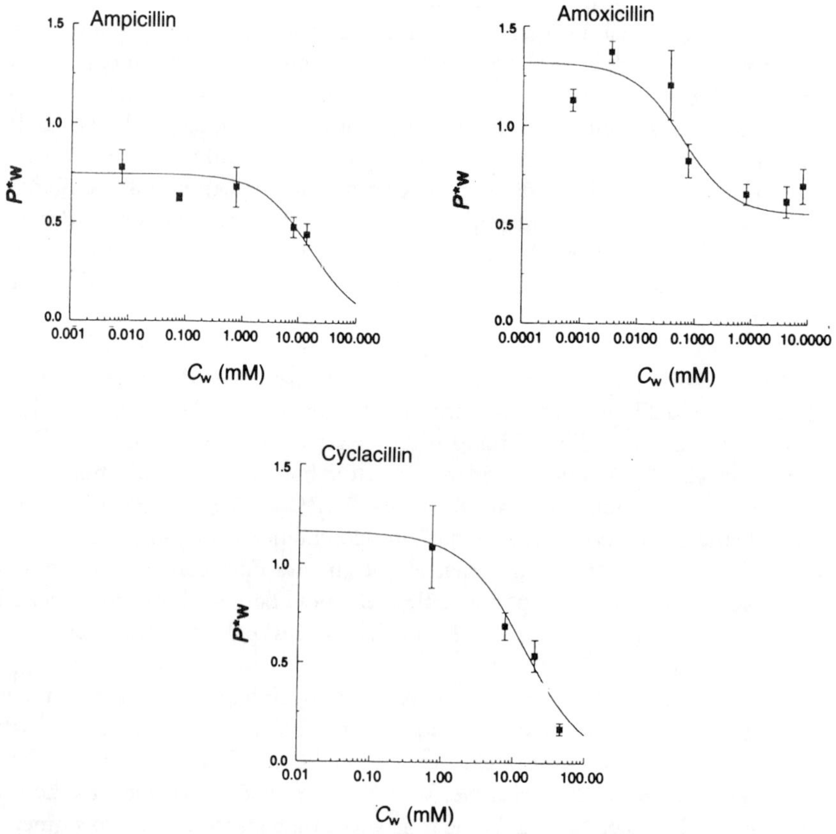

Figure 1A. Intrinsic wall permeability (P_w^*) versus wall concentration (C_w) of aminopenicillins (ampicillin, amoxicillin, and cyclacillin) in rat jejunum. Permeabilities are reported as the mean ± SE; (—) best fit line. (Reproduced with permission from reference 70. Copyright 1992 Elsevier.)

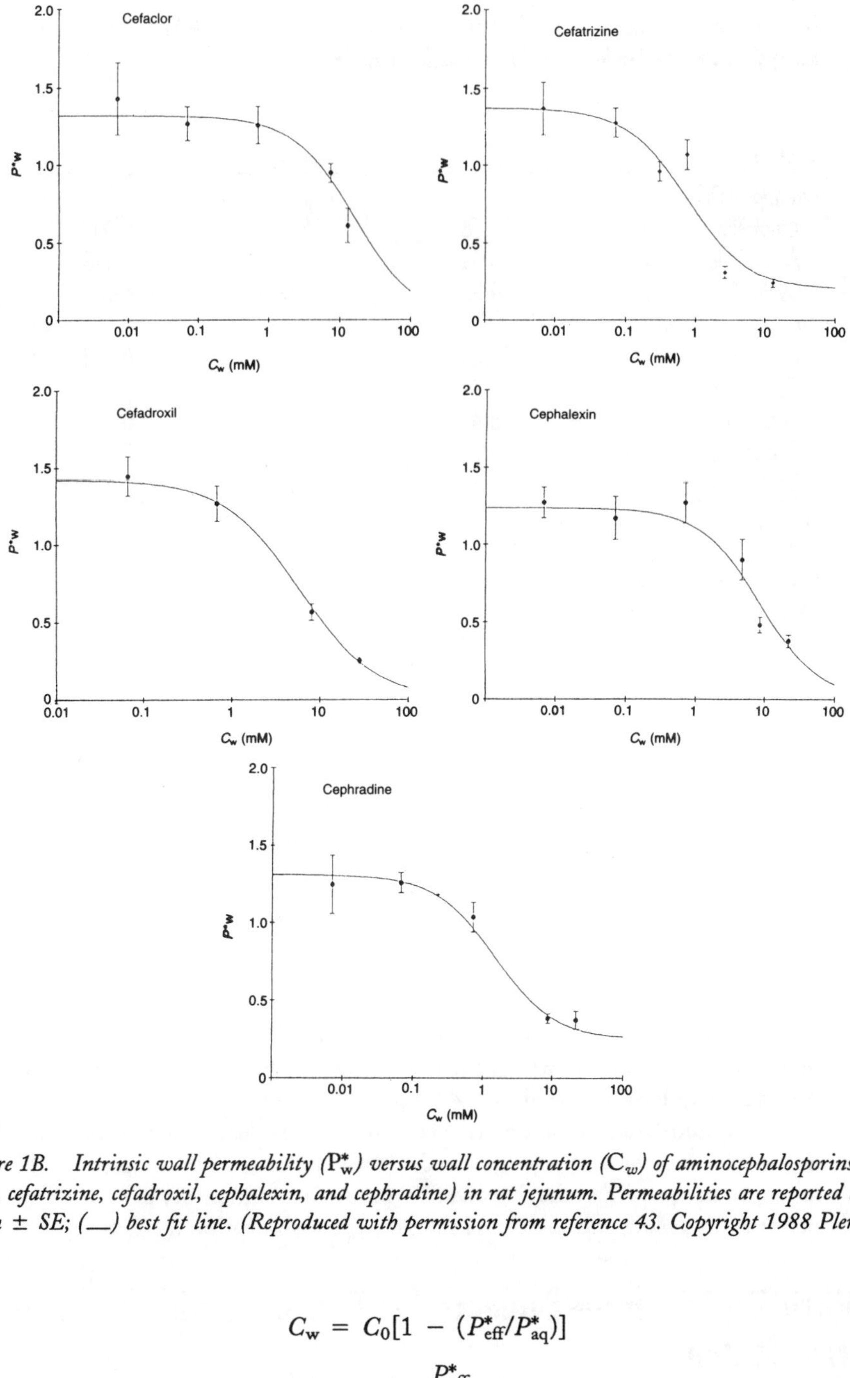

Figure 1B. Intrinsic wall permeability (P^*_w) versus wall concentration (C_w) of aminocephalosporins (cefaclor, cefatrizine, cefadroxil, cephalexin, and cephradine) in rat jejunum. Permeabilities are reported as the mean ± SE; (—) best fit line. (Reproduced with permission from reference 43. Copyright 1988 Plenum.)

$$C_w = C_0[1 - (P^*_{eff}/P^*_{aq})] \quad (2)$$

$$P^*_w = \frac{P^*_{eff}}{[1 - (P^*_{eff}/P^*_{aq})]} \quad (3)$$

Assuming that the difference between the rates of mass flow into and out of the intestine is equal to the rate of mass absorbed, the dimensionless effective wall permeability (P^*_{eff}) and the dimensionless aqueous permeability (P^*_{aq}) were calculated from the inlet (C_0) and outlet (C_m) perfusate concentrations and the Graetz number (Gz) by eqs 4–6:

Table I. Intrinsic Membrane (Jejunum) Absorption Parameters Derived from Nonlinear Regression Analysis with the Model of Wall Permeability in Equation 1

Compound	J^*_{max} (mM)	K_m (mM)	P^*_m	P^{*a}_c
Aminopenicillins				
Ampicillin	11.78	15.80	NDO[b]	0.75
Amoxicillin	0.044	0.058	0.558	0.757
Cyclacillin	16.30	14.00	NDO	1.14
Aminocephalosporins				
Cefaclor	21.30	16.1	0 (0.36)[c]	1.3
Cefatrizine	0.7	0.6	0.2	1.3
Cefadroxil	8.4	5.9	0	1.4
Cephalexin	9.1	7.2	0	1.3
Cephradine	1.6	1.5	0.3	1.1

NOTE: Reported values are fitted values.
[a] $P^*_c = J^*_{max} K_m$.
[b] NDO, not different from zero.
[c] Value in parentheses for colon.
SOURCE: Reproduced with permission from references 43 and 70. Copyright 1988 and 1992.

$$P^*_{\text{eff}} = \frac{1 - (C_{\text{m}}/C_0)}{4\ Gz} \tag{4}$$

$$P^*_{\text{aq}} = (A\ Gz^{1/3})^{-1} \tag{5}$$

$$Gz = \frac{\pi D\ L}{2\ Q} \tag{6}$$

where D denotes the diffusion coefficient, L is the length of the intestine perfused, and Q represents the fluid flow rate. The details of the method to obtain the values of the constants A and D have been reported (*42, 43, 52, 70*).

Amino-β-lactam antibiotic perfusion studies demonstrated that jejunal absorption in the rat occurs by a carrier-mediated process, judging from the presence of the saturable kinetic term in eq 1 and the detection of competitive–inhibitory effects, whereas the colonic permeability is low and involves a simple passive absorption mechanism.

In Vitro Experimental Evidence for Transport via the Peptide Carrier System

What kind of transport system participates in the saturable intestinal absorption of amino-β-lactam antibiotics? For the elucidation of the transport mechanism and the driving force of the transport, in vitro techniques with everted intestinal sacs, a cultured epithelial cell line, or BBMVs are more suitable than the in situ intestinal perfusion or in situ loop methods.

We have determined (*26*) the initial uptake rate of amino-β-lactam antibiotics in

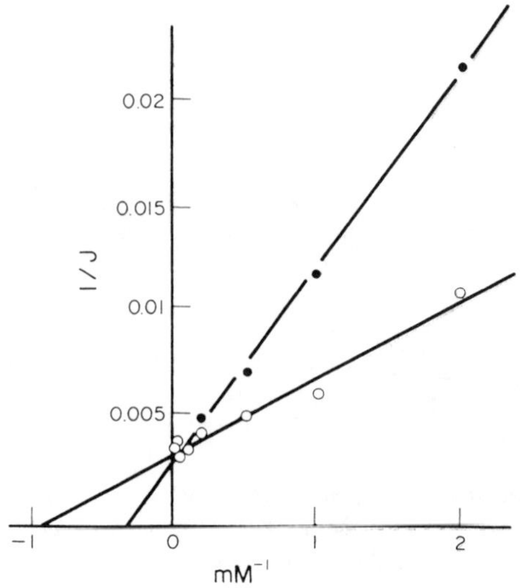

Figure 2. Lineweaver–Burk plots of uptake of cyclacillin in the absence (○) and in the presence (●) of cephalexin (10 mM). Values are corrected for non-mediated uptake. The line for the control experiment was calculated using the parameters K_t = 1.15 (mM) and J_{max} = 316 (nmol min^{-1} g^{-1} wet tissue). The line for uptake in the presence of cephalexin is a weighted least-squares regression line. (Reproduced with permission from reference 26. Copyright 1984 Pergamon.)

rat everted jejunum segments (4-cm length), which were mounted on polyethylene tubing (outer diameter, 4 mm; length, 4.5 cm) to avoid uptake from the serosal side. Lineweaver–Burk plots for the inhibitory effect of cephalexin (Figure 2) or glycylglycine (Figure 3A) on the uptake of cyclacillin and for the inhibitory effect of cyclacillin on the uptake of glycylglycine (Figure 3B) clearly show competition between cyclacillin and cephalexin and between cyclacillin and glycylglycine for transport. The results in Table II show that the uptakes of cyclacillin and cefadroxil were significantly inhibited by several dipeptides. Also, the uptake of glycylglycine was significantly inhibited by several amino-β-lactam antibiotics (Table III). These results indicate the involvement of a common transport system in the intestinal uptakes of peptides and amino-β-lactam antibiotics. The apparent activation energies of the intestinal uptake of cyclacillin and cephalexin at 1 mM concentration were in the 23–25 kcal/mol range, which is higher than that of passive diffusion (3–5 kcal/mol). Furthermore, the remarkable reduction of the uptake rates of both antibiotics by metabolic inhibitors (sodium azide, 2,4-dinitrophenol, and sodium cyanide) strongly indicates that cyclacillin and cephalexin are transported in an active manner.

However, whether this transport is primary or secondary active transport has not been clarified by the in vivo everted sac mathod. Evidence has accumulated to show that small peptides such as di- or tripeptides are actively transported into intestinal epithelial cells by a carrier system that is different from those involved in the transport of amino acids. Recent studies by Ganapathy and Leibach (*77*) and Takuwa et al. (*78*) with BBMVs prepared from the rabbit small intestine have demonstrated that Na^+ does not play a direct role in the transport of dipeptides (glycyl-L-proline, glycylsarcosine, and glycylglycine), but an inwardly directed H^+ gradient is the driving force for the transport of dipeptides. To elucidate if amino-β-lactam antibiotics can be transported by the H^+–peptide cotransporter existing in the intestinal brush-border membrane, the transport characteristics of these antibiotics have been studied with intestinal BBMVs. These BBMVs possess considerable

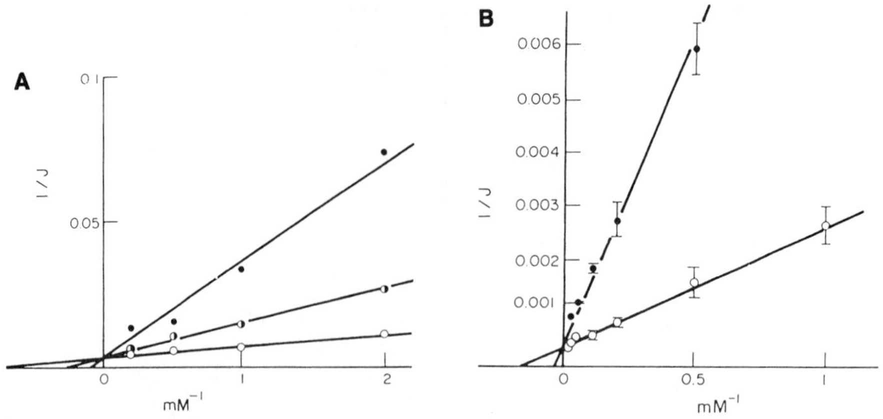

Figure 3. A. Lineweaver–Burk plots of uptake of cyclacillin in the absence (○) or in the presence of 5 mM (◑) and 10 mM (●) glycylglycine. Values are corrected for non-mediated uptake. The line for the control experiment was calculated using the parameters K_t = 1.15 (mM) and J_{max} = 316 (nmol min^{-1} g^{-1} wet tissue). The lines for uptake in the presence of glycylglycine are weighted least-squares regression lines. Each point is the mean of at least three determinations. B. Lineweaver–Burk plots of uptake of glycylglycine in the absence (○) or in the presence (●) of 5 mM cyclacillin. The values are corrected for non-mediated untake. The line for the control experiment was calculated using the parameters K_t = 6.02 (mM) and J_{max} = 2770 (nmol min^{-1} g^{-1} wet tissue). The line for uptake in the presence of glycylglycine is a weighted least-squares regression line. Each point is the mean of at least three determinations. (Reproduced with permission from reference 26. Copyright 1984 Pergamon.)

Table II. Effects of Dipeptides (30 mM) on Uptake of Cyclacillin (1 mM) and Cefadroxil (1 mM) Determined in the Everted Intestine of Rats[a]

	J (nmol min^{-1} g^{-1} wet tissue)	
Inhibitor	*Cyclacillin*	*Cefadroxil*
None	156.5 ± 15.1	95.9 ± 14.7
Gly-Leu	6.4 ± 5.5[a]	7.2 ± 12.4[a]
Gly-Sar	19.0 ± 2.3[a]	26.5 ± 7.3[a]
Leu-Gly	2.7 ± 9.5[a]	13.4 ± 6.8[a]
Phe-Gly	9.6 ± 10.6[a]	22.8 ± 9.0[a]
Carnosine	31.3 ± 9.7[a]	21.6 ± 17.3[a]

NOTE: Each result is the mean ± SE of three determinations: Gly-leu (glycylleucine), Gly-Sar (glycylsarcosine), Leu-Gly (leucylglycine), and Phe-Gly (phenylalanylglycine).

[a] Significantly different from the control uptake studied in the absence of inhibitors ($p < 0.05$), as assessed by the test.

SOURCE: Reproduced with permission from reference 26. Copyright 1984 Pergamon.

Table III. Effects of Antibiotics (30 mM) on the Uptake of Glycylglycine (1 mM) Determined in the Everted Intestine of Rats

Inhibitor	*J (nmol min^{-1} g^{-1} wet tissue)*
None	400.0 ± 20.1 (5)
Cyclacillin	44.0 ± 1.2 (3)[a]
Cefadroxil	181.8 ± 6.6 (3)[a]
Cefroxadine[c]	358.4 ± 24.8 (3)
Cephalexin	239.1 ± 13.6 (3)[a]
Cephradine	227.2 ± 13.6 (3)[a]

NOTE: Each value is the mean ± SE, with the number of experiments in parentheses.

[a] Significantly different from the control uptake studied in the absence of inhibitors ($p < 0.05$), as assessed by the *t* test.

[b] Cefroxadine (5 mM) was used because of its limited solubility.

SOUCE: Reproduced with permission from reference 26. Copyright 1984 Pergamon.

advantages over the intact tissue preparation for transport studies; most notably, control over the supply of energy and the absence of complicating factors such as intracellular metabolism. BBMVs can be isolated and purified without contamination by basolateral membrane from animal or human small intestine by the calcium precipitation method of Kessler et al. (*79*). The uptake of drugs, in general, is measured with a rapid filtration technique. If there is significant adsorption of the drug used for the transport study on the filters (e.g., Millipore filter with pore size of 0.45 mm), the transport rate of the drug is difficult to measure. Fortunately, most cephalosporins and aminopenicillins are adsorbed to a negligible extent and are suitable for study by the membrane vesicle method.

Okano et al. (*33*) demonstrated clearly that the uptake of cephradine by BBMVs isolated from rabbit small intestine was stimulated by the countertransport effect of dipeptide (L-carnosine and glycylsarcosine), which indicates the existence of a carrier-transport system. An inward H^+-gradient (pH_{in} = 6.0–8.4, pH_{out} = 6) induced a clear overshoot uptake of cephradine (Figure 4A and 4c), and this uptake was remarkably reduced in the presence of the protonophore carbonylcyanide *p*-trifluoromethoxyphenyl hydrazone (FCCP; Figure 4B). A valinomycin-induced K^+ diffusion potential (interior-negative) stimulated the H^+-gradient-dependent uptake of cephradine (Figure 4D). The uptake of other aminocephalosporins (cefadroxil, cefaclor, cephalexin) was also stimulated in the presence of an inward H^+-gradient, whereas the uptake of the parenteral cephalosporins without an α-amino group (cefazolin cefotiam) was not changed in the presence or absence of the H^+-gradient, (Figure 5). Very similar observations were made in rat intestinal BBMVs (*32*). On the contrary, Iseki (*50, 51*) and Sugawara et al. (*63–65, 67*) have recently concluded, based on similar membrane vesicle studies, that there is little contribution of H^+/peptide cotransporter because little or no overshoot of amino-β-lactam antibiotic uptake was observed in rats and humans, despite the clear overshoot observed in rabbits.

In contrast to the brush-border membranes, there is little information as to whether the orally active β-lactam antibiotics are transferred across the intestinal basolateral membranes via a specific process or by simply diffusion. Considering their low solubility in lipids, these antibiotics are unlikely to cross the basolateral membranes easily by passive diffusion.

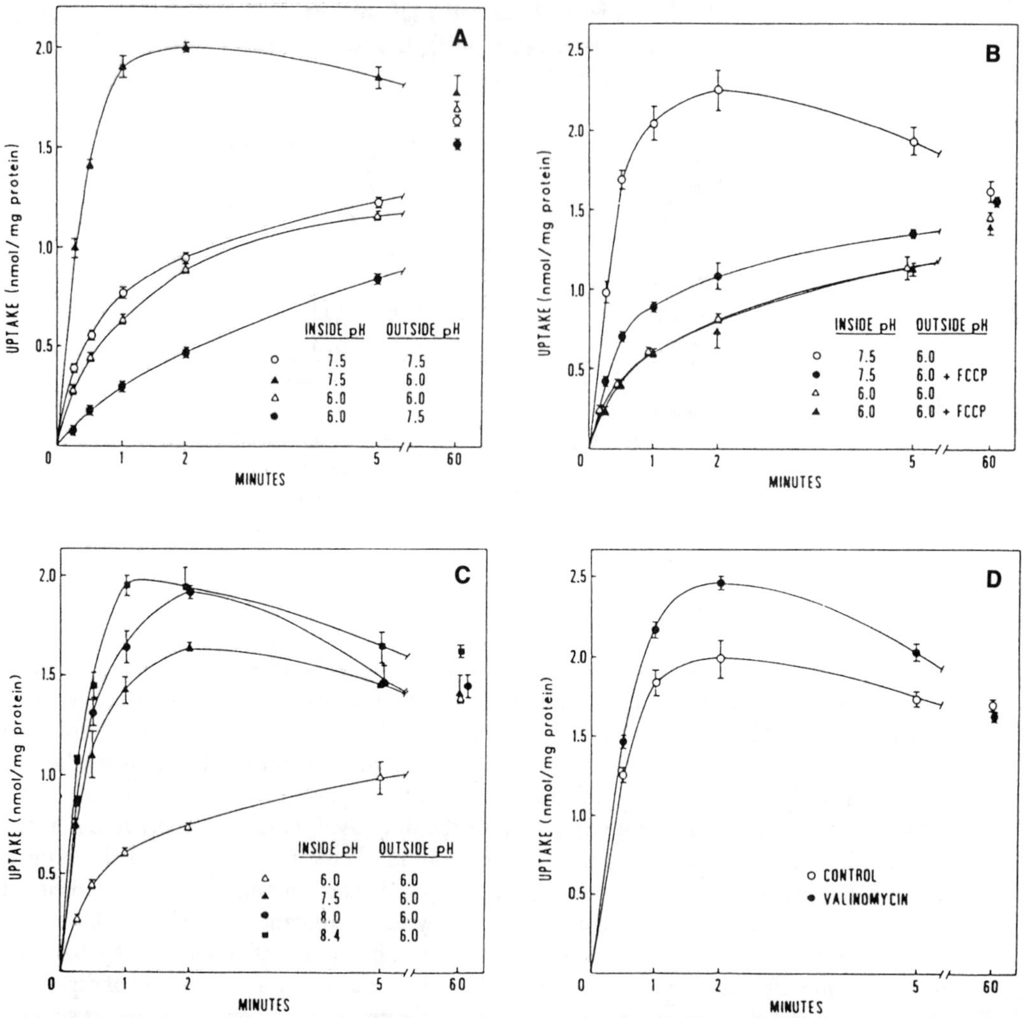

Figure 4. A. Effect of pH on cephradine uptake by intestinal BBMVs. Membrane vesicles (20 μL), suspended in 100 mM mannitol, 100 mM KCl, and either 10 mM Hepes (pH 7.5) (○, ▲) or 10 mM Mes (pH 6.0) (●, △), were incubated with the substrate mixture (200 μL) comprising 100 mM mannitol, 100 mM KCl, 1 mM cephradine, and either 10 mM Hepes (pH 7.5) (○, ●) or 10 mM Mes (pH 6.0) (△, ▲). Each point represents the mean ± SE of three determinations. B. Effect of FCCP on H^+-gradient-dependent cephradine uptake by intestinal BBMVs. Membrane vesicles (20 μL), suspended in 100 mM mannitol, 100 mM KCl, and either 10 mM Hepes (pH 7.5) (○, ●) or 10 mM Mes (pH 6.0) (△, ▲), were incubated with the substrate mixture (200 μL) comprising 100 mM mannitol, 100 mM KCl, 10 mM Mes (pH 6.0), 1 mM cephradine, and 0.5% ethyl alcohol in the presence (●, ▲) or absence (○, △) of 50 μM FCCP. Each point represents the mean ± SE of three determinations. C. Effect of various inward H^+-gradients on cephradine uptake by intestinal BBMVs. Membrane vesicles (20 μL), suspended in 100 mM mannitol, 100 mM KCl, and either 10 mM Hepes (pH 8.4) (■), 10 mM Hepes (pH 8.0) (●), 10 mM Hepes (pH 7.5) (▲), or 10 mM Mes (pH 6.0) (△), where incubated with the substrate mixture (200 μL) comprising 100 mM mannitol, 100 mM KCl, 10 mM Mes (pH 6.0), and 1 mM cephradine. The medium pH was adjusted with KOH. Each point represents the mean ± SE of three determinations. D. Effect of membrane potential on H^+-gradient-dependent cephradine uptake. Membrane vesicles (20 μL), suspended in 100 mM mannitol, 67 mM K_2SO_4, and 10 mM Hepes (pH 7.5), were incubated with the substrate mixture (200 μL) comprising 100 mM mannitol, 67 mM Na_2SO_4, 10 mM Mes (pH 6.0), 1 mM cephradine, and 0.2% ethyl alcohol in the presence (●) or absence (○) of 9 μM valinomycin (17 μg/mg protein). Each point represents the mean ± SE of three determinations. (Reproduced with permission from reference 33. Copyright 1986.)

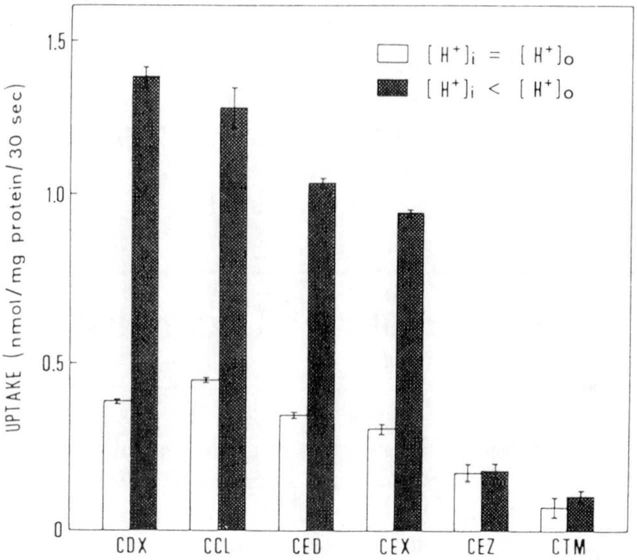

Figure 5. Effect of H^+-gradient on the uptake of various cephalosporin antibiotics by intestinal brush-border membranes. Membrane vesicles (20 μL), suspended in 100 mM mannitol, 100 mM KCl, and either 10 mM Hepes (pH 7.5) (dotted columns) or 10 mM Mes (pH 6.0) (open columns), were incubated with the substrate mixture (200 μL) comprising 100 mM mannitol, 100 mM KCl, 10 mM Mes (pH 6.0), and 1 mM cephalosporins. Key: (CDX) cefadroxil; (CCL) cefaclor; (CED) cephradine; (CEX) cephalexin; (CEZ) cefazolin; (CTM) cefotiam. Each value represents the mean ± SE of four determinations. (Reproduced with permission from reference 33. Copyright 1986.)

Recently, Dyer et al. (*80*) reported, in the intestinal basolateral membranes of rabbit enterocytes, the presence of a system of transporting glycyl-L-proline that exhibits H^+-dependent characteristics like the brush-border peptide transporter. To reach a better understanding of the intestinal transport mechanisms of these antibiotics as well as those of dipeptides, epithelial transport studies with intact intestinal epithelia are required. Recently, the human colon adenocarcinoma cell line Caco-2 has been used as a model for studying the function of the intestinal epithelial cells (*80*). Caco-2 cells sponantaneously differentiate into polarized cell monolayers with microvilli on their apical surface and they form tight junctions after attaining confluency, exhibiting transporting epithelia (*81–83*). The transport systems for sugars (*84–86*), amino acids (*87*), and bile acids (*88*) normally found in the small intestine are expressed in the Caco-2 cells. Recently, Thwaiters et al. (*89*) and Saito and Inui (*90*) demonstrated, by measuring the transepithelial transport of the biologically stable dipeptide, glycylsarcosine (*89*) or the dipeptide-like anticancer agent bestatine (*90*) across Caco-2 cell monolayers, the presence of H^+-coupled small peptide transporters at both apical and basolateral membranes, but that the basolateral small peptide transporter is distinct from the apical H^+-coupled transporter. These observations are consistent with previous studies with BBMVs (*77, 78*) and basolateral membrane vesicles (*80*). With respect to β-lactam antibiotics, Dantzig and Bergin (*62*), Dantzig et al. (*68*), and Inui et al. (*72*) showed that orally active aminocephalosporins (cephalexin, cefaclor, and cephradine) accumulate in the Caco-2 cell monolayer via the H^+/peptide cotransport system localized in the apical membranes, in accordance with the transport mechanism established in BBMVs (*32, 33*). Inui et al. (*72*) demonstrated that the H^+/dipeptide transport system at the apical membrane contributes to the net transepithelial transport of these antibiotics and that a specific peptide transport system, which has a functional sulfhydryl group, is involved in the efflux of these antibiotics across the basolateral membranes.

Overall (with some exceptions), the accumulated evidence suggests that amino-β-lactam antibiotics are actively absorbed from the small intestine because, as structural analogs of tripeptides, they share the intestinal H^+-gradient-dependent transport system for small peptides.

Transport Mechanism of Dicarboxylic Acid Type Cephalosporins

β-Lactam antibiotics can be absorbed from the gastrointestinal tract to variable extents depending on their chemical structures. Until recently, the intestinal carrier-mediated transport of β-lactam antibiotics has been believed to be specific for derivatives with an α-amino group in the side chain at the 6-position in penicillins and 7-position in cephalosporins (*see* structures in Chart I). Recently, new orally active cephalosporins without an α-amino group have been found; among them, the α-amino group was replaced by a carboxymethoxyimino group in cefixime (FK027) and FK089 and by a carboxybutenoyl group in ceftibuten (*see* structures in Chart I). These new cephalosporins are well absorbed after oral administration, despite their low lipid solubility, with bioavailabilities of 30–50% in animals and humans for cefixime (*34, 36, 40*) and almost 100% in every species for ceftibuten (*54, 55*).

Studies (*34, 36*) in our laboratory with the rat everted sac technique indicated that cefixime and FK089 (*see* structures in Chart I) are absorbed via peptide transport systems like amino-β-lactam antibiotics. The intestinal uptake of cefixime and FK089 was apparently pH and partially Na^+ dependent, with maximal rates around pH 5.0 and rates 3–5 times lower at pH 7.0 in in vitro intestinal everted sacs. Studies with similar in vitro intestinal tissue preparations indicated that peptide transport is reduced by replacement of Na^+ (*77*), as observed for the intestinal tissue upake of cefixime (*36*). However, studies (*77, 78*) with intestinal BBMVs have shown that the intestinal transport of small peptides is entirely independent of the presence of a Na^+-gradient and is dependent on an H^+-gradient. Therefore, clarification of the roles of Na^+- and H^+-gradients in the case of cefixime appears to be necessary to determine if cefixime can be transported via the dipeptide transport system(s).

Recent studies with intestinal BBMVs revealed identical transport mechanisms in rats (*37, 41, 91*) and rabbits (*48, 75*): Cefixime is transported in an inward H^+-gradient-dependent and Na^+-independent manner via a peptide carrier. This mechanism is evident from the results shown in Figures 6A and 6B. An inward H^+-gradient (pH_{in} = 7.5, pH_{out} = 5.0) induced an overshoot of cefixime uptake, and uptake was reduced in the presence of the protonophore FCCP. Cefixime uptake was trans-stimulated by glycylsarcosine (*48*), glycyl-L-proline (*75*), cephradine (*48*), and cephalexin (*75*); countertransport effect of glycylsarcosine and cephradine, see Figure 7) and cis-inhibited (*37, 41, 75, 91*) by dipeptides and aminocephalosporins. However, one group demonstrated no overshoot uptake of cefixime (*64*) and cephradine (*51*) even in the presence of H^+-gradient in rat and human intestinal BBMVs. Although the reason for such discrepancies between investigators is obscure, treatment of rabbit intestinal BBMVs with diethylpyrocarbonate led to a complete loss of the H^+-gradient-dependent transport activity for cefixime and cephalexin (*75*), indicating the possible participation of histidyl residues in membrane proteins for intestinal

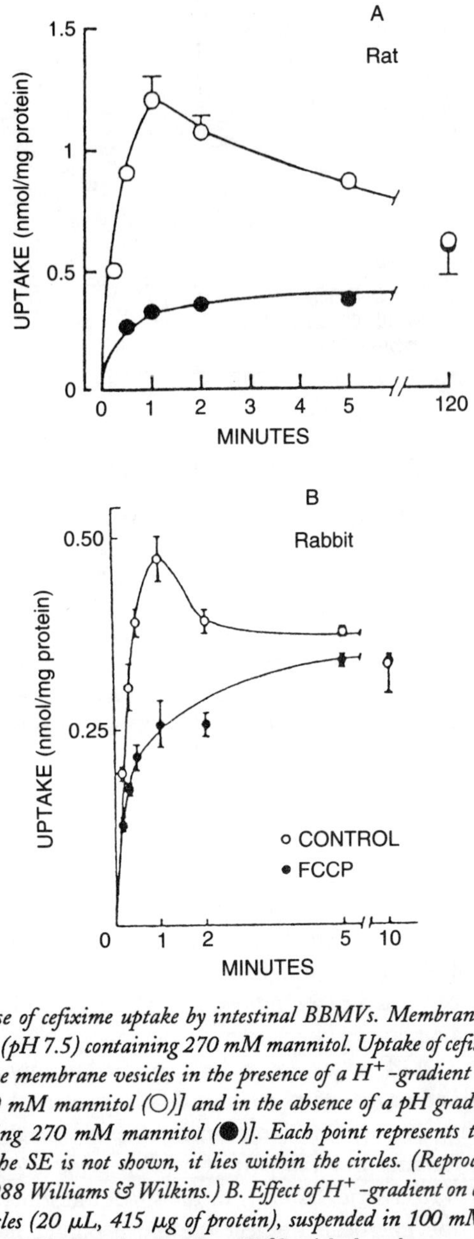

Figure 6. A. Time course of cefixime uptake by intestinal BBMVs. Membrane vesicles were preloaded in 10 mM Tris/Hepes buffer (pH 7.5) containing 270 mM mannitol. Uptake of cefixime (1 mM) was measured at 37 °C by incubating the membrane vesicles in the presence of a H^+-gradient [10 mM Tris/citrate buffer (pH 5.0), containing 270 mM mannitol (○)] and in the absence of a pH gradient [10 mM HEPES/Tris buffer (pH 7.5), containing 270 mM mannitol (●)]. Each point represents the mean ± SE of three to five experiments. When the SE is not shown, it lies within the circles. (Reproduced with permission from reference 91. Copyright 1988 Williams & Wilkins.) B. Effect of H^+-gradient on cefixime uptake by intestinal BBMVs. Membrane vesicles (20 μL, 415 μg of protein), suspended in 100 mM mannitol, 100 mM KCl, and 10 mM HEPES (pH 7.5), were incubated at 37 °C with the substrate mixture (200 μL) consisting of 100 mM mannitol, 100 mM KCl, 10 mM Mes (pH 5.0), 0.2 mM cephalosporins, and 0.5% ethanol in the presence (●) or absence (○) of 50 μM FCCP. Each point represents the mean ± SE of three determinations. (Reproduced with permission from reference 48. Copyright 1988 Williams & Wilkins.)

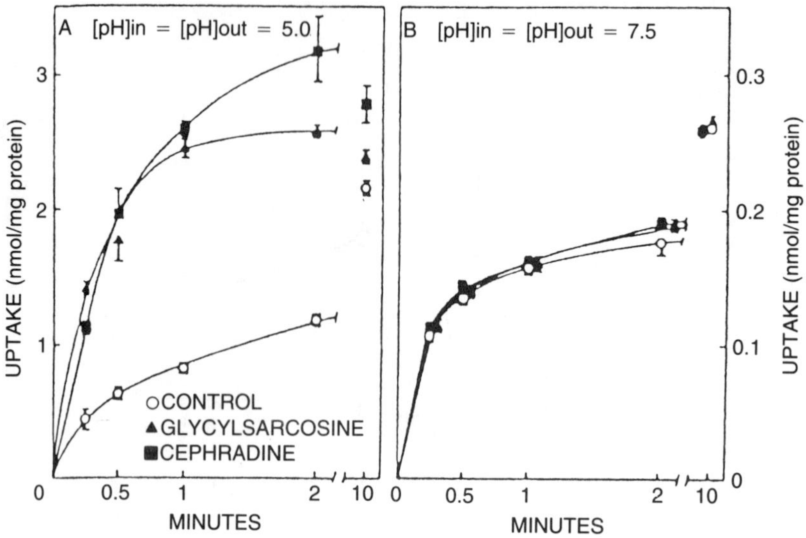

Figure 7. Countertransport effect of glycylsarcosine and cephradine on cefixime uptake by intestinal BBMVs at pH 5.0 (A) and pH 7.5 (B). A. Membrane vesicles were preincubated at 1 h in 100 mM mannitol, 100 mM KCl, and 10 mM Mes (pH 5.0), with 10 mM glycylsarcosine or 10 mM cephradine, and then aliquots (20 μL, 248 μg of protein) were incubated at 37 °C with the substrate mixture (200 μL) consisting of 100 mM mannitol, 100 mM KCl, 10 mM Mes (pH 5.0), and 1 mM cefixime. B. Membrane vesicles were preincubated for 1 h in 100 mM mannitol, 100 mM KCl, and 10 mM HEPES (pH 7.5), with 10 mM glycylsarcosine or 10 mM cephradine, and then aliquots (20 μL, 292 μg of protein) were incubated at 37 °C with the substrate mixture (200 μL) consisting of 100 mM mannitol, 100 mM KCl, 10 mM HEPES (pH 7.5) and 1 mM cefixime. Key: (○) control; (▲) glycylsarcosine preloaded; and (■) cephradine preloaded. Each point represents the mean ± SE of three determinations. (Reproduced with permission from reference 48. Copyright 1988 Williams & Wilkins.)

transport of both cephalosporins. Taking into account previous observations (*37, 41, 48, 75, 91*), we conclude that a dianionic cephalosporin, cefixime, is transported via the carrier-medicated system shared by small peptides and amino-β-lactam antibiotics.

The partial dependence of cefixime transport on Na^+ observed in intact tissue preparations (*36*) may be related to an indirect role of the Na^+–H^+-antiport system existing in the brush border membrane. The presence of an inward Na^+-gradient, which can be maintained by Na^+, K^+-ATPase present in the basolateral membrane, stimulates the Na^+–H^+-antiport system and thus maintains an inward H^+-gradient. Such a discrepancy with respect to Na^+ dependency between intact intestinal tissue and membrane vesicles has been demonstrated in recent years for the transport of small peptides (*77, 78, 92*) and other amino-β-lactam antibiotics (*16, 28, 30, 32, 33*). In the small intestine, particularly the proximal and middle jejunum, the microclimate near the brush-border membrane of the enterocyte is acidic compared with the bulk of the luminal solution (*77, 78*). Thus, a sufficient H^+-gradient exists across the brush-border membrane to energize the transport of cefixime and amino-β-lactam antibiotics as well as that of small peptides.

For ceftibuten, a similar transport mechanism was reported by the Shionogi research group (*54, 55*) in rat intestinal BBMVs. Interestingly, as shown in Figure 8, the H^+-gradient-dependent transport via the dipeptide (or small peptide) transport system occurs only for the cis-isomer but not the trans-isomer. Recently, Sugawara et al. (*67*) also demon-

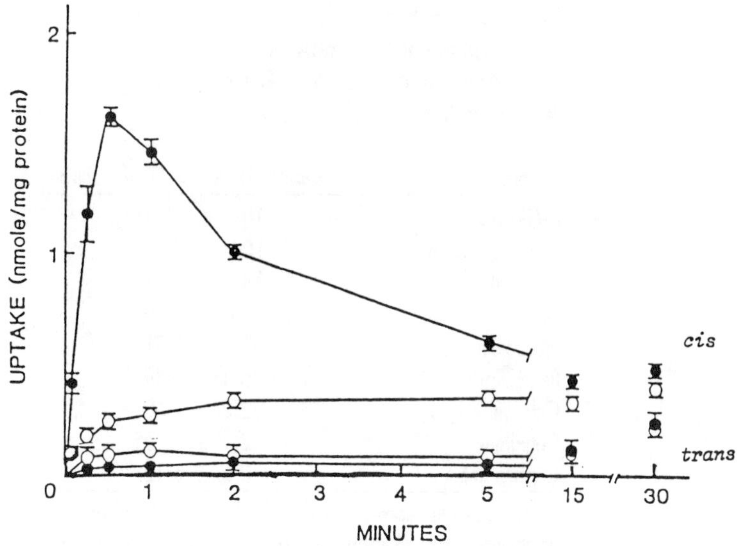

Figure 8. Effect of H^+-gradient on ceftibuten uptake by intestinal BBMVs. Membrane vesicles were suspended in either (○) HEPES buffer (pH 7.5) or (●) Mes buffer (pH 5.5) and then transport was studied. (Reproduced with permission from reference 54. Copyright 1989 Plenum.)

strated the H^+-gradient-dependent transport of ceftibuten in rat, rabbit, and human intestinal BBMVs.

Transport Mechanism of Cefdinir

The newly developed cephalosporin antibiotic cefdinir (*see* structure in Chart I), which is a monocarboxylic acid-type cephalosporin, has a broad antibacterial spectrum that is especially effective against gram-positive bacteria such as *Staphylococcus aureus*. The bioavailability after oral administration of cefdinir varies among animal species; 12–15% (lowest) in mice and rats, 75% (highest) in dogs, and 37% (moderate) in rabbits (*93*). Although cefdinir is structurally similar to cefixime, intestinal absorption of these two derivatives is different. The absorption of cefixime from the large intestine is negligible, whereas absorption of cefdinir from this region is comparable with that from the upper small intestine (*94*). These observations suggest involvement of a specialized transport function in the intestinal absorption of cefdinir in contrast to cefixime and amino-β-lactam antibiotics.

We have recently found (*95–97*), in studies with intestinal BBMVs, that monocarboxylic acids such as acetic acid and nicotinic acid are transported via an intestinal H^+-cotransport system and/or H^+-dependent bicarbonate exchange system. Very similar results were reported by Harig et al. (*98*) for propionate in human ileal BBMVs. To determine if the monocarboxylic acid transport system can transport β-lactam antibiotics, we measured inhibitory effects of various types of β-lactam antibiotics on the uptake of [^{3}H]acetic acid in the presence of an inward H^+-gradient (pH_{in} = 7.5, pH_{out} = 6.0) in rabbit intestinal

Table IV. Inhibitory Effects of Various Compounds on the Uptake of [^{3}H]Acetic Acid (2 μM) by Rabbit Intestinal Brush-Border Membrane Vesicles

Inhibitor	*Acetic Acid Uptake (% of Control)*
D-Glucose	103 ± 15.3
Glycyl-L-proline	101 ± 4.81
Acetic Acid	34.4 ± 5.91[a]
Dicloxacillin	33.6 ± 2.96[a]
Propicillin	61.7 ± 5.91[a]
Penicillin V	75.8 ± 0.390[a]
Cefdinir	22.2 ± 9.50[a]
Cefixime	72.5 ± 5.75[a]
Cephradine	60.6 ± 6.15[a]

NOTE: Each value represents the mean ± SE of three to eight experiments; the uptake was measured in the presence of inwardly directed H^+-gradient (pH_{in} = 6.0, pH_{out} = 7.5); concentration of inhibitor was 10 mM.

[a] Significantly different from the control uptake studied in the absence of inhibitors ($p < 0.05$), as assessed by the *t* test.

SOURCE: Reproduced with permission from reference 99. Copyright 1993 Royal Pharmaceutical Society of Great Britain.

BBMVs. As shown in Table IV, neither D-glucose nor dipeptide (glycyl-L-proline) had a significant effect on the uptake of [^{3}H]acetic acid, indicating that the monocarboxylic acid transport system is independent of the H^+-dependent peptide transport system. On the contrary, cefdinir, dicloxacillin, unlabeled acetic acid, cephradine, propicillin, cefixime, and penicillin V at concentrations of 10 mM significantly reduced, in this order, the uptake of [^{3}H]acetic acid. The inhibitory effects of β-lactam antibiotics suggest that cefdinir and dicloxacillin, monocarboxylic acid-type β-lactam antibiotics, have higher affinities for the monocarboxylic acid transport system than cephradine and cefixime.

We then assessed the relative contribution of the monocarboxylic acid transport system and of the dipeptide transport system to cefdinir absorption by measuring the uptake of cefdinir in BBMVs prepared from rabbit small intestine (*99*). The initial uptake of cefdinir was pH dependent, with increased uptake at acidic pH, and was not influenced by either a Na^+-gradient or membrane potential difference. Cefdinir uptake was saturable, with an apparent Michaelis constant of 8.1 mM and a maximal rate of 0.534 nmol/30 s/mg protein. As shown in Table V, dipeptides (glycyl-L-proline and glycylsarcosine) and cephalosporins (cephradine and cefixime) inhibited cefdinir uptake in the presence of an inward H^+-gradient (pH_{in} = 7.5, pH_{out} = 6.0). Furthermore, the H^+-gradient-dependent uptakes of cefixime and cephradine were significantly inhibited by cefdinir. These results, including mutual inhibition of cedfinir and substrates of the peptide carrier, suggest that cefdinir is transported at least partly via the carrier-mediated system for peptides.

As shown in Table V L-lactic acid and acetic acid inhibited cefdinir uptake without any significant effect on the uptake of cefixime or cephradine. The mutual inhibition observed between cefdinir and acetic acid (Tables IV and V) strongly suggests that cefdinir

Table V. Inhibitory Effect of Various Compounds on the Uptake of Cefdinir (2 mM), Cefixime (2 mM), and Cephradine (2 mM) by Rabbit Intestinal Brush-Border Membrane Vesicles

Inhibitor	*Cefdinir*[a]	*Cefixime*	*Cephradine*
Acetic acid	73.1 ± 3.09[b,c]	94.6 ± 3.25[c]	100 ± 7.40[c]
L-Lactic acid	73.4 ± 7.66[b,c]	93.0 ± 6.24[c]	97.1 ± 4.28[c]
Nicotinic acid	—[d]	104 ± 1.91[c]	112 ± 11.7[b,c]
Glycyl-L-proline	67.7 ± 4.05[b,e]	—	—
Glycylsarcosine	83.1 ± 1.81[b,e]	58.2 ± 4.60[b,f]	72.1 ± 3.73[b,f]
Penicillin V	53.0 ± 0.578[b,f]	—	—
Cefdinir	—	66.5 ± 7.42[b,f]	82.5 ± 7.06[b,f]
Cefixime	80.8 ± 4.28[b,c]	—	—
Cephradine	73.0 ± 5.63[b,f]	—	—
D-Glucose	—	98.8 ± 7.65[c]	90.1 ± 5.11[c]

NOTE: Each value is percent of control and represents the mean ± SE of three to six experiments.
[a] The uptake was determined in the presence of inwardly directed H^+-gradient: $pH_{in} = 6.0$, $pH_{out} = 7.5$ for cefdinir and cephradine; $pH_{in} = 5.5$, $pH_{out} = 7.5$ for cefixime.
[b] Significantly different from the control uptake studied in the absence of inhibitors ($p < 0.05$), as assessed by the *t* test.
[c] Concentration of inhibitor was 10 mM.
[d] —, Not determined.
[e] Concentration of inhibitor was 50 mM.
[f] Concentration of inhibitor was 20 mM.
SOURCE: Reproduced with permission from reference 99. Copyright 1993 Royal Pharmaceutical Society of Great Britain.

can be transported via both the H^+-dependent monocarboxylic acid carrier and the H^+-dependent peptide carrier. In view of the structural similarity of dicloxacillin, propicillin, and penicillin V to monocarboxylic acids and the significant inhibitory effects of these drugs on the uptake of [^{3}H]acetic acid, these penicillins may also share transport carriers with monocarboxylic acids. On the contrary, cephradine and cefixime are not transported by monocarboxylic acid carrier despite having affinity to the carrier, because their uptake was not reduced by monocarboxylates (*see* Table V).

The preceding results (*99*) obtained in our laboratory strongly suggest that some β-lactam antibiotics are transported by a H^+-gradient-dependent monocarboxylic acid carrier as well as a H^+-dependent peptide transporter in addition to simple diffusion. The relative contributions of these two distinct carrier systems (peptide and monocarboxylate transport systems) to cefdinir transport were estimated from the inhibitory effects of dipeptides and acetic acid on cefdinir transport. It was concluded that peptide and monocarboxylate transporters make similar contributions of 30% of total uptake to cefdinir absorption, with the remaining 40% due to diffusion. The apparent differences in intestinal absorption behaviors, including animal species dependence and intestinal region dependence, between cefdinir and cefixime may be ascribed to the difference of the functioning carriers between them. This conclusion is supported by the observation (*94*) of significant absorption of cefdinir from the large intestine, because monocarboxylic acids, like short-chain fatty acids, are absorbed from the large intestine as well as from the small intestine (*100*).

In conclusion, β-lactam antibiotics may be taken up via a H^+/peptide cotransporter and a H^+/monocarboxylate cotransporter into the intestinal enterocyte, as summarized in Figure 9.

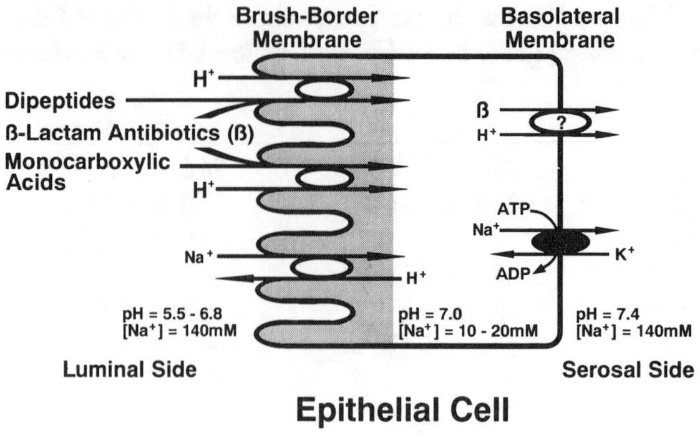

Figure 9. A model for intestinal absorption of β-lactam antibiotics.

Characterization and Identification of the Intestinal Transporter for Peptides and β-Lactam Antibiotics

Despite the importance of the intestinal peptide transporter in the maintenance of protein nutrition in animals and humans, little is known about the transporter at the molecular level. Only recently has evidence of a membrane polypeptide (molecular weight, 127 kD) been presented by Kramer et al. (*45–47, 57–61, 69, 75*). The polypeptide was identified by direct photoaffinity labeling of rabbit small intestinal BBMVs with [^{3}H]benzylpenicillin. This peptide specifically binds dipeptides and β-lactam antibiotics, including aminopenicillins, aminocephalosporins, and dianionic cephalosporins such as cefixime, in a competitive manner and is different from the brush-border-membrane-bound peptidases aminopeptidases N and dipeptidase V. Most recently, the same authors (*69*) have succeeded in the preparation of liposomes reconstituted with the binding protein for β-lactam antibiotics and oligopeptides, which were purified to >95% homogeneity from the rabbit brush-border membrane (*see* Figure 10). Transport experiments with proteoliposomes containing the purified 127 kD binding protein revealed H^+-dependent and stereospecific transport activity. As we observed previously in rat intestinal uptake experiments (*35*), D-cephalexin was taken up into these proteoliposomes more efficiently than the L-enantiomer, in a manner very similar to that observed for D- and L-cephalexin in rabbit intestinal BBMVs (*62*). The results of these extensive studies by Kramer et al. (*45–47, 57–61, 69, 75*) indicate that the binding of the 127 kD protein is by the intestinal H^+-dependent transport system for oligopeptides and β-lactam antibiotics.

Isolation of proteins that are responsible for the transport of various organic solutes across animal cell membranes by conventional protein purification procedures has generally been found to be difficult. This has led Hediger et al. (*101*) to devise an alternate strategy that has proven to be successful in the identification and purification of the mRNA encoding the intestinal Na^+/glucose cotransporter. This strategy, now commonly known as "expression cloning", involves the use of *Xenopus laevis* oocytes to express a particular transport system in a functionally competent form following microinjection of exogenous mRNA into the oocytes. *Xenopus laevis* oocytes are known to be capable of not only translating injected

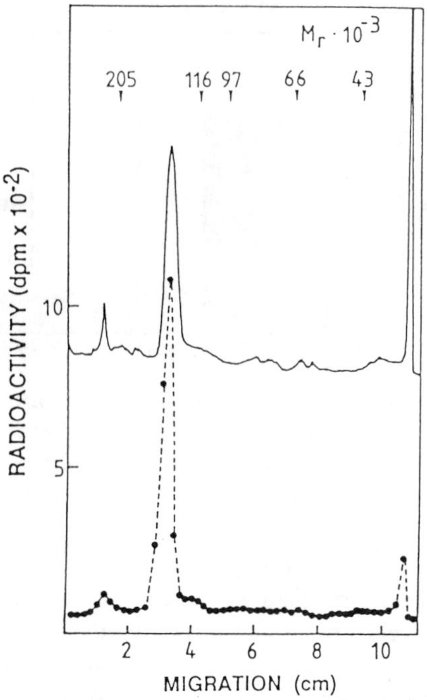

Figure 10. Photoaffinity labeling of proteoliposomes prepared from purified 127-kDa binding protein for β-lactam antibiotics/oligopeptides. Purified 127-kDa binding protein for β-lactam antibiotics/oligopeptides (300 μg protein) was reconstituted into liposomes (80 mg asolectin). Then, 10 μL of the final liposome suspension, equilibrated with buffer B, was inclubated at 20 °C for 5 min in the dark with 1.35 μM (5 μCi) [^{3}H]benzylpenicillin in buffer F and subsequently irradiated at 254 nm for 120 s. Key: (—) densitogram of stained polypeptides; (●----●) distribution of radioactivity. (Reproduced with permission from reference 69. Copyright 1992 Springer.)

mRNA from any eucaryotic source, but also performing posttranslational modification and targeting as appropriate for each protein. As a first step in cloning the gene for the intestinal H^+/dipeptide cotransporter as a common transport system for β-lactam antibiotics we have attempted to express the carrier in *Xenopus leavis* oocytes and to identify the size of the mRNA encoding the 127 kD protein. In our laboratory, expression of the rat intestinal Na^+/uridine cotransporter has been achieved (*102*). Our most recent results on the expression of size-fractionated mRNA encoding the rat intestinal H^+/peptide cotransporter in *Xenopus laevis* oocytes are as follows (*103*). RNA was isolated from the rat intestinal mucosa by a guanidium thiocyanate extraction method followed by cesium chloride gradient centrifugation. Poly(A)$^+$RNA was then selected by oligo(dT)-cellulose chromatography and was injected into oocytes. The expression of the H^+/peptide cotransporter in the oocyte plasma membrane was assayed by measuring the H^+-dependent uptake of ceftibuten. As shown in Figure 11, ceftibuten uptake was greater in oocytes injected with poly(A)$^+$RNA than in those injected with water and was remarkably enhanced in the presence of an inwardly directed H^+-gradient only in oocytes injected with poly(A)$^+$RNA. As illustrated in Figure 12, ceftibuten uptake in poly(A)$^+$RNA-injected oocytes was significantly inhibited by the presence of glycyl-L-proline, glycylglycine, cyclacillin, cephradine, cefixime, and benzylpenicillin, but was not inhibited by proline or glycine. The uptake of ceftibuten was significantly trans-stimulated by preloading glycycl-L-proline (countertransport effect) only in poly-

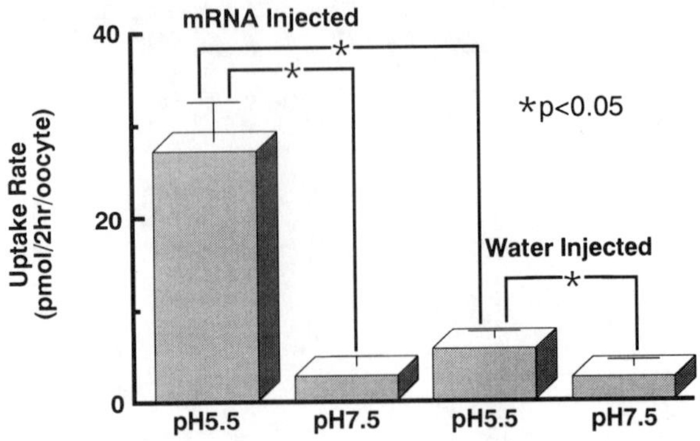

Figure 11. H^+-Dependent ceftibuten uptake at 25 °C by mRNA- or water-injected Xenopus laevis *oocytes. The uptake values represent the mean ± SE of 4–6 determinations. Significant differences ($p <$ 0.05) from the uptake indicated, as assessed by the* t *test are denoted with an asterisk (*). (Reproduced with permission from reference 103. Copyright 1994 Pergamon.)*

(A)$^+$RNA-injected oocytes (Figure 13A) and the uptake of the cis-isomer was significantly enhanced in the presence of an inward H^+-gradient and cis-inhibited by glycyl-L-proline (Figure 13B); these results are consistent with the observations (*54, 55*) in rat intestinal BBMVs. These properties of the expressed ceftibuten transporter strongly suggest that the intestinal H^+/peptide cotransporter can be expressed in *Xenopus laevis* oocytes and can transport ceftibuten. Miyamoto et al. (*104*) and Saito et al. (*105*) reported the functional expression of H^+/peptide cotransporter in *Xenopus laevis* oocytes injected with poly-(A)$^+$RNA prepared from rabbit intestinal mucosal cells and from Caco-2 cells, respectively. Both groups reported that the expressed peptide transporters, assayed by measuring the uptake of [^{14}C]glycylsarcosine and [^{3}H]bestatin (a dipeptide anticancer agent), respectively, retained their substrate specificities and had affinity for dipeptides or cephradine but no

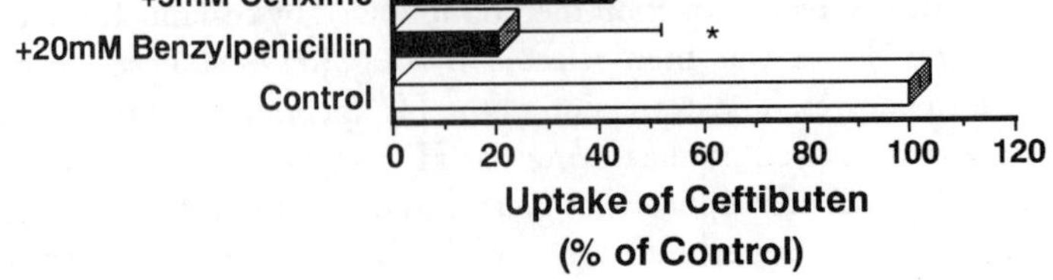

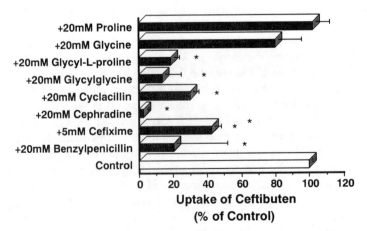

Figure 12. Effect of various compounds of H^+-dependent ceftibuten uptake at 25 °C and at pH 5.5 by mRNA- or water-injected Xenopus laevis *oocytes. The uptake values represent the mean ± SE of 3–9 determinations. Significant differences from the control uptake in the absence of inhibitors ($p <0.05$), as assessed by the* t *test are denoted with an asterisk (*). (Reproduced with permission from reference 103. Copyright 1994 Pergamon.)*

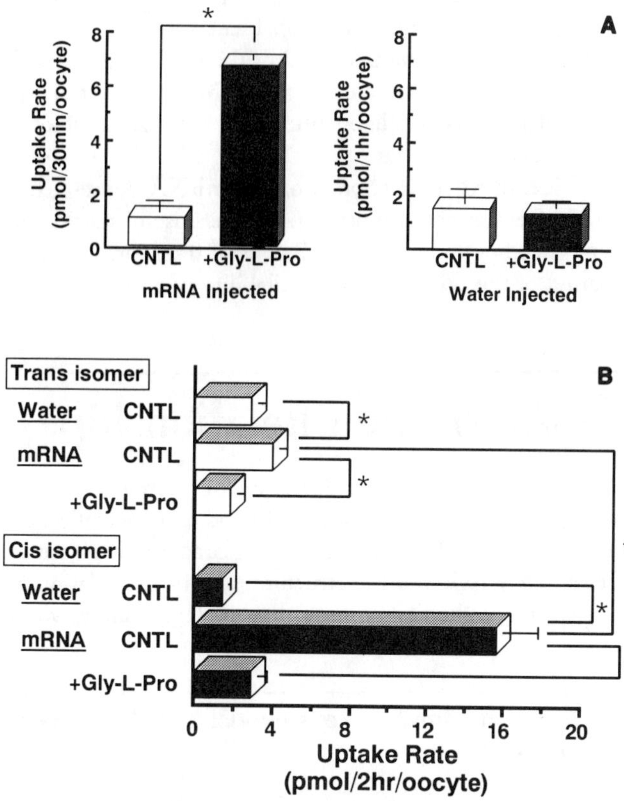

Figure 13. A. Countertransport effect of glycyl-L-proline on ceftibuten uptake at 25 °C in the presence of H^+-gradient by mRNA- or water injected Xenopus laevis *oocytes. Oocytes were preincubated for 2 h in the uptake medium of pH 5.5 with 10 mM glycylglycine. The uptake values represent the mean ± SE of 4–8 determinations. Significant difference (p <0.05) from the uptake indicated as assessed by the* t *test are denoted with an asterisk (*). B. Stereospecific uptake of ceftibuten at 25 °C in the presence of H^+-gradient and inhibitory effects of 25 mM glycyl-L-proline (gly-L-pro) on their uptakes at 25 °C and at pH 5.5 by mRNA- or water-injected* Xenopus laevis *oocytes. The uptake values represent the mean ± SE of 4–8 determinations. Significant difference from the uptake indicated (p < 0.05), as assessed by the* t *test are denoted with an asterisk (*). (Reproduced with permission from reference 103. Copyright 1994 Pergamon.)*

affinity for the free amino acids. The expressed peptide transporters showed pH-dependent maximal uptakes of [^{14}C]glycylsarcosine and [^{3}H]bestatin around pH 5.5.

Recent findings from our laboratory showed very clear expression of the transport systems for ceftibuten and aminocephalosporins (cephalexin, cephradine, and cefadroxil) in *Xenopus laevis* oocytes injected with poly(A)$^+$RNA prepared from rat, rabbit, and human intestines (unpublished observation), indicating that orally active β-lactam antibiotics are transported, as the common route among species, via carrier proteins existing in the intestinal mucosa.

To screen a cDNA library using expression in the *Xenopus laevis* oocyte as an assay, it is desirable to minimize testing of the clones by fractionating the poly(A)$^+$RNA before cDNA synthesis. The total mRNA prepared from rat intestinal mucosa was fractionated by centrifugation through a sucrose gradient, and each fraction was assayed for its ability to induce the transport activity of ceftibuten and cefadroxil. As shown in Figure 14, the mRNA encoding the H^+/dipeptide transporter was found in fraction 2, corresponding to the size range 2.20–3.75 kilobases (kb). Hence, assuming an average molecular weight of 100–130 for amino acids, subfraction 2 of the mRNA could well account for expression of the 127 kD polypeptide that was suggested by Kramer et al. (*69*) to be directly involved in H^+-dependent uptake of oligopeptides and β-lactam antibiotics into the intestinal brush-border membrane.

The identified subfraction 2 of the mRNA component for the expression of the intestinal H^+/peptide cotransporter should be a useful starting point for expression cloning of the gene encoding this important intestinal transport protein without the need for large scale isolation of the protein or the availability of an antibody.

Structure–Absorption Relationship

There are two main strategies for development of orally active cephalosporins: one is based on screening cephalosporin derivatives for good oral absorption and good antibacterial activity and the other is based on making prodrugs of a candidate with good antibacterial activity. Sakane et al. (*93*) reviewed the structure–absorption relationship in the development of cefdinir according to the first strategy. Tables VI–XI show, with regard to 3-vinyl cephalo-

Table VI. Antibacterial Activity of Cephalosporins 3 and Urinary Recovery after Oral Administration in Rats

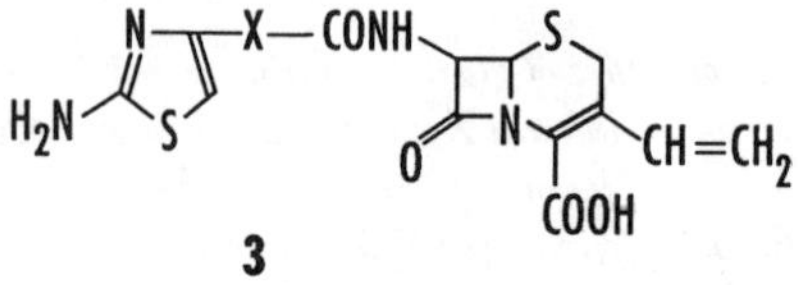

3

Compound	*X*	MIC (μg/mL), 10^6 cfu/mL: S. aureus *209P JC-1*	E. coli *NIHJ JC-2*	K. pneumoniae *12*	P. mirabilis *1*	Proteus vulgaris *49*	*Urinary recovery, 0–24 h (%)*[a]
3a	$-C(=O)-$	1.56	3.13	0.78	0.39	100	12.0
3b	$-CH(NH_2)-$	1.56	12.5	6.25	12.5	>100	22.5
3c	$-CH(OH)-$	0.78	3.13	1.56	1.56	50	13.8
3d	$-CH(CH_2OH)-$	3.13	6.25	6.25	3.13	>100	8.7
3e	$-CH_2-$	0.78	25.0	3.13	1.56	>100	11.2
Cefixime	$-C(=N-OCH_2COOH)-$	25.0	0.20	0.10	≦0.025	≦0.025	34.0

[a] Dose, 100 mg/kg.

SOURCE: Reproduced with permission from reference 93. Copyright 1992.

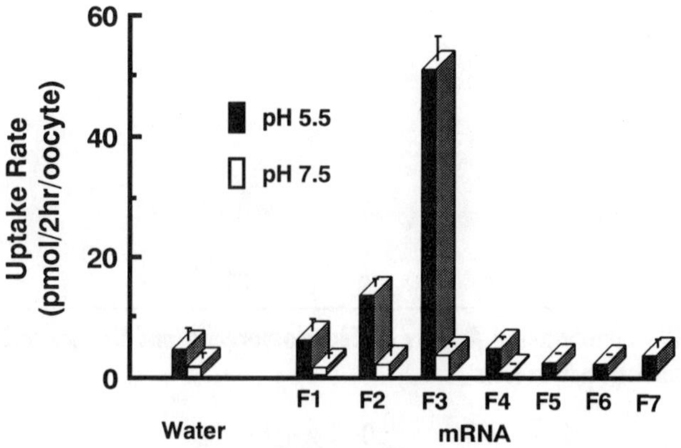

Figure 14. Expression of ceftibuten uptake at 25 °C in the presence (closed columns; pH 5.5) or absence (open columns; pH 7.5) of a H^+-gradient by Xenopus laevis *oocytes injected with size-fractionated mRNA. The uptake values represent the mean ± SE of 4–10 determinations. (Reproduced with permission from reference 103. Copyright 1994 Pergamon.)*

sporins with various 7-acyl side chains and cefdinir derivatives with various 3-substituents, the antibacterial activities against gram-positive and gram-negative bacteria and the urinary and/or biliary recoveries after oral dosing (20 and/or 100 mg/kg) in rats. Compound **3b** and cefixime show relatively good absorption (Table VI). Compound **4g** (cefdinir) and cefixime show higher absorption (Table VII). Only cefdinir exhibits good antibacterial

Table VII. Antibacterial Activity of Cephalosporins 4 and Urinary Recovery after Oral Administration in Rats

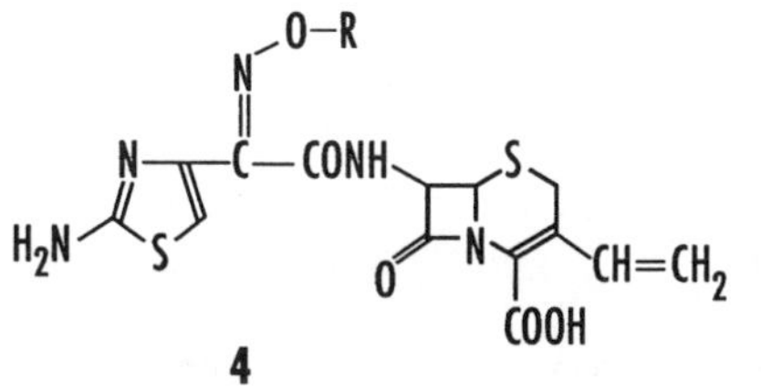

4

		MIC (μg/mL), 10^6 cfu/mL					Urinary recovery, 0–24 h (%)[a]
Compound	*R*	S. aureus *209P JC-1*	E. coli *NIHJ JC-2*	K. pneumoniae *12*	P. mirabilis *1*	P. vulgaris *49*	
4a	$-(CH_2)_2COOH$	100	1.56	0.78	0.10	0.10	6.5
4b	$-(CH_2)_3COOH$	12.5	1.56	0.39	0.05	0.39	12.3
4c	$-CH(CH_3)COOH$	50.0	0.78	0.39	0.05	0.05	7.3
4d	$-C(CH_3)_2COOH$	12.5	0.78	0.78	0.10	0.10	2.7
4e	$-CH_2COOC_2H_5$	3.13	0.78	1.56	0.20	0.39	5.7
4f	$-CH_2PO(OH)_2$		0.78	0.39	0.05	0.05	4.2
4g	$-H$	10	0.39	0.20	0.10	0.78	18.0
Cefixime	$-CH_2COOH$	.0	0.20	0.10	≦0.025	≦0.025	34.0

[a] Dose, 100 mg/kg.

SOURCE: Reproduced with permission from reference 93. Copyright 1992.

Table VIII. Antibacterial Activity of Cephalosporins 5 and Urinary and Biliary Recoveries after Oral Administration in Rats

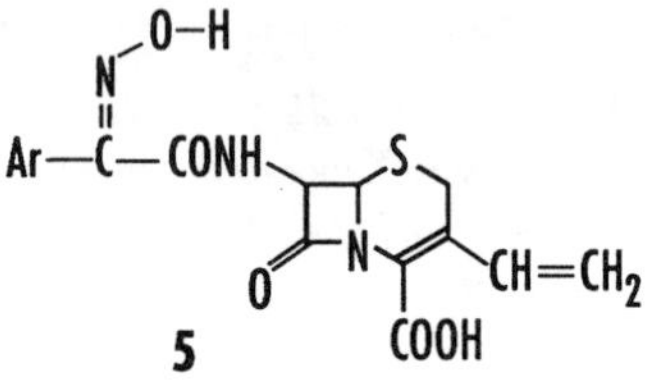

5

Compound	Ar	MIC (μg/mL), 10^6 cfu/mL					Recovery 0–24 h (%)[a]	
		S. aureus 209P JC-1	E. coli NIHJ JC-2	K. pneumoniae 12	P. mirabilis 1	P. vulgaris 49	Urine	Bile
5a	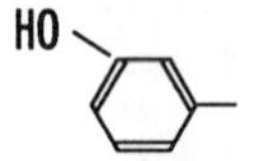	0.39	3.13	6.25	0.78	25	16.0	1.6
5b	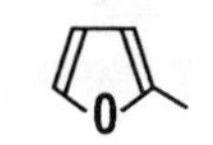	0.20	25	50	6.25	>100	32.0	N.D.[b]
5c	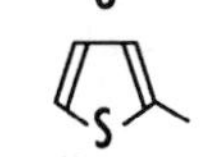	0.78	50	>100	12.5	>100	15.4	N.D.
5d	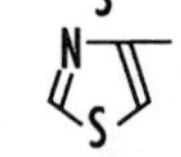	0.39	6.25	12.5	3.13	100	60.2	0.4
5e	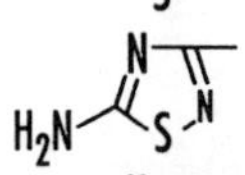	0.78	0.78	0.78	1.56	6.25	6.9	1.4
4g (Cefdinir)	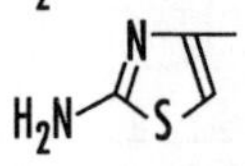	0.10	0.39	0.20	0.10	0.78	31.6	0.9

[a] Dose, 20 mg/kg.
[b] N.D., Not detected.
SOURCE: Reproduced with permission from reference 93. Copyright 1992.

Table IX. Antibacterial Activity of Cephalosporins 6 and Urinary and Biliary Recoveries after Oral Administration in Rats

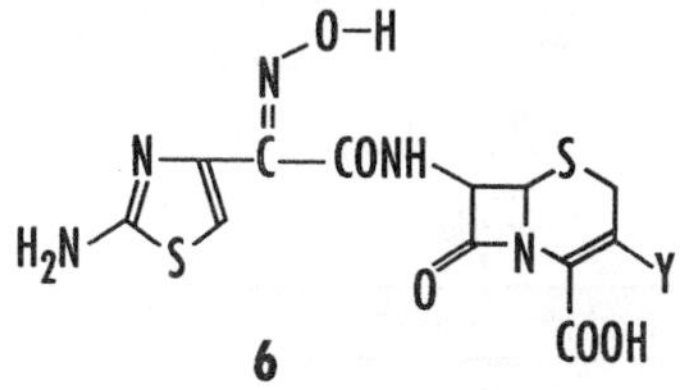

6

		MIC (μg/mL), 10^6 cfu/mL						Recovery 0–24 h (%)[a]	
		S. aureus			Bacillus subtilis	E. coli			
Compound	Y	209P JC-1	6	32	ATCC 6633	NIHJ JC-2	K. pneumonia 12	Urine	Bile
6a	H	0.39	0.78	0.78	6.25	0.05	0.05	10.7	0.7
6b	CH_3	6.25	6.25	6.25	50.0	1.56	0.78	12.5	0.4
6c	C_2H_5	1.56	3.13	3.13	6.25	1.56	1.56	1.1	3.2
6d	C≡CH	3.13	3.13	3.13	3.13	1.56	1.56	3.6	1.9
6e	(Z) $CH{=}CHCH_3$	0.39	0.78	0.78	1.56	0.20	0.20	11.2	0.9
6f	OCH_3	3.13	6.25	12.5	25.0	3.13	1.56	9.7	0
6g	SCH_3	0.39	0.78	1.56	0.78	0.78	0.39	11.4	0
6h	SC_2H_5	0.20	0.39	0.39	0.39	0.39	0.39	7.6	0.9
6i	$SCH{=}CH_2$	0.20	0.39	0.78	0.20	0.39	0.39	7.5	0.7
4g (Cefdinir)	$CH{=}CH_2$	0.20	0.39	0.39	1.56	0.20	0.20	31.6	0.9

[a] Dose, 20 mg/kg.

SOURCE: Reproduced with permission from reference 93. Copyright 1992.

Table X. Influence of Substitution in the Aryl Moiety of Cephalosporins 5 and 7 on Urinary and Biliary Recoveries after Oral Administration in Rats

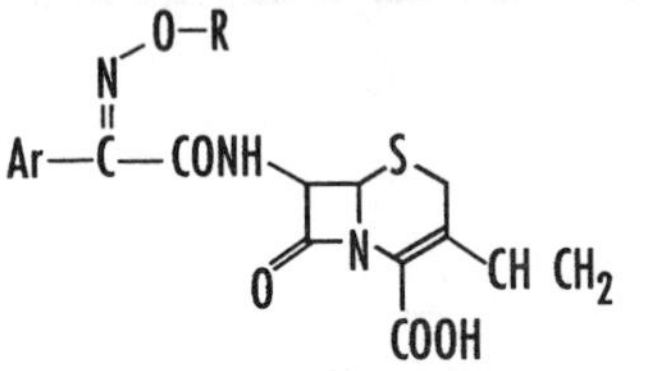

R=H **(5)**

$R{=}CH_2CO_2H$ **(7)**

	R = H **(5)**[a]			R = $CH_2CO_2CO_2H$ **(7)**[b]		
Ar	*Compound*	*Urinary Recovery (%)*	*Biliary Recovery (%)*	*Compound*	*Urinary Recovery (%)*	*Biliary Recovery (%)*
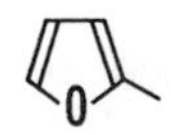	**5b**	32.0	N.D.[c]	**7b**	43.8	26.9
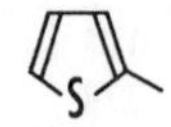	**5c**	15.4	N.D.	**7c**	66.7	30.2
	**5d**	60.2	0.4	**7d**	43.1	22.1
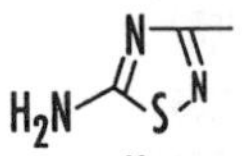	**5e**	6.9	1.4	**7e**	28.8	2.7
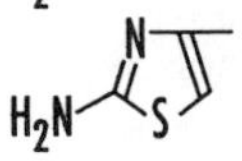	Cefdinir	31.6	0.9	Cefixime	34.0	18.2

[a] Dose, 20 mg/kg.
[b] Dose, 100 mg/kg.
[c] N.D., Not detected.
SOURCE: Reproduced with permission from reference 93. Copyright 1992.

Table XI. Influence of Substitution in the Oxime Moiety of Cephalosporins 6 and 8 on Urinary and Biliary Recoveries after Oral Administration in Rats

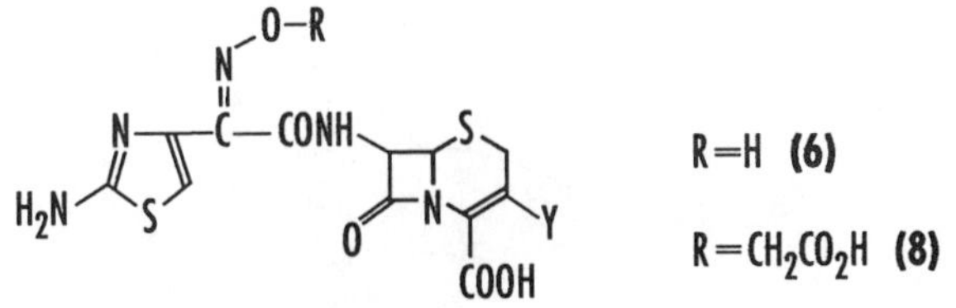

	R=H (6)		R=CH_2CO_2H (8)	
	Recovery (%)[a]		Recovery (%)[b]	
Y	Urine	Bile	Urine	Bile
—H	10.7	0.7	41.0	3.8
—CH_3	12.5	0.4	25.7	9.7
—OCH_3	9.7	0	41.6	18.0
—SCH_3	11.4	0	51.6	41.1
—C≡CH	3.6	1.9	20.0	7.5
—CH=CH_2	31.6	0.9	34.0	18.2

[a] Dose, 20 mg/kg.
[b] Dose, 100 mg/kg.
SOURCE: Reproduced with permission from reference 93. Copyright 1992.

activity and moderate absorption, although **5b** and **5d** also exhibit good absorption characteristics (Tables VIII and IX). Therefore, cefdinir was selected as a fourth generation orally active cephalosporin with excellent antibacterial activity against *Staphylococcus aureus* and enhanced absorption properties compared with its analogues. As summarized in Tables X and XI, oral availability is higher in 2-carboxymethoxyimino derivatives (compounds **7** and **8**) than in 2-hydroxyimino derivatives (compounds **5** and **6**). Interestingly, among compounds **5** and **6**, only cefdinir shows good absorbability.

Conclusions

The amino-β-lactam antibiotics cefixime and ceftibuten are absorbed from the small intestine via a H^+/peptide cotransporter and cefdinir is absorbed via both H^+/peptide and H^+/monocarboxylate cotransporters (*see* Figure 9). As shown in Figure 15, aminocephalosporin

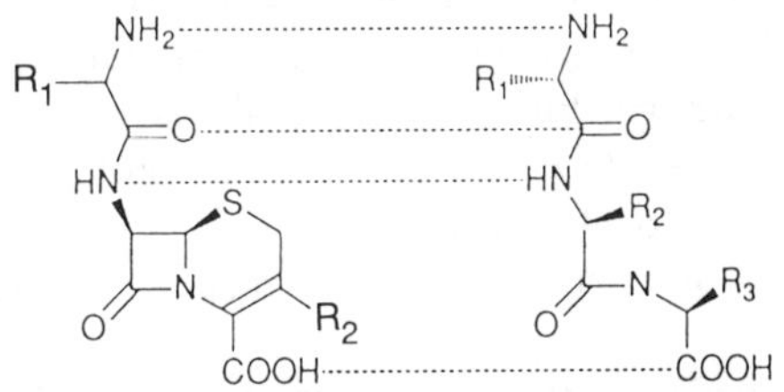

Figure 15. Structural similarity between aminocephalosporin and tripeptide.

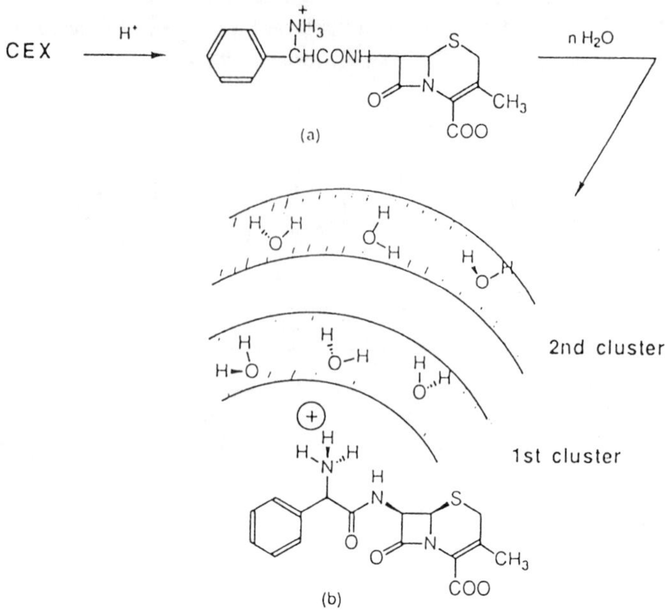

Figure 16. Zwitterionic form of cephalexin and its hydration. (Reproduced with permission from reference 93. Copyright 1992.)

and aminopenicillin structurally resemble tripeptides. Therefore, we can easily rationalize that amino-β-lactam antibiotics are absorbed via a peptide transporter. However, how can we explain the structural recognition of cefixime, ceftibuten, and cefdinir, which have no α-amino group, by the peptide transporter? Because the pK_a value of the α-ammonium group of amino-β-lactam antibiotics is 6–8 at the microclimate pH (5.5–6.8) of the intestinal brush-border membrane, amino-β-lactam antibiotics exist in the zwitterionic form because of protonation of the α-amino group. In general, ammonium compounds such as amino-β-lactam antibiotics tend to be hydrated in aqueous solution and seem to be surrounded by two clusters of water, as illustrated in Figure 16. It seems likely that the peptide carrier recognizes and transports amino-β-lactam antibiotics such as cephalexin as the hydrated structure (*93*). By computer simulation, Sakane et al. (*93*) showed that the structures of cefixime and ceftibuten (cis-isomer) can be well superimposed on the structure of cephalexin clustered with two molecules of water (Figure 17). There is a slightly less good superposition of cefdinir with one water molecule at the oxime terminal upon cephalexin with two water molecules (Figure 18). The reason why the trans-isomer of ceftibuten is poorly absorbed can be easily understood on the basis of this model approach. Although there is no experimental evidence to support the hypothesis proposed by Sakane et al. (*93*), the hypothesis provides a rationale for the apparent structural similarity of tripeptides with the amino-β-lactam antibiotics cefixime, ceftibuten, and cefdinir.

The molecular structure and biological characteristics of the intestinal H^+/peptide cotransporter should soon be elucidated with the cloning of the gene coding for this transport protein. Thereafter, rational drug design should be possible for development of orally active β-lactam antibiotics and peptide-like drugs that can be taken up into intestinal enterocytes by this transporter.

After completion of this chapter, the intestinal oligopeptide transporters have been successfully cloned as PepT1 from rabbit small intestine (707-amino acid protein with 12

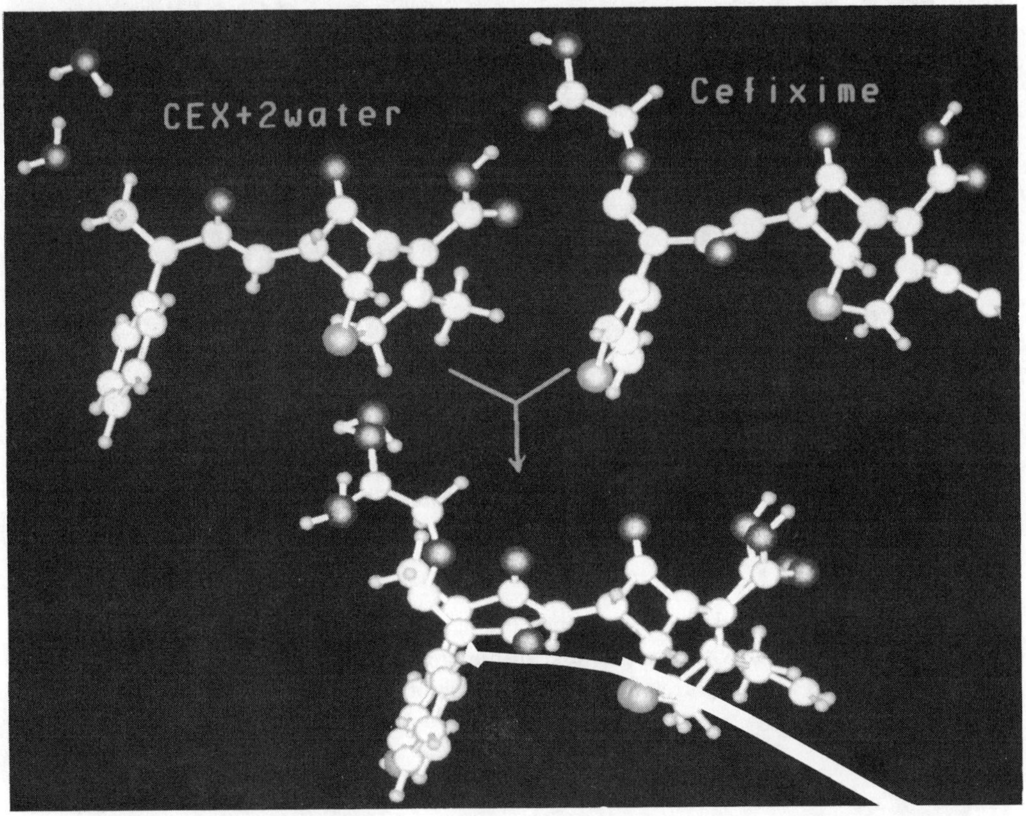

Figure 17. Cefixime superimposed on cephalexin clustered with two water molecules. (Reproduced with permission from reference 93. Copyright 1992.)

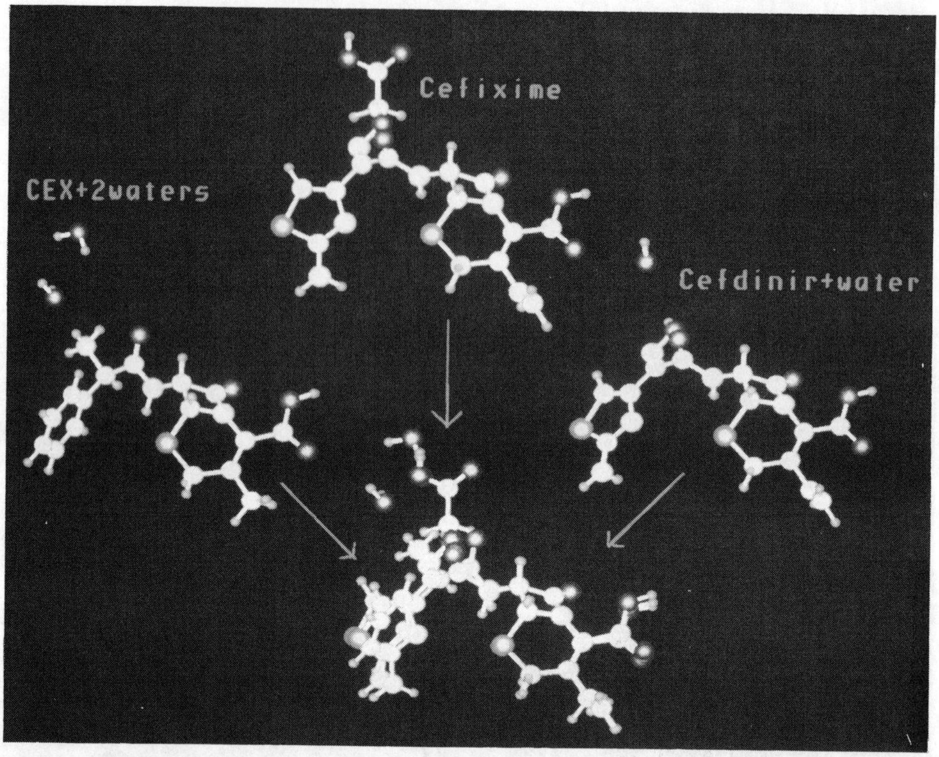

Figure 18. Cefdinir clustered with one water molecule superimposed on cephalexin clustered with two water molecules and one cefixime. (Reproduced with permission from reference 93. Copyright 1992.)

membrane-spanning domains, reference 106) and HPT-1 from human Caco-2 cells (832-amino acid protein with 2 membrane-spanning domains, reference 107), both of which were reported to transport amino β-lactam antibiotics.

Acknowledgments

This research was supported in part by Grants-in-Aid for Scientific Research from the Ministry of Education, Science and Culture, Japan. I am grateful to Dr. Takao Takaya, New Drug Research Laboratories, Fujisawa Pharmaceutical Company, Ltd., Osaka, for helpful discussions and for the use of the computer graphics in Figures 17 and 18. I am indebted to a number of scientists for their collaboration during the course of the studies described, and I thank Hitomi Takanaga for her help in the preparation of the figures and tables.

References

1. Quay, J. F. *Physiologist* **1972,** *15,* 241.
2. Addison, J. M.; Burston, D.; Dalrymple, J. A.; Matthews, D. M.; Payne, J. W.; Sleisenger, M. H.; Wilkinson, S. *Clin. Sci. Mol. Med.* **1975,** *45,* 313–322.
3. Penzotti, S. C.; Poole, J. W. *J. Pharm. Sci.* **1974,** *63,* 1803–1806.
4. Tsuji, A.; Nakashima, E.; Kagami, I.; Honjo, N.; Yamana, T. *J. Pharm. Pharmacol.* **1977,** *29,* 707–708.
5. Miyazaki, K.; Ogino, O.; Nakano, M.; Arita, T. *Chem. Pharm. Bull.* **1977,** *25,* 246–252.
6. Dixon, C.; Mizen, L. W. *J. Physiol.* **1977,** *269,* 549–559.
7. Yasuhara, M.; Miyoshi, Y.; Yuasa, A.; Kimura, T.; Muranishi, S.; Sezaki, H. *Chem. Pharm. Bull.* **1977,** *25,* 675–679.
8. Kimura, T.; Endo, H.; Yoshikawa, M.; Muranishi, S.; Sezaki, H. *J. Pharmacobio-Dyn.* **1978,** *1,* 262–267.
9. Tsuji, A.; Nakashima, E.; Kagami, I.; Asano, T.; Nakashima, R.; Yamana, T. *J. Pharm. Pharmacol.* **1978,** *30,* 508–509.
10. Tsuji, A.; Miyamoto, E.; Hashimoto, N.; Yamana, T. *J. Pharm. Sci.* **1978,** *67,* 1705–1711.
11. Shindo, H.; Fukuda, K.; Kawai, K.; Tanaka, K. *J. Pharmacobio-Dyn.* **1978,** *1,* 310–323.
12. Umeniwa, K.; Ogiso, O.; Miyazaki, K.; Arita, T. *Chem. Pharm. Bull.* **1979,** *27,* 2177–2182.
13. Tsuji, A.; Nakashima, E.; Asano, T.; Nakashima, R.; Yamana, T. *J. Pharm. Pharmacol.* **1979,** *31,* 718–720.
14. Tsuji, A.; Miyamoto, E.; Kubo, O.; Yamana, T. *J. Pharm. Sci.* **1979,** *68,* 812–816.
15. Halpin, T. C.; Perkins, R. L.; Greenberger, N. *J. Pharmacol. Res. Commun.* **1980,** *12,* 725–738.
16. Kimura, T.; Kobayashi, H.; Sezaki, H. *Life Sci.* **1980,** *27,* 1667–1672.
17. Stewart, H. E.; Jackson, M. J. *J. Pharmacol. Exp. Ther.* **1981,** *218,* 453–458.
18. Tsuji, A.; Nakashima, E.; Kagami, I.; Yamana, T. *J. Pharm. Sci.* **1981,** *70,* 768–772.
19. Tsuji, A.; Nakashima, E.; Kagami, I.; Yamana, T. *J. Pharm. Sci.* **1981,** *70,* 772–777.
20. Tsuji, A.; Miyamoto, E.; Terasaki, T.; Yamana, T. *J. Pharm. Sci.* **1982,** *71,* 403–406.
21. Miyazaki, K.; Ohtani, K.; Umeniwa, K.; Arita, T. *J. Pharmacobio-Dyn.* **1982,** *5,* 555–563.
22. Miyazaki, K.; Iseki, K.; Arita, T. *J. Pharmacobio-Dyn.* **1982,** *5,* 593–602.
23. Kimura, T.; Yamamoto, T.; Mizuno, M.; Suga, Y.; Kitade, S.; Sezaki, H. *J. Pharmacobio-Dyn.* **1983,** *6,* 246–253.
24. Yamashita, S.; Yamazaki, Y.; Mizuno, M.; Masada, M.; Nadai, T.; Kimura, T.; Serzaki, H. *J. Pharmacobio-Dyn.* **1984,** *7,* 227–233.
25. Nakashima, E.; Tsuji, A.; Kagatani, S.; Yamana, T. *J. Pharmacobio-Dyn.* **1984,** *7,* 452–464.
26. Nakashima, E.; Tsuji, A.; Mizuo, H.; Yamana, T. *Biochem. Pharmacol.* **1984,** *33,* 3345–3352.
27. Iseki, K.; Iemura, A.; Sato, H.; Sunada, K.; Miyazaki, K.; Arita, T. *J. Pharmacobio-Dyn.* **1984,** *7,* 768–775.
28. Nakashima, E.; Tsuji, A. *J. Pharmacobio-Dyn.* **1985,** *8,* 623–632.
29. Naskashima, E.; Tsuji, A.; Nakamura, M.; Yamana, T. *Chem. Pharm. Bull.* **1985,** *33,* 2098–2106.
30. Kimura, T.; Yamamoto, T.; Ishizuka, R.; Sezaki, H. *Biochem. Pharmacol.* **1985,** *34,* 81–84.
31. Yamashita, S.; Yamazaki, Y.; Masada, M.; Nadai, T.; Kimura, T.; Sezaki, H. *J. Pharmacobio-Dyn.* **1986,** *9,* 368–374.

32. Okano, T.; Inui, K.; Takano, M.; Hori, R. *Biochem. Pharmacol.* **1986,** *35,* 1781–1786.
33. Okano, T.; Inui, K.; Maegawa, H.; Takano, M.; Hori, R. *J. Biol. Chem.* **1986,** *261,* 14130–14134.
34. Tsuji, A.; Hirooka, H.; Tamai, I.; Terasaki, T. *J. Antibiot.* **1986,** *39,* 1592–1597.
35. Tamai, I.; Ling, H.-Y.; Simanjuntak, M. T.; Nishikido, J.; Tsuji, A. *J. Pharm. Pharmacol.* **1987,** *40,* 320–324.
36. Tsuji, A.; Hirooka, H.; Terasaki, T.; Tamai, I.; Naskashima, E. *J. Pharm. Pharmacol.* **1987,** *39,* 272–277.
37. Tsuji, A.; Tamai, I.; Hirooka, H.; Terasaki, T. *Biochem. Pharmacol.* **1987,** *36,* 565–567.
38. Iseki, K.; Mori, K.; Miyazaki, K.; Arita, T. *Biochem. Pharmacol.* **1987,** *36,* 1843–1846.
39. Iseki, K.; Mori, K.; Miyazaki, K.; Arita, T. *Biochem. Pharmacol.* **1987,** *36,* 1837–1842.
40. Tsuji, A. In *Frontiers of Antibiotic Research;* Umezawa, H., Ed.; Academic: Tokyo, Japan, 1987; pp 255–268.
41. Tsuji, A.; Terasaki, T.; Tamai, I.; Hirooka, H. *J. Pharmacol. Exp. Ther.* **1987,** *241,* 594–601.
42. Sinko, P. J.; Hu, M.; Amidon, G. L. *J. Control. Rel.* **1987,** *6,* 115–121.
43. Sinko, P. J.; Amidon, G. L. *Pharm. Res.* **1988,** *5,* 645–654.
44. Dantzig, A. N.; Bergin, L. *Biochim. Biophys. Res. Commun.* **1988,** *155,* 1082–1087.
45. Kramer, W.; Leioe, I.; Petzoldt, E.; Girbig, F. *Biochim. Biophys. Acta* **1988,** *939,* 167–172.
46. Kramer, W.; Girbig, F.; Petzoldt, E.; Leipe, I. *Biochim. Biophys. Acta* **1988,** *943,* 288–296.
47. Kramer, W.; Girbig, F.; Leipe, I.; Petzoldt, E. *Biochem. Pharmacol.* **1988,** *37,* 2427–2435.
48. Inui, K.; Okano, T.; Maegawa, H.; Kato, M.; Takano, M.; Hori, R. *J. Pharmacol. Exp. Ther.* **1988,** *247,* 235–241.
49. Hori, R.; Okano, T.; Kato, M.; Maegawa, H.; Inui, K. *J. Pharm. Pharmacol.* **1988,** *40,* 646–647.
50. Iseki, K.; Sugawara, M.; Saitoh, H.; Miayazaki, K.; Arita, T. *J. Pharm. Pharmacol.* **1988,** *40,* 701–705.
51. Iseki, K.; Sugawara, M.; Saitoh, H.; Miayazaki, K.; Arita, T. *J. Pharm. Pharmacol.* **1989,** *41,* 628–632.
52. Sinko, P. J.; Amidon, G. L. *J. Pharm. Sci.* **1989,** *78,* 723–727.
53. Pico, A. S.; Peris-Ribera, J.-E.; Toledano, C.; Torres-Molina, F.; Casabo, V.-G.; Martin-Villodre, A.; Pla-Delfina, J. M. *J. Pharm. Pharmacol.* **1989,** *41,* 179–185.
54. Yoshikawa, T.; Muranushi, N.; Yoshida, M.; Oguma, T.; Hirano, K.; Yamada, H. *Pharm. Res.* **1989,** *6,* 302–307.
55. Muranushi, N.; Yoshikawa, T.; Yoshida, M.; Oguma, T.; Hirano, K.; Yamada, H. *Pharm. Res.* **1989,** *6,* 308–312.
56. Kato, M.; Maegawa, H.; Okano, T.; Inui, K.; Hori, R. *J. Pharmacol. Exp. Ther.* **1989,** *251,* 745–749.
57. Kramer, W.; Girbig, F.; Gutjahr, U.; Leipe, I. *J. Chromatogr.* **1990,** *521,* 199–210.
58. Kramer, W.; Girbig, F.; Gutjahr, U.; Kleemann, H.; Leipe, I.; Urbach, H.; Wagner, A. *Biochim. Biophys. Acta* **1990,** *1027,* 25–30.
59. Kramer, W.; Dechent, C.; Girbig, F.; Gutjahr, U.; Neubauer, H. *Biochim. Biophys. Acta* **1990,** *1030,* 41–49.
60. Kramer, W.; Gutjahr, U.; Girbig, F.; Leipe, I. *Biochim. Biophys. Acta* **1990,** *1030,* 50–59.
61. Kramer, W.; Durckheimer, W.; Girbig, F.; Gutjah, U.; Leipe, I.; Oekonomopulos, R. *Biochim. Biophys. Acta* **1990,** *1028,* 174–182.
62. Dantzig, A. H.; Bergin, L. *Biochim. Biophys. Acta* **1990,** *1027,* 211–217.
63. Sugawara, M.; Saitoh, H.; Iseki, K.; Miyazaki, K.; Arita, T. *J. Pharm. Pharmacol.* **1990,** *42,* 314–318.

64. Sugawara, M.; Iseki, K.; Miyazaki, K. *J. Pharm. Pharmacol.* **1991,** *43,* 433–435.
65. Sugawara, M.; Iseki, K.; Miyazaki, K.; Shiroto, H.; Kondo, Y.; Uchino, J. *J. Pharm. Pharmacol.* **1991,** *43,* 882–884.
66. Westphal, J. F.; Deslandes, A.; Brogard, J. M.; Carbon, C. *J. Antimicrob. Chemother.* **1991,** *27,* 647–654.
67. Sugawara, M.; Toda, T.; Iseki, K.; Miyazaki, K.; Shiroto, H. *J. Pharm. Pharmacol.* **1992,** *44,* 968–972.
68. Dantzig, A. H.; Tabas, L. B.; Bergin, L. *Biochim. Biophys. Acta* **1992,** *1112,* 167–173.
69. Kramer, W.; Girbig, F.; Gutjahr, U.; Kowalewski, S.; Adam, F.; Schiebler, W. *Eur. J. Biochem.* **1992,** *204,* 923–930.
70. Oh, D-M.; Sinko, P. J.; Amidon, G. L. *Int. J. Pharm.* **1992,** *85,* 181–187.
71. Inui, K.; Tomita, Y.; Katsura T.; Okano, T.; Takano, M.; Hori, R. *J. Pharmacol. Exp. Ther.* **1992,** *260,* 482–486.
72. Inui, K.; Yamamoto, M.; Saito, H. *J. Pharmacol. Exp. Ther.* **1992,** *261,* 195–201.
73. Wang, H.-P.; Bair, C.-H.; Huahg, J.-D. *J. Pharm. Pharmacol.* **1992,** *44,* 1027–1029.
74. Morita, E.; Mizuno, N.; Nishikata, M.; Takahashi, K. *J. Pharm. Sci.* **1992,** *81,* 337–340.
75. Kramer, W.; Gutjahr, U.; Kowalewski, S.; Girbig, F. *Biochem. Pharmacol.* **1993,** *46,* 542–545.
76. Saito, H.; Ishi, T.; Inui, K. *Biochem. Pharmacol.* **1993,** *45,* 776–779.
77. Ganapathy, V.; Leibach, F. H. *Am. J. Physiol.* **1985,** *249,* G153–G160.
78. Takuwa, N.; Shimada, T.; Matsumoto, H.; Himukai, M.; Hoshi, T. *Jpn. J. Physiol.* **1985,** *35,* 629–642.
79. Kessler, M.; Acuto, O.; Storelli, C.; Murer, H.; Muller, M. *Biochim. Biophys. Acta* **1978,** *506,* 136–154.
80. Dyer, J.; Beechey, R. B.; Gorvel, J. P. ; Smith, R. T.; Wootton, R.; Shirazi-Beechey, S. P. *Biochem. J.* **1990,** *978,* 565–571.
81. Pinto, M.; Robine-Leon, S.; Appay, M. D.; Kedinger, M.; Triadou, N.; Dussaulx, E.; Lacroix, B.; Simon-Assmann, P.; Haffen, K.; Fogh, J.; Zweibaum, A. *Biol. Cell* **1983,** *47,* 323–330.
82. Hidalgo, I. J.; Raub, T. J.; Borchardt, R. T. *Gastroenterology* **1989,** *96,* 736–749.
83. Hilgers, A. R.; Conradi, R. A. Burton, P. S. *Pharm. Res.* **1990,** *7,* 902–910.
84. Blais, A.; Bissonnette, A.; Berteloot, A. *J. Membr. Biol.* **1987,** *99,* 113–125.
85. Harris, D. S.; Slot, J. W.; Geuze, H. J.; James, D. E. *Proc. Natl. Acad. Sci. U.S.A.* **1992,** *89,* 7556–7560.
86. Mahraoui, L.; Rousset, M.; Dussaulx, E.; Darmoul, D.; Zweibaum, A.; Brot-Laroche, E. *Am. J. Physiol.* **1992,** *263,* G312–G318.
87. Hidalgo, I. J.; Borchardt, R. T. *Biochim. Biophys. Acta* **1990,** *1028,* 25–30.
88. Hidalgo, I. J.; Borchardt, T. T. *Biochim. Biophys. Acta* **1990,** *1035,* 97–103.
89. Thwaiters, D. T.; Brown, C. D. A.; Hirst, B. H.; Simmons N. L. *J. Biol. Chem.* **1993,** *268,* 7640–7642.
90. Saito, H.; Inui, K. *Am. J. Physiol.* **1993,** *265,* G289–G294.
91. Tamai, I.; Tsuji, A.; Kin, Y. *J. Pharmacol. Exp. Ther.* **1988,** *246,* 338–344.
92. Hoshi, T. *Jpn. J. Physiol. (Engl.)* **1985,** *35,* 179–191.
93. Sakane, K.; Inamoto, Y.; Takaya, T. *Jpn. J. Antibot. (Jpn.)* **1992,** *45,* 909–925.
94. Mine, Y.; Sakamoto, H.; Hirose, T.; Shibayama, F.; Kikuchi, H.; Kuwahara, S. *Abstracts of Papers,* 27th Interscience Conference on Antimicrobial Agents and Chemotherapy, New York; American Society for Microbiology: Los Angeles, CA, 1987; Abstract 654.
95. Simanjuntak, M. T.; Tamai, I.; Terasaki, T.; Tsuji, A. *J. Pharmacobio–Dyn.* **1990,** *13,* 301–309.

96. Tsuji, A.; Simanjuntak, M. T.; Tamai, I.; Terasaki, T. *J. Pharm. Sci.* **1990,** *79,* 1123–1124.

97. Simanjuntak, M. T.; Terasaki, T.; Tamai, I.; Tsuji, A. *J. Pharmacobio–Dyn.* **1991,** *14,* 501–508.

98. Harig, J. M.; Soergel, K. H.; Barry, J. A.; Ramaswamy, K. *Am. J. Physiol.* **1991,** G776–G782.

99. Tsuji, A.; Tamai, I.; Nakanishi, M.; Terasaki, T.; Hamano, S. *J. Pharm. Pharmacol.* **1993,** *45,* 996–998.

100. Bugaut, M. *Comp. Biochem. Physiol.* **1987,** *86B,* 439–472.

101. Hediger, M. A.; Coady, M. J.; Ikeda, T. S.; Write, E. M. *Nature (London)* **1987,** *330,* 379–381.

102. Terasaki, T.; Kadowaki, A.; Higashida, H.; Nakayama, K.; Tami, I.; Tsuji, A. *Biol. Pharm. Bull. (continuation of J. Pharmacobio–Dyn.)* **1993,** *16,* 493–496.

103. Tamai, I.; Tomizawa, N.; Kadowaki, A.; Terasaki, T.; Nakayama, K.; Higasida, H.; Tsuji, A. *Biochem. Pharmacol.* **1994,** *48,* 881–888.

104. Miyamoto, Y.; Thompson, Y. G.; Howard, E. F.; Ganapathy, V.; Leibach, F. H. *J. Biol. Chem.* **1991,** *266,* 4742–4745.

105. Saito, H.; Ishi, T.; Inui, K. *Biochem. Pharmacol.* **1993,** *45,* 776–779.

106. Fei, Y-J.; Kanai, Y.; Nussberger, S.; Ganapathy, V.; Leibach, F. H.; Romero, M. F.; Singh, S. K.; Boron, W. F.; Hediger, M. A. *Nature (London)* **1994,** *368,* 563–566.

107. Dantzig, A. H.; Hoskins, J.; Tabas, L. B.; Bright, S.; Shepard, R. L.; Jenkins, I. L.; Duckworth, D. C.; Sportsman, J. R.; Mackensen, D.; Rosteck, P. R., Jr.; Skatrud, P. L. *Science (Washington, DC)* **1994,** *264,* 430–433.

RECEIVED for review July 12, 1993. ACCEPTED revised manuscript January 10, 1994.

CHAPTER 6

Oral Absorption of Angiotensin-Converting Enzyme Inhibitors and Peptide Prodrugs

Shiyin Yee
and Gordon L. Amidon*

Orally active angiotensin-converting enzyme (ACE) inhibitors represent one of the newest and most significant classes of cardiovascular and antihypertensive agents. These drugs inhibit ACE by preventing the conversion of the prohormone angiotensin I to the active hormone angiotensin II. The ACE inhibitors also inhibit the degradation of the vasodilator bradykinin (*1*). The clinical effectiveness of these agents is well established, but the oral absorption and systemic availability of ACE inhibitors range from 25 to 71%, with high intra- and intersubject variability (Table I). Recent results concerning the absorption mechanism and first-pass metabolism of the ACE inhibitors are reviewed in this chapter, and approaches for improving the oral bioavailability of these drugs through the use of prodrugs are discussed.

Captopril was the first orally active sulfhydryl Ala-Pro derivative approved for use in the treatment of hypertension and congestive heart failure (Chart I). Further synthesis

* Corresponding author.

Table I. Pharmacokinetic Parameters of ACE Inhibitors

Compound	*Bioavailability (%)*	*Excretion in Feces %*	*Fe in Urine (%)*	*Biliary Elimination*[a]	*Ref.*
Captopril	71	18	67	NS	2
Lisinopril	25	71	29	N	5
Fosinopril	25	73 (62% diacid)	13 (10% diacid)	N	39
Fosinoprilat	NA[b]	46 (iv)	44 (iv)	Y	39
Enalapril	36–44	33 (27% diacid)	61 (43% diacid)	Y (50%)	2, 5
Quinapril	35–40	37 (27–37% diacid)	62	Y	42
Benazepril	35–40	60 (27% diacid)	37 (17% diacid)	Y (>50%)	40, 41
Enalaprilate	3	—[c]	—	—	11

[a] NS, Not significant; Y, yes; N, no.
[b] Not applicable.
[c] —, Not determined.

was directed toward compounds structurally related to captopril and led to ACE inhibitors such as fentiapril (SA446), pivalopril, and zofenopril (SQ 26991) (*2*). However, adverse effects associated with the sulfhydryl moiety led to a search for ACE inhibitors that do not contain thiol and have a longer duration of action.

The first approved, orally active, ACE inhibitor that did not contain a sulfhydryl group was enalapril (Chart I), a substituted *N*-carboxylmethyl dipeptide ethyl ester (*1, 3*). The affinity of enalaprilat, the active dicarboxylic acid of enalapril, is 5–10 times higher than the affinity of captopril for ACE (*4*). However, because the bioavailability of enalaprilat is very low (*4*), the monoethyl ester prodrug enalapril was developed to improve oral effec-

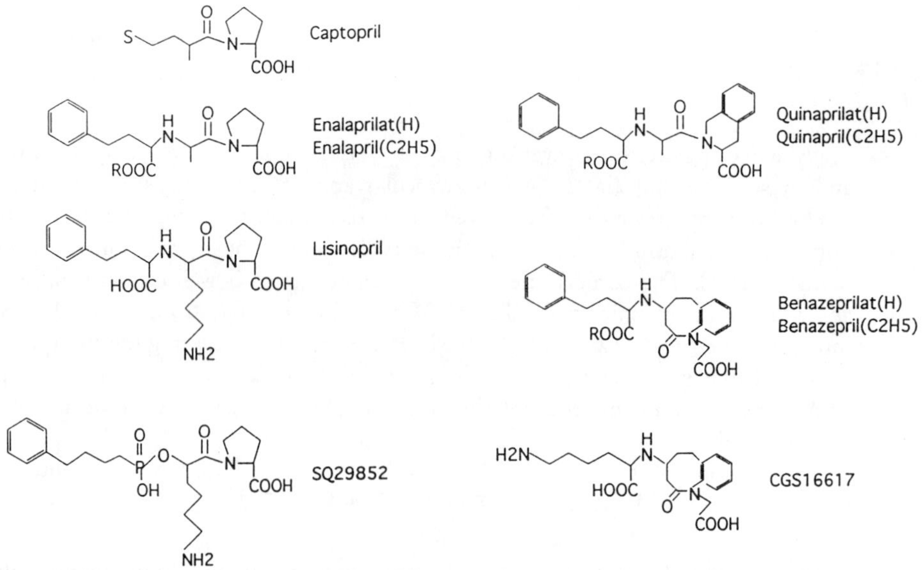

Chart I. ACE inhibitors.

tiveness. Although enalapril has low activity as an ACE inhibitor, it is fairly well absorbed from the gut and subsequently hydrolyzed to the active diacid form by esterases in the gut, liver, blood, and other tissues (*5*). Therefore, enalapril became the prototype for the development of structurally related prodrug ACE inhibitors such as indolapril (CI-907), quinapril (CI-906) (Chart I), benazepril (CGS 14824A), and ramipril (HOE 498). Ramipril has a greater bioavailability and a higher potency against ACE in plasma and tissues following oral administration compared with enalapril (*6*).

Other bicyclic, prodrug ACE inhibitors include perindopril (S 9490-3) and pentopril (CGS 13945); perindopril is the most active compound of this group. The active diacid of perindopril, perindoprilat, has a bioavailability and an affinity for ACE similar to those of ramipril and ramiprilat, respectively (*7*). Non-prodrug ACE inhibitors without a sulfhydryl group include indalapril [REV 6000 A(SS)], lisinopril (MK 521), and libenzapril (CGS 16617) (Chart I). Lisinopril, a Lys-Pro peptide analogue, has an affinity for ACE similar to that of enalaprilat. The potential advantage of lisinopril over enalapril is that lisinopril has a better (but still low) systemic availability despite being a dicarboxylic acid. Consequently, lisinopril is not a prodrug because it does not have to be metabolized to the active compound in vivo. Another synthetic approach to develop nonsulfhydryl ACE inhibitors resulted in the replacement of the sulfhydryl group with a phosphorus group. Fosinopril (SQ 28555) is a prodrug ACE inhibitor with an in vitro inhibitory potency similar to that of captopril but with a longer duration of action than captopril following oral administration (*8*).

In Situ Intestinal Membrane Permeability

The intrinsic wall permeability (P_w^*) at various drug concentrations correlates with the fraction of dose absorbed (FA) of cephalosporins and other carrier-mediated and passively absorbed compounds (*9, 10*) in humans. The parameter P_w^* can be measured by the single-pass rat perfusion technique, which is simple and involves the cannulation of a segment of the intestine ~10 cm long, perfusion of the drug solution in a suitable buffer (pH 6.5), and collection of the perfusate at the outlet under steady-state conditions (after ~30–45 min). The water-flux-corrected ratios of the outlet-to-inlet concentrations (peak heights) are then used to calculate P_w^*.

A summary of the estimated intrinsic membrane absorption parameters of some ACE inhibitors obtained from the single-pass intestinal perfusion technique is shown in Table II. The measurement of P_w^* in the presence or absence of metabolic inhibitors provides additional insight into the mechanism of oral absorption of ACE inhibitors. The permeability and concentration results from perfusion experiments [Figure 1 (*11*)] can be used in the following functional expression (*12*) to calculate the intestinal wall permeability, P_w:

$$P_w = J_{max}/(K_m + C) + P_m \tag{1a}$$

$$P_w = P_c/(1 + K_m/C) + P_m \tag{1b}$$

where J_{max} and K_m are the maximal flux and apparent carrier affinity, respectively, and P_m and P_c (= J_{max}/K_m) are the passive and carrier-mediated components of the transport

Table II. Estimated Intestinal Membrane Absorption Parameters of Some ACE Inhibitors

Compound[a]	J^*_{max} (mM)[b]	K_m (mM)[c]	P^*_c[d]	P^*_m[e]	Ref.
Quinapril	0.31 (0.01)	0.55 (0.02)	0.57 (0.002)	0.89	Unpublished
Quinaprilat	—[f]	—[f]	—[f]	0	Unpublished
Benazepril	0.07 (0.003)	0.07 (0.001)	0.96 (0.04)	0.75 (0.02)	Unpublished
Benazeprilat	—[f]	—[f]	—[f]	—[f]	Unpublished
CGS 16617	—[f]	—[f]	—[f]	0.27 (0.04)	Unpublished
Lisinopril	0.032 (0.004)	0.08 (0.003)	0.39 (0.03)	0	11
SQ29,852	0.16 (0.04)	0.08 (0.01)	2 (0.2)	0.25 (0.07)	11
Enalapril	0.13	0.07	1.9	0.35	14
Enalapril	0.10 (0.01)[g]	0.02 (0.0)	4.58 (0.05)[h]	1.62 (0.07)[h]	Unpublished
Enalaprilat	—[f]	—[f]	—[f]	0.1	14
Captopril	12	6	2	0.75	13
Fosinopril	—[f]	—[f]	—[f]	>4	14

NOTE: Parameters are mean fitted values. The numbers in parentheses are standard error.
[a] All results are from in situ rat perfusion, except for enalapril, which is from in vitro rabbit BBMVs.
[b] Maximal flux (*11, 12*).
[c] Michaelis constant.
[d] Dimensionless (*11, 12*) carrier permeability.
[e] Dimensionless (*11,12*) passive permeability.
[f] —, Not determined.
[g] Values are nmol mg^{-1} min^{-1}.
[h] Values are μL mg^{-1} min^{-1}.

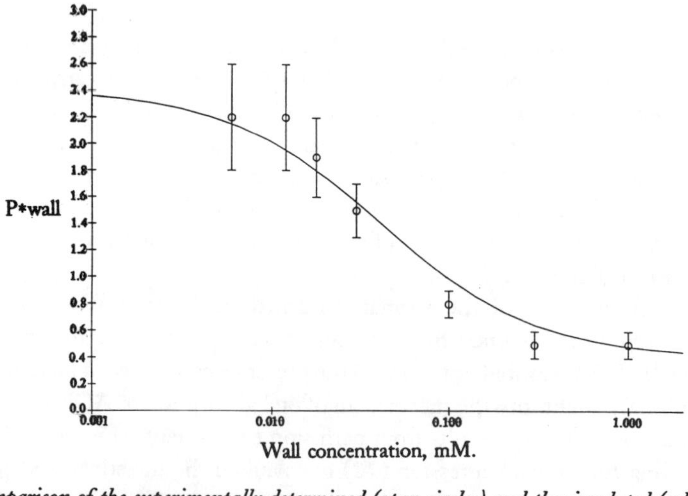

Figure 1. Comparison of the experimentally determined (open circles) and the simulated (solid line) of the dimensionless wall permeability (P^*_w) *as a function of wall concentration* (C_w) *for enalapril. Mean [± standard deviation (SD)] of four to eight rats at each concentration* (11).

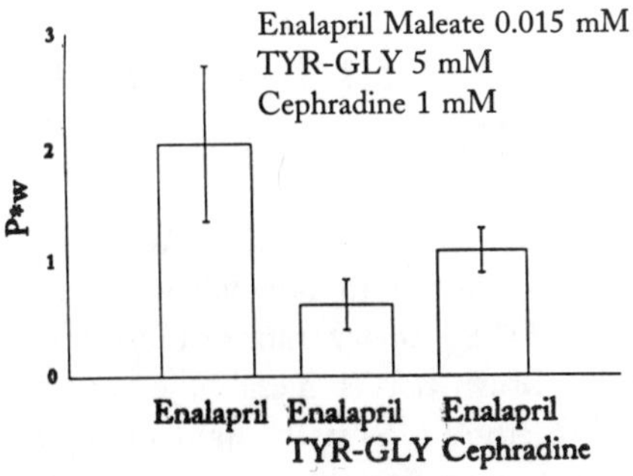

Figure 2. Influence of Tyr-Gly (5 mM) and cephradine (1 mM) on P^*_w of 0.015 mM enalapril. Results are the mean of four to six rats ± SD (11).

process, respectively. In the presence of a competitive inhibitor (or substrate), the analysis of the steady-state perfusion results replaces K_m in eq 1 with

$$K_m = (1 + [I]/K_I)K_m$$

where [I] is the inhibitor concentration in the perfusate and K_I is the apparent inhibitor binding constant [Figure 2 (*11*)].

Results obtained by this methodology show that several ACE inhibitors are transported in part by the carrier-mediated transport system for small peptides (*11, 13, 14*) that is also shared by many β-lactam antibiotics (*15*), some renin inhibitors (*16*), alafosfalin (*17*), and thyrotropin-releasing hormone (*18*). Most of the ACE inhibitors have a passive component to their intestinal membrane transport (i.e., passive permeability, P^*_m; Table II), but evidence also suggests a carrier-mediated component. For example, the intestinal permeability of captopril is both pH- and concentration-dependent and is decreased by co-perfusion with 2,4-dinitrophenol, cephradine, or di- and tripeptides (*13*). This finding is the first demonstration that a peptide carrier may be involved in the transport of ACE inhibitors and further supports the suggestion that the peptide carrier system can transport a substrate without an *N*-terminal nitrogen atom. The results indicate that whereas enalapril, ceronapril (SQ29,852), and lisinopril have similar K_m values, the carrier permeability of lisinopril ($P^*_c = J^*_{max}/K_m$) is ~10-fold lower than those of enalapril and ceronapril, a result that agrees with the relatively low oral absorption of lisinopril in humans. In addition, unlike other ACE inhibitors, lisinopril is transported only by a carrier mechanism. In contrast fosinopril, a nonpolar prodrug with a large P_m, appears to be transported by passive diffusion. The diacids benazeprilat, quinaprilat, enalaprilat, and libenzapril have low permeabilities, ranging from 0 to 0.27 (Table II), which are consistent with their poor oral absorption. Consequently, on the basis of the observations that the oral bioavailability of lisinopril is threefold higher than that of enalaprilat and the in vitro activities of lisinopril and enalaprilat are similar, the structural requirements for the binding to the mucosal cell peptide carrier and to ACE are dissimilar. The introduction of the lysyl residue seems to promote a low but significant carrier-mediated transport for lisinopril. The difference in permeabilities between enalapril and enalaprilat suggests that the carboxylic moiety is important for the interaction with ACE but hinders the interaction of the drug with the peptide transporter. Hence, esterification of the carboxyl group improves the oral absorption of the ACE inhibitors, in part, by making the drug more peptide-like.

Mechanistic Studies in Intestinal Brush-Border-Membrane Vesicles

Several processes involved in the transport of ACE inhibitors and peptide prodrugs across intact intestinal tissue, in addition to metabolism in tissue, cytosol, and lumen, can complicate the identification of the primary translocation step across the brush border membrane of the epithelial cells. Brush-border-membrane vesicles (BBMVs) were prepared by the calcium precipitation method (*19*) with slight modifications. Briefly, the preparation involved scraping and homogenizing the mucosal layer from jejunum of male New Zealand rabbits, precipitating with $CaCl_2$, centrifuging, and resuspending the pellet in suitable buffer (pH 6 or 7.5) containing mannitol. The BBMVs were then quantitated for protein content, alkaline phosphatase activity, and glucose uptake. Uptake was measured by a rapid filtration technique (*20*). Typically, uptake was initiated by incubating the membrane preparation with substrate, and the reaction was terminated at the desired time by diluting the reaction mixture with ice-cold stop solution. The membrane was immediately collected by vacuum filtration.

The measured K_m of 0.02 mM for enalapril uptake in the vesicles is consistent with the results from the isolated perfused intestinal segment experiments, a finding suggesting no fundamental difference in the substrate–carrier association mechanism between the two species (*21*). Uptake of enalapril was significantly inhibited by other ACE inhibitors, cephalosporins, and di- and tripeptides, a result suggesting a common transport system for these compounds.

Countertransport experiments to confirm the involvement of a carrier in the transport of enalapril were conducted with vesicles in which the concentration of the preloaded substrate at the *trans* face of the membrane was set well above the K_m at that face and the uptake of radioactive enalapril from the *cis* face was measured. *Trans* stimulation occurred in vesicles preloaded with the aminocephalosporins, cephradine, and cefadroxil, but *trans* inhibition was found with the ACE inhibitors enalapril and lisinopril. In contrast, enalaprilat had no *trans* effect (Figure 3a). The phenomenon of *trans* stimulation and *trans* inhibition of enalapril uptake can be explained by Figure 3b (*22*) and the following equation for a simple cotransport system:

$$(\nu_{12}^{S})_2^{P}/(\nu_{12}S)^0 = [K^P + P_2]/[K^P + P_2(V_{max_{00}}/V_{max_{21}}^{P})]$$

where the left side of the equation represents the ratio of the flux of label substrate S in the presence of a *trans* substrate P_2 $[(\nu_{12}^{S})_2^{P}]$ to the flux in the absence of *trans* substrate $[(\nu_{12}^{S})^0]$, K^P is the dissociation constant of substrate P, and $V_{max_{00}}/V_{max_{21}}^{P}$ is the ratio of the rate of conformational change of the free carrier compared with the maximal velocity of transport of substrate P_2 in the 2-to-1 direction. In *trans* stimulation by the cephalosporins, the substrate–carrier complex on the *trans* side reorients to the *cis* side faster than the free carrier. The increase in free carrier on the *cis* side accelerates enalapril uptake in the BBMV. The inhibition by lisinopril or enalapril on the *trans* side suggests that either the rate of reorientation or dissociation of the ACE inhibitor–carrier complex is the rate-limiting step for the ACE inhibitor uptake. These results correlate well with the observation that the cephalosporins (*23, 24*) have a 100-fold higher maximal flux (J_{max}) compared with the ACE inhibitors (*11, 14*). The ACE inhibitors have a 100-fold higher K_m compared with the cephalosporins. Hence, the ACE inhibitors have a high-affinity but low-capacity mechanism of transport by the carrier, whereas the cephalosporins have a low-affinity but high-capacity transport. Stimulation of enalapril transport in the presence of cephradine or cefa-

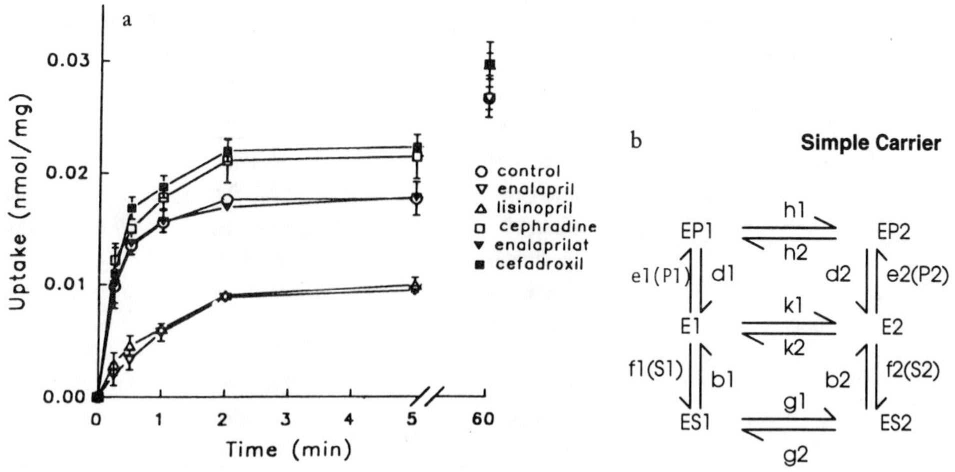

Figure 3. (a) Countertransport effect of various compounds on enalapril uptake in BBMVs. The vesicles were pre-equilibriated with either 10 mM of drug in loading buffer (300 mM mannitol, 20 mM Mes–Tris, pH 6.0) or loading buffer alone (control) for 30 min at 25 °C. Uptakes at various time points were conducted by incubating 10 μL of this vesicle at 25 °C with 500 μL of labeled 1 μM enalapril in Mes–Tris buffer (pH 6.0) containing 300 mM mannitol. Values are mean ± standard error of three determinations. (b) Kinetic model for countertransport where 1 is outside and 2 is inside the membrane with indicated rate constants b, d, e, g, h, *and* k. *S is a substrate and P is a different substrate. (Adapted from reference 22.)*

droxil on the *trans* side further indicates that the uptake of enalapril involves a carrier system and not channels (*22*). Enalaprilat on the *trans* side did not affect enalapril uptake at the concentration studied, and this finding suggests that it does not bind to the carrier on the *trans* side. These studies suggest that the peptide carrier is asymmetric in terms of rate of reorientation and substrate affinity. Results obtained from *trans* studies for cephradine uptake in the rabbit vesicle also show *trans* stimulation by cefadroxil, *trans* inhibition by enalapril or lisinopril, and no effect by enalaprilat (*23*, Table III). Enalaprilat inhibited the

Table III. Inhibition (K_i) and Michaelis (K_m) Constants of Various Compounds

Compound	K_i *(mM)*[b]	K_m *(mm)*
Lisinopril	0.11 + 0.01	0.08[a]
Enalaprilat	5.76 + 0.17[b]	—[c]
Quinapril	0.89 + 0.02	0.55[d]
Cephradine	4.79 + 0.11	4.30[e]

NOTE: The *cis* inhibition of enalapril uptake by these compounds was studied in the presence of an inward proton gradient and ethylenediaminetetraacetic acid at 25 °C for 15 s; values are represented as mean + standard error. All inhibition is competitive type.

[a] Reference 14.

[b] Comparable to k_i of enalaprilat obtained from cephradine uptake in the rabbit BBMV (K_i = 5.4; reference 23).

[c] Not determined.

[d] Unpublished results from rat perfusion experiments.

[e] Rabbit intestinal vesicle study (reference 23).

Table IV. *cis* and *trans* Effects of Various Compounds of Enalapril and Cephradine Uptake in Rabbit Intestinal BBMVs

Analyte	*Compound*	*cis Effect*	*trans Effect*
Enalapril	Enalaprilat	Inhibition	No effect
	Lisinopril	Inhibition	Inhibition
	Enalapril	Inhibition	Inhibition
	Cephradine	Inhibition	Stimulation
	Cefadroxil	Inhibition	Stimulation
Cephradine[a]	Enalaprilat	Inhibition	No effect
	Lisinopril	ND[b]	Inhibition
	Enalapril	Inhibition	Inhibition
	Cefaroxil	Inhibition	Stimulation
	L-Gly-Pro	Inhibition	ND

[a] Reference 23.
[b] ND, Not determined.

transport of enalapril in *cis* experiments (Table IV); most likely enalaprilat is a true competitive inhibitor of the peptide transporter.

Systemic Availability Considerations

Prodrug Considerations

A very common prodrug strategy is to use simple esters of a polar parent drug to increase the hydrophobicity of the drug (*25*). A complimentary strategy to increase the oral absorption of water-insoluble hydrophobic drugs is to use amino acid derivatives targeted at intestinal brush border membrane enzymes (*26–28*). Enzyme-catalyzed hydrolytic cleavage is the most common reaction for conversion. Active drug species containing –OH or –COOH groups can often be converted to prodrug esters from which the active forms are regenerated by esterases within the body. The poorly absorbed active ACE inhibitors were designed as ester prodrugs to potentiate their biological in vivo activity (*2*).

Comparison of the oral efficacy and intrinsic activity profiles of, for example, enalapril and enalaprilat indicates that although enalaprilat binds tightly to ACE, it has low efficacy for the peptide carrier. The ester prodrug enalapril, on the other hand, has a higher affinity for the peptide carrier. These studies of the intestinal absorption of the ACE inhibitors indicate that esterification of one of the carboxylic acid groups results in a more peptide-like compound that is transported by the peptide carrier (*14*). This conclusion is consistent with other results with β-lactam antibiotics that suggest a free N-terminal nitrogen is not an absolute requirement for peptide carrier-mediated transport (*29*). The peptide carrier thus appears to have a broad specificity, and use of prodrugs designed to exploit this broad specificity may be applicable to other drugs in the di- and tripeptide size. A recent example of this strategy for polar drugs, based on the peptide carrier, is illustrated in Figure 4 for the poorly absorbed drug α-methyldopa (*30*). Various dipeptides of α-methyldopa were synthesized and shown to have a significantly higher membrane permeability than the parent

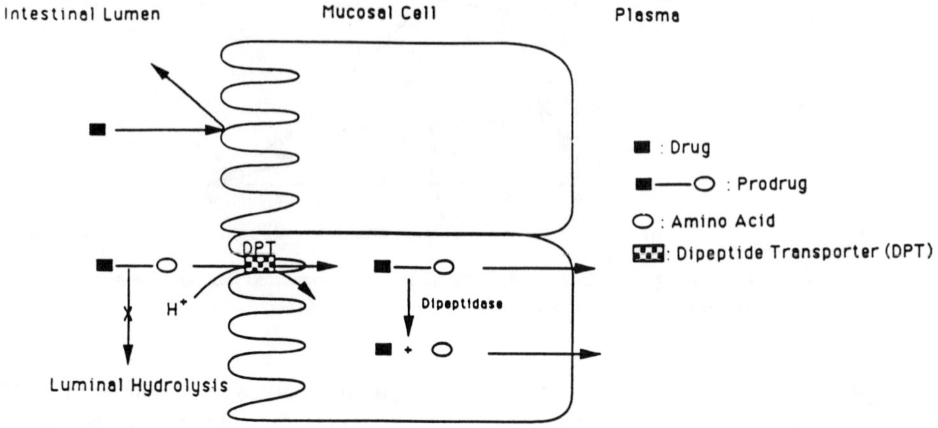

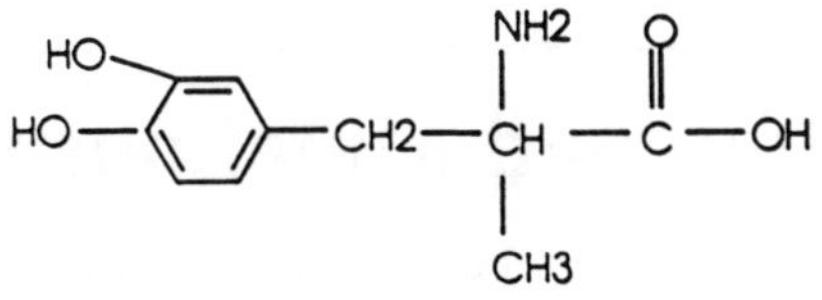

L-Alpha-methyldopa

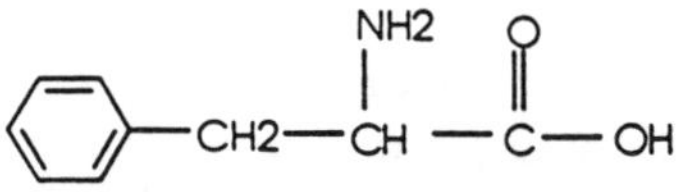

L-Phenylalanine

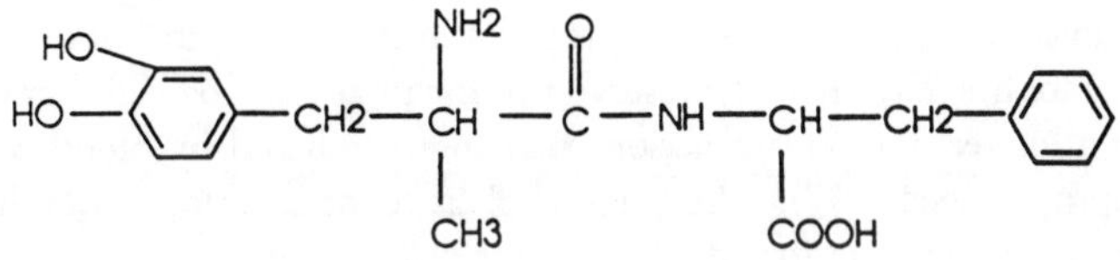

L-Alpha-methyldopa-phe

Figure 4. A peptide prodrug strategy for improving oral absorption.

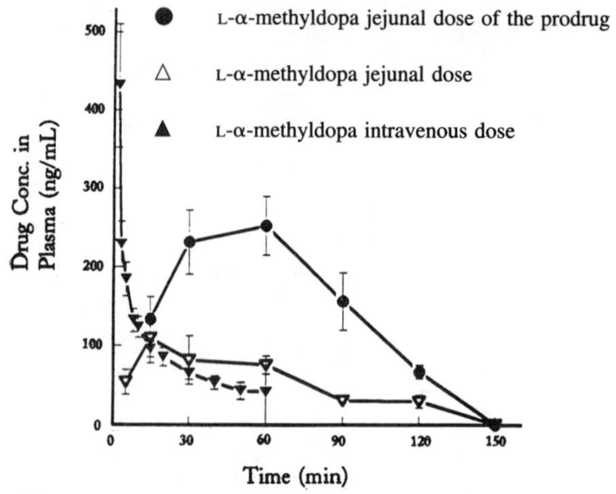

Figure 5. Plasma profile of L-α-methyldopa following intravenous dose of L-α-methyldopa and jejunal dose of L-α-methyldopa-phenylalanine and L-α-methyldopa (n = 6–7).

drug and to be metabolized to the active drug in mucosal tissue. Recent results (*31*) showed that the use of prodrugs increased the in vivo systemic availability of α-methyldopa to nearly 100% [Figure 5 (*31*)].

First-Pass Extraction and Metabolism

The ACE inhibitor prodrugs need to be converted via deesterification to the active drug to elicit pharmacological activity. The esterases required for the conversion are concentrated mainly in the liver (*32, 33*) and to some extent in the gastrointestinal tract (*34*) and blood. The 27–37% recovery of diacids in feces after an oral dose of the prodrug indicates esterolysis of the prodrug in the gut lumen and/or hepatic extraction and excretion of the diacid in the bile (*35*). High doses of enalapril do not saturate hepatic biotransformation of the prodrug to the active diacid enalaprilat (*13*), and transfer of enalaprilat across rat liver cell membranes is barrier-limited (*36*). Other than the bioactivation to the diacid, very little metabolism of the prodrugs to other identifiable metabolites is detected (*2*). Lisinopril also undergoes very little metabolism. Captopril, on the other hand, is oxidized in plasma to form the disulfide dimer and mixed disulfides with endogenous sulfhydryl compounds (e.g., glutathione and cysteine). Captopril and its metabolites undergo interconversions, and the metabolite may serve as a reservoir of the pharmacologically active moiety, thus contributing to a longer duration of action than that predicted by blood concentrations of unchanged captopril itself (*37*). These metabolism considerations complicate the development of a predictive scheme for ACE-inhibitor absorption.

Correlation of FA with P_w^*

The mean permeability, percent predicted dose absorbed, and re-calculated FA for some ACE inhibitors are shown in Table V. The mean intrinsic wall permeabilities ($P_w^{*\prime}$) and predicted FA were calculated from the absorption parameters obtained from rat perfusion data using the following equations (*9*):

Table V. Properties of Some ACE Inhibitors

Compound	Dose (mg)	MW	Mean P_w^*	Predicted FA (%)	Re-estimated FA (%)	Ref.
Captopril	100	216	2.44	99	60–75[a]	2
Enalapril	10	493	1.53	95	61–88[a]	2, 5
Enalaprilat	10	384	0.1	0.18	3	43
Quinapril	20	475	1.37	94	62–89[a]	42
Benazepril	20	461	1.20	91	37–64[a]	40, 41
Fosinopril	20	579	4.00	100	13–75[a]	39
Lisinopril	20	377	0.17	29	29	2, 5

[a] Corrected for biliary elimination and gastrointestinal metabolism.

$$P_w^{*\prime} = P_m^* + P_c^*(K_m/C_0)\ln(1 + C_0/K_m)$$
$$FA = 1 - \exp(-\beta P_w^{*\prime}) \qquad (2)$$

where C_0 is the initial concentration (dose/volume, mM), volume is 0.2 L, and $\beta = 2$, but can be a fitted parameter (*9, 10*). The predicted FA was estimated by eq 2 with the $P_w^{*\prime}$ values determined from the disappearance of drug from the intestinal perfusing solution and therefore does not account for any intestinal metabolism or first-pass extraction and biliary excretion. The calculated FA for the diacid lisinopril, which is eliminated unchanged mainly by the kidney (*5, 38;* Tables I and II) is in good agreement with its predicted value.

In contrast captopril and the prodrugs have a high intestinal permeability in the rat, and their low systemic availability indicates that systemic exposure of these drugs is limited by presystemic first-pass metabolism, which may include pre-absorptive gut metabolism as well as post-absorptive biliary elimination of the free acid. Results from in vitro studies with intestinal BBMVs indicate that the diacid formed from the de-esterification of the prodrug is poorly absorbed and may compete with the prodrug for peptide carrier binding. Although eliminated mainly by the kidney, captopril, which contains a free sulfhydryl moiety, undergoes rapid metabolism with thiol-containing endogenous compounds in vivo to form various mixed disulfides and dimer products that may be poorly absorbed (*2, 13*). This in vivo metabolism causes <100% systemic availability of captopril in humans, even though captopril has good permeability.

The FA for the prodrugs quinapril, benazepril, enalapril, and fosinopril can be only roughly estimated. If the fecal excretion of free diacid is due to biliary excretion, the fecal excretion can be added to the renal elimination of the diacid to estimate total absorption. For fosinopril, 62% of an oral dose is excreted in the feces as the diacid and 13% is excreted in the urine, giving a FA of 75% (*38, 39*). For the prodrugs enalapril (*2, 7*), benazepril (*40, 41*), and quinapril (*42*), ~27% of the oral dose is eliminated in the feces as their diacids. Adding this amount to the reported oral absorption for enalapril, benazepril, and quinapril (61, 62, and 37%, respectively) gives estimates of total absorption of 88, 89, and 64% for enalapril, quinapril, and benazepril, respectively. These amounts suggest good absorption of these prodrugs, as would be expected given their permeabilities. However, the systemic availability of these compounds is limited by biliary elimination. In addition it cannot be ruled out that the prodrugs are hydrolyzed in the gut lumen and/or intestinal mucosa and subsequently excreted in the feces as suggested by the finding that 27% of an intravenous dose of enalaprilat is excreted in the feces (*2*).

Summary

The ACE inhibitors are transported in part by the peptide carrier pathway across the intestinal epithelial cell brush-border membrane. Esterification of the active diacid ACE inhibitors results in a more peptide-like structure and better fit for the peptide carrier, thereby increasing the effective membrane permeability of the ACE inhibitors. Results of countertransport studies suggest that the peptide carrier is asymmetric in terms of rate of reorientation and substrate affinity. Although the prodrugs have better oral bioavailability than the diacids, their systemic availability is limited by pre-absorptive gut metabolism and/or post-absorptive biliary excretion. The results of this investigation into the mechanism of ACE inhibitor transport add to the existing knowledge of structural requirements for peptide transport. ACE inhibitors and other small peptide-type drugs can be designed with improved systemic availability by taking these transport mechanism results into consideration early in the drug design program.

References

1. Unger, T.; Gokhale, P.; Gruber, M. G. In *Handbook of Experimental Pharmacology;* Ganten, D.; Mulrow, P. J., Eds.; Springer-Verlag: New York, 1990; pp 377–482.
2. Kostis, J. B.; DeFelice, E. A. In *Angiotensin-Converting Enzyme Inhibitors;* Alan R. Liss: New York, 1987.
3. Patchett, A. A.; Harris, E.; Tristram, E. W.; Wyvratt, M. J.; Wu, M. T.; et al. *Nature (London)* **1980,** *288,* 280–283.
4. Sweet, C. S. *Fed. Proc.* **1983,** *42,* 167–170.
5. Ulm, E. H.; Hichens, M.; Gomez, H. J.; Till, A. E.; Hand, E.; Vassil, T. C.; Biollaz, J.; Brunner, H. R. *Br. J. Clin. Pharmacol.* **1982,** *14,* 357–362.
6. Unger, T.; Ganten, D.; Lang, R. E.; Scholkens, B. A. *J. Cardiovasc. Pharmacol.* **1984,** *6,* 872–880.
7. Unger, T.; Moursi, M.; Ganten, D.; Hermann, H.; Lang, R. E. *J. Cardiovasc. Pharmacol.* **1986,** *8,* 276–285.
8. Powell, J. R.; DeForrest, J. M.; Cushman, D. W.; Rubin, B.; Petrillo, E. W. *Fed. Proc.* **1984,** *43,* 733.
9. Amidon, G. L.; Sinko, P. J.; Fleisher, D. *Pharm. Res.* **1988,** *5,* 651–654.
10. Sinko, P. J.; Leesman, G. D.; Amidon, G. L. *Pharm. Res.* **1991,** *8,* 979–988.
11. Friedman, D. I.; Amidon, G. L. *Pharm. Res.* **1989,** *6,* 1043–1047.
12. Johnson, D. A.; Amidon, G. L. *J. Theor. Biol.* **1988,** *131,* 93–106.
13. Hu, M.; Amidon, G. L. *J. Pharm. Sci.* **1988,** *77,* 1007–1011.
14. Friedman, D. I.; Amidon, G. L. *J. Pharm. Sci.* **1989,** *78,* 995–998.
15. Nakashima, E.; Tsuji, A.; Kagatani, S.; Yamana, T. *J. Pharmacobio–Dyn.* **1984,** *7,* 452–464.
16. Kramer, W.; Girbig, F.; Gutjahr, U.; Kleemann, H.; Leipe, I.; Urbach, H.; Wagner, A. *Biochim. Biophys. Acta* **1990,** *1027,* 25–30.
17. Allen, J. G.; Havas, L.; Leicht, E.; Lenox-Smith, J.; Nisbet, L. J. *Antimicrob. Agents Chemother.* **1979,** *16,* 306–313.
18. Yokohama, S.; Yoshioka, T.; Yamashita, K.; Kitamori, N. *J. Pharmacobio–Dyn.* **1984,** *7,* 445.

19. Kessler, M.; Acuto, O.; Storelli, C.; Murer, H.; Muller, M.; Semenza, G. *Biochim. Biophys. Acta* **1978,** *506,* 136–154.

20. Malathi, P.; Preiser, H.; Fairclough, P.; Mallet, P.; Crane, R. K. *Biochim. Biophys. Acta* **1979,** *554,* 259–263.

21. Yee, S.; Yuasa, H.; Amidon, G. L. Submitted for publication in *Pharm. Res.*

22. Stein, W. D. In *Transport and Diffusion Across Cell Membranes;* Harcourt Brace Jovanovich: London, 1986.

23. Yuasa, H.; Amidon, G. L.; Fleisher, D. *Pharm. Res.* **1991,** *8,* S-221.

24. Sinko, P. J.; Amidon, G. L. *Pharm. Res.* **1988,** *5,* 645–650.

25. Banerjee, P. K.; Amidon, G. L. In *Design of Prodrugs;* Bundgaard, H., Ed.; Elsevier Science, B.V.: New York, 1985; pp 93–133.

26. Amidon, G. L.; Stewart, B. H.; Pogany, S. *J. Controlled Rel.* **1985,** *2,* 13–26.

27. Amidon, G. L.; Sinko, P. J.; Hu, M.; Leesman, G. D. In *Novel Drug Delivery;* Prescott, L. F.; Nimmo, W. S., Eds.; John Wiley & Sons: London, 1989; pp 45–56.

28. Stewart, B. H.; Amidon, G. L.; Brabec, R. K. *J. Pharm. Sci.* **1986,** *75,* 940–945.

29. Bai, P.; Subramanian, P.; Mosberg, H. I.; Amidon, G. L. *Pharm. Res.* **1991,** *8,* 593–599.

30. Hu, M.; Subramanian, P.; Mosberg, H. I.; Amidon, G. L. *Pharm. Res.* **1989,** *6,* 66–70.

31. Asgharnejad, M.; Amidon, G. L. *Pharm. Res.* **1992,** *9,* S-248.

32. Pang, K. S.; Barker, F. D., III; Cherry, W. F.; Goresky, C. A. *J. Pharmacol. Exp. Ther.* **1991,** *257(1),* 294–301.

33. Larmour, I.; Jackson, B.; Cubela, R.; et al. *Br. J. Clin. Pharmacol.* **1985,** *19,* 701–704.

34. Pang, K. S.; Cherry, W. F.; Elm, E. H. *J. Pharmacol. Exp. Ther.* **1985,** *233,* 788–795.

35. Todd, P. A.; Heel, R. C. *Drugs* **1986,** *31,* 198–248.

36. Schwab, A. J.; Barker, F., III; Goresky, C. A.; Pang, K. S. *Am. J. Physiol.* **1990,** *258,* G461–475.

37. Duchin, K. L.; McKinstry, D. N.; Cohen, A. I.; Migdalof, B. H. *Clin. Pharmacokinet.* **1988,** *14,* 241–259.

38. Carr, R. D.; Cooper, A. E.; Hutchinson, R.; Mann, J.; O'Connor, S. E.; Robinson, D. H.; Wells, E. *Br. J. Pharmacol.* **1990,** *100,* 90–94.

39. Singhvi, S. M.; Duchin, K. L.; Morrison, R. A.; Willard, D. A.; Everett, D. W.; Frantz, M. *Br. J. Clin. Pharmacol.* **1988,** *25,* 9–15.

40. Waldmeier, F.; Schmid, K. *Arzneim.-Forsch.* **1989,** *39,* 62–67.

41. Waldmeier, F.; Kaiser, G.; Ackerman, R.; Faigle, J. W.; Wagner, J.; Barner, A.; Lasseter, K. C. *Xenobiotica* **1991,** *21,* 251–261.

42. Olsen, S. C.; Horvath, A. M.; Michniewicz, B. M.; Sedman, A. J.; Colburn, A. W.; Welling, P. G. *Angiology* **1989,** *40,* 351–358.

43. Irwin, J. D.; Till, A. E.; Vlasses, P. H.; et al. *Clin. Pharmacol. Ther.* **1984,** *33,* 248.

RECEIVED for review November 9, 1993. ACCEPTED revised manuscript March 22, 1994.

CHAPTER 7

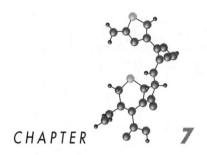

The Intestinal Oligopeptide Transporter

Molecular Characterization and Substrate Specificity

Werner Kramer,
Frank Girbig,
Ulrike Gutjahr,
and Simone Kowalewski

The oral route is by far the most desirable way of drug administration for the treatment of chronic diseases. New methodologies for the random synthesis of peptides and intelligent screening strategies (*1*) will lead to the discovery of highly biologically active peptides. Therefore, the importance of peptides as drugs of the future might increase tremendously. However, the medicinal importance of peptides as drugs is limited because peptides have two principle disadvantages compared with most other drugs; namely, high susceptibility to hydrolysis of the peptide bonds (*2, 3*) and poor intestinal absorption (*4, 5*).

To improve the bioavailability of peptides after oral application, many strategies, such as coating of peptides with polymers (*6*), use of penetration enhances (*5, 7*), or administration of peptidase inhibitors (*7, 8*), have been applied. Whereas these approaches focus on the protection of peptides against the intestinal hydrolytic milieu and on the nonselective increase of intestinal permeability with enhancers, in principle two more specific ways to increase the intestinal permeability of peptides are possible: adaption of the peptide structure to the (hitherto unknown) structure–activity relationships (SARs) of the intestinal oligopeptide transport systems for optimal molecular recognition as a transportable substrate, and "smuggling in" of the peptides through a specific endogenous uptake mechanism by coupling to a natural ligand, either by receptor-mediated endocytosis (RME) as with vitamin B_{12} (*9*) or by a carrier mechanism as with bile acids (*10, 11*). The possibility of intestinal absorption of intact peptides was generally viewed as a curiosity until 1970, but independent results from several laboratories (for review, *see* reference 4) gave evidence that di- and tripeptides are actively absorbed into intestinal cells by specific transport systems that are distinct from those involved in the absorption of amino acids. In contrast to the transport of amino acids, D-glucose, or bile acids, the uptake of dipeptides is not energized by a Na^+-gradient but by a H^+-gradient (*12*).

Figure 1 summarizes the present view of peptide absorption. According to this model, di- and tripeptides are cotransported across the enterocyte brush border membrane with protons. The proton gradient across the brush border membrane (*13, 14*) is generated by a combined action of a Na^+–H^+-exchanger located in the brush border membrane (*15*) and a Na^+–K^+-ATPase in the basolateral membrane of the enterocyte (*16*). Because the uptake of di- and tripeptides occurs by a carrier-mediated H^+-dependent transport system and intestinal absorption of peptides is a prerequisite for oral drug treatment, we investigated the intestinal peptide transport system on a molecular level. An understanding of the transport mechanism and the SARs of compounds for enteral absorption through the peptide

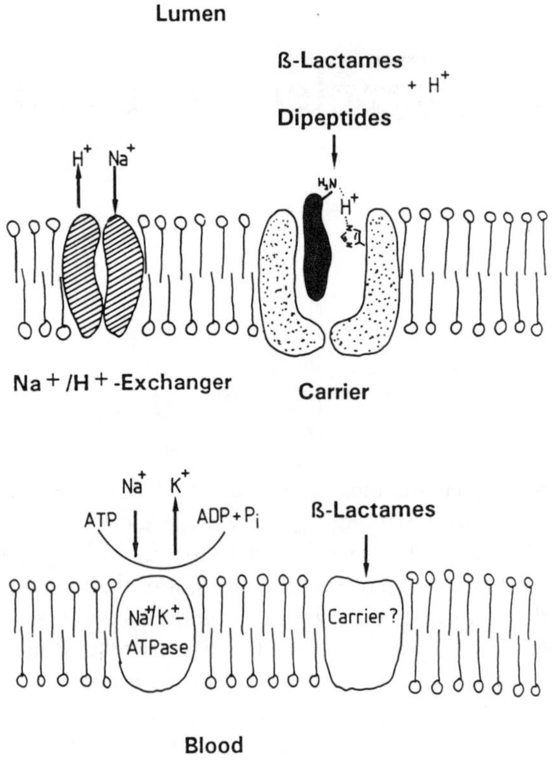

Figure 1. Model for the molecular mechanism of intestinal oligopeptide absorption. The upper membrane represents the brush border membrane and the lower membrane represents the basolateral membrane of the enterocyte.

transporter is necessary for the rational design of orally active peptide drugs. Such a characterization on the molecular level involves

1. the identification of driving forces and kinetic characterization of the transport systems
2. the identification of amino acids essential for transport and molecular recognition
3. the identification of the protein components of the transport system
4. the purification of the protein subunits of the transport system
5. the functional reconstitution of the transport system
6. the immunohistochemical localization of the transporter
7. the identification of the drug binding domain
8. the sequencing, cloning, and crystal structure determination of the lipid-embedded transporter

Characterization of the Oligopeptide Transport System in Intestinal Brush-Border-Membrane Vesicles

For the elucidation of peptide transport across the brush border membrane of small intestinal enterocytes on a molecular level, several physiological and methodological aspects are of major importance to obtain clear results and to avoid misinterpretations of experimental results. Net absorption measurement of peptides by complex biological models, such as in vivo intestinal perfusion and everted intestinal rings or sacs, are difficult to interpret for two major reasons. First, intestinal absorption measurements of peptides and drugs using such models are the sum of several overlapping processes that cannot be separated easily; for examples, transport across the enterocyte brush-border membrane, transcellular transport from the brush-border membrane to the basolateral pole of the enterocyte, transport processes into intracellular organelles and compartments, intracellular metabolism, transport across the basolateral membrane, secretory transport processes of the small intestine in the opposite direction (*17, 18*). Second, the disappearance of a peptide from the intestinal lumen in these models does not always reflect intestinal absorption of the peptide. Also, no detection of intact peptides in mesenteric blood does not demonstrate lack of peptide absorption, because hydrolysis may have occurred in the cytosol of the enterocyte or in the plasma. Because several lines of evidence indicate that the rate-limiting step during intestinal absorption is the transfer across the brush border membrane (*19*), we and others (*12, 20–24*) used brush-border-membrane vesicles (BBMV) to characterize the intestinal transport system for oligopeptides for several reasons. First, there is no superimposition by other transport and metabolic processes during intestinal absorption. Second, the investigation of substrate specificity and ion dependence of the transport process independently on the overall absorption process is a prerequisite for the evaluation of clear SARs. Third, after kinetic characterization of a peptide transport system in BBMV, these preparations, being 15–20-fold enriched over intestinal homogenate, are a promising starting material for the identification of the protein components of the peptide transporter and its purification.

Another prerequisite for the characterization and identification of a putative peptide transport system is the use of metabolically stable substrates of this particular transport system. The instability of peptides, which is dependent on their amino acid composition and amino acid sequence, against hydrolysis by membrane-bound peptidases of the brush-border membrane may account for the conflicting results concerning the absorption of peptides greater than tripeptides and the ion dependence of peptide transport (*4*). Whereas it is well documented that di- and tripeptides can reach the blood in intact form (*25–28*), controversial results about the intestinal absorption of tetrapeptides have been reported. Most studies argue against the uptake of tetrapeptides by intestinal mucosal cells (*29–32*), whereas others give evidence for transport in intact form (*33, 34*). The reason for these conflicting results predominantly lies in the susceptibility of peptides to brush-border hydrolysis (*30, 35*). Using radiolabeled peptides as substrates and measuring uptake of these compounds by determination of the radioactivity accumulated in the BBMV may give misleading results because several radiolabeled compounds can be formed from the original oligopeptide (amino acids, dipeptides, and tripeptides) by hydrolysis. Therefore, the accumulated radioactivity could be the result of uptake of these fragments rather than be the uptake of the intact oligopeptide.

Carrier-Mediated Uptake of β-Lactam Antibiotics as Surrogates for Di- and Tripeptides

Orally active α-amino-β-lactam antibiotics share the intestinal uptake system for di- and tripeptides (*20, 36*) and are taken up by human small intestine nearly quantitatively (*36–38*). The UV absorption characteristics of α-amino-β-lactam antibiotics allow a very sensitive determination of these compounds by high pressure liquid chromatography (HPLC), enabling determination of the actual amount of these compounds as well as detection of possible hydrolytic products. Use of HPLC avoids the aforementioned difficulties and misinterpretations with radiolabeled substrates. Therefore, we and others (*20–22, 39*) used α-aminocephalosporins as prototype substrates for the characterization and identification of the oligopeptide transport system in the small intestine of several species. The uptake of α-aminocephalosporins such as cephalexin by BBMV isolated from the small intestine of rabbit, rat, pig, or human cells is stimulated by an inwardly directed H^+-gradient (Figure 2) that reflects the acidic microclimate at the surface of the intestinal cells (*13, 40*) and shows saturation kinetics (*20–22, 36, 41, 42*). In vitro experiments with everted intestinal sacs or human intestinal cells (Caco-2) as well as in vivo measurements support the findings of an active H^+-dependent, saturable uptake system for α-amino-β-lactam antibiotics and oligopeptides (*43–53*).

The uptake of both dipeptides such as Gly-Pro or Gly-Gly (*54, 55*) and of α-aminocephalosporins exhibits a clear pH-optimum of the H^+-stimulative uptake (*21, 44, 55, 56*) into BBMV. Furthermore, both the uptake of dipeptides (*57–59*) and of α-aminocephalosporins (*60, 61*) shows a profound stereospecificity, indicating a specific molecular interaction of these substrates with the protein components of the respective intestinal uptake systems. Tsuji et al. (*39*) concluded "that all types of β-lactam antibiotics share the brush border transport mechanism for peptides of dietary origin and that the uptake of β-lactam antibiotics possessing sufficient lipid solubility, such as benzylpenicillin, phenoxymethylpenicillin, propicillin, or dicloxacillin, may occur via either the peptide transport carrier or the lipid membrane barrier". Figure 3 shows the time-dependent uptake of benzylpenicillin into rabbit BBMV, indicating that the uptake is independent of a Na^+-gradient and is stimulated by a H^+-gradient. This pH-gradient stimulation of benzylpenicillin uptake may result either from a stimulation of the oligopeptide transport system or by an increase

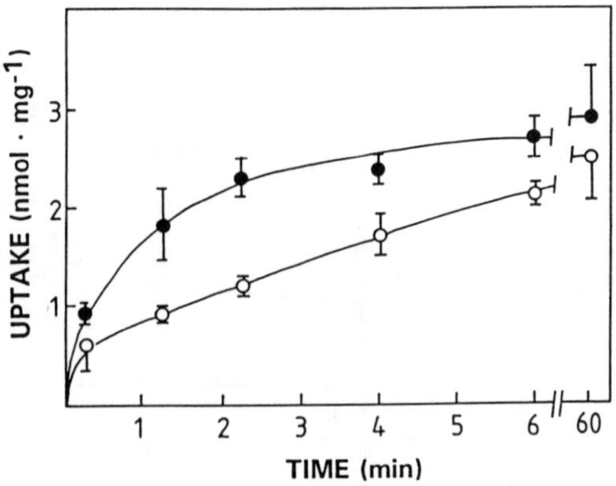

Figure 2. Uptake of D-cephalexin into BBMV from rabbit small intestine. The BBMV (100 μg of protein, 20 μL) preloaded with 10 mM Tris–Hepes buffer (pH 7.4)/300 mM mannitol were mixed at 30 °C with 180 μL of buffer containing 2 mM D-cephalexin, and uptake was measured by HPLC after the indicated time points. Key: (●) uptake in 10 mM citrate–Tris (pH 6.0)–140 mM KCl; (○) uptake in 10 mM potassium phosphate (pH 7.4)–140 mM KCl.

in the percentage of undissociated benzylpenicillin leading to increased diffusional uptake. To discriminate between these possibilities, the uptake of cephalexin and benzylpenicillin into BBMV, which were pretreated at 70 °C, was studied. Figure 4 shows that the uptake of cephalexin as well as that of benzylpenicillin were significantly inhibited after thermal pretreatment of membrane vesicles. Equilibrium uptake measurements into control and heat-pretreated vesicles dependent on the medium osmolarity did not indicate increased

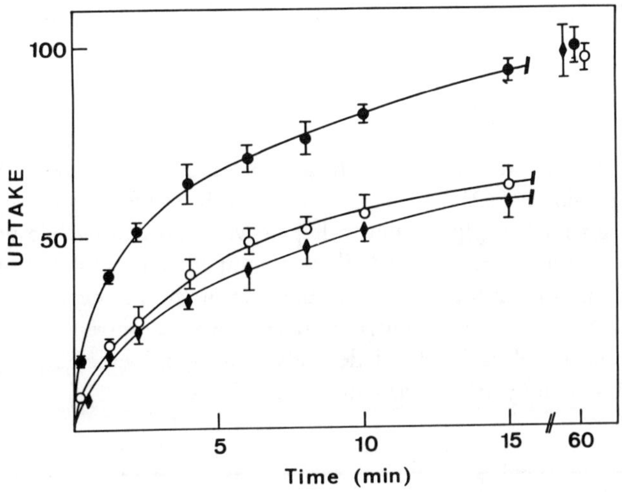

Figure 3. Uptake of benzylpenicillin into BBMV from rabbit small intestine. The BBMV (100 μg of protein, 20 μL) preloaded with 10 mM Tris–Hepes buffer (pH 7.4)–300 mM mannitol were mixed at 30 °C with 180 μL of buffer containing 100 μM (1 μCi) [³H]benzylpenicillin, and uptake was measured by liquid scintillation counting after the indicated time points and expressed as percentage of uptake of the respective equilibrium. Key: (●) uptake in 10 mM citrate–Tris-buffer (pH 6.0)–140 mM KCl; (○) uptake in 10 mM Tris–Hepes buffer (pH 7.4)–140 mM KCl; (◆) uptake in 10 mM Tris–Hepes buffer (pH 7.4)–140 mM NaCl.

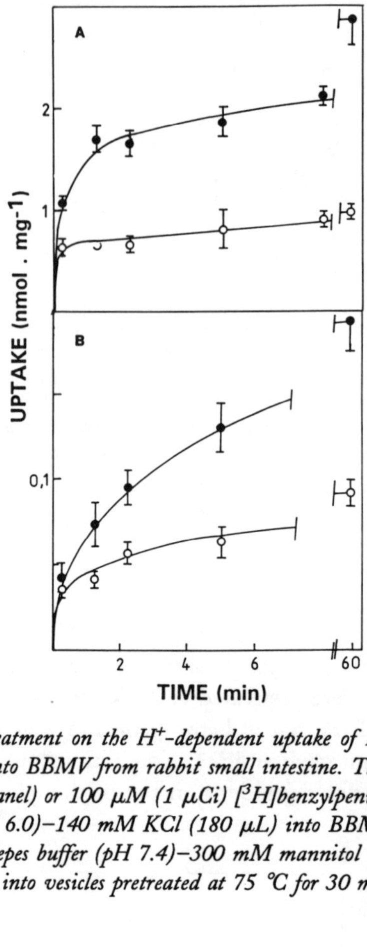

Figure 4. Effect of heat pretreatment on the H^+-dependent uptake of D-cephalexin (upper panel) and benzylpenicillin (lower panel) into BBMV from rabbit small intestine. The time-dependent uptake either of 2 mM D-cephalexin (upper panel) or 100 μM (1 μCi) [^{3}H]benzylpenicillin (lower panel) dissolved in 10 mM citrate–Tris buffer (pH 6.0)–140 mM KCl (180 μL) into BBMV (100 μg of protein, 20 μL) preloaded with 10 mM Tris–Hepes buffer (pH 7.4)–300 mM mannitol was measured. Key: (●) uptake into control vesicles; (○) uptake into vesicles pretreated at 75 °C for 30 min.

binding, and efflux studies did not indicate an increased leakage of heat-treated vesicles. These results demonstrate that benzylpenicillin, like cephalexin, is transported across the intestinal brush border membrane by a carrier-mediated process rather than by diffusion as thought previously (*62*). This is also supported by the finding of carrier-mediated transport mechanisms for benzylpenicillin in hepatocytes and choroid plexus (*63–65*). The uptake of cephalexin into intestinal BBMV was competitively inhibited by benzylpenicillin (*66*), indicating that benzylpenicillin is a substrate of the H^+-dependent uptake system shared by di- and tripeptides and orally active α-aminocephalosporins. This finding was a prerequisite for the use of radiolabeled derivatives of benzylpenicillin to probe the oligopeptide transport system by affinity-labeling techniques.

Identification of Putative Protein Components of the H^+–Oligopeptide Transporter

A method is needed for the identification of the putative transport proteins for oligopeptides and β-lactam antibiotics that is able to fix the reversible interactions between the ligands and their specific binding proteins and to change these transient noncovalent bonds into a

$$L + R \rightleftharpoons L \cdot R$$
$$L' + R \rightleftharpoons L' \cdot R$$
$$L' \cdot R \xrightarrow{UV} [L'^{*} \cdot R] \Rightarrow L' - R$$

Scheme I. Reversible, noncovalent interactions between radioactively labeled ligand (L) or a photolabile derivative of L (L′) and the receptor (R). Irradiation with UV light transforms the noncovalent interaction between ligand and receptor into a covalent one by an immediate nonselective chemical reaction between the receptor and the ligand.

covalent attachment of the ligand to the transport protein. Such an approach is photoaffinity labeling (*67, 68*), a method that is generally useful for the identification of molecular interactions.

The radioactively labeled ligand L (Scheme I) or a photolabile derivative L′ thereof specifically binds to its receptor R; this noncovalent interaction between ligand and receptor is transformed into a covalent one by an immediate nonselective chemical reaction between the receptor and the ligand that is triggered by irradiation with UV light. A prerequisite is that either highly reactive radicals that are capable of covalent binding with the ligand are generated in the binding protein or the ligand carries a chemically inert but photolabile group that can be converted by light into a highly reactive carbene, nitrene, or radical intermediate able to react indiscriminately with its immediate vicinity. The photoprobes also must be radiolabeled with a high specific radioactivity to allow the identification of specifically labeled binding proteins in a complex biological probe composed of hundreds of proteins such as the intestinal brush-border membrane. For the identification of the protein components of the transport system shared by oligopeptides and β-lactam antibiotics, we used photoreactive derivatives of cephalorins, penicillins, and dipeptides (Chart I).

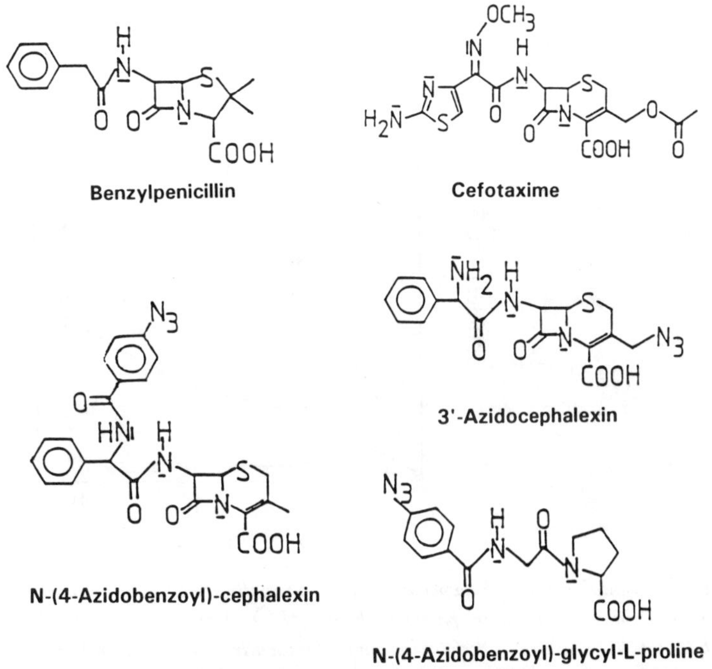

Chart I. Structure of photoaffinity probes for the H^{+}–oligopeptide transporter.

Photoaffinity Probes for the H^+–Oligopeptide Transporter

Penicillins

[^{3}H]Benzylpenicillin of high specific radioactivity is commercially available and, because of its aromatic ring system, can be activated to highly reactive radicals by irradiation at 254 or 300 nm, leading to the covalent attachment to proteins. Investigation of the photochemical mechanism of cross-linking revealed, however, that photoaffinity labeling of penicillin-binding proteins presumably proceeds via photoactivation of the binding protein (*66*).

Cephalosporins

The α-aminocephalosporin cephalexin was used in most studies as a prototype substrate for the peptide transport system. Consequently, photolabile derivatives of cephalexin were synthesized and used for photoaffinity labeling. [^{3}H]3′-Azidocephalexin is a competitive inhibitor of cephalexin uptake (*69*) with similar enteral absorption and antibacterial activity as cephalexin (*70*) and an aliphatic nitrene-generating azido group.

N-(4-Azido[3,5-^{3}H]benzoyl)-cephalexin (*42*) is a cephalexin analogue with a nitrene-generating aromatic azido function that can be activated by UV irradiation at 254 nm with a half-life of only 3.5 s for the azido derivative. This short half-life allows very short irradiation times, avoiding possible impairment of the structure or function of the brush-border-membrane proteins.

[^{3}H]Cephalexin itself and [^{14}C]cefotaxime (a cephalosporin for parenteral use), because their α, β-unsaturated carbonyl functions in the cephem nucleus, can be activated by irradiation at 300–320 nm in a n ⇒ π* transition to a reactive carbonyl radical leading to covalent labeling of respective cephalosporin binding proteins such as albumin (Figure 5).

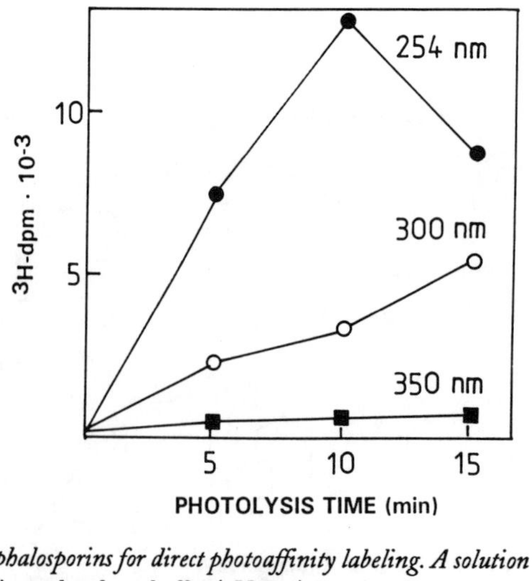

Figure 5. Suitability of cephalosporins for direct photoaffinity labeling. A solution of human serum albumin (1 mg/mL) in 50 mM sodium phosphate buffer (pH 7.4) was incubated for 1 min in the dark at 20 °C with 0.46 mM (1 μCi) [^{14}C]cefotaxime and subsequently irradiated in a Rayonet RPR 100 photochemical reactor at 254, 300, or 350 nm. After the indicated irradiation times, 20-μg aliquots were removed, and the incorporation of radioactivity into albumin was measured after SDS gel electrophoresis.

Dipeptides

N-(4-Azido[3,5-^{3}H]benzoyl)-glycyl-L-proline (*42*) is a photoreactive analogue of the hydrolysis-resistant dipeptide Gly-Pro (*71*).

Conditions for Photoaffinity Labeling

For the identification of binding proteins for these ligand analogues by photoaffinity labeling, the following conditions must be fulfilled to obtain conclusive results and to avoid misinterpretations.

Stability of the Photoprobes

The photoreactive derivatives used must be resistant to brush border hydrolysis. Otherwise, labeling of proteins other than the dipeptide- and β-lactam antibiotic-binding proteins would occur. For example, opening of the β-lactam ring in benzylpenicillin did not result in a specific labeling of the oligopeptide transporter. The photoprobes just described were stable for more than 20 min after incubation with BBMV from rat, rabbit, or pig (*42*) and thus it is ensured that binding proteins identified with these probes are indeed binding proteins for dipeptides, cephalosporins, and penicillins, respectively.

Physiological Behavior of the Photoprobes

Prior to the identification of transport proteins by photoaffinity labeling it must be shown that the photolabile derivatives used interact with the respective transport systems and behave physiologically and pharmacologically as the "natural" ligands. Transport measurements with BBMV showed that benzylpenicillin and the azido derivatives of cephalexin and Gly-Pro share the transport system for oligopeptides and α-aminocephalosporins both in the intestine and the kidney (*42, 66, 69, 72*).

Stability of the Transport System upon Irradiation

The photoactivation of benzylpenicillin, 3′-azidocephalexin, or the α,β′-unsaturated carbonyl function in cephalexin and cefotaxime requires relatively harsh irradiation conditions with wavelengths of 254 or 300 nm and irradiation times of 2–5 min. Because the irradiation of membrane proteins with UV light leads to damage of protein structure and function (*73*), a compromise between duration of photolysis and extent of photoaffinity labeling has to be found. The H^+-dependent uptake of cephalexin was not significantly affected by irradiation of BBMV at 254, 300, or 350 nm for up to 5 min, demonstrating a remarkable photochemical stability of the oligopeptide transporter (data not shown). This finding minimizes the possibility that a photolabeled binding protein is an artifact by labeling of damaged proteins or photolytic fragments of the transporter protein.

Identification of the Putative H^+–Oligopeptide Transporter

For the identification of the protein components presumably involved in the uptake of oligopeptides and β-lactam antibiotics, BBMV isolated from the small intestine of rat, rabbit, or pigs were incubated with the respective radioactively labeled photoprobes in the dark. Subsequently, the BBMV were irradiated, and then the membrane proteins were separated by denaturating SDS-gel electrophoresis to detect radiolabeled binding proteins. With benzylpenicillin, [^{3}H]cephalexin, and the axido derivatives of Gly-Pro and cephalexin,

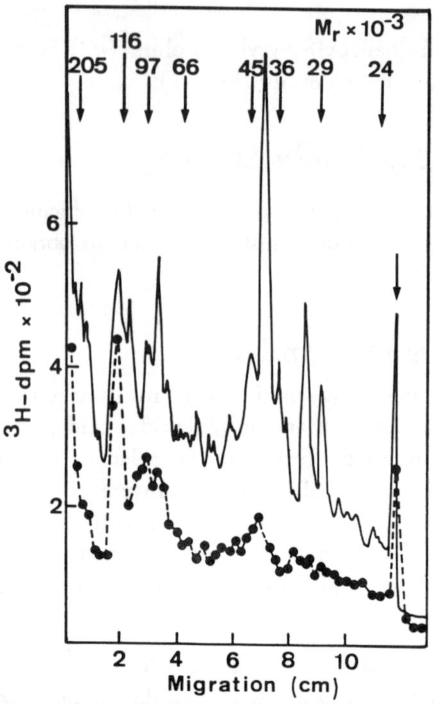

Figure 6. Incorporation of radioactivity into membrane proteins after direct photoaffinity labeling of BBMV from rat small intestine with [^{3}H] benzylpenicillin. Rat small intestinal BBMV (200 μg of protein) were incubated with 1.7 μM (3 μCi) [^{3}H]benzylpenicillin in 10 mM citrate–Tris buffer (pH 6.0)/140 mM KCl for 5 min in the dark at 30 °C and subsequently irradiated for 2 min at 254 nm in a Rayonet RPR 100 photochemical reactor. After SDS gel electrophoresis, these gels were sliced into 2-mm pieces, and radioactivity was measured after digestion of proteins by liquid scintillation counting. The upper curve shows the distribution of Serva Blue R 250-stained polypeptides, and the lower curve shows the distribution of radioactivity.

similar labeling patterns were obtained within a given species. In rat BBMV, all derivatives led to a specific labeling of membrane proteins with apparent M_r values of 127,000, 100,000, 94,000 and 86,000, with the 127-kDa protein being the most labeled (*42*) (Figure 6). In BBMV isolated from rabbit or pig small intestine, the photoprobes were incorporated nearly exclusively into a membrane polypeptide of M_r 127,000 (*42, 66, 70*). The distribution of radioactivity along the gel tracks is quite different from the distribution of stained proteins, indicating specific labeling of the abovementioned proteins rather than nonspecific hydrophobic labeling. Cefotaxime is a cephalosporin for parenteral use that is only poorly absorbed. Photoaffinity labeling of rat BBMV with [^{14}C]cefotaxime led to the labeling of many membrane proteins without a profound labeling of the 127-kDa binding protein (data not shown), indicating a low affinity of cefotaxime to the 127-kDa protein. In contrast, direct photoaffinity labeling with [^{3}H]cephalexin led, as it did with the azido derivatives of cephalexin, to a high and specific incorporation of the photoprobe into the 127-kDa protein.

The significance of the 127-kDa protein as the specific binding protein for oligopeptides and β-lactam antibiotics was also indicated by labeling BBMV with fluoresceineisothiocyanate (FITC). Chemical modification of BBMV with FITC in the presence of cephalexin led to a specific inhibition of fluorescence labeling of a membrane protein of M_r 127,000

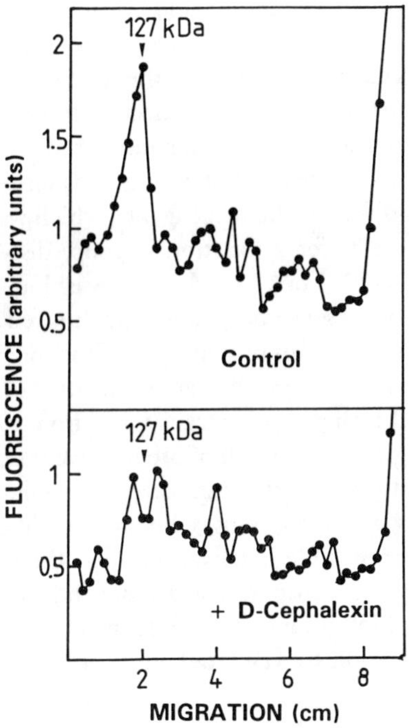

Figure 7. Labeling of BBMV from rat small intestine with FITC in the absence and presence of D-*cephalexin. The BBMV from rat small intestine (1 mg/mL protein) were incubated with 50 μM FITC in 50 mM Tris–HCL buffer (pH 9.2)–2 mM EDTA–100 mM NaCl–<5% dimethylformamide for 30 min at 20 °C either in the absence or the presence of 10 mM* D-*cephalexin. Subsequently, the vesicles were diluted 30-fold with ice-cold buffer (*see *above) and centrifuged at 34,000 × g for 20 min. The resulting pellets were submitted to sodium dodecyl sulfate (SDS) gel electrophoresis on 9% gels. The gels were sliced into 2-mm pieces, and protein was eluted with 1 mL of the aforementioned buffer and analyzed by fluorimetry.*

(Figure 7). Similar results were obtained by a combined phenylisothiocyanate (PITC)–FITC-labeling technique that has been used successfully for the identification of the D-glucose transporter from rabbit small intestine (*74*).

Identification of Functional Amino Acids of the H⁺–Oligopeptide Transporter Essential for Transport Function

The protein-mediated transport of oligopeptides and α-amino-β-lactam antibiotics across the enterocyte brush border membrane with its H^+-dependence, stereospecificity, and substrate specificity implies a specific molecular recognition of the transported ligand by the transport protein. Specificity and energization of the transport system depend on specific interactions and contacts between distinct epitopes of the ligand molecule and defined amino acid side chains of the transport protein. Knowledge of these molecular interactions between ligands and transport proteins on a molecular level is essential for understanding the transport mechanism and the SARs for binding and transport. A valuable approach to the identification of amino acids of the transport system essential for transport activity, binding affinity, and energization by a chemical gradient is the systematic chemical modification of functional amino acid side chains with respective selective chemical agents and subsequent measurement of the transporter activity and substrate affinity.

To identify the involvement of amino acids for transport activity, membrane vesicles were incubated with different chemical agents, which were selective or specific for a distinct amino acid side chain or for a functional group, dissolved in appropriate buffers in the concentration range of 10^{-6} to 10^{-2} M. After washing the vesicles, initial uptake of cephalexin in the absence or presence of an inwardly directed H^+-gradient was measured. Of the different amino acids present in proteins with functional side chains, only the modification of histidine and tyrosine residues led to an inactivation of the oligopeptide transporter (Table I). The behavior of the oligopeptide transporter from rabbit small intestine upon chemical modification differs greatly from that of other active transporters (Table I). The Na^+-dependent uptake systems for bile acids, D-glucose, and L-alanine showed a strict dependence on the presence of intact cysteine groups. Chemical modification of thiol groups with a variety of agents led to an irreversible inactivation of all three Na^+-dependent transport systems, whereas the oligopeptide transporter was independent of functional cysteinyl groups. The absence of essential thiol groups of the oligopeptide transporter explains the resistance of this system to inactivation by UV irradiation in contrast to the Na^+-dependent bile acid

Table I. Effect of Amino Acid Modification on H^+–Oligopeptide Transport, Na^+–Bile Acid Transport, Na^+–D-Glucose Transport, and Na^+–L-Alanine Transport

Functional Amino Acid (Group)	*Reagent*	*H^+–Oligopeptide*	*Na^+–Bile Acid*	*Na^+–D-Glucose*	*Na^+–L-Alanine*
Cysteine	$HgCl_2$	0	+ + +	+ + +	+ + +
	PCMB	0	+ + +	+ + +	+ + +
	PMS	(+)	+ +	+ + +	+ +
	NEM	0	+ +	+ + +	+ +
	DTNB	0	+	+	+ +
	Iodoacetamide	0	(+)	(+)	n.d.
Vicinal cysteines	PAO	0	+ +	+ +	+ +
Disulfide	DTT	0	0	0	(+)
	2-Mercaptoethanol	0	0	0	n.d.
Amino group	NBD-chloride	0	+ + +	+ + +	+ + +
	TNBS	0	(+)	(+)	n.d.
	FITC	0	+ +	+	+
	PITC	0	+ +	+	+
Hydroxyl group	PMSF	0	0	0	0
	DFP	0	n.d.	0	0
Carboxyl group	DCCD	0	0	0–(+)	+
Arginine	PGO	0	0	0	0
Tyrosine	NAI	+	0	(+)	0
Histidine	DEPC	+ + +	0	+ + +	+

[a] Key to effects: (0) no inhibition; (+) moderate inhibition; (+ +) strong inhibition; (+ + +) very strong inhibition; ((+)), tendency to inhibition; (n.d.) not determined.

transporter (*73*). Modification of amino groups, carboxyl groups, hydroxyl groups, or the basic amino acid arginine had no significant influence on the H^+–oligopeptide transport system. Treatment of BBMV with the histidine-modifying agent diethylpyrocarbonate (DEPC) completely abolished the H^+-dependence of the oligopeptide transporter (*56, 75*). Kato et al. (*75*) reported an apparent noncompetitive inhibition of the oligopeptide transporter after DEPC treatment, whereas in our hands (*41*) histidine modification led to an apparent competitive inhibition of cephalexin uptake with an increase in the Michaelis constant (K_m). This latter finding is in accordance with the results from photoaffinity labeling studies; after pretreatment of BBMV with DEPC, photoaffinity labeling with the photolabile α-aminocephalosporin 3-[*phenyl*-4-^{3}H]azidocephalexin resulted in a decrease in the extent of labeling of the 127-kDa binding protein compared with controls. This result indicates a decreased affinity that is reflected by an increase in the K_m value for binding of cephalexin to the oligopeptide transporter (*69*). The different behavior of the oligopeptide and α-amino-β-lactam antibiotic transporters in the small intestine and the kidney cortex with respect to energization by a H^+-gradient (*72*), inactivation by DEPC (noncompetitive in kidney, competitive in intestine) (*20, 24*), and susceptibility to *N*-acetylimidazol (*71*) and thiol-group modification (*24, 71, 75*) support the hypothesis of different transport systems for oligopeptides in the small intestine and the kidney despite similar molecular weights of the protein components (*72*). The findings that the uptake of cephalosporins without an α-amino group, such as cefotiam and cefoperazone, was not stimulated by a H^+-gradient (*20*) and that the uptake of such compounds lacking an α-amino group, like cefoperazone or cephaloridine, was not influenced by DEPC treatment of BBMV (*41*) led us to the model for the molecular mechanism of the H^+-dependent oligopeptide transporter outlined in Scheme II. Optimal uptake of cephalexin occurs under pH gradient conditions of pH_{out} = 6.0 and pH_{in} = 7.4. The pK_a values of the α-amino group in orally active α-amino-β-lactam antibiotics are in the range 6.89–7.35 (*53, 76*) and that of histidine residues in proteins is around pH 7.0. These pK_a values suggest a proton donor–acceptor relationship between the α-amino group of a transportable substrate and a histidine residue of the transporter protein leading to a stimulation of translocation of the substrate across

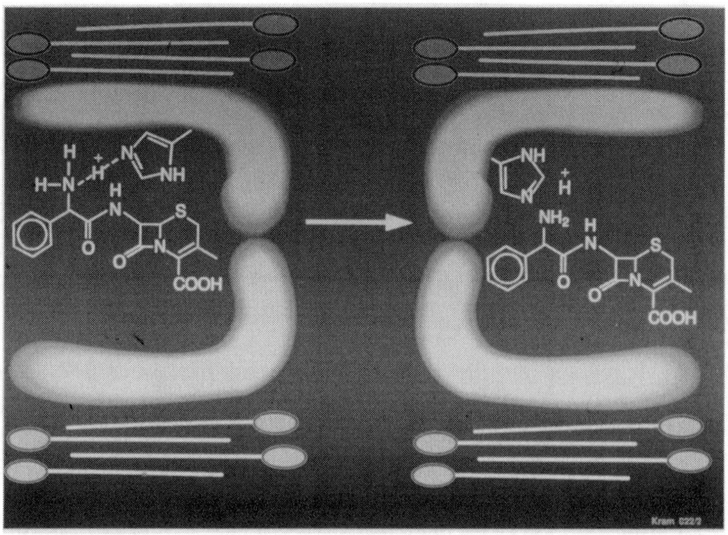

Scheme II. Hypothetical mechanism of H^+-dependent uptake of α-aminocephalosporins by the small intestinal oligopeptide transporter.

the brush border membrane. This concept is in accordance with the finding that the efflux of α-amino-β-lactam antibiotics from BBMV is strongly inhibited by histidine carboxyethylation in contrast to substrates without an α-amino group (*41*). Such a proton donor–acceptor function between the α-amino group and a histidine residue in the protein is also disturbed after acylation of the α-amino group in the ligand molecule. This explains the lack of a significant effect of DEPC treatment of BBMV on photoaffinity labeling of the 127-kDa protein by *N*-(4-azido[3,5-^{3}H]benzoyl)-cephalexin in contrast to the α-amino-cephalosporin 3-[*phenyl*-4-^{3}H]azidocephalexin (*70*) or the weakening of the affinity of dipeptides for the transporter by methylation, acetylation, or other substitution of the N-terminal amino group (*4, 25, 77, 78*). The molecular involvement of the tyrosine residues of the transporter protein is not clear now, but it is possible that the histidine and tyrosine residues act with the primary α-amino group of a transportable ligand as a charge-relay system.

Purification and Reconstitution of the H$^+$–Oligopeptide Transporter

For the elucidation of the mechanism of molecular recognition of a substrate by the intestinal oligopeptide transporter, the purification, reconstitution, sequencing, and cloning of the respective transport protein is necessary. Intestinal absorption of proteins occurs by a sequential process initiated by hydrolysis of proteins to oligopeptides and amino acids by secreted peptidases in the intestinal lumen, followed by membrane hydrolysis to di- and tripeptides and amino acids by ectoenzymes like aminopeptidase N or dipeptidylpeptidase IV. Because these enzymes are located in the enterocyte brush-border membrane, a direct role of these peptidases in the absorption process of oligopeptides was proposed (*79*). With different approaches involving inactivation of dipeptidylpeptidase IV and aminopeptidase N with group-specific agents and antibodies against purified aminopeptidase N, we demonstrated that peptidases of the brush-border membrane are not directly involved in the intestinal uptake process for small peptides and are not constituents of the H$^+$–oligopeptide cotransport system (*80, 81*). This finding was supported by the observation that BBMV obtained from Japanese Fisher 344 rats do not possess the dipeptidylpeptidase IV protein but exhibit normal peptide transport activity (*82*).

Attempts to purify the putative oligopeptide transport protein of M_r 127,000 imply the solubilization of brush-border-membrane proteins with nonionic detergents. We found that solubilization with nonionic detergents such as Triton X-100, CHAPS, or *n*-octylglycoside greatly decreased the affinity of the 127-kDa protein for binding of its substrates, and therefore affinity chromatography with immobilized cephalexin-derivatives was not very successful (*83*). For the purification of the oligopeptide transporter the following strategy was attempted: photoaffinity labeling of the 127-kDa protein in intact BBMV to introduce a measurable marker into the desired protein, solubilization of BBMV and purification of the radiolabeled 127-kDa protein to homogeneity, demonstration of the involvement of the 127-kDa protein with antibodies against the purified protein, and reconstitution of the purified 127-kDa protein into artificial lipid membranes and demonstration of H$^+$-dependent stereospecific transport of cephalexin.

Purification to >95% homogeneity was achieved by lectin-affinity chromatography with wheat germ lectin agarose followed by ion-exchange chromatography on (diethylamino)ethyl-sephacel (*83*) or Mono S HR 5/5 columns (*84*). Antibodies prepared against the purified protein were able to inhibit photoaffinity labeling of the 127-kDa protein in intact BBMV and the transport of cephalexin (*83*). A direct role of the 127-kDa protein as the intestinal oligopeptide transporter was made probable by reconstitution into liposomes (*61, 85, 86*): (1) The uptake of D-cephalexin into liposomes prepared from brush border membrane proteins was stimulated by an inwardly directed H^+-gradient, whereas pure liposomes without incorporated proteins did not show a H^+-dependent D-cephalexin uptake; (2) The uptake of cephalexin into BBMV was strictly stereospecific for the D-enantiomere (*60, 61*). This strict stereospecificity of uptake was also found after reconstitution into liposomes with enriched protein fractions and the purified protein (*61*); (3) Whereas the affinity of the 127-kDa protein for binding of ligands is greatly decreased after solubilization (*83*), the binding affinity was restored in the reconstituted liposomes showing the same substrate specificity for binding and transport as an intact BBMV (*61*); and (4) Dipeptides like L-carnosine or Gly-Pro, as well as β-lactam antibiotics like cephalexin, were able to inhibit photoaffinity labeling of the purified reconstituted 127-kDa protein, whereas D-glucose, amino acids (such as glycine) or bile acids (such as taurocholate) had no influence on the labeling (Figure 8).

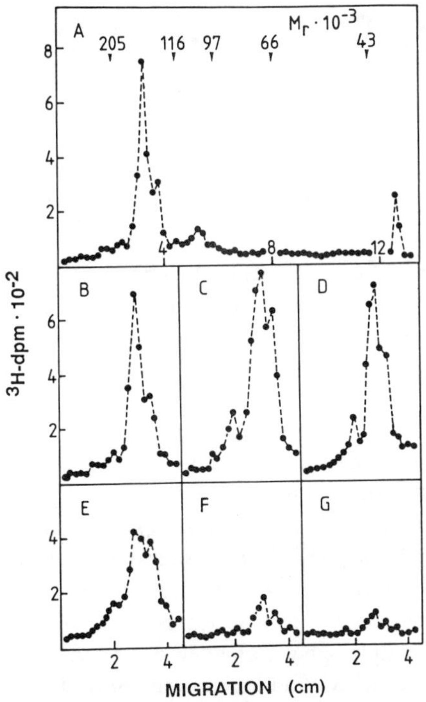

Figure 8. Differential photoaffinity labeling of proteoliposomes prepared from purified 127-kDa binding proteins for oligopeptides–β-lactam antibiotics. Liposomes were prepared from purified 127-kDa binding protein (300 μg of protein) and 80 mg of asolectin (59). The resulting liposomes were resuspended in 500 μL of 10 mM Tris–Hepes buffer (pH 7.4)/300 mM mannitol. Then, 60 μL of this liposome suspension were incubated for 5 min in the dark with 0.9 μM (5 μCi) [³H]benzylpenicillin in 20 mM citrate–Tris buffer (pH 6.0)/140 mM KCl in the absence of inhibitors (A) or in the presence of 5 mM glycine (B), 10 mM D-glucose (C), 500 μM taurocholate (D), 10 mM Gly-Pro (E), 5 mM L-carnosine (F), or 100 μM D-cephalexin (G), and subsequently irradiated for 2 min at 254 nm. After washing, the distribution of radioactivity was analyzed by SDS gel electrophoresis on 7.5% gels by liquid scintillation counting after slicing of the gels into 2-mm slices.

Attempts to clone the intestinal peptide transporter are under way. Using *Xenopus laevis* oocytes, a functional expression of H^+–dipeptide transport was achieved after injection of rabbit intestinal poly(A)$^+$ mRNA (*87*) or poly(A)$^+$ mRNA obtained from the human colon carcinoma cell line Caco-2 (*88*). However, the identification of the cDNA of the oligopeptide transporter has not yet been reported.

By chemical and enzymatic fragmentation of the purified 127-kDa binding protein, we obtained and sequenced several peptide fragments with 9–13 amino acids and did not find any sequence homology to hitherto known transport proteins. With antibodies prepared against these peptides, the tissue expression, topology and gene structure of the intestinal H^+–oligopeptide transport system will be investigated. By immunohistochemical methods, the 127-kDa protein was localized to the jejunum and the proximal tubule of the kidney, whereas no immunostaining was found in the colon. In the jejunum and the kidney, the peptide transporter was exclusively expressed in the brush-border membrane of the epithelial cells and was completely absent from the basolateral and lateral membranes (*89*).

Substrate Specificity of the Intestinal Uptake System for Oligopeptides

For evaluation of the SARs of ligands necessary for molecular recognition by the intestinal oligopeptide transport system, the following two synergistic approaches were followed.

1. The influence of compounds (to be investigated for interaction with the oligopeptide transport system) on the H^+-dependent initial uptake of cephalexin into BBMV was measured. No effect on cephalexin uptake excludes a direct molecular interaction of the respective ligand with the oligopeptide transporter, whereas a concentration-dependent inhibition of cephalexin uptake may indicate a direct molecular interaction with the oligopeptide transporter provided that other effects like dissipation of the H^+-gradient or changes in the membrane fluidity can be excluded. A subsequent kinetic analysis of inhibition gives insight into the mode of interaction of the respective ligand with the oligopeptide transporter. An inhibition of the oligopeptide transport system by a given ligand indicates binding of the respective ligand to the oligopeptide transporter but does not give any information about if the ligand itself is also transported across the brush-border membrane. Uptake of an inhibitory ligand on cephalexin uptake has to be demonstrated by direct measurement of transport of this ligand into BBMV and measurement of the influence of substrates of the oligopeptide transporter on the uptake of this ligand.
2. In a carrier-mediated transport process, binding of a ligand to the respective carrier protein is the initial step of transport. Therefore, valuable information about the molecular interaction of a substrate or inhibitor of the oligopeptide transport system is available by determination of the effect of such compounds on photoaffinity labeling of the specific binding protein of M_r 127,000, which is presumably a component of the oligopeptide transporter. Subsequently, common structural elements of the different ligands interacting with the oligopeptide transporter may be identified as a first important step for the design of orally active drugs that use the oligopeptide transporter pathway.

β-Lactam Antibiotics

Investigation by several groups of the substrate specificity of the intestinal absorption system for β-lactam antibiotics with everted sacs or in vivo jejunal perfusion led to an inconsistent picture. Nakashima et al. (*47*) found by in situ experiments in rat small intestine that the α-aminocephalosporins (cefadroxil or cephalexin) and the α-aminopenicillin cyclacillin share a common transport system with di- and tripeptides like L-carnosine and Phe-Ala-Gly, respectively. In contrast, a diversity of transport pathways for α-aminopenicillins and α-aminocephalosporins was found by others (*46, 48, 90, 91*) with inconclusive specificities. These findings are difficult to interpret because intact intestinal tissues were used and the superimposition of a variety of transport and metabolic processes as just outlined may have led to the inconsistent specificity interrelationships. Use of BBMV gave a much clearer picture. The initial H^+-dependent uptake of the α-aminocephalosporin cephalexin into BBMV from rabbit small intestine was measured in the absence and presence of the indicated substrates. From the results in Table II it is evident that all β-lactam antibiotics (penicillins and cephalosporins), whether they are enterally absorbed or not, inhibit the uptake of cephalexin to a different extent depending on their chemical structure, thus indicating a direct interaction with the oligopeptide transport system. Furthermore, all β-lactam antibiotics tested so far (in addition to those listed in Table II, >30 different cephalosporins of different structure and both orally active and inactive) decreased the labeling of the 127 kDa binding protein, no matter whether the respective β-lactam antibiotics are enterally absorbed or not. The low inhibitory activity of cefotaxime or ceftizoxime on cephalexin uptake does not correlate well with their strong inhibitory effect on photoaffinity labeling of the 127-kDa binding protein (*66*), but can be explained by the low labeling of the 127-kDa protein by [^{14}C]cefotaxime. Obviously all β-lactam antibiotics are able to bind to the oligopeptide transporter from the luminal side of the brush-border membrane with different affinity, but further molecular events are necessary for a translocation of a carrier-bound substrate across the brush-border membrane. The inactivation of the oligopeptide transporter by DEPC treatment could be greatly inhibited or even prevented only by β-lactam antibiotics and oligopeptides carrying a free α-amino group (*41*), whereas other compounds that are rather strong inhibitors of cephalexin uptake and photoaffinity labeling of the 127-kDa protein, such as cephalothin or cefotiam, without an α-amino group could not protect from inactivation. The dianionic orally active cephalosporin cefixime was also able to protect the oligopeptide transporter from inactivation by DEPC, suggesting that the carboxymethoxyimino function in position 7 of the cephem nucleus can interact with the essential histidine residue of the transporter protein. This finding also strongly argues for a direct role of the α-amino group in β-lactam antibiotics and oligopeptides in the translocation process of a carrier-bound substrate across the brush-border membrane. Tsuji et al. (*39, 92, 93*), using the dianionic orally active cephalosporin cefixime as substrate, came to the same conclusion "that all types of β-lactam antibiotics share a common transport system with peptides in the intestinal brush-border membrane" (*39*). In contrast, Inui et al. (*55*) suggested the existence of two different peptide transport systems both shared by α-aminocephalosporins, whereas cefixime should be transported only by the acidic pH-preferring peptide transport system (Type II). Kinetic analysis of cephalexin uptake showed that benzylpenicillin (*66*), dipeptides like L-carnosine (Figure 9A), α-amino-β-lactam antibiotics like cefadroxil (Figure 9B), or the dianionic cephalosporin cefixime (Figure 9C) competitively inhibited cephalexin uptake as well as photoaffinity labeling of the 127-kDa protein. These results indicate an identical molecular interaction of these different substrates with the oligopeptide carrier system. The uptake system for α-amino-β-lactam antibiotics was markedly impaired upon heat pretreatment (*see* Figure 4), whereas the uptake of cefixime remained unchanged (*56,*

Table II. Effect of Substrates on H^+-Dependent Uptake of Cephalexin Into Rabbit Small Intestinal BBMV and on Photoaffinity Labeling of the Oligopeptide Transporter Protein

Inhibitor	*Effect on Uptake*[a] *Percent Control (%) or IC_{50} (mM)*	*Inhibition of Photolabeling of 127-kDa Protein*[b]
Orally active β-lactam antibiotics		
Cephradine (25 mM)	60.8 ± 0.9% (2)	+ + +
Cefadroxil (12.5 mM)	38.4 ± 9.1% (1)	+ + +
Amoxicillin (5 mM)	81.6 ± 8.8% (1)	+ + +
Ceftibuten (5 mM)	39.8 ± 2.5% (1)	+ + +
Cefixime (4 mM)	40.4 ± 6.3% (1)	+ + +
Penicillins for parental use		
Benzylpenicillin (12.5 mM)	56.1 ± 8.1% (1)	+ + +
Carbenicillin (25 mM)	76.8 ± 5.4% (2)	+ +
Dicloxacillin (10 mM)	53.0 ± 9.2% (1)	+ +
Cephalosporins for parental use		
Cefapirine (25 mM)	88.5 ± 9.4% (2)	+ + +
Ceftazidine (25 mM)	28.1 ± 3.2% (2)	+ + +
Ceftizoxime (25 mM)	94.9 ± 9.1% (2)	+ +
Cefotaxime (12.5 mM)	97.6 ± 7.4% (1)	+ +
Cefoperazone (25 mM)	60.6 ± 9.7% (2)	+ + +
Cefoperazone (5 mM)	81.0 ± 12.6% (1)	+ +
Cefotiam (5 mM)	77.7 ± 7.48% (1)	+ +
Cephaloridine (5 mM)	96.9 ± 3.00% (1)	+ +
Cephaloridine (25 mM)	70.5 ± 7.3% (2)	+ +
Cephalothin (12.5 mM)	54.9 ± 3.6% (1)	+ +
7-α-Aminocephalosporanic acid	9 mM (2)	+ + +
Penems		
HR 664	No inhibition (2)	+
SUN 5555	No inhibition (2)	+
Amino acids		
L-Proline (25 mM)	95.6 ± 6.5% (2)	0
L-Alanine (25 mM)	100.9 ± 7.1% (2)	0
Glycine (12.5 mM)	103 ± 9.6% (1)	0
Peptides		
Glycyl-L-proline, (12.5 mM)	47.2 ± 1.3% (1)	+ +
Glycyl-L-proline (25 mM)	51.4 ± 8.5% (2)	+ +
L-Prolyl-glycine (25 mM)	69.8 ± 6.7% (2)	+ +
L-Carnosine (12.5 mM)	46.5 ± 0.3% (1)	+ +
L-Carnosine (25 mM)	55.3 ± 4.9% (2)	+ +
Triglycine (10 mM)	73.2 ± 3.9% (1)	+ +
Tetraglycine (10 mM)	73.1 ± 8.5% (1)	+
Pentaglycine (10 mM)	85.4 ± 5.6% (1)	(+)
Bestatin (10 mM)	35.3 ± 2.20% (2)	+ +
HOE 427 (15 mM)	67.8 ± 2.3% (2)	+ + +

Table II. Continued

Inhibitor	*Effect on Uptake[a] Percent Control (%) or IC_{50} (mM)*	*Inhibition of Photolabeling of 127-kDa Protein*
Peptide		
HOE 427 (30 mM)	49.3 ± 1.2% (2)	+ + +
Cyclosporin A	n.d.	0
Somatostatin	n.d.	+ + +
Phalloidin	n.d.	+ + +
Glutathione	n.d.	+ + +
p-Aminohippurate (12.5 mM)	36.1 ± 6.9% (1)	+ + +
Renin inhibitors		
S 2863	3 mM (2)	
S 86 2033	2.5 mM (2)	+ +
S 86 3390	1.1 mM (2)	+ +
S 0094	>10 mM (2)	+ +
S 2119	3.75 mM (2)	+ +
S 1078	>10 mM (2)	(+)
S 2586	>4.5 mM (5–8) (2)	+
S 1045	3 mM (2)	+
ACE Inhibitors		
Ramipril	2.6 mM (2)	+ +
Ramiprilate	6 mM (2)	+ +
Enalapril	4.3 mM (2)	+ +
Enalaprilate	5 mM (2)	+ +
Other compounds		
Taurocholate (1 mM)	99.4 ± 1.8% (2)	0
Nicotinamide (25 mM)	109.9 ± 2.8% (2)	0
DIDS (0.5 mM)	80 ± 7.8% (2)	0
Streptomycin (10 mM)	96.7 ± 4.2% (2)	n.d.
Lincomycin (10 mM)	84.8 ± 4.18% (2)	n.d.
Fusidic acid (2.5 mM)	39.3 ± 6.59% (2)	n.d.
Erythromycin (5 mM)	68.0 ± 4.04% (2)	n.d.
Polymyxin (10 mM)	94.3 ± 3.1% (2)	n.d.
Cythidine (10 mM)	83.9 ± 9.2% (2)	n.d.
Thiamine (10 mM)	54.5 ± 5.16% (2)	n.d.
(−) Penicillamine (10 mM)	113.1 ± 1.8% (2)	n.d.
(+) Penicillamine (10 mM)	85.4 ± 17.5% (2)	n.d.
Nalidixic acid (0–20 mM)	No inhibition (2)	n.d.
Ofloxacin (0–20 mM)	No inhibition (2)	n.d.

[a] The concentration of D-cephalexin is given in parentheses; n.d., not determined.

[b] Key to symbols: (0), no inhibition; (+), moderate inhibition; (+ +), strong inhibition; (+ + +), very strong inhibition; ((+)), tendency to inhibition.

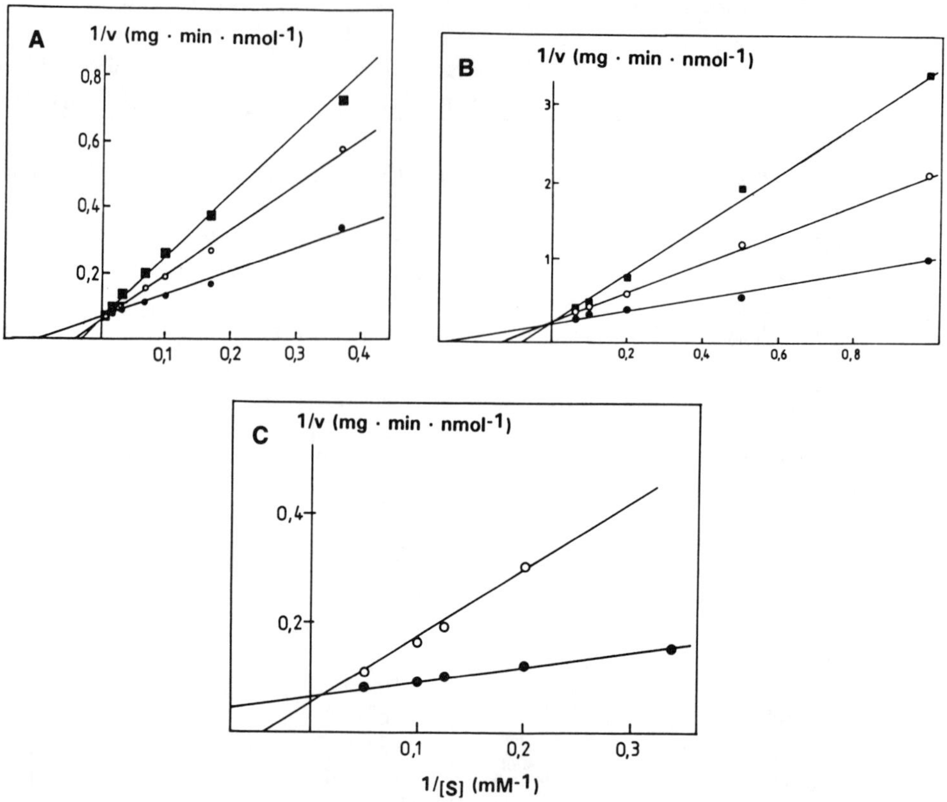

Figure 9. Inhibition of H^+-dependent D-cephalexin uptake into rabbit small intestinal BBMV by L-carnosine (A), cefadroxil (B), and cefixime (C). The BBMV (100 μg of protein, 20 μL) loaded with 10 mM Tris–Hepes buffer (pH 7.4)/300 mM mannitol, were mixed at 30 °C with 180 μL of 10 mM citrate–Tris buffer (pH 6.0)/140 mM KCl containing D-cephalexin and the respective inhibitors. Uptake of D-cephalexin was measured for 1 min. (A) Uptake of 2.5, 5, 10, 15, 20, and 25 mM D-cephalexin in the absence (●) and presence of 10 (○) and 20 (■) mM L-carnosine. (B) Uptake of 1, 2, 5, 10, and 15 mM D-cephalexin in the absence (●) and presence of 10 (○) and 20 (■) mM cefadroxil. (C) Uptake of 3, 5, 7.5, 10, and 20 mM D-cephalexin in the absence (●) and presence (○) of 4 mM cefixime.

94). These findings, the different behavior against the anion transport inhibitor DIDS with inhibition of cefixime uptake and no inhibition of cephalexin uptake (*55*), together with further evidence (inhibition of cefixime but not of cephalexin uptake by fatty acids) suggest that the transport of dianionic cephalosporins occurs despite a direct molecular interaction with the oligopeptide transporter by a second unidentified transport system.

Penems are a new class of β-lactam antibiotics and therefore their interaction with the intestinal oligopeptide transporter was also investigated. The two penems selected, HR 664 and SUN 5555, did not inhibit cephalexin uptake up to a concentration of 20 mM (Figure 10). In contrast, to 7-aminocephalosporanic acid exhibited a concentration-dependent inhibition of the oligopeptide transporter. However, the uptake of the penem HR 664 was not influenced by substrates of the oligopeptide transporter such as cephalexin or L-carnosine. These findings demonstrate that penems do not interact with the intestinal oligopeptide transporter. The main structural difference between penems and 7-aminocephalosporanic acid or α-amino-β-lactam antibiotics is a hydroxyl group instead of a basic

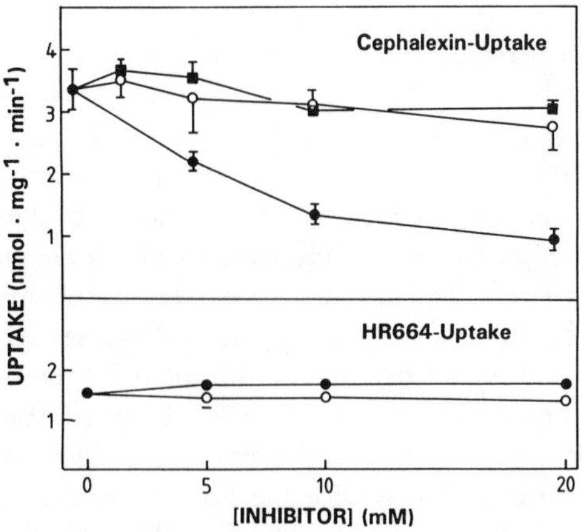

Figure 10. Effect of penems and 7-aminocephalosporanic acid on the H^+-dependent uptake of D-cephalexin (upper panel) and of cephalexin and L-carnosine on the uptake of the penem HR 664 (lower panel). Upper panel: The uptake of 4 mM D-cephalexin dissolved in 180 μL of 10 mM citrate–Tris buffer (pH 6.0)–140 mM by rabbit BBMV (20 μL, 100 μg of protein) preloaded with 10 mM Tris–Hepes buffer (pH 7.4)–300 mM mannitol was measured at 30 °C for 1 min in the absence and in the presence of the indicated concentrations of 7-aminocephalosporanic acid (●), or the penems HR 664 (○) or SUN 5555 (■). Lower panel: The uptake of 4 mM of the penem HR 664 dissolved in 180 μL of 10 mM citrate–Tris buffer (pH 6.0)–140 mM KCl by rabbit BBMV (20 μL, 100 μg of protein) preloaded with 10 mM Tris–Hepes buffer (pH 7.4)–300 mM mannitol was measured at 30 °C for 1 min in the absence and in the presence of the indicated concentrations of L-carnosine (●) or D-cephalexin (○).

amino function at position 6 or 7. The results with penems support an essential role of the α-amino function of β-lactam antibiotics and oligopeptides in the molecular recognition of a substrate by the oligopeptide transporter.

Peptides

Various dipeptides, such as Gly-Pro, Pro-Gly, or L-carnosine, are competitive inhibitors of cephalexin uptake and also inhibit photoaffinity labeling of the 127-kDa binding protein, suggesting a common binding site and transport system for dipeptides, α-aminocephalosporins, and α-aminopenicillins in accordance with results reported by other investigators (*21, 38, 47, 50*). Bestatin, a synthetic dipeptide-derived protease inhibitor, also acts as a competitive inhibitor of cephalexin uptake (*95*). Tri-, -tetra-, and pentaglycine inhibit cephalexin uptake and photoaffinity labeling of the 127-kDa protein, the inhibitory effect decreasing with increasing number of amino acids. Using BBMV, the uptake of Gly-Pro was competitively inhibited by a wide variety of dipeptides such as His-Pro, Leu-Gly, carnosine, Gly-Leu (*96*), or Leu-Pro, Gly-Phe, Gly-Gly (*97*). Also, the uptake of Gly-Leu was inhibited by Leu-Gly, Ala-Gly, Gly-Met, Gly-Ala, and Leu-Gly (*98*). These results indicate that all dipeptides are transported by a common transport system for dipeptides with a very wide substrate specificity, a conclusion also made probable by the results of experiments by Das and Radhakrishnan (*99*) in monkey small intestine. Whereas the results obtained with BBMV strongly argue for a single transport system for oligopeptides, conflicting results from in vivo and in vitro experiments with whole organs and tissues lead to the assumption of the existence of more than one brush-border transport system for dipeptides (*78, 100–103*) despite the fact that, with one exception (*78*), no biphasic kinetic plot that would indicate the presence of more than one peptide transport system has been observed. A major difficulty in these experiments with intact tissue is that the interpretation of the results with competitive uptake by two different peptides is overshadowed by different intracellular and brush-border membrane hydrolysis rates of the respective peptides. Furthermore, the affinity of di- and tripeptides to the intestinal peptide transport systems greatly depends on the peptide sequence, making a conclusive interpretation of competition kinetics even more doubtful. The presence of long lipophilic amino acid side chains enhances affinity for peptide transport (*104*), whereas the introduction of negative charges or more than one positive charge into the side chain of a dipeptide greatly decreases affinity (*105*). It was suggested, that tetrapeptides and higher peptides are unsuited for carrier-mediated peptide transport by the small intestine (*28, 31, 104, 106–108*). However, biologically active peptides like Tyr-D-Ala-Gly-Phe (*34*), penta- and tetragastrin (*109*), morphiceptin (*110*), Tyr-Pro-Phe-ProNH$_2$, the latter obviously using the dipeptide transporter (*111*), the renin-inhibiting nonapeptide RI-61 (Pro-His-Pro-Phe-His-Leu-Phe-Val-Phe) (*112*), or thyrotropin releasing factor (Pyro Glu-His-ProNH$_2$), are absorbed in intact form by the small intestine. The hexapeptide HOE 427 [ebiratide, H-Met(0)-Glu-His-Phe-D-Lys-Phe-NH(CH$_2$)$_8$NH$_2$] is a strong inhibitor of cephalexin uptake (Figure 11). Kinetic analysis reveals a competitive- to mixed-type inhibition of cephalexin uptake by HOE 427. Photoaffinity labeling of the 127-kDa protein was more strongly inhibited by HOE 427 than by dipeptides like Gly-Pro (Figure 12); we can not exclude a small rate of hydrolysis of HOE 427 to di- or tripeptides during the transport or photoaffinity labeling experiments, but the stronger inhibitory effect on photoaffinity labeling of the 127-kDa protein of a given concentration of

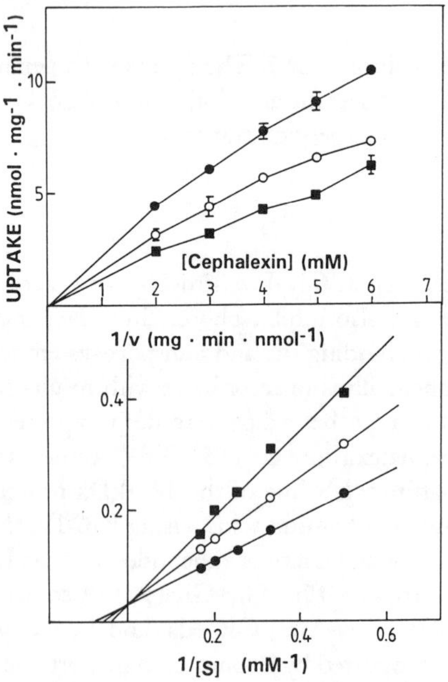

Figure 11. Inhibition of H⁺-dependent D-cephalexin uptake by the hexapeptide-derivative HOE 427 and by Gly-Pro. The uptake of the indicated concentrations of D-cephalexin dissolved in 180 μL of 10 mM citrate–Tris buffer (pH 6.0)–140 mM KCl into rabbit BBMV (20 μL, 100 μg of protein) loaded with 10 mM Tris–Hepes buffer (pH 7.4)–300 mM mannitol was measured for 1 min at 30 °C in the absence (●) and presence of 20 mM HOE 427 (○) or 20 mM Gly-Pro (■). Upper panel: Concentration-dependent uptake of D-cephalexin. Lower panel: 1/v versus 1/[s] diagram.

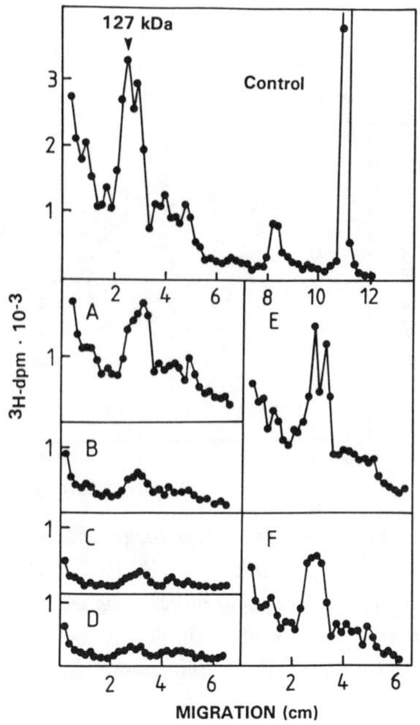

Figure 12. Effect of the hexapeptide derivative HOE 427 on photoaffinity labeling of rabbit BBMV. Rabbit BBMV (250 μg of protein) loaded with 10 mM Tris–Hepes buffer (pH 7.4)–300 mM mannitol were incubated with 1.38 μM (5 μCi) [^{3}H]benzylpenicillin in 10 mM citrate–Tris buffer (pH 4.0)–140 mM KCl for 5 min in the dark either in the absence or in the presence of HOE 427 or Gly-Pro and subsequently irradiated at 254 nm for 2 min. After SDS gel electrophoresis with a 7.5% gel, the distribution of radioactivity was determined by liquid scintillation counting after slicing the gel into 2-mm pieces. A, +1 mM HOE 427. B, +5 mM HOE 427. C, +10 mM HOE 427. D, +20 mM HOE 427. E, +1 mM Gly-Pro, and F, +10 mM Gly-Pro.

HOE 427 in comparison with an identical concentration of a dipeptide like Gly-Pro or L-carnosine make a direct molecular association of HOE 427 with the oligopeptide transporter probable.

Other linear oligopeptides, like decaplanin or encastines, also inhibited cephalexin uptake in a concentration dependent manner, with IC_{50} values >20 mM. The linear peptides glutathione or somatostatin were very strong inhibitors of photoaffinity labeling of the 127 kDa protein. Vice versa, the transport of glutathione across everted sacs of rat small intestine was also strongly inhibited by di- and tripeptides, whereas amino acids exerted no effects (*113*). This finding supports the view that glutathione shares a common intestinal oligopeptide transport system and that it makes the hypothesis that the γ-glutamyl cycle is responsible for membrane transport of peptides and amino acids (*114*) unlikely. Among cyclic peptides that are resistant to hydrolysis, like phalloidin or cyclosporin A, phalloidin in the concentration range 0.5–5 mM inhibited photoaffinity labeling of the 127-kDa protein in a concentration-dependent manner, whereas cyclosporin A had no effect, presumably because of its solubility limit of 100 μM. *p*-Aminohippurate, which can formally be considered a dipeptide formed between 4-aminoenzoic acid and glycine, is a strong competitive inhibitor of the 127-kDa protein; photoaffinity labeling with photoreactive derivatives of *p*-aminohippurate results in photoaffinity labeling of 127-kDa protein, the labeling of which was prevented by dipeptides and β-lactam antibiotics.

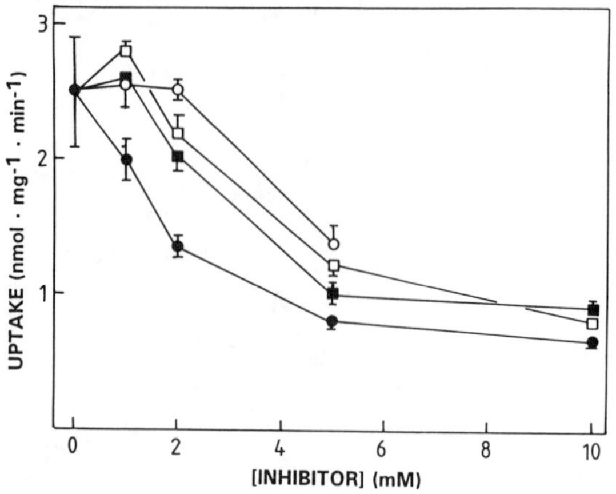

Figure 13. Inhibition of H^+-dependent D-cephalexin uptake by angiotensin-converting enzyme inhibitors. The uptake of 2 mM D-cephalexin dissolved in 180 μL of 10 mM citrate–Tris buffer (pH 6.0)–140 mM KCl into rabbit BBMV (20 μL, 100 μg of protein) loaded with 10 mM Tris–Hepes buffer (pH 7.4)–300 mM mannitol was measured for 1 min at 30 °C in the absence of inhibitor and in the presence of the indicated concentrations of ramipril (●), ramiprilate (○), enalapril (■), and enalaprilate (□).

Angiotensin-Converting Enzyme Inhibitors

The dipeptide-derived drug captopril, an inhibitor of angiotensin-converting enzyme (ACE), is a competitive inhibitor of the intestinal dipeptide transport system (*115*). Absorption of non-sulfhydryl-containing ACE inhibitors, like lisinopril or SQ 29.852, also occurs via the dipeptide transport pathway (*116*). The uptake of cephalexin into rabbit BBMV was also inhibited in a concentration-dependent manner by the anionic ACE inhibitors ramipril or enalapril (both by their monoanionic prodrug form) as well as by the active drugs enalaprilate and ramiprilate, which are dianions (Figure 13). Photoaffinity labeling of the 127-kDa binding protein by these ACE inhibitors was also inhibited in a concentration-dependent manner, suggesting that peptide-derived ACE inhibitors are substrates of the intestinal oligopeptide transporter. The higher inhibitory effect of ramipril and enalapril, which contain a basic NH group and one carboxyl group, compared with their poorly absorbed prilates, which contain an additional carboxyl group, correlates well with the findings of Wooton and Hazelwood (*105*) that the affinity of peptides to the intestinal oligopeptide transporter greatly decreases by introduction of negative charges in the amino acid side chains.

Renin Inhibitors

The aspartic proteinase renin (EC 3.4.23.15) is the first and rate-limiting enzyme in the renin angiotensin cascade, which is significantly involved in the regulation of blood pressure and fluid volume. Therefore, specific inhibitors of renin are a promising new class of antihypertensive agents if they possess good intestinal absorption leading to systemic circulation and dose-related lowering of blood pressure when given orally. Most of the peptide-based renin inhibitors containing a dipeptide core have poor oral bioavailability (<10% in rats and 5% in primates) (*117*). Therefore, the search for orally active renin inhibitors remains a challenge to make valuable drugs using the principle of renin inhibition.

Recently, a novel peptide-derived renin inhibitor (A-72517) with sufficient bioavailability in several species was reported (*118*). Because renin inhibitors are peptide-derived drugs, we investigated the interaction of several renin inhibitors with the intestinal H^+-dependent transport system. The renin inhibitors used all inhibited the oligopeptide transporter in a concentration-dependent manner, with IC_{50} values ranging from 1 to >10 mM, indicating different affinities of these compounds for the oligopeptide transporter. The inhibition constants in the millimolar range are similar to the K_m value of cephalexin for the intestinal peptide uptake system (*41*), indicating affinities of these renin inhibitors comparable to those of orally active α-amino-β-lactam antibiotics. Kinetic analysis of the inhibitors S 86 2033 and S 86 3390 revealed a competitive inhibition of cephalexin uptake (*119*), showing that these renin inhibitors compete with cephalexin for binding to the transport site of the intestinal oligopeptide transporter. Because these renin inhibitors are peptide derivatives, the possibility was excluded that dipeptides formed by brush border hydrolysis caused the competitive inhibition. After incubation with BBMV, the compounds remained intact for periods up to 2 h. The renin inhibitor S 2586 showed antihypertensive activity after introduodenal application to monkeys (*120*) despite a relatively weak affinity for the intestinal oligopeptide transporter, as indicated by concentrations of 2–3 mM for half-maximal inhibition of photoaffinity labeling of the 127-kDa protein as compared with 500 and 800 μM for S 86 2033 and S 86 3390, respectively. Because of the pentahydroxy moiety at the C-terminus of S 2586, an interaction with the Na^+–D-glucose transport system seemed possible. At a D-glucose concentration of 20 μM, the Na^+-dependent uptake of D-[^{14}C]glucose was inhibited in a concentration-dependent manner by S 2586, with an IC_{50} value of 3–5 mM. Measurement of the uptake of S 2586 into BBMV (concentration 4 mM) showed that D-glucose up to a concentration of 20 mM had no inhibitory effect on uptake, whereas cephalexin strongly inhibited S 2586 uptake, with an IC_{50} value of 5–8 mM. These results indicate that this renin inhibitor is probably taken up by the oligopeptide transporter.

To obtain clear evidence that peptide-derived renin inhibitors share the intestinal uptake system for dipeptides and α-amino-β-lactam antibiotics, the uptake of the renin inhibitor S 86 3390 into BBMV from rabbit small intestine was measured. The uptake of this inhibitor was stimulated by an inwardly directed H^+-gradient with transient accumulation compared to equilibrium (*119*). The uptake of S 86 3390 was also inhibited in a concentration-dependent manner by cephalexin. These experimental findings suggest that the uptake of renin inhibitors of the chemical structure described earlier occurs by the H^+–oligopeptide transport system and that orally active renin inhibitors should fulfill the structural requirements of a substrate for the intestinal peptide transporter.

The renin inhibitors described earlier all contain a histidyl residue at the so-called P_2-site of the molecule (*118, 120*). Keeping the structural elements at the amino terminus, the P_3-site, and the carboxy terminus constant, the substitution of the histidyl residue at the P_2-site greatly influences intestinal absorption of a renin inhibitor. In a series of polar renin inhibitors containing a C-terminal oxazolidinone structure, intestinal absorption (as measured by drug levels in portal blood) was stimulated six- to sevenfold by substitution of the imidazolyl residue by a thiazolyl-4-yl radical (*121*). Substitution of the imidazol-4-yl by 1-methylimidazol-4-yl and by thiophen-2-yl resulted in 3.3- and 2.8-fold better intestinal absorption, respectively, whereas a pyrazol-3-yl residue decreased intestinal absorption by a factor of 10. Conserving the structural elements at the N-terminus and the P_3-site and replacement of the carboxy terminus by a less polar glycol residue slightly changed the influence of substitution at the P_2-site. 1-Methylimidazol-4-yl, *N*-methylmethyloxy, thiazol-4-yl, and pyrazol-3-yl substitution yielded 20-, 18-, 7-, and 4-fold better intestinal absorption compared with the imidazol-4-yl (His) compound. These studies indicate that weakly basic residues lacking an exchangeable proton result in portal and systemic drug levels that are significantly higher than those observed for the parent histidine derivatives (*122*).

Other Compounds

From the results just described, it is evident that the intestinal oligopeptide transporter shows a broad specificity for binding of ligands. Common to all these compounds is the existence of a peptide core, with a free C-terminal carboxyl group in most compounds. An N-terminal amino group or a weakly basic group at the N-terminus seem to be essential for the translocation process of a carrier-bound substrate across the brush border membrane. We therefore investigated whether other anionic, zwitterionic, or structurally unrelated compounds had an influence on the oligopeptide transporter. All these compounds did not interfere with the oligopeptide transporter with the exception of erythromycin, fusidic acid, and thiamine (Table II). The inhibitory effect of erythromycin and fusidic acid may be caused by intrinsic characteristics of these compounds acting as ionophores or detergents. By chemical modification of the cephalexin molecule, we learned that attachment of amino acid or peptide radicals to the 3-position retained affinity for binding to the 127-kDa protein but led to a significant drop in transport across the brush-border membrane. By substitution of the D-phenylglycine moiety of cephalexin with D-phenylalanine, an α-amino-cephalosporin was found that was transported even better than cephalexin itself by the oligopeptide transporter.

Conclusions

From the SARs found, we hypothesized the essential structure of a peptide substrate to be recognized and transported by the intestinal oligopeptide transporter (*see* Chart II). The SARs of renin inhibitors for enteral absorption do not completely fit into this concept, maybe because these compounds also interfere with transporters other than the oligopeptide transporter. The finding that modification of the amino acid occupying the P_2-site affects intestinal absorption is in accordance with the proposed structure, and it may be that the C-terminal carboxyl group can be substituted by other polar group such as pentols, amino groups, or pyridine rings, which are able to form hydrogen bonds with amino acid side chains of the transporter protein. From the SARs described, a greater understanding of the molecular specificity of the intestinal oligopeptide transport system could clearly provide

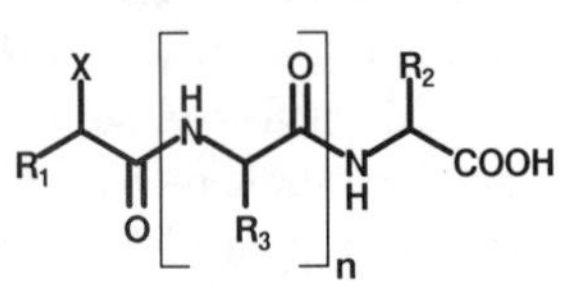

Chart II. Hypothetical peptide substrate to be recognized and transported by the intestinal oligopeptide transporter. Key: (n) *0 or 1; (R_2) only a small, electrically neutral substituent like H, CH_3, vinyl; (X) a group capable of accepting protons in a proton donor–acceptor relationship with a histidyl residue of the transporter (the pK_a value of X should be in the range of about 6.5 to 7.5 such as with NH_2, imidazolyl, and thiol; (R_1, R_3) the allowed structural variety of R_1 and R_3 is still unclear, but R_3 must not be a negatively charged group.*

important leads for the synthesis of orally active peptides and peptide-derived drugs. Furthermore, the oligopeptide transporter may possibly be used as a shuttle system to increase intestinal absorption of poorly absorbable drugs by covalent attachment of these drugs to a dipeptide or tripeptide backbone as expressed in cephalexin. Such a drug delivery approach, using substrates of endogenous nutrient transporters of the small intestine, has recently been demonstrated by us with modified bile acids to achieve liver selective drug targeting and intestinal absorption of peptides and drugs (*10, 11*). Because of the similarities of the intestinal and the renal oligopeptide transporters (*24, 72*), such compounds may also be reabsorbed in the proximal kidney tubule, leading to prolonged plasma levels and duration of action of the corresponding drugs.

Note Added in Proof

During the reviewing process of this submitted manuscript, two manuscripts describing the expression cloning of the H^+–oligopeptide transporter from rabbit small intestine (*123*) and Caco-2 cells (*124*) were published. Both proteins have a different amino acid sequence and are not related to each other. Both proteins also have no sequence homology to the 127-kDa polypeptide we describe, a finding that suggests the existence of at least three different H^+–oligopeptide transporters in small intestinal brush-border membranes.

Acknowledgment

We thank Susanne Winkler for excellent secretarial assistance and for preparation of the manuscript.

References

1. Holmes, C. P.; Adams, C. L.; Fodor, S. P. A.; Yu-Yang, P. In *Perspectives in Medicinal Chemistry;* Testa, B.; Kyburz, E.; Fuhrer, W.; Geiger, R. Eds.; Verlag Helvetica Chimica Acta: Basel, Switzerland, 1993; pp 489–500.
2. Smith, P. L.; Wall, D. A.; Gochoco, C. H.; Wilson, G. *Adv. Drug Deliv. Rev.* **1992,** *8,* 253–290.
3. Lee, V. H. L. *CRC Crit. Rev. Ther. Drug Carrier Syst.* **1988,** *5,* 69–97.
4. Matthews, D. M. In *Protein Absorption: Development and Present State of the Subject;* Wiley-Liss: New York, 1991; pp 235–315.
5. Lee, V. H. L.; Yamamoto, A. *Adv. Drug Deliv. Rev.* **1990,** *4,* 171–207.
6. Saffran, M.; Kumar, G. S.; Savariar, C.; Burnham, J. C.; Williams, F.; Neckers, D. C. *Science (Washington, DC)* **1986,** *233,* 1081–1084.
7. Fujii, S.; Yokoyama, T.; Ikegaya, K.; Sato, F.; Yokoo, N. *J. Pharm. Pharmacol.* **1985,** *37,* 545–549.
8. Ziv, E.; Kidron, M.; Berry, E.; Baron, H. *Life Sci.* **1981,** *29,* 803–809.

9. Russell-Jones, G. J.; Aizpurna, H. J. *Proceedings of the International Symposium Controlled Release Bioactive Materials;* Controlled Release Society, Inc.: Dearfield, IL, 1988; Vol. 15, pp 142–143.
10. Kramer, W.; Wess, G.; Schubert, G.; Bickel, M.; Girbig, F.; Gutjahr, U.; Kowalewski, S.; Baringhaus, K.-H.; Enhsen, A.; Glombik, H.; Müllner, S.; Neckermann, G.; Schulz, S.; Petzinger, E. *J. Biol. Chem.* **1992,** *267,* 18598–18604.
11. Kramer, W.; Wess, G.; Schubert, G.; Bickel, M.; Baringhaus, K.-H.; Enhsen, A.; Glombik, H.; Hoffmann, A.; Müllner, S.; Neckermann, G.; Schulz, S.; Petzinger, E. In *Bile Acids and the Biliary System;* Paumgartner, G.; Stiehl, A.; Gerok, W., Eds.; Kluwer Academic: Hingham, MA, 1993; pp 161–176.
12. Ganapathy, V.; Leibach, F. H. *J. Biol. Chem.* **1983,** *258,* 14189–14192.
13. Lucas, M. L.; Cooper, B. T.; Lei, F. H.; Johnson, T.; Holmes, G. K. T.; Blair, J. A.; Cooke, W. T. *Gut* **1978,** *19,* 735–742.
14. Lucas, M. L. *Gut* **1983,** *24,* 734–739.
15. Murer, H.; Hopfer, U.; Kinne, R. *Biochem. J.* **1976,** *154,* 597–604.
16. Murer, H.; Burckhardt, G. *Rev. Physiol. Biochem. Pharmacol.* **1983,** *96,* 1–51.
17. Lauterbach, F. *Arzneim.-Forsch.* **1975,** *25,* 479–488.
18. Lauterbach, F. In *Pharmacology of Intestinal Permeation II;* Csáky, T. Z., Ed.; Springer Verlag: Berlin, Germany, 1984; pp 271–299.
19. Shindo, H.; Fukuda, K.; Kawai, K.; Tanaka, K. *J. Pharmacobio-Dyn.* **1978,** *1,* 310–323.
20. Okano, T.; Inui, K.; Maegawa, H.; Takano, M.; Hori, R. *J. Biol. Chem.* **1986,** *261,* 14130–14134.
21. Okano, T.; Inui, K.; Takano, M.; Hori, R. *Biochem. Pharmacol.* **1986,** *35,* 1781–1786.
22. Kimura, T.; Yamamoto, T.; Ishizuka, R.; Sezaki, M. *Biochem. Pharmacol.* **1985,** *34,* 81–84.
23. Takuwa, N.; Shimada, T.; Matsumoto, H.; Himukai, M.; Hoshi, T. *Jpn. J. Physiol.* **1985,** *35,* 629–642.
24. Miyamoto, Y.; Ganapathy, V.; Leibach, F. H. *J. Biol. Chem.* **1986,** *261,* 16133–16140.
25. Addison, J. M.; Burston, D.; Dalrymple, J. A.; Matthews, D. M.; Payne, J. W.; Sleisenger, M. H.; Wilkinson, S. *Clin. Sci. Mol. Med.* **1975,** *49,* 313–322.
26. Adibi, S. A.; Mercer, D. W. *J. Clin. Invest.* **1973,** *52,* 1586–1594.
27. Adibi, S. A. *J. Clin. Invest.* **1971,** *50,* 2266–2275.
28. Boullin, D. J.; Crampton, R. F.; Hedding, C. E.; Pelling, D. *Clin. Sci. Mol. Med.* **1973,** *45,* 849–858.
29. Boyd, C. A. R.; Ward, M. R. *J. Physiol.* **1982,** *324,* 411–428.
30. Adibi, S. A.; Morse, E. L. *J. Clin. Invest.* **1977,** *60,* 1008–1016.
31. Smithson, K. W.; Gray, G. M. *J. Clin. Invest.* **1977,** *60,* 665–674.
32. Kerschner, G. A.; Geary, L. E. *J. Pharmacol. Exp. Ther.* **1983,** *226,* 33–38.
33. Chung, Y. C.; Silk, D. B. A.; Kim, Y. S. *Clin. Sci.* **1979,** *57,* 1–11.
34. Rogers, C. S.; Heading, C. E.; Wilkinson, S. *IRCS Med. Sci. (Biochem.)* **1980,** *8,* 648–649.
35. Kania, R. K.; Santiago, N. A.; Gray, G. M. *J. Biol. Chem.* **1977,** **252,** 4929–4934.
36. Nakashima, E.; Tsuji, A.; Mizuo, H.; Yamana, T. *Biochem. Pharmacol.* **1984,** *33,* 3345–3352.
37. Bergan, T. *Scand. J. Infect. Dis. (Suppl.)* **1984,** **42,** 83–98.
38. Nakashima, E.; Tsuji, A.; Kagatani, S.; Yamane, T. *J. Pharmacobio-Dyn.* **1984,** *7,* 452–464.
39. Tsuji, A.; Tamai, I.; Hirooka, H.; Terasaki, T. *Biochem. Pharmacol.* **1987,** *36,* 565–567.
40. Lucas, M. L.; Blair, J. A.; Cooper, B. T.; Cooke, W. T. *Biochem. Soc. Trans.* **1976,** *4,* 154–156.
41. Kramer, W.; Girbig, F.; Petzoldt, E.; Leipe, I. *Biochim. Biophys. Acta* **1988,** *943,* 288–296.
42. Kramer, W. *Biochim. Biophys. Acta* **1987,** *905,* 65–74.

43. Dantzig, A. H.; Bergin, L. *Biochem. Biophys. Res. Commun.* **1988,** *155,* 1082–1087.

44. Dantzig, A. H.; Bergin, L. *Biochim. Biophys. Acta* **1990,** *1027,* 211–217.

45. Yamashita, S.; Yamazaki, Y.; Masada, M.; Nadai, T.; Kimura, T.; Sezaki, H. *J. Pharmacobio-Dyn.* **1986,** *9,* 368–374.

46. Iseki, K.; Iemura, A.; Sato, H.; Sunada, K.; Miyazaki, K.; Arita, T. *J. Pharmacobio-Dyn.* **1984,** *7,* 768–775.

47. Nakashima, E.; Tsuji, A.; Kagatani, S.; Yamana, T. *J. Pharmacobio-Dyn.* **1984,** *7,* 452–464.

48. Yamashita, S.; Yamazaki, Y.; Mizuno, M.; Masada, M.; Nadai, T.; Kimura, T.; Sezaki, H. *J. Pharmacobio-Dyn.* **1984,** *7,* 227–233.

49. Kimura, T.; Yamamoto, T.; Mizuno, M.; Suga, Y., Sumiko, K.; Sezaki, H. *J. Pharmacobio-Dyn.* **1983,** *6,* 246–253.

50. Kimura, T.; Endo, H.; Yoshikawa, M.; Muranishi, S.; Sezaki, H. *J. Pharmacobio-Dyn.* **1978,** *1,* 262–267.

51. Tsuji, A.; Nakashima, E.; Asano, T.; Nakashima, R.; Yamana, T. *J. Pharm. Pharmacol.* **1979,** *31,* 718–720.

52. Tsuji, A.; Nakashima, E.; Kagami, I.; Asano, T.; Nakashima, R.; Yamana, T. *J. Pharm. Pharmacol.* **1978,** *30,* 508–509.

53. Tsuji, A.; Nakashima, E.; Kagami, I.; Yamana T. *J. Pharm. Sci.* **1981,** *70,* 768–772.

54. Ganapathy, V.; Leibach, F. H. *J. Biol. Chem.* **1983,** *258,* 14189–14192.

55. Inui, K.; Okano, T.; Maegawa, H.; Kato, M.; Takano, M.; Hori, R. *J. Pharmacol. Exp. Ther.* **1988,** *247,* 235–241.

56. Kramer, W.; Girbig, F.; Gutjahr, U.; Kowalewski, S. *Biochem. Pharmacol.* **1993,** *46,* 542–546.

57. Asatoor, A. M.; Chadha, A.; Milne, M. D.; Prosser, D. *J. Clin. Sci. Mol. Med.* **1973,** *45,* 199–212.

58. Burston, D.; Addison, J. M.; Matthews, D. M. *Clin. Sci.* **1972,** *43,* 907–911.

59. Cheeseman, C. I.; Smyth, D. H. *J. Physiol. (Lond.)* **1973,** *229,* 45P–46P.

60. Tamai, I.; Ling, H.-Y.; Timbal, S.; Nishikido, J.; Tsuji, A. *J. Pharm. Pharmacol.* **1988,** *40,* 320–324.

61. Kramer, W.; Girbig, F.; Gutjahr, U.; Kowalewski, S.; Adam, F.; Schiebler, W. *Eur. J. Biochem.* **1992,** *204,* 923–930.

62. Steward, H. E.; Jackson, M. J. *J. Pharmacol. Exp. Ther.* **1981,** *218,* 453–458.

63. Suzuki, H.; Sawada, Y.; Sugiyama, Y.; Iga, T.; Hanano, M. *J. Pharmacol. Exp. Ther.* **1987,** *243,* 1147–1152.

64. Suzuki, H.; Sawada, Y.; Sugiyama, Y.; Iga, T.; Hanano, M. *J. Pharmacol. Exp. Ther.* **1987,** *242,* 660–665.

65. Tsuji, A.; Terasaki, T.; Tamai, I.; Nakashima, E.; Takanosu, K. *J. Pharm. Pharmacol.* **1985,** *37,* 55–57.

66. Kramer, W.; Girbig, F.; Leipe, I.; Petzoldt, E. *Biochem. Pharmacol.* **1988,** *37,* 2427–2435.

67. Bayley, H. In *Laboratory Reagents in Biochemistry and Molecular Biology;* Work, T. S.; Burdon, R. H., Eds.; Elsevier: Amsterdam, The Netherlands, 1983; pp 1–187.

68. Schuster, D. I.; Probst, W. C.; Ehrlich, G. K.; Singh, G. *Photochem. Photobiol.* **1989,** *49,* 785–804.

69. Kramer, W.; Dürckheimer, W.; Girbig, F.; Gutjahr, U.; Leipe, I.; Oekonomopulos, R. *Biochim. Biophys. Acta* **1990,** *1028,* 174–182.

70. Willner, D.; Holdredge, C. T.; Baker, S. R.; Cheney, L. C. *J. Antibiot.* **1972,** *25,* 64–67.

71. Ganapathy, V.; Mendicino, J. F.; Pashley, D. H.; Leibach, F. M. *Biochem. Biophys. Res. Commun.* **1980,** *97,* 1133–1139.

72. Kramer, W.; Leipe, I.; Petzoldt, E.; Girbig, F. *Biochim. Biophys. Acta* **1988,** *939,* 167–172.

73. Burckhardt, G.; Kramer, W.; Kurz, G.; Wilson, F. A. *J. Biol. Chem.* **1983,** *258,* 3618–3622.

74. Peerce, B. E.; Wright, E. M. *J. Biol. Chem.* **1984,** *259,* 14105–14112.

75. Kato, M.; Maegawa, H.; Okano, T.; Inui, K.; Hori, R. *J. Pharmacol. Exp. Ther.* **1989,** *251,* 745–749.

76. Tsuji, A.; Nakashima, E.; Yamana, T. *J. Pharm. Sci.* **1979,** *68,* 308–311.

77. Addison, J. M.; Matthews, D. M.; Burston, D. *Clin. Sci. Mol. Med.* **1974,** *46,* 707–714.

78. Rubino, A.; Field, M.; Schwachman, H. *J. Biol. Chem.* **1971,** *246,* 3542–3548.

79. Ganapathy, V.; Leibach, F. H. *Life Sci.* **1982,** *30,* 2137–2146.

80. Kramer, W. *Naunyn–Schmiedebergs Arch. Pharmacol.* **1989,** *339,* R 42.

81. Kramer, W.; Dechent, C.; Girbig, F.; Gutjahr, U.; Neubauer, H. *Biochim. Biophys. Acta* **1990,** *1030,* 41–49.

82. Tiruppathi, C.; Ganapathy, V.; Leibach, F. H. *J. Biol. Chem.* **1990,** *265,* 14870–14874.

83. Kramer, W.; Gutjahr, U.; Girbig, F.; Leipe, I. *Biochim. Biophys. Acta* **1990,** *1030,* 50–59.

84. Kramer, W.; Girbig, F. Gutjahr, U.; Leipe, I. *J. Chromatogr.* **1990,** *521,* 199–210.

85. Kramer, W.; Schiebler, W. *Naunyn–Schmiedebergs Arch. Pharmacol.* **1991,** *340,* R 78.

86. Kramer, W. *Naunyn–Schmiedebergs Arch. Pharmacol.* **1991,** *343,* R 48.

87. Miyamoto, Y.; Thompson, Y. G.; Howard, E. F.; Ganapathy, V.; Leibach, F. H. *J. Biol. Chem.* **1991,** *265,* 4742–4745.

88. Saito, H.; Ishii, T.; Inui, K. *Biochem. Pharmacol.* **1993,** *45,* 776–779.

89. Kramer, W.; Girbig, F.; Gutjahr, U.; Kowalewski, S.; Adam, F.; Langer, K.-H.; Tripier, D.; Schiebler, W. In *Recent Advances in Chemotherapy;* Adam, D.; Lode, H.; Rubinstein, E., Eds.; Futuramed Publishers: Munich, Germany, 1992; pp 1694–1695.

90. Miyazaki, K.; Ohtani, K.; Umeniwa, K.; Arita, T. *J. Pharmacobio–Dyn.* **1982,** *5,* 555–563.

91. Dixon, C.; Mizen, L. W. *J. Physiol. (Lond.)* **1977,** *269,* 549–559.

92. Tsuji, A. In *Frontiers of Antibiotic Research;* Umezawa, H.; Ed.; Academic: Orlando, FL, 1987; pp 253–268.

93. Tsuji, A. *Adv. Biosci.* **1987,** *65,* 125–131.

94. Kramer, W.; Girbig, F.; Gutjahr, U. *Naunyn–Schmiedebergs Arch. Pharmacol.* **1992,** *345,* R 65.

95. Tomita, Y.; Katsura, T.; Okano, T.; Inui, K.-I.; Hori, R. *J. Pharmacol. Exp. Ther.* **1990,** *252,* 859–862.

96. Ganapathy, V.; Mendicino, J. F.; Leibach, F. H. *J. Biol. Chem.* **1981,** *256,* 118–126.

97. Wooton, R. *Biochem. Soc. Trans.* **1986,** *14,* 1192–1193.

98. Sigrist-Nelson, K. *Biochim. Biophys. Acta* **1975,** *394,* 220–226.

99. Das, M.; Radhakrishnan, A. N. *Biochem. J.* **1975,** *146,* 133–137.

100. Burston, D.; Wapnir, R. A.; Taylor, E.; Matthews, D. M. *Clin. Sci.* **1982,** *62,* 617–626.

101. Matthews, D. M. *Biochem. Soc. Trans.* **1982,** *11,* 808–810.

102. Matthews, D. M.; Burston, D. *Clin. Sci.* **1983,** *65,* 177–184.

103. Matthews, D. M.; Burston, D. *Clin. Sci.* **1984,** *67,* 541–549.

104. Matthews, D. M.; Adibi, S. A. *Gastroenterology* **1976,** *71,* 151–161.

105. Wooton, R.; Hazelwood, R. *Biochem. Soc. Trans.* **1989,** *17,* 691–692.

106. Addison, J. M.; Burston, D.; Payne, J. W.; Wilkinson, S.; Matthews, D. M. *Clin. Sci. Mol. Med.* **1975,** *49,* 305–312.

107. Matthews, D. M.; Craft, I. L.; Geddes, D. M.; Wise, I. J.; Hyde, C. W. *Clin. Sci.* **1968,** *35,* 415–424.

108. Adibi, S. A.; Morse, E. L. *J. Clin. Invest.* **1977,** *60,* 1008–1016.

109. Jennewein, H. M.; Waldeck, F.; Konz, W. *Arzneim.-Forsch.* **1974,** *24,* 1225–1228.

110. Mahé, S.; Tome, D.; Dumentier, A. M.; Desjeux, J. F. *Peptides* **1989,** *10,* 45–52.

111. Kimura, T. *Pharm. Int.* **1984,** 75–83.

112. Takaori, K.; Donowitz, M.; Burton, J. *Peptides: Structure, Function, Proceedings of the American Peptide Symposium,* 9th ed.; Debe, C. M.; Hruby, V. J.; Kopple, K. D., Eds.; Pierce Chemical Company: Rockford, Ill., 1985; pp 767–770.

113. Hunjan, M. K.; Evered, D. F. *Biochim. Biophys. Acta* **1985,** *815,* 184–188.

114. Meister, A. *Trends Biochem. Sci.* **1981,** *6,* 231–234.

115. Hu, M.; Amidon, G. L. *J. Pharm. Sci.* **1988,** *77,* 1007–1011.

116. Friedman, D. I.; Amidon, G. L. *J. Pharm. Sci.* **1989,** *78,* 995–998.

117. Greenlee, W. J. *Pharm. Res.* **1987,** *4,* 364–374.

118. Kleinert, H. D.; Rosenberg, S. H.; Baker, W. R.; Stein, H. H.; Klinghofer, V.; Barlow, J.; Spina, K.; Polakowski, J.; Kovar, P.; Cohen, J.; Denisseu, J. *Science (Washington, DC)* **1992,** *257,* 1940–1943.

119. Kramer, W.; Girbig, F.; Gutjahr, U.; Kleemann, H.-W.; Leipe, I.; Urbach, H.; Wagner, A. *Biochim. Biophys. Acta* **1990,** *1027,* 25–30.

120. Kleemann, H.-W.; Heitsch, H.; Henning, R.; Kramer, W.; Kocher, W.; Lerch, U.; Linz, W.; Nickel, W.-U.; Ruppert, D.; Urbach, H.; Utz, R.; Wagner, A.; Weck, R.; Wiegand, F. *J. Med. Chem.* **1992,** *35,* 559–567.

121. Rosenberg, S. H.; Spina, K. P.; Woods, K. W.; Polakowski, J.; Martin, D. L.; Yao, Z.; Stein, H. H.; Cohen, J.; Barlow, J. L.; Egan, D. A.; Tricarico, K. A.; Baker, W. R.; Kleinert, H. D. *J. Med. Chem.* **1993,** *36,* 449–459.

122. Rosenberg, S. H.; Spina, K. P.; Condon, S. L.; Polakowski, J.; Yao, Z.; Kovar, P.; Stein, H. H.; Cohen, J.; Barlow, J. L.; Klinghofer, V.: Egan, D. A.; Tricarico, K. A.; Perun, T. J.; Baker, W. R.; Kleinert, H. D. *J. Med. Chem.* **1993,** *36,* 460–467.

123. Fei, Y. J.; Kanai, Y.; Nussberger, S.; Ganapathy, V.; Leibach, F. H.; Romero, F. H.; Singh, S. K.; Boron, W. F.; Hediger, M. A. *Nature* (*London*) **1994,** *368,* 563–566.

124. Dantzig, A. H.; Hoskinns, J.; Tabas, L. B.; Bright, S.; Shepard, R. L.; Jenkins, J. L.; Duckworth, D. C.; Sportsman, J. R.; Mackensen, D.; Rosteck, P. R., Jr.; Skatrud, P. L. *Science* (*Washington, DC*) **1994,** *264,* 430–433.

RECEIVED for review July 12, 1993. ACCEPTED revised manuscript September 24, 1993.

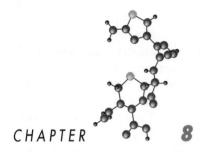

CHAPTER 8

Oral Delivery of Therapeutic Proteins and Peptides by the Vitamin B_{12} Uptake System

Gregory J. Russell-Jones

Many pharmaceuticals, such as antibiotics (tetracycline, penicillin, etc.) and lipophilic hormones (testosterone, estrogen, etc.), can be successfully administered to the patient orally and subsequently manifest their action systemically. The majority of peptide (greater than four amino acids) and protein pharmaceuticals, however, are almost completely ineffective when given orally. Thus, the efficacy of small hormones such as calcitonin, vasopressin, insulin, and luteinizing hormone releasing hormone (LHRH) is reduced by at least two orders of magnitude following oral administration. Larger proteins, such as erythropoietin, granulocyte colony stimulating factor, and porcine somatotrophin are almost completely ineffectual when given orally or must be given in such large doses that oral delivery of these proteins is an economic unreality. The major obstacles to the oral delivery of peptide and protein pharmaceuticals are the inability of the intestinal mucosal cells to absorb these compounds, the breakdown of these substances by various physiological agents encountered in the harsh environment of the intestinal milieu, and the relatively rapid transit time of the pharmaceuticals through the small intestine.

To overcome the problem of degradation, a number of encapsulation methods that enable the encapsulated material to bypass both the gastric acidity and pepsin-mediated proteolysis encountered within the lumen of the stomach have been employed. Enteric coatings of these capsules, with materials such as eudragit, have been used to deliver the material to the lower small intestine and the upper bowel. Despite these protective devices, the uptake of these peptide pharmaceuticals remains low. It appears, therefore, that the major obstacle to peptide and protein uptake in the intestine is the inability of these pharmaceuticals to enter and cross the intestinal epithelial cell barrier, which remains an almost impenetrable barrier to the majority of peptides larger than a dipeptide in size.

However, two molecules do overcome this intestinal barrier and are transported in significant quantities across the intestinal barrier of the mature vertebrate; namely, iron and vitamin B_{12} (VB_{12}). In both cases, a specific binding protein is released into the intestine and binds to its ligand in the lumen of the gut. During iron uptake, transferrin is released from the stomach and binds to iron, and the [transferrin–iron] complex is in turn bound by a receptor on the duodenal mucosa. The [receptor–transferrin–iron] complex is then taken up by receptor-mediated endocytosis. The uptake of VB_{12} occurs by a similar process and is the subject of this review.

Uptake of VB_{12} in the Small Intestine

The naturally occurring dietary molecule VB_{12} is actively absorbed from the intestine. During this process, VB_{12} is first released from food substances by the action of pepsin in the stomach. In the stomach, VB_{12} is then complexed to a specific binding protein secreted in saliva, haptocorrin (Hc), which has a higher affinity for VB_{12} than intrinsic factor (IF) at the acid pH of the stomach. The [VB_{12}–Hc] complex leaves the stomach and enters the duodenum where the Hc is degraded by the action of trypsin and chymotrypsin. The IF produced in and released from the parietal cells of the stomach then binds to VB_{12} to form an [IF–VB_{12}] complex that passes down the small intestine until it reaches the ileum. Here, the complex binds to a specific IF receptor (IFR) located on the villous epithelium. The [VB_{12}–IF–IFR] complex then enters the cell via receptor-mediated endocytosis (Figure 1). In a poorly understood process, the VB_{12} is released from IF and binds to another VB_{12} binding protein, transcobalamin II (TcII), which completes the process of transcytosis of the VB_{12}. The VB_{12} subsequently enters the circulation complexed to TcII. The whole process takes several hours as can be seen by the delayed systemic appearance of orally administered [^{57}Co]VB_{12} (Figure 2). In contrast, the uptake of VB_{12} differs significantly from the rapid nonspecific uptake seen following oral administration of peptides or proteins (Figure 3). The high percentage of ^{125}I seen after feeding ^{125}I-labelled bovine serum albumin ([^{125}I]BSA) presumably represents nonspecific uptake of degraded peptide fragments or deiodination of peptide in the gut.

Allen and Majerus (*1, 2*) showed that it was possible to modify VB_{12} chemically, to couple it to a resin, and to use the [VB_{12}–resin] complex to affinity purify IF. Recently, we have shown that it is possible to chemically link various substances (including peptide hormones and proteins) to VB_{12} in a manner that does not interfere with [VB_{12}–IF] complex formation. Oral administration of these [VB_{12}–drug] complexes allows these molecules to be cotransported from the intestinal lumen to the circulation while complexed to VB_{12} (*3*).

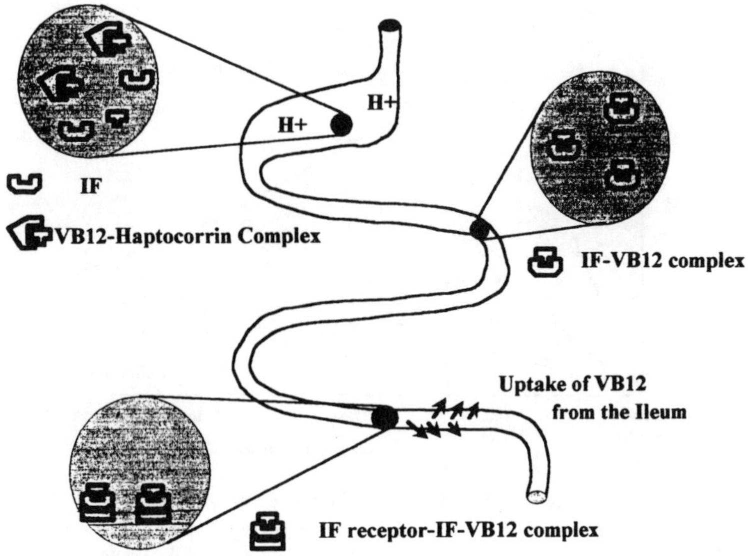

Figure 1. Uptake of VB_{12} from the gut. VB_{12} is initially released from food and binds to Hc in the stomach. The [VB_{12}:Hc] complex passes out of the stomach and the Hc is degraded by intestinal proteases and thereby releases VB_{12}. The VB_{12} is then bound by IF. The [VB_{12}–IF] complex passes down the intestine until it reaches the ileum (in most species). Here it binds to an IFR. Once bound, the [VB_{12}–IF–IFR] complex is internalized by receptor-mediated endocytosis. Several hours later, the VB_{12} appears in serum complexed to TcII (not shown).

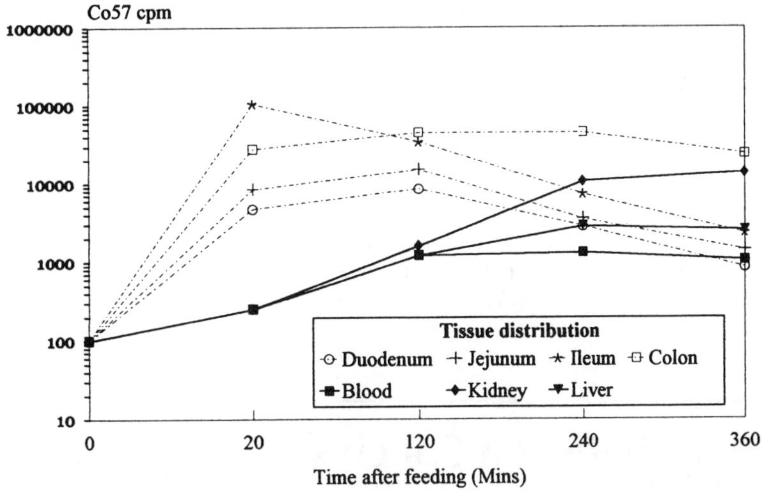

Figure 2. Time course of uptake of [^{57}Co]VB_{12} orally administered to mice. Female outbred Swiss mice (n = 5) received a 5-ng dose of [^{57}Co]VB_{12} administered in 0.1% VB_{12}-free BSA–saline. At various times after feeding, the mice were sacrificed and their tissues removed for counting. Intestinal sections containing their contents were counted without washing. Results are expressed as total [^{57}Co]VB_{12} counts per tissue including blood.

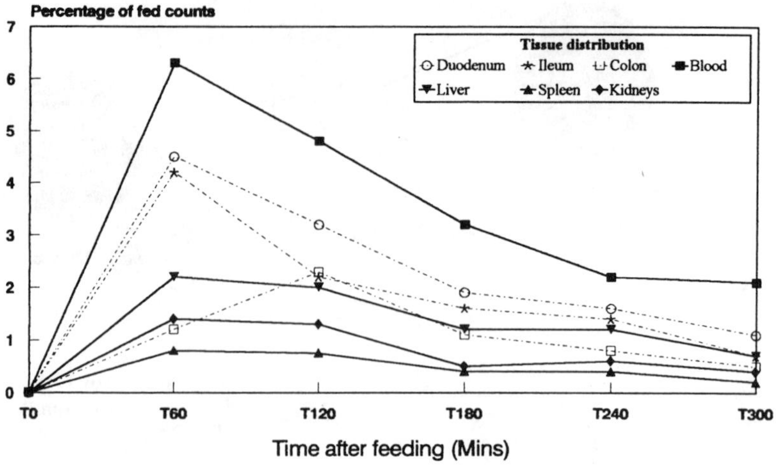

Figure 3. Time course of uptake of [^{125}I]BSA orally administered to mice. Female outbred Swiss mice received a 5-ng dose of [^{125}I]BSA administered in 0.1% VB_{12}-free BSA–saline. At various times after feeding, the mice were sacrificed and their tissues removed for counting. Results are expressed as percentage of fed ^{125}I counts per tissue.

Capacity of the Uptake System

The capacity of the VB_{12} uptake system varies from species to species. In humans, ~1 nmol of VB_{12} is taken up per oral dose (*4*), whereas in rats and mice, 20–30 pmol are taken up per feed (*5–7*). To achieve therapeutic levels of some peptide and protein pharmaceuticals in humans, repeated dosing (2–3 times per hour) over a 24-h period may be required. Experiments by the author and others (*6*) have shown that it is possible to repeat dosing this frequently; thus, a potential daily uptake in humans of 48 to 72 nmol per day can be achieved. This level of uptake is sufficient to achieve therapeutic concentrations of a number of peptide and protein pharmaceuticals, such as calcitonin analogues, vasopressin, oxytocin, LHRH agonists, and erythropoietin.

Variation between Species

Whereas the general method of uptake of VB_{12} is similar in different species, a number of significant differences in the site of release of IF and the site of uptake of the [VB_{12}–IF] complex between species exists. In the dog, the majority of IF is synthesized in the pancreatic duct epithelium (*8, 9*) and secreted into the bile, with subsequent release into the duodenum. In contrast, the site of synthesis of IF in the rat is not in the parietal cells of the fundus of the stomach, but rather in the chief cells (*10*). The rat also varies from the majority of animals in that the site of uptake of VB_{12} is not in the ileum, but rather in the jejunum (*11, 12*).

Methods for Conjugation of VB_{12} to Peptides and Proteins

Preparation of Analogues Suitable for Conjugation

Monocarboxylic Acid Derivatives of VB_{12}

Native VB_{12} does not contain any suitable functional groups for conjugation to peptides or proteins, so it must first be modified to provide such groups. The monocarboxylic acid derivative of VB_{12} can readily be prepared by hydrolyzing native VB_{12} in 0.4 M HCl at room temperature (*13*). Hydrolysis of the propionimide side chains of the A, B, and C rings of the VB_{12} molecule (*see* structure 1) results in the formation of the "b", "d", and "e" monocarboxylic acids of VB_{12}, respectively. The reaction is stopped by neutralization with NaOH, whereafter the material is desalted with a suitable molecular weight cut-off membrane in a Pollicon flat-bed dialysis machine. The desalted material is then passed down a Dowex 1X2 resin (Bio-Rad Labs, Richmond, CA), where the monocarboxylic acid isomers are separated by elution with 50 mM acetate buffer pH 5.5. Further purification can be achieved by reversed-phase HPLC (RP-HPLC) with a linear gradient of 5–80% acetonitrile in 0.1% trifluoroacetic acid. Eluted material can be lyophilized and stored indefinitely for use in conjugation.

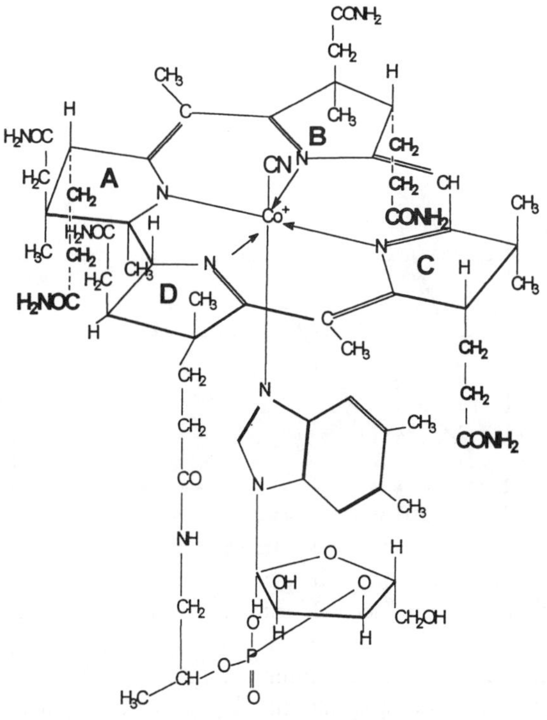

Structure 1. Vitamin B_{12}.

The monocarboxylic acid derivatives of VB_{12} can be directly conjugated to amino groups of the peptide or protein with a suitable carbodiimide (*3, 13*). If these groups do not exist on the peptide or protein or if modification of these groups leads to inactivation of the proteins, an alternative strategy is to make the amino derivative of VB_{12} by reaction of the monocarboxylic acid with a diamine (diaminoethane, diaminohexane, etc.) and then reacting the amino functionality with a carboxylate on the peptide or protein by carbodiimide chemistry.

Preparation of Axial Ligands of VB_{12}

A second method of conjugation of peptides to VB_{12} is by axial substitution of functional groups into the Co atom of the corrin ring. In this method, the axial CN ligand of VB_{12} can be replaced with a functionalized alkyl chain (*13*). This substituted functional group can then be used for conjugation to a peptide or protein using similar chemistry to that just outlined. One major disadvantage of this method, however, is that the resultant conjugate contains a light-sensitive Co—C bond. Thus, care must be taken not to expose solutions of the alkylcobalamins to visible light.

Affinity of IF for Various VB_{12} Isomers and Their Conjugates

The "b", "d", and "e" monocarboxylic acid isomers of VB_{12} vary greatly in their affinity for IF. The "e" acid has the greatest affinity for IF, followed by the "b" and "d" isomers (Figure 3). The affinity of the "e" isomer of VB_{12} for IF (Figure 4) can generally be maintained during information of [VB_{12}–drug] complexes; however, in some instances, the affinity is greatly reduced following conjugation. In this situation, the affinity can be restored by the

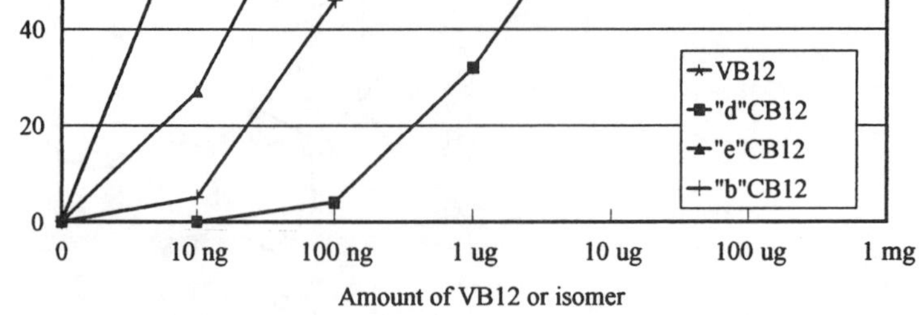

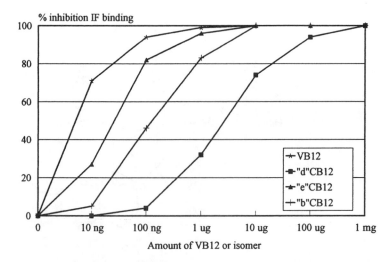

Figure 4. Relative binding of VB_{12} carboxylic acid isomers to porcine IF. Dilutions of VB_{12} or "b", "d," or "e" monocarboxylic acid isomers, dissolved in 0.1% VB_{12}-free BSA in phosphate-buffered saline (PBS), were mixed with a constant amount of [^{57}Co]VB_{12}. One unit of porcine IF (Sigma Chemical Company) was then added to the mixtures and incubated for 20 min. The [^{57}Co]VB_{12} not bound to IF was removed by addition of activated charcoal. Results are expressed as the percentage inhibition of [^{57}Co]VB_{12} binding to IF (N.B. one unit of IF binds 1 ng of VB_{12}).

Table I. Relative IF Affinity of VB_{12} Analogues

Spacer	*Relative Affinity, %[a]*
Diaminoethane	48
Diaminohexane	91
Diaminododecane	74
Adipyl hydrazine	44
Dithiopyridylethylamine	7
Dithiopyridylhexylamine	6
Dithiopyridyldodecylamine	11

[a] The affinity of "e" VB_{12} analogues was determined in the standard IF assay. Dilutions of VB_{12} or analogue were mixed with a constant amount of [^{57}Co] VB_{12} and added to a constant amount of IF. After 20 min, activated charcoal was added to the mixture, which was then centrifuged. The percentage binding was determined by the relative percentage of counts remaining in the supernatant. The relative affinity was determined from the concentration of analogue that could displace 50% of the [^{57}Co] VB_{12} relative to native VB_{12}.

use of suitable spacers during conjugation (*see* Table I). Thus, conjugation of "e" VB_{12} to diaminododecane increased the affinity of the resultant aminododecyl "e" VB_{12} to 75% of the native VB_{12}. This is in comparison to the 48% relative IF affinity seen with the diaminoethane derivative. The affinity of the "b" and "d" isomers for IF is generally greatly reduced during conjugation to ligands (Figure 5) and cannot be restored with spacers.

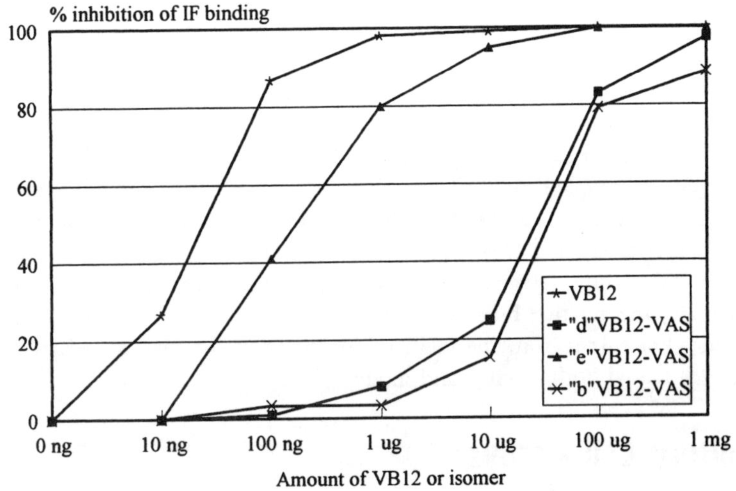

Figure 5. Relative binding of different vasopressin–VB_{12} carboxylic acid isomer conjugates to porcine IF. Arg_8-vasopressin was conjugated to the "b", "d", or "e" VB_{12} monocarboxylic acid isomers. Dilutions of the vasopressin conjugates to the "b", "d", or "e" VB_{12} monocarboxylic acid isomers, dissolved in 0.1% VB_{12}-free BSA in PBS, were mixed with a constant amount [^{57}Co]VB_{12}. One unit of IF was then added to the mixtures and incubated for 20 min. The [^{57}Co]VB_{12} not bound to IF was then removed by addition of activated charcoal. Results are expressed as the percentage inhibition of [^{57}Co]VB_{12} binding to IF.

Intrinsic Factor Assay

The affinity of various VB_{12} analogues and VB_{12} conjugates for IF can readily be determined by one of two assays. In the first, a competitive binding assay, dilutions of unlabeled VB_{12} or VB_{12} analogue or conjugate are mixed with a constant amount of [^{57}Co]VB_{12}. A constant amount of IF is then added to the mixture and allowed to incubate. Material that does not bind to the IF is removed by the addition of activated charcoal, followed by centrifugation. The relative number of counts in the supernatant is then used to determine the relative affinity of IF for the material tested.

In the second assay, a constant amount of the VB_{12} analogue or conjugate is mixed with increasing amounts of [^{57}Co]VB_{12}. Then, IF is added and the percent of counts bound to the IF is used to produce a Scatchard plot to determine absolute affinity of the analogue for IF.

Oral Delivery of an LHRH Analogue by VB_{12}: A Case Study

As stated previously, a number of hormones, such as estrogen and progesterone, are actively absorbed upon oral administration. However, many other hormones have little effect when given orally. Notable among these hormones is the peptide hormone LHRH or gonadotrophin releasing hormone (GnRH). This hormone is normally secreted by the anterior pituitary and is responsible for the control of release of luteinizing hormone (LH) and follicle stimulating hormone (FSH). Parenteral injections of LHRH have previously been shown to be effective in stimulating FSH and LH release; however, orally presented LHRH has little effect even at doses 100 times that given parenterally. Many studies have been performed on varying the sequence of LHRH, with the result that a number of agonists and antagonists have now been identified. One of the agonists, D-Lys_6-LHRH, contains a suitable functional group (viz. the ε-amino group of lysine) for conjugation to the "e" carboxylic acid isomer of VB_{12} ("e"VB_{12}) by carbodiimide chemistry.

D-Lys_6-LHRH was synthesized by us and purified by RP-HPLC. The purified analogue was coupled to "e"VB_{12} with EDAC [1-ethyl-3-(dimethylaminopropyl)carbodiimide], and the conjugated product was purified by chromatography using Sephadex G-25 in 10% acetic acid and then analyzed by RP-HPLC.

Before oral delivery of the [VB_{12}–D-Lys_6-LHRH] conjugate, or any other VB_{12} conjugate, the relative bioactivity of the conjugate must be compared with the starting peptide. For this reason, the relative bioactivity of the [VB_{12}–D-Lys_6-LHRH] conjugate was ascertained both in vitro and in vivo.

Rat Pituitary Cell Assays

The assessment of the relative LH releasing potential of LHRH analogues can readily be obtained by the addition of the analogues to rat pituitary cell cultures. Following addition of the analogues to the cultures, the amount of LH released can be quantitated by a suitable radioimmunoassay. As can be seen from the results in Table II, the [VB_{12}–D-Lys_6-LHRH] conjugate had a greatly reduced in vitro activity compared with the native analogue D-Lys_6-LHRH and significantly less activity than native LHRH.

Table II. Relative LH Releasing Activity of LHRH and Its Analogues in the Rat Pituitary Cell Culture

Analogue	Relative Activity[a]
LHRH	1.0
D-Lys_6-LHRH	6.0
VB_{12}: D-Lys_6-LHRH	0.264

[a] The relative stimulatory activity of LHRH (given a nominal value of 1.0), D-Lys_6-LHRH, or the VB_{12}: D-Lys_6-LHRH conjugate is given following additon of the LHRH or analogue to male rat pituitary cell cultures during 4 h according to the method of Farnworth et al. (29). The quantity of LHRH to elicit the release of a fixed amount of LH during a 4-h culture with LHRH is compared with the quantity of analogue or VB_{12} conjugate required for the same level of release.

Demonstration of Intravenous Potency in Rats

The data on in vitro potency of the [VB_{12}–D-Lys_6-LHRH] conjugate show that the analogue had <4% of the activity of the unconjugated agonist and was less active than native LHRH. Subsequent intravenous (iv) injection of the conjugate into rats showed the VB_{12} conjugate to be more active than the native LHRH, but still less active than the D-Lys_6-LHRH agonist (Figure 6).

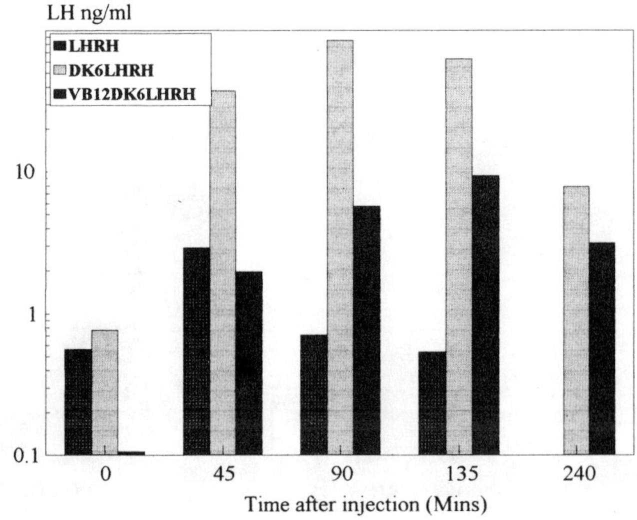

Figure 6. Stimulation of LH following iv injection of LHRH, DK_6-LHRH, or VB_{12}–D-Lys_6-LHRH conjugate. Male Wistar rats (n - 4) were injected iv with 1 μg of LHRH or DK_6-LHRH, or 2 μg of VB_{12}–D-Lys_6-LHRH conjugate ([VB_{12}–DK6LHRH]). Immediately prior to injection and at various times after injection, the rats were bled from the tail vein and sera collected for analysis of LH levels by RIA. Data is expressed as the mean level (n = 4) of LH measurable in serum at the various time points. Low but detectable levels of LH were seen in all rats at time 0.

Mouse Ovulation Studies

Relative Potency of D-Lys$_6$-LHRH and VB$_{12}$–D-Lys-LHRH Conjugates in Inducing Ovulation after Parenteral Administration to Mice

One of the major uses for LHRH and its agonists is in the treatment of secondary amenorrhoea. In this condition, the hypothalamus does not produce enough LHRH to stimulate the pituitary to release LH and FSH to cause ovulation. Addition of LHRH in a pulsatile fashion to women can often result in stimulation of LH and FSH secretion with subsequent ovulation. The addition of LHRH to pregnant mare serum gonadotrophin (PMSG)-primed mice has a similar effect in eliciting increased ovulation frequencies. This model of ovulation induction was used to test the true biological potency of D-Lys$_6$-LHRH and the [VB$_{12}$–D-Lys$_6$-LHRH] conjugate (*14*). Thus, D-Lys$_6$-LHRH and the [VB$_{12}$–D-Lys$_6$-LHRH] were given at various doses to PMSG-primed mice. The following day, the mice were killed and the ovulation frequency was determined.

A similar pattern of stimulation of ovulation was seen with both the D-Lys$_6$-LHRH and the [VB$_{12}$–D-Lys$_6$-LHRH] conjugate given iv (Figure 7). Maximal induction of ovulation occurred at a 10-ng dose of analogue itself or VB$_{12}$ conjugate. In other experiments (not shown) increased ovulation frequency could be demonstrated with doses of as little as 100 pg of the analogue. Doses of 100 ng or greater induced down-regulation of ovulation, as was expected (*15*).

Oral Delivery of D-Lys$_6$-LHRH by the VB$_{12}$ Uptake System

Having demonstrated that the [VB$_{12}$–D-Lys$_6$-LHRH] conjugate was active both in rats by inducing LH release and in mice by inducing ovulation following parenteral administration, the VB$_{12}$ conjugate was examined for its ability to stimulate ovulation in PMSG-primed mice following oral administration.

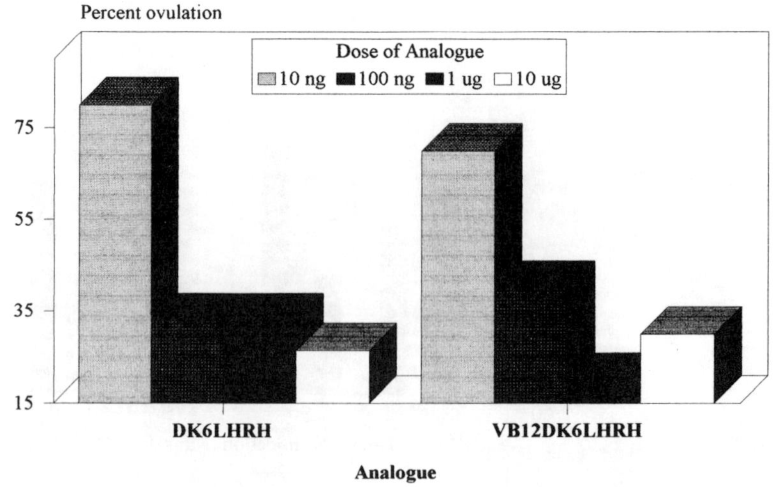

Figure 7. Stimulation of ovulation by iv administered LHRH, D-Lys$_6$-LHRH, or VB$_{12}$ conjugate. Female Swiss mice (n = 20) were injected intraperitoneally with 10 IU of PMSG. Then, 48 h later, the mice were injected iv with various doses of D-Lys$_6$-LHRH (DK6LHRH) or VB$_{12}$ (b) isomer–D-Lys$_6$-LHRH conjugate ([VB12–DK6LHRH]). After a further 24 h, the mice were sacrificed and examined for the presence of corpora hemorrhagica (CH). Mice with two or more CH were deemed to have ovulated.

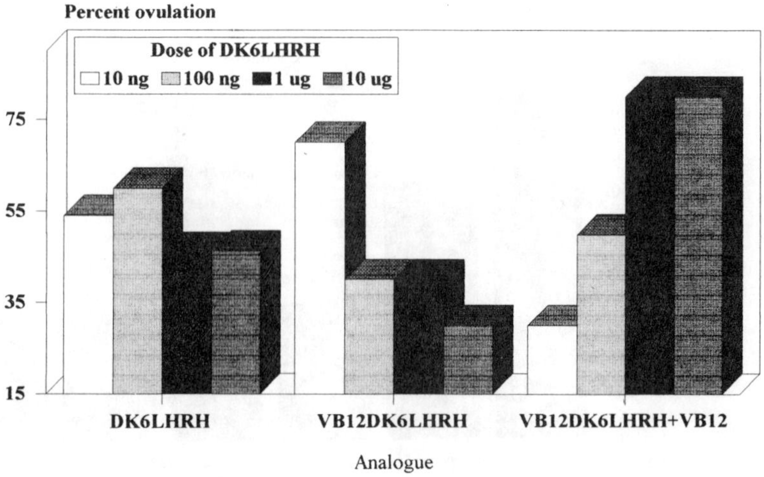

Figure 8. Stimulation of ovulation by orally administered LHRH, D-Lys$_6$-LHRH, or VB$_{12}$ conjugate. Female Swiss mice (n = 20) were injected intraperitoneally with 10 IU of PMSG. Then, 48 h later, the mice were fed with various doses of D-Lys$_6$-LHRH (DK6LHRH) or VB$_{12}$ (b) isomer conjugated to D-Lys$_6$-LHRH ([VB12–DK6LHRH]). A further group of mice received VB$_{12}$–D-Lys$_6$-LHRH that had been premixed with 25 μg of free VB$_{12}$ ([VB12–DK6LHRH] + VB12). After a further 24 h, the mice were sacrificed and examined for the presence of corpora hemorrhagica (CH). Mice with two or more CH were deemed to have ovulated. Control mice showed a spontaneous ovulation frequency of 35%.

The data show that the [VB$_{12}$–D-Lys$_6$-LHRH] conjugate elicited ovulation over a similar range when administered orally or iv (Figure 8). Thus, a 10-ng dose showed maximal stimulation of ovulation when given by either route; doses above this level resulted in down-regulation regardless of route. In contrast, when D-Lys$_6$-LHRH was given orally, it was much less effective than when administered iv: The moderate effect observed was presumably due to a low level of oral uptake combined with the high sensitivity of the system (ovulation having been induced by this LHRH analogue at doses as low as 100 pg).

Uptake of the [VB$_{12}$–D-Lys$_6$-LHRH] conjugate was mediated via the VB$_{12}$ uptake system, because cofeeding an excess of free VB$_{12}$ vastly increased the dose of conjugate required to produce ovulation. Again, the stimulation of ovulation produced by high doses of the [VB$_{12}$–D-Lys$_6$-LHRH] conjugate in the presence of excess VB$_{12}$ was presumably due to the 0.1–1% level of nonspecific uptake generally observed following the oral administration of polypeptides. Higher doses of the [VB$_{12}$–D-Lys$_6$-LHRH] conjugate given orally elicited down-regulation of ovulation.

Effect of Route of Administration on the Stimulation of Ovulation Achieved by D-Lys$_6$-LHRH or the [VB$_{12}$–D-Lys$_6$-LHRH] Conjugate

To compare the relative efficiency of VB$_{12}$ conjugates given iv or orally, a 10-ng dose of D-Lys$_6$-LHRH or VB$_{12}$ conjugate (weight of analogue) was administered to PMSG-primed mice. The oral or iv administration of a 10-ng dose of LHRH analogue given as a VB$_{12}$ conjugate stimulated the ovulation of ~2 follicles per primed mouse, which was significantly

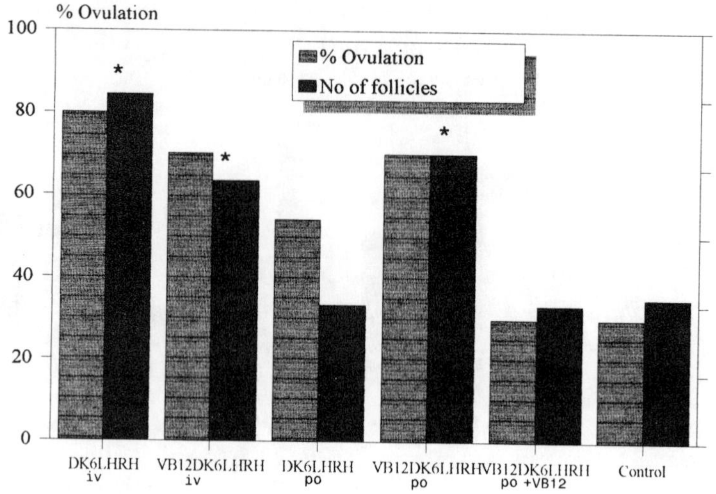

Figure 9. Stimulation of ovulation by iv or oral administration of D-Lys$_6$-LHRH or its VB_{12} isomer conjugate. PMSG-primed mice (n = 10) were administered a 10-ng dose of D-Lys$_6$-LHRH (DK6LHRH) or VB_{12}–D-Lys$_6$-LHRH conjugate ([VB12–DK6LHRH]) either by iv injection or by oral feeding (po). A further group received 10 ng of VB_{12}–D-Lys$_6$-LHRH conjugate mixed with 5 μg of free VB_{12} ([VB12–DK6LHRH]+ VB12). Twenty-four hours later, the mice were killed and corpora hemorrhagica (CH) were counted. Results are expressed as the percentage of mice with more than one CH, or as the mean number of CH per mouse.

higher ($p < 0.05$, student t test) than in control mice (Figure 9). This level of ovulation did not differ significantly from that observed following the iv administration of D-Lys$_6$-LHRH alone. Oral administration of the 10-ng dose of the analogue alone or the VB_{12}–D-Lys$_6$-LHRH conjugate administered in the presence of excess VB_{12} did not produce an ovulation frequency significantly above that observed in control (PMSG-primed) mice.

The data presented in the experiments just described confirm that the VB_{12} uptake system can deliver conjugates of D-Lys$_6$-LHRH to the circulation effectively following oral administration, and the conjugates stimulate ovulation in PMSG-primed mice when given orally. Furthermore, the data demonstrate that the uptake is a specific VB_{12}-mediated phenomenon, as the addition of excess VB_{12} was able to abolish the specific uptake mechanism.

Models of Uptake

Caco-2 Cell Cultures

Two cell lines have been isolated that have been demonstrated to bind, internalize, and transcytose VB_{12} in an IF-dependent fashion: the colon carcinoma cell line Caco-2 (*16–18*) and the HT29 human colon carcinoma cell lines (*19, 20*). The most widely studied of these cell lines is the Caco-2 cell line. This cell line has been shown to form a confluent monolayer that polarizes itself during growth to form apical and basal surfaces. Significant tight junction formation can also be demonstrated for these cells. The cells can be grown on 0.4- or 0.22-μm membranes, which allow the assessment of transport of VB_{12} from the apical to the

Table III. VB$_{12}$-Mediated Transport of a 20-kDa Protein in the Caco-2 (Clone DF2) Monolayer Culture

Conjugate	*Cross-Linking Agent*	*Reducible (Y/N)*	*Transport in 24 h (fmol)*
P20	—	N	13.5 ± 10
VB_{12}:P20-1	SPDP	Y	290 ± 70
VB_{12}:P20-2	EDAC/cystamine	Y	180 ± 90
VB_{12}:P20-3	EDAC/hydrazine	N	195 ± 65

NOTE: Monolayers were derived from the Caco-2 dilution clone DF2 cultured for 23 days on Falcon cell inserts. All inserts were pretested with [^{14}C]inulin for leakiness, and those that were intact were then tested for their ability to transport [^{57}Co]VB_{12}. Inserts were selected for their ability to transport >1 fmol of [^{57}Co]VB_{12}; these inserts were then randomized before being used to test for P20 transcytosis. P20 is a 20-kDa protein currently being studied at AMGEN research laboratories. The protein was conjugated to VB_{12} at Biotech Australia Party Ltd. The basolateral medium was analyzed for the presence of P20 after 24 h of incubation by an ELISA specific for P20 (*21*).

basolateral chamber. The Caco-2 cell line has been shown to bind, internalize, and transport VB_{12} bound to IF; however, the transport capacity for VB_{12} is low. Furthermore, these cells are not normal cells and do not possess full properties of adult enterocytes. Ramanujam et al. (*17*) have been able to measure 250 femptomoles (fmol) of IF-dependent [^{57}Co]VB_{12} transport in this cell line in 4 h. The transport capacity of this cell line is variable and depends on the source of the cells and the culture method. Thus, while Ramanujam et al. (*17*) were able to demonstrate 250 fmol of transport in 4 h of culture, other groups have only been able to demonstrate transport of 40 fmol [^{57}Co]VB_{12} in 18–21 h (*16, 18*), despite the fact that Dix et al. (*16*) could demonstrate the release of 600 fmol of VB_{12} binding material into the basolateral chamber of the cell culture at this time.

Habberfield et al. (*21*) have recently used the Caco-2 cell culture system to show that these cells are capable of transporting 180–300 fmol of a 20-kDa protein (P20) linked to VB_{12} ([VB_{12}–P20]) from the apical to basolateral chamber during 24 h of culture. Control wells in which the unconjugated protein was placed alone showed 14 fmol of transport (Table III).

Future Directions

Biodegradable Bonds

One of the major disadvantages of the VB_{12} uptake system is the need to form a covalent linkage between the pharmaceutical to be delivered and VB_{12}. This noncleavable covalent linkage can sometimes result in a dramatic loss in activity of the parent pharmaceutical, to the extent that the increased uptake seen with the conjugate is negated by the loss of bioactivity seen with the conjugate. For instance, covalent linkage of VB_{12} to vasopressin (either Arg_8-vasopressin, or Lys_8-vasopressin) leads to almost complete loss of bioactivity of the parent compound (activity, <1%). Just as obvious, although less dramatic, is the 20-fold loss of activity of VB_{12}–D-Lys_6-LHRH conjugates seen when these conjugates are administered iv to rats (Figure 3). (For some reason, a similar loss of activity was not seen in the mouse ovulation studies.)

The observed loss of activity just described is presumably the result of steric hindrance by the bulky VB_{12} group attached to small peptides such as vasopressin or LHRH. This problem is not as obvious in large protein molecules, but it would still be desirable to link the VB_{12} to the peptide to be delivered by a biodegradable bond that would be cleaved following transport from the intestine to the circulation and preferably release the native analogue. A limited number of biodegradable linkages present themselves. One with perhaps the most utility would be the conjugation of the VB_{12} to an exposed free thiol present on the peptide itself. This can readily be achieved by the formation of a dithiopyridyl derivative of VB_{12} with reagents such as *N*-succinimidyl 3-(2-pyridyldithio)propionate (SPDP) or long-chain *N*-succinimidyl 3-(2-pyridyldithio)propionate (LC-SPDP) (Pierce). The disulfide bond so formed would be cleavable by systemic glutathione and would regenerate the native peptide intact and unaltered following transport from the intestine into the circulation.

Conjugation of VB_{12} to Peptides with Thiol-Cleavable Linkages

One of the most potent antagonists of LHRH is the decapeptide ANTIDE (*22, 23*). The analogue itself does not possess any functional groups suitable for conjugation to VB_{12}, but the insertion of either D-Lys_6 or Lys_8 into the peptide results in the incorporation of an ε-amine on lysine that is suitable for conjugation. Direct conjugation to D-Lys_6 or Lys_8 residues in ANTIDE-1 or ANTIDE-2, respectively, greatly reduces the activity of each analogue both in vitro and in vivo. Extension of the spacer used for conjugation between VB_{12} and the two ANTIDE analogues increases the activity of the two conjugates when tested in vitro; however, their activity is still very low following in vivo injection into castrated rats. Conjugates of VB_{12} to D-Lys_6- or Lys_8-ANTIDE, formed with a thiol-cleavable linker, show increased activity both in vitro and in vivo to levels comparable to that of the native analogue ANTIDE (Tables IV and V) (*24*).

Amplification of the Uptake Capacity

One disadvantage of the VB_{12} uptake system is the limited capacity that it has for uptake of pharmaceuticals. In humans, there is a limited of 1 nmol per dose that can be taken up, per feed, whereas in mice, the uptake capacity is ~20 pmol per dose. This level of uptake would be sufficient to deliver molecules such as vasopressin, calcitonin, D-Lys_6-LHRH, or erythropoietin (EPO) but the uptake capacity is not sufficient for the delivery of pharmaceutical doses of peptides and proteins such as ANTIDE, insulin, granulocyte colony stimulating factor (G-CSF) or granulocyte monocyte colony stimulating factor (GM-CSF). For this reason it would be desirable to amplify the uptake system.

Amplification of the VB_{12} Uptake System with Polymers

Significant amplification of the VB_{12} uptake system could be achieved by the covalent linkage of peptides and VB_{12} to a polymer backbone, such as dextran or *N*-(2-hydroxypropyl)methacrylamide (HPMA). In this way the peptide–protein to be delivered can be linked to a multisubstituted polymer backbone, and a single VB_{12} molecule could then be linked to the polymer backbone. Using this technique a 20- or 100-fold amplification possibly could be achieved with molecules such as Dextran T70 (Sigma Chemical Co.) or HMPA, respectively, when these polymers are used as the polymer backbone. Similar systems have been proposed and tested by workers such as Kopecek and co-workers (*25–28*), who have linked reactive groups such as daunomycin to HPMA using peptide spacers cleavable in the lysosomal vacuoles. Experiments are currently being undertaken to determine whether

Table IV. In Vitro Bioassay for Blockade of LHRH-Stimulated Release of LH from Rat Pituitary Cells by ANTIDE, Its Analogues, and Various VB_{12} Conjugates

Analogue	IC_{50}
ANTIDE	4.9 ± 2.8
ANTIDE-1	2.5 ± 0.5
ANTIDE-2	1.2 ± 0.6
VB_{12}:ANTIDE-1 (direct)	88 ± 47
VB_{12}:ANTIDE-2 (direct)	36 ± 20
VB_{12}:EGS*:ANTIDE-1	8.9 ± 3.5
VB_{12}:EGS:ANTIDE-2	3.4 ± 2.4
VB_{12}:SS:ANTIDE-1	11.0 ± 4.0
VB_{12}:SS:NH^+:ANTIDE-2	7.4 ± 0.8

NOTE: Median inhibitory concentrations of ANTIDE, ANTIDE analogues, and various VB_{12}–ANTIDE conjugates for antagonizing GnRH-stimulated release of LH from rat anterior pituitary cell cultures during 4 h of incubation. Results for conjugates are given as nanogram of incorporated analogue per milliliter (final concentation in culture well; see Farnworth et. al., *29*). Conjugates between VB_{12} and ANTIDE analogues were formed by direct conjugation to carboxylic acid isomer of VB_{12} with EDAC (direct), by the use of the hydrophilic spacer ethylene glycol*bis*(succinimidylsuccinate) (EGS*), or via a thiol cleavable spacer (SS) (*24*).

Table V. Reduction in Serum LH Levels in Castrated Rats Following Injection of ANTIDE, ANTIDE Analogues, or Various VB_{12} Conjugates

	Mean LH ± Std (ng/mL)		
Analogue	*100 μg*	*25 μg*	*6.25 μg*
Control	6.40 ± 1.87		
ANTIDE	0.29 ± 0.06	0.29 ± 0.05	4.16 ± 2.26
ANTIDE-1	6.37 ± 1.00		
ANTIDE-2	0.27 ± 0.05	2.22 ± 0.66	5.99 ± 2.19
VB_{12}:ANTIDE-1 (direct)	6.37 ± 1.00		
VB_{12}:ANTIDE-2 (direct)	5.81 ± 0.22		
VB_{12}:EGS:ANTIDE-1	8.40 ± 3.00		
VB_{12}:EGS:ANTIDE-2	4.33 ± 1.41		
VB_{12}:SS:ANTIDE-1		0.94 ± 0.40	
VB_{12}:SS:ANTIDE-2		2.33 ± 0.68	

NOTE: The in vivo effect of subcutaneous injection of ANTIDE, ANTIDE analogues, or VB_{12}–ANTIDE conjugates was examined in 6-week-old castrated rats. The rats were killed and trunk blood was collected for analysis of serum LH levels by RIA 24 h after injection of the analogues; *see* reference 30. The VB_{12} conjugates were formed as described in Table IV (*24*).

the [VB_{12}–polymer–drug] complex can be taken up and transported via the IF–VB_{12} transport system and subsequently can release the peptide in a pharmacologically active form.

Microspheres

Whereas the use of polymer backbones provides the potential to amplify the VB_{12} uptake system some 20–100-fold, it does not protect the peptide–protein from proteolysis in the intestine. Furthermore, some peptides, such as vasopressin, cannot be modified for conjugation to the polymer backbone without loss of activity. Other large molecular weight proteins also are unsuitable for conjugation, as their sheer bulk prevents sufficient numbers of molecules from being conjugated to the polymer for amplification. One technology that would overcome both these limitations is the incorporation of the protein to be delivered into biodegradable microspheres. The microspheres could in turn be linked to VB_{12}. The whole complex could then be administered orally and be taken up by IF-mediated endocytosis. Once the VB_{12} microspheres had entered the circulation, the microspheres would have to degrade rapidly to release their internal pharmaceutical load. One advantage of this technology would be that the molecule released would not have been chemically linked to the VB_{12} at any stage and so should retain full bioactivity and should be no more immunogenic than the natural pharmaceutical. Currently, the size of particle that can be internalized and transported by the VB_{12} system is unknown.

Summary

One major obstacle to the oral delivery of peptide and protein pharmaceuticals is their inability to cross the intestinal mucosa once they have been administered to the intestine. It is possible to overcome this obstacle by covalently linking these pharmaceuticals to VB_{12}. The natural mechanism of VB_{12} uptake can then be used to transport these molecules from the lumen of the intestine to the circulation. The importance of these findings lie in the potential to use the VB_{12} uptake mechanism as a specific carrier mechanism to deliver highly potent hormones, peptides, and protein pharmaceuticals via the oral route of administration rather than via parenteral injection, which is costly and inconvenient. A number of approaches with polymers and microspheres are currently being developed in an attempt to overcome the limited capacity of the VB_{12} uptake system.

Acknowledgments

Thanks is given to J. Kanellos, S. W. Westwood, and H. de Aizpurua, and to A. Jarvis and A. Phillips for their contribution to the preparation of conjugates, data, and affinities reproduced in this manuscript. Acknowledgment is also given to J. Findlay, P. Farnworth, and H. Burger (Prince Henry's Institute of Medical Research, Melbourne, Australia) for their analysis of LHRH analogues in the pituitary cell assay and for the analysis of ANTIDE analogues in the castrated rat model. The work on ANTIDE was supported in part by a grant from the National Health and Medical Research Council of Australia, Grant No. 920244.

References

1. Allen, R. H.; Majerus, P. W. *J. Biol. Chem.* **1972,** *247,* 7702–7708.
2. Allen, R. H.; Majerus, P. W. *J. Biol. Chem.* **1972,** *247,* 7709–7717.
3. de Aizpurua, H. J.; Burge, G. L.; Howe, P. A.; Russell-Jones, G. J. "Oral Delivery Systems"; International Patent Application PCT/AU86/0299.
4. Chanarian, J. *The Megaloblastic Anaemias;* Blackwell Scientific Publication; Oxford, England, 1969.
5. Russell-Jones, G. J. Biotech Australia Pty. Ltd., unpublished results.
6. Roberton, J. A.; Gallagher, N. D. *Gastroenterology* **1985,** *88,* 908–912.
7. Gallagher, N. D.; Foley, K. *Gastroenterology* **1971,** *61,* 332–338.
8. Vaillant, C.; Horadagoda, N. U.; Batt, R. M. *Cell Tissue Res.* **1990,** *260,* 117–122.
9. Simpson, K. W.; Alpers, D. H.; de Wilde, J.; Swanson, P.; Farmer S.; Sherding, R. G. *Am. J. Physiol.* **1993,** *265,* G179–G188.
10. Hoedmaiker, P. J.; Abels, J.; Wachters, J. J.; Arends, A.; Nieweg, H. O. *Lab. Invest.* **1964,** *12,* 1394.
11. Booth, C. C.; Chanarin, I.; Anderson, B. B.; Mollin, D. L. *Br. J. Haematol.* **1957,** *3,* 253–261.
12. Okuda, K. *Am. J. Physiol.* **1960,** *199,* 84–90.
13. Schneider, Z. In *Comprehensive B_{12};* Schneider, Z.; Stroinski, A., Eds.; Walter de Gruyter & Co.: Berlin, Germany, 1987; pp 17–43.
14. Russell-Jones, G. J.; de Aizpurua, H. J. *Proc. Int. Symp. Control. Rel. Bioact. Mater.* **1988,** *15,* 142–143.
15. Heber, D.; Bhasin, S.; Steiner, B.; Swerdloff, R. S. *J. Clin. Endocrinol.* **1984,** *58,* 1084–1088.
16. Dix, D. J.; Hassan, I. F.; Obray, H. Y.; Shah, R.; Wilson, G. *Gastroenterology* **1990,** *88,* 1272–1279.
17. Ramanujam, K. S.; Seetharam, S.; Ramassamy, M.; Seetharam, B. *Am. J. Physiol.* **1991,** *260,* 6416–6422.
18. Wilson, G.; Hassan, I. F.; Dix, C. J.; Williamson, I.; Shah, R.; Mackay, M.; Artusson, P. *J. Controlled Rel.* **1990,** *11,* 25–40.
19. Guéant, J.-L.; Masson, D.; Schohn, J.; Girr, M.; Saunier, M.; Nicolas, J. P. *FEBS Lett.* **1992,** *297,* 229–232.
20. Schohn, H.; Guéant, J.-L.; Girr, M.; Nexø, E.; Baricault, L.; Zweibaum, A.; Nicolas, J.-P. *Biochem. J.* **1991,** *280,* 427–430.
21. Habberfield, A., AMGEN, 1000 Oaks, CA; Westwood, S. W.; Russell-Jones, G. J., Biotech Australia Pty. Ltd., Roseville, Australia, personal communications, 1992.
22. Srivastava, R. K.; Sridaran, R. *Contraception* **1991,** *42,* 309–316.
23. Leal, J. A.; Gordon, K.; Williams, R. F.; Danforth, D. R.; Roh, S. I.; Hodgen, G. D. *Contraception* **1989,** *40,* 623–633.
24. Russell-Jones, G. J.; McInerny, B.; Findlay, J.; Farnworth, P.; Burger, H., 1993, Biotech Australia P/L, Roseville, Australia; Prince Henry's Institute of Medical Research, Melbourne, Australia, unpublished results.
25. Ulbrich, K.; Strohaml, J.; Kopecek, J. *Biomaterials* **1983,** *3,* 150–154.
26. Solovskij, M. V.; Ulbrich, K.; Kopecek, J. *Biomaterials* **1983,** *4,* 44–48.
27. Cartlidge, S. A.; Duncan, R.; Lloyd, J. B.; Kopecková-Rejomanová, P.; Kopecek, J. *J. Controlled Rel.* **1987,** *4,* 254–278.

28. Duncan, R.; Yardley, H. J.; Ulbrich, K.; Kopecek, J. *Proc. Int. Symp. Controlled Rel. Bioact. Mater.* **1989,** *16,* 140–141.

29. Farnworth, P. G.; Robertson, D. M.; de Krester, D. M.; Burger, H. G. *J. Endocrinol.* **1988,** *119,* 233–241.

30. Puénte, M.; Catt, K. J. *Steroid Biochem.* **1986,** *25,* 917–925.

RECEIVED for review July 12, 1993. ACCEPTED revised manuscript December 21, 1993.

CHAPTER 9

Prodrug Approaches for Improving Peptidomimetic Drug Absorption

Barbra H. Stewart and Michael D. Taylor*

The potent and selective biological properties of many naturally occurring peptides make them an irresistible target for drug design (*1*). However, oral drug delivery of peptide and protein drugs remains a major obstacle to their successful development (*2, 3*). By starting from a peptide lead, medicinal chemists have approached the challenges of peptide-based drug design and delivery principally by a process of gradually and sequentially reducing the peptide character of each generation of derivatives. This approach eventually leads to "peptidomimetics" that retain the biological properties of the lead peptide with much less peptide-like structure. The first step in this process is often dissecting the peptide lead to identify a minimum fragment that retains activity. Next, peptides are designed that contain modifications, such as unnatural amino acids, peptide-bond replacements, terminal modifications, or conformational restrictions. Subsequently, a larger number of small modifications or installation of large nonpeptide fragments may be introduced. After several generations, the resulting compounds may have little in common with peptides by way of structure but still will mimic selected pharmacological attributes of the naturally occurring peptide leads.

* Corresponding author.

This process of rational design, while certainly nontrivial, is usually straightforward (*4*). In contrast, *de novo* design of relatively small nonpeptide ligands of peptide receptors or enzyme inhibitors remains extremely challenging. Indeed, the discovery of nonpeptide ligands of peptide receptors has been largely the result of empirical screening programs (*5*). This "brute force" approach has been revitalized in recent years with the advent of modern automated high-volume screening technology.

Peptidomimetics may be considered as intermediate in their degree of peptide structure between modified peptides and nonpeptides. Beyond such a general characterization, they are not easily classified, because they include a broad range of structural types that have varying degrees of peptide character. Even though they contain mostly nonpeptide fragments, peptidomimetics often incorporate peptide (amide) bonds, natural or unnatural amino acids, and functional group side chains that create a peptide-like motif. A common target for modification is the peptide bond. Peptide bond isosteres have been used extensively in both modified peptides and peptidomimetics (*6*). Such substitutions will stabilize the peptidomimetic compound to proteolysis and modify the polarity, hydrogen bonding ability, and conformation of the backbone. Peptide bond isosteres may consist of changes in oxidation state (CH_2NH), substitution by hydroxyl-containing methylene chains ($CHOHCH_2$), or involve substantial changes such as heterocyclic rings (*7*).

Incorporation of unnatural amino acids may also provide metabolic stability. The use of monocyclic or bicyclic structures as rigid frameworks has been a particularly successful strategy that is common to both naturally occurring peptidomimetics such as penems or asperlicin, and synthetic compounds such as angiotensin-containing enzyme (ACE) inhibitors. Large ring structures formed by cyclization of the backbone or connecting side chains of the peptide may also provide a degree of conformational constraint that may lead to improved potency or stability, especially to proteinases (*8, 9*), or may be designed specifically to mimic peptide turn structures (*10*).

Some investigators employ a functional rather than structural definition and would include nonpeptide compounds that interact with peptide receptors under the peptidomimetic banner as well. For example, Morgan and Gainor (*11*) define peptidomimetics as "structures which serve as appropriate substitutes for peptides in interactions with receptors and enzymes."

Several classes of peptidomimetic drugs have been the focus of intense drug discovery research. Among these are ACE inhibitors (*12*), renin inhibitors (*13*), cholecystokinin antagonists (*14*), and human immunodeficiency virus (HIV) protease inhibitors (*15*). The objective of this chapter is to briefly review the drug-delivery obstacles of peptidomimetics and describe selected approaches for improving delivery properties. We will emphasize prodrug strategies that were used to improve the intestinal absorption of peptidomimetic drugs.

Delivery Problems of Peptidomimetic Drugs

The challenge of peptide-based drug design is twofold: pharmacological–structural and biopharmaceutical. Medicinal and peptide chemists have had great success in rapidly designing potent and selective modified peptides based on naturally occurring peptide leads (*16*), but these compounds rarely have good oral activity. Drugs that can be delivered successfully by oral administration must overcome a series of physical and enzymatic barriers not presented to drugs given by a parenteral route. These barriers include dissolution from a solid form into aqueous media, extremes of pH that may cause hydrolysis, pancreatic and intestinal proteolytic enzymes, absorption of the drug from the gastrointestinal tract, and "first-pass" extraction by the liver. Peptidomimetic drugs may possess advantages relative to peptides and proteins in stability or size, for example, that improve oral activity. Nonetheless, because they mimic the structural attributes of peptides, these agents also share many of the physicochemical properties of peptides that result in poor absorption (*17*), rapid clearance (*18, 19*), or metabolism. These properties occur as a gauntlet of absorption barriers, proteinases, and clearance mechanisms that exist to control the production, degradation, and elimination of peptides and proteins in organisms.

Poor stability in the gastrointestinal tract (GIT) has been addressed successfully by chemical modification of peptide-based drugs. Instability in the GIT generally results from a susceptability to hydrolysis of peptide or amide bonds of the backbone, rather than oxidation or conjugation of functional groups. Therefore, relevant modifications focus on the amide bond itself or sites adjacent to it that may affect enzyme affinity (*see* Chapter 17 for a more detailed treatment of this topic). Amide N-alkylation, peptide-bond replacement with a suitable isostere, or substitution with D-amino acids are among the commonly employed strategies. Synthetic modifications can increase chemical stability and protect against hydrolysis by pancreatic enzymes and peptidases of the intestinal brush-border membrane (*20*). Even after absorption, stability of peptides in plasma and serum must be controlled (*21*).

Other obstacles are less easily resolved without profoundly altering physicochemical properties and, consequently, biological activity. The interplay of polarity, lipophilicity, and molecular weight on dissolution of the dosage form and biological transport may pose several rate limitations to effective drug absorption from the intestine into the systemic circulation. Slow intestinal absorption, rather than instability in the GIT, was shown to be a factor in the low bioavailability of a cyclic hexapeptide of somatostatin (*22*).

Poor aqueous solubility, which results in a low lumenal drug concentration, reduces the driving force for partitioning of drug from gastrointestinal lumen into lipid membranes. This effect is exacerbated when poor water solubility is combined with a slow dissolution rate. Although much justifiable emphasis is placed on the drug partition coefficient, the solubility of drug in water and in lipid must be sufficient as well. Lumenal drug concentration may be increased by formulation approaches, such as the use of cosolvents or complexing agents, but these approaches are not straightforward. Because many peptidomimetic drugs are poorly soluble in water, enhancing aqueous solubility has been a common objective in the quest to improve oral activity. HIV protease inhibitors are a recent and typical example. The first examples of highly potent and selective agents were poorly soluble except at low pH, where aqueous solubility sometimes improved dramatically. As a result, drug design efforts were initiated in several laboratories to increase aqueous solubility at physiological pH starting from agents that were potent and selective, but poorly soluble and inactive in vivo. An example is the symmetric inhibitors (*23*). Multiple structural modifications were required to improve aqueous solubility from the prototypes so that concentrations exceeded 1 μg/mL. Although aqueous solubility could be improved while retaining enzyme inhibitor potency with certain modifications in structure, other modifications resulted in a reduction in activity. Overall, aqueous solubility was not strongly correlated with bioavailability for these HIV protease inhibitors, a finding that suggests that aqueous stability is but one of several factors. Renin inhibitors with improved aqueous solubility have been designed, but like the HIV protease inhibitors, no direct relationship between increased water solubility and oral bioavailability was found (*24–26*).

The partition coefficient, determined most often between water and octanol, is generally considered to be predictive of the tendency of simple compounds to associate with a lipid membrane. Peptidomimetic molecules are often relatively large molecules composed of diverse components similar to amino acids and capable of more complex interactions with solvent and membrane than simpler molecules. Even though the partition coefficient is still a useful measure of lipophilicity, its predictability for partitioning into membranes is less direct. Partition coefficients on the order of 100–1000 ($\log P = 2–3$) are required for good intestinal absorption. Compounds with partition coefficients greater than 1000 may be limited by poor aqueous solubility. We studied (*27*) a series of renin inhibitors and structurally related compounds in a rat-intestinal perfusion model and found intestinal permeability from solution increased with increasing lipophilicity up to a $\log P$ of approximately 4. Increasing lipophilicity is generally associated with increasing extent of biliary clearance, but only within a series of compounds of similar structure. Structural differences between series may overwhelm physicochemical determinants of clearance and may suggest a complex relationship between physicochemical properties and hepatic uptake, metabolism, and secretion into bile (*28*).

Like peptides, peptidomimetic drugs can form multiple hydrogen bonds with the aqueous environment. Breaking these drug–solvent bonds is thermodynamically unfavorable but is required for partitioning the drug into the membrane. The extent of solvation to water via hydrogen-bond donating and acceptor groups is a major determinant of intestinal absorption of model peptides and analogues (*29, 30*). Permethylation of hydrogen-bonding sites lowers desolvation energy without major effects on lipophilicity and improves absorption at a given concentration. Similar modifications increase the permeability of small peptide molecules across Caco-2 cells (*31*). However, structural modifications that decrease desolvation may also decrease aqueous solubility. The benefit of increasing membrane permeability through lowering desolvation energy must be weighed against the resultant decrease in aqueous solubility as a driving force for absorption.

The large size of many biologically active peptides and proteins often is cited as a key determinant of their poor pharmacokinetics. The impact of increasing molecular size on passive membrane transport is manifest primarily in lowered drug diffusivity (*32*). Yet even peptides as small as tripeptides may be absorbed to the extent of only a few percent, despite relatively good stability to metabolic enzymes (*33, 34*). Because peptidomimetics are usually designed to be smaller than the peptides they mimic, membrane transport properties are often improved relative to the peptides. However, size is no less important for peptidomimetics in determining absorption. In one study, in situ intestinal permeability decreased with increased molecular weight for a series of renin inhibitors (*27*).

Carrier proteins for amino acids and small peptides are present in the small intestine and are essential for the absorption of dietary protein (*35*). See Chapters 1 and 5–8 for detailed treatment of carriers and their use in drug absorption. Peptidomimetic drugs of low molecular weight may be substrates for these carrier proteins. These transporters are attractive targets for enhancing absorption of charged or polar compounds for which passive transcellular permeation is low or when size exceeds the limits of the paracellular pathway. A number of ACE inhibitors and β-lactam antibiotics are substrates for the peptide carrier. However, with few exceptions, such as specific renin inhibitors that are transported by the peptide transporter (*36*), other peptidomimetic drugs are not substrates for these carriers. Thus, most peptidomimetic drugs are restricted to absorption by passive processes.

Even when compounds are well-absorbed, rapid clearance or metabolism may be limiting. Proteolysis of peptide (amide) bonds may be involved, although often metabolism due to proteolysis can be eliminated through the modification of susceptible peptide bonds. However, other clearance mechanisms such as hepatic extraction and excretion into bile can overwhelm even the most efficiently absorbed compound. Hepatic extraction has been a particularly acute problem for renin inhibitors, and biliary excretion has been implicated in short duration of action of the inhibitors (*37, 38*). Other tissues also efficiently extract compounds of this class (*39*). Chemical approaches to reducing clearance include glycosylation, which in the case of renin inhibitors appears to decrease hepatic extraction (*40*).

Therefore, rate and extent of absorption, and to a lesser degree clearance, are driven by lipophilicity, aqueous solubility, solvation, and molecular weight. The optimization of these parameters for drug delivery while maintaining a high degree of intrinsic biological activity has been shown in many instances to be a herculean task. Certain classes of peptidomimetic drugs are extremely resistant to design strategies that attempt to enhance oral activity while retaining intrinsic biological activity. Large, resource-intensive efforts that combine drug design, synthesis, and in vivo screening may be necessary to identify peptidomimetic drugs with such a profile (*41, 42*).

This high degree of difficulty prompts many researchers to seek approaches that target the delivery problem specifically and do not compromise the intrinsic biological activity of the drug candidate. Prodrug strategies constitute one approach that has proven successful in a number of cases.

Prodrug Approaches to Enhancing Absorption of Peptidomimetics

Design Considerations

Peptidomimetics include a diversity of structures. As a result, a variety of prodrug approaches can be employed depending on the particular chemical and delivery requirements for the drug of interest. Because both peptide and nonpeptide fragments are usually incorporated into peptidomimetic structures, prodrug strategies may be applied that were first developed for either nonpeptide or peptide drugs. Prodrugs of peptides and proteins were used to overcome physicochemical limitations (*43*) or to limit the metabolism of peptide derivatives by proteinases (*44*) (*see* Chapter 18). Prodrugs of amino acids and their analogues are another class that may yield prodrug strategies that are directly applicable to peptidomimetics.

For peptidomimetic drugs, most reported prodrug applications have used traditional methods of increasing lipophilicity or aqueous solubility through bioreversible functional-group modification, such as esterification of carboxylic acid groups. Ester prodrugs have been one of the most widely used and successful approaches to reversibly modifying the physicochemical properties of drugs to improve delivery (*45*). The approach has several advantages. The synthesis of esters is straightforward and many different functionalities can be introduced easily. Esters are relatively stable to aqueous solution at physiological pH. The rate of hydrolysis, either chemical or enzymatic, can usually be controlled by appropriate substitution of the ester function. Esters are substantially more lipophilic than the precursor acids, a characteristic that often results in an increased membrane permeability (*46*). Esterases are ubiquitous in biological systems and provide readily accessible enzyme targets for reconversion of the prodrugs following administration.

Carboxylic acids also may be converted to amide prodrugs. Amines, particularly at the N-terminus, can be acylated to form amides. The use of amide prodrugs is hampered by their greater stability relative to esters (*47*). Disulfide bonds linking a drug that contains thiols to a prodrug auxilliary is another successful strategy. This strategy uses endogenous disulfide reductases to unmask the active moiety (*48*).

The following examples illustrate a variety of approaches and chemical designs for applying prodrug strategies to peptidomimetic drugs and closely related compounds.

Examples

Angiotensin-Converting Enzyme Inhibitors

The era of ACE inhibitor therapy in the treatment of hypertension arrived with the discovery of captopril **1** (*49*), a sulfhydryl analogue of L-Ala-L-Pro. Highly efficacious and producing a lower incidence of side effects than other agents, ACE inhibitors rapidly became first-line therapy for hypertension as well as a preferred treatment for congestive heart failure (*50*). Replacement of the sulfhydryl group and prolonging the duration of action were the main objectives for second-generation ACE inhibitors. These objectives were realized in enalaprilat **2** several years after the introduction of captopril (*51*). The oral absorption of the free acid, enalaprilat, was less than 5%; however, the fraction absorped of the ethyl ester prodrug, enalapril **3**, ranged from 50 to 73% (*52*). Reconversion of enalapril to its parent is catalyzed by nonspecific esterases in the liver and result in a bioavailability of 36–44% as enalaprilat (*53*). This large increase in oral absorption is somewhat the result of the higher lipophilicity of the prodrug and greater degree of membrane partitioning. More important, enalapril, but not enalaprilat, is a substrate for the peptide transporter in the small intestine (*54*). The improved oral bioavailability of enalapril is due to greatly enhanced absorption because the hepatic extraction of the prodrug is 16-fold greater than enalaprilat (*55*).

Of 14 new ACE inhibitors in development in the late 1980s, 12 were prodrugs (*50*), an acknowledgment of the generally improved absorption characteristics of the prodrug form. In all cases, the prodrug forms are more lipophilic than the parent compounds (*56*). For some compounds, this greater lipophilicity is directly responsible for improved absorption. One example is fosinopril **4**, where passive diffusion rather than carrier-mediated transport is the principal absorption pathway. The fosinopril ester structure uses the double ester concept, which will be discussed in detail in the section on antibiotic prodrugs.

As explained by Friedman and Amidon (*54*), the diacid class of ACE inhibitor prodrugs exemplifies how the prodrug approach can address the conflicting structural requirements of absorption and biological activity. In this class, the two acid moieties are essential for good enzyme-inhibitor activity, but absorption by carrier-mediated processes or passive diffusion is poor. Monoesterification reduces enzyme-inhibitory activity but greatly improves carrier-mediated intestinal absorption for agents such as enalapril and fosinopril. Other carboxyalkyl ACE inhibitors such as quinapril **5** (*57*) and perindopril **6** (*58*) are also absorbed by carrier-mediated processes.

The transporter mechanism is a particularly important target when the drug or prodrug is polar or charged and the passive transcellular absorption would be negligible. Although the peptide transporters implicated in these processes are specific protein receptors in the drug design context, the variety of structures that are substrates suggests that it is a versatile target. Design of drugs or prodrugs that can be transported by such carrier systems is a potentially powerful approach to improving the absorption of the small to moderately sized peptide and peptidomimetic drugs.

Neutral Endopeptidase and Aminopeptidase N Inhibitors

Neutral endopeptidase (NEP) is also known as neutral metalloendopeptidase, endopeptidase-24.11, enkephalinase, and kidney brush-border neutral proteinase. This enzyme is responsible for the hydrolysis of atrial natriuretic factor (ANF) and other peptides such as

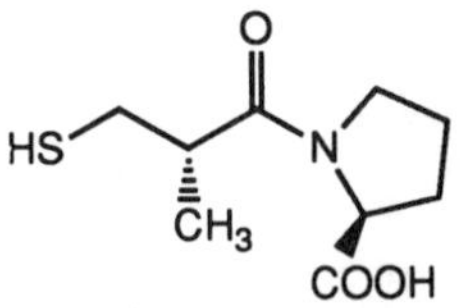

1

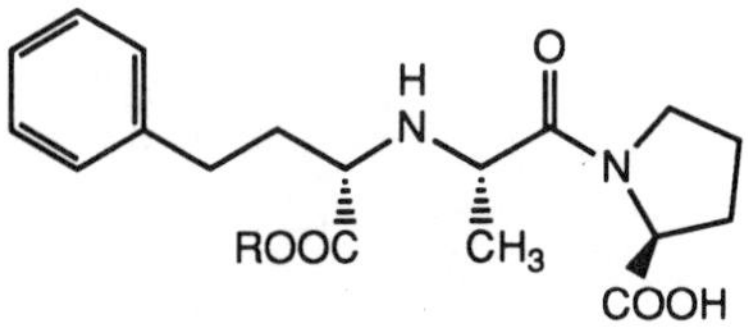

2 R = H
3 R = Et

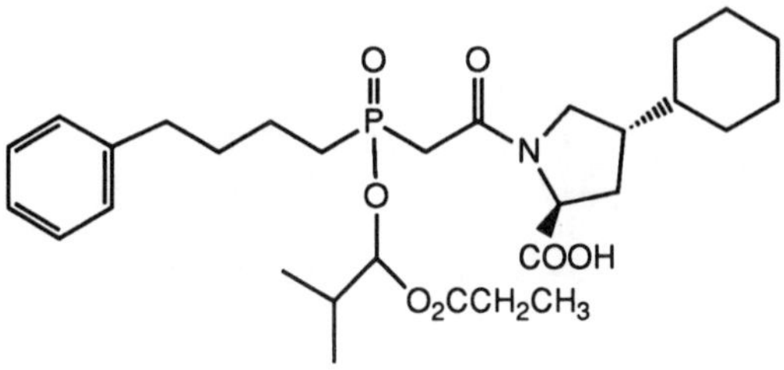

4

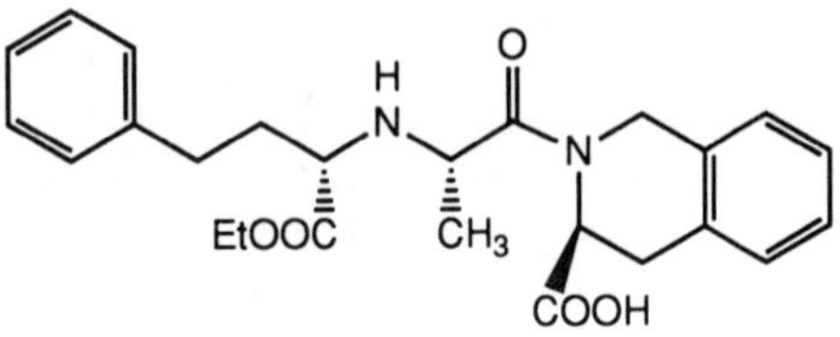

5

cholecystokinin, substance P, neurotensin, angiotensin, and oxytocin (*59*). The combined action of NEP and aminopeptidase N (AP-N) is responsible for the degradation of endogenous enkephalin. Specific NEP inhibitors are highly desirable for a number of therapeutic indications such as analgesia (potentiation of endogenous or administered enkephalin) and hypertension and heart failure (stabilization of circulating ANF). NEP inhibitors are similar to ACE inhibitors in that both target enzymes are metalloproteinases. As a result, NEP inhibitors resemble ACE inhibitors in their structure, and similar approaches have been used to improve their delivery properties.

Thiorphan **7** is the prototypic NEP inhibitor with nanomolar affinity in vitro and good in vivo activity after intracerebroventricular injection (*60*). Short duration of action, limited selectivity, and lack of oral activity are impediments to its development as a therapeutic agent. Acetorphan **8** is a bifunctionalized prodrug of thiorphan in which both carboxylic acid and mercaptan groups are masked (*59*). These modifications greatly increase lipophilicity relative to the parent drug and improve membrane transport. No reports have suggested that acetorphan is active after oral administration; however, Gros et al. (*61*) reported protection of ANF after oral dosing of acetorphan to human volunteers as shown by increased levels of circulating ANF, diuresis, and natriuresis. Most NEP inhibitors are diacids, and like ACE inhibitors, their monoesters such as candoxatril (**9**; UK 79,300) (*62*) and SCH 34826 **10** (*63*) show improved oral activity (*64*). Specific NEP inhibitors or their prodrugs have not been reported to be substrates for carrier-mediated transport.

The degradation of enkephalin is the result of the combined actions of NEP and AP-N; therefore, interest is keen for development of a mixed inhibitor. Kelatorphan **11** is a bidentate, hydroxamate-containing, multiple-enzyme inhibitor with nanomolar binding to NEP and dipeptidylaminopeptidase but only micromolar affinity for AP-N in vitro (*65*). Fournie-Zaluski et al. (*65*) demonstrated that kelatorphan produced stronger antinociceptive effects than those induced by administration of NEP or AP-N inhibitors alone. An improvement in AP-N affinity was achieved by Xie et al. (*66*) in RB 38A **12**. This compound is also hydroxamate-containing and has the same disadvantages as kelatorphan of high hydrophilicity and limited penetration through the blood–brain barrier. These disadvantages

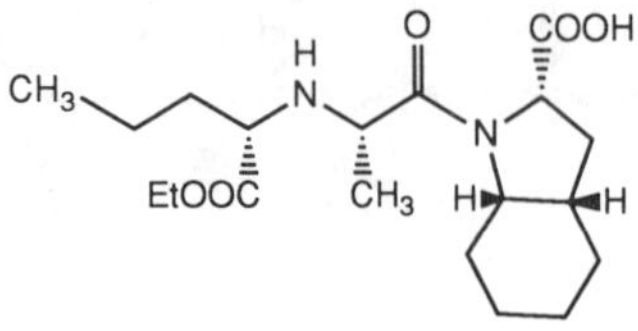

6

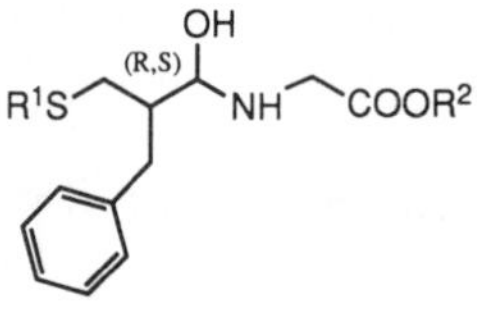

7 $R^1 = H; R^2 = H$

8 $R^1 = CH_3CO; R^2 = CH_2Ph$

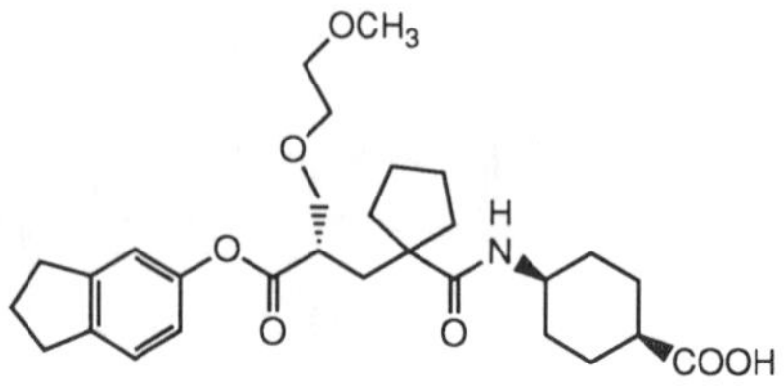

9

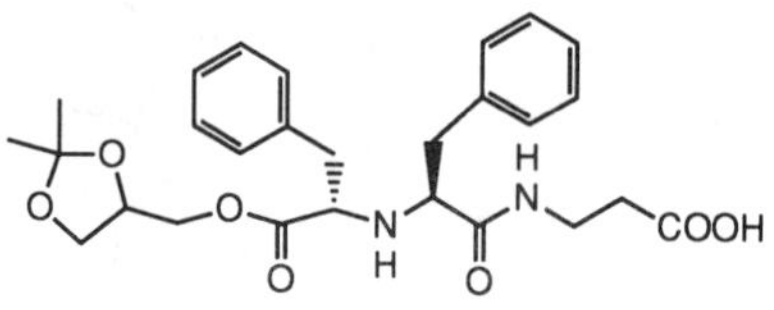

10

make it inactive even after intravenous administration. In a recent approach (*48*) combining the concept of mixed inhibitor and prodrug, RB 101 **13**, a disulfide prodrug of NEP and AP-N inhibitors, was synthesized. RB 101 is sufficiently lipophilic to cross the blood–brain barrier, and although stable in serum, prodrug is converted to the active compounds in brain homogenates containing the prodrug. Administration of RB 101 by intravenous, intraperitoneal, or subcutaneous routes produced strong antinociceptive effects in mice. Intravenous administration of the corresponding disulfide forms of the constituent NEP and AP-N inhibitors had minimal antinociceptive effect at concentrations twofold higher than RB 101 (*67*). AP-N inhibitors containing free thiols were not antinociceptive after systemic administration. The disulfide prodrugs prepared from *β*-aminothiol-type AP-N inhibitors effectively crossed the blood–brain barrier (*68*).

β-Lactam Antibiotics

β-Lactam antibiotics inhibit the synthesis of bacterial cell walls by irreversible inhibition of transpeptidases, which catalyze the incorporation of acetylmuramic peptides. β-Lactam antibiotics are peptidomimetic analogues of di- and tripeptides that mimic D-Ala–D-Ala in the transpeptidase reaction (*69*). Although profoundly effective as antibiotic agents, most are not orally active. Those that exhibit oral activity are usually transported from the intestine by a carrier-mediated process (*see* Chapter 5). For other poorly absorbed β-lactam antibiotics prodrug forms have been useful in enhancing oral activity.

Prodrug esters of ampicillin **14** increase absorption by increasing lipophilicity and reducing charge (*70*). In contrast, lipophilic amides of several penicillins and cephalosporins failed to improve their antibacterial activity apparently because of the metabolic stability of the amide linkage between parent drug and prodrug auxilliary (*71*). Such metabolic and chemical stability is not unusual and is increased by the sterically crowded environment near the penem nucleus. Even esters of penicillins may be too metabolically stable to be useful prodrugs.

The problem is avoided by increasing the distance between the enzymatically susceptible ester linkage and the ring via a chemically unstable linking group. Double ester prodrugs (*72*) use a hydroxymethyl ester as the linking group to a second ester function. Once enzymatic cleavage of the distal ester function occurs, the residual hydroxymethyl ester rapidly decomposes to yield the parent drug. By varying the distal ester moiety or the linking group, acetaldehyde versus formaldehyde for example, one can modulate the rate of hydrolysis. This general approach has resulted in several commercially available prodrugs of ampicillin, including pivampicillin **15,** bacampicillin **16,** and talampicillin **17,** all of which have improved oral bioavailability relative to the parent antibiotic. This approach continues to be used with newer antibiotics, such as FCE 22891 **18,** which is an orally active prodrug of FCE 22101 **19.** Hydrolysis of FCE 22891 is so rapid that the prodrug is not observed in plasma, and presystemic hydrolysis is the principal cause of variability in plasma drug levels (*73*).

The double ester prodrug approach has been integrated with the mixed-inhibitor concept to combine ampicillin with the lactamase inhibitor sulbactam. The resulting prodrug, sultamicillin **20,** has 200–250% of the bioavailability of ampicillin and sulbactam and is more than 80% absorbed. Hydrolysis of 20 produces equimolar amounts of antibiotic and lactamase inhibitor in plasma (*74*). Many examples of cepham antibiotic prodrugs were reviewed by Dürckheimer et al. (*75*).

The relationship between oral absorption and physicochemical properties of cephalosporin ester prodrugs was studied by Yoshimura et al. (*76*). For a series of 7-*O*-acylmandelamido-3-methyl-3-cephem-4-carboxylic acids **21,** hydrolysis rates of the 7-*O*-acyl groups were less important than lipophilicity in determining extent of absorption (*76*). Lipophilicity was parabolically correlated to bioavailability and was optimal in the range of log P = 1.2–2.1. This modification left the free 4-carboxyl group, which improved aqueous solubil-

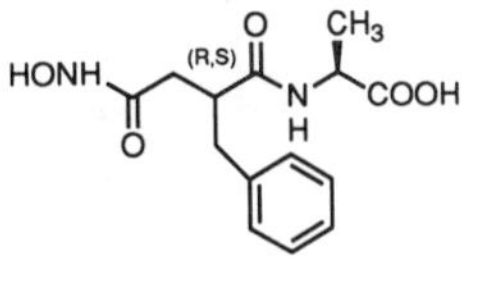

11

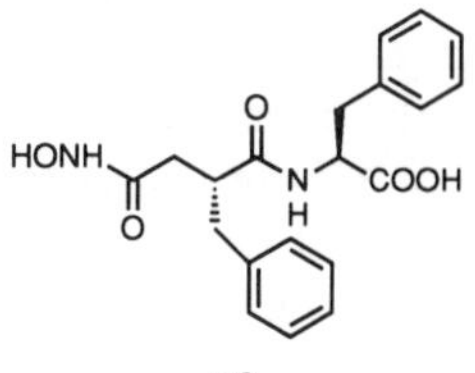

12

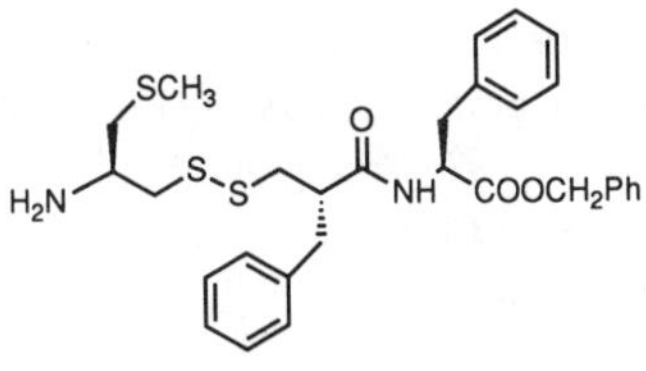

13

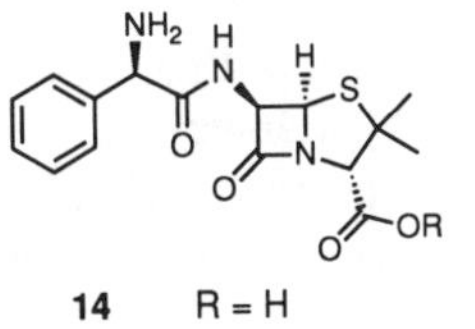

14 R = H

15 R =

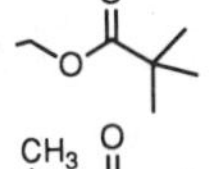

16 R =

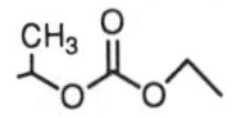

17 R =

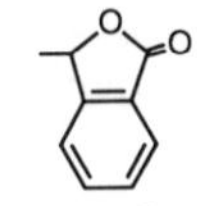

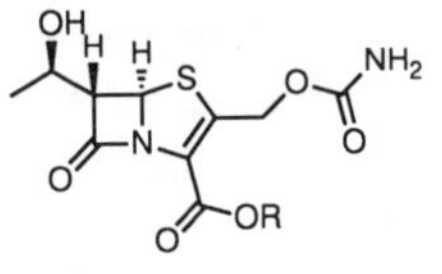

18 R = CH2OCOCH3

19 R = H

20 R =

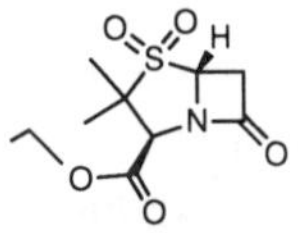

ity. A similar parabolic relationship was observed for the π value of the substituent on acyloxyethyl esters of cefotiam **22** (*77*). However, evaluation of the aqueous solubility, lipophilicity, and hydrolysis rates of pivaloyloxymethyl esters for 10 parenteral antibiotics versus bioavailability found that water solubility was the principal distinguishing factor and bioavailability improved with solubility. No correlation was seen for log *P*, although all but one of the esters appeared to be sufficiently lipophilic (log $P > 1.2$) and provided good absorption (*78*). Acyloxyalkyl esters of 3-(2-propenyl)cephem increased absorption relative to the parent as much as fourfold (*79*).

Prodrug approaches have also been used to overcome stability problems that contribute to poor oral activity of certain β-lactams. Carbenicillin **23** is unstable in gastric juice. Esterification of the α-carboxyl group overcomes this liability and increases bioavailability 30-fold in the case of the phenyl ester carfecillin (*80*). Peptide derivatives of C_4β-aminoalkyl carbapenem **24** are bioreversible in vivo and overcome a chemical instability of the compound that leads to rearrangements of the penem nucleus (*81*).

Amino Acid Prodrugs

Even though they are not peptidomimetics *per se,* prodrugs of amino acids and analogues provide useful templates for prodrug designs that may be applicable to peptidomimetic drugs. Amino acid transporters are highly specific and tolerate few modifications. However, dipeptides, including dipeptide prodrugs of L-α-methyl DOPA **25,** are transported efficiently by the intestinal peptide carrier (*32, 83*). Once transported, reconversion to the amino acid can be accomplished by peptidases. For example, the structure–activity relationships developed from a series of proline dipeptide analogues suggest that prolidase may be a useful target for reconversion of proline-containing prodrugs after transport into the intestinal mucosal cell by the peptide carrier (*84*).

Other prodrugs of L-DOPA include NB-355 **26,** which is not actively transported but is slowly absorbed from the small intestine (*85*). GluDopa **27** is metabolized in the kidney by γ-glutamyl transpeptidase and aromatic L-amino acid decarboxylase to release dopamine in the kidney and produce the consequent renovascular effects (*86*). Alafosfalin **28** is an antibacterial prodrug that is actively transported by the peptide transporter in animals. Once absorbed, it is converted to the toxic amino acid mimetic 1-aminoethylphosphonate (*87*).

Milacemide **29,** an anticonvulsant prodrug of glycine, crosses the blood–brain barrier to a much greater extent than glycine or glycinamide (*88, 89*). Reconversion of milacemide is accomplished by monoamine oxidase B (*88*). Prodrugs of thiol-containing amino acids include GCE **30,** an ethyl ester prodrug of glutathione (*90*), and RibCys **31,** a thiazolidine prodrug of L-cysteine, which increases glutathione levels in vivo (*91, 92*).

Amino Acid Ester Prodrugs of Renin Inhibitors

Prodrug approaches have been used sparingly in more recent peptidomimetic drug classes, such as cholecystokinin antagonists, HIV proteinase inhibitors, and renin inhibitors. Nonetheless, prodrug approaches would seem to be a worthwhile strategy to directly address the lack of oral activity of most of the prototype compounds. We explored (*93–95*) a targeted prodrug approach applied to poorly water-soluble renin inhibitors with the objective of improving oral activity through enhanced intestinal absorption.

Prodrugs are often used to improve the aqueous solubility of parent drugs with poor solubility properties. These prodrugs typically contain auxilliaries with charged or highly polar groups to increase aqueous solubility. However, absorption of the prodrugs is not

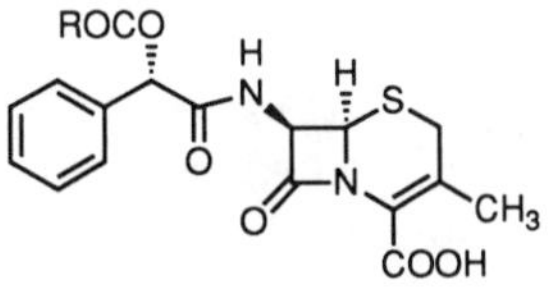

21

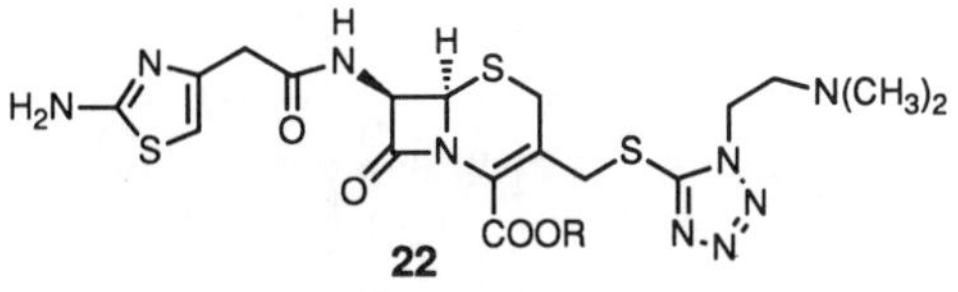

22

cefotiam R = H

prodrugs R = CHR^1OCOR^2

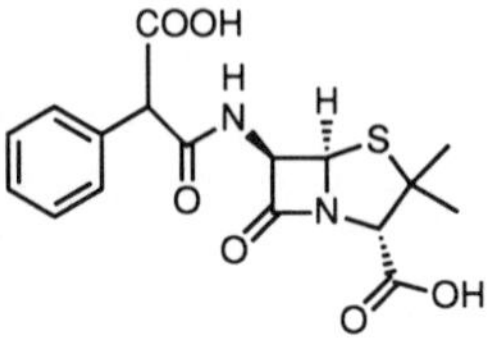

23

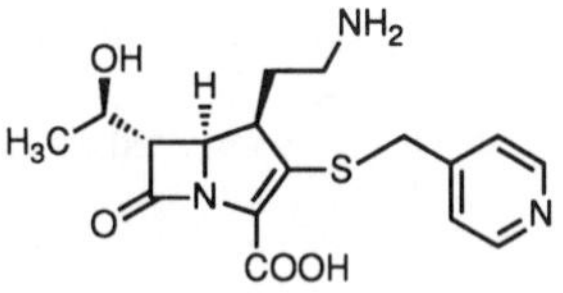

24

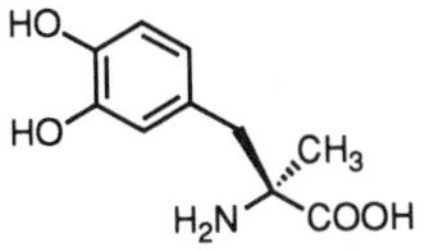

25

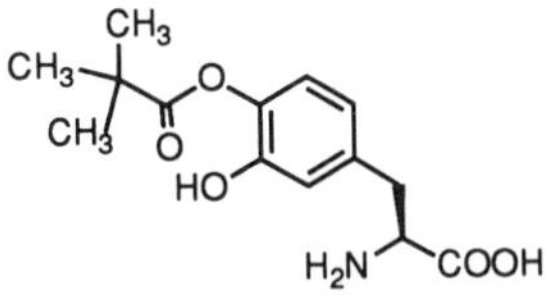

26

always improved because the added polarity and charges will reduce the ability of the modified drug to partition into lipid membranes (*96*). This trade-off between aqueous solubility and permeability is shown in the relationship between the flux across the intestinal membrane, *J*, and membrane permeability, *P*, and the concentration gradient between gut lumen and blood, $\Delta C_{gl \rightarrow b}$. Under sink conditions in the blood, $\Delta C_{gl \rightarrow b}$ is maximal at the aqueous solubility, C_s, of the compound in the lumen:

$$J = P \times (C_{gl} - C_b) \rightarrow P \times C_s$$

The flux of highly lipophilic renin inhibitors across the intestinal membrane is not generally limited by permeability, but by aqueous solubility. The goals in this project were to produce prodrugs that were more soluble in water than their parent drugs by at least an order-of-magnitude and preferably several orders-of-magnitude and that were highly selective for hydrolysis by intestinal brush-border membrane (BBM) hydrolases (*97, 98*); therefore, absorption at a permeability (rate) comparable to the parent drug would be facilitated.

Renin inhibitors **32** and **38** were designed to contain a hydroxyl group at a subsite that could be functionalized with solubilizing amino acids to form esters; these subsites were designated "P_2" and "P_4," respectively, for **32** and **38.** The "P" subsite nomenclature relates amino acid residues or mimics in the inhibitor to corresponding residues in the natural substrate, angiotensinogen, relative to the cleavage site (*99*). Installation of the hydroxyl groups at these sites was consistent with good renin-inhibitory activity. Results

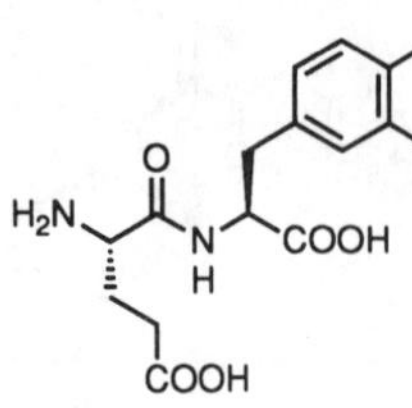

27

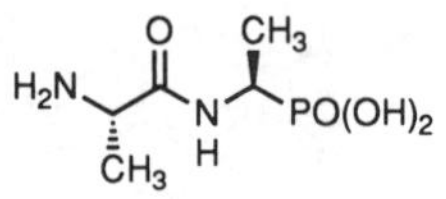

28

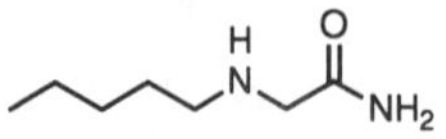

29

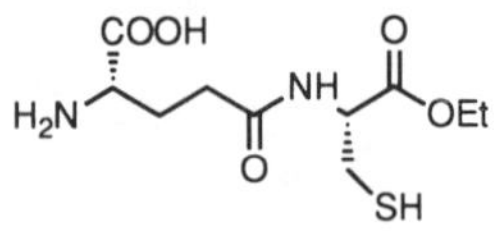

30

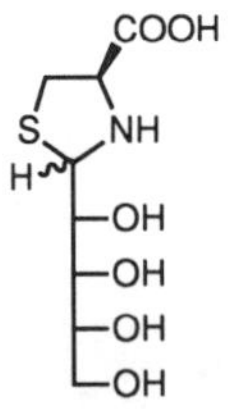

31

summarized in Table I show that aqueous solubility was enhanced relative to the clinical candidate renin inhibitor Cl-992 **40** (*100*), first by installation of the primary hydroxy groups and then by esterification with various amino acids. The P_2 series (**32–37**), in which the amino acid progroup was varied, displayed a range of modest increases in water solubility over the parent compound that ranged from slightly greater than fourfold (Asp ester, **37**)

Table I. Solubility of Renin Inhibitors and Prodrugs in MES Buffer (pH 6.5)

Compound	*Amino Acid Ester*	*Solubility (mM)*
P_2-Series		
32	Parent (H)	0.597
33	Lys	16.7
34	Ala	9.03
35	Gln	4.33
36	Glu	4.72
37	Asp	2.50
P_4-Series		
38	Parent (H)	0.121
39	Asp	1.60
40	Cl-992	0.011

NOTE: MES is 2-morpholinoethanesulfonic acid.

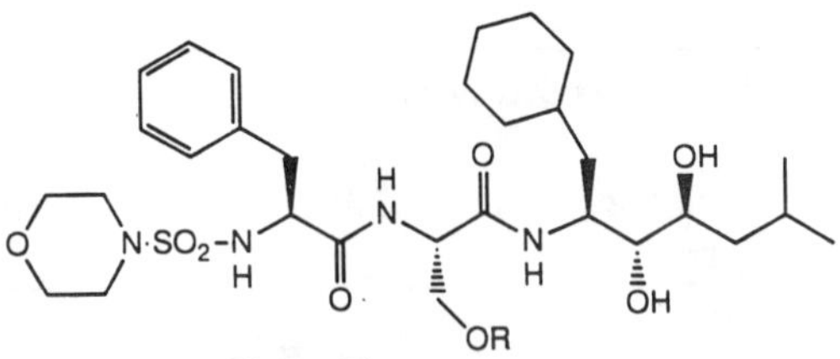

32 R = H
33 R = Lys
34 R = Ala
35 R = Gln
36 R = Glu
37 R = Asp

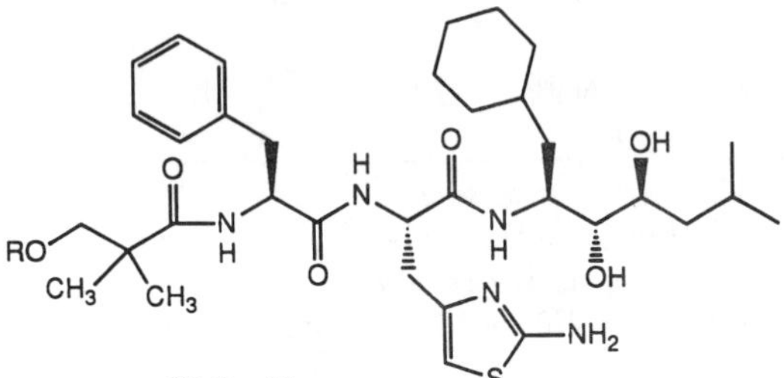

38 R = H
39 R = Asp

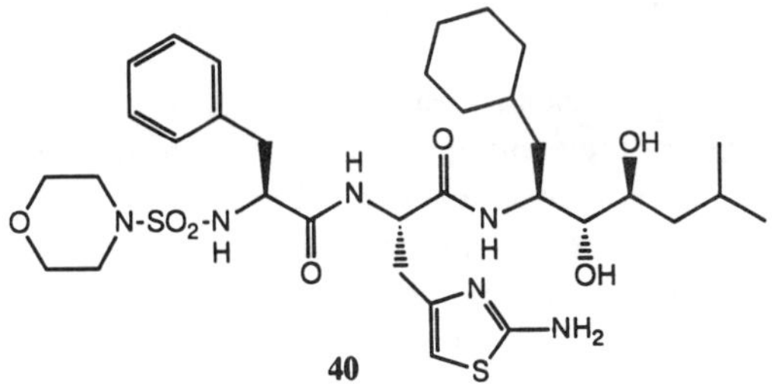

40

to 28-fold (Lys ester, **33**). At pH 6.5, solubility of the P_4-Asp prodrug **39** was approximately 13-fold greater than the parent compound **38.** From the simplified equation previously defined, the increase in driving force for absorption of the prodrugs is proportional to the increase in aqueous solubility, provided the permeability is unchanged.

Hydrolysis of prodrug to parent drug in the intestinal lumen could result in precipitation of the parent drug and a compromise in the aqueous solubility enhancement; hydrolysis at the intestinal wall should maximize the absorption advantages of solubility enhancement by presenting a high drug concentration at the site of absorption. Targeted reconversion should, therefore, be a function of both enzyme distribution and activity (*101*). Stability studies (*102*) were conducted in intestinal perfusate generated in rat jejunum and in suspensions of purified BBM from rat jejunum. The selectivity ratio was defined as the prodrug half-life in intestinal perfusate divided by the half-life in BBM suspensions (Table II). Within the P_2 series, the selectivity ratio increased from one for the −Ala, −Gln, and −Lys esters to approximately two for the acidic amino acid esters, −Asp, and −Glu. The absolute stability of the acidic amino acid prodrugs was also greatest (388−576 min). Subsequently, the aspartate ester was installed at the N-terminus region of the renin inhibitor, a position

Table II. Selectivity of Producing Reconversion

Compound	Amino Acid Ester	BBMV $t_{1/2}$ (min)	Perfusate $t_{1/2}$ (min)	Selectivity Ratio[a]
P_2-Series				
32	Parent (H)	S[b]	S	—
33	Lys	54.1	50.2	0.93
34	Ala	72.6	69.7	0.96
35	Gln	36.9	36.4	0.99
36	Glu	235	388	1.7
37	Asp	291	576	2.0
P_4-Series				
38	Parent (H)	248	S	—
39	Asp	14	>171[c]	>12

NOTE: BBMV is brush-border membrane vesicles.
[a] Selectivity ratio = (Perfusate $t_{1/2}$)/(BBMV $t_{1/2}$)
[b] Less than 10% of the prodrug was degraded in 4 h.
[c] For this value $n = 4$; $t_{1/2}$ = 220 min, 122 min, S and S.

favoring rapid hydrolysis by intestinal BBM aminopeptidases. The prodrug **39** was relatively stable in perfusate but was rapidly hydrolyzed to parent drug in the BBM suspensions.

Intestinal permeabilities for selected compounds were determined by single-pass rat-gut perfusion (Figure 1). All intestinal permeabilities were high (i.e., $°P^*_{eff}$ greater than 1) and indicated that permeability was not rate-limiting to the absorption of these renin inhibitors. Thus, the flux of renin inhibitor across the intestinal membrane should increase proportionately with increases in aqueous solubility.

These studies have identified prodrugs for evaluation in vivo. Prodrug **39,** for example, was highly selective for reconversion by BBM hydrolases, relatively stable against backbone hydrolysis by pancreatic enzymes, and over 10-fold more soluble in water than its

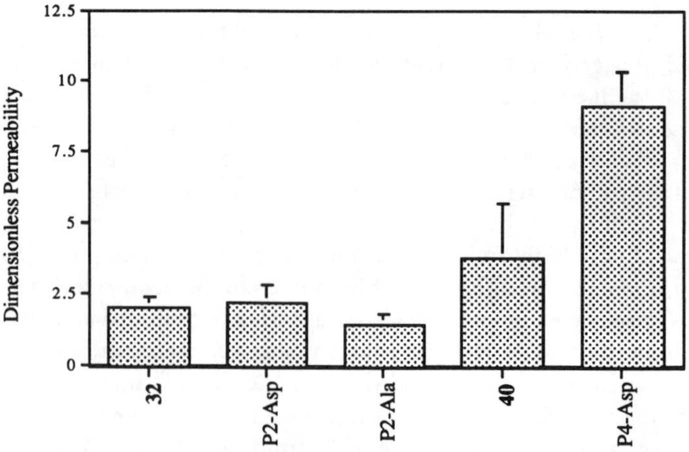

Figure 1. Effective permeabilities for renin inhibitors ***32*** *and* ***40*** *and their amino acid ester prodrugs. Dimensionless permeabilities were calculated using the method of Amidon et al.* (103).

parent, **38.** Intestinal perfusion results indicated that permeability was not rate-limiting to absorption.

Future Directions and Potential for Peptidomimetic Prodrugs

Development of prodrugs has been a successful strategy for improving the absorption of several classes of peptidomimetic drugs, especially ACE inhibitors and β-lactam antibiotics. Although less widely used for other classes such as renin or HIV protease inhibitors, the commonly encountered problems of poor aqueous solubility and limited absorption suggest that prodrug strategies will be useful in these classes as well. However, because hepatic clearance rather than poor intestinal absorption more often limits bioavailability for these classes, prodrug approaches designed to improve absorption have been given less attention. Prodrugs that directly target clearance by the liver may have great potential for future investigation.

The number of chemical approaches mentioned in this and related chapters certainly provide a wealth of techniques to apply to peptidomimetic drugs of diverse structure. At the same time, the increasing number of in vitro models and the understanding of the individual components of drug delivery, absorption, distribution, metabolism, and clearance provide much needed tools to identify specific delivery obstacles and to design approaches that address these obstacles.

References

1. Portoghese, P. S. *J. Med. Chem.* **1991,** *34,* 1757–1762.
2. Lee, V. L.; Dodda-Kashi, S.; Grass, G. M.; Rubas, W. In *Peptide and Protein Drug Delivery*; Lee, V. H. L., Ed.; Marcel Dekker: New York, 1991; pp 691–738.
3. Smith, P. L.; Wall, D. W.; Gochoco, C. H.; Wilson, G. *Adv. Drug Deliv. Rev.* **1992,** *8,* 253–290.
4. Fauchère, J.-L. *Adv. Drug Res.* **1986,** *15,* 29–69.
5. Freidinger, R. *Trends Pharmacol. Sci.* **1989,** *10,* 270–274.
6. Kaltenbronn, J. S.; Hudspeth, J. P.; Lunney, E. A.; Michneiwicz, B. M.; Nicolaides, E. D.; Repine, J. T.; Roark, W. H.; Stier, M. A.; Tinney, F. J.; Woo, P. K. W.; Essenburg, A. D. *J. Med. Chem.* **1990,** *33,* 838–845.
7. Gordon, T.; Hansen, P.; Morgan, B.; Singh, J.; Balzman, E.; Ward, S. *Biorg. Med. Chem. Lett.* **1992,** *3,* 915–920.
8. Sham, H. I.; Bolis, G.; Stein, H. H.; Fesik, S. W.; Marcotte, P. A.; Plattner, J. J.; Rempel, C. A.; Greer, J. *J. Med. Chem.* **1988,** *31,* 284–295.
9. Yee, G. C. *Pharmacotherapy* **1991,** *11,* 130S–134S.
10. Su, T.; Nakanishi, H.; Xue, L.; Chen, B.; Tuladhar, S.; Johnson, M. E.; Kahn, M. *Biorg. Med. Chem. Lett.* **1993,** *3,* 835–840.
11. Morgan, B. A.; Gainor, J. A. *Annu. Rep. Med. Chem.* **1989,** *24,* 243–252.

12. *Angiotensin Converting Enzyme Inhibitors*; Kostis, J. B.; DeFelice, E. A., Eds.; Alan R. Liss: New York, 1987.

13. Greenlee, W. J. *Med. Res. Rev.* **1990,** *10,* 173–236.

14. Horwell, D. C.; Hughes, J.; Hunter, J. C.; Pritchard, M. C.; Richardson, R. S.; Roberts, E.; Woodruff, G. N. *J. Med. Chem.* **1991,** *34,* 404–414.

15. Getman, D. P.; DeCrescenzo, G. A.; Heintz, R. M.; Reed, K. L.; Talley, J. J.; Bryant, M. L.; Clare, M.; Houseman, K. A.; Marr, J. J.; Mueller, R. A.; Vazquez, M. L.; Shieh, H. S.; Stallings, W. C.; Stegeman, R. A. *J. Med. Chem.* **1993,** *36,* 288–291.

16. Hruby, V.; Sharma, S. D. *Curr. Opin. Biotech.* **1991,** *2,* 599–605.

17. Burston, D.; Matthews, D. M. *Clin. Sci.* **1990,** *79,* 267–272.

18. Greenfield, J. C.; Cook, K. J.; O'Leary, I. A. *Drug Metab. Disp.* **1989,** *17,* 518–525.

19. Stewart, B. H.; Lu, R. H. H.; Chan, O. H. *Pharm. Res.* **1992,** *9,* S295.

20. Erickson, R. H.; Kim, Y. S. *Annu. Rev. Med.* **1990,** *41,* 133–139.

21. Powell, M. F. *Annu. Rep. Med. Chem.* **1993,** *28,* 285–294.

22. Veber, D. F.; Saperstein, R.; Nutt, R. F.; Freidinger, R. M.; Brady, S. F.; Curley, P.; Perlow, D. S.; Paleveda, W. J.; Colton, C. D.; Zacchei, A. G.; Tocco, D. F.; Hoff, D. R.; Vandlen, R. I.; Gerich, J. E.; Hall, L.; Mandarino, L.; Cordes, E. H.; Anderson, P. S.; Hirschmann, R. *Life Sci.* **1984,** *34,* 1371–1378.

23. Kempf, D. J.; Codacovi, L.; Wang, X. C.; Kohlbrenner, W. E.; Wideburg, N. E.; Saldivar, A.; Vasavanonda, S.; March, K. C.; Bryant, P.; Sham, H. L.; Green, B. E.; Betebenner, D. A.; Erickson, J.; Norbeck, D. W. *J. Med. Chem.* **1993,** *36,* 320–330.

24. Thaisrivongs, S.; Pals, D. T.; DuCharme, D. W.; Turner, S. R.; DeGraaf, G. L.; Lawson, J. A.; Couch, S. J.; Williams, M. V. *J. Med. Chem.* **1991,** *34,* 633–642.

25. Rosenberg, S. H.; Woods, K. W.; Sham, H. L.; Kleinert, H. D.; Martin, D. L.; Stein, H. *J. Med. Chem.* **1990,** *33,* 1962–1969.

26. Bundy, G. L.; Pals, D. T.; Lawson, J. A.; Couch, S. J.; Lipton, M. F.; Mauragis, M. A. *J. Med. Chem.* **1990,** *33,* 2276–2283.

27. Chan, O. H.; Hamilton, H. W.; Steinbaugh, B.; Taylor, M. D.; Stewart, B. H. *Pharm. Res.* **1992,** *9,* S347.

28. Ruwart, M. J.; Wilkinson, K. F.; Rush, B. D.; Thaisrivongs, S.; Bundy, G. L.; Vidmar, T. J. *Pharm. Res.* **1992,** *9,* S205.

29. Karls, M. S.; Rush, B. D.; Wilkinson, K. F.; Vidmar, T. J.; Burton, P. S.; Ruwart, M. J. *Pharm. Res.* **1991,** *8,* 1477–1481.

30. Conradi, R. A.; Hilgers, A. R.; Ho, N. F. H.; Burton, P. S. *Pharm. Res.* **1991,** *8,* 1453–1460.

31. Conradi, R. A.; Hilgers, A. R.; Ho, N. F. H.; Burton, P. S. *Pharm. Res.* **1992,** *9,* 435–439.

32. Cussler, E. L. In *Diffusion: Mass Transfer in Fluid Systems*; Cambridge University: Cambridge, England, 1984; p 118.

33. Hichens, M.; *Drug Metab. Rev.* **1983,** *14,* 77–98.

34. Yokohama, S.; Yamashita, K.; Toguchi, H.; Takeuchi, J.; Kitamori, N. *J. Pharm. Dyn.* **1984,** *7,* 101–111.

35. Burston, D.; Matthews, D. M. *Clin. Sci.* **1990,** *79,* 267–272.

36. Kramer, W.; Girbig, F.; Gutjahr, U.; Kleeman, H.-W.; Leipe, I.; Urbach, H.; Wagner, A. *Biochim. Biophys. Acta* **1990,** *1027,* 25–30.

37. Adedoyin, A.; Perry, P. R.; Wilkinson, G. R. *Drug Metab. Dispos. Biol. Fate Chem.* **1993,** *21,* 184–188.

38. Cumin, F.; Schnell, C.; Richert, P.; Raschdorf, F.; Probst, A.; Schmid, K.; Wood, J. M.; de Gasparo, M. *Drug Metab. Dispos. Biol. Fate Chem.* **1990,** *18,* 831–835.

39. Greenfield, J. C.; Cook, K. J.; O'Leary, I. A. *Drug Metab. Disp.* **1989,** *17,* 518–525.

40. Fisher, J. F.; Harrison, A. W.; Bundy, G. L.; Wilkinson, K. F.; Rush, B. D.; Ruwart, M. J. *J. Med. Chem.* **1991,** *34,* 3140–3143.

41. Rosenberg, S. H.; Kleinert, H. D.; Stein, H. H.; Martin, D. L.; Chekal, M. A.; Cohen, J.; Egan, D. A.; Tricarico, K. A.; Baker, W. R. *J. Med. Chem.* **1991,** *34,* 469–471.

42. Rosenberg, S. H.; Spina, K. P.; Condon, S. L.; Polakowski, J.; Yao, Z.; Kovar, P.; Stein, H. H.; Cohen, J.; Barlow, J. L.; Klinghofer, V.; Egan, D. A.; Tricaro, K. A.; Perun, T. J.; Baker, W. R.; Kleinert, H. D. *J. Med. Chem.* **1993,** *36,* 460–467.

43. Oliyai, R.; Stella, V. J. *Annu. Rev. Pharmacol. Toxicol.* **1993,** *32,* 521–544.

44. Kahns, A. H.; Friis, G. J.; Bundgaard, H. *Bioorg. Med. Chem. Lett.* **1993,** *3,* 809–812.

45. Bundgaard, H. *Med. Actual Drugs Today* **1983,** *19,* 499–538.

46. Kahns, A. H.; Buur, A.; Bundgaard, H. *Pharm. Res.* **1993,** *10,* 68–74.

47. Bundgaard, H. In *Design of Prodrugs*; Bundgaard, H., Ed.; Elsevier: Amsterdam, The Netherlands, 1985; pp 1–92.

48. Fournie-Zaluski, M.-C.; Coric, P.; Turcaud, S.; Lucas, E.; Noble, F.; Maldonado, R.; Roques, B. P. *J. Med. Chem.* **1992,** *35,* 2473–2481.

49. Ondetti, M. A.; Rubin, B.; Cushman, D. W. *Science (Washington, DC)* **1977,** *196,* 441–444.

50. DeFelice, E. A.; Kostis, J. B. In *Angiotensin Converting Enzyme Inhibitors*; Kostis, J. B.; DeFelice, E. A., Eds.; Alan R. Liss: New York, 1987; pp 213–261.

51. Williams, G. H. *N. Engl. J. Med.* **1988,** *319,* 1517–1525.

52. Kostis, J. B.; Raia, J. J., Jr.; DeFelice, E. A.; Barone, J. A.; Deeter, R. G. In *Angiotensin Converting Enzyme Inhibitors*; Kostis, J. B.; DeFelice, E. A., Eds.; Alan R. Liss: New York, 1987; pp 19–54.

53. Irvin, J. D.; Till, A. E.; Vlasses, P. H.; Hichens, M.; Rotmensch, H. H.; Harris, K. E.; Merrill, D. D.; Ferguson, R. K. *Clin. Pharmacol. Ther.* **1984,** *35,* 248.

54. Friedman, D. I.; Amidon, G. L. *Pharm. Res.* **1989,** *6,* 1043–1047.

55. Pang, K. S.; Cherry, W. F.; Terrell, J. A.; Ulm, E. G. *Drug Metab. Disp.* **1984,** *12,* 309–313.

56. Ranadive, S. A.; Serajuddin, A. T. M.; Chen, X.; Wadke, D. A. *Pharm. Res.* **1989,** *6,* S149.

57. Olson, S. C.; Horvath, A. M.; Michniewicz, B. M.; Sedman, A. J.; Posvar, E. *Angiology* **1989,** *40,* 351–359.

58. Devissaguet, J. P.; Ammoury, N.; Devissaguet, M.; Perret, L. *Fundam. Clin. Pharmacol.* **1990,** *4,* 175–189.

59. Lecomte, J.-M; Costentin, J.; Vlaiculescu, A.; Chaillet, P.; Marcais-Collado, H.; Llorens-Cortes, C.; Leboyer, M.; Schwartz, J.-C. *J. Pharmacol. Exp. Ther.* **1986,** *237,* 937–944.

60. Roques, B. P.; Fournie-Saluski, M. C.; Soroca, E.; Lecomte, J. M.; Malfroy, B.; Llorens, C.; Schwartz, J.-C. *Nature (London)* **1980,** *288,* 286–288.

61. Gros, C.; Souque, A.; Schwartz, J.-C.; Duchier, J.; Cournot, A.; Baumer, P.; Lecomte, J.-M. *Proc. Natl. Acad. Sci. U.S.A.* **1989,** *86,* 7580–7584.

62. O'Connell, J. E.; Jardine, A. G.; Davidson, G.; Connell, J. M. C. *J. Hypertens.* **1992,** *10,* 271–277.

63. Chipkin, R. E.; Berger, J. G.; Billard, W.; Iorio, L. C.; Chapman, L. C.; Barnett, A. *J. Pharmacol. Exp. Ther.* **1988,** *245,* 829–838.

64. Richards, M.; Espiner, E.; Frampton, C.; Ikram, H.; Yandle, T.; Sopwith, M.; Cussans, N. *Hypertension* **1990,** *16,* 269–276.

65. Fournie-Zaluski, M.-C.; Chaillet, P.; Bouboutou, R.; Coulaud, A.; Cherot, P.; Waksman, G.; Costentin, J.; Roques, B. P. *Eur. J. Pharmacol.* **1984,** *102,* 525–528.

66. Xie, J.; Soleilhac, J.-M.; Schmidt, C.; Peyroux, J.; Roques, B. P.; Fournie-Zaluski, M.-C. *J. Med. Chem.* **1989,** *32,* 1497–1503.

67. Noble, F.; Soleilhac, J. M.; Soroca-Lucas, E.; Turcaud, S.; Fournie-Zaluski, M. C.; Roques, B. P. *J. Pharmacol. Exp. Ther.* **1992,** *261,* 181–189.

68. Fournie-Zaluski, M. C.; Coric, P.; Turcaud, S.; Bruetschy, L.; Lucas, E.; Noble, F.; Roques, B. P. *J. Med. Chem.* **1992,** *35,* 1259–1266.

69. Albert, A. In *Selective Toxicity,* 6th ed.; Chapman and Hall: London, 1979; pp 449–453.

70. Jones, K. H.; Langley, P. F.; Lees, L. J. *Chemotherapy* **1978,** *24,* 217–226.

71. Toth, I.; Hughes, R. A.; Ward, P.; Baldwin, M. A.; Welham, K. J.; McColm, A. *Int. J. Pharm.* **1991,** *73,* 259–266.

72. Bundgaard, H. *Drugs Fut.* **1991,** *16,* 443–456.

73. Efthymiopoulos, C.; Benedetti, M. S.; Sassella, D.; Boobis, A.; Davies, D. *Antimicrob. Agents Chemother.* **1992,** *36,* 1958–1963.

74. Bush, K. *Clin. Microbiol. Rev.* **1988,** *1,* 109–123.

75. Dürckheimer, W.; Adam, F.; Fischer, G.; Kirrstetter, R. *Adv. Drug Res.* **1988,** *17,* 189–234.

76. Yoshimura, Y.; Hamaguchi, N.; Kakeya, N.; Yashiki, T. *Int. J. Pharmaceut.* **1985,** *26,* 317–328.

77. Yoshimura, Y.; Hamaguchi, N.; Yashiki, T. *J. Antibiot.* **1986,** *39,* 1329–1342.

78. Yoshimura, Y.; Hamaguchi, N.; Yashiki, T. *Int. J. Pharmaceut.* **1985,** *23,* 117–129.

79. Kim, W. J.; Ko, K. Y.; Jung, M. H.; Kim, M.; Lee, K. I.; Kim, J. H. *J. Antibiot.* **1991,** *44,* 1083–1087.

80. Clayton, J. P.; Cole, M.; Elson, S. W.; Hardy, K. D.; Mizen, L. W.; Sutherland, R. *J. Med. Chem.* **1975,** *18,* 172–177.

81. Rao, V. S.; Fung-Tomc, J. C.; Desiderio, J. V. *J. Antibiot.* **1993,** *46,* 167–176.

82. Hu, M.; Subramanian, P.; Mosberg, H. I.; Amidon, G. L. *Pharm. Res.* **1989,** *6,* 66–70.

83. Bodor, N.; Sloan, K. B.; Higuchi, T. *J. Med. Chem.* **1977,** *20,* 1435–1445.

84. Bai, J. P.-F.; Hu, M.; Subramanian, P.; Mosberg, H. I.; Amidon, G. L. *J. Pharm. Sci.* **1992,** *81,* 113–116.

85. Hisaka, A.; Kasamatsu, S.; Takenaga, N.; Ohtawa, M *Drug. Metab. Dispo. Biol. Fate Chem.* **1990,** *18,* 621–625.

86. Barthelmebs, M.; Caillette, A.; Ehrhardt, J. D.; Velly, J.; Imbs, J. L. *Kidney Int.* **1990,** *37,* 1414–1422.

87. Allen, J. G.; Havas, L.; Leicht, E.; Lenox-Smith, J.; Nisbet, L. J. *Antimicrob. Agents Chemother.* **1979,** *18,* 897.

88. Yadid, G.; Zinder, O.; Youdim, M. B. H. *Br. J. Pharmacol.* **1991,** *104,* 760–764.

89. Youdim, M. B. H.; Kerem, D.; Duvdevani, Y. *Eur. J. Pharmacol.* **1988,** *150,* 381–384.

90. Kobaysahi, H.; Kurokawa, T.; Kitahara, S.; Nonami, T.; Harada, A.; Nakao, A.; Sugiyama, S.; Ozawa, T.; Takagi, H. *Transplantation* **1992,** *54,* 414–418.

91. Roberts, J. C.; Francetic, D. J. *Toxicol. Lett.* **1991,** *59,* 245–251.

92. Roberts, J. C.; Francetic, D. J.; Zera, R. T. *Cancer Chemother. Pharmacol.* **1991,** *28,* 166–170.

93. Wright, J. L.; Stewart, B. H.; Kugler, A. R.; Taylor, M. D. Presented at the 204th National Meeting of the American Chemical Society, Washington, DC, August 1992.

94. Stewart, B. H.; Massey, K. D.; Kugler, A. R.; Wright, J. L.; Repine, J. T.; Taylor, M. D. *Pharm. Res.* **1992,** *9,* S177.

95. Taylor, M. D.; Wright, J. L.; Repine, J. T.; Cheng, X. M.; Stewart, B. H.; Massey, K. D.; Kugler A. R. *J. Cell. Biochem.* **1993,** *Suppl. 17C,* L328.

96. Yoshimi, A.; Hashizume, H.; Tamaki, S.; Tsuda, H.; Fukata, F.; Nishimura, K.; Yata, N. *Pharmacobiodynamics* **1992,** *15,* 339–345.

97. Amidon, G. L.; Stewart, B. H.; Pogany, S. *J. Contr. Rel.* **1985,** *2,* 13–26.

98. Stewart, B. H.; Amidon, G. L.; Brabec, R. K. *J. Pharm. Sci.* **1986,** *75(10),* 940–945.

99. Schecter, I.; Berger, A. *Biochem. Biophys. Res. Commun.* **1967,** *27,* 157.

100. Patt, W. C.; Hamilton, H. W.; Ryan, M. J.; Painchaud, C. A.; Taylor, M. D.; Rapundalo, S. T.; Batley, B. L.; Connolly, C. J. C.; Taylor, D. G., Jr. *Med. Chem. Res.* **1992,** *2,* 10–15.

101. Fleisher, D.; Johnson, K. C.; Stewart, B. H.; Amidon, G. L. *J. Pharm. Sci.* **1986,** *75(10),* 934–939.

102. Tsuji, A.; Terasaki, T.; Tamai, I.; Hirroka, H. *J. Pharmacol. Exp. Ther.* **1987,** *241,* 594–601.

103. Amidon, G. L.; Sinko, P. J.; Fleisher, D. *Pharm. Res.* **1988,** *5,* 651–654.

RECEIVED for review September 9, 1993. ACCEPTED revised manuscript January 24, 1994.

Liver Peptide Transport

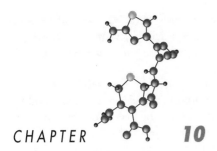

CHAPTER 10

Hepatic Processing of Peptides

David L. Marks,
Gregory J. Gores
and Nicholas F. LaRusso*

Many peptides, including hormones, xenobiotics, and drugs, are rapidly cleared by the liver. Also, some peptides that are extracted by the liver are excreted into bile often after hepatic metabolism. The cellular and molecular mechanisms by which the liver transports peptides from plasma to bile, however, are rarely identified in pharmacologic studies. In addition, even though the metabolism of xenobiotics by microsomal enzymes has been studied extensively, the mechanisms by which the liver metabolizes or degrades peptides are not well-recognized. Therefore, we will review the underlying mechanisms by which peptides are processed in the liver. We use the term "hepatic processing" to include extraction and internalization by liver cells, intracellular transport, and metabolism and excretion of peptides (*1, 2*).

* Corresponding author.

Background

Structure of the Liver

The microarchitecture of the liver has been reviewed extensively (*3, 4*). We will focus on several aspects of the organization of the liver that are especially relevant to processing of peptides; these aspects are blood flow to the liver, the cells present in the liver, and the organization of liver cells into acini or functional units.

The liver is the only organ with a dual blood supply. The liver receives nutrient-rich, oxygen-poor blood from the intestines via the portal vein and oxygen-rich blood via the hepatic artery (*5*). As a consequence of portal circulation, the liver is the first organ to interact with substances that have been absorbed into the blood from the small intestine, a feature that is important in the hepatic clearance of many xenobiotics and orally administered drugs (*6*). The liver also is responsible for the rapid clearance of many circulating hormones or peripherally injected drugs (e.g., exogenous insulin), which can enter the liver via either the hepatic artery or the portal vein. The relative importance of the liver in uptake of a circulating peptide depends not only on the mode of entry into the liver (i.e., portal vs. systemic) but also on the specificity and distribution of uptake mechanisms (e.g., receptors and carrier proteins) for that peptide. For example, radiolabeled transforming growth factor-β (TGF-β) was taken up similarly the the rat liver (approximately 50% of total dose at 20 min after injection) when injected portally or systemically (*7*). In contrast, radiolabeled epidermal growth factor (EGF) accumulated in the liver to a greater degree when injected portally rather than systemically [80 vs. 45% of total dose at 20 min (*7*)]. These data suggest that when circulating systemically, EGF is cleared extensively by receptors at other sites before reaching the liver, whereas TGF-β is not.

Blood moving through the both portal vein and hepatic artery eventually flows into terminal portal venules that feed the sinusoids (specialized capillaries) of the liver (*3*). Most liver cells interact with the blood in the sinusoids and clear some substances while secreting others. The cellular organization of the sinusoid is shown in Figure 1. Hepatic sinusoids

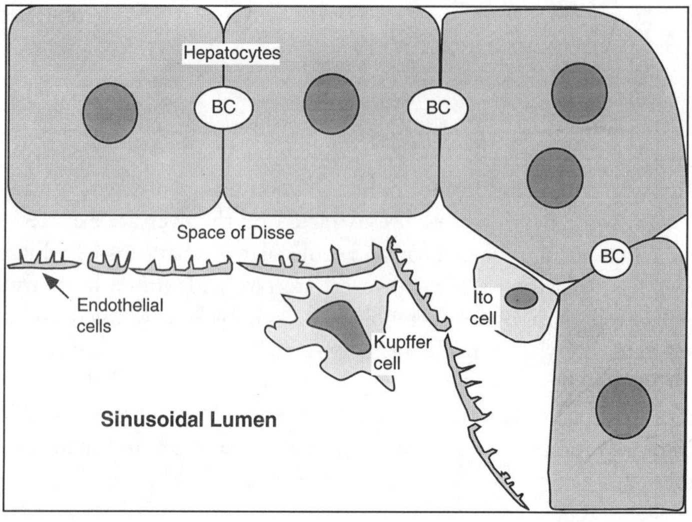

Figure 1. Cellular organization of the hepatic sinusoids: BC is bile canaliculus.

are unique structures lined by specialized endothelial cells and Kupffer cells [hepatic macrophages (*4*)]. Just outside of the sinusoidal-lining cells are occasional perisinusoidal, myofibroblast-like cells referred to variously as fat-storing, stellate, or Ito cells (*8*). Sinusoidal endothelial cells have no basement membrane and possess fenestrae (holes) through their cell bodies that allow blood solutes to pass into the space of Disse, where they come into direct contact with hepatocytes, the predominant cell type in the liver (*3*). Hepatocytes are polarized epithelial cells with specialized basolateral plasma membranes facing the sinusoids and specialized apical membranes that form bile canaliculi (*4*). Bile secreted into the canaliculi by hepatocytes is channeled into bile ducts, which are lined by intrahepatic bile duct epithelial cells called cholangiocytes (*9, 10*). Bile then moves into successively larger intraheptic ducts and eventually flows into extrahepatic bile ducts that carry bile to the common bile duct (*4*).

Most of the cell types of the liver have known or potential roles in peptide processing. Hepatocytes play multiple roles, including the clearance, metabolism, and biliary excretion of many hormones, drugs, and xenobiotics (*11, 12*). Kupffer cells are well known for their phagocytic capabilities to internalize particulates [e.g., bacteria and colloids (*13*)]. This feature of Kupffer cells makes them potential sites for clearance of very large molecular weight drug vehicles such as liposomes and colloids. Kupffer cells and endothelial cells possess receptors for several types of proteins [e.g., mannosylated glycoproteins and acetylated and formaldehyde-altered proteins (*13–15*)] that could potentially be involved in the processing of peptides. Cholangiocytes possess receptors (presumably basolateral) for the peptide hormones, EGF, secretin, and hepatocyte growth factor and absorb substances across their lumenal plasma membrane (*16–20*). Thus, cholangiocytes possess mechanisms that suggest that these cells may also play a role in the metabolism of peptides.

Blood flowing through the sinusoids is eventually collected in terminal hepatic venules (central veins) and returned to the general circulation via the hepatic vein (*21*). Blood flow in the liver sinusoids is predominantly unidirectional (*5*). Thus, the liver can be considered to be organized into functional units or acini originating at the terminal portal venule and ending with the terminal hepatic venules (Figure 2) (*4*). Based on this concept, the liver acinus can be divided (somewhat arbitrarily) into three functional zones: periportal (zone 1), intermediate (zone 2), and perivenous (zone 3) (*22*). Zone 1 receives the blood when it is most rich in oxygen and nutrients, whereas zone 3 receives the least nutrient-rich blood. Methods available for differential study of the functions and metabolic capabilities of hepatocytes within zones 1 and 3 (*see* next section) have shown that hepatocytes located in the periportal zone differ functionally from those in the perivenous zone (*22–24*). This functional heterogeneity may have major consequences for the location of hepatic processing of different peptides. For example, bile acid excretion and glutathione conjugation occur at a higher level in hepatocytes in the periportal area, but cytochrome P-450 and uridine diphosphate (UDP) glucuronyl transferase are more active in perivenous hepatocytes (*22*). In addition, the acinar zones may be affected differently by different types of liver damage (*25*).

Models Used for Study of Peptide Processing

Although human studies are eventually a necessity in drug development, for studies of mechanisms of hepatic processing, ethical and monetary considerations have led most researchers to use the rat liver. We previously discussed (*1, 26*) in detail some advantages and disadvantages of different rat-liver models used for the study of hepatic processing of proteins and peptides. In addition, in vitro models of hepatic processing are discussed elsewhere in this volume (*See* Chapter 22). We wish to stress, in this chapter the multiple

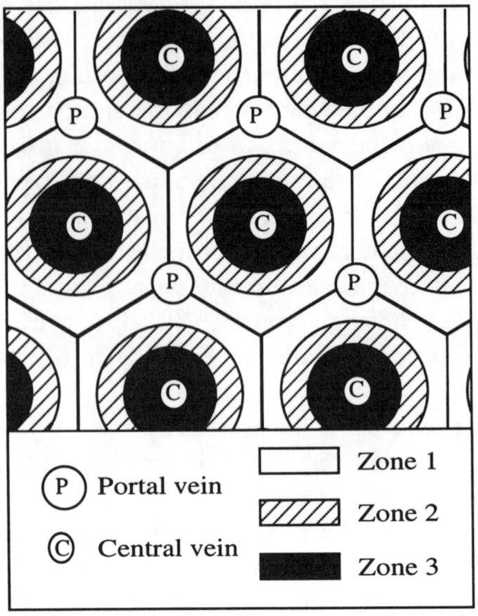

Figure 2. Schematic representation of the acinar zones of the liver. Blood enters the hepatic acinus mainly via the portal vein, flows past hepatocytes via the sinusoids through zones 1, 2, and 3, and is eventually collected in the central vein. As blood moves through the different zones, the composition (e.g., oxygen, nutrients, and metabolites) of blood is altered. Hepatocytes in the three zones possess different functional specializations.

models available and their uses for answering specific questions about hepatic processing (Table I). For example, intact animals and bile-duct-cannulated models (*7, 27*) will show if the liver is an important site for hepatic elimination of a compound, but results can be confounded by effects due to other organs or blood constituents. The isolated rat perfused liver (IPRL) does not suffer from these disadvantages and is widely used to quantify hepatic clearance. This model can be used to identify specific mechanisms of hepatic uptake and metabolism when appropriate inhibitors are known (*7, 28, 29*) but has the disadvantages of relatively rapid loss of viability (1–2 h) and long setup times for each experiment. A number of manipulations are possible with the isolated liver that are not possible in the intact animal. Perfusate composition (e.g., buffers, proteins and inhibitors), time of ligand interaction with the liver, and temperature can be controlled precisely. Both liver effluent and bile can be collected. Perfusion of the liver can be single pass, recirulating, or even retrograde via the hepatic vein instead of the portal vein (*24, 26, 30, 31*). The liver can be perfused either via the portal vein or hepatic artery or via both vessels (*32*). This manipulation may be especially useful for the study of peptide processing by cholangiocytes because they receive blood only from the hepatic artery, whereas most other liver cell types are fed by sinusoidal blood that flows from both the portal vein and the hepatic artery (*3, 33*). Application of several of these IPRL models has increased our understanding of mechanisms of peptide processing.

For studies of hepatic metabolism, liver slices may be used (*34*); however, such studies do not differentiate between effects of hepatic ectoenzymes or intracellular enzymes. To determine if a peptide is internalized, to identify the liver cell types involved in internalization, or to help define the mechanisms involved in uptake (e.g., endocytic or carrier-mediated transport), isolated liver cells may be the model of choice (*35, 36*). Methods are

Table I. Models Used for Study of Hepatic Processing

Model	*Uses*	*Advantages*	*Limitations*
Intact rat	Demonstration of uptake	Most physiological	No bile sampling; influences of other organs and blood elements
Bile fistula rat	Demonstration of biliary excretion	Near physiological	Influences of other organs and blood elements, interrupted enterohepatic circulation
Isolated perfused liver	First pass clearance; preparation of liver cells	Intact architecture; control of perfusate composition and temperature	Short viability; extensive time and equipment required, usually 1 data point per animal
Liver slices	Metabolism	Intact architecture	Do not make bile; do not distinguish intracellular from extracellular events
Isolated cells	Cell binding, uptake and metabolism	Convenient; many data points per animal	Short viability; loss of phenotype
Isolated subcellular fractions	Mechanisms of transport and metabolism some proteins	Defined materials	Loss of interacting mechanisms; inactivation of some proteins
Purified proteins	Transport and metabolic capabilities of proteins	Defined materials	Loss of interacting mechanisms; inactivation of some proteins

available for the isolation of relatively pure hepatocytes and endothelial, Kupffer, and Ito cells (*37–40*). We recently developed methods (*41, 42*) for the isolation of pure preparations of cholangiocytes from both normal and bile-duct ligated rats. However, some characteristics present in vivo may be lost upon cell isolation when using isolated cell models. For example, some cell surface receptors are reduced or absent in freshly isolated cells (*13*); and in long-term cultures of hepatocytes, extensive changes in phenotype, such as loss of cell polarity and loss of bile acid transporters, are observed (*34, 43–45*).

To provide more conclusive evidence of which components in the liver are involved in the processing of a substance, liver fractions enriched in specific cellular compartments (e.g., plasma membrane, endoplasmic reticulum, lysosomes, or cytosol) may be used. The use of hepatic microsomes for studies of metabolism are well-known (*46, 47*). The development of methods for the separate isolation of membrane vesicles enriched in either sinusoidal or canalicular domains of hepatocytes has been essential for characterization of the carrier-mediated systems that transport substances into or out of hepatocytes (*48–52*). Finally, as specific proteins are identified as playing roles in hepatic transport or metabolism, the use of purified hepatic proteins is becoming increasingly important in understanding hepatic processing (*53–55*). In general, because each model has its limits, multiple models are required to elucidate the multiple mechanisms involved in the haptic processing of a given peptide.

Table II. Pathways in Hepatic Processing

Stage of Processing	*Vesicular Pathway*	*Nonvesicular Pathway*
Internalization	Endocytosis	Carrier-mediated
Intracellular compartmentalization	Vesicles	Cytosol, binding proteins
Intracellular transport	Microtubules	Actin-myosin, diffusion
Site of metabolism	Lysosomes	Cytosol, endoplasmic reticulum
Externalization	Exocytosis	Carrier-mediated

Stages in Hepatic Processing

The focus of this section is on peptides that are internalized by the liver. The degradation of peptides or some dipepties (*56*) by hepatic ectoenzymes without internalization (hepatic inactivation as opposed to hepatic clearance) will not be discussed. We have defined four stages of hepatic processing.

1. internalization
2. intracellular transport
3. metabolism or degradation
4. externalization or excretion (*1*)

Conceptually, we envision two major pathways by which peptides are processed by the liver, a vesicular and a nonvesicular route (Table II) (*2*). Peptides that enter the cell within vesicles tend to stay in vesicles, whereas substances delivered to the cytosol rarely become packaged in vesicles. Undoubtedly, the pathways overlap. A lipophilic peptide may be inserted into the plasma membrane by diffusion, internalized into vesicles by invagination of the membrane, and finally metabolized by microsomal enzymes to a more polar form that is then excreted into bile by a nonvesicular mechanism. This overlap may be the case for cyclosporin A, which probably enters the liver cell by diffusion into the membrane, is extensively metabolized by cytochrome P-450 enzymes, and may be excreted into bile by the multidrug resistant gene product carrier system (*57–61*).

Mechanisms of Hepatic Uptake of Peptides

The first stage in the hepatic processing of a peptide is internalization of the peptide by liver cells. The passage of a substance across the plasma membrane into the cell may occur by several mechanisms that can be categorized as either passive or active (Figure 3). Because of the large surface area of the liver in contact with the blood, passive diffusion can play a major role in the clearance of some lipophilic substances (e.g., long-chain fatty acids), which are inserted rapidly into the membranes of liver cells. Many hydrophobic organic anions and cations, however, circulate in plasma strongly bound to proteins (albumin or α_1-acid glycoprotein, respectively) and are less likely to diffuse across the liver cell membrane (*12*). For these protein-bound substances as well as less hydrophobic molecules such as peptide growth factors and insulin, active mechanisms are usually required for transport across the

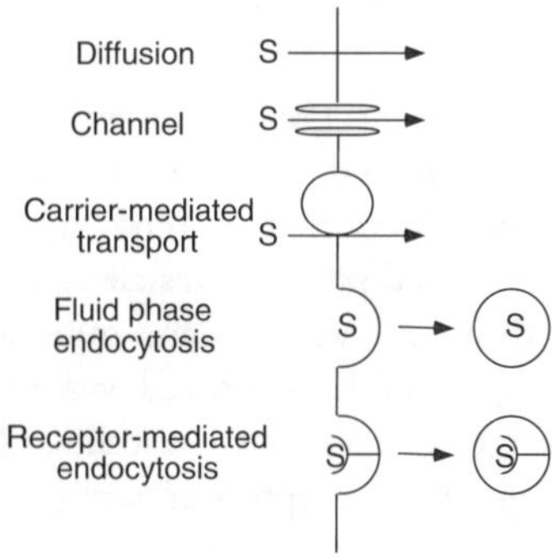

Figure 3. Mechanisms by which solutes(s) are internalized by cells. In diffusion, solutes pass through the plasma membrane due to their solubility characteristics without the aid of membrane proteins. Channels allow the passage of a solute through a pore in the membrane. In carrier-mediated transport, solutes are transported into the cell cytosol by a carrier protein in the plasma membrane without internalization of the carrier protein. In fluid-phase endocytosis, solutes are internalized into vesicles without prior binding to plasma membrane receptors. In receptor-mediated endocytosis, solutes bind to a receptor, which is then internalized into vesicles along with the solute.

plasma membrane of liver cells. These active processes can be divided into two classes, endocytosis and carrier-mediated transport (Figure 3).

Endocytosis

In endocytosis, water and dissolved substances in the extracellular milieu are internalized into intracellular vesicles by the invagination and pinching off of part of the plasma membrane (*62, 63*). Peptides may enter liver cells by fluid-phase or adsorptive endocytosis. In fluid-phase endocytosis, which is a constitutive process, dissolved substances enter the cell without prior binding to the plasma membrane. In adsorptive endocytosis, substances bind to the plasma membrane either nonspecifically (e.g., by electrostatic charge) or specifically, by interaction with cell surface receptors in receptor-mediated endocytosis (*63, 64*). Hepatocytes apparently have a huge capacity for fluid-phase endocytosis and membrane turnover. Cultured hepatocytes were estimated to internalize the equivalent of ≥20% of their volume per hour (*65*). In rats, the half-life for plasma clearance of [^{125}I]labeled bovine serum albumin, which is presumably cleared mainly by the liver via fluid-phase endocytosis, is approximately 200 min (*66*). Thus, most water-soluble peptides that are not cleared by specific mechanisms are probably internalized into liver cells by fluid-phase endocytosis to some degree. Similarly, hydrophobic peptides that diffuse into the hepatocyte plasma membrane may be internalized constitutively into vesicles during fluid-phase endocytosis.

Peptide hormones such as insulin, glucagon, EGF, and TGF-β are cleared by liver cells via receptor-mediated endocytosis (*7, 67–69*). The term "receptor" (*63*) is used to refer to membrane proteins that are internalized by endocytosis along with their ligand (Figure 3). Membrane proteins that are responsible for the internalization of ligands but that are not internalized will be referred to as carrier proteins or transporters (Figure 3). The process of receptor-mediated endocytosis requires the presence of a specific receptor on the plasma membrane, and many of these receptors have been characterized (*63*). The details of the internalization of a receptor and its ligand were reviewed elsewhere (*63, 64, 70*), but the process is described briefly as follows:

1. Receptors move to or are already present in coated pit regions when ligands bind to them.
2. The plasma membrane invaginates and pinches off to form a coated vesicle containing the receptor and its ligand. Specific amino acid sequences in the cytosolic tail of some receptors were identified as internalization signals that may serve to regulate internalization of coated pits into vesicle coats (*71–73*). The vesicle coats consist of the proteins clathrin and adaptins and other associated proteins (*74*).
3. Coated vesicles are rapidly uncoated via an uncoating adenosine triphosphatase (ATPase) and targeted to various intracellular locations (e.g., bile canaliculus or lysosomes) depending on the receptors present (*11, 64*) (*See* also Intracellular Transport).

Exogenous peptides with structures similar to native ligands that are cleared by the liver via receptor-mediated endocytosis are likely to be cleared by the same mechanism. Peptides could also be bound to larger molecules that are then cleared by receptor-mediated endocytosis. For example, TGF-β and interleukin-1β bind to α-2 macroglobulin, which is then rapidly cleared by α-2 macroglobulin receptors in the liver (*75, 76*). One report (*77*) suggested that cyclosporin A was complexed to low density lipoprotein (LDL) and cleared by the LDL receptor. However, in a study (*78*) using IPRL, the presence or absence of LDL in perfusate had no significant effect on the clearance of cyclosporin A. Peptide inhibitors of plasma enzymes could also be cleared by receptor-mediated endocytosis along with their target enzyme. The plasma enzyme renin is cleared by the liver via receptors on hepatic sinusoidal-lining cells that recognize mannose-terminated glycoproteins (*27*). Given the great interest in renin-inhibitory peptides, studies to determine if any of these peptides are cleared by receptor-mediated endocytosis when bound to renin would be useful. Also, researchers may want to know if any of these peptides affect the receptor-mediated clearance of renin from the blood. Finally, hepatocytes possess receptors for galactose- or *N*-acetyl-galactosamine-terminated asialo-glycoproteins. These receptors have been used as targets to deliver drugs specifically to the liver by coupling of the drug to asialoglycoproteins (*79*).

Carrier-Mediated Transport

Carrier-mediated transport is probably the most important mechanism by which small, hydrophobic peptides are rapidly internalized by liver cells. Over the past several years, a great deal was learned about the mechanisms of internalization of organic ions by hepatocytes. Even though these studies focused mainly on nonpeptides the identification of specific carrier systems and the principles by which these systems function are relevant of the liver because carrier-mediated transport was shown to be involved in the heaptic uptake by the liver of several classes of peptides.

General Concepts of Carrier-Mediated Transport

In hepatocytes, carrier-mediated transport is involved in both the hepatic uptake of organic ions at the basolateral membrane and biliary excretion of organic ions at the canalicular membrane (*80*). Thus, the mechanisms reviewed here are relevant to both hepatic uptake and biliary excretion. In general, carrier-mediated transport is the movement of a solute into or out of a cell against a concentration or electrochemical gradient (*81*). Carrier-mediated transport can be classified into several different types based on the source of energy

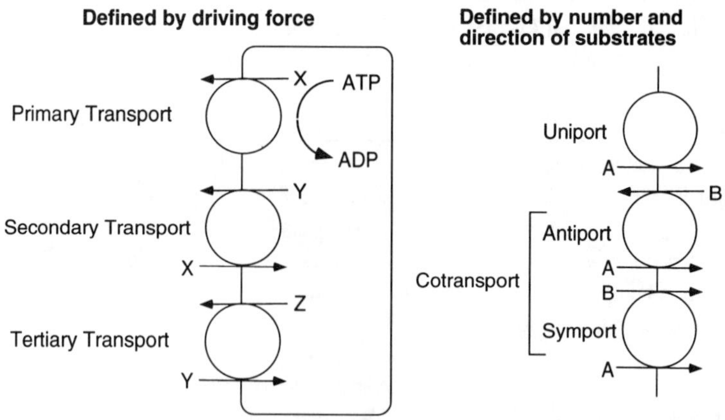

Figure 4. Terminology used to describe carrier-mediated transport processes. Transporters defined by driving force. Primary transporters are driven by ATP hydrolysis. Secondary transporters are driven by a gradient produced by the action of a primary transporter (shown above, the transport of ion Y is driven by the ion X gradient produced by the primary transporter). Tertiary transport is driven by a gradient produced by a secondary transporter (shown above, the transport of ion Z is driven by ion Y). Transporters defined by number and direction of substrates. Transporters may move only one substrate (uniport) or more than one substrate (cotransport). Cotransporters may move more than one substance in the same direction (symport) or in opposite directions (antiport).

used to move substances across the plasma membrane (Figure 4) (*81*). Primary transporters use ATP directly for energy to transport one or more ion [e.g., Na^+ and K^+-ATPase (*82*)]. Secondary transporters use the favorable electrochemical gradient generated by the transport of one solute across the cell membrane to drive transport of another substance against its concentration gradient (*83*). Similarly, tertiary transporters use a gradient generated by a secondary transport of an ion to drive transport of a further ion (*81*). Examples of ATP-dependent primary transporters will be discussed in the section on biliary excretion of peptides. Most studies indicate that organic ions are carried across the hepatocyte basolateral membrane by secondary or tertiary transporters that are driven, either directly or indirectly, by the inwardly directed Na^+ gradient maintained by the Na^+, K^+-ATPase on the basolateral membrane (*80*). One study (*84*) reports that purified human liver basolateral vesicles are able to transport the bile acid taurocholate via an ATP-dependent mechanism; however, the physiological significance of this observation is unknown.

The hepatocyte basolateral membrane possesses a number of secondary or tertiary organic-ion transport systems, one or more of which may be involved in peptide transport. Although research in this area is active, some of the findings regarding these carrier systems disagree on which specific membrane proteins are the actual carriers. Nevertheless, the hepatocyte basolateral membrane is thought to possess separate transporters for bile acids, bilirubin–bromosulphophthalein (BSP), fatty acids, amino acids, and organic cations (Table III) (*7, 12, 67–69, 80, 85, 86*). Of greatest relevance here is that several classes of peptides have been shown to be transported by multispecific transporters that appear to be related to bile transporters.

The carriers responsible for sinusoidal bile acid transport have been the subject of extensive investigation. Although questions remain, bile acids are thought to be internalized by hepatocytes, predominantly via an Na^+-dependent carrier that has a greater affinity for conjugated than unconjugated bile acids (*85*). When photoaffinity probes based on bile acids are bound to rat-liver plasma membranes, characteristic polypeptides of 48–54 kDa

Table III. Organic Ion Carriers of the Hepatocyte

Carrier	Substrates	Driving Force	Gene Cloned	Reference
Sinusoidal Membrane Importers				
Bilirubin/BSP transporter	Bilirubin, BSP, rifampicin	Cl^{-} [a]	Yes	(*85, 89*)
Amino acid carriers	Amino acids	Na^{+}, Na^{-} [b]	Yes	(*86, 182*)
Fatty acid transporter	Fatty acids	Na^{+}	No	(*183*)
Organic cation transporter	Thiamin-HCl, quaternary Ammonia compounds		No	(*12, 184, 185*)
Bile acid transporter(s)	Bile acids, steroids	Na^{+} [c]	Yes	(*52, 67, 69*)
	Peptides			(*91, 93, 102*)
Canalicular Membrane Exporters				
Bile acid transporter(s)	Monovalent bile acids, some β-lactams	ATP	No	(*57, 151*)
Multispecific organic anion transporter	Divalent bile acids, glutathione-disulfides, bilirubin conjugates	ATP	No	(*57, 82*)
P-Glycoprotein (MDR gene product)	Colchicine, vinblastine, daunomycin, cyclosporin A	ATP	Yes	(*57, 151, 156, 158*)

NOTE: ATP is adenosine 5′-triphosphate.

[a] Some controversy exists as to the driving force and identity of the BSP–bilirubin carrier. The recently cloned protein that supports transport (*85*) is Cl^{-}-dependent.

[b] Several amino acid carrier systems have been characterized. Most are reported to be Na^{+}-dependent.

[c] The primary carrier responsible for bile-acid transport is Na^{+}-dependent. Linear peptide transport and a portion of bile-acid transport are Na^{+}-independent.

are labeled (*87*). Recently, Hagenbuch et al. (*88*) demonstrated the functional expression of a Na^{+}-dependent, bile-acid carrier from rat liver. Expression of the carrier was accomplished by injecting rat liver poly $(A)^{+}$ mRNA into *Xenopus laevis* oocytes and screening for Na^{+}-dependent taurocholate uptake. The gene for this transporter was cloned from a rat liver cDNA library using the same expression and screening techniques (*67*). Sequence analysis of the cDNA of the positive clone indicated that the transporter protein had a polypeptide core of approximately 39 kDa and five potential glycosylation sites. In vitro translation of the mRNA for the transporter yielded a glycoprotein of 41 kDa (34 kDa after deglycosylation). This 41-kDa polypeptide is not known to be related to the 50-kDa polypeptide identified by photoaffinity labeling with bile-acid proteins (*87*). The functional mass of the Na^{+}-dependent bile-acid transporter was estimated by radiation-inactivation analysis to be 170 kDa (*68*). Bile acids are also transported across the hepatocyte sinusoidal membrane by an Na^{+}-independent mechanism that has a higher affinity for unconjugated than conjugated bile acids (*85*). The identity of the Na^{+}-independent transporter is still in question (*85*); however, Na^{+}-independent transport of bile acids by the cloned Cl^{-}-dependent BSP–bilirubin transporter was reported (*89*).

Multispecific Bile-Acid Transporter

Over the past decade, Ziegler, Frimmer, and others have reported on the concept of bile-acid carriers as multispecific transporters. In most of these studies, isolated hepatocytes were incubated with radiolabeled peptides under various conditions, and the cell was centrifuged through oil to remove unbound radioactivity (*36*). Cyclic peptides were among the first peptides identified as substrates of bile-acid carriers (*90–93*). Cyclic somatostatins were shown to competitively inhibit the uptake by isolated hepatocytes of radiolabeled cholate and phalloidin (*90, 92*). After exposure to light, a photoaffinity derivative of a synthetic, cyclic somatostatin analogue irreversibly inhibited the uptake of taurocholate by hepatocytes and labeled proteins of the hepatocyte plasma membrane. These labeled proteins included those in the size range (48–54 kDa) of bile acid carriers (*91*). Munter et al. (*92*) compared the effectiveness of closely related fungal peptides as competitive inhibitors of radiolabeled phalloidin uptake by hepatocytes. They concluded that the common transport system for these substances preferred substrates of low molecular weight (400–1200 Da), high lipophilicity, and rigid ring structure (*92*).

Another cyclic peptide that may be cleared by the multispecific transporter is microcystin, a toxic heptapeptide from the blue–green alga *Microcystis aeruginosa.* Hooser et al. (*94*) showed that tritiated microcystin was taken up by rat hepatocytes and by the IPRL. Uptake by the IPRL was inhibited by low temperature and by rifampicin, a substrate for the bilirubin carrier. Tissue and cell fractionation revealed that the majority (70%) of recovered microcystin was present in the liver cytosol rather than associated with membranes. These results suggest that microcystin is internalized via the multispecific transporter rather than receptor-mediated endocytosis. Thus, a number of cyclic peptides are all apparently internalized by hepatocytes by the same mechanism. Not all cyclic peptides, however, are transported by this carrier. Cyclosporin, which inhibits uptake of ligands by this carrier, is apparently not transported into the cell by this mechanism (*61, 95*).

Our own research in the area of hepatic clearance of small peptides began with studies of the mechanism of hepatic clearance of cholecystokinin (CCK) peptides. CCK, an important gastrointestinal peptide hormone, is secreted into the blood from the small intestine (*96*). In the blood, CCK is present as several different biologically active forms, the most important of which are probably CCK-33, which has a sulfated tyrosine at amino acid 27, and CCK-8, which consists of the eight amino acids at the C-terminus of CCK-33 (*96*). Previous studies (*97*) suggested that the concentration of CCK-8, but not CCK-33, was higher in portal than systemic venous blood. Therefore, we investigated the hepatic clearance of several forms of CCK using the nonrecirculating IPRL and [^{125}I]Bolton–Hunter-labeled CCK peptides (*98*). We first compared the extraction by the IPRL of radiolabeled CCK-33, CCK-8, desulfated CCK-8 (CCK-8-DS), and two tetrapeptides (CCK-4 and CCK-26-29-DS). CCK-4 and CCK-26-29-DS make up the C-terminus and the N-terminus of CCK-8-DS, respectively. When injected into the IPRL as a bolus, the first pass extraction (percent of injected peptide found in the liver) after 20 min was similar for CCK-8, CCK-8-DS, and CCK-4 (approximately 30%) and much lower for CCK-33 (3%) and CCK-26-29-DS (10%). These results suggested that rapid hepatic uptake of CCK forms was determined by the CCK-4 portion of the molecule, but that the larger size of CCK-33 probably prevented its uptake. We further showed that the hepatic uptake of CCK-8 could be inhibited by lectins, a finding that indicated the potential importance of hepatic membrane glycoproteins for uptake, and oxidation of CCK, a finding that suggested CCK structure was an important determinant of uptake (*99*).

The specific mechanism of hepatic uptake of [^{125}I]Bolton–Hunter-labeled CCK-8 was investigated using isolated liver cells (*36*). By using centrifugation through oil to separate bound from free radioactivity, we showed the following:

1. CCK-8 was taken up by hepatocytes but not by Kupffer or endothelial cells.
2. Uptake was time- and temperature-dependent.
3. Uptake was saturable by the addition of unlabeled CCK-4.
4. Uptake was inhibitable by the addition of taurocholate, BSP, metabolic inhibitors, and 4,4′-diisothiocyanostilbene-2-2′-disulfonic acid (an inhibitor of organic anion transport.
5. Uptake was not Na^+ dependent but did require the presence of extracellular ions.

These results suggest that CCK-8 is internalized by hepatocytes via a similar or identical system to the multispecific transporter identified by Frimmer, Ziegler, and co-workers (*90, 91, 100*). A concern in these studies is that the Bolton–Hunter label, which is itself hydrophobic, may target CCK-8 to the multispecific transporter; however, inhibition uptake of radiolabeled CCK-8 by unlabeled CCK-4 suggests that the Bolton–Hunter reagent is not required for the targeting of CCK forms to the transporter.

We investigated the determinants for hepatic extraction of oligopeptides by using the IPRL model to measure the first pass clearance of 13 synthetic, radiolabeled tetrapeptides, some of which were related to CCK-4 (*101*). Peptide uptake by the liver 20 min after injection ranged from 4 to 86%. The magnitude of uptake of a peptide was significantly positively related to its hydrophobicity (r + 0.96; $P < 0.05$). A subset of the peptides that are most related to CCK-4 are shown in Figure 5. Peptide charge was apparently not very important in determining extent of uptake; Trp–Met–Asp–Phe–NH_2, Trp–Met–Asn–Phe–NH_2, Trp–Met–Arg–Phe–NH_2, Trp–Met–Asp–Phe–OH have charges of −1, 0, +1, and −2, respectively, and were all cleared to a similar degree. Amino acid sequence, however, was apparently an important determinant because the reversal of the position of two amino acids (Met–Asp to Asp–Met) doubled the extent of hepatic uptake. One caveat in the interpretation of these results is that we did not demonstrate that each of these peptides is cleared by the same transporter and would mutually compete for uptake. Thus, the possibility exists that some of the peptides were cleared by a different mechanism such as the organic-cation transporter.

Our studies showed that in addition to cyclic peptides, linear peptides may be cleared by the multispecific transporter. Further evidence for clearance of linear peptides by the multispecific transporter came from studies of linear peptide-based renin inhibitors. Because of the important role of renin in hypertension, a great deal of effort has been put into the development of orally active renin inhibitors. Many compounds that were developed had good inhibitory properties and were absorbed in the intestine but had very short half-lives in the blood. Once these inhibitors were known to be cleared by the liver, Bertrams and Ziegler (*35, 102*) attempted to determine the mechanism of hepatic clearance. They showed that several renin inhibitors competitively inhibit the uptake of cholate and taurocholate by isolated hepatocytes (*102*). In contrast, uptake of substrates for other hepatic carriers (bilirubin, amino acids, fatty acids, and organic cations) was not inhibited. The uptake by hepatocytes of one renin inhibitor (EMD 51921) was studied in detail (*35*). Although a portion of the uptake of this compound could be ascribed to diffusion, most of its uptake was saturable and inhibitable by metabolic poisons, absence of O_2, and low temperature (*35*). Uptake was inhibited in a competitive manner by cholate, taurocholate, and the cyclo-somatostatin analogue described earlier. The question of Na^+-dependence of the carrier process was further complicated by the results of these studies. The uptake of EMD 51921 was Na^+-independent, but the same compound competitively inhibited the Na^+-dependent uptake of taurocholate (*35, 102*). Finally, Ziegler and Sänger (*103*) identified hepatocyte plasma membrane proteins that bind to renin inhibitors by passing detergent liver-solubilized plasma membranes over a renin inhibitor affinity column and eluting with taurocholate.

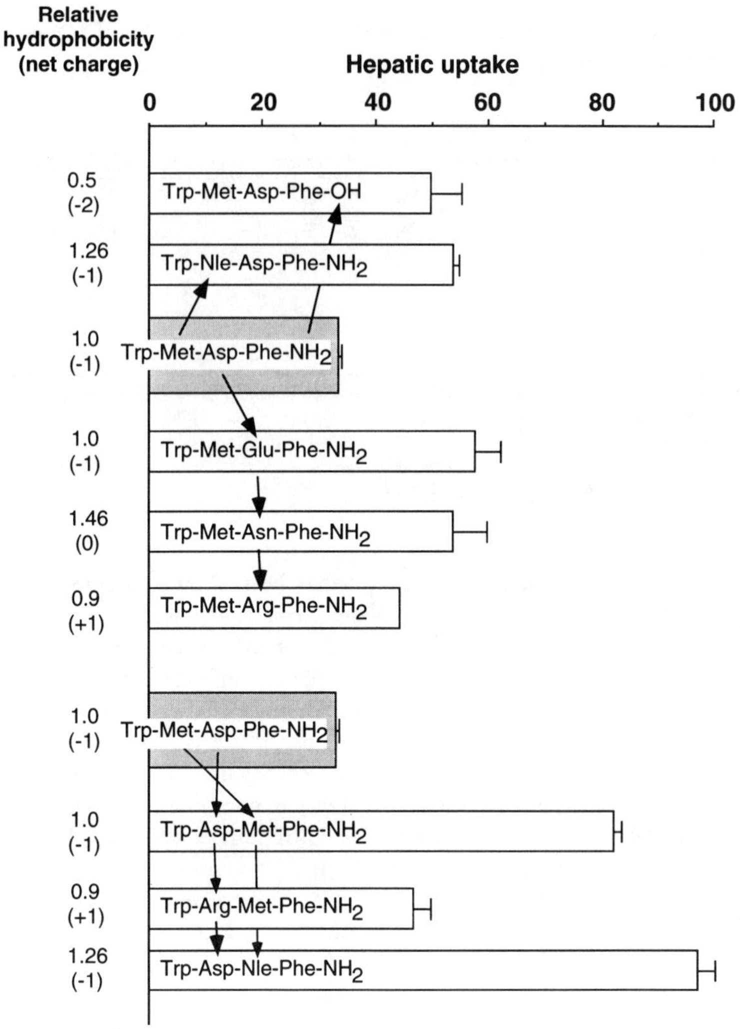

*Figure 5. Uptake of CCK-4 and related peptides by the isolated perfused rat liver: influence of amino acid sequence, hydrophobicity, and charge. The length of bars represents mean (± standard errors) of first-pass clearance of each [^{125}I]BH-labeled peptide by the perfused liver expressed as a percent of total infused radioactivity. Numbers at the upper left of each bar are calculated relative hydrophobicity of each peptide. Numbers in parentheses at the lower left of each bar are net charges of the peptide. Shaded bars indicate the parent peptide (CCK-4) (*101*).*

The profiles of purified proteins were similar to those identified using photoaffinity probes for other substrates of the bile-acid transporter (46–54 kDa) and provided further evidence that the renin inhibitors are transported into hepatocytes by the same carrier (*87, 91, 103*).

Our data show that a number of different cyclic and linear peptides may be transported by a multispecific bile-acid transporter. Uptake of the previously discussed peptides is apparently saturable, inhibitable in a mutually competitive manner, and energy- and temperature-dependent. A comparison of characteristics of the transport into hepatocytes of CCK-8, renin inhibitors, taurocholate, and BSP (Table IV) shows that some major features of uptake are similar for all four substrates. The identity of the peptide carriers,

Table IV. Comparison of Characteristics of Uptake by Hepatocytes of CCK-8, Renin Inhibitor EMD 51921, Taurocholate, and BSP

		Inhibitor				*Replacement of Cl^-*			
Substrate	*Saturable*	*Taurocholate*	*BSP*	*Energy-Dependent*	*Na^+-Dependent*	*SO_4^-*	*Gluconate*	*NO_3^-*	*SCN^-*
CCK-8	Yes	Yes[a]	Yes[a]	Yes	No	↓	↓	—[b]	↑
Renin inhibitor (EMD 51921)	Yes	C	NC	Yes	No	↓	—[b]	↑	↑
Taurocholate	Yes		NC	Yes	Yes	↓	—[b]	↑	—[b]
BSP	Yes	NC		Cl^-[c]	No	↓	↓	↑	—[b]

NOTES: C is competitive; NC is noncompetitive. Up and down arrows denote increased or decreased uptake activity in the presence of the ion.
[a] Competitiveness of inhibition was not determined.
[b] Dashes indicate that no data were available.
[c] Energy requirements of the BSP and bilirubin transporter are controversial.
SOURCE: Data are from references *36, 49, 67, 93, 102,* and *186.*

however, is still in question. Phalloidin has been shown to competitively inhibit Na^+-dependent taurocholate uptake by isolated hepatocyte basolateral-membrane vesicles (*52*). Conversely, phalloidin uptake by isolated hepatocytes is competitively inhibitable by taurocholate. These results strongly suggest that phalloidin is cleared by the Na^+-dependent bile-acid transporter. Although the uptake of both renin inhibitors and CCK-8 by hepatocytes is inhibitable by taurocholate, uptake of these peptides was Na^+-independent (*35, 36*). Thus, a different, previously unidentified transporter may be responsible for hepatic uptake. With the recent cloning and expression in *Xenopus* oocytes of the Na^+-dependent bile-acid carrier and the Cl^--dependent bilirubin–BSP carrier (*67, 88, 89*), the ability of these transporters to support uptake of peptides will be determined conclusively. When one considers the varied structures of peptides cleared by carrier-mediated transport the characteristics that allow recognition of these substances by a transporter that excludes some other organic anions are difficult to pinpoint. At present, we can surmise that these molecules probably circulate in the blood bound to albumin, are relatively hydrophobic, have molecular weights between 200–1400 Da, and are mainly negatively charged at physiological pH. Also, many of these compounds contain rigid ring structures and aromatic amino acids (*92*).

Fate of Internalized Peptides

Once internalized into liver cells, peptides are directed to an intracellular compartment such as vesicles, or cytosol. Peptides are then transported to an appropriate location for metabolism and excretion. The following sections summarize current knowledge about the hepatic intracellular compartmentalization, metabolism, and biliary excretion of peptides.

Intracellular Compartmentalization

Substances internalized into cells via an endocytic mechanism are immediately packaged in vesicles. In contrast, substances internalized by carrier-mediated transport are initially delivered to the cytosol. One method for the localization of internalized substances to a specific intracellular compartment is subcellular fractionation, in which cells or tissues are disrupted and fractions enriched in various intracellular compartments are isolated by centrifugational techniques (*7, 27*). For example, after uptake of radiolabeled TGF-β, CCK-8, microcystin, and cyclosporin by liver or hepatocytes, the relative association of radiolabel with vesicles versus cytosol was quantified (*7, 36, 61, 94*).

These results are summarized in Figure 6. Twenty minutes after TGF-β was internalized into liver cells by endocytosis, radiolabel was predominantly (55%) associated with the membrane fraction (*7*). This membrane-associated radiolabel was localized within hepatic vesicles rather than bound to the membrane itself because the radiolabel was easily releasable from the membrane with a low concentration of Triton X-100 (*7*). In contrast, after internalization of CCK-8 and microcystin, which are apparently internalized via carrier-mediated transport, >76% of recovered radiolabel was found in the cytosol (*36, 94*). Finally, for cyclosporin A, radiolabel was distributed equally between membranes and cytosol (*61*). This result is consistent with other observations on the hepatic processing of cyclosporin that showed that cyclosporin is apparently internalized into liver cells by diffusion and not by carrier-mediated transport (*61*) and that cyclosporin binds to both membrane proteins and cytosolic proteins (*60*).

Peptides internalized into the cytosolic compartment may be bound to cytosolic binding proteins, as are nonbile acid organic ions such as bile acids, BSP, and fatty acids (*104–106*). Three families of cytosolic, organic ion binding proteins have been identified: glutathione S-transferases; oxidoreductases, referred to as Y′ bile-acid binders; and fatty acid binding proteins (*107–109*). Although some of these proteins have enzyme activities toward the organic ions that bind to them, the physiologic function of the binding of these proteins to internalized organic ions is unknown. One possibility is that these proteins act as a sink for lipophilic compounds lower the free cytosolic concentrations of their ligands, and consequently facilitate internalization or minimize toxicity of organic ions (*106, 110*).

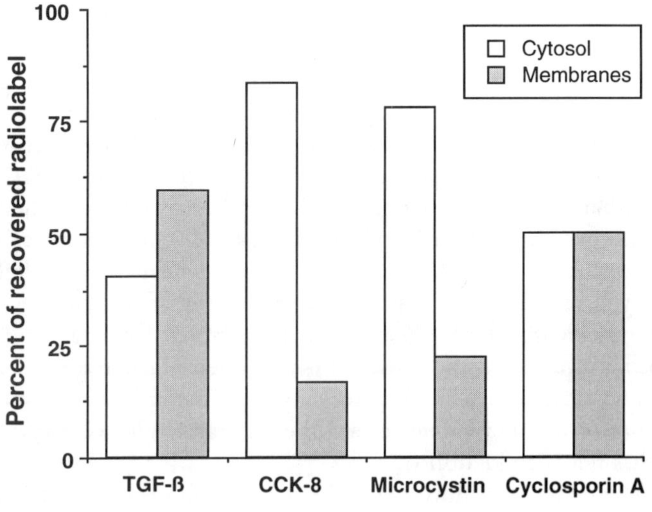

Figure 6. Hepatic subcellular compartmentalization of internalized radiolabeled peptides. Cytosol and membranes were prepared by ultracentrifugation of homogenized rat liver or rat hepatocytes at 5 min for [CCK-8 (data from reference 36) and cyclosporin A (data from reference 100)], 20 min for TGF-β (data from reference 7) or 45 min for microcystin (data from reference 94)] after internalization of radiolabeled peptides.

To our knowledge, the binding of hydrophobic peptides to organic binding proteins has not been explored. However, microcystin internalized by rat liver cell was found in the cytosol bound mainly to cytosolic proteins rather than in the free state (*94*), a finding that suggests binding of microcystins to the above mentioned or other cytosolic proteins. Cyclosporin A is known to bind to intracellular proteins called cyclophilins, which are the probable target molecules for the immunosuppressive effects of cyclosporin (*111*). In hepatocytes, a cyclosporin photoaffinity probe was shown to bind to a 17-kDa cytosolic polypeptide that may be similar to cyclophilin (*60*).

Intracellular Transport

The mechanisms by which vesicles are targeted to specific locations are areas of great research interest [for review, *see* (*2, 64, 112*)]. Ligands internalized into vesicles are transported intracellularly through a series of compartments called endosomes. These compartments serve to sort out ligands from their receptors and segregate substances bound for different intracellular locations (e.g., the sinusoidal plasma membrane, lysosomes, or bile canaliculus) (*113, 114*). Most substances that have undergone endocytosis are eventually transported to the lysosomal compartment for degradation (*7, 27, 115*). For example, we showed (*7*) that 20 min after internalization of radiolabeled TGF-β by the liver, radioactivity was concentrated in lysosomes (*7*). Some substances cleared by the liver via endocytosis, however, may not be targeted to lysosomes. For example, polymeric immunoglobulin A, which is cleared by a receptor on the sinusoidal membrane of hepatocytes in rats, largely bypasses the lysosomal compartment and is delivered to the area of the bile canaliculus intact (*11*). Even for ligands that are normally targeted to the lysosomes, a small percent is sometimes secreted unmetabolized into bile, presumably via the same transcytotic mechanism (*7, 116*). The specific sorting signals that target vesicles to different locations have not been identified but probably reside in receptors and other integral membrane proteins.

Substances internalized into vesicles undergo intracellular transport mainly via the microtubule cytoskeleton. Early evidence (*29, 117, 118*) for this pathway came from indirect studies where inhibitors of microtubule function (e.g., colchicine, vinblastine, and taxol) inhibited the transport of a ligand across the hepatocyte. Recent studies (*119, 120*) demonstrated the movement of vesicles along microtubules in vitro in the presence of ATP. The means by which vesicle move along microtubules are under study in a number of laboratories. Two proteins have been identified that probably provide the motive force for movement along microtubules. These proteins, kinesin and cytoplasmic dynein, are mechanochemical ATPases that transduce the energy of ATP hydrolysis into movement (*121, 122*). Kinesin and dynein, which are both found in liver (*54, 123*), move vesicles in opposite directions along microtubules (*119, 124, 125*). In general, dynein has been linked to the movement of endocytic vesicles and kinesin to the movement of secretory vesicles (*119, 126, 127*). Dynein was identified on endocytic vesicles isolated from rat liver (*128*). Thus, dynein is probably involved in the inward-directed intracellular transport of vesicles containing endocytotic substances at the hepatocyte plasma membrane. The motor enzyme, which is responsible for the movement toward the bile canaliculus of transcytotic vesicles or lysosomes, has not been identified.

The mechanisms involved in the transport of substances delivered to the cytosol are obscure. Transport through the cytosol may occur by diffusion, as has been suggested for bile acids (*110*). Organic ions bound to cytosolic proteins could be transported via the actin-myosin microfilament cytoskeleton. Evidence for transport of organic ions via the actin cytoskeleton is largely indirect and comes from studies showing that inhibitors of actin polymerization (e.g., phalloidin and cytochalasins) interfere with the biliary excretion of bile acids (*118, 129, 130*). Alternatively, the suggestion was made (*131*) that microfilament poisons inhibit the actin-myosin network present around the canaliculus and therefore, inhibit the necessary contractions of the canaliculus for bile secretion. Microtubules are probably not involved in the intracellular transport of internalized substances that are compartmentalized in the cytosol. In our studies (*99*) of CCK processing, we showed that 20 min after injection of radiolabeled CCK-8, more than 75% of the radiolabel extracted by the liver was excreted into bile. Delivery of radiolabel to bile was not inhibited by microtubule-disrupting agents, a finding that indicated a lack of involvement of microtubules in CCK transport (*99*). Under basal conditions, microtubule inhibitors had little effect on hepatic transport of bile acids (*132*). Under certain conditions, however, organic ions may be transported via the microtubule cytoskeleton. In experiments (*132, 133*) where rats were depleted of bile acids by biliary diversion and then reinfused with bile acids, biliary excretion of bile acids and bilirubin conjugates became inhibitable by colchicine. Further, under these conditions, bile acids could be localized to the endoplasmic reticulum and Golgi by autoradiography and immunocytochemistry (*109, 134*). Under the conditions of high bile-acid flux through the hepatocyte, bile acids and bilirubin conjugates accumulate in membrane-bound compartments and may be excreted into bile by exocytosis (*110*). No evidence exists for the extension of this mechanism to the transport of hydrophobic peptides.

Metabolism of Internalized Peptides

As summarized in Table II, substances internalized into vesicles usually are degraded in lysosomes, whereas substances taken up by carrier-mediated transport are likely to be metabolized in the cytosol. Many endocytotic proteins and peptides are transported to the lysosomal compartment of the cell, which is usually near the cell center (*11, 135, 136*). There, the ligand-containing vesicles fuse with or mature into lysosomes (*64, 137*). Lysosomes are specialized acidic vesicles that contain a vast array of hydrolases with acid pH optima capable of degrading most or all biological macromolecules (*138*). We showed (*7*) that [^{125}I]TGF-β accumulated in hepatic lysosomes after injection into bile fistula rats. In addition, after uptake of radiolabeled peptide by the liver, radiolabeled degradation products of TGF-β were rapidly excreted into bile. When rats were pretreated with chloroquine and leupeptin, (which inhibit lysosomal proteolysis) before internalization of TGF-β, the delivery of radiolabeled metabolites to bile was significantly reduced (*7*). These results indicate that TGF-β is degraded within hepatic lysosomes.

For peptides internalized by carrier-mediated transport into the cytosol, the potential for metabolism by microsomal enzymes or cytosolic peptidases exists. The well-known metabolism of xenobiotics by microsomal cytochrome P-450 monooxygenases usually involves the activation of a substance followed by the conjugation to a polar molecule (*139, 140*). This process is important because it serves to target the substance for biliary excretion by canalicular carrier proteins. Although peptides are not normally thought to be substrates for cytochrome P-450 enzymes, cyclosporin A is metabolized by enzymes of the P-450IIIA family to a series of hydroxyl and *N*-methyl derivatives that retain the cyclic peptide nucleus of cyclosporin (*141, 142*). No glucuronide conjugates of cyclosporin have been found, although a sulfate conjugate was reported (*143*). Other peptidase-resistant peptides are probably metabolized by cytochrome P-450 enzymes or conjugated to polar forms by the liver, but no other data are available.

A number of cytosolic enzymes have been described, including proteases with broad specificity (e.g., calpains) and peptidases with more specific cleavage sites (e.g., prolyl peptidase and some enkephalinases), that could be responsible for the degradation of internalized peptides (*144–146*). Our studies with [^{125}I]CCK-8 showed that it was degraded in the liver

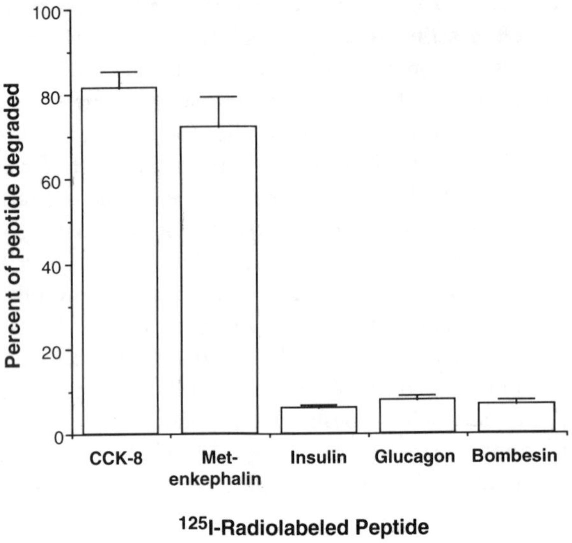

Figure 7. Degradation of radiolabeled peptides by CCK-8 peptidase purified from rat liver. [^{125}I]Bolton–Hunter-labeled peptides were incubated with 70 ng of purified CCK-8 peptidase in 25 mM Tris-maleate, 1 mM Ca^{2+}, pH 6.0, for 15 min at 37 °C. Degradation was monitored by aluminum silicate binding for CCK-8, Met- enkephalin, and bombesin and by trichloroacetic acid precipitation for insulin and glucagon.

and its metabolites were excreted into bile; however, the lysosomotropic agents leupeptin and chloroquine had no effect on the release of metabolites into bile, and these results suggest that metabolism did not occur in lysosomes (*36*). These results plus the demonstration that CCK-8 was compartmentalized mainly in the cytosol and not in vesicles (*99*) suggest that CCK-8 is degraded by a cytosolic enzyme. Thus, we tried to identify the enzyme responsible for this degradative activity. We first developed a method for rapid quantitation of CCK-8 degrading activity (*147*). This method took advantage of the fact that intact CCK-8 binds to aluminum silicate but that degraded CCK-8 products do not. Using this technique for screening, we purified a thermostable metallopeptidase from rat liver cytosol that degrades CCK-8, enkephalins, and neurotensin but not insulin, glucagon, or bombesin (Figure 7) (*55*). Resistance of this peptidase to phosphoramidon, an inhibitor of enkephalinase (EC 24.11), and other properties suggested that the peptidase may be a novel enzyme that is involved in the degradation of oligopeptides delivered to the hepatic cytosol.

Hepatic Excretion of Peptides

Consistent with the pathway shown in Table II, substances internalized by the liver within vesicles are often excreted from the liver via the vesicular process of exocytosis. Exocytosis is the functional reverse of endocytosis in that it requires the fusion of a vesicle membrane with the plasma membrane and the extrusion of the vesicle contents into the extracellular medium. As mentioned previously, some materials are transported to the pericanalicular area via the transcytotic pathway (*11*). From this area, the contents of these vesicles can be released into bile by exocytosis (*148, 149*). Also, the contents of hepatocyte lysosomes can be delivered eventually to the bile canaliculus and excreted by exocytosis (*138, 150*). Thus, both intact and metabolized peptides could be delivered into bile via exocytosis. Alternatively, lysosomal contents, including degraded proteins or peptides, may also be released from liver cells into the blood. This release would be the case if the substance of interest was internalized and degraded by the cells of the liver other than hepatocytes. For example, we showed (*27*) that a significant fraction of breakdown products of radiolabeled renin, which was degraded largely in lysosomes of hepatic sinusoidal endothelial cells and Kupffer cells, was released back into the blood and eventually cleared by the kidneys.

Substances that enter the liver via carrier-mediated transport typically are excreted into bile by a transporter. A number of carrier systems have been identified in the hepatocyte canalicular membrane. As was the case for the basolateral carriers that imported organic ions, canalicular exporting carrier systems have been studied intensively but are only partially characterized. At present, evidence for a number of multispecific, canalicular carrier systems exists, such as transporters for bile acids, glutathione conjugates, and xenobiotics (Table III) (*48, 57, 80*). Plasma-membrane vesicles enriched in canalicular membranes possess both electrogenic and ATP-dependent, monovalent, bile-acid transporting capabilities (*82*). The electrogenic system was reported to transport some β-lactam antibiotics (*151*). A second carrier system was identified and has broad specificity for organic anions. This system, called the multispecific organic anion transporter (MOAT) is a primary ATP-dependent carrier that transports BSP, oxidized glutathione, cysteinyl leukotrienes, and a number of glucuronide conjugates (*57, 82, 151*). The ability of this system to export oxidized glutathione (a peptide) and glucuronides suggests a potential role in the biliary excretion of peptide-related drugs.

The third canalicular carrier system with potential relevance to peptide transport is the multidrug resistance (MDR) gene product, also known as P-glycoprotein or P-170 (*57, 152*). MDR was first identified in cultured cells where treatment with colchicine induced resistance to this drug and other unrelated drugs (*153*). Resistance was correlated with lower levels of uptake of the drugs by the resistant cells (*154, 155*). The MDR phenotype was linked to the expression of P-glycoprotein in the plasma membrane of these cells (*156*). By using molecular techniques, MDR was found (*57, 156*) to be a family of proteins consisting of three members in rodents and two members in humans. Transfection studies (*157, 158*) showed that P-glycoprotein was an ATP-dependent transporter with the ability to export a broad range of drugs. Immunocytochemical studies showed that P-glycoprotein is present in normal tissues, including liver. In hepatocytes, P-glycoprotein is localized almost exclusively to the canalicular membrane (*159*). P-Glycoprotein is able to transport many hydrophobic, cationic compounds with molecular masses between 300–1000 Da, such as colchicine, taxol, vinca alkaloids, doxorubicin, and daunomycin; however, its natural substrates are unknown (*160, 161*). Peptides may be the physiologic substrates for P-glycoprotein based on the following (*57, 58*):

- P-glycoprotein homology with peptide transporters from yeast and T-cells
- the ability of *N*-acetyl-leucyl-leucyl-norleucinal to induce the MDR phenotype
- the ability of cyclosporin to competitively inhibit transport of drugs by P-glycoprotein

After excretion into bile, substances may follow several pathways. First, peptides could be degraded by canalicular ectopeptidases [e.g., dipeptidyl peptidase and aminopeptidases (*57*)]. The amino acid breakdown products of the peptides could then be reabsorbed by hepatocytes or cholangiocytes via amino acid carriers, as is the case for glutathione (*57, 151*). It is possible that some biliary peptides can be reabsorbed by cholangiocytes. We showed (*18*) that cholangiocytes are capable of absorption of proteins by fluid-phase endocytosis at their lumenal membrane. Further, because of the hypercholeretic properties of the unconjugated bile acid ursodeoxycholate (UDC), the proposal was made that UDC is reabsorbed from bile by cholangiocytes, either by diffusion or a carrier system (*162*). UDC could then be released from cholangiocytes back into blood, reabsorbed by hepatocytes, and resecreted into bile in a so called "cholehepatic shunt" (*162*). Isolated cholangiocytes are able to metabolize and conjugate some bile acids (*163*). Thus, cholangiocytes may possess mechanisms that would allow the reabsorption of peptides from bile.

If no further reabsorption of an excreted substance occurs in the intrahepatic biliary tree, the substance is delivered by the common bile duct to the duodenum. From here, the compound can be excreted in feces or reabsorbed into the blood through the intestinal wall, either with or without further metabolism by colonic bacteria. For bile acids, all of these possibilities occur (*164*). After reabsorption into blood, bile acids are taken up once again by the liver. This cycle of hepatic clearance, biliary excretion, and intestinal reabsorption is called the "enterohepatic circulation" (*164*). Evidence from patients treated with cyclosporin and from animal models suggests that metabolites of cyclosporin, but very little of the parent compound, are excreted into bile (*141, 165*). The quantities of cyclosporin metabolites that are eventually detected in blood and the increasing numbers of these metabolites detected over time (*141, 166*) suggest that cyclosporin metabolites may pass through several entero-hepatic cycles before excretion.

Alterations of Hepatic Processing

Numerous pathophysiologic and pharmacologic alterations are known to affect various stages of hepatic processing; however, little is known about specific effects of these perturbations on the processing of peptides.

Pathophysiologic Alterations

Several genetic defects interfere with the normal processing of substances by the liver. One example of a defect in hepatic internalization by endocytosis is familial hypercholesterolemia, in which the expression of the LDL receptor is absent, reduced, or aberrant (*63*). The effect of reduced clearance of LDL on the processing of peptides that bind to LDL is unknown. Hereditary defects in intracellular metabolism include enzymes for both the vesicular and nonvesicular pathways. Several lysosomal-storage diseases in humans are due to the decreased or defective expression of a glucosidase (e.g., Tay–Sachs disease, mannosidosis, and fucosidosis). These diseases cause the accumulation of oligosaccharide fragments in the lysosomes (*167*). Genetic deficiencies in microsomal UDP-glucuronyl transferases, in which individuals are unable to hydrolyze bilirubin and other organic ions via glucuronidase include Crigler–Najjar syndrome type I and Gilbert's syndrome in humans and the Gunn rat model (*168*). Genetic polymorphisms in cytochrome P-450 enzymes are well-documented and may be responsible for some individual differences in sensitivity to drugs (*140*). A final class of genetic disorders that can affect processing are defects in excretion. Three strains of rats were identified that clear bilirubin glucuronides normally but are unable to excrete them (*82, 169*). The defect in these rats is apparently in the expression of the organic-ion carrier MOAT. A similar defect is seen in humans in Dubin–Johnson syndrome (*169*).

Hepatobiliary diseases clearly affect the ability of the liver to process drugs at multiple stages. For example, in humans, hepatic disease impairs the clearance of cyclosporin (*59*). Obviously, diseases that cause cholestasis will impair biliary secretion of drugs; however, more specific effects of liver disease are known. For example, in cholestasis, bile acids accumulate in hepatocytes and blood (*170*). Multiple consequences of these high levels of bile acids for the processing of peptides include competitive inhibition of bile-acid carriers on both the basolateral and canalicular membrane, noncompetitive inhibition of other carriers, and inhibition of microsomal cytochrome P-450 enzymes and other metabolizing enzymes (*35, 92, 152, 171*). Several lines of evidence using cirrhotic rat models suggest that hepatic processing is impaired in cirrhosis. First, hepatocytes isolated from cirrhotic rat livers had reduced binding sites for asialoglycoproteins and EGF compared with hepatocytes from control rats (*172*). Second, using the IPRL model, livers from cirrhotic rats showed reduced uptake and metabolism of propranolol compared with controls (*173*). Cirrhotic rats had increased lysosomal enzyme activity and greater numbers of lysosomes in the pericanalicular region of hepatocytes, a finding that suggested impaired biliary excretion of lysosomal contents into bile (*174*). Reduced transcellular transport of a marker for fluid-phase endocytosis was also described in experimental cirrhosis (*174*). Finally, in humans, a study comparing the concentration of type III, procollagen, aminoterminal propeptide in arterial versus hepatic venous blood showed a reduced extraction of this peptide in patients with alcoholic cirrhosis compared with normal patients (*175*).

Pharmacologic Alterations

Transporter proteins of the basolateral and canalicular membranes are multispecific. Thus, drugs that are substrates of these carrier systems can competitively inhibit the transport of other substrates of the same carrier. Examples include the inhibition of the basolateral bilirubin carrier by BSP and inhibition of the basolateral, multispecific anion transporter by antamanide (*85, 92*). Some drugs that are not substrates also inhibit these carrier systems [e.g., the inhibition of P-glycoprotein by progesterone (*176*)]. Drugs such as polycyclic aromatics, dexamethazone, and ethanol alter the expression or activity of microsomal P-450 enzymes (*139, 140*), whereas drugs such as some steroids, 3-methylcholanthrene, and clofibrate affect UDP-glucuronyl transferases (*177–179*). Phenobarbitol increases the activity of both P-450 enzymes and UDP-glucuronyl transferases (*139, 178, 179*). The intracellular movement of peptides that are transported intracellularly in vesicles may be inhibited by microtubule poisons used in chemotherapy (e.g., vinca alkaloids, and taxol); transport of peptides processed via the nonvesicular pathway may be affected by microfilament inhibitors (*see* Intracellular Transport). Finally, cyclosporin A apparently can inhibit multiple stages of hepatic processing. Cyclosporin A noncompetitively inhibits the basolateral multispecific anion transporter, inhibits the activity of cytochrome P-450IIIA enzymes, and binds to and is possibly a substrate for P-glycoprotein (*47, 58, 95*). Cyclosporin also inhibits ATP-dependent transport by the canalicular bile-acid transporter and MOAT in isolated canalicular membrane vesicles. This ability suggests a possible mechanism for cyclosporin-induced cholestasis (*180, 181*). Thus, expectations of the ability of a patient to process a peptide drug must take into account the coadministration of other drugs.

Summary and Conclusions

The mechanisms by which peptides are processed by the liver are still obscure. The mechanisms of hepatic uptake of peptides are best-understood and involve uptake by receptor-mediated endocytosis or carrier-mediated transport by a basolateral multispecific transporter. With the exception of peptides cleared by endocytosis and degraded in lysosomes, little is

Table V. Strategies To Decrease Hepatic Clearance of Peptide Drugs

Strategy	*Limitations*
Structural Alterations	
Alter structure of drug to reduce uptake	Determinants of uptake not fully known. Alterations may decrease biological activity or intestinal absorption.
Alter structure of drug to reduce metabolism	Alterations may decrease biological activity or intestinal absorption.
Adjuvent Drugs	
Inhibit transporter	May inhibit transport of other substances (e.g., bile acids) and induce cholestasis.
Inhibit metabolizing enzymes	Identity of enzymes not always known. Specific inhibitors may not be available.

known about the specific mechanisms of intracellular transport and biliary excretion of peptides. By analogy with bile acids and other organic anions, short, hydrophobic peptides that are cleared rapidly by the liver are probably transported in a protein-bound state; they are bound to albumin in the blood, transported into the hepatocyte cytosol by basolateral carriers where they are bound to intracellular binding proteins, and are finally exported into bile by canalicular carriers. On the basis of our limited knowledge of the mechanisms of hepatic peptide processing, several strategies can be considered for reducing the hepatic elimination of peptide-based drugs (Table V). First, within the constraints of biological effectiveness, peptide structures could be altered either to reduce hepatic uptake (e.g., by decreasing hydrophobicity) or hepatic metabolism (e.g., by using reduced peptide bonds). These strategies are hampered by our lack of knowledge of the characteristics of hepatic carrier systems and peptide metabolizing enzymes. Structural modifications to reduce hepatic uptake may also decrease the intestinal absorption of orally administered drugs, because the hepatic and intestinal carrier systems appear to have similar determinants. A second general approach to reducing hepatic elimination could be to use adjuvant drugs to reduce uptake or metabolism. Again, such efforts are hindered by lack of information about the specificities of carriers and metabolizing enzymes. Further, inhibition of uptake of hepatic carriers may decrease uptake of endogenous substances such as bile acids, and bilirubin and cause cholestasis. Only with further research to identify and characterize the mechanisms involved in hepatic peptide-based drug uptake and metabolism will such drug design strategies become practical. With the availability of multiple liver models and the use of molecular biological techniques to identify hepatic proteins involved in processing, answers to many of the questions regarding the hepatic transport and metabolism of peptides may now be forthcoming.

References

1. Marks, D. M.; Gores, G. J.; LaRusso, N. F. In *Perfused Liver: Clinical and Basic Applications;* Ballet, F. A., Thurman, R. G., Eds.; John Libby: London, 1991; pp 209–234.
2. Marks, D. L.; LaRusso, N. F. In *Hepatic Transport and Bile Secretion: Physiology and Pathophysiology;* Tavoloni, N.; Berk, P.-D., Eds.; Raven: New York, 1993; pp 513–530.

3. McCuskey, R. S. In *Hepatic Transport and Bile Secretion: Physiology and Pathophysiology;* Tavoloni, N.; Berk, P. D., Eds.; Raven: New York, 1993; pp 1–10.

4. Jones, A. L. In *Hepatology, a Textbook of Liver Disease;* Zakim, D.; Boyer, T. D., Eds.; W.B. Saunders: Philadelphia, PA, 1990; vol. 1, pp 3–30.

5. Gumuncio, J. J.; Guibert, E. E. In *Hepatic Transport and Bile Secretion: Physiology and Pathophysiology;* Tavoloni, N.; Berk, P. D., Eds.; Raven: New York, 1993; pp 71–82.

6. Saville, B. A.; Gray, M. R.; Tam, Y. K. *Drug Metab. Rev.* **1992,** *24,* 49–88.

7. Coffey, R. J.; Kost, L. J.; Lyons, R. M.; Moses, H. L.; LaRusso, N. F. *J. Clin. Invest.* **1987,** *80,* 750–757.

8. Ramadori, G.; Rieder, H.; Knittel, T. In *Hepatic Transport and Bile Secretion: Physiology and Pathophysiology;* Tavoloni, N.; Berk, P. D., Eds.; Raven: New York, 1993; pp 83–102.

9. Tarsetti, F.; Lenzen, R.; Salvi, R.; Schuler, E.; Dembitzer, R.; Tavoloni, N. In *Hepatic Transport and Bile Secretion: Physiology and Pathophysiology;* Tavoloni, N.; Berk, P. D., Eds.; Raven: New York, 1993, pp 619–635.

10. LaRusso, N. F.; Ishii, M.; Vroman, B. T. In *Transactions of the American Clinical Climatology Association;* Waverly: Baltimore, MD, 1991; Vol. CII, pp 245–259.

11. Jones, A. L.; Burwen, S. J. *Semin. Liver Dis.* **1985,** *5,* 136–174.

12. Meijer, D. K. F. *J. Hepatol.* **1987,** *4,* 259–268.

13. Jones, E. A. *Hepatology* **1983,** *3,* 259–266.

14. Horiuchi, S.; Takata, K.; Morino, Y. *J. Biol. Chem.* **1985,** *260,* 482–488.

15. Nagelkerke, J. F.; Barto, K. P.; van Berkel, T. J. C. *J. Biol. Chem.* **1983,** *258,* 12221–12227.

16. Alpini, G.; Ulrich, C. D.; Pham, L. D.; Miller, L. J.; LaRusso, N. F. *Hepatology* **1992,** *16,* 98A.

17. Ishii, M.; Vroman, B.; LaRusso, N. F. *Gastroenterology* **1990,** *98,* 1284–1291.

18. Ishii, M.; Vroman, B.; LaRusso, N. F. *J. Histochem. Cytochem.* **1990,** *38,* 515–524.

19. Joplin, R.; Hishida, T.; Tsuboushi, H.; Daikuhara, Y.; Ayers, R.; Neuberger, J. M.; Strain, A. J. *J. Clin. Invest.* **1992,** *90,* 1284–1289.

20. Kato, A.; Gores, G.; LaRusso, N. F. *J. Biol. Chem.* **1992,** *267,* 15523–15529.

21. Withrington, P. G.; Richardson, P. D. I. In *Hepatology, a Textbook of Liver Disease;* Zakim, D.; Boyer, T. D., Eds.; W.B. Saunders: Philadelphia, PA, 1990; Vol. 1, pp 30–48.

22. Katz, N.; Jungermann, K. In *Hepatic Transport and Bile Secretion: Physiology and Pathophysiology;* Tavoloni, N.; Berk, P. D., Eds.; Raven: New York, 1993; pp 55–70.

23. Lamers, W. H.; Hilberts, A.; Furt, E.; Smith, J.; Jonges, G. N.; Van Noorden, C. J. F.; Gaasbeek Janzen, J. W.; Charles, R.; Moorman, A. F. M. *Hepatology,* **1989,** *10,* 72–76.

24. Van der Sluijs, P.; Braakman, I.; Meijer, D. K. F.; Groothuis, G. M. M. *Hepatology* **1988,** *8,* 1521–1529.

25. Branch, R. A.; Cotham, R.; Johnson, R.; Porter, J.; Desmond, P. V.; Schenker, S. *J. Lab. Clin. Med.* **1983,** *102,* 805–812.

26. Gores, G. J.; Kost, L. J.; LaRusso, N. F. *Hepatology* **1986,** *6,* 511–517.

27. Marks, D. L.; Kost, L. J.; Kuntz, S. M.; Romero, J. C.; LaRusso, N. F. *Am. J. Physiol.* **1991,** *261,* G349–G358.

28. Dennis, P. A.; Aronson, N. N., Jr. *Biochim. Biophys. Acta* **1984,** *798,* 14–20.

29. Mullock, B. M.; Jones, R. S.; Peppard, J.; Hinton, R. H. *FEBS Lett.* **1980,** *120,* 278–282.

30. Dunn, W. A.; Wall, D. A.; Hubbard, A. L. *Meth. Enz.* **1983,** *98,* 225–241.

31. Scholmerich, J.; Kitamura, S.; Miyai, K. *Res. Exp. Med.* **1986,** *186,* 379–405.

32. Pang, K. S.; Goresky, C. A.; Schwab, A. J. In *Perfused Liver: Clinical and Basic Applications;* Ballet, F. A.; Thurman, R. G., Eds.; John Libby: London, 1991; pp 259–302.

33. Campra, J. L.; Reynolds, T. B. In *The Liver, Biology and Pathobiology;* Arias, I. M.; Jacoby, J. W.; Popper, H.; Schachter D.; Schafritz, D. A. Eds.; Raven: New York, 1988; pp 911–930.

34. Ballet, F. A.; Thurman, R. G. In *Perfused Liver: Clinical and Basic Applications;* Ballet, F. A.; Thurman, R. G., Eds.; John Libby: London, 1991; pp 1–20.

35. Bertrams, A.; Ziegler, K. *Biochim. Biophys. Acta* **1991,** *1091,* 337–348.

36. Gores, G. J.; Kost, L. J.; Miller, L. J.; LaRusso, N. F. *Am. J. Physiol.* **1989,** *257,* G242–G-248.

37. Zahlten, R. N.; Hagler, H. K.; Nejtex, M. E.; Day, C. J. *Gastroenterology* **1978,** *75,* 80–87.

38. Seglan, P. O. *Methods Cell Biol.* **1976,** *13,* 29–83.

39. Berry, M. S.; Friend, D. S. *J. Cell Biol.* **1969,** *43,* 506–529.

40. Knook, D. L.; Seffalaar, A. M.; de Leeuw, A. M. *Exp. Cell Res.* **1982,** *139,* 468–471.

41. Ishii, M.; Vroman, B.; LaRusso, N. F. *Gastroenterology* **1989,** *97,* 1236–1247.

42. Alpini, G.; Pham, L. D.; Ulrich, C. D.; Miller, L. J.; LaRusso, N. F. *FASEB J.* **1993,** *7,* A496.

43. Reid, L. M.; Jefferson, D. M. *Hepatology* **1984,** *4,* 548–559.

44. Dich, J.; Vind, C.; Grunnet, N. *Hepatology* **1988,** *8,* 39–45.

45. Dunn, J. C. Y.; Tompkins, R. G.; Yarmush, M. L. *J. Cell Biol.* **1992,** *116,* 1043–1053.

46. Shah, I. A.; Whiting, P. H.; Omar, G.; Thomson, A. W.; Burke, M. D. *Transplant. Proc.* **1991,** *23,* 2783–2785.

47. Moochhala, S. M.; Lee, E. J. D.; Earnest, L.; Wong, J. Y. Y.; Ngoi, S. S. *Transplant. Proc.* **1991,** *23,* 2786–2788.

48. Suchy, F. J. In *Hepatic Transport and Bile Secretion: Physiology and Pathophysiology;* Tavoloni, N.; Berk, P. D., Eds.; Raven: New York, 1993; pp 211–223.

49. Inoue, M.; Kinne, R.; Tran, T.; Arias, I. M. *Hepatology* **1982,** *2,* 572–579.

50. Inoue, M.; Kinne, R.; Tran, T.; Biempica, L.; Arias, I. M. *J. Biol. Chem.* **1983,** *258,* 5183–5188.

51. Kamimoto, Y.; Gatmaitan, Z.; Hsu, J.; Arias, I. M. *J. Biol. Chem.* **1989,** *264,* 11693–11698.

52. Zimmerli, B.; Valantinas, J.; Meier, P. J. *J. Pharm. Exp. Ther.* **1989,** *250,* 301–308.

53. Tiribelli, C.; Lunazzi, G. C.; Sottocasa, G. L. In *Hepatic Transport and Bile Secretion: Physiology and Pathophysiology;* Tavoloni, N.; Berk, P. D., Eds.; Raven: New York, 1993; pp 235–244.

54. Marks, D. L.; LaRusso, N.F.; McNiven, M. A. *Hepatology* **1992,** *16,* 125A.

55. Janas, R.; Marks, D. L.; LaRusso, N. F. *Gastroenterology* **1991,** *100,* 646.

56. Lombardo, Y. B.; Morse, E. L.; Adibi, S. A. *J. Biol. Chem.* **1988,** *263,* 12920–12926.

57. Arias, I. M.; Che, M.; Gatmaitan, Z.; Leveille, C.; Nishida, T.; St. Pierre, M. *Hepatology* **1993,** *17,* 318–329.

58. Foxwell, B. M. J.; Mackie, A.; Ling, V.; Ryfell, B. *Mol. Pharmacol.* **1989,** *36,* 543–546.

59. Yee, G. C. *Pharmacotherapy* **1991,** *11,* 130S–134S.

60. Ziegler, K.; Frimmer, M. *Biochim. Biophys. Acta* **1986,** *855,* 147–156.

61. Ziegler, K.; Polzin, G.; Frimmer, M. *Biochim. Biophys. Acta* **1988,** *938,* 44–50.

62. Silverstein, S. C.; Steinman, R. M.; Cohn, Z. A. *Annu. Rev. Biochem.* **1977,** *46,* 669–722.

63. Goldstein, J. L.; Brown, M. S.; Anderson, R. G. W.; Russell, D. W.; Schneider, W. J. *Annu. Rev. Cell Biol.* **1985,** *1,* 1–39.

64. Mellman, I.; Howe, H.; Helenius, A. *Curr. Top. Membr. Transp.* **1987,** *29,* 255–288.

65. Scharschmidt, B. F.; Lake, J. R.: Renner, E. L.; Licko, V.; Van Dyke, R. W. *Proc. Natl. Acad. Sci. U.S.A.* **1986,** *83,* 9488–9492.

66. Young, G. P.; Rose, I. S.; Cropper, S.; Seetharam, S.; Alpers, D. H. *Am. J. Physiol.* **1984,** *247,* G419–G426.

67. Hagenbuch, B.; Stieger, B.; Foguet, M.; Lübbert, H.; Meier, P. J. *Proc. Natl. Acad. Sci. U.S.A.* **1991,** *88,* 10629–10633.

68. Elsner, R. H.; Ziegler, K. *Biochim. Biophys. Acta* **1989,** *983,* 113–117.

69. Premont, R. T.; Iyengar, R. In *Peptide Hormone Receptors;* Kalimi, M. Y.; Hubbard, J. R., Eds.; Walter de Gruyter: New York, 1987; pp 129–177.

70. Wileman, T.; Harding, C.; Stahl, P. *Biochem. J.* **1985,** *232,* 1–14.

71. Trowbridge, I. S. *Curr. Opinion Cell Biol.* **1991,** *3,* 634–641.

72. Canfield, W. M.; Johnson, K. F.; Ye, R. D.; Gregory, W.; Kornfeld, S. *J. Biol. Chem.* **1991,** *266,* 5682–5688.

73. McCune, B. K.; Prokop, C. A.; Earp, H. S. *J. Biol. Chem.* **1990,** *265,* 9715–9721.

74. Pearse, B. M. F.; Robinson, M. S. *Annu. Rev. Cell. Biol.* **1990,** *6,* 151–171.

75. LaMarre, J.; Hayes, M. A.; Wollenberg, G. K.; Hussaini, I.; Hall, S. W.; Gonias, S. L. *J. Clin. Invest.* **1991,** *87,* 39–44.

76. Borth, W.; Luger, T. A. *J. Biol. Chem.* **1989,** *264,* 5818–5825.

77. Gupta, S. K.; Manfro, R. C.; Tomlanovich, S. J.; Gambertoglio, J. G.; Garovoy, M. R.; Benet, L. Z. *J. Clin. Pharmacol.* **1990,** *30,* 643–653.

78. Prueksaritanont, T.; Hoener, B.; Benet, L. Z. *Drug Metab. Dispos.* **1992,** *20,* 547–552.

79. Jansen, R. W.; Kruijt, J.; van Berkel, T. J. C.; Meijer, D. K. F. *Hepatology* **1992,** *16,* 142A.

80. Nathanson, M. H.; Boyer, J. L. *Hepatology* **1991,** *14,* 551–566.

81. Graf, J. In *Hepatic Transport and Bile Secretion: Physiology and Pathophysiology;* Tavoloni, N.; Berk, P. D., Eds.; Raven: New York, 1993; pp 155–170.

82. Zimniak, P.; Awasthi, Y. C. *Hepatology* **1993,** *17,* 330–339.

83. Fitz, G. In *Hepatic Transport and Bile Secretion: Physiology and Pathophysiology;* Tavoloni, N.; Berk, P. D., Eds.; Raven: New York, 1993; pp 281–295.

84. Wolters, H.; Fuipers, F.; Stooff, M. J. H.; Vonk, R. J. *J. Clin. Invest.* **1992,** *90,* 2321–2326.

85. Stremmel, W.; Tiribelli, C.; Vyska, K. In *Hepatic Transport and Bile Secretion: Physiology and Pathophysiology;* Tavoloni, N.; Berk, P. D., Eds.; Raven: New York, 1993; pp 225–233.

86. Moseley, R. D. In *Hepatic Transport and Bile Secretion: Physiology and Pathophysiology;* Tavoloni, N.; Berk, P. D., Eds.; Raven: New York, 1993; pp 337–349.

87. Ziegler, K.; Frimmer, M.; Müllner, S.; Fasold, H. *Biochem. Biophys. Acta* **1989,** *980,* 161–168.

88. Hagenbuch, B.; Lubbert, H.; Stieger, B.; Meier, P. J. *J. Biol. Chem.* **1990,** *265,* 5357–5360.

89. Jacquemin, E.; Hagenbuch, B.; Stieger, B.; Wolkoff, A. W.; Meier, P. J. *Hepatology* **1992,** *16,* 89A.

90. Ziegler, K.; Frimmer, M.; Kessler, H.; Damm, I.; Eiermann, V.; Koll, S.; Zarbock, J. *Biochim. Biophys. Acta* **1985,** *845,* 86–93.

91. Ziegler, K.; Frimmer, M.; Kessler, H.; Haupt, A. *Biochim. Biophys. Acta* **1988,** *945,* 263–272.

92. Munter, K.; Mayer, D.; Faulstich, H. *Biochim. Biophys. Acta* **1986,** *860,* 91–98.

93. Petzinger, E.; Joppen, C.; Frimmer, M. *Naunyn-Schmiedebergs Arch. Pharmacol.* **1983,** *322,* 174–179.

94. Hooser, S. B.; Kuhlenschmidt, M. S.; Dahlem, A. M.; Beasley, V. R.; Carmichael, W. W.; Haschek, W. M. *Toxicon* **1991,** *6,* 589–601.

95. Ziegler, K.; Frimmer, M. *Biochim. Biophys. Acta* **1986,** *855,* 136–142.

96. Mutt, V. In *Gastrointestinal Hormones;* Glass, G. B. H., Ed.; Raven: New York, 1980; pp 169–221.

97. Bottcher, W.; Eysselein, V. E.; Kauffman, G. L.; Walsh, J. H. *Gastroenterology* **1983,** *84,* 1112.

98. Gores, G. J.; LaRusso, N. F.; Miller, L. J. *Am. J. Physiol.* **1986,** *250,* G344–G349.

99. Gores, G. J.; Miller, L. J.; LaRusso, N. F. *Am. J. Physiol.* **1986,** *250,* G350–G356.

100. Frimmer, M.; Ziegler, K. *Biochim. Biophys. Acta* **1988,** *947,* 75–99.

101. Hunter, E. B.; Powers, S. P.; Kost, L. J.; Pinon, D. I.; Miller, L. J.; LaRusso, N. F. *Hepatology* **1990,** *12,* 76–82.

102. Bertrams, A.; Ziegler, K. *Biochim. Biophys. Acta* **1991,** *1073,* 213–220.

103. Ziegler, K.; Sänger, U. *Biochim. Biophys. Acta* **1992,** *1103,* 219–228.

104. Wolkoff, A. W. In *Liver and Biliary Tract Physiology I;* Javitt, N. B., Ed.; International Review of Physiology, Vol. 21; University Park: Baltimore, MD, 1980; pp 151–169.

105. Theilmann, L.; Stollman, Y. R.; Arias, I. M.; Wolkoff, A. W. *Hepatology* **1984,** *4,* 923–926.

106. Litowsky, I. In *Hepatic Transport and Bile Secretion: Physiology and Pathophysiology;* Tavoloni, N.; Berk, P. D., Eds.; Raven: New York, 1993; pp 397–405.

107. Takikawa, H.; Stolz, A.; Sugiyama, Y.; Yoshida, H.; Yamanaka, M.; Kaplowitz, N. *J. Biol. Chem.* **1990,** *265,* 2132–2136.

108. Stolz, A.; Rahimi-Kiani, M.; Ameis, D.; Chen, E.; Ronk, M.; Shively, J. *J. Biol. Chem.* **1991,** *266,* 15253–15257.

109. Erlinger, S. In *Hepatic Transport and Bile Secretion: Physiology and Pathophysiology;* Tavoloni, N.; Berk, P. D. Eds.; Raven: New York, 1993; pp 467–475.

110. Crawford, J. M.; Gollan, J. L. In *Hepatic Transport and Bile Secretion: Physiology and Pathophysiology;* Tavoloni, N.; Berk, P. D., Eds.; Raven: New York, 1993; pp 447–465.

111. Handschumachaer, R. E.; Harding, M. W.; Rice, J.; Drugge, R. J.; Speicher, D. W. *Science Washington, DC)* **1984,** *226,* 544–547.

112. Gruenberg, J.; Howell, K. E. *Annu. Rev. Cell. Biol.* **1989,** *5,* 453–481.

113. Geuze, H. J.; Slot, J. W.; Strous, G. J. A. M.; Lodish, H. F.; Schwartz, A. L. *Cell* **1983,** *32,* 277–287.

114. Geuze, H. J.; Slot, J. W.; Strous, G. J. A. M.; Peppard, J.; van Figura, K.; Hasilik, A.; Schwartz, A. L. *Cell* **1984,** *37,* 195–204.

115. Renaud, G.; Hamilton, R. L.; Havel, R. J. *Hepatology* **1989,** *9,* 380–392.

116. Schiff, J. M.; Fisher, M. M.; Underdown, B. J. *J. Cell Biol.* **1984,** *98,* 79–89.

117. Terris, S.; Hofmann, C.; Steiner, D. F. *Can. J. Biochem.* **1979,** *57,* 459–468.

118. Kacich, R. L.; Renston, R. H.; Jones, A. L. *Gastroenterology* **1983,** *85,* 385–394.

119. Schnapp, B. J.; Reese, T. S.; Bechtold, R. *J. Cell Biol.* **1992,** *119,* 389–399.

120. Urrutia, R.; McNiven, M. A.; Albanesi, J. P.; Murphy, D. B.; Kachar, B. *Proc. Natl. Acad. Sci. U.S.A.* **1991,** *88,* 6701–6705.

121. Schroer, T. A.; Sheetz, M. P. *Annu. Rev. Physiol.* **1991,** *53,* 629–652.

122. Vallee, R. B.; Shpetner, H. S. *Annu. Rev. Biochem.* **1990,** *59,* 909–932.

123. Collins, C. A.; Vallee, R. B. *Cell Motil. Cytoskeleton* **1989,** *14,* 491–500.

124. Vale, R. D.; Scholey, J. M.; Sheetz, M. P. *Trends Biochem. Sci.* **1986,** *11,* 464–468.

125. Schnapp, B. J.; Reese, T. S. *Proc. Natl. Acad. Sci. U.S.A.* **1989,** *86,* 1548–1552.

126. Lin, S. X. L.; Collins, C. A. *J. Cell Sci.* **1992,** *101,* 125–137.

127. Leopold, P. L.; McDowall, A. W.; Pfister, K. K.; Bloom, G. S.; Brady, S. T. *Cell Motil. Cytoskeleton* **1992,** *23,* 19–33.

128. Goltz, J. S.; Wolkoff, A. W.; Novikoff, P. M.; Stockert, R. J. *Proc. Natl. Acad. Sci. U.S.A.* **1992,** *89,* 7026–7030.

129. Dubin, M.; Maurice, M.; Feldmann, G.; Erlinger, S. *Gastroenterology* **1980,** *70,* 646–654.

130. Vonk, R. J.; Yousef, I. M.; Corriveau, J. P.; Tuchweber, B. *Liver* **1982,** *2,* 133–140.

131. Phillips, M. J.; Poucell, S.; Oda, M. *Lab. Invest.* **1986,** *54,* 593–608.

132. Crawford, J. M.; Berken, C. A.; Gollan, J. L. *J. Lipid Res.* **1988,** *29,* 144–156.

133. Crawford, J. M.; Gollan, J. L. *Am. J. Physiol.* **1988,** *255,* G121–G131.

134. Suchy, F. J.; Balistreri, W. F.; Hung, J.; Miller, P.; Garfield, S. A. *Am. J. Physiol.* **1983,** *245,* G681–G689.

135. Hubbard, A. S. H. *J. Cell Biol.* **1979,** *83,* 65–81.

136. Dunn, W. A.; LaBadie, J. H.; Aronson, N. N., Jr. *J. Biol. Chem.* **1979,** *254,* 4191–4196.

137. Stoorvogel, W.; Strous, G. J.; Geuze, H. J.; Oorschot, V.; Schwartz, A. L. *Cell* **1991,** *65,* 417–427.

138. LaRusso, N. F. In *Handbook of Physiology. The Gastrointestinal System III.;* Forte, J. G., Ed.; American Physiological Society: Bethesda, MD, 1989; pp 677–691.

139. Vessey, D. A. In *Hepatology, a Textbook of Liver Disease;* Zakim, D.; Boyer, T. D., Eds.; W.B. Saunders: Philadephia, PA, 1990; vol. 1, pp 196–234.

140. Winters, D. K.; Cedarbaum, A. I. In *Hepatic Transport and Bile Secretion: Physiology and Pathophysiology;* Tavoloni, N.; Berk, P. D., Eds.; Raven: New York, 1993; pp 407–420.

141. Maurer, G.; Lemaire, M. *Transplant. Proc.* **1986,** *18(Suppl 5),* 25–34.

142. Kronbach, T.; Fischer, V.; Meyer, U. A. *Clin. Pharmacol. Ther.* **1988,** *43,* 630–635.

143. Johansson, A.; Henricsson, S.; Moller, E. *Transplantation* **1990,** *49,* 619–622.

144. Melloni, E.; Pontremoli, S.; Salamino, F.; Sparatore, B.; Michetti, M.; Horecker, B. L. *Arch. Biochem. Biophys.* **1984,** *232,* 505–512.

145. Ogawa, W.; Shii, K.; Yonezawa, K.; Baba, S.; Yokono, K. *J. Biol. Chem.* **1992,** *267,* 1310–1316.

146. Wilk, S. *Life Sci.* **1983,** *33,* 2149–2157.

147. Janas, R. M.; Marks, D. L.; LaRusso, N. F. *Anal. Biochem.* **1992,** *206,* 6–11.

148. Lorenzini, I.; Sakisaka, S.; Meier, P. J.; Boyer, J. L. *Gastroenterology* **1986,** *91,* 1278–1288.

149. Renston, R. H.; Maloney, D. G.; Jones, A. L.; Hradek, G. T.; Wong, K. Y.; Goldfine, I. D. *Gastroenterology* **1980,** *78,* 1373–1388.

150. Sewell, R. B.; Barham, S. S.; Zinsmeister, A. R.; LaRusso, N. F. *Am. J. Physiol.* **1984,** *246,* G8–G15.

151. Meier, P. J. In *Hepatic Transport and Bile Secretion: Physiology and Pathophysiology;* Tavoloni, N.; Berk, P. D., Eds.; Raven: New York, 1993; pp 587–591.

152. Arias, I. M. *Hepatology* **1990,** *12,* 159–165.

153. Ling, V.; Thompson, L. H. *J. Cell Physiol.* **1974,** *83,* 103–116.

154. Riehm, H.; Biedler, J. L. *Cancer Res.* **1972,** *32,* 1195–1200.

155. Dano, K. *Cancer Chemother. Rep.* **1972,** *56,* 701–708.

156. Juliano, R. L.; Ling, V. *Biochim. Biophys. Acta* **1976,** *455,* 152–162.

157. Ueda, K.; Cardarelli, C.; Gottesman, M. M.; Pastan, I. *Proc. Natl. Acad. Sci. U.S.A.* **1987,** *84,* 3004–3008.

158. Gros, P. N. Y. B.; Croop, J. M.; Housman, D. E. *Nature (London)* **1986,** *322,* 728–731.

159. Thiebault, F.; Tsuruo, T.; Hamada, H.; Gottesman, M. M.; Pastan, I.; Willingham, M. C. *Proc. Natl. Acad. Sci. U.S.A.* **1987,** *84,* 7735–7738.

160. Pastan, I.; Willingham, M. C.; Gottesman, M. *FASEB J.* **1991,** *5,* 2523–2528.

161. Gottesman, M. M.; Pastan, I. *J. Biol. Chem.* **1988,** *263,* 12163–12166.

162. Yoon, Y. B.; Hagey, L. R.; Hofmann, A. F.; Gurantz, D.; Michelotti, E. L.; Steinbach, J. H. *Gastroenterology* **1986,** *90,* 837–852.

163. Hylemon, P. B.; Bohdan, P. M.; Sirica, A. E.; Heuman, D. M.; Vlahcevic, Z. R. *Hepatology* **1990,** *11,* 982–988.

164. Vlahcevic, Z. R.; Heuman, D. M.; Hylemon, P. B. In *Hepatology, a Textbook of Liver Disease;* Zakim, D.; Boyer, T. D., Eds.; W.B. Saunders: Philadelphia, PA, 1990; Vol. 1, pp 341–377.

165. Wagner, O.; Schreier, E.; Heitz, F.; Maurer, G. *Drug Metab. Dispos.* **1987,** *15,* 377–383.

166. Venkataramanan, R.; Starzl, T. E.; Yang, S.; Burckart, G. J.; Ptachcinski, R. J.; Shaw, B. W.; Iwatsuki, S.; Van Thiel, D. H.; Sanghvi, A.; Seltman, H. *Transplant. Proc.* **1985,** *17,* 286–289.

167. Aronson, N. N.; Kuranda, M. J. *FASEB J.* **1989,** *3,* 2615–2622.

168. Roy Chowdhury, J.; van Es, H. H. G.; Roy Chowdhury, N. In *Hepatic Transport and Bile Secretion: Physiology and Pathophysiology;* Tavoloni, N.; Berk, P. D., Eds.; Raven: New York, 1993; pp 713–719.

169. Jansen, P. L. M.; Oude, E., R. P. J. In *Hepatic Transport and Bile Secretion: Physiology and Pathophysiology;* Tavoloni, N.; Berk, P. D., Eds.; Raven: New York, 1993; pp 721–731.

170. Ostrow, J. D. In *Hepatic Transport and Bile Secretion: Physiology and Pathophysiology;* Tavoloni, N.; Berk, P. D., Eds.; Raven: New York, 1993; pp 673–712.

171. Reichen, J. In *Hepatic Transport and Bile secretion: Physiology and Pathophysiology;* Tavoloni, N.; Berk, P. D., Eds.; Raven: New York, 1993; pp 665–672.

172. D'Arville, C. N.; Le, M.; Kloppel, T. M.; Simon, F. R. *Hepatology* **1989,** *9,* 6–11.

173. Fenyves, D.; Gariepy, L.; Villeneuve, J. *Hepatology* **1993,** *17,* 301–306.

174. Dufour, J.; Gehr, P.; Reichen, J. *Hepatology* **1992,** *16,* 997–1006.

175. Bentsen, K. D.; Henriksen, J. H.; Bendtsen, F.; Hørslev-Petersen, K.; Lorenzen, I. *Hepatology* **1990,** *11,* 957–963.

176. Yang, C. P.; Depinho, S. G.; Greenberg, L. M.; Arceci, R. J.; Horwitz, S. B. *J. Biol. Chem.* **1989,** *264,* 782–788.

177. Burchell, B. In *Hepatic Transport and Bile Secretion: Physiology and Pathophysiology;* Tavoloni, N.; Berk, P. D., Eds.; Raven: New York, 1993; pp 489–499.

178. Blanckaert, N.; Jevery, J. In *Hepatology;* Zakim, D.; Boyer, T. D., Eds.; W.B. Saunders: Philadelphia, PA, 1990; pp 254–302.

179. Vanstapel, F.; Blanckaert, N. In *Hepatic Transport and Bile Secretion: Physiology and Pathophysiology;* Tavoloni, N.; Berk, P. D., Eds.; Raven: New York, 1993; pp 477–488.

180. Arias, I. M. *Gastroenterology* **1993,** *104,* 1558–1560.

181. Kadmon, M.; Klünemann, C.; Böhme, M.; Ishikawa, T.; Gorgas, K.; Otto, G.; Herfarth, C.; Keppler, D. *Gastroenterology* **1993,** *104,* 1507–1514.

182. Kong, C.; Yet, S.; Lever, J. E. *J. Biol. Chem.* **1993,** *268,* 1509–1512.

183. Potter, B. J.; Berk, P. D. In *Hepatic Transport and Bile Secretion: Physiology and Pathophysiology;* Tavoloni, N.; Berk, P. D., Eds.; Raven: New York, 1993; pp 253–267.

184. Moseley, R. H.; Morrissette, J.; Johnson, T. R. *Am. J. Physiol.* **1990,** *259,* G973–G982.

185. Neef, C.; Keulemans, K. T. P.; Meijer, D. K. F. *Biochem. Pharmacol.* **1984,** *33,* 3977–3990.

186. Wolkoff, A. W.; Samuelson, A. C.; Johansen, K. L.; Nakata, R.; Withers, D. M.; Sosiak, A. *J. Clin. Invest.* **1987,** *79,* 1259–1268.

RECEIVED for review July 12, 1993. ACCEPTED revised manuscript November 3, 1993.

CHAPTER 11

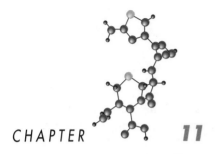

Approaches To Modulating Liver Transport of Peptide Drugs

Mary J. Ruwart

Peptides represent a class of potentially useful therapeutic drugs, including analgesics (*1*), renin inhibitory peptides RIPs; (*2*), and human immunodeficiency virus (HIV) protease inhibitors (*3*). The development of peptides and proteins has been impaired, however, by a number of factors, including rapid removal from the systemic circulation by the liver. In this chapter, we explore the factors that influence hepatic clearance (e.g., size and hydrophobicity) so that this parameter may be more readily addressed by those interested in the development of peptidic drugs.

Influence of Peptide Size on Hepatic Clearance

Generally, peptides with chain lengths of three to six amino acids are readily extracted by the liver, whereas hepatic clearance is usually, but not always, minimal for peptides and proteins outside this range. Unfortunately, the smaller peptides, which are most readily synthesized, are also rapidly extracted. Thus, successful pharmaceutical development of synthetic peptides may require a thorough understanding of why the liver exercises this selectivity.

Di- and Tripeptides

Natural Configurations

Dipeptides usually undergo rapid enzymatic hydrolysis at the hepatocyte membrane (*4, 5*) so that uptake of the intact peptide never occurs. However, even small peptides resistant to such hydrolysis are poorly extracted by the liver. For example, two-thirds of cyclo(His-Pro) was excreted unchanged into the urine after intravenous (iv) administration to rats, suggesting that the liver poorly extracted and hydrolyzed this peptide (*6*). Less than 3% of an administered dose of intact cyclo(His-Pro) was excreted into the bile, although the presence of radiolabeled fragments in the liver suggested that some of the extracted peptide had been metabolized. Another dipeptide, Leu-Tyr, which did not appear to be hydrolyzed by the liver, was also poorly extracted (5%) after iv injection to rats (*7*). Like most dipeptides, tripeptides (e.g., Met-Leu-Tyr, Gly-Leu-Tyr, and Leu-Leu-Tyr) undergo <10% hepatic extraction when given iv to rats (*7*). Thus, the liver probably does not effectively transport di- or tripeptides with natural amino acid configurations.

Modified Peptides

However, a different pattern is observed for the unnatural "D" amino acids or peptides with N-terminal modifications. For example, ~50% of a dimer and several methylated trimers of D-phenylalanine appeared in the bile of rats given iv injections (*8*). The unnatural configuration of the peptides were likely responsible for hepatic uptake because liver extraction increased by 5–10% with the substitution of a single "D" amino acid in other naturally configured tripeptides (*7*).

In addition, the D-phenylalanine oligomers just described were acetylated. Increased hepatic extraction of small peptides was reported after acetylation of the N-terminus, as well as after the addition of formyl, propionyl, carbobenzoxy, or Bolton–Hunter acyl groups (*7, 9*). Thus, only minor structural modifications are required to change peptides that are routinely excluded from hepatic extraction to peptides that are efficiently cleared by the liver (30–90%).

Peptides with Four or More Amino Acids

The liver often fails to extract peptides with molecular weights of 1000 Da or greater, just as it leaves most di- and tripeptides in the circulation. For example, gastrin analogues greater than eight amino acids long were neither extracted nor metabolized by the liver (*10, 11*), and only 5% of gastrin-8 (G-8) was taken up by the liver (*10*). The unextracted G-8 nevertheless experienced some metabolism, because only 86% of the peptide in the venous effluent was intact. Thus, the liver is apparently capable of metabolizing peptides without targeting them for biliary excretion. In interpreting the disappearance of peptides from the circulation, therefore, it might be necessary to distinguish the impact of liver metabolism from liver extraction.

In contrast to the poor liver uptake experienced by G-8 and larger analogues, almost half of the G-6 and 95% of G-4–cholecystokinin (CCK)-4 was extracted by the liver. Only 17% of the unextracted G-6 and none of the G-4–CCK-4 remained intact (*10*). Thus, the liver most likely can effectively remove intact peptide from the circulation by a combination of metabolism and extraction.

Although most investigators agree that larger peptides are usually poorly extracted by the liver, they disagree on the exact clearance estimate for each peptide. For example, one laboratory reported that <10% of CCK-7, -8, and -12 were taken up by perfused rat liver, whereas 83% of CCK-6 was extracted (*12*). In another laboratory, considerably more extraction was observed: 59% of CCK-8 was removed in a single pass through rat liver whereas only 3% of CCK-33 was extracted (*13*).

Likewise, whereas one laboratory reported that somatostatin-14 and -28 were <10% extracted by hepatic clearance in perfused rat liver (*11*), other investigators reported extraction as high as 36% and 17% for the same peptides in a similar model (*14, 15*). However, even investigators who report rapid clearance rates for somatostatin-14 agree that somatostatin-28 is extracted by the liver only half as well as the small analogue. Thus, within a given series of peptides, the preponderance of evidence suggests that elongating a peptide causes it to undergo less hepatic extraction.

Possibility of Carrier-Mediated Transport

The type of molecular weight dependence just described might be expected if peptides undergo carrier-mediated transport. As described later, some peptides in the intermediate molecular weight range can compete with bile acids for hepatic uptake, presumably because they share the same carrier for transport. However, preliminary evidence suggests that at least one peptide, ditekiren, undergoes carrier-mediated uptake that is largely independent of bile acid transporters (*16*).

Multiple uptake processes would be expected to complicate the search for a structure–activity relationship (SAR) in the liver extraction of intermediate size peptides. Both hydrophobicity, a determinant of passive diffusion, and conformational aspects, which might dictate carrier-mediated processes, could become important. Indeed, in examining attempts to elicit a pattern of hepatic uptake, a complex SAR is suggested.

Impact of Hydrophobicity on Carrier-Mediated Transport

Many of the SARs for hepatic transport of intermediate size peptides (three to six amino acids) were derived from competition experiments with the presumed carrier, ignoring any potential input from passive diffusion. For example, cyclolinopeptide A protects hepatocytes against the deleterious effects of phalloidin, presumably by competing for the bile acid transport carrier that phalloidin also uses. Four analogues of cyclolinopeptide A were tested for their ability to compete with cholate uptake in hepatocytes as a measure of this "cytoprotection". Replacing one of the two proline moieties with the more hydrophobic alanine decreased the binding affinity slightly, suggesting that hydrophobicity was not the major determinant for uptake by that route. Replacing the phenylalanine-3 with a less hydrophobic alanine caused a sevenfold increase in the 50% inhibitory concentration (IC_{50}), whereas replacing phenylalanine-4 with the same amino acid produced a fourfold increase (*17*). If the position of the replaced phenylalanine does not influence hydrophobicity, this factor apparently is of little importance in binding to the bile acid receptor.

However, some data suggest that carrier-mediated transport can involve different carriers and that binding might be facilitated in a secondary way by increased lipophilicity. For example, the somatostatins are another class of cyclic peptides that can protect liver cells against phalloidin poisoning. Photoaffinity labeling of hepatocytes with a labeled somatostatin hexapeptide and analogue was comparable with binding observed with labeled bile acids (*18*). One somatostatin analogue ("008") was transported by hepatocytes but not by a hepatoma cell line known to be unable to transport bile acids and phalloidin (*19*). Unlike cyclosporin A, 008 binding was not compromised by rifampicin, but binding was decreased by sulfobromophthalein, suggesting that 008 was transported by a somewhat different group of carriers than cyclosporin (*19*). Obviously, multiple carriers for different peptides obscures any attempt to establish SARs governing such transport.

Several somatostatin analogues were compared with respect to their cytoprotective IC_{50} against hepatocyte damage due to phalloidin. The extent of cytoprotection was used as a reflection of binding to the transporter (*20*). Peptides differing only in the substitution of the more hydrophobic phenylalanine for proline decreased IC_{50} almost fivefold, a finding consistent with a role for hydrophobicity.

Furthermore, comparison of the IC_{50} of a number of cyclic peptides indicated that inhibition of cholate uptake in hepatocytes was clearly not related to size (*20*). When the sum of the "consensus" estimation of hydrophobicity for the functional amino acid groups (*21*) was used as a surrogate for the log of the octanol–water partition coefficient (log *P*), the correlation between IC_{50} and lipophilicity (Figure 1) was not significant ($r^2 = 0.50$, $p > 0.1$), although the trend was in that direction (i.e., lower IC_{50} at higher hydrophobicities).

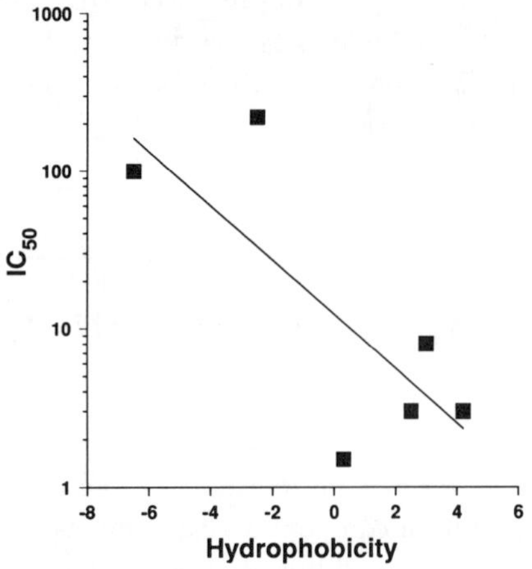

Figure 1. Hydrophobicity versus IC_{50} for inhibition of cholate uptake by cyclic peptides.

Thus, the findings with somatostatin suggests that hydrophobicity might only play a secondary role for carrier-mediated transport.

Impact of Hydrophobicity on Hepatic Clearance

If carrier-mediated transport plays a predominant role in hepatic clearance and if this transport is not primarily driven by the lipophilic character of the peptide, hydrophobicity should not be the primary factor in the liver uptake of peptides. However, a peptide binding to a carrier does not necessarily ensure efficient transport of that peptide. Therefore, examination of whole liver transport of peptides is necessary to determine the balance between passive diffusion and carrier-mediated transport. Thus, whereas hydrophobicity clearly is not the dominant factor in peptide binding to the phalloidin–bile acid transporter, it might still be important in hepatic extraction if passive diffusion plays a significant role.

For example, sulfating the tyrosine residue of CCK-8 did impair its extraction from perfused liver as might be expected if hydrophobicity were important (*22*). When a series of CCK-4 analogues was examined in perfused rat liver, a significant sigmoidal relationship was seen between hepatic extraction and the logit transform of the calculated hydrophobicity (*23*), a fact that suggests a potentially large passive diffusion component. However, two peptides with identical amino acids but with different sequences (i.e., the two internal amino acids were switched) had significantly different ($p < 0.0001$) extraction rates (29% versus 73%). In another study (*24*) with HIV protease inhibitors, a C-terminal pyridyl group greatly affected liver clearance depending on the position (2, 3, or 4) of the nitrogen. Stereospecific alterations (*R*, *R* to *R*, *S* or *S*, *S*) in the insert also affected clearance greatly (*24*), once again suggesting that conformation, rather than hydrophobicity, is the primary determinant for clearance. Unless the measured hydrophobicities of these compositionally identical analogues were very different, it is unlikely that hydrophobicity-driven passive diffusion was truly the primary factor of hepatic extraction.

One difficulty in assembling more in vivo data of this type is the paucity of assays available to conduct such studies. Unless peptides are radiolabeled, quantification usually requires high-pressure liquid chromatography (HPLC), which is tedious when large numbers of compounds are tested. Therefore, we turned to a structurally similar series of RIPs (structure 1) that could be assessed by an activity assay (*25*) to gain understanding of the role of hydrophobicity in hepatic extraction. Rats were given iv injections of RIPs, and their fractional recovery in bile (i.e., biliary recovery divided by the sum of biliary and urinary

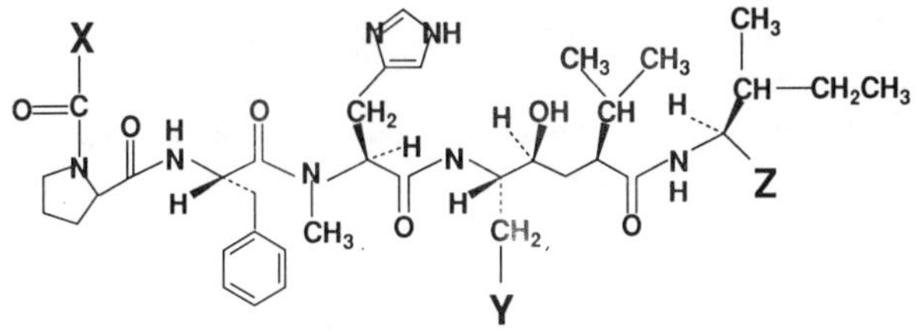

Structure 1. Parent structure of the RIPs studied, where X = γ GluPro, γ Glu, or βVal; Y = 5S-amino-6-cyclohexy-4S-hydroxy-2S-isopropylhexanoy (CVA) or 5S-amino-4S-hydroxy-2S-isopropyl-7-methyloctanoyl (LVA); Z = H, 2-pyridylmethylamine (AMP), AMP → O, or εLys.

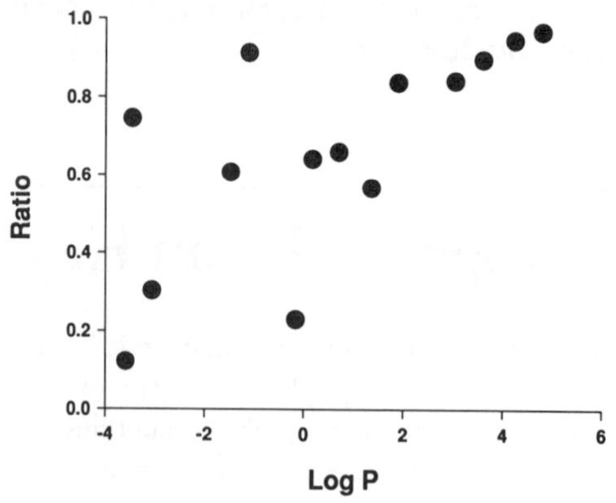

Figure 2. Ratio (fractional biliary clearance) versus log P *for all renin inhibitors tested.*

recovery) was used as an indicator of hepatic clearance. This parameter is useful when the peptides studied are not metabolized to a significant extent and complete recovery of an injected dose can be observed, as seen with the renin inhibitor ditekiren (*26*), a structurally similar peptide to the peptides studied.

When the data were viewed in their entirety, no correlation between log *P* and fractional biliary recovery was observed (Figure 2). When the C-terminus was systematically varied while the rest of the structure remained constant, however, the predicted sigmoidal relationship between log *P* and biliary excretion began to emerge. When the N-terminus was altered in addition to these systematic C-terminal substitutions, a series of parallel sigmoidal curves appeared that shifted according to the moiety on the N-terminus (Figure 3). These data suggested that hydrophobicity is a determinant of liver extraction, but is not the predominant one. Thus, the impact of hydrophobicity may only be observed when closely related peptides are studied.

Unfortunately, such an analysis is only as good as the method of quantifying the peptides. If substantial metabolism occurred and was not uniform between the liver and the kidneys, for example, the fractional biliary recovery determined by this method would be meaningless. Members of chemical families differing only by substitution in one part of the molecule might be metabolized similarly and thus allow for the sigmoidal relationship to appear even if recoveries were not correct in the absolute sense as implied in Figure 3. Therefore, we reexamined data from other laboratories (*7, 8*) to see if they too observed better correlations with closely related chemical families.

Indeed, other laboratories (*7*) reported data consistent with a role for hydrophobicity. For example, studies with di- and tripeptides (*7*) that were esterified to enhance their hepatic extraction showed little correlation between their biliary excretion and measured partition coefficients ($r = 0.205$, $p < 0.386$) unless the closely related subset of *N*-acetyl derivatives are examined ($r = 0.642$, $p < 0.005$), a fact that suggests that hydrophobicity play a predominant role in this closely related peptide series. Another study with a D-phenylalanine trimer and its methylated analogues found a good correlation between calculated log *P* and biliary excretion after iv injection to rats (*8*).

This relationship with hydrophobicity may be a secondary requirement of carrier-mediated transport rather than an indicator of passive diffusion. For example, when renin inhibitors are assigned IC_{50}s based on their ability to inhibit cholate uptake in hepatocytes (*27*) instead of hepatic clearance, a dependence on lipophilicity is observed. These IC_{50}s correlated reasonably well with experimentally derived octanol–water log *P* ($r = 0.74$)

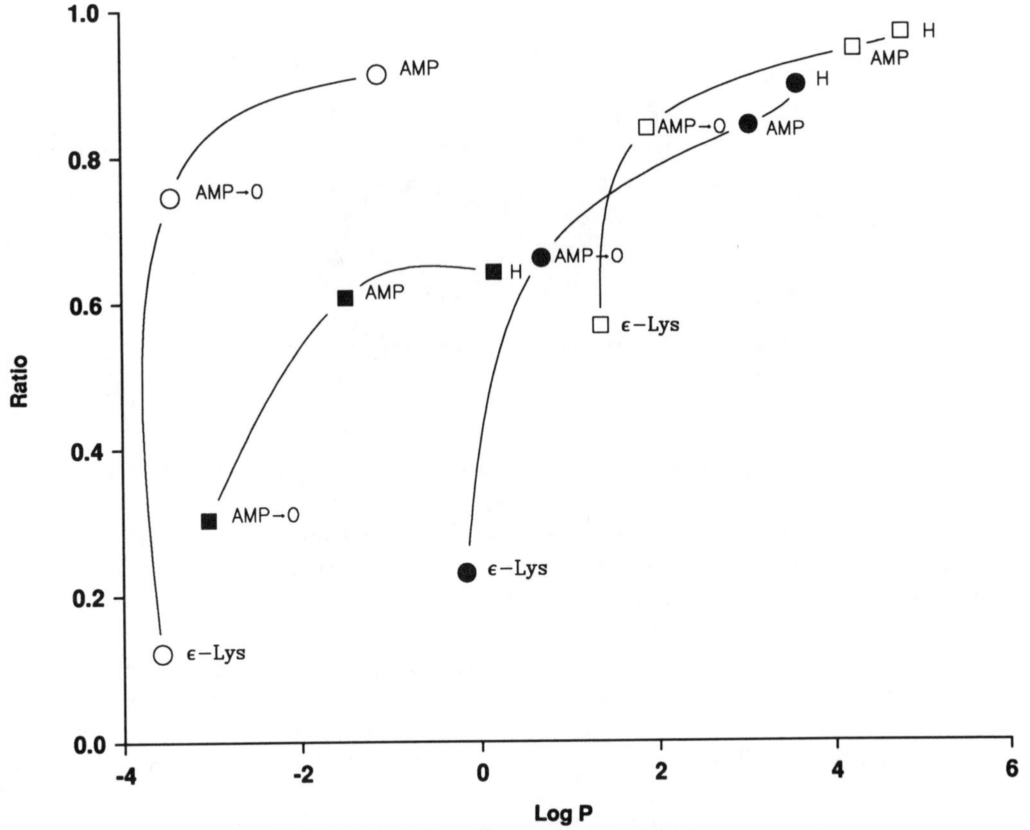

Figure 3. Ratio (fractional biliary clearance) versus log P *for four families of renin inhibitors, where X = γ GluPro, Y = LVA (■); X = βVal, Y = CVA (□); X = βVal, Y = LVA (●); and X = γ Glu, Y = LVA (○). For each family, Z = H, AMP, AMP → O, or εLys.*

(Figure 4). However, when only compounds from the family Boc-L-Phe-L-His-ACHP-Z (where Z is the only variant) were similarly examined, $r = 0.91$ (Figure 5). Just as in the in vivo studies, the role of hydrophobicity in carrier-related studies was most pronounced within a structurally similar family grouping, a fact that suggests that this secondary factor is not necessarily due to passive diffusion.

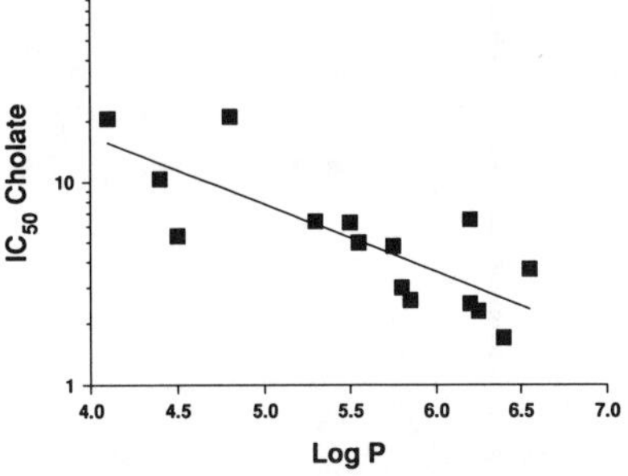

Figure 4. IC_{50} versus log P *for all renin inhibitors tested.*

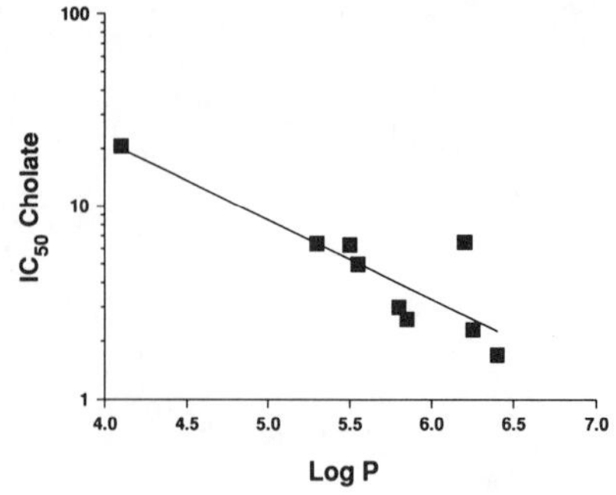

Figure 5. IC_{50} versus log P *for one family of renin inhibitors.*

Thus, in examining the relationship between hydrophobicity and biliary excretion of peptides, evidence suggests a primary dependence on peptide conformation and a secondary dependence on hydrophobicity. However, we did not find any correlation with hydrophobicity and serum peptide levels when comparing closely related analogues of HIV peptidic protease inhibitors. Fasted, anesthetized rats were given iv injections of various peptide analogues, and blood from the orbital sinus was sampled. A previously reported activity assay was used to quantify peptide remaining in the sera (*28*). Because each peptide had a different limit of detection dependent on its activity, areas-under-the-curves (AUCs) were not strictly comparable between the compounds. Therefore, we examined the 20-min sera levels that were easily detectable for all peptides. No correlation with log *P* was observed (Figure 6).

Because the HIV protease inhibitors were somewhat diverse in their chemical structure, metabolic transformations, volume of distribution, and factors other than hepatic extraction influencing the AUC might have been quite different for each analogue. Naturally, this would make global correlations such as those with log *P*, suspect. However, by comparing analogues that differed in a single amino acid, the chance of differences in metabolic or pharmacokinetic handling would hopefully be eliminated. Based on that assumption, the more lipophilic analogue should consistently give lower sera values if hydrophobicity plays a major role in clearance.

Indeed, HIV protease inhibitors with the internal histidine moiety resulted in lower

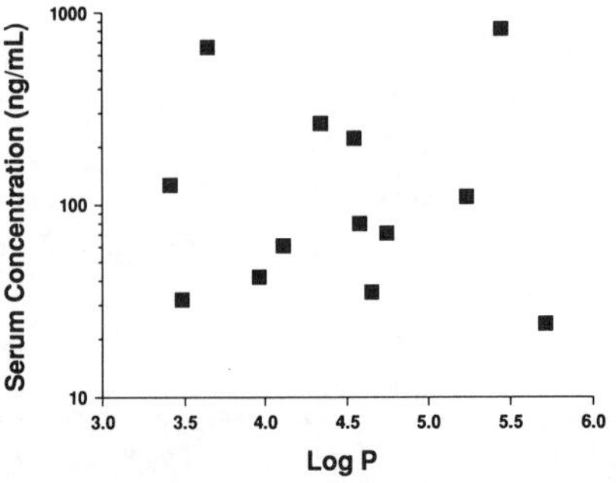

Figure 6. Log serum concentration versus log P *for all HIV protease inhibitors tested.*

Table I. Sera Levels of Peptides with Different Amino Acid Substitutions

Parent Structure	R	Log P	U No.	Level ng/mL
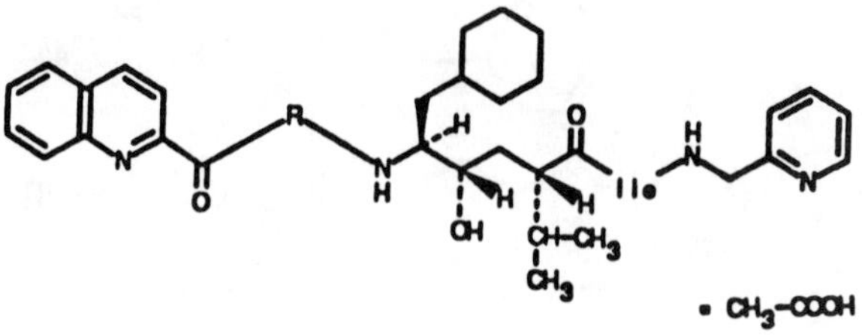	Asp	3.413	89855E	127 ± 8
	His	4.113	91318E	61 ± 12
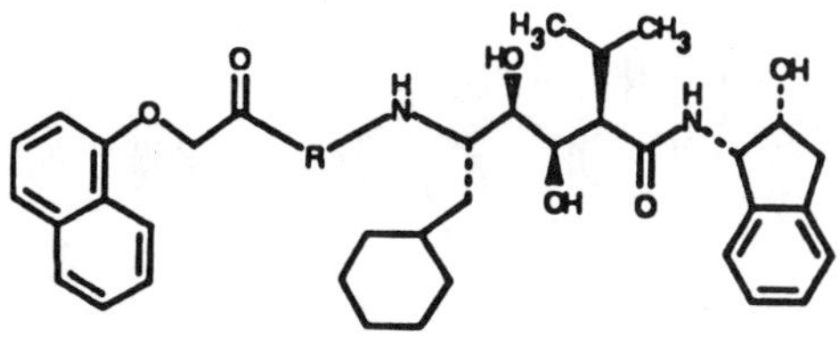	Ser	4.342	94791	265 ± 45
	Thr	4.651	94778	35 ± 2
	His	4.577	95630	80 ± 6

sera levels compared with analogues with the less hydrophobic asparagine (Table I). In the same series of peptides, however, the threonine analogue exhibited lower sera levels than the histidine analogue, which in turn had lower sera levels than the serine analogue. All three of the analogues, however, had comparable log *P* values, so that discrimination in sera levels had to reflect a different parameter (e.g., conformation).

Replacing the N-terminal pyridine carboxylic acid in one of the peptide structures with an acetyl group resulted in elevation of sera levels, whereas a phenoxyethoxyproprionic acid substitution lowered sera levels (Table II). These changes were consistent with a role for hydrophobicity (i.e., more hydrophobic compounds exhibited lower sera levels). In other pairings (Table II), however, the relationship between sera levels and log *P* was in the opposite direction.

Of course, sera levels at a single time point may not be reliable indicators of hepatic extraction, even when the compounds are as closely related as the pairs of peptides examined herein. Indeed, as the studies described below indicate, sera-based pharmacokinetics appear to be quite complex and worthy of careful interpretation.

Pharmacokinetic Studies in Rats, Dogs, Monkeys, and Humans

Several structurally related renin inhibitors were given to chronically cannulated, unanesthetized rats to determine the sera pharmacokinetic profile as quantified by an activity assay (*29*). As C-terminal modifications became increasing hydrophilic (i.e., AMP, AMP → O,

Table II. Sera Levels of Peptides with Different N-Terminal Substitutions

Parent Structure	R	Log P	U No.	Level ng/mL
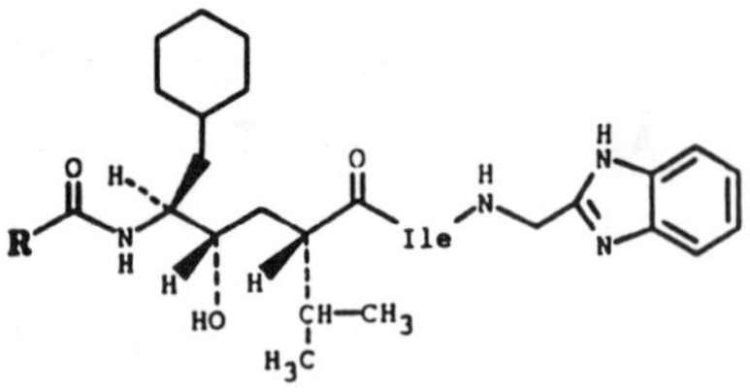	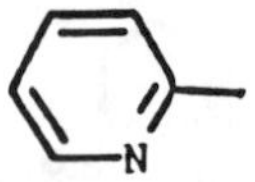	4.543	93464	222 ± 85
	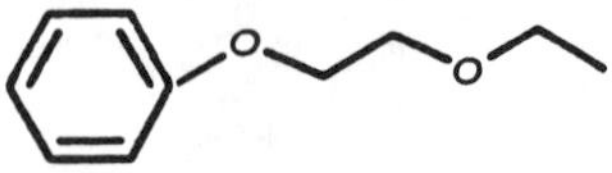	5.715	93281	24 ± 2
	—CH_3	3.643	91797E	658 ± 87
	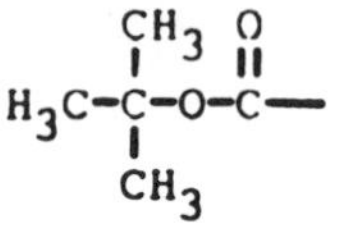	5.446	82222	820 ± 49
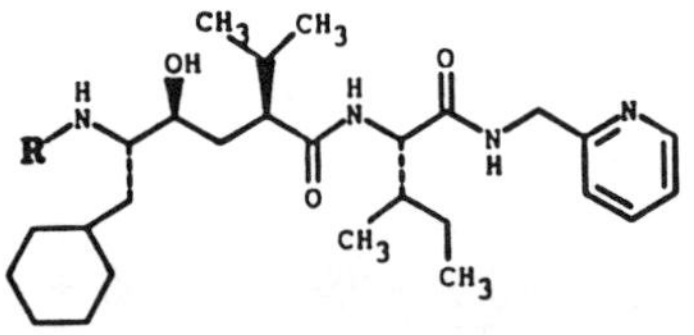	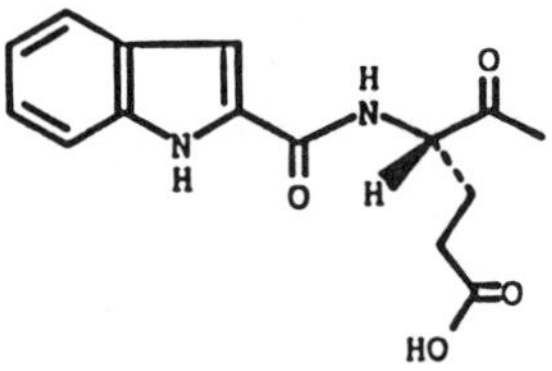	3.483	80532	32 ± 6
	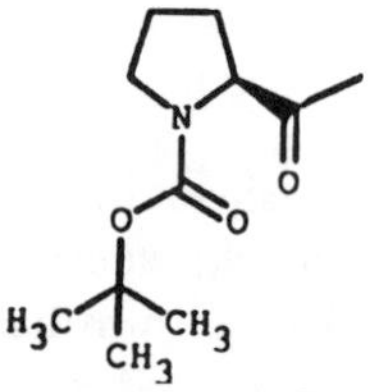	5.238	93205	110 ± 10
	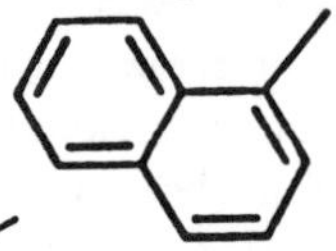	3.965	75875	42 ± 6
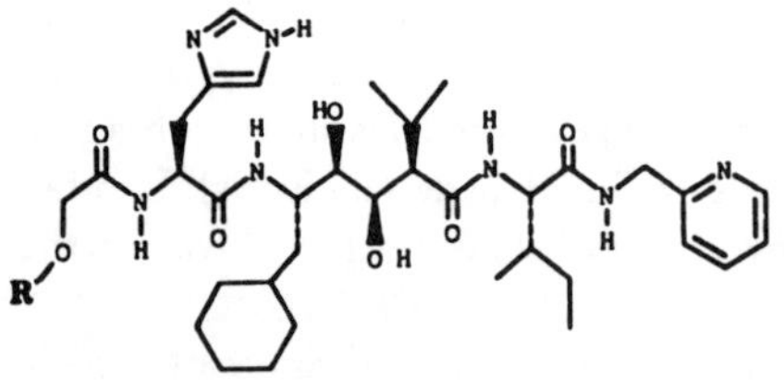	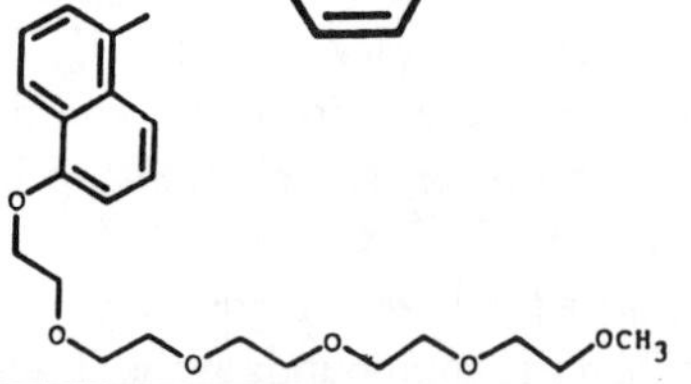	4.745	94023	71 ± 8

and ε-lysine substitutions), total body clearance (Cl_T) decreased from 33 to 5.9 mL/min/kg. Volume of distribution at steady state (V_{ss}) remained unchanged when analogues differing only in their C-terminal modification were compared.

However, N-terminal glucosamine substitution not only decreased clearance, but V_{ss} as well. Thus, trying to infer clearance differences between analogues with synthetic modifications at opposite ends of the molecule without a full pharmacokinetic workup could be misleading. Even closely matched pairs, such as those utilized in the HIV protease inhibitors work just described, might be subject to discrepant interpretation should such modifications be accompanied by alterations in V_{ss} as well as in changes in clearance.

In the study with renin inhibitors just described, no correlation was seen between the calculated log P and the pharmacokinetic clearance when all tested analogues were

Table III. Pharmacokinetic Parameters of Peptides on Different Species

Compound	Species	α (min^{-1})	$t_{1/2\alpha}$ (min)	β (min^{-1})	$t_{1/2\beta}$ (min)
U-76393	Rhesus[a]	[b]	[b]	0.020	35.6
	Rat	0.54	1.3	0.074	10.2
U-71038	Human[c]	[b]	[b]	0.006	126.0
	Rhesus[a]	[b]	[b]	0.006	126.0
	Rat	0.22	3.3	0.043	16.9
U-75875	Rhesus[d]	0.69	1.1	0.019	35.0
	Rat	0.41	1.8	0.025	27.7
	Mouse	0.21	3.4	0.007	95.0

[a] Lakings D. B.; Friis, J. M.; Bruns, M. B.; Jones, B. W.; Forbes, A. D., The Upjohn Company, Kalamazoo, MI, personal communication, 1987.
[b] No data available.
[c] Terry, J. J.; La Kings, D. B.; Closson, S. K.; Ludere, J. R., The Upjohn Company, Kalamazoo, MI, personal communication, 1990.
[d] Kakuk, T. J.; Rush, B. D.; Zaya, R. M.; Wilkinson, K. F.; McShane, M. M.; Cole, S. L.; Rieden, C. L., The Upjohn Company, Kalamazoo, MI, personal communication, 1993.

considered. Unlike the HIV protease inhibitors described herein, however, the clearance in each of the four closely related pairs appeared to increase as lipophilicity did. The failure of the HIV protease inhibitors to exhibit similar behavior probably was caused by alterations in the volume of distribution that were unaccounted for by looking only to 20-min sera levels as an estimation of hepatic extraction.

Not only are peptides difficult to assay in sera, but most methods must also be retooled for the sera of each individual species. This barrier, as well as the larger quantities of peptide needed for studies in monkey, dog, and humans, has resulted in minimal pharmacokinetic comparison across species. Another problem is that the most reproducible parameters of the pharmacokinetic profile are distribution phase (α), terminal elimination phase of a two-compartment model (β), terminal half-life ($t_{1/2\beta}$), and distribution half-life ($t_{1/2\alpha}$). The AUC, V_{ss}, and Cl_T are more dependent on the completeness of the sampling profile and are useful primarily when comparing several compounds in the same experiment.

Some of the available data suggest that rats may clear small peptides more rapidly than other species. For example, one report examining the properties of the HIV protease inhibitor A-77003 found that the disposition half-life ($t_{1/2\beta}$) increased with increasing size of animal (rat, dog, and monkey). For the renin inhibitor U-76394, $t_{1/2\beta}$ values were higher for monkey than rat (Table III). U-71038 had 10-fold higher $t_{1/2\beta}$ values in monkey than in rat; human values were accurately predicted by the monkey data. However, animal size does not appear to consistently predict clearance rate. For example, U-75875, a dual protease and renin inhibitor had similar $t_{1/2\beta}$ values for rat and monkey, but substantially higher values for mouse (Table III). These data suggest peptides in monkey and humans frequently, but not always, have longer $t_{1/2\beta}$ values than in rats.

Receptor-Mediated Transport of Proteins

As stated earlier, most peptidic compounds with more then eight amino acids do not undergo substantial hepatic clearance. However, peptide length is not the only determinant for hepatic uptake, because 43% of vasoactive intestinal peptide (28 amino acids long) was extracted by perfused liver in laboratories reporting 9% hepatic clearance for gastrin-17 and 6% for gastrin-27 (*11*).

Indeed, at least some of the peptidic molecules, notably insulin, glucagon, interleukin-2 (IL-2), tumor necrosis factor α, growth hormone, epidermal growth factor, some glycoproteins, lipoproteins, and immunoglobulins appear to undergo receptor-mediated endocytosis in addition to the nonselective pinocytosis by which other protein molecules are taken up by the hepatocyte (for a good review, *see* reference 30). Although nonselective pinocytosis appears to have low capacity, receptor-mediated endocytosis can be quite efficient because the receptor is recycled to the membrane surface after internalization. Estimates of hepatic transport of intravenously administered proteins that undergo receptor-mediated endocytosis range from 41% for IL-2 (*31*) to 64% for r-erythropoietin (*32*). Little structure–activity information on the binding of larger peptide molecules to these receptors is available, however, because systematic synthesis of analogues has not been performed. For glycoproteins such as erythropoietin, however, removing sialic acid residues that block receptor access to galactosamine can enhance hepatic uptake (*33, 34*).

Rapid clearance of proteins, especially of those that undergo receptor-mediated endocytosis, poses considerable problems for development of a long-acting pharmaceutical preparation. Therefore, attempts have been made to increase the serum residence time of these macromolecules by conjugating them to dextran, albumin, poly(vinylpyrrolidone), or poly(ethylene glycol) (for a recent review, *see* reference 35). In general, such conjugation decreases clearance of the proteins from systemic circulation, but can impair their activity to some extent. Conjugation to anionic and neutral macromolecules delay hepatic uptake best, whereas cationic conjugates are more rapidly cleared (*36*).

Summary

Hepatic clearance of peptides is complex. Current evidence suggests that the liver efficiently extracts peptides of three to six amino acids in length through carrier-mediated processes. The role of passive diffusion is difficult to assess, because at least one of the carrier mechanisms may exhibit a secondary dependence on lipophilicity, the property believed to drive passive transport across membranes. As might be expected, clearance properties of peptides differ across species, but not always in a predictable manner.

Peptides and proteins greater than six amino acids in length are generally efficiently extracted by the liver only if they exhibit receptor-mediated endocytosis. Unmasking carbohydrate residues that bind to these receptors or masking the binding moeity by conjugating the proteins to other marcromolecules can influence hepatic uptake accordingly. All peptidic molecules probably undergo some degree of nonspecific pinocytosis. Any strategies for modifying hepatic uptake of peptides should take into account the apparent dependence on some type of carrier–receptor mechanism for efficient transport.

References

1. Morley, J. S. *Annu. Rev. Pharmacol. Toxicol.* **1980,** *20,* 81–106.
2. Greenlee, W. J. *Pharm. Res.* **1987,** *4,* 364–372.
3. Johnson, M. I.; Allaudeen, H. S.; Sarver, N. *TIPS* **1989,** *10,* 305–307.

4. Lochs, H.; Morse, E. L.; Adibi, S. A. *J. Biol. Chem.* **1986,** *261,* 14976–14981.

5. Lombardo, Y. B.; Morse, E. L.; Adibi, S. A. *J. Biol. Chem.* **1988,** *263,* 12920–12926.

6. Koch, Y.; Battini, F.; Peterkofsky, A. *Biochem. Biophys. Res. Commun.* **1982,** *104,* 823–829.

7. Anderson, R. P.; Butt, T. J.; Chadwick, V. S. *Dig. Dis. Sci.* **1992,** *37,* 248–256.

8. Karls, M. S.; Rush, B. D.; Wilkinson, K. F.; Vidmar, T. J.; Burton, P. S.; Ruwart, M. J. *Pharm. Res.* **1991,** *8,* 1477–1481.

9. Dennis, P. A.; Aronson, N. N. *Arch. Biochem. Biophys.* **1985,** *240,* 768–776.

10. Doyle, J. W.; Wolfe, M. M.; McGuigan, J. E. *Gastroenterology* **1984,** *87,* 60–68.

11. Wolfe, M. M.; Doyle, J. W.; McGuigan, J. E. *Life Sci.* **1987,** *40,* 335–342.

12. Doyle, J. W.; Wolfe, M. M.; McGuigan, J. E. *Methods Enzymol.* **1983,** *98,* 225–241.

13. Gores, G. J.; La Russo, N. F.; Miller, L. J. *Am. J. Physiol.* **1986,** *250,* 344–349.

14. Sacks, H.; Terry, L. C. *J. Clin. Invest.* **1981,** *67,* 419–429.

15. Seno, M.; Seino, Y.; Takemura, Y.; Nishi, S.; Ishida, H.; Kitano, N.; Imura, H. *Can. J. Physiol. Pharmacol.* **1985,** *63,* 62–67.

16. Kim, R. B.; Perry, P. R.; Wilkinson, G. R. *Hepatology* **1993,** *18(4),* 298A.

17. Zanotti, G.; Rossi, F.; Di Blasio, B.; Pedon, C.; Benedetti, E.; Ziegler, K.; Tancredi, T. In *Peptides;* Rivier, J. E.; Marshall, G. K., Eds.; ESCOM: Leiden, The Netherlands, 1990; pp 118–119.

18. Kessler, H.; Haupt, A.; Frimmer, M.; Ziegler, K. *Int. J. Pept. Prot. Res.* **1987,** *29,* 621–628.

19. Ziegler, K.; Lins, W.; Frimmer, M. *Biochim. Biophys. Acta* **1991,** *1061,* 287–296.

20. Kessler, H.; Gerhke, M.; Haupt, A.; Klein, M.; Muller, A.; Wagner, K. *Klin. Wochenschr.* **1986,** *64,* 74–78.

21. Eisenberg, D. *Annu. Rev. Biochem.* **1984,** *53,* 595–623.

22. Gores, G. J.; Kost, L. J.; Miller, L. J.; La Russo, N. F. *Am. J. Physiol.* **1989,** *257,* G242–G248.

23. Hunter, E. B.; Powers, S. P.; Kost, L. J.; Pinon, D. I.; Miller, L. J.; La Russo, N. F. *Hepatology* **1990,** *12,* 76–82.

24. Kempf, D. J.; Marsh, K. C.; Paul, D. A.; Knigge, M. F.; Norbeck, D. W.; Kohlbreener, W. E.; Codacovi, L.; Vasavanonda, S.; Bryant, P.; Wang, X. C.; Wideburg, N. E.; Clement, J. J.; Plattner, J. J.; Erickson, J. *Antimicrob. Agents Chemother.* **1991,** *35,* 2209–2214.

25. Ruwart, M. J.; Sharma, S. K.; Harris, D. W.; Lakings, D. B.; Rush, B. D.; Wilkinson, K. F.; Cornette, J. C.; Evans, D. B.; Friis, J. M.; Cook, K. J.; Johnson, G. A. *Pharmacol. Res.* **1990,** *7(4),* 407–410.

26. Greenfield, J. C.; Cook, K. J.; O'Leary, I. A. *Drug Metab. Dispos.* **1989,** *17(5),* 518–525.

27. Bertrams, A.; Ziegler, K. *Biochim. Biophys. Acta* **1991,** *1073,* 213–220.

28. Wilkinson, K. F.; Rush, B. D.; Sharma, S. K.; Evans, D. B.; Ruwart, M. J.; Friis, J. M.; Bohanon, M. J.; Tomich, P. K. *Pharm. Res.* **1991,** *10(4),* 562–566.

29. Rush, B. D.; Wilkinson, K. F.; Lakings, D. B.; Ruwart, M. J. *Pharm. Res.* **1992,** *9(10),* S-282.

30. Bhaskar, U.; Lee, V. H. L. In *Peptide and Protein Drug Delivery;* Lee, V. H. L., Ed.; Marcel Dekker: New York, 1991; pp 391–484.

31. Sands, H.; Loveless, S. E. *Int. J. Immunopharmacol.* **1989,** *11,* 411–416.

32. Fu, J. S.; Lertora, J. J. L.; Brookins, J.; Rice, J. C.; Fisher, J. W. *J. Lab. Clin. Med.* **1988,** *111,* 669–676.

33. Sasaki, H.; Bothner, B.; Dell, A.; Fukuda, M. *J. Biol. Chem.* **1987,** *262,* 12059–12079.

34. Dordal, M. S.; Wang, F. F.; Goldwasser, E. *Endocrinology* **1985,** *116(6),* 2293–2299.

35. Davis, F. F.; Kazo, G. M.; Nucci, M. L.; Abuchowski, A. In *Peptide and Protein Drug Delivery;* Lee, V. H. L., Ed.; Marcel Dekker: New York, 1991; pp 831–864.

36. Nishida, K.; Mihara, K.; Takino, T.; Nakane, S.; Takakura, Y.; Hashida, M.; Sezaki, H. *Pharm. Res.* **1991,** *8(4),* 437–444.

RECEIVED for review July 12, 1993. ACCEPTED revised manuscript December 6, 1993.

Blood–Brain Barrier Peptide Transport

CHAPTER 12

Blood–Brain Barrier Peptide Transport and Peptide Drug Delivery to the Brain

William M. Pardridge

Peptides are defined as polypeptides comprised of two or more amino acids. In discussing transport of peptides to the brain, no distinction is made between peptides of 8, 80, or 800 amino acids in length, because the same general principles apply. For example, the cell biology of plasma membrane binding and receptor-mediated endocytosis of cholecystokinin (CCK) (eight amino acids), insulin (~50 amino acids), or transferrin (~800 amino acids), is comparable (*1–3*).

Peptides do not freely enter the brain from blood because of the minimal in vivo free diffusion of these hydrophilic molecules through the brain capillary endothelial wall, which makes up the blood–brain barrier (BBB). Peptides are important neuromodulator compounds in brain (*4*), and peptide analogues may prove to be highly efficacious neuropharmaceuticals should these molecules be made transportable through the BBB (*5*). In this chapter, I will review existing strategies for peptide drug delivery to the brain. However, prior to reviewing these strategies, it is necessary to discuss methodologic considerations related to BBB transport of peptides, given the often conflicting statements in the literature in this area. For example, a recent review (*6*) on BBB transport of peptides lists 22 different peptides that are purported to readily cross the BBB intact. If such ready transport of peptides actually occurs, then no need to develop strategies for peptide drug delivery to the brain exists, because peptides should exert pronounced central nervous system (CNS) pharmacologic effects in vivo following systemic administration. Although in some cases the systemic administration of massive doses of lipidized peptides (e.g., >10 mg/kg) results in CNS effects (*7*), the systemic administration of nonlipidized peptides at doses of <5–10 mg/kg rarely results in CNS pharmacologic effects.

The permeability of the BBB to peptides is low and on the order of that reported for other centrally active compounds, such as glutamic acid, an excitatory neurotransmitter, or γ-aminobutyric acid (GABA), an inhibitory neurotransmitter (*8*). However, CNS effects following the administration of submassive doses of these neurotransmitters is not observed (*9, 10*). That is, the transport of glutamic acid or GABA through the BBB is not physiologically or pharmacologically significant. To develop practical models of neurotransmitter or peptide transport through the BBB that is acceptable to all investigators it is necessary to critically evaluate the existing methodologies employed for the study of peptide transport through the BBB.

Methodologic Considerations

Experimental Variables

The experimental variables that should be considered in the study of peptide transport through the BBB include the differentiation of peptide transcytosis through the BBB from brain capillary endothelial binding or endocytosis, the elimination of "uptake" artifacts due to peptide metabolism, and definition of the various mechanisms of peptide transport through the brain capillary endothelium (Table I).

Peptide Transcytosis and Measurement of Volume of Distribution

Peptide transcytosis through the BBB involves complete transendothelial passage of the peptide from blood into brain interstitial fluid (ISF). Conversely, peptides may simply be bound or sequestered to the lumenal membrane of the brain capillary endothelium or may undergo binding and endocytosis into the intracellular endothelial compartment, but not

Table I. Peptide Transcytosis through the BBB

Experimental Variables	*Class*	*Technique*
Differentiate transcytosis versus endothelial binding–endocytosis	Morphologic	autoradiography peroxidase histochemistry gold/silver enhancement
Eliminate artifacts due to peptide metabolism	Physiologic	carotid artery perfusion capillary depletion technique dialysis fiber
Mechanism of transport receptor mediated absorptive mediated free diffusion	In vitro brain endothelial monolayers	

undergo net transport into brain ISF. The importance of distinguishing between transcytosis versus endothelial binding–endocytosis in the study of BBB transport of peptides has only recently been appreciated. In the past, this distinction was not made in the study of BBB transport of peptides (*6, 11*).

It has generally been concluded that a given peptide is transportable through the BBB when the brain volume of distribution (V_D) of the peptide is experimentally observed to exceed the V_D of a blood volume marker, such as albumin or sucrose (*11*). However, as described later, a high brain V_D is readily detected following the intracarotid infusion of acetylated low-density lipoprotein (LDL) (*12*), a protein that undergoes receptor-mediated endocytosis into cells (*13*). In another example, the brain V_D of β-endorphin greatly exceeds that of a blood volume marker following a 10-min carotid artery infusion of the peptide (*14*). However, in neither case is the peptide actually transported through the BBB (*12, 14*), as reviewed later in the discussion of the capillary depletion technique. The techniques available for differentiating transcytosis from endothelial binding–endocytosis include morphologic, physiologic, and in vitro brain endothelial monolayers (Table I); these experimental approaches are briefly reviewed in a subsequent section.

Peptide Metabolism

A second experimental variable that often confounds the interpretation of BBB transport of peptides is the rapid endothelial metabolism of peptides, often by ectoenzymes that project from the lumenal membrane of the capillary endothelium into the plasma compartment. This enzymatic metabolism may release radiolabeled amino acids that are then transported through the BBB by specific BBB amino acid transporters (*15*). Thus, if no attempts are made to perform a chromatographic analysis of brain extracts following carotid artery perfusion of labeled peptides, it is difficult to evaluate the extent to which the peptide actually undergoes a net transcellular movement into the brain ISF in intact form.

Mechanism of Peptide Transport and Saturation of Transport

A third experimental variable that is of importance to the study of BBB transport of peptides is saturability because this variable pertains to the mechanism of peptide transport through the brain capillary endothelium (Table I). If the apparent brain "uptake" of a peptide is nonsaturable, then the peptide is either undergoing free diffusion through the brain capillary endothelium or is simply being nonspecifically bound to the brain capillary endothelium. The distinction between actual free diffusion of the peptide into brain ISF from nonspecific binding to the brain vascular endothelium cannot be made unless additional methodologies (Table I) are used to evaluate net peptide transport into the brain ISF.

The extent to which a given peptide may undergo free diffusion through the BBB may be approximated by determining either the octanol–saline partition coefficient (log *P*) or the hydrogen bond number on the peptide. Determination of hydrogen bond number may be estimated by simply visual inspection of the structure of the peptide, using the following rules for computation of hydrogen bond numbers: a hydroxyl group forms two hydrogen bonds, an amide group forms two hydrogen bonds, and an amino- or carboxyl-terminus each forms two hydrogen bonds (*16*). Thus, a simple dipeptide, such as glycylglycine, forms six hydrogen bonds, and a dipeptide, such as serine–serine, forms 10 hydrogen bonds with solvent water. The extent to which solutes form hydrogen bonds with solvent water predicts the membrane permeability for the given solute (*17*). In the case of the BBB, previous studies (*18*) with steroid hormones have shown that BBB permeability decreases a log order of magnitude with the addition of each pair of hydrogen bonds. When the hydrogen bond number exceeds 10, the BBB transport of the solute is minimal (*18*). From these considerations, it can be seen that peptides as large as three or even two amino acids have hydrogen bond numbers in excess of 10 and would not be expected to undergo significant transport through the BBB by free diffusion.

In addition to the hydrophilicity of peptides, another factor that minimizes peptide transport through the BBB is the molecular size of the molecule. Studies (*19, 20*) with the peptide cyclosporin, which has a molecular weight of ~1200 Da, showed that this highly lipid–soluble peptide is transported through the BBB several log orders of magnitude less than that predicted by its lipid solubility, even after normalization for molecular weight. Similarly, the BBB transport of alkaloids with molecular weights of ~800 Da is much lower than that predicted from the octanal–saline log *P* (*21*). This divergence between lipid solubility and BBB permeation rates when the molecular weight of a compound exceeds 800–1000 Da has been ascribed to steric hindrances (*5*) that restrict solute partitioning through pores or "kinks" in the fatty acyl side chains of the membrane lipid bilayer (*22*). This apparent molecular weight threshold of 800–1000 Da further restricts the free diffusion of peptides through the BBB beyond the restriction from the inherent hydrophilicity of these molecules.

Experimental Approaches That Differentiate between Transcytosis and Endothelial Binding–Endocytosis

Morphologic Techniques

Thaw-mount autoradiography was initially used to demonstrate transcytosis of ^{125}I-insulin into brain following carotid arterial infusion of the labeled peptide for 10 min (*23*). Radioactive silver grains were found well within brain parenchyma and were removed from the vasculature. Reversed-phase high-performance liquid chromatography (HPLC) was used to demonstrate that the uptake of radioactivity in the brain parenchyma was not due to systemic degradation of circulating insulin. The disadvantage of thaw-mount autoradiography is that it often takes several weeks to achieve adequate exposure of the emulsion. A faster technique is the use of horseradish peroxidase (HRP)–peptide conjugates and peroxidase histochemistry. A wheat germ agglutinin–HRP conjugate was used to document the endothelial subcellular organelles involved in transcytosis of this conjugate through the brain capillary endothelium (*24*). A method comparable to HRP histochemistry has recently been used to study protein transcytosis through the BBB, wherein 5-nm gold conjugates are infused into the internal carotid artery and distribution of gold conjugate is detected by immunogold silver staining (IGSS) techniques (*25*). A disadvantage of both HRP histochemistry and of the IGSS technique is that they require electron microscopy to assess transcytosis and therefore are labor intensive and time consuming.

Physiologic Techniques

BBB transcytosis of peptides may be quantified in a single day with internal carotid artery perfusion coupled with either a capillary depletion or a dialysis fiber technique. The carotid artery perfusion method allows for determination of the brain V_D of the peptide (*8, 20*). The capillary depletion (*12*) or dialysis fiber technique (*26*) allows for quantitation of transcytosis of the peptide through the BBB. The disadvantage of the capillary depletion technique is that radiolabeled solute that is nonspecifically bound to the microvasculature during the brain perfusion may dissociate from the vessels during preparation of the homogenate.

This dissociation causes an artifactual "distribution" of the solute into the postvascular supernatant that exceeds the uptake of the blood volume marker (e.g., sucrose or albumin).

Chromatographic analysis of either postvascular supernatant (as with the capillary depletion technique) or with the dialysate (as with the dialysis fiber technique) is important to document transport of undegraded peptide into brain ISF. Dialysis fiber sampling, and not cerebrospinal fluid (CSF) sampling, is the preferred method for measuring the composition of brain ISF. As discussed later, CSF sampling allows for measurement of choroid plexus permeability of peptides but does not allow for measurement of brain capillary or BBB permeability of peptides.

Specific examples of BBB transport of peptides are now reviewed within the context of how the use of the capillary depletion technique or the dialysis fiber technique, in conjunction with chromatographic analysis of brain extracts, allows for a quantitative assessment of whether the peptide actually undergoes transport through the BBB in intact form.

Acetylated LDL. The acetylation of LDL–lysine residues with acetic anhydride causes this protein to be a substrate for the Type I scavenger receptor depicted in Figure 1A. The scavenger receptor is widely expressed on endothelium (*13*) and allows for the receptor-mediated endocytosis of acetylated LDL (acetylLDL) and other modified proteins into endothelial cells. The acetylLDL receptor is present on brain capillary endothelium, and its presence mediates the brain uptake of circulating acetylLDL (Figure 1B). When ^{125}I-acetylLDL is infused into the internal carotid artery of anesthetized rats, the brain V_D reaches 30 μL/g by 10 min (*12*). However, when the capillary depletion technique is performed, virtually all of this distribution of acetylLDL into brain is confined to the capillary endothelium. The acetylLDL–albumin V_D ratio in the postcapillary supernatant does not exceed one during the 10-min perfusion (Figure 1B), a fact that indicates essentially all of the apparent "uptake" of acetylLDL following carotid artery perfusion is confined to the endothelium. That is, this protein undergoes only binding and endocytosis within the brain capillary endothelium, not transcytosis into the brain ISF. If the capillary depletion technique had not been performed in these studies, a BBB permeability surface (PS) area product of ~3 μL min^{-1} g^{-1} would be calculated and it would be concluded, erroneously, that the BBB is readily permeable to acetylLDL. The lumenal side of the brain capillary endothelium is readily permeable to acetylLDL but the BBB is not.

β-Endorphin. When the D-Ala^2 analogue of [^{125}I]β-endorphin is infused into the internal carotid artery of anesthetized rats, a brain V_D of ~40 μL/g is measured following 10 min of perfusion (*14*). Without performing the capillary depletion technique, a BBB PS product of ~4 μL min^{-1} g^{-1} would be calculated and it would be concluded, erroneously, that the peptide readily undergoes transport through the BBB. However, when the capillary depletion technique is performed, the V_D of [D-Ala^2]β-endorphin is not significantly different from zero in the postvascular supernatant. These data indicate β-endorphin does not cross the BBB.

Adrenocorticotropin (ACTH) Analogues. A novel ACTH analogue which is a hexapeptide that is extended at the carboxyl-terminus by a diaminooctane residue, was developed (*26*). This conversion of the carboxyl-terminus into an extended amino-terminus confers a highly cationic charge on the peptide and raises the isoelectric point (pI) to ~10. The internal carotid artery perfusion technique, coupled with either the capillary depletion or the dialysis fiber technique, has been used to show that this cationic peptide undergoes transcytosis through the BBB (*26*). Other studies (*27*) have shown that the brain capillary uptake of the analogue, called ebiratide, is inhibited by a polycatonic compound, such as

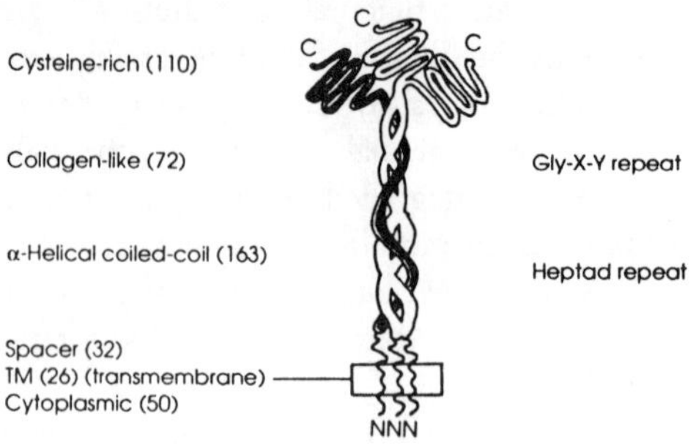

A

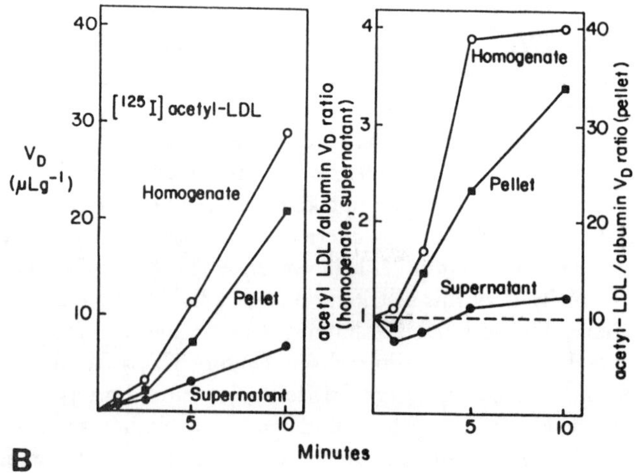

B

Figure 1. (A) Model of type I acetylLDL or scavenger receptor predicted from deduced amino acid sequence. More than 85% of the receptor is predicted to project into the extracellular space with a small cytoplasmic domain. (Reproduced with permission from references 92. Copyright 1990 American Association for the Advancement of Science.) (B) V_D *of* ^{125}*I-acetylated LDL in rat brain homogenate, vascular pellet, and postvascular supernatant is shown on the left side of the graph. The data are replotted as a ratio of acetylLDL* V_D *to albumin* V_D *in the homogenate, vascular pellet, and supernatant. The values for the homogenate and postvascular supernatant* V_D *ratios are shown on the left* y*-axis, and the* V_D *ratios for the vascular pellet are shown on the right* y*-axis. Graphs are drawn from data previously reported by Triguero et al. (*12*).*

protamine or polylysine. Thus, the mechanism of transport of ebiratide through the BBB would appear to be absorptive-mediated transcytosis, which was observed for a number of polycationic proteins (*5*). In this study (*26*), the appropriate experimental methodologies were performed for differentiating transcytosis from endothelial binding–endocytosis, and additional evidence is given to characterize the mechanism of peptide transport through the BBB.

Delta Sleep-Inducing Peptide (DSIP). The BBB transport of ^{125}I-labeled DSIP, a nonapeptide, was studied by the internal carotid artery perfusion technique (*28*). The brain V_D of DSIP reached a value, corrected for the sucrose blood volume, of 9 μL/g following a 10-min perfusion, and the calculated BBB PS product, also called K_{in}, was 0.93 μL/min/

g for the parietal cortex (*28*). The brain uptake of ^{125}I-DSIP was completely suppressed to the level of sucrose uptake by the inclusion of 7 μM unlabeled DSIP in the perfusate. Thus, based on these studies, little if any DSIP is transported through the BBB through a nonsaturable mechanism. This finding contrasts with earlier studies showing that the apparent brain distribution of DSIP is nonsaturable (*11*). DSIP has a hydrogen bond number well in excess of 10 and is a highly water-soluble compound that would not be expected to undergo free diffusion through the BBB. Thus, if DSIP does undergo transport through the BBB, this transport is likely mediated by either a specific carrier or receptor system, and support for this mechanism is found in the saturability of DSIP transport through the BBB (*28*). However, the apparent brain uptake of [^{125}I]-DSIP is also greatly suppressed by the inclusion of 1 mM L-tryptophan in the carotid artery perfusate (*28*). This observation raises the question as to whether the brain uptake of DSIP is artifactual due to microvascular metabolism of DSIP and to release of [^{125}I]tyrosine from the peptide. Both tyrosine and tryptophan share affinity for a large neutral amino acid transporter within the BBB (*15*), and the inhibition of DSIP "uptake" by tryptophan suggests that metabolism is playing an important role in the apparent brain uptake of this peptide. The transport of DSIP through the BBB essentially cannot be evaluated on the basis of the existing data. What is needed is a chromatographic analysis of the postvascular supernatant using the capillary depletion technique or of the brain dialysate using the dialysis fiber technique to document the transport of undegraded DSIP into the brain ISF.

Arginine Vasopressin. Recently, the transport of [^{3}H]arginine vasopressin through the BBB has been documented by Zlokovic et al. (*29*). This study is an important advance in the field of BBB transport of vasopressin analogues in at least three respects. First, a ^{3}H-labeled vasopressin analogue was used instead of an iodinated analogue that was used in previous studies. Iodinated vasopressin analogues are not biologically active (*30*), and thus the significance of measurements of brain uptake of iodinated vasopressin analogues is not apparent. Second, a capillary depletion technique was added to the carotid artery perfusion technique in these studies and, third, chromatographic analysis of the postvascular supernatant was performed (*29*). These studies showed that intact arginine vasopressin was measurable in the postvascular supernatant following a 3-min carotid artery perfusion. However, following a 10-min carotid artery perfusion, the amount of intact arginine vasopressin in the postvascular supernatant was only ~10% of the total brain radioactivity (*29*). Nevertheless, the finding of intact arginine vasopressin following 3 min carotid artery perfusion in the postvascular supernatant suggests that intact vasopressin distributes into the brain ISF following transcytosis through the BBB. Given that brain capillary endothelium is endowed with vasopressin receptors (*31*), the transcytosis of arginine vasopressin through the BBB may be receptor mediated.

In Vitro Brain Endothelial Monolayers

The cultivation of isolated brain capillary endothelial cells in tissue culture on filters that can be fitted into side-by-side chambers allows for the potential establishment of in vitro models for measuring BBB peptide transcytosis (*32, 33*). For example, either passaged brain capillary endothelial cell lines, which have a cuboidal morphology (Figure 2A), or primary cultures of brain capillary endothelial cells, which have a spindle-shape morphology (Figure 2B), may be grown in tissue culture. Following cultivation for 10 to 14 days, the filters are positioned in side-by-side diffusion chambers and the transport of various radiolabeled drugs through the endothelial monolayer is measured (*19*). From the rate of transport or clearance from the donor chamber to the acceptor chamber, the PS product for the monolayer may

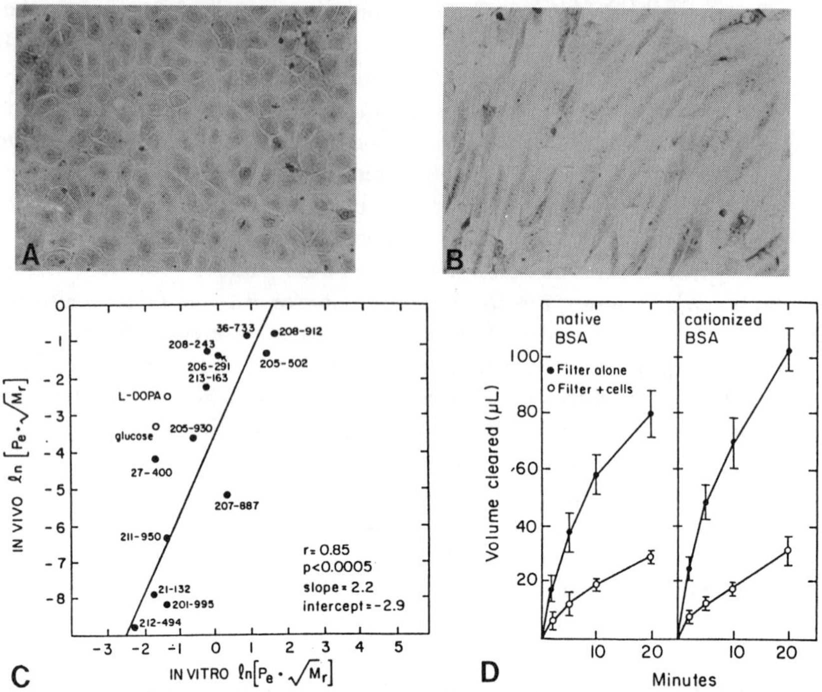

Figure 2. (A) Cuboidal morphology of bovine brain capillary endothelium grown as a continuous line of culture, as described by Estrada et al. (93). (B) Spindle-shaped morphology of bovine brain capillary endothelium grown in primary tissue culture. Cells are lightly stained with Mayer's hematoxylin. (C) The in vivo ln BBB permeability coefficient (P_e) normalized for molecular weight (M_r) is plotted versus the same function for the in vitro BBB model for 13 different drugs (●) to generate the slope and intercept of the linear regression analysis shown in the Figure. L-Dopa and glucose (○) were not used in the regression analysis because these molecules traverse the BBB in vivo by carrier-mediated transport processes independent of lipid solubility. (A, B, and C are reproduced with permission from reference 19. Copyright 1990 Williams & Wilkins Company.) (D) The transport of ^{3}H-labeled native BSA (left) or ^{3}H-labeled cationized BSA (right) across a monolayer of bovine brain capillary endothelium in a continuous line in culture (A) is plotted versus time of incubation at 37 °C. Data are mean ± SE (n = 3 wells per point). Transport of native or cationized BSA across a collagen–fibronectin-coated polycarbonate filter alone (○) or across a filter covered by a monolayer of the endothelial cells (●) is shown. The transport of ^{3}H-labeled native BSA or ^{3}H-labeled cationized BSA across primary cultures of bovine brain capillary endothelium (B) were also measured, and the data are identical to those observed with the endothelial line. No significant difference between the rate of transport of either native or cationized BSA across the endothelial monolayer in vitro is observed. (D is reproduced with permission from reference 5. Copyright 1991 Raven.)

be calculated. The PS product may be converted into P_e, which is a permeability coefficient, by knowing the surface area of the Transwell monolayer (e.g., 0.33 cm^2) (*19*). The in vitro P_e values are plotted on the x-axis in Figure 2C for a bovine brain endothelial cell line. Bovine primary cultures and the cell line (Figures 2A and 2B) were studied; the cell line provided a tighter barrier to solute diffusion and yielded P_e values about threefold lower than those determined with the primary cultures. The same radiolabeled drugs studied with the in vitro BBB model were then infused into the internal carotid artery and the rate of BBB transcytosis of the drugs was quantified with the capillary depletion technique (*19*). Knowing the rate of cerebral blood flow and the unidirectional extraction for each drug, the in vivo BBB PS values may be computed. Knowing the surface area of the BBB in vivo (100 cm^2/g) (*34*), the in vivo P_e values were calculated from the PS values and are plotted on the y-axis of Figure 2C. These data show a linear correlation between in vivo and in vitro models of BBB transport. However, the in vitro P_e values are on average 150-fold greater than the in vivo values based on the slope of the relationship in Figure 2C. Thus, the in vitro barrier for these molecules that undergo transport through the BBB by free diffusion is 150-fold leakier than that found in vivo (*19*). Conversely, for molecules that undergo carrier-mediated transport through the BBB, the in vitro model underestimates by several log orders of magnitude the P_e values for nutrients, such as glucose or L-dopa (Figure 2C). This underestimation of P_e values for compounds that undergo carrier-mediated transport through the BBB arises due to the profound down-regulation of these nutrient transporters when brain capillary endothelial cells are grown in tissue culture. For example, the mRNA for the GLUT1 glucose transporter is severely down-regulated in cultures of brain capillary endothelial cells (*35*). The reason for this down-regulation is presumably due to the loss in vitro of astrocyte trophic factors, which are normally present in vivo (*5*).

In parallel to the down-regulation of the glucose or L-dopa transporter in brain capillary endothelial cells grown in culture, a down-regulation in the peptide transcytosis pathway occurs. Although initial studies showed enhanced rates of transcytosis through the in vitro BBB of cationized bovine serum albumin (BSA) relative to native BSA (*36*), this enhanced rate of transport could not be confirmed in our laboratory. As shown in Figure 2D, no difference between the rate of transcytosis through the endothelial monolayer of either native or cationized BSA occurs. However, a marked difference in the rate of transcytosis of these two compounds at the BBB in vivo occurs (*5*), wherein cationized albumin is readily transcytosed through the BBB with no transport of native albumin (*see* later). These data indicate that the in vitro endothelial monolayer model deviates to such a great extent from the conditions of the BBB in vivo that an accurate assessment of peptide transport through the BBB is not yet possible with this in vitro model. Further advances in the understanding of the molecular and cell biology of BBB-specific gene expression may, in the future, refine the in vitro BBB model so that it more closely approximates the BBB in vivo.

A model that may approximate the in vivo BBB is a coculture model wherein rat brain astrocytes and bovine brain endothelial cells are grown in a heterologous system on opposite sides of a porous filter (*37*). The P_e values in this system are three- to eightfold lower than when endothelial cells are grown alone (*19*). However, the P_e values were measured in the presence of 100% human serum (*37*). Because serum proteins bind many of the lipophilic amine drugs studied in the coculture system (*37*), the in vitro P_e is reduced regardless of a tighter BBB in the in vitro model.

Brain-to-Blood Transport of Peptides

Two Barrier Systems in Brain

As depicted in Figure 3, two major barrier systems occur in brain, the choroid plexus, or blood–CSF barrier, and the brain capillary, or BBB. The choroid plexus segregates the CSF from blood and the BBB separates brain ISF from blood. The CSF compartment may be accessed experimentally through lumbar or cisternal puncture and the brain ISF compartment may be accessed experimentally with dialysis fibers. Although no anatomic barrier segregates the CSF and the ISF, a physiologic barrier arises from the continuous absorption of CSF at the arachnoid villi and the return of CSF constituents to the circulating

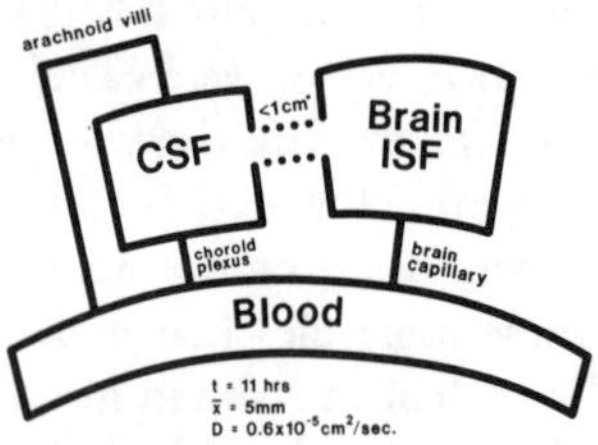

Figure 3. The two central extracellular compartments of brain, that is, CSF and brain ISF are segregated from blood by the choroid plexus, or blood–CSF barrier, and by the brain capillary, or BBB, respectively. The two extracellular fluid compartments are separated by a functional barrier preventing complete equilibration between these two fluid compartments. This "barrier" exists because CSF is constantly formed by the choroid plexus and drained into the peripheral bloodstream at the arachnoid villi and is completely cleared in the human brain approximately every 4–5 h. Conversely, a small molecule such as glucose, with a diffusion coefficient of 0.6 $\times$ 10^{-5} cm^2/s, would take ~11 h to diffuse 5 mm, assuming no enzymatic metabolism or tissue sequestration of the compound occurs. (Reproduced with permission from reference 5. Copyright 1991 Raven.)

systemic blood compartment. The CSF in the human brain turns over completely every 4–5 h. In contrast to the relatively rapid rate of CSF bulk flow from brain to blood, the rate of solute diffusion from CSF into brain parenchyma is relatively slow. For example, given a diffusion coefficient of 0.6 $\times$ 10^{-5} cm^2/s, it may be calculated that it takes ~11 h for a solute the size of glucose to diffuse 5 mm (*5*). Conversely, a peptide such as myoglobin (molecular weight = 17,500 Da) has a diffusion coefficient of 0.11 $\times$ 10^{-5} cm^2/s, and diffuses 5 mm in 65 h (*5*).

Owing to the rapidity of CSF bulk flow relative to the slowness of peptide diffusion, the distribution of peptides into brain ISF is minimal when these molecules are injected into CSF. For example, insulin has been injected into the ventricular compartment followed by immunocytochemical detection of insulin in brain tissue (*38*). However, insulin was localized only to the ventricular surfaces of the brain. In another example, when 5 μg of CCK was injected into the ventricular compartment of rats, it was readily found in the peripheral blood stream and achieved a blood concentration of 2 nM within 30 min of intraventricular injection (*39*). In addition to the rapid distribution of peptide into peripheral blood and to the minimal distribution of the peptide into brain parenchyma following ventricular administration, another caveat in this area is the rapid metabolism of peptides by choroid plexus following intraventricular administration. For example, the half-life of angiotensin II in the ventricular compartment is only 25 s following ventricular administration, due to the rapid degradation of this peptide by choroid plexus enzymes (*40*). Similarly, dynorphin analogues are also rapidly metabolized within seconds following intraventricular administration of the peptide (*41*).

An example of the differential permeability of the blood–CSF barrier and the BBB for azidothymidine (AZT) is shown in Figure 4. AZT readily distributes into CSF following oral administration and achieves a peak CSF concentration that is ~40% of the corresponding plasma concentration. Conversely, when radiolabeled AZT is administered systemically and its distribution to various organs is measured by autoradiography, the compound readily distributes into virtually all organs except for brain and spinal cord (Figure 4). Similarly, its parent compound, thymidine, readily undergoes transport through the blood–CSF barrier, but not through the BBB (*5*). The differential transport of either thymidine or AZT through these two barriers is due to the presence of thymidine or AZT transport systems

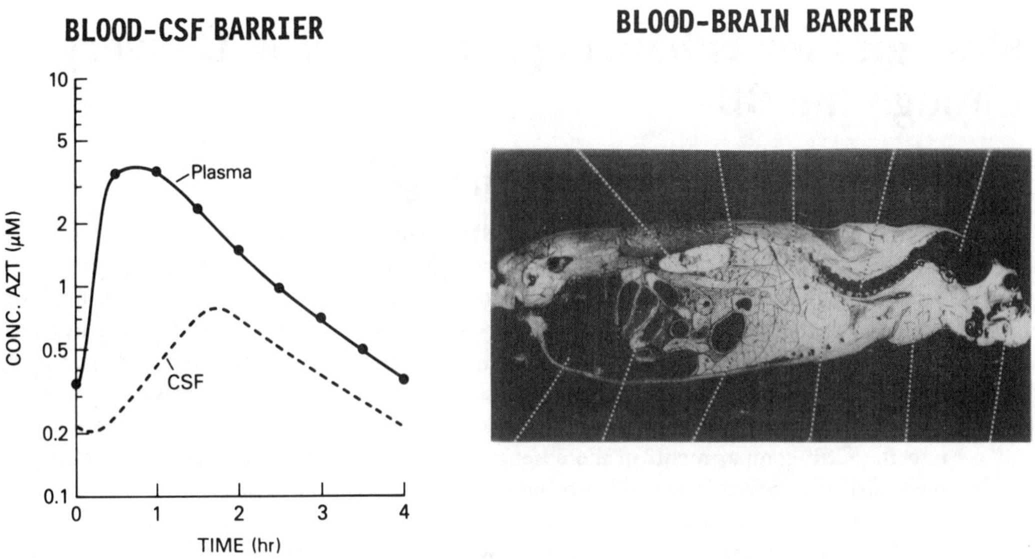

Figure 4. Left: The concentration of AZT in human CSF or plasma is plotted versus time after oral drug administration. The drug readily distributes into CSF because of the presence of a pyrimidine nucleoside transport system at the choroid plexus. (Reproduced with permission from reference 94. Copyright 1987 New England Journal of Medicine.) Right: Autoradiography of a rat following intravenous administration of radiolabeled AZT shows no distribution of the drug into the brain or spinal cord because of the absence of a pyrimidine nucleoside transport system at the BBB. (Reproduced with permission from reference 95. Copyright 1991.)

in the choroid plexus and the absence of these transport systems at the BBB. Thus, BBB permeability to drugs or peptides is not reflected in CSF sampling following peripheral administration. A corollary of this principle is that BBB permeability, on the brain side of the BBB, is not reflected by blood sampling following intraventricular administration.

Peptide Efflux at the BBB

Recently, peptide transport in the direction of brain-to-blood has been measured to characterize the permeability of brain capillaries on the ablumenal endothelial membrane (*6, 11*). These studies (*6, 11*) led to the hypothesis that several so-called peptide transport systems (PTS) exist and are localized on the BBB; these systems were even subdivided into PTS-1 through PTS-5. However, the studies (*6, 11*) characterizing the ablumenal endothelial membrane permeability to peptides involved intraventricular peptide administration and measurement of peptide distribution from CSF to blood. Therefore, the principle factors affecting these experimental measurements include rates of CSF bulk flow, rates of peptide transport at the choroid plexus epithelium, and choroidal peptide metabolism. These factors related to choroid plexus and CSF flow are so dominant in controlling the rate of solute movement from CSF to blood (*42, 43*) that the role played by the transport properties of intraparenchymal capillaries is minimal. The permeability of the ablumenal membrane of the BBB is difficult to quantify experimentally, but may be measurable by injecting peptides into dialysis fibers implanted within brain parenchyma.

Strategies for Enhancing Peptide Drug Delivery through the BBB

Overview of Brain Drug Delivery Strategies

Brain drug delivery strategies may be generally classified into the following three areas: (1) neurosurgical-based strategies, (2) pharmacologic-based strategies, and (3) physiologic-based strategies. Neurosurgical-based strategies include intraventricular drug infusion. The advantages and disadvantages of this approach were recently reviewed (*5*). In the consideration of how well drugs penetrate into brain parenchyma following intraventricular drug infusion, it is useful to recall the basic physiology of convection and diffusion within the extracellular compartments of brain as depicted in Figure 3. When peptides are injected into the CSF compartment, in the absence of axoplasmic flow mechanisms operating at the ventricular surface, very little distribution occurs beyond the surface of the brain because the peptide is rapidly exported to the peripheral circulation. The concentration of drug within brain parenchyma decreases steeply with each millimeter of distance removed from the surface of the brain (*42*). Paradoxically, drug may distribute to brain parenchyma following intraventricular administration through the following route: injection into the ventricular compartment, followed by bulk flow and exodus into the systemic circulation, followed by transport directly across the brain capillary or BBB, followed by entry into brain parenchymal ISF. Such a pathway accounts for the fact that the concentration of barbiturate that induces sleep is essentially the same whether the drug is administered intravenously or into the ventricles (*43*). In this respect, intraventricular administration may be fundamentally viewed as an intravenous injection. For example, after radiolabeled ascorbic acid and radiolabeled mannitol are coinjected into the ventricular compartment of rabbits, the distribution of ascorbic acid into the brain parenchyma, several millimeters removed from the ventricular surface, is selectively increased for the ascorbic acid relative to the mannitol (*44*), although mannitol is not taken up by brain cells. These observations suggest that ascorbic acid is exported from the CSF to the systemic circulation followed by reentry into brain parenchyma through direct transport across the BBB.

Pharmacologic-based strategies include liposomes and peptide lipidization. Recent advances in liposome technology include the development of multivesicular liposomes (MVLs) that have a diameter of 40 μM (*45*), which is a diameter approximately five times that of an erythrocyte. When dideoxycytidine (ddC) is incorporated in MVLs and is injected into the ventricular compartment, the half-life of drug distribution in the CSF is 23 h compared with the half-life of free drug distribution in CSF, which is 1.1 h (*45*). Because ddC exits the ventricular compartment through bulk flow, the rate of exodus of free ddC should be identical to the rate of exodus of ddC incorporated into 40-μM MVLs. That is, exodus by bulk flow is independent of molecular weight. The fact that the half-life of drug in the ventricular compartment is increased so dramatically with the use of MVLs suggests that these large aggregated liposomes are somehow trapped within the CSF compartment. These particles may be too large to freely exit the CSF flow tracks through absorption at the arachnoid villi.

The use of conventional liposomes for drug delivery across the brain capillary recently was reviewed (*5*). Briefly, even liposomes with a diameter as small as 100 nm are too large to undergo free diffusion through the BBB. However, it is possible that small unilamellar vesicles (SUVs) coupled to brain drug transport vectors (*see* later) may be transported through the BBB by receptor-mediated or absorptive-mediated transcytosis. Similarly, cationic liposomes recently were developed (*46*), and these structures may undergo absorptive-mediated endocytosis into cells. Whether cationic liposomes successfully undergo absorptive-mediated transcytosis through the BBB has not yet been determined.

New strategies for peptide lipidization recently were developed for enkephalin peptides. 1-Adamantane derivatives of [D-Ala^2]leucine enkephalin were prepared (*7*). When the amino-terminus of the enkephalin pentapeptide is derivatized with the 1-adamantane, loss of biological activity of the opioid peptide occurs. These findings are consistent with earlier observations that opioid peptides require a free amino terminal tyrosine residue to be biologically active (*47*). Conversely, when the 1-adamantane derivative is coupled to the carboxyl-terminus of the leucine enkephalin analogue, analgesia is observed following systemic administration of the peptide. However, the transport of this lipidized leucine enkephalin analogue across the BBB may not be substantially enhanced with this strategy because massive doses (5–50 mg/kg) of subcutaneously administered peptide are required to induce centrally mediated analgesia (*7*). The reason for this apparent poor transport of the lipidized enkephalin may be that this form of lipidization results in only a small reduction in the number of hydrogen bond-forming functional groups on the enkephalin molecule. For example, conjugation of a carboxyl-terminus reduces two hydrogen bonds on the peptide nucleus, but this only results in a reduction from 14 to 12 of the total hydrogen bonds formed with solvent water by the leucine enkephalin analogue. A lipidization strategy should reduce the hydrogen bond number to <10 to allow for effective transport of the peptide through the BBB, providing the molecular weight of the peptide is under the threshold of 800–1000 Da.

Another lipidization strategy for enkephalin analogues recently was reported by Bodor et al. (*48*). However, again, very large doses of enkephalin peptide (5–20 mg/kg) were required to induce centrally mediated analgesia. The reasons for the large dose of peptide required to induce a CNS pharmacologic effect in this system may be at least threefold. First, the molecular weight of the lipidized enkephalin adduct is in excess of 1000 Da and thus exceeds the apparent threshold of 800–1000 Da for lipid-mediated transfer through the BBB. Second, the lipidization strategy requires blockade of the free amino terminal tyrosine moiety, which markedly reduces opioid peptide biological activity (*47*). Third, the analogue is rapidly taken up by all organs and has poor pharmacokinetics; the blood level of the analogue is immeasurably low within 15 min following intravenous administration. Finally, in this study, it is assumed that the analgesia, as measured with the tail flick test, represents centrally activated opioid receptors. However, reversibility with naloxone was not demonstrated. The analgesia was also not reversible with time, and this may be because the tail was initially subjected to a 50-s heat exposure (*48*). Another factor that may have contributed to non-opioid-mediated toxicity was the fact that toxicologic doses of dimethyl sulfoxide and ethanol were coadministered as the vehicle with the lipidized enkephalin analogue. The doses that were administered (e.g., ~1 g/kg) were shown in previous studies to cause solvent-mediated opening of the BBB (*49, 50*).

Physiologic-based strategies for drug delivery through the BBB involve the use of chimeric peptides (*5, 51*). Chimeric peptides are formed when a nontransportable drug is covalently attached to a drug transport vector. A drug transport vector is a protein that undergoes receptor-mediated or absorptive-mediated transcytosis through the BBB. The prototypic vector that undergoes receptor-mediated transcytosis through the BBB is the OX26 monoclonal antibody to the transferrin receptor, and the prototypic vector that undergoes absorptive-mediated transport through the BBB is cationized albumin.

Chimeric Peptides for Brain Drug Delivery

Cell Biology of Endothelial Transcytosis

Transcytosis of brain drug transport vectors through the BBB involves three steps: binding at the lumenal or blood side of the brain capillary endothelium, movement through the 0.3 μm of endothelial cytoplasm, and exocytosis at the ablumenal or brain side of the BBB. Generally, transcytosis through capillary endothelium cannot be inferred in the absence of ultrastructural confirmation of this pathway because solutes or vectors may be transported across endothelial barriers through paracellular pathways. However, essentially no paracellular pathway in the brain capillary endothelium exists (*52*) in the absence of opening of BBB tight junctions. Therefore, the only route for delivery across the brain capillary endothelium is the transcellular pathway, which involves transcytosis. Transcytosis is generally a term that refers to a cell biological phenomenon that is initiated by binding at the plasma membrane and endocytosis of the ligand into an intracellular tubulovesicular compartment that is segregated from the cell cytosol (*53*). This segregation of the transcytotoic pathway from the cell cytosol is important in the delivery of chimeric peptides through the BBB. Chimeric peptides are generally formed through a disulfide linkage, and the rapid reduction of the disulfide forming the chimeric peptide within the capillary endothelial cytosol would greatly restrict the delivery of drug to brain ISF. However, the intracellular membrane limited compartments are generally devoid of disulfide reductase activity (*54*). Indeed, when isolated brain capillaries are exposed to disulfide-based chimeric peptides, no measurable reduction of the disulfide-based linkage is observed (*55*), although rapid reduction (e.g., 50% within 60 s) in a cell-free system of brain cytosol occurs (*14*).

The presence of an active transcytotic pathway of brain capillary endothelium was for many years thought not to exist because of several observations. First, unlike capillary endothelia in peripheral tissues, a paucity of pinocytosis or vesicular transport of fluid across the brain capillary endothelium occurs (*52*). However, studies (*56*) showed that brain capillary endothelium is, in fact, endowed with an extensive smooth surface tubular system that may be involved not in bulk flow fluid movement, as is observed in endothelia in peripheral capillaries, but may participate in active receptor-mediated transcytosis. A second puzzling observation with respect to transcytosis across brain endothelium was noted by Broadwell et al. (*57*); this observation pertains to the minimal endocytosis observed at the ablumenal membrane of the brain capillary endothelium. For example, the systemic administration of a wheat germ agglutinin–HRP conjugate results in endocytosis and transcytosis across the brain capillary endothelium of the blood-borne conjugate (*24*). Conversely, when the wheat germ agglutinin–HRP conjugate is injected in the ventricular compartment, no measurable endocytosis of the conjugate on the ablumenal membrane of capillary endothelium in brain exposed to the conjugate occurs (*57*). However, this polarization of the transcytotic pathway at the brain capillary endothelium is in accordance with transcytosis pathways observed in peripheral epithelial barrier systems. For example, epidermal growth factor (EGF) undergoes receptor-mediated transcytosis across some epithelial barriers. However, this transcytosis pathway is polarized and occurs only in the basolateral-to-apical direction (*58*). The basis for the polarization of the transcytosis pathway may be that no receptor on the contralateral side of the cellular barrier system exists. For example, if minimal endocytosis at the ablumenal side of the brain capillary endothelium occurs, this minimal endocytosis may relate to the minimal deposition of receptor on the ablumenal membrane. In such a setting, the receptor would not participate in the exocytosis arm of the transcytosis pathway. This model is in accord with observations in peripheral tissues, where the EGF receptor mediates endocytosis, but not exocytosis of EGF across epithelial barrier systems (*58*).

Finally, another consideration in the cell biology of vector transcytosis through the BBB is the role of the Golgi apparatus in the subcellular transcytotic pathway. Studies by Broadwell et al. (*24*) showed that, although a wheat germ agglutinin–HRP conjugate is transcytosed through the brain capillary endothelium following systemic administration, this transcytosis occurs slowly over several hours. In contrast, physiologic measurements of transcytosis rates showed that brain drug delivery vectors achieve distribution within brain ISF relatively rapidly and within 10 min of internal carotid artery perfusion (*12*). The delay in wheat germ agglutinin–HRP conjugate transcytosis through the brain capillary endothelium may be caused by transport of this conjugate through the Golgi apparatus of the brain capillary endothelium, which was observed previously (*24*). This observation is in accord with previous studies (*60*) performed in peripheral transcytotic systems wherein it was shown that a wheat germ agglutinin–ferritin conjugate did not undergo transfer through the Golgi apparatus, whereas the wheat germ agglutinin–HRP conjugate was delayed within the Golgi apparatus (*60*). That is, HRP conjugates may selectively target the Golgi apparatus. In other systems of transcytosis (e.g., viral membrane glycoproteins), no association of the ligand with the Golgi apparatus is observed (*61*).

With respect to the cell biology of vector transcytosis through the BBB, the following is a summary of observations to date (these observations are provisional, as a great deal more work needs to be performed on the subcellular localization of BBB transcytosis pathways). First, BBB transcytosis may be asymmetric and may predominate in the direction of lumenal-to-ablumenal membrane. Second, the transcytosis pathway may involve a smooth surface tubular system rather than a vesicular system that involves extensive invagination of lumenal membrane. Third, the lumenal membrane receptor may be separated from the ligand relatively early in the transcytosis pathway and may not participate in exocytosis. Fourth, passage through the Golgi apparatus may be ligand dependent and may be restricted to certain ligands, such as HRP.

Brain Drug Transport Vectors

OX26 Monoclonal Antibody. The brain capillary endothelium is endowed with high concentrations of transferrin receptor. This was demonstrated by Jefferies et al. (*62*) with a mouse monoclonal to the rat transferrin receptor called the OX26 monoclonal antibody. This antibody was shown by Friden et al. (*63*) to be a highly efficacious vector for drug delivery through the BBB. The antibody may be conveniently prepared with hybridoma technology as described in Figure 5. The OX26 hybridoma was generously provided by A. F. Williams of Oxford University, and several million hybridoma cells were grown in tissue culture laboratory (Figure 5, upper left). Ascites was prepared in BALB/c mice (Figure 5, upper right), and the OX26 monoclonal antibody was purified to homogeneity by sodium dodecyl sulfate polyacrylamide gel electrophoresis using protein G affinity chromatography (Figure 5). The OX26 monoclonal has two light chains (Figure 5), which may be because the initial myeloma line used to prepare the hybridoma was an IgG-secreting line.

The OX26 monoclonal antibody binds to an extracellular epitope on the transferrin receptor that is removed from the transferrin ligand binding site. Thus, binding of the OX26 monoclonal antibody to the receptor does not interfere with transferrin binding (*64, 65*). The transcytosis of the OX26 monoclonal antibody through the BBB was demonstrated with the capillary depletion technique (*65*), and recent experiments with gold conjugates of the OX26 monoclonal antibody began to identify the subcellular organelles involved in transcytosis through the brain capillary endothelium (25).

The transcytosis of the OX26 monoclonal antibody through the BBB parallels the receptor-mediated transcytosis of transferrin through the BBB as demonstrated by Fishman

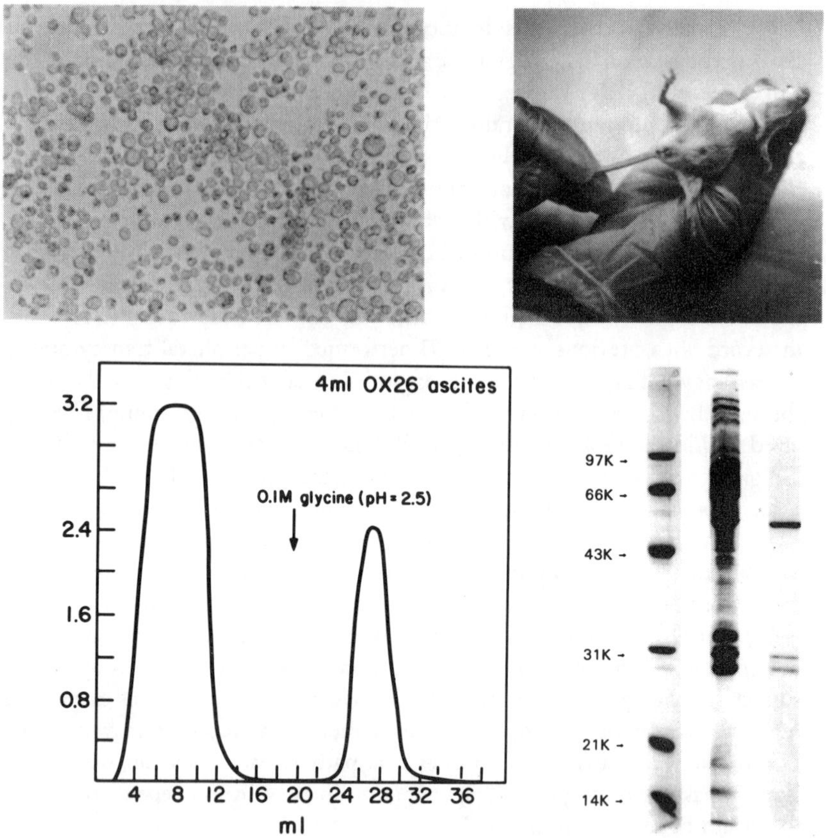

Figure 5. Upper left: Hybridoma cells grown in tissue culture. Upper right: Hybridoma ascites generated in BALB/c mice. Lower left: Protein G affinity chromatography of ascites with elution of the IgG peak by 0.1 M glycine (pH 2.5). Lower right: Coomassie blue stain of sodium dodecyl sulfate polyacrylamide gel electrophoresis of molecular weight standards (lane 1), nonpurified mouse ascites (lane 2), and protein G-purified OX26 monoclonal antibodies (lane 3). The molecular weights of the standards are shown in the Figure. (Lower panels are reproduced with permission from reference 78. Copyright 1992 Williams & Wilkins Company.)

et al. (*66*). However, the transcytosis of transferrin through the BBB has recently been questioned on the basis of two different observations. First, no transferrin receptor can be identified on the ablumenal membrane of the brain capillary endothelium (*67*). However, as just reviewed, it may not be expected that the transferrin receptor is localized on the ablumenal membrane if it does participate in transferrin exocytosis at the ablumenal membrane. Second, the brain distribution of Fe-59 exceeds the brain distribution of ^{125}I-labeled transferrin (*68*). However, it is probably not appropriate to compare transferrin radiolabeled with Fe-59 with iodinated transferrin. The oxidative iodination damages the transferrin protein, and, in our hands, adequate binding of [^{125}I]holo-transferrin to the BBB transferrin receptor is not observed when the iodinated protein is used for >12–24 h following the iodination reaction (*69*). The denaturation of large plasma proteins and the inability of these plasma proteins to bind to receptors following iodination was observed for a number of different plasma proteins (*70*).

Cationized Albumin. Native albumin does not undergo significant transcytosis through the BBB and no binding of gold native albumin conjugates to the lumenal membrane of the brain capillary endothelium is detectable (*56*). However, when the pI of albumin is chemically raised from the acidic to the basic range by cationization, the cationized albumin undergoes absorptive-mediated transcytosis through the BBB (Figure 6). The concern with cationized albumin as a transport vector relates to the known pathogenicity of cationic proteins (*71*). When administered to laboratory animals, cationized albumin is nephrotoxic, because of the deposition of immune complexes within the kidney. Moreover, other studies (*72*) showed that cationization enhances the underlying immunogenicity of albumin. Presumably, cationization causes absorptive-mediated endocytosis of albumin into cells of the

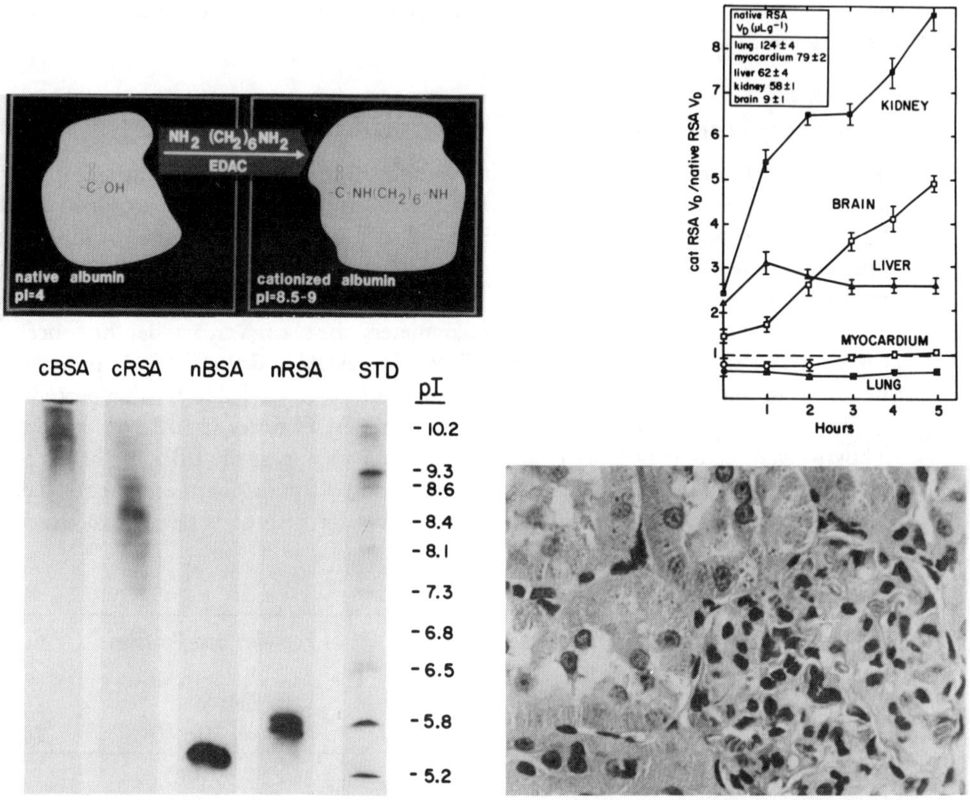

Figure 6. Upper left: Conversion of carboxyl groups on native albumin to extended primary amino groups using hexamethylenediamine and carbodiimide (EDAC), which results in an elevation of the pI of the native albumin from ~4 to ~8.5–9. Lower left: Coomassie blue stain of polyacrylamide gel isoelectric focusing study of cationized bovine serum albumin (cBSA), cationized rat serum albumin (cRSA), native BSA, native RSA, and isoelectric focusing standard (STD). The pI values of the isoelectric standards are shown. The pI value of native BSA is raised to a greater extent than is the pI value for RSA following cationization under identical conditions because of the greater number of surface carboxyl groups on bovine albumin as compared with rat albumin. (Reproduced with permission from reference 5. Copyright 1991 Raven.) Upper right: The ratio of V_D *of* ^3H*-labeled cRSA (catRSA) to* ^{125}I*-labeled native RSA in brain and other organs at various times after single intravenous injection in anesthetized rats is shown. Data are mean ± SE (*n = *3 rats each time point). The* V_D *of native RSA reached equilibrium by 1–2 h in all organs, and the equilibrium* V_D *values for native RSA are shown in the inset. Lower right: Hematoxylin and eosin stain of rat renal cortex after 8 weeks of administration of cRSA at a daily dose of 1 mg/kg subcutaneously. The study shows glomeruli and renal tubules of normal histology and no evidence of immune complex formation. (Reproduced with permission from reference 75. Copyright 1990 Williams & Wilkins Company.)*

immune system, which induces a marked immune response to this antigen. However, it is important to recall that cationic proteins, per se, are not pathogenic when administered in nontoxic doses. For example, protamine is a highly cationic and toxic protein when administered in high doses (*73*). However, protamine is administered daily to insulin-requiring diabetic subjects (*74*). The second point to emphasize is that cationic protein may not be particularly immunogenic when used in a homologous system. When cationized rat albumin was administered daily to rats in subcutaneous doses of 1 mg/kg for up to 8 weeks, no toxic sequellae were observed after reviewing organ histologies for kidney (Figure 6) and all other major organs and after reviewing 20 different serum chemistries related to renal, hepatic, and metabolic function tests (*75*). Similarly, the rat immune response to cationized rat albumin was minimal. Thus, cationized human albumin may prove to be a useful brain drug delivery vector for humans (*5*).

Pharmacokinetics of Brain Drug Transport Vectors

The amount of drug delivered to brain is a function of the delivery of the transport vector, and this delivery may be quantified with the determination of the percent injected dose (ID) per gram brain. This parameter is listed in Table II for a number of different BBB drug delivery vectors and for plasma blood volume markers, such as native rat serum albumin or the mouse IgG2a isotype control for the OX26 monoclonal antibody. The percent ID per gram brain is a function of two other parameters: the plasma area under the concentration curve (AUC) and the BBB PS product (*59*). The AUC and the BBB PS product for the first 60 min following an intravenous injection are listed in Table II. The percent ID per gram brain is equal to the product of the plasma AUC times its BBB PS product (*59*). Thus, a given vector may enjoy a relatively high permeability at the BBB, but have a relatively low percent injected dose delivered to brain because of poor pharmacokinetics (i.e., a low plasma AUC). For example, the BBB PS product for histone is nearly two-thirds of the

Table II. Pharmacokinetic Parameters of BBB Transport of Peptides and Proteins

Protein	*Plasma AUC*[b] *[(%ID)·min·mL^{-1}]*	*BBB PS*[c] *(μL·min^{-1}·g^{-1})*	*%ID (g^{-1})*[d]
[^{3}H]OX26	168 ± 76	1.56 ± 0.13	0.262
[^{3}H]Biotin–avidin–OX26	72 ± 18	1.35 ± 0.16	0.100
[^{3}H]catRSA	387 ± 74	0.16 ± 0.02	0.062
[^{3}H]catbIgG[a]	43 ± 14	0.63 ± 0.12	0.027
[^{3}H]rCD4	223 ± 49	0.11 ± 0.01	0.025
[^{125}I]Histone[a]	18 ± 10	0.91 ± 0.11	0.016
[^{3}H]Biotin–avidin	13	0.30 ± 0.10	0.004
[^{14}C]nRSA	533 ± 55	0	0
[^{125}I]mIgG$_{2a}$	440 ± 11	0	0

[a] Estimates based on whole blood analysis; all other studies performed on serum measurements.
[b] Plasma AUC = $[A_1(1 - e^{-K_1 t})/K_1] + [A_2(1 - e^{-K_2 t})/K_2]$, where t = 60 min after single injection.
[c] Permeability-surface (PS) area product = $[(V_D - V_0)C_p(T)]$–AUC, where V_0 = plasma volume, $C_p(T)$ = plasma concentration at t = 60 min after intravenous injection.
[d] Product of [AUC/1000] × PS, where %ID = percent of injected dose delivered to brain at 60 min after injection.
SOURCE: Data are from reference 59.

BBB PS product for the OX26 monoclonal antibody. However, the histone AUC is only about 10% of that for the OX26 AUC. Consequently, the amount of histone delivered to brain is <10% of the amount of the OX26 antibody delivered to brain. Thus, poor pharmacokinetics limit histone as a BBB delivery vector. The point to emphasize is that the pharmacokinetics of a vector are as important as the permeability of the BBB to the vector in ultimately determining the amount of drug delivered to brain. The AUC is also a measure of the extent to which the vector is delivered to nonbrain organs. The greater the uptake of the vector by major systemic organs, such as liver or kidney, the lower the AUC and the lower percent injected dose of drug delivered to brain. These considerations illustrate the importance of evaluating the pharmacokinetics in developing brain drug delivery vectors.

Coupling Strategies for Chimeric Peptide Synthesis

Chemical Coupling Strategies

Drugs may be coupled to brain drug delivery vectors with disulfide-linked chemical coupling agents, such as *N*-succinimidyl-3-(2-pyridylthio)propionate (SPDP) as shown in Figure 7A. SPDP was used to couple ^{3}H-β-endorphin, a drug that is normally not transported through the BBB, to cationized albumin, which was used as a BBB drug delivery vector (*76*). The β-endorphin conjugated to the cationized albumin was separated from unconjugated β-endorphin and from high molecular weight aggregates with Sepharose CL-6B (agarose) gel filtration chromatography in the presence of 6 M guanidine (Figure 7C). The guanidine subsequently was removed by extensive dialysis, and this cationized albumin proved to be just as efficacious a drug delivery vector as the cationized albumin unexposed to 6 M guanidine. This observation is in accord with previous studies (*77*) showing that proteins refold following removal of denaturing agents, such as guanidine. The 6 M guanidine was used in the chromatography to present nonspecific binding of unconjugated peptide to the cationized albumin transport vector.

The [^{3}H]β-endorphin coupled to the cationized albumin vector was added to isolated bovine brain capillaries, which are shown in a scanning electron micrograph in Figure 7B. The β-endorphin conjugate was extensively taken up by the isolated brain capillaries, but no significant uptake of the unconjugated ^{3}H-β-endorphin occurred (Figure 7D). Other studies show that the ^{3}H-β-endorphin coupled to the cationized albumin chimeric peptide entered an endothelial space that was resistant to mild acid wash and thus represented endocytosed chimeric peptide (*76*). The data in Figure 7C, which shows the relatively low yield in coupling peptides to drug transport vectors using disulfide-based coupling reagents, such as SPDP, is important. Only about a 10–15% yield in the coupling efficiency was achieved (*76*). This low coupling efficiency was suitable for testing the chimeric peptide model with respect to the physiology of brain uptake. However, higher yields of coupling were desired for subsequent pharmacologic reduction to practice of the chimeric peptide model. This higher yield of coupling was achieved with avidin–biotin technology (*5, 78*).

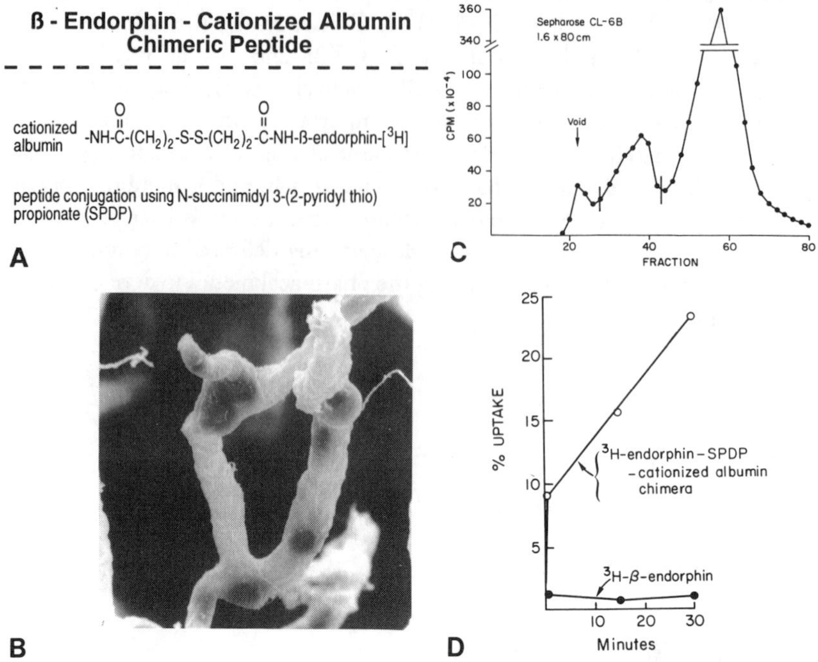

Figure 7. (A) Structure of β-endorphin-cationized albumin chimeric peptide. (B) Scanning electron micrograph of isolated bovine brain capillaries. (C) Elution profile of ^{3}H-β-endorphin covalently coupled to cationized bovine albumin (pI 8.5–9) using the heterobifunctional cross-linking reagent SPDP (A). The conjugate was applied to 1.6 × 80-cm column of Sepharose CL-6B, using an elution buffer containing 0.1 M Na_2HOP_4 and 6 M guanidine. A small aggregate peak eluted at the void volume, and the monomeric form of the β-endorphin-cationized albumin chimeric peptide migrated as a single peak between fractions 28 and 42. This peak was pooled, dialyzed against water, and concentrated prior to incubation of the ^{3}H-labeled chimeric peptide with the bovine brain capillaries shown in B. These data are plotted in D. (C is reproduced with permission from reference 76. Copyright 1987 The American Society for Biochemistry & Molecular Biology.) (D) The percent uptake of [^{3}H]β-endorphin SPDP-cationized albumin chimeric peptide or unconjugated [^{3}H]β-endorphin by isolated bovine brain capillaries is plotted versus incubation time. The ^{3}H-chimeric peptide taken up by the brain capillaries represents both membrane-bound and endocytosed peptide because >35% of the radioactivity associated with the capillaries was resistant to a mild acid wash. (D is reproduced with permission from reference 55. Copyright 1987 Academic.)

Use of Avidin–Biotin Technology To Facilitate Drug Coupling to Brain Drug Transport Vectors

Avidin as a Vector

Avidin binds the water-soluble vitamin biotin with very high affinity. Avidin is a 64,000 Da homotetramer that has four biotin binding sites. The K_D of biotin binding is 10^{-15} M and the dissociation half-life is 89 days (*79*). Therefore, the avidin–biotin bond is one of the tightest noncovalent bonds in nature. Unlike the bacterial homologue of avidin, the slightly acidic protein streptavidin, the avidin protein is a highly cationic protein with a pI of ~10 (*5*). This cationic nature afforded the avidin protein the ability to undergo absorptive-mediated endocytosis into brain capillary endothelial cells in vitro and absorptive-mediated

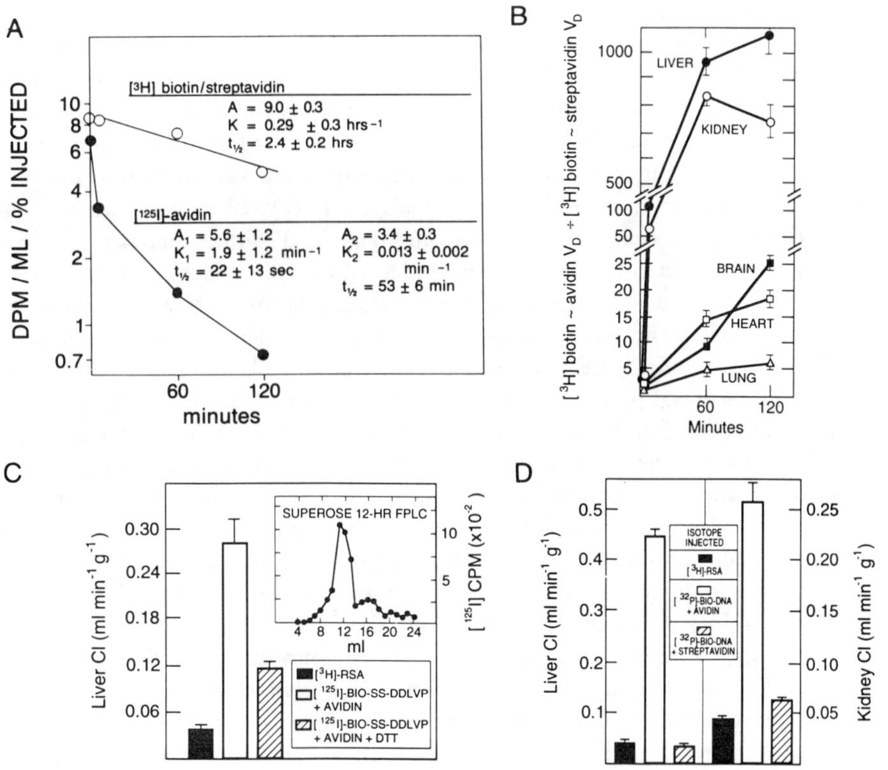

Figure 8. (A) ^{3}H-Biotin–streptavidin and ^{125}I-avidin were coinjected, and serum radioactivity was measured for 2 h after injection and plotted as a percentage of injected disintegrations per milliliter of serum. The serum radioactivity was fit to either a monoexponential or a biexponential decay function, using a nonlinear regression analysis to yield the intercept (A) and slope (K) *of the clearance curve. The half-life of clearance (Cl) was calculated from ln 2/K. Data are means (*n = *3 rats per point); SE are omitted for clarity and range from 10–25% of the mean. (B) The ratio of ^{3}H biotin* V_D *following coinjection with avidin relative to the ^{3}H biotin* V_D *following coinjection with streptavidin is measured in five different organs for up to 2 h following single intravenous injection of the isotope. (C) The liver Cl values (5 min after intravenous injection) for plasma marker, ^{3}H-labeled rat serum albumin (RSA), for ^{125}I-labeled bio-SS-DDLVP coinjected with an excess of avidin, and for ^{125}I-labeled bio-SS-DDLVP coinjected with an excess of avidin following a 60-min incubation of the peptide with 25 mM DTT are shown. Data are means ± SE (*n = *4 rats per point). (Inset) Serum radioactivity obtained 5 min after injection of ^{125}I-labeled bio-SS-DDLVP coinjected with avidin was applied to Superose 12HR FPLC gel filtration column; the majority of the radioactivity eluted at 12 mL, which is identical to the elution volume of the ^{125}I avidin control. The biotinylated ^{125}I-DDLVP elutes at 16 mL in the absence of avidin (DDLVP is desamino Cys^{1}, D-Lys^{8} vasopressin). (D) The liver and kidney Cl values (5 min after intravenous injection), determined with the external organ technique, are shown for a plasma marker, ^{3}H-RSA, for 5′-bio[^{32}P]antisense oligonucleotide (DNA) coinjected with avidin, or for the labeled DNA coinjected with streptavidin. Data are means ± SE (*n = *3 rats per point). (A, B, C, and D are reproduced with permission from reference 81. Copyright 1992 Academic.)*

transport across the BBB in vivo (*80, 81*). Figure 8A shows the pharmacokinetics of avidin and streptavidin clearance from rat blood in vivo. Streptavidin radiolabeled by binding with [^{3}H]biotin was cleared from rat plasma monoexponentially with a half-time of 2.4 ± 0.2 h. Conversely, [^{125}I]avidin was rapidly cleared with an initial half-time of 22 ± 13 s. When avidin is indirectly labeled by binding with [^{3}H]biotin, the tritium radioactivity disappears from plasma even faster than does the ^{125}I-avidin because of the rapid dissociation of ^{3}H-biotin from the avidin binding site by peripheral tissues, such as liver and kidney (*81*). Figure 8B shows this rapid uptake of ^{3}H-biotin bound to avidin, which is a plot of the V_D of biotin following intravenous injection of avidin relative to the V_D of [^{3}H]biotin following intravenous injection of streptavidin. Avidin also was used to enhance the delivery of biotinylated peptide (Figure 8C) or antisense oligonucleotide (Figure 8D) to tissues.

[DesaminoCys^{1}, D-Lys^{8}]lysine vasopressin (DDLVP) was iodinated and biotinylated with sulfosuccinimidyl 2-(biotinamido)ethyl-, 3′-dithiopropionate (NHS-SS-biotin), and the biotinylated DDLVP was coupled to avidin and intravenously injected into anesthetized rats. As shown in Figure 8C, the hepatic V_D of the DDLVP was increased manyfold by the coadministration of avidin. However, the hepatic V_D was suppressed when the DDLVP was initially treated with dithiothreitol (DTT), which cleaved DDLVP from the biotin moiety and prevented the binding of the DDLVP peptide to the avidin vector. The hepatic uptake of the biotinylated DDLVP represented uptake of the avidin peptide complex as evidenced by the migration of DDLVP on a gel filtration column at the elution volume characteristic of the 64,000 Da avidin, and not at the elution volume characteristic of the low molecular weight peptide (Figure 8C, inset). Similarly, an antisense oligonucleotide that hybridizes to the GLUT1 glucose transporter mRNA was biotinylated (*82*), bound to either avidin or streptavidin, and injected intravenously into anesthetized rats. The V_D in liver or kidney of the antisense oligonucleotide bound to either avidin or streptavidin was measured (Figure 8D). The avidin, but not the streptavidin, resulted in marked enhancement of the liver or kidney uptake of the antisense oligonucleotide; the V_D was increased severalfold above that for a plasma volume marker, such as ^{3}H-native RSA.

The results in Figure 8 indicate that avidin is a useful delivery vector for tissues such as liver or kidney because of the rapid uptake of this protein by these organs. However, the usefulness of avidin as a brain drug delivery vector is limited by the pharmacokinetic considerations. As shown in Table II, the AUC of ^{3}H-biotin bound to avidin is very low and this accounts for a low %ID delivered to brain, despite a reasonable permeability of the BBB to the avidin protein. However, avidin may still be used in brain drug delivery with the production of avidin conjugates or avidin fusion proteins, whereby the avidin is joined to a brain drug delivery vector, such as the OX26 antibody or cationized albumin. Following the initial synthesis of the avidin–vector conjugate, this hybrid protein may deliver almost any biotinylated therapeutic substance to the brain.

Avidin–Vector Conjugates

An avidin–OX26 monoclonal antibody conjugate was prepared by coupling avidin through a thioether linkage to the antibody, as described previously (*78*). The conjugate was purified from unconjugated avidin through gel filtration fast-protein liquid chromatography (FPLC). The retention of the bifunctionality of the hybrid protein was measured to insure that the conjugate still transcytosed through the BBB and still bound biotin with high affinity. The biotin binding capacity of the conjugate was measured with a filtration technique (*78*). The biotin binding capacity was 75 ± 2 nM (Figure 9) in the presence of 32 nM conjugate. Therefore, 2.3 biotin binding sites existed on the conjugate protein. Because the conjugation may destroy one biotin binding site, the residual two to three biotin binding sites were functionally active. The avidin–OX26 conjugate was also equally active as a brain drug delivery vector. As shown in Figure 9, the brain V_D of the [^{3}H] biotin coupled to the avidin–OX26 conjugate actually exceed that of the [^{3}H]OX26 antibody (*78*). This increased brain V_D reflects pharmacokinetic differences, however, and computation of the BBB PS

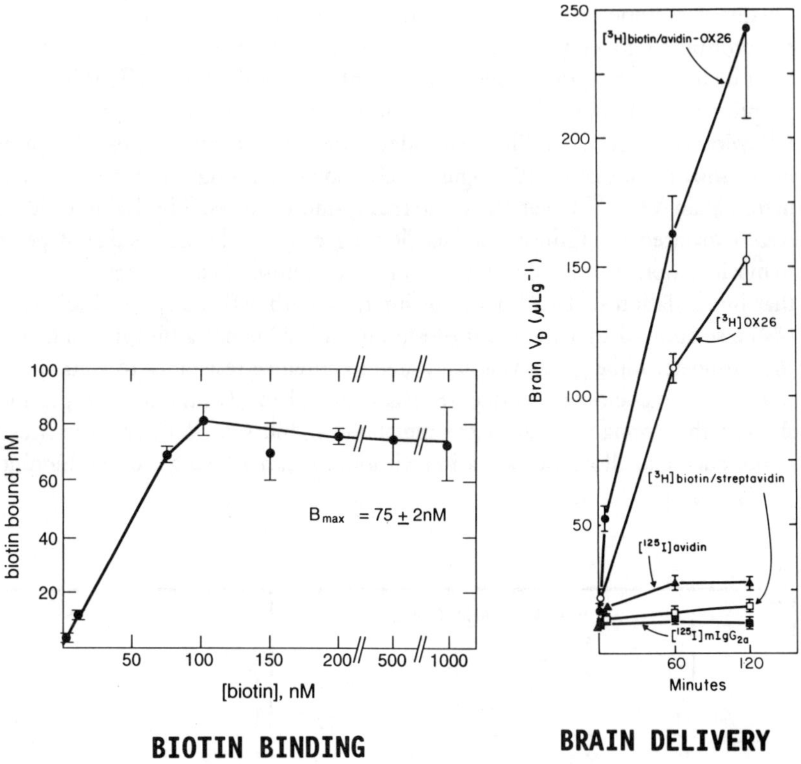

Figure 9. Left: ^{3}H-Biotin binding to avidin–OX26 conjugate. Data are mean ± SE (n = *3). B_{max} is calculated to be 75 ± 2 nM in he presence of 32 nM conjugate. Right: The* V_D *of ^{3}H-OX26, ^{3}H-biotin–avidin–OX26 conjugate, ^{125}Iavidin, ^{3}H-biotin–streptavidin, and ^{125}I-labeled mouse IgG2a is plotted over a 120-min time period after intravenous bolus injection of isotopes in anesthetized rats. (Reproduced with permission from reference 78. Copyright 1992 Williams & Wilkins Company.)*

product showed that no significant difference between the avidin–OX26 conjugate and the unconjugated OX26 exists (Table II).

In Vivo CNS Pharmacologic Effects of Systemically Administered Peptide Drugs

Peptide Design and Structure

A vasoactive intestinal peptide (VIP) analogue was developed to illustrate the design strategies necessary to insure retention of biological activity of a peptide drug following coupling and release from a BBB drug transport vector (*83*). Two important criteria must be satisfied in this design strategy. First, the peptide drug must be subject to only monobiotinylation, as higher degrees of biotinylation would lead to the formation of high-molecular weight aggregates, because of the multivalency of avidin binding of biotin. Second, the biotinylation, usually of a lysine residue, must occur at a site on the peptide drug that is removed from receptor binding so it will not interfere with biological activity of the peptide as it is released from its transport vector through cleavage of the disulfide bond. This is because a mercapto-proprionate molecular adduct remains attached to the ϵ-amino a group of the lysine residue following cleavage (*14*). These considerations are illustrated in the design of the VIP analogue shown in Figure 10. Figure 10A shows the single-letter amino acid sequence of mammalian VIP. VIP has three internal lysine residues. The lysine residues at 20 and 21 were converted to arginine residues, leaving only the lysine residue at position 15 subject to biotinylation. The residue at position 15 was chosen because previous studies (*84*) showed that biotinylation of Lys^{15} does not interfere with VIP receptor binding. Previous studies (*84*) also showed that the amino-terminus of VIP is not a biotinylation site. Nevertheless, the amino-terminal group was acetylated to enhance resistance to aminopeptidase degradation. The isoleucine at residue 26 was changed to alanine because previous studies (*85*) showed this change enhances the metabolic stability of VIP, and the Met^{17} was changed to norleucine to allow for iodination without oxidative changes of methionine sulfur groups.

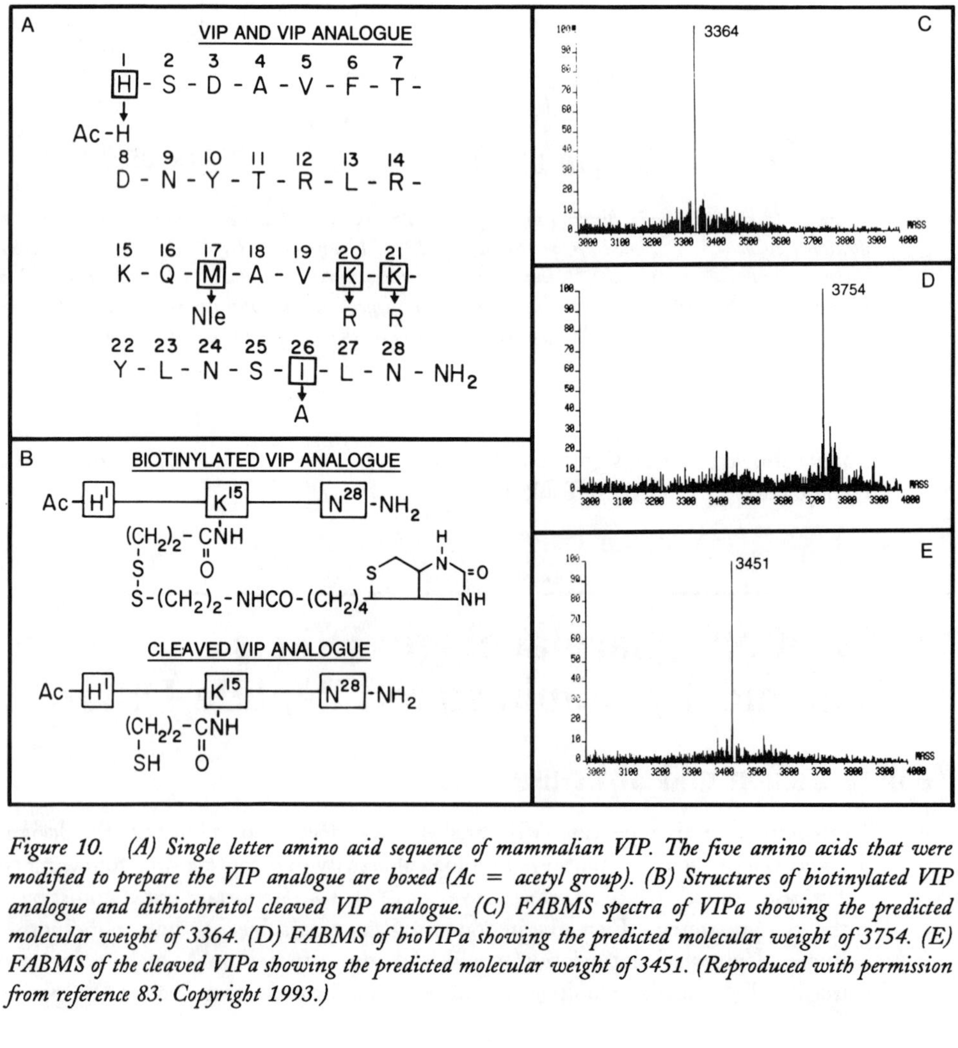

Figure 10. (A) Single letter amino acid sequence of mammalian VIP. The five amino acids that were modified to prepare the VIP analogue are boxed (Ac = acetyl group). (B) Structures of biotinylated VIP analogue and dithiothreitol cleaved VIP analogue. (C) FABMS spectra of VIPa showing the predicted molecular weight of 3364. (D) FABMS of bioVIPa showing the predicted molecular weight of 3754. (E) FABMS of the cleaved VIPa showing the predicted molecular weight of 3451. (Reproduced with permission from reference 83. Copyright 1993.)

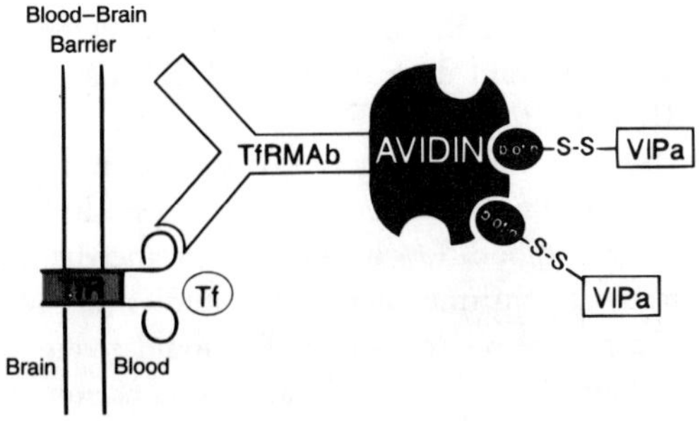

Figure 11. Model of VIPa drug delivery to the brain using a conjugate of avidin and the OX26 Anti-TfRMAb. The VIPa was biotinylated with a disulfide linker, as shown in Figure 10B. (Reproduced with permission from refernce 83. Copyright 1993.)

The VIP analogue (VIPa) was biotinylated with NHS-SS-biotin (the structure of this analogue is shown in Figure 10B). The biotinylated VIP was cleaved with DDT and the structure of the cleaved VIP analogue is also shown in Figure 10B.) The structures of the VIPa, its biotinylated derivative, and the desbiotinylated or cleaved derivative were confirmed by fast-atom-bombardment mass spectrometry spectra (FABMS), as shown in Figures 10C, 10D, and 10E, respectively. The binding of the VIPa, its biotinylated derivative, and the cleaved or desbio analogue to the mammalian VIP receptor was assessed by using a VIP radioreceptor assay with rat lung membranes. These studies (*83*) showed that the affinity of the three analogues for the VIP receptor was comparable to that of mammalian VIP. Thus, the mercaptopropionate group at Lys^{15} of the VIPa did not substantially interfere with VIP receptor binding.

In Vivo BBB Transport

The biotinylated VIPa was completely coupled to the avidin–OX26 conjugate (the scheme of VIP drug transport through the BBB using this delivery system is shown in Figure 11). The VIP was joined through disulfide linkage to the biotin moiety, which was attached to avidin that was further attached through a noncleavable thioether linkage to the OX26 antitransferrin receptor monoclonal antibody (TfRMAb). The OX26 antibody bound to the BBB transferrin receptor (TfR) at an epitope removed from the transferrin (Tf) ligand binding site. The transport of the iodinated, biotinylated VIPa through the BBB with this delivery system was documented with the internal carotid artery perfusion–capillary depletion technique and was comparable to the BBB delivery of the unconjugated OX26 antibody. Conversely, unconjugated VIP was found to undergo no measurable transcytosis through the BBB (*83*).

In Vivo CNS Pharmacologic Assay

The in vivo CNS pharmacologic effect of VIP was assessed by measurement of cerebral blood flow. Smooth muscle cells of intraparenchymal cerebral arterioles are studded with VIPergic nerve endings, and VIP is believed to be the principle cortical vasodilator of brain (*86*). When VIP was topically applied to pial vessels, vasodilation was observed (*87*). However, no vasodilation was observed when VIP was infused into the carotid artery (*88, 89*), because of the failure of VIP to cross the BBB and the localization of smooth muscle

VIP receptors beyond the BBB. The enhancement of cerebral blood flow would be expected if VIP is coupled to a brain drug delivery vehicle that could allow for transport of VIP through the BBB and into brain ISF.

The cerebral blood flow assay was developed in nitrous oxide-ventilated rats. Other forms of anesthesia were not used because previous studies (*90*) showed that pharmacologic modulation of cerebral blood flow is inhibited by anesthetics, such as barbiturates or ketamine, and this fact was confirmed in our own studies (unpublished observations). However, cerebral blood flow may be pharmacologically modulated in nitrous oxide-ventilated animals because this model more closely approximates the conscious condition. Accordingly, animals were intubated, immobilized with curare, and ventilated with 70% nitrous oxide and 30% oxygen (*83*). The arterial blood gases were maintained constant for 30 min prior to initiation of the cerebral blood flow assay (the mean pO_2, pCO_2, and arterial pH values are shown in Figure 12B). Cerebral blood flow was measured with an external organ technique, using ^{3}H-diazepam as a fluid microsphere to monitor cerebral blood flow and ^{14}C-sucrose to measure cerebral blood volume. These two isotopes were injected rapidly into a femoral vein, and the data in Figure 12A show that the isotopes complete a full circulation within 10 s following rapid injection. Therefore, femoral arterial blood was collected for 10 s following the rapid intravenous injection of these two isotopes; this was followed by decapitation of the animals and measurement of cerebral radioactivity. The measurement of the ratio of brain-to-femoral arterial plasma radioactivity allowed for determination of cerebral blood flow and cerebral blood volume.

In the cerebral blood flow protocol, the various drug solutions were infused into the internal carotid artery for 10 min following stabilization of arterial blood gases and prior to measurement of cerebral blood flow or blood volume (*83*). The various drug solutions included saline (control), biotinylated VIPa (bioVIPa), avidin–OX26 conjugate (AV–OX26), and bioVIP coupled to the avidin–OX26 conjugate (Figure 12D). The biological activity of bioVIPa was documented (Figure 12C); that is, it was shown that the carotid artery infusion of bioVIPa resulted in a doubling of thyroid blood flow. No drug delivery system for VIP delivery to thyroid vessels was required, as thyroid capillaries have no barrier to small peptides such as VIP. VIP has been shown in previous studies (*91*) to augment thyroid blood flow. Therefore, the enhancement of thyroid blood flow with bioVIPa demonstrated the biological activity of this analogue and confirmed the radioreceptor assay, which showed high affinity binding of bioVIPa to the mammalian VIP receptor. Conversely, when biotinylated VIP was infused into the carotid artery, no enhancement of cerebral blood flow was observed. Similarly, when the avidin–OX26 conjugate was infused, in the absence of VIP, no alteration of cerebral blood flow was observed. However, when bioVIPa coupled to the avidin–OX26 delivery vector was infused, a highly statistically significant 65% increase in cerebral blood flow was observed following the systemic administration of relatively low doses (12 μg/kg) of bioVIP peptide drug. These results are the first demonstration of an in vivo brain pharmacologic effect following the systemic delivery of low doses of neuropeptide coupled to a brain drug delivery system (*83*).

Summary

The VIP study represents a paradigm that may be applied to numerous other peptide-based drugs for the induction of in vivo CNS pharmacologic effects in brain following systemic administration of peptides. Within this paradigm, several design features of the drug delivery strategy are important. First, the peptide-based drug should be designed specifically, usually

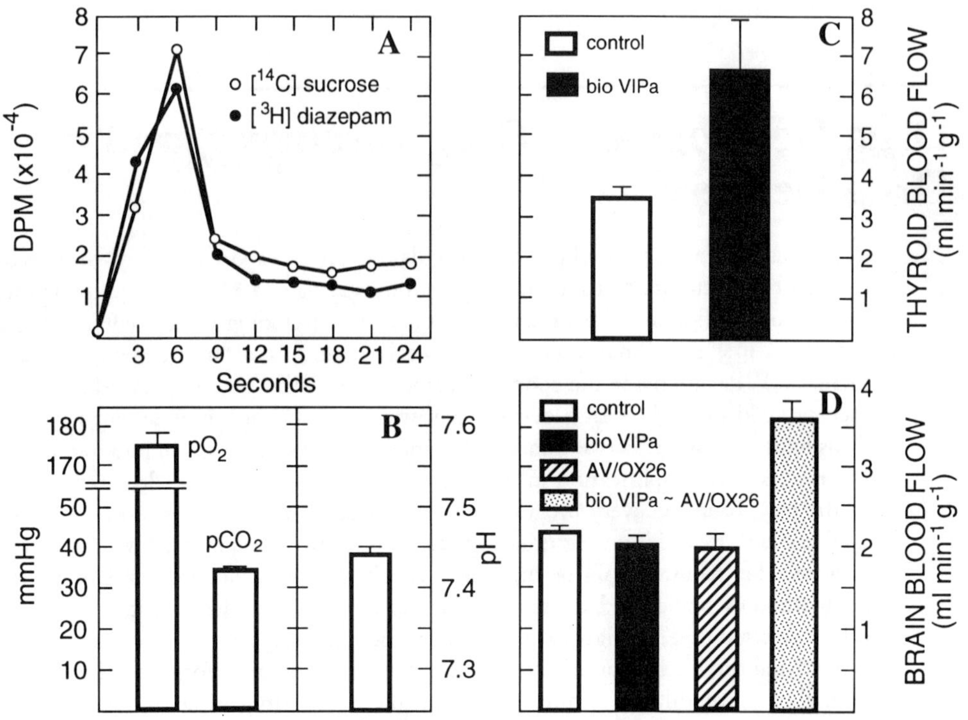

Figure 12. Measurement of tissue blood flow after a 10-min carotid artery infusion of vehicle (control), of bio-VIPa (12 μg/kg), avin–OX26 (430 μg/kg), or bio-VIPa–avidin–OX26 (12 μg/kg of bio-VIPa and 430 μg/kg of OX26) in N_2O anesthetized rats. (A) Arterial blood radioactivity of ^{3}H-diazepam or ^{14}C-sucrose was determined with sampling periods of 3 s following intravenous bolus injection. (B) No significant differences in the arterial blood gas measurements of the four treatment groups listed in D were observed; the means ± SE for all animals (n = 16) are shown. (C) The rate of thyroid blood flow is increased by the systemic administration of bio-VIPa over the 10-min infusion period. (D) Brain blood flow is increased 65% (p < 0.0025, Students t-test versus control) by the systemic administration of bio-VIPa coupled to the AV–OX26 vector, whereas administration of either the bio-VIPa or the AV–OX26 alone causes no change in brain blood flow relative to the control infusion. No differences in brain blood flow were observed between the right and left hemispheres. Data are mean ± SE (n = 4 rats per group, body weight 270–280 g). (Reproduced with permission from reference 83. Copyright 1993.)

at the level of solid-phase synthesis, to ensure both monobiotinylation and biotinylation at a site of the peptide that is not actively involved in receptor binding. Therefore, the peptide maintains its biological activity following cleavage of the disulfide linker bond. Second, the peptide must be efficiently coupled to the brain drug delivery vector so that high yields of coupling may be achieved. This is facilitated by the use of avidin–biotin technology in brain drug delivery. Third, a brain drug transport vector, such as the OX26 monoclonal antibody or cationized albumin, is used to allow for the construction of the chimeric peptides. These chimeric peptides have several different functional domains, as depicted in Figure 11. These functional domains include the transport vector, which has an affinity for specific binding sites on the lumenal membrane of the BBB, an avidin moiety, which facilitates biotin binding, a biotin group, which is linked to the drug via a disulfide linkage, and the peptide-based drug.

Drug Development, Drug Discovery, and Drug Delivery

Peptides are highly water-soluble compounds that are transported poorly through the BBB through free diffusion. Thus, brain drug delivery strategies are required, and one such strategy is the use of chimeric peptides. Subsequent to developing a transcellular drug delivery system, such as the chimeric peptide approach, a number of known peptide-based drugs, such as a VIPa, are made transportable through the BBB. These peptide-based drugs may then proceed through a drug delivery pathway toward drug development (Figure 13). A focus on developing effective transcellular drug delivery systems, as opposed to drug discovery, results in an amplification in the drug development process, as depicted in Figure 13. In the past, drug discovery programs used the traditional "trial-and-error" process; lipid-soluble drugs, that did not require drug delivery systems, were invariably selected by in vivo bioassays that required drug permeation through biological membranes. Conversely, when the method of new drug discovery changes to "rational drug design", it is likely that drugs will be selected that are highly water soluble. Rational drug design results in drug discovery by studying the interaction of drug candidates with drug receptors, and this interaction is invariably hydrophilic in nature. Consequently, a new drug discovery program based on rational drug design requires a parallel transcellular drug delivery program as a final common pathway to drug development. Conversely, the diagram in Figure 13 depicts the concept that drug development may be achieved by focusing only on transcellular drug delivery. The development of a novel transcellular drug delivery system allows for the successful application of known drugs that were previously poor drug candidates because of the poor permeation of these drugs through biological membranes. The concept that drug development may proceed in the absence of a new drug discovery program is, of course, an exaggeration, but is made to emphasize the necessity of attaching equal significance to both drug delivery and drug discovery in the overall drug development program.

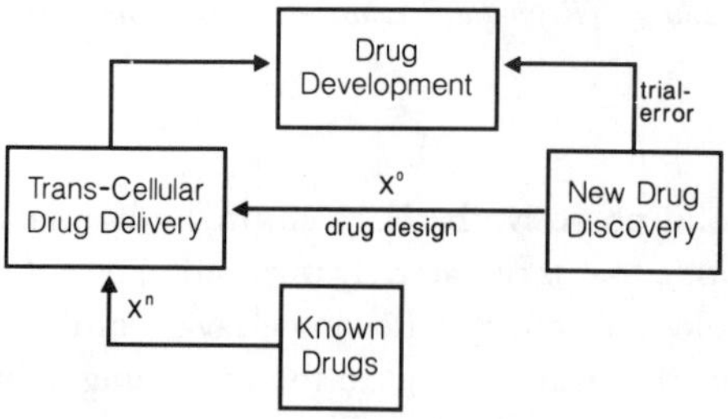

Figure 13. As the "trial and error" approach used in the past for new drug discovery is replaced by rational drug design, which is based on information on drug–receptor interaction at the molecular level, new drugs that are discovered will invariably be water-soluble compounds that have poor cell membrane permeation rates. Therefore, these new drugs will require a transcellular drug delivery vehicle. When drug development is initiated by drug discovery, drugs are developed individually; conversely, when drug development is initiated by transcellular drug delivery, many known drugs may be developed simultaneously. (Reproduced with permission from reference 59. Copyright 1993.)

Acknowledgments

The work in my laboratory is supported by NIH Grant R01–DA06748. I am indebted to Ulrich Bickel, Takayoshi Yoshikawa, and Young-Sook Kang for many valuable discussions.

References

1. Matozaki, T.; Göke, B.; Tsunoda, Y.; Rodriguez, M.; Martinez, J.; Williams, J. A. *J. Biol. Chem.* **1990,** *265,* 6247–6254.
2. Czech, M. P.; Klarlund, J. K.; Yagaloff, K. A.; Bradford, A. P.; Lewis, R. E. *J. Biol. Chem.* **1988,** *263,* 11017–11020.
3. McClelland, A.; Kühn, L. C.; Ruddle, F. H. *Cell* **1984,** *39,* 267–274.
4. Barchas, J. D.; Elliott, G. R. In *Neuropeptides in Neurologic and Psychiatric Disease;* Martin, J. B.; Barchas, J. D., Eds.; Raven: New York, 1986, pp 287–307.
5. Pardridge, W. M. *Peptide Drug Delivery to the Brain;* Raven: New York, 1991; pp 1–357.
6. Banks, W. A.; Kastin, A. J. *Am. J. Physiol.* **1990,** *259,* E1–E10.
7. Tsuzuki, N.; Hama, T.; Hibi, T.; Konishi, R.; Futaki, S.; Kitagawa, K. *Biochem. Pharmacol.* **1991,** *41,* R5–R8.
8. Zlokovic, B. V.; McComb, J. G.; Perlmutter, L.; Weiss, M. H.; Davson, H. *NIDA Monographs* **1992,** *120,* 26–42.
9. Perry, T. L.; Hansen, S. *J. Neurochem.* **1978,** *30,* 679–684.
10. Himwich, W. A.; Petersen, J. C.; Allen, M. L. *Neurology* **1957,** *7,* 705–710.
11. Banks, W. A.; Kastin, A. J.; Barrera, C. M. *Pharm. Res.* **1991,** *8,* 1345–1350.
12. Triguero, D.; Buciak, J. B.; Pardridge, W. M. *J. Neurochem.* **1990,** *54,* 1882–1888.
13. Voyta, J. C.; Via, D. P.; Butterfield, C.; Zetter, B. R. *J. Cell Biol.* **1984,** *99,* 2034–2040.
14. Pardridge, W. M.; Triguero, D.; Buciak, J. L. *Endocrinology* **1990,** *126,* 977–984.
15. Pardridge, W. M. *Annu. Rev. Physiol.* **1983,** *45,* 73–82.
16. Stein, W. D. *The Movement of Molecules across Cell Membranes;* Academic: Orlando, FL, 1967.
17. Diamond, J. M.; Wright, E. M. *Proc. Roy. Soc.* **1969,** *172,* 273–316.
18. Pardridge, W. M.; Mietus, L. J. *J. Clin. Invest.* **1979,** *64,* 145–154.
19. Pardridge, W. M.; Triguero, D.; Yang, J.; Cancilla, P. A. *J. Pharmacol. Exp. Ther.* **1990,** *253,* 884–891.
20. Begley, D. J. *Handbook of Experimental Pharmacology;* Bradbury, M. W. B., Ed.; Springer-Verlag: Berlin, Germany, 1992; Vol. 103, pp 151–203.
21. Greig, N. H.; Soncrant, T. T.; Shetty, H. U.; Momma, S.; Smith, Q. R.; Rapoport, S. I. *Cancer Chemother. Pharmacol.* **1990,** *26,* 263–268.
22. Träuble, H. *J. Membr.* **1971,** *4,* 193–208.
23. Duffy, K. R.; Pardridge, W. M. *Brain Res.* **1987,** *420,* 32–38.
24. Broadwell, R. D.; Balin, B. J.; Salcman, M. *Proc. Natl. Acad. Sci. U.S.A.* **1988,** *85,* 632–636.
25. Pardridge, W. M.; Farrell, C. L. *J. Histotechnol.* **1993,** *16,* 249–257.
26. Shimura, T.; Tabata, S.; Terasaki, T.; Deguchi, Y.; Tsuji, A. *J. Pharm. Pharmacol.* **1992,** *44,* 583–588.

27. Shimura, T.; Tabata, S.; Ohnishi, T.; Terasaki, T.; Tsuji, A. *J. Pharmacol. Exp. Ther.* **1991,** *258,* 459–465.

28. Zlokovic, B. V.; Susic, V. T.; Davson, H.; Begley, D. J.; Jankov, R. M.; Mitrovic, D. M.; Lipovac, M. N. *Peptides* **1989,** *10,* 249–254.

29. Zlokovic, B. V.; Banks, W. A.; Kadi, H. E.; Erchegyi, J.; Mackic, J. B.; McComb, J. G.; Kastin, A. J. *Brain Res.* **1992,** *590,* 213–218.

30. Schiffrin, E. L.; Genest, J. *Endocrinology* **1983,** *113,* 409–411.

31. van Zwieten, E. J.; Ravid, R.; Swaab, D. F.; v.d. Woude, Tj. *Brain Res.* **1988,** *474,* 369–373.

32. Audus, K. L.; Borchardt, R. T. *Pharm. Res.* **1986,** *3,* 81–87.

33. van Bree, J. B. M. M.; de Boer, A. G.; Verhoef, J. C.; Danhof, M.; Breimer, D. D. *J. Pharmacol. Exp. Ther.* **1989,** *249,* 901–905.

34. Bradbury, M. *The Concept of a Blood–Brain Barrier;* John Wiley & Sons: New York, 1979.

35. Boado, R. J.; Pardridge, W. M. *Mol. Cell. Neurosci.* **1990,** *1,* 224–232.

36. Smith, K. R.; Borchardt, R. T. *Pharm. Res.* **1989,** *6,* 466–473.

37. Dehouck, M.-P.; Jolliet-Riant, P.; Bree, F.; Fruchart, J.-C.; Cecchelli, R.; Tillement, J.-P. *J. Neurochem.* **1992,** *58,* 1790–1797.

38. Baskin, D. G.; Woods, S. C.; West, D. B.; van Houten, M.; Posner, B. I.; Dorsa, D. M.; Porte, D., Jr. *Endocrinology* **1983,** *113,* 1818–1825.

39. Crawley, J. N.; Fiske, S. M.; Durieux, C.; Derrien, M.; Roques, B. P. *J. Pharmacol Exp. Ther.* **1991,** *257,* 1076–1080.

40. Harding, J. W.; Yoshida, M. S.; Dilts, R. P.; Woods, T. M.; Wright, J. W. *J. Neurochem.* **1986,** *46,* 1292–1297.

41. Molineaux, C. J.; Ayala, J. M. *J. Neurochem.* **1990,** *55,* 611–618.

42. Blasberg, R. G.; Patlak, C.; Fenstermacher, J. D. *J. Pharmacol. Exp. Ther.* **1975,** *195,* 73–83.

43. Aird, R. B. *Exp. Neurol.* **1984,** *86,* 342–358.

44. Spector, R. *Exp. Neurol.* **1981,** *72,* 645–653.

45. Kim, S.; Scheerer, S.; Geyer, M. A.; Howell, S. B. *J. Infect. Diseases* **1990,** *162,* 750–752.

46. Rose, J. K.; Buonocore, L.; Whitt, M. A. *BioTechniques* **1991,** *10,* 520–525.

47. Bewley, T. A.; Li, C. H. *Biochemistry* **1983,** *22,* 2671–2675.

48. Bodor, N.; Prokai, L.; Wu, W.-M.; Farag, H.; Jonalagadda, S.; Kawamura, M.; Simpkins, J. *Science* (*Washington, DC*) **1992,** *257,* 1698–1700.

49. Brink, J. J.; Stein, D. G. *Science* (*Washington, DC*) **1967,** *158,* 1479–1480.

50. Hanig, J. P.; Morrison, J. M., Jr.; Krop, S. *Eur. J. Pharm.* **1972,** *18,* 79–82.

51. Pardridge, W. M. *Endocrine Rev.* **1986,** *7,* 314–330.

52. Brightman, M. W. *Exp. Eye Res.* **1977,** *25,* 1–25.

53. Simionescu, N. *Adv. Inflammation Res.* **1979,** *1,* 61–70.

54. Hwang, C.; Sinskey, A. J.; Lodish, H. F. *Science (Washington, DC)* **1992,** *257,* 1496–1502.

55. Pardridge, W. M.; Kumagai, A. K.; Eisenberg, J. B. *Biochem. Biophys. Res. Commun.* **1987,** *146,* 307–315.

56. Simionescu, M.; Ghinea, N.; Fixman, A.; Lasser, M.; Kukes, L.; Simionescu, N.; Palade, G. E. *J. Submicrosc. Cytol. Pathol.* **1988,** *20,* 243–261.

57. Broadwell, R. D.; Balin, B. J.; Salcman, M.; Kaplan, R. S. *Proc. Natl. Acad. Sci. U.S.A.* **1983,** *80,* 7352–7356.

58. Brändli, A. W.; Adamson, E. D.; Simons, K. *J. Biol. Chem.* **1991,** *266,* 8560–8566.

59. Bickel, U.; Yoshikawa, T.; Pardridge, W. M. *Adv. Drug Delivery Rev.* **1993,** *10,* 205–245.

60. Gonatas, N. K.; Stieber, A.; Hickey, W. F.; Herbert, S. H.; Gonatas, J. O. *J. Cell Biol.* **1984,** *99,* 1379–1390.

61. Pesonen, J.; Bravo, R.; Simons, K. *J. Cell. Biol.* **1984,** *99,* 803–809.

62. Jefferies, W. A.; Brandon, M. R.; Hunt, S. V.; Williams, A. F.; Gatter, K. C.; Mason, D. Y. *Nature (London)* **1984,** *312,* 162–163.

63. Friden, P. M.; Walus, L.; Musso, G. F.; Taylor, M. A.; Malfroy, B.; Starzyk, R. M. *Proc. Natl. Acad. Sci. U.S.A.* **1991,** *88,* 4771–4775.

64. Jefferies, W. A.; Brandon, M. R.; Williams, A. F.; Gatter, K. C.; Mason, D. Y. *Immunology* **1985,** *54,* 333–341.

65. Pardridge, W. M.; Buciak, J. L.; Friden, P. M. *J. Pharmacol. Exp. Ther.* **1991,** *259,* 66–70.

66. Fishman, J. B.; Rubin, J. B.; Handrahan, J. V.; Connor, J. R.; Fine, R. E. *J. Neurosci. Res.* **1987,** *18,* 299–304.

67. Roberts, R.; Sandra, A.; Siek, G. C.; Lucas, J. J.; Fine, R. E. *Annu. Neurol.* **1992,** *32,* S43–S50.

68. Morris, C. M.; Keith, A. B.; Edwardson, J. A.; Pullen, R. G. L. *J. Neurochem.* **1992,** *59,* 300–306.

69. Pardridge, W. M.; Eisenberg, J.; Yang, J. *Metabolism* **1987,** *36,* 892–895.

70. Tack, B. G.; Dean, J.; Eilat, D.; Lorenz, P. E.; Schecter, A. N. *J. Biol. Chem.* **1980,** *255,* 8842–8847.

71. Huang, J. T.; Mannik, M.; Gleisner, J. *J. Neuropathol. Exp. Neurol.* **1984,** *43,* 489–499.

72. Muckerheide, A.; Apple, R. J.; Pesce, A. J.; Michael, J. G. *J. Immunol.* **1987,** *138,* 833–837.

73. Vehaskari, V. M.; Chang, T.-C.; Stevens, J. K.; Robson, A. M. *J. Clin. Invest.* **1984,** *73,* 1053–1061.

74. Nell, L. J.; Thomas, J. W. *Diabetes* **1988,** *37,* 172–176.

75. Pardridge, W. M.; Triguero, D.; Buciak, J. L.; Yang, J. *J. Pharmacol. Exp. Ther.* **1990,** *255,* 893–899.

76. Kumagai, A. K.; Eisenberg, J.; Pardridge, W. M. *J. Biol. Chem.* **1987,** *262,* 15214–15219.

77. Tandon, S.; Horowitz, P. M. *J. Biol. Chem.* **1987,** *262,* 4486–4491.

78. Yoshikawa, T.; Pardridge, W. M. *J. Pharmacol. Exp. Ther.* **1992,** *263,* 897–903.

79. Green, N. M. *Methods Enzymol.* **1990,** *184,* 51–67.

80. Pardridge, W. M.; Boado, R. J. *FEBS Lett.* **1991,** *288,* 30–32.

81. Pardridge, W. M.; Boado, R. J.; Buciak, J. L. *Drug Delivery* **1993,** *1,* 43–50.

82. Boado, R. J.; Pardridge, W. M. *Bioconj. Chem.* **1992,** *3,* 519–523.

83. Bickel, U.; Yoshikawa, T.; Landaw, E. M.; Faull, K. F.; Pardridge, W. M. *Proc. Natl. Acad. Sci. U.S.A.* **1993,** *90,* 2618–2622.

84. Andersson, M.; Marie, J.-C.; Carlquist, M.; Mutt, V. *FEBS Lett.* **1991,** *282,* 35–40.

85. O'Donnell, M.; Garippa, R. J.; O'Neill, N. C.; Bolin, D. R.; Cottrell, J. M. *J. Biol. Chem.* **1991,** *266,* 6389–6392.

86. Lee, T. J.-F.; Berezin, I. *Science (Washington, DC)* **1984,** *224,* 898–900.

87. Yaksh, T. L.; Wang, J.-Y.; Go, V. L. W. *J. Cereb. Blood Flow Metabol.* **1987,** *7,* 315–326.

88. McCulloch, J.; Edvinsson, L. *Am. J. Physiol.* **1980,** *238,* H449–H456.

89. Wilson, D. A.; O'Neill, J. T.; Said, S. I.; Traystman, R. J. *Circ. Res.* **1981,** *48,* 138–148.

90. Berntman, L.; Carlsson, C.; Hägerdal, M.; Siesjö, B. K. *Acta Physiol. Scand.* **1976,** *97,* 264–266.

91. Huffman, L. J.; Connors, J. M.; Hedge, G. A. *Am. J. Physiol.* **1988,** *254,* E435–E442.

92. Kodama, T.; Freeman, M.; Rohrere, L.; Zabrecky, J.; Matsudaira, P.; Krieger, M. *Nature (London)* **1990,** *343,* 531–535.

93. Estrada, C.; Bready, J.; Berliner, J.; Pardridge, W. M.; Cancilla, P. A. *J. Neuropath. Exp. Neurol.* **1990,** *49,* 539–549.

94. Yarchoan, R.; Broder, S. *N. Engl. J. Med.* **1987,** *316,* 557–564.

95. Ahmed, A. E.; Jacob, S.; Loh, J.-P.; Sama, S. K.; Nokta, M.; Pollard, R. B. *J. Pharmacol. Exp. Ther.* **1991,** *257,* 479–486.

RECEIVED for review July 12, 1993. ACCEPTED revised manuscript January 21, 1994.

CHAPTER 13

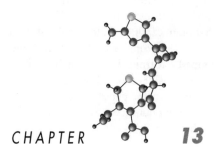

Oligopeptide Drug Delivery to the Brain

Importance of Absorptive-Mediated Endocytosis and *P*-Glycoprotein Associated Active Efflux Transport at the Blood–Brain Barrier

Tetsuya Terasaki
and Akira Tsuji

During the past decade, a number of central nervous system (CNS)-acting peptides have been discovered in the brain (*1*). Although these neuropeptides could have many uses as neuropharmaceuticals, clinical application has not yet been completely accomplished. One of the most crucial problems to be solved in the development of CNS-acting peptide drugs is to increase the cerebral availability of the peptide drug after its administration. To overcome this obstacle, at least four different approaches can be considered: (1) modification of a small peptide (e.g., the active fragment of a native peptide) to increase the stability (resistance to enzymatic metabolism) and to increase the basicity so that it can efficiently enter the brain through absorptive-mediated endocytosis (*2, 3*), (2) modification of a native peptide to increase blood–brain barrier (BBB) transport activity (*4*), (3) conjugation of a native peptide and vector peptide that can be transported at the BBB (*5, 6*) and (4) transfer of a gene encoding the neuropeptide to the brain. Possibilities 2 and 3 were discussed in

Chapter 12. Regarding gene therapy, more studies are needed to elucidate the ultimate strategy for therapy of genetic disorders in the brain. An experimental brain tumor has been successfully treated by the in vivo transfer of a gene with a retrovirus (*7*). Further progress in molecular cell biology may make it possible to deliver a specific gene to the brain.

In this chapter, we discuss strategy 1 for the development of a CNS-acting peptide drug (i.e., design of oligopeptides) that can be transported through the BBB. One of the most interesting topics regarding BBB transport is that brain capillary endothelial cells have an active efflux transport system that is ascribed to *P*-glycoprotein. We found that cyclosporin A, a cyclic lipophilic peptide drug, is pumped out from the endothelial cells to the circulating blood side by *P*-glycoprotein. Therefore, consideration of this active efflux system would be important in the development of a strategy for peptide delivery to the brain. Accordingly, in the latter part of this chapter, we present our recent findings regarding the *P*-glycoprotein-associated active efflux system at the BBB.

Factors Affecting Pharmacokinetics of CNS-Acting Peptide Drugs

The cerebral availability of peptide drugs depends on the route of administration. Intracerebral, intraventricular, or intracarotid arterial administration provides high concentrations in the cerebrospinal fluid (CSF), and diffusion of the peptide drug from CSF through the interstitial fluid (ISF) is significantly limited. Oral administration is a more convenient route for the daily use of a drug, but small intestinal absorption and hepatic first-pass effects both affect drug availability at the site of action. Intravenous or subcutaneous administration is a practical route of administration for CNS-acting peptide drugs. Factors affecting the pharmacokinetics of a peptide drug after systemic administration include (1) elimination in the systemic circulation (e.g., hepatic uptake, enzymatic metabolism in the blood and liver, and renal excretion), (2) plasma protein binding, (3) transport at the BBB, (4) enzymatic stability in the brain ISF, and (5) diffusion in the brain ISF. Because it would be difficult to solve all of these complicated problems at one time, traditional trial and error approaches are inefficient. To improve the pharmacokinetic characteristics of peptide drugs, it is necessary to understand the biochemical mechanisms of the aforementioned distribution processes. These distribution processes could be studied by chemically modifying native peptides and synthesizing analogue peptides, as has been accomplished with the development of synthetic antibiotics (e.g., β-lactam antibiotics that are di- or tripeptide drugs).

BBB as a Major Route of Drug Transfer into the Brain

In contrast to the endothelial cells in capillaries of peripheral organs, brain capillary endothelial cells are connected by tight junctions that restrict the paracellular permeation of substrates from the circulating blood to the brain ISF (*8*). By increasing the lipophilicity of a peptide drug, the passive diffusion rate for small peptides at the BBB is increased, the permeability of peripheral organ plasma membrane would also increase simultaneously. Influx and efflux of nutrients, which are highly hydrophilic substrates, between the capillary lumen and brain ISF are well known to be efficiently regulated by several kinds of transport systems at the brain capillary endothelial cells (*9*). At least eight independent carrier-mediated transport systems for nutrients and hormones (i.e., hexoses, neutral amino acids, basic amino acids, monocarboxylic acids, amines, purine bases, nucleosides, thyroid hormones) have been found to play important roles in normal brain function (*9*). Carrier-mediated transport systems for peptides have also been described (*see* following discussion). Moreover, receptor-mediated transcytosis has been found to be responsible for peptide transfer from the circulating blood to the brain. Recent findings (*10, 11*) have revealed that brain capillary endothelial cells can transfer insulin, insulin-like growth factor, and transferrin by an independent receptor-mediated endocytosis system. These specific transport systems at the BBB may be useful for peptide delivery to the brain.

To study specific delivery of peptides to the brain, it is important to determine the similarities and differences of membrane transport systems between brain capillary endothelial plasma membranes and peripheral organ plasma membranes. Some synthetic peptides, such as somatostatin analogue peptide, are efficiently taken up by the liver through a multispecific bile acid transport system (*12*). Thus, to increase the biological half-life of peptide drugs after their systemic administration, it is necessary to take this hepatic transport into consideration. As the transporter would require a strict structure–substrate relationship to bind the substrate, three-dimensional computer graphic analysis would provide a rational way to design the structure of the synthetic peptide to fit the carrier or receptor protein at the BBB.

Like the BBB, the blood–cerebrospinal fluid barrier (BCSFB), which is located at circumventricular organs such as the choroid plexus and median eminence in the brain, is also known to transfer substrates between blood and brain (*13*). Although the surface area of the BCSFB is only one five thousandth that of the BBB, a carrier-mediated transport system at the BSCFB may play an important role in drug transfer into CSF. For example, as most acquired immunodeficiency syndrome (AIDS) patients have CNS disorders and a high viral burden in the brain, effective chemotherapy for HIV infection requires the efficient penetration of anti-HIV drugs such as 3′-azido-3′-deoxythymidine (AZT). Thymidine, the parent compound of AZT, does not have an affinity for the carrier-mediated transport system of nucleosides or purine bases at the BBB (*14*). Moreover, thymidine and AZT are highly polar compounds that do not penetrate the endothelial cell membrane through passive diffusion. Interestingly, the CSF concentration of AZT has been reported to be more than half that of plasma after administration of AZT to AIDS patients (*15*), whereas no transport of AZT at the BBB has been demonstrated in rats (*16*). The dual findings of a lack of permeation of AZT through the BBB (*16*) and of a high CSF concentration (*15*) could be explained by the evidence that the BCSFB can transport thymidine rapidly (*17*). These findings indicate that CSF and brain ISF are not in equilibrium and that the concentration of drug in CSF may not necessarily reflect BBB transport but rather transport through the BCSFB (*16*).

Recent studies (*18*) have revealed that the BCSFB also acts as an active efflux pump for the transport of β-lactam antibiotics from brain to blood. The active efflux pump for β-lactam antibiotics is an anion exchange transporter located at the epithelial plasma membrane of the choroid plexus (*18*). Because this transport system recognized β-lactam antibiotics that are di- or tripeptides, the system may also recognize some neuropeptides, thereby facilitating (along with the bulk flow of the CSF) their elimination from the CNS.

Mechanism of Peptide Transport at the BBB

Because brain capillaries are connected by tight junctions, many researchers believed until the mid-1980s that hydrophilic peptides with large molecular weight cannot cross the BBB. However, significant studies reviewed by Pardridge (*10*) and Banks et al. (*19, 20*) have opened up new research fields regarding peptide transport at the BBB. Although the permeability rate was limited, a good correlation between lipophilicity and the BBB permeability rate was reported for 18 peptides (*21*). Moreover, nonsaturable transport has been reported for delta sleep-inducing peptide (DSIP) at the BBB (*22*). These reports (*22*) suggest that passive diffusion through the capillary endothelial cell membrane is one possible mechanism for peptide transport at the BBB. In addition, as reported (*23*) previously for vasopressin-like peptide transport, paracellular transport is another possible mechanism for transport at the BBB.

Although it has been clarified that the BBB has several carrier-mediated transport systems for nutrients (*9*), these systems are not promising candidates for the efficient transport of peptide drugs. Recently, Banks and Kastin and co-workers (*9, 20, 24*) suggested that a family of transport systems, termed protein transport system-1, -2, -3, and -4 (PTS-1, PTS-2, PTS-3, and PTS-4), may exist to transfer several kinds of endogenous peptides, such as opiate enkephalins, antiopiates like tyrosine melanocyte-stimulating hormone inhibitory factor 1 (Tyr-MIF-1), and oxytocin. Although the biochemical mechanism requires further clarification, this family of transport systems would be a potentially useful pathway in the brain. In addition to these findings, Kaplowitz and co-workers (*25*) found that reduced glutathione (GSH), a tripeptide and a substrate for glutathione *S*-transferase that is responsible for the hepatic conjugation of drugs, can be taken up at the BBB by a carrier-mediated transport system. The brain uptake index (BUI), a useful parameter to evaluate BBB permeability, was almost 20% in the rat, indicating significant transport of GSH at the BBB. Moreover, the transport of GSH was saturable, with an apparent Michaelis constant (K_m) of 5.84 ± 1.92 mM. Because γ-glutamyltranspeptidase (γ-GTP) is localized at the luminal membrane of brain capillary endothelial cells, GSH could be significantly metabolized and the resulting metabolite (i.e., ^{35}S-methione) would be taken up. However, the BUI values of GSH with and without an irreversible γ-glutamyl transpeptidase inhibitor, acivicin, were similar (*25*). Also, 83% of the intact form of GSH was found in the brain 15 s after the intracarotid artery injection of GSH to the acivicin-pretreated rat. No inhibitory effect on the GSH uptake into the brain was shown by amino acids, amino acid analogues, cystein, phenylalanine, glutathione disulfide, γ-glutamylglutamate, γ-glutamyl *p*-nitroanilide, or 2-aminobicyclo(2,2,1)heptane-2-carboxylic acid (BCH). These results demonstrate that GSH is transported through the BBB in an intact form by a specific system (*25*). The brain contains significant amounts of GSH for antioxidant defense; therefore, the transport system would play an important role in supplying cerebral GSH from the circulating blood. Although no uptake of oxidized glutathione (GSSG) at the BBB has been demonstrated, unlike the renal and intestinal transport system of GSH, it would be very interesting to clarify the transport mechanism of GSH and the structure–recognition relationship of this carrier system. The GSH transport system may also be a possible pathway to deliver small peptides (e.g., di- or tripeptide drugs) to the brain.

Pardridge and co-workers found that positively charged peptides, such as cationized albumin (*5*) and cationized immunoglobulin, can cross the BBB (*4*). The electrostatic interaction between the positively charged moiety of the peptide and the negatively charged region of capillary endothelial membrane, such as the sialic acid of membrane glycoprotein, is considered to be a trigger of this absorptive-mediated endocytosis system. Moreover, the isoelectric point seems to be the most important physicochemical property to regulate the organ specificity of peptide transfer (*26*). To date, cationized rat albumin, the isoelectric point of which is 8.5, has been demonstrated to be selectively delivered to the brain (*26*).

Absorptive-Mediated Transcytosis at the BBB as a Useful System for Synthetic Peptide Drug Delivery to the Brain

E-2078, an Analogue Peptide of Dynorphin$_{1-8}$

Dynorphin$_{1-8}$ and dynorphin$_{1-13}$ are naturally occurring opioid peptides that are probably involved in the regulation of synaptic transmission related to the κ-type opioid receptor. Although these peptides are known to have high affinity for opioid receptors, no significant analgesic effect after systemic administration is observed. This lack of analgesia is because, like other opioid peptides, dynorphin is very susceptible to enzymatic degradation and shows very low permeability through the BBB. To overcome the disadvantage of rapid degradation of dynorphin$_{1-8}$, a novel analogue of dynorphin$_{1-8}$, E-2078, was developed. Interestingly, E-2078 (*see* structure 1) exhibits several times greater analgesic effect than morphine after systemic administration (*3, 27*). Therefore, it can be assumed that this synthetic peptide crosses the BBB. Because E-2078 has high affinity for the κ-type opioid receptor, is a positively charged peptide (the isolectric point is 10) at a physiological pH of 7.4, and is also an octapeptide with an N-terminal tyrosine, several transport mechanism have been hypothesized; for examples, receptor-mediated transcytosis (*11*), absorptive-mediated transcytosis (*5*), and carrier-mediated transport of small peptides with an N-terminal tyrosine (*19*). To prove the hypothesis that E-2078 crosses the BBB and to clarify the transport mechanism involved, the following in vivo and in vitro experimental systems have been established: in vitro isolated brain capillaries (*28*), carotid artery injection method (*28*), and brain microdialysis (*29*).

Figure 1 illustrates the HPLC chromatogram of E-2078 in bovine brain capillaries after incubation with a capillary suspension for 30 min at 37 °C. In contrast to endogenous opioid peptides, such as dynorphin and enkephalin, E-2078 was fairly resistant to enzymatic degradation by the peptidases present in the brain capillaries. This advantageous property

E-2078 (Dynorphin analogue, DLAP)

CH_3 (on Arg)

H_3C-Tyr-Gly-Gly-Phe-Leu-Arg-Arg-(D)Leu-NHC_2H_5

M.W. 1035 pI 10

Ebiratide (ACTH analogue)

Met(O_2)-Glu-His-Phe-(D)Lys-Phe-NH(CH_2)$_8$$NH_2$

M.W. 997 pI 10

Structure 1.

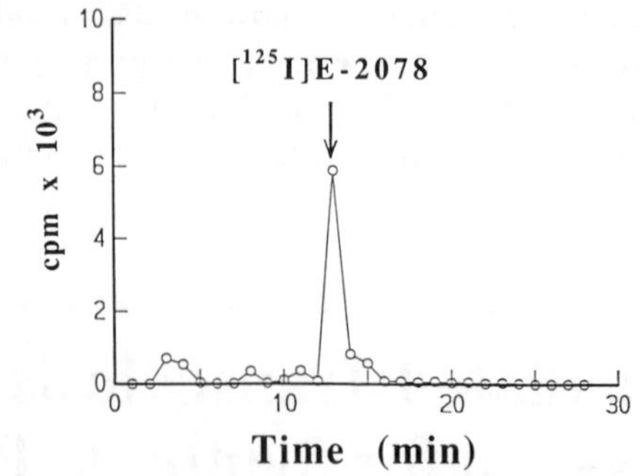

Figure 1. HPLC chromatogram of [^{125}I]E-2078 after incubation with isolated bovine brain capillaries for 30 min at 37 °C. The supernatant of the incubation mixture was added to the reversed-phase HPLC column. An HPLC chromatogram of a standard sample of [^{125}I]E-2078 is illustrated in the inset. (Reproduced with permission from reference 28. Copyright 1989 Pergamon.)

of E-2078 can be attributed to suitable chemical modifications of enzyme-susceptible sites of native dynorphin [i.e., *N*-methyl groups at the N-terminal of Tyr^1 and at Arg^7, D-Leu^8 instead of L-Leu^8, and an ethylamine group at the C-terminal of D-Leu^8 (*see* structure 1)]. E-2078 is bound and internalized by isolated bovine brain capillaries, as indicated by an acid-wash procedure. The surface binding of E-2078 to brain capillaries was concentration dependent, with a half-saturation concentration (K_d) of 4.62 ± 0.59 μM (Figure 2). The K_d value was 5.8-fold greater than that of cationized albumin (0.8 ± 0.1 μM) (*5*). The acid-resistant binding of E-2078 to brain capillaries was completely inhibited by phenylarsine oxide, which is known to be an endocytosis inhibitor of epidermal growth factor and insulin (*30*). Moreover, significant temperature and osmotic pressure dependencies were observed for the acid-resistant binding of E-2078 to brain capillaries. This evidence supports the conclusion that the acid-resistant binding of E-2078 is due to internalization into bovine brain capillaries through an endocytosis process.

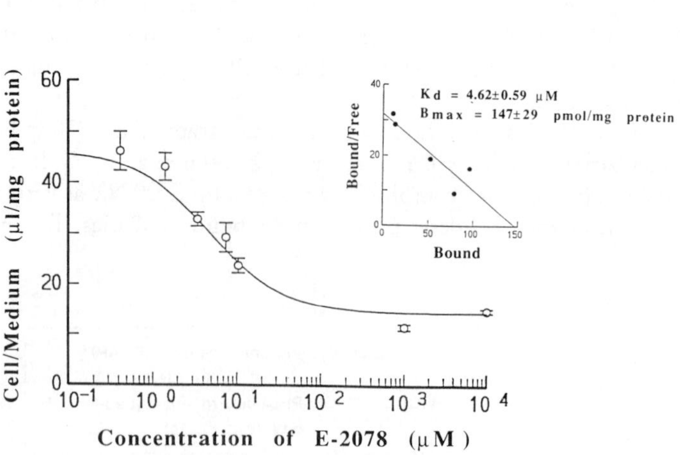

Figure 2. Concentration dependence of the surface binding of [^{125}I]E-2078 to isolated bovine brain capillaries. The inset presents the Scratchard plot of the saturable surface binding of [^{125}I]E-2078, where saturable surface binding was calculated from the surface binding minus nonsaturable surface binding that was determined in the presence of 10 mM unlabeled E-2078. Each point represents the means ± SE of five experiments (B_{max} is the maximal binding capacity). (Reproduced with permission from reference 28. Copyright 1989 Pergamon.)

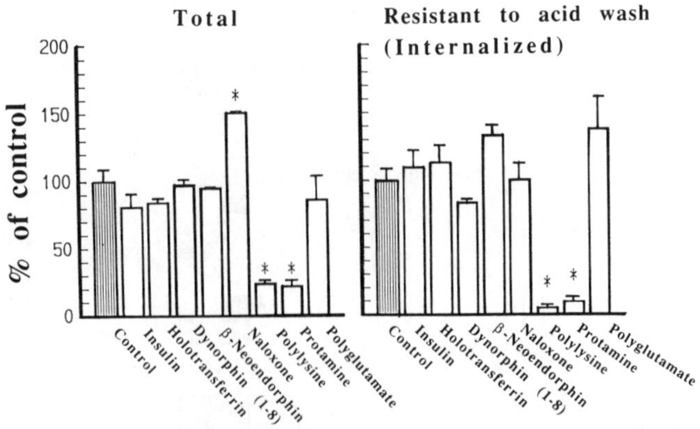

Figure 3. Effects of several inhibitors on the total and acid-resistant bindings of [^{125}I]E-2078 to isolated bovine brain capillaries. Brain capillaries were preincubated for 5 min at 37 °C with either 10 μM insulin, 10 μM holotransferrin, 1 mM dynorphin$_{1-8}$, 1 mM β-neoendorphin, 10 μM naloxone, 300 μM polylysine, 300 μM protamine, or 300 μM polyglutamate. Each bar represents the mean ± SE of five experiments. (Reproduced with permission from reference 28. Copyright 1989 Pergamon.)

Inhibition studies (*28*) have been conducted to characterize the endocytosis system of E-2078 into the brain capillaries. Poly-*L*-lysine (300 μM) and protamine (300 μM) significantly inhibited both total and acid-resistant binding of E-2078 (Figure 3). In contrast, porcine insulin (10 μM), human holotransferrin (10 μM), porcine dynorphin$_{1-8}$ (1 mM), β-neoendorphin (1 mM), and polyglutamate (300 μM) did not change the total or acid-resistant bindings of E-2078. Although the acid-resistant binding of E-2078 was not changed by the opioid antagonist naloxone, the total binding of E-2078 was significantly enhanced. The allosteric effect of naloxone on the surface binding of E-2078 to the capillary plasma membrane cannot be explained at the present time. The significantly different effects of polycationic peptide and polyanionic peptide indicate that electrostatic interactions of E-2078 with the anionic surface region of the capillary plasma membrane play a dominant role in the surface binding and subsequent internalization of E-2078 into the brain capillary endothelial cells. As these inhibitory effects of polycationic peptide are very similar to those of cationized albumin (*5*), cationized IgG (*4*), and histone (*31*), absorptive-mediated endocytosis is thought to be the BBB transport mechanism of E-2078.

Although the molecular mechanism of absorptive-mediated transcytosis of basic peptides has not yet been clarified, Raub and Audus (*32*) reported interesting results that luminal membrane glycoproteins labeled with a lectin, *Ricinus communis* I (RCAI) that binds carbohydrates containing β-D-galactosyl or β-D-*N*-acetylgalactosaminyl residues are internalized by a selective endocytosis mechanism and sorted to the *trans*-Golgi network (TGN) in primary cultured bovine brain capillary endothelial cells. The movement of membrane proteins through the TGN is well known to be involved in the cellular protein recycling pathway in many cells (*32–36*). Thus, a basic peptide bound to the luminal membrane glycoproteins could be transferred into the TGN. Subsequently, the transferred basic peptide in the TGN may also bind to the other glycoproteins that will move to the abluminal membrane by the exocytosis mechanism (*32*). Figure 4 shows a hypothetical mechanism for absorptive-mediated transcytosis. Although the physiological meaning of the absorptive-mediated transcytosis has not been completely clarified, this cellular traffic of membrane glycoprotein that can bind basic peptides may be one of the possible mechanisms of peptide transport to the brain.

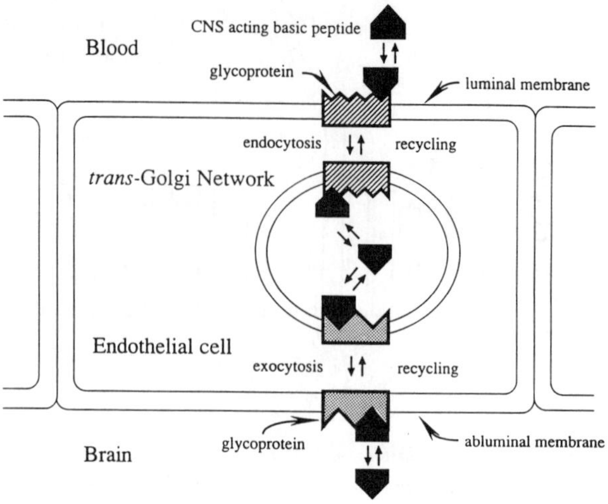

Figure 4. Hypothetical mechanism of absorptive-mediated transcytosis at the BBB.

In vitro uptake studies with isolated brain capillaries have several advantages, such as direct proof of BBB transport and clarification of the biochemical mechanism of transport. However, it is necessary to separate surface binding and internalization of peptide into brain capillaries in in vitro studies by acid-washing techniques (*5, 29*). It is difficult to distinguish whether the substrate is taken up at the luminal or albuminal plasma membrane of brain capillaries when the uptake study is performed in vitro with isolated brain capillaries. To overcome this problem, the in vivo carotid artery injection method and in vivo brain microdialysis have been used to study the brain uptake of E-2078. Because the brain extraction was expected to be small, the extracellular marker sucrose was used for the internal reference in the in vivo brain uptake studies. The BUI of E-2078 was 368 ± 55% (*28*). Moreover, the presence of unlabeled E-2078 (1 mM) significantly reduced the BUI of E-2078 to 185 ± 16% (*28*), suggesting that a specific transport system, probably an absorptive-mediated endocytosis system, occurs at the luminal side of the brain capillary endothelial plasma membrane (*28*).

As shown in Figure 4, at least three steps are involved in peptide transport through the BBB: (1) endocytosis at the luminal membrane, (2) intracellular movement, and (3) exocyotis directed to the albuminal membrane. By using two techniques, such as in vitro uptake by isolated brain capillaries and in vivo brain uptake by carotid artery injection, the endocytosis of E-2078 has been confirmed. Several techniques to obtain direct evidence for an exocytosis process, including in vivo brain microdialysis (*29, 37, 38*), in vivo capillary depletion (*38, 39*), and in vitro uptake by primary cultured bovine endothelial cells (*40–42*), are available. Brain microdialysis has been developed to collect neurotransmitters directly in the brain ISF by implanting a semipermeable dialysis fiber in the brain (*37*). As shown in Figure 5, a brain microdialysis probe of the transcranial type, which was constructed from a renal dialysis fiber with molecular cutoff of 12.5 kDa, inner diameter of 200 μm, and stainless steel tubing, was implanted in the hippocampus of rats. Because the in vivo permeability of E-2078 appeared to be small, in vivo brain perfusion was performed simultaneously with the brain microdialysis. The implantation of dialysis fiber in the brain may cause significant damage to the brain capillaries, resulting in overestimation of transendothelial transport of substrate. Therefore, the capillary nonpermeable marker sucrose was used to evaluate the nonspecific leakage of E-2078 from the circulating blood through the damaged

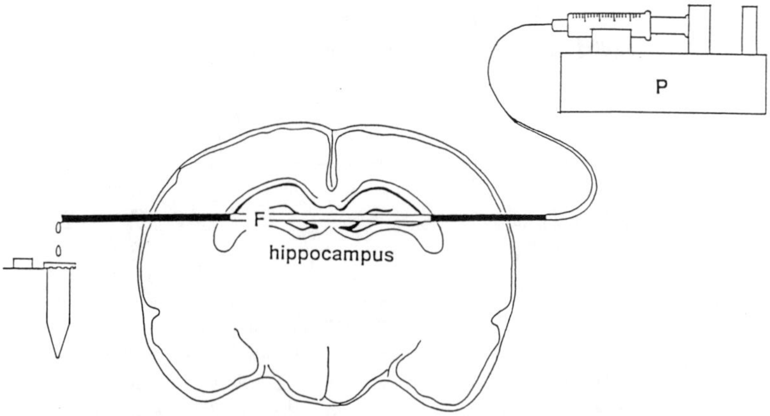

Figure 5. Schematic representation of the brain microdialysis apparatus (F and P represent the microdialysis fiber and the constant infusion pump, respectively). (Reproduced with permission from reference 29. Copyright 1991 Plenum.)

capillaries. The concentration ratio of sucrose in the brain ISF and the brain perfusate was $2.71 \times 10^{-3} \pm 1.43 \times 10^{-3}$. This significantly low value of sucrose suggests that nonspecific permeation of the perfusate through the brain capillaries is negligible after the implantation. Interestingly, the concentration ratio of E-2078 in the brain ISF and the brain perfusate was 0.296 ± 0.026 (~100-fold greater than that of sucrose), providing direct evidence for significant in vivo transport of E-2078 from the vascular lumen to the brain ISF. In contrast, the concentration ratio of E-2078 in the CSF and the brain perfusate was $2.96 \times 10^{-3} \pm 0.26 \times 10^{-3}$, which was not significantly different from that of sucrose ($1.31 \times 10^{-3} \pm 0.43 \times 10^{-3}$), demonstrating that the transport of E-2078 into the brain can be ascribed mainly to transport across the BBB but not across the BCSFB. In addition to this evidence, no metabolite of E-2078 was found in the brain ISF by HPLC analysis of brain dialysate, indicating the absence of significant metabolism of E-2078 during the transcytosis process through the BBB and diffusion through the brain ISF (*29*).

Ebiratide, an Analogue Peptide of Adrenocorticotropic Hormone

Adrenocorticotropic hormone (ACTH) and its analogue peptides are known to restore the avoidance learning caused by removal of the pituitary gland (*43, 44*) and also to be effective for inhibitory (passive) avoidance conditioning (*45*). Ebiratide, an analogue peptide of $ACTH_{4-9}$, has significant neurotrophic effects in cultured cerebral neurons (*46*), increases acetylcholine release in the rat brain (*47*), and is remarkably effective in improving learning and memory impairments in rats and mice (*2*). Thus, ebiratide is considered to be a potential memory-enhancing peptide and is expected to be developed as a drug for the treatment of Alzheimer-type dementia. These CNS effects were demonstrated after the systemic administration of ebiratide (*2*), so it is assumed that ebiratide crosses the BBB. Ebiratide consists of six amino acids in which the following three amino acids of $ACTH_{4-9}$ have been modified: an oxidized methionine residue in position 4, a D-lysine residue in position 8 instead of arginine, and the replacement of tryptophan by phenylalanine octylamide in position 9 (*see* structure 1). It is noteworthy that ebiratide is also a basic peptide drug, having the same isoelectric point as that of E-2078 [i.e., 10 (*48*)], whereas the primary structure of ebiratide is quite different from that of E-2078 (*see* structure 1). Therefore, the mechanism of BBB transport of ebiratide, like that of E-2078, may be absorptive-mediated endocytosis.

Table I. Effect of Protease Inhibitors on the Metabolism of [^{125}I] Ebiratide in Bovine Brain Capillary Suspension

Inhibitor	*Percentage of Intact [^{125}I]Ebiratide in the Incubation Medium*
Control	8.8 ± 2.5
Bacitracin (1 mM)	37.1 ± 2.0
Aprotinin (1000 U/mL)	12.9 ± 6.2
EDTA (10 mM)	59.8 ± 0.5

NOTE: [^{125}I]Ebiratide was incubated with 600 μg of bovine brain capillaries for 30 min at 37 °C in the presence or absence of each inhibitor. The incubation medium was analyzed by HPLC. Each value represents the mean ± SE of three or four experiments.
SOURCE: Reproduced with permission from reference 48. Copyright 1991 Pergamon.

The BBB transport mechanism of ebiratide has been investigated with isolated bovine brain capillaries (*48*). In contrast to E-2078, ebiratide was metabolized during incubation with a suspension of bovine brain capillaries. Although the metabolism of ebiratide was not completely inhibited by protease inhibitors, significant inhibition was observed in the presence of EDTA (10 mM) in the suspension of brain capillaries (Table I). These results suggest that some metaloenzyme may be responsible for the metabolism of ebiratide. In the same transport study, in the presence of EDTA, the acid-resistant binding of ebiratide increased with time and equilibrium was attained at 15 min. The acid-resistant binding of ebiratide at steady state (30 min) (i.e., the internalized amount represented by the cell-to-medium ratio) was 5.00 ± 0.18 μL/mg protein, which is one-fourth that of E-2078 (19.1 ± 0.6 μL/mg protein). The uptake of ebiratide was also inhibited by the endocytosis inhibitor dansylcadaverine (500 μM) and the basic peptides poly-L-lysine (300 μM) and protamine (300 μM). These results suggest that ebiratide is also internalized by absorptive-mediated endocytosis into the brain capillary endothelial cells, although there has been some criticism that EDTA may stimulate the endocytosis of ebiratide. Therefore, we have tried to establish another experimental transport system to clarify this point.

Interestingly, no metabolite of ebiratide was observed in the incubation medium after incubating ebiratide for 60 min at 37 °C with the primary cultured bovine brain capillary endothelial cells (BCEC) (*42*). Moreover, the lack of a significant inhibition effect of 10 mM histidine on the acid-resistant binding of ebiratide to primary cultured BCEC suggests that the ^{125}I radioactivity of the acid-resistant fraction could not be attributed to the carrier-mediated transport of [^{125}I]histidine, an amino acid labeled with the ^{125}I of ebiratide (Table II). Significant metabolism of ebiratide was observed during its incubation in blood constituents, indicating that the metabolism in the isolated brain capillary suspension may be ascribed to contamination by blood in the capillary lumen. As shown in Table II, the effects of several peptides on the acid-resistant binding of ebiratide to primary cultured BCEC were very similar to those effects on the acid-resistant binding of ebiratide to isolated bovine brain capillaries. Transferrin significantly enhanced the acid-resistant binding of ebiratide. A possible explanation for this enhancement is an allosteric interaction of transferrin with ebiratide at the luminal membrane of BCEC. In in vitro transport studies with isolated brain capillaries, it is difficult to distinguish between the endocytosis transport systems present at the luminal and albuminal sides of the plasma membrane of brain capillaries. However, one of the advantages of uptake studies with primary cultured BCEC is that transport characteristics can be ascribed to the luminal side of the brain capillary endothelial

Table II. Effect of Histidine and Several Peptides on the Acid-Resistant Binding of [^{125}I]Ebiratide to Cultured Monolayer of BCEC

Inhibitor	*Concentration (mM)*	*Acid-Resistant Binding (% of Control)*[a]
Control	—	100 ± 4.0
Histidine	10.0	98.5 ± 5.5
Poly-L-lysine	0.30	29.6 ± 2.4[b]
Protamine	0.30	12.4 ± 1.1[b]
Poly-L-glutamic acid	0.30	95.5 ± 13.6
Histone	0.10	68.7 ± 5.7[c]
ACTH	0.10	76.1 ± 5.4[c]
Insulin	0.010	123 ± 21.0
Transferrin	0.010	188 ± 36.0

[a] Cells were incubated at 37 °C for 60 min in the presence or absence of several inhibitors. Each value represents the percentage of control value (mean ± SE of three or four experiments).
[b] Significantly different than control ($P < 0.001$) as determined by the *t* test.
[c] Significantly different than control ($P < 0.05$) as determined by the *t* test.
SOURCE: Reproduced with permission from reference 42. Copyright 1992 Pergamon.)

cells (*32, 40*). Moreover, the primary cultured BCEC exhibit high viability, whereas isolated bovine brain capillaries can be stained with trypan blue. As shown in Table III, the acid-resistant binding of ebiratide was significantly inhibited by dancylcadaverine and phenylarsine oxide. Phenylarsine oxide is known to be an endocytosis inhibitor that interacts with the SH group in the cell membrane (*30*). Dansylcadaverin is also known to be an endocytosis

Table III. Effect of Temperature, Osmolarity, Endocytosis Inhibitors, and a Metabolic Inhibitor on the Acid-Resistant Binding of [^{125}I]Ebiratide to Cultured Monolayers of BCEC

Condition	*Concentration (mM)*	*Acid-Resistant Binding (% of Control)*
Control	[a]	100 ± 16.1
Low temperature (4 °C)	[a]	11.8 ± 1.8[b]
Hypertonic (1600 mOsm)		47.1 ± 3.6[c]
Dansylcadaverine	0.50	20.5 ± 4.9[b]
2,4-DNP	1.0	42.3 ± 2.4[c]
Control	[a]	100 ± 23.2[d]
Phenylarsine oxide	0.10	35.9 ± 6.1[c,d]

NOTE: After the preincubation of cells for 20 min under several conditions, [^{125}I]ebiratide was added to initiate the uptake. Acid-resistant binding was determined after the 60 min incubation. Each value represents the percentage of control value (mean ± SE of three or four experiments).
[a] Not applicable.
[b] Significantly different than control ($P < 0.05$) as determined by the *t* test.
[c] Significantly different than control ($P < 0.05$) as determined by the *t* test.
[d] Phenylarsine oxide was dissolved in 0.1% dimethyl sulfoxide (DMSO), so the control experiment was performed in the presence of 0.10% DMSO.
SOURCE: Reproduced with permission from reference 42. Copyright 1992 Pergamon.

inhibitor that inhibits a transglutaminase in the cell membrane, resulting in the suppression of coated pit formation (*49*). In addition to these effects, the significant inhibitory effect of the metabolic inhibitor 2,4-dinitrophenol (2,4-DNP) on the acid-resistant binding of ebiratide strongly supports the presence of an endocytosis system at the luminal side of BCEC.

To prove the transcytosis of ebiratide across BCEC, a brain microdialysis study and capillary depletion study were conducted (*38*). After infusion of ebiratide or sucrose into the internal carotid artery for 10 min, the rat brain hemisphere was isolated and treated by the capillary depletion method. As ebiratide is significantly metabolized in vivo, HPLC analysis was performed to determine the unmetabolized ebiratide in the brain parenchyma and capillary fractions. The apparent distribution volume (Vd_{app}) of unmetabolized ebiratide in the brain parenchyma fraction (167.8 ± 62.2 μL/g brain) was about sevenfold greater than that of sucrose (24.9 ± 4.0 μL/g brain) and was also 35-fold greater than that of sucrose in the brain capillary fraction (6.2 ± 1.8 μL/g brain); however, the standard error of Vd_{app} in the brain parenchyma fraction for ebiratide was almost 40% of the mean value. Although the capillary depletion method is a useful technique to use when evaluating the in vivo transfer of drugs from circulating blood to brain parenchyma (*39*), it may underestimate the in vivo transport through the BBB of a peptide drug that is susceptible to metabolism because the peptide in the brain ISF could be metabolized by the cellular enzymes during and after homogenization. One of the advantages of brain microdialysis is that the substrate can be directly analyzed in the brain ISF (*29, 50*). Thus, brain microdialysis study was simultaneously performed (*38*). A microdialysis fiber was implanted into the hippocampus, which plays an important role in learning and memory functions. Benveniste and Diemer (*51*) reported that normal neuropiles and only occasional hemorrhages were observed in the region surrounding the microdialysis fiber within the first 2 days, whereas astrocytes close to the dialysis fiber hypertrophied at 3 days. Accordingly, the dialysis was performed 48 h after the implantation to minimize the influence of the cellular reactions and impairment (*38*). The brain dialysate was analyzed by HLPC, and unmetabolized ebiratide was determined to account for >80% of total ^{125}I radioactivity in the dialysate of the brain microdialysis, supporting the contention that ebiratide crosses the BBB in an intact form.

The ISF concentration was estimated from the concentration of dialysate by a method that uses the correction factor determined by the in vitro dialysis study. However, this in vitro recovery method underestimated the ISF concentration. Several extrapolation methods have been proposed to solve the problem of underestimation (*52–57*), but, in practice, no promising approach has been developed. Unlike this approach, our proposed in vivo reference method is simple and reliable because the effective dialysis coefficient (R_d) for the reference compound (i.e., an in vivo internal standard for microdialysis) is used to correct the difference in the permeability rate constant between in vivo and in vitro dialysis conditions (*58–60*). The R_d was defined by the ratio between the in vivo and the in vitro permeability constants of the test drug or reference compound. As only a slight difference in the ratio of the permeability rate constants was determined in the erythrocyte suspension and the buffer solution among several substrates with different molecular weights [e.g., [^{3}H]water, antipyrine, sucrose, and ebiratide (*58*)], no difference in R_d values between a test drug and a reference compound (e.g., antypirine) that crosses the BBB rapidly by passive diffusion was assumed. Thus, the following equation was derived (*58, 60*):

$$C_{isf,u,drug} = C_{d,vivo,drug}/\{1 - \exp(-R_{d,ref}PA_{vitro,drug}/F)\} \quad (1)$$

where $C_{isf,u,drug}$ is the unbound concentration of the test drug in the ISF, $C_{d,vivo,drug}$ is the concentration of the test drug in the dialysate, $R_{d,ref}$ is the ratio of permeability rate constant of reference compound between in vivo brain dialysis condition and in vitro dialysis against buffer, $PA_{vitro,drug}$ is the in vitro permeability rate constant of the test drug, and F is the dialysis rate. The capillary permeability of small molecules in the peripheral tissues is significantly rapid, so a lack of a significant difference in unbound drug concentrations between circulating blood and tissue ISF can be assumed. The reliability of this in vivo reference method was confirmed for the microdialysis study in the muscle, lung, and liver (*59*). The applicability of equation 1 was also confirmed by the brain microdialysis of [^{3}H]water, aminopyrine, and caffeine (*60*). Based on our in vivo reference method, the concentrations of ebiratide and sucrose in the brain ISF were extrapolated from those of the dialysate (*38*). The concentration ratio between brain ISF and the internal carotid artery was $4.24 \times 10^{-2} \pm 0.38 \times 10^{-2}$ for ebiratide, which was significantly greater than that of sucrose ($7.64 \times 10^{-3} \pm 1.57 \times 10^{-3}$). These results demonstrate that ebiratide permeated through the brain capillaries in an intact form. Furthermore, the BBB could be the major pathway for the transfer of ebiratide from the circulating blood into the CNS, as no significant amount of ebiratide was found in the CSF even 30 min after intravenous infusion (*38*). These results for ebiratide and E-2078 indicate that the absorptive-mediated transport system is a very useful pathway for peptide delivery through the BBB. Chemical modifications to increase the basicity of a peptide may increase its affinity to the positively charged plasma membrane surface of BCEC. Although only limited information is currently available, the isoelectric point of the peptide would be the most important physicochemical parameter in regulating the association to the membrane surface of BCEC endothelial cells (*4, 5, 28, 31, 48, 61*). Some steric effect may also play an important role in the association process, as K_d differs significantly between E-2078 and ebiratide (*28, 42, 48*). In addition to the association process (Figure 4), intracellular peptide traffic, such as sorting in the TGN and secretion to the abluminal membrane, may also influence the efficiency of the transcytosis rate (*32*).

P-Glycoprotein as an Active Efflux Pump of Cyclosporin A and Multidrug Resistant Sensitive Drugs at the BBB

Lipophilicity is a well-known physicochemical property affecting the BBB permeability of drugs, and fairly good correlation between the lipophilicity and BBB permeability rate has been reported (*62*). However, some highly lipophilic drugs, such as cyclosporin A (CsA) (*63*), vincristine (VCR) (*64*), and doxorubicin (Adriamycin) (*64*), are also known to have significantly low BBB permeability. The molecular weight of these drugs is >500, so kinks model or pore theory has been proposed as a possible mechanism to explain their low permeability at the BBB (*61, 65*). To find a rational strategy to increase the BBB permeability of these drugs, it is very important to clarify the membrane transport mechanism. In multidrug-resistant (MDR) tumor cells the apparent cellular membrane permeability of antitumor drugs, such as doxorubicin and VCR, is known to be significantly reduced by the function of *P*-glycoprotein as an active efflux pump. It is also interesting that CsA, a peptide drug, also can be recognized by *P*-glycoprotein of MDR tumor cells (*66*). Moreover, immunohistochemical studies have revealed that *P*-glycoprotein was found in some of the normal cells (*67, 68*) including BCEC, biliary canalicular surface of liver cells, brush border

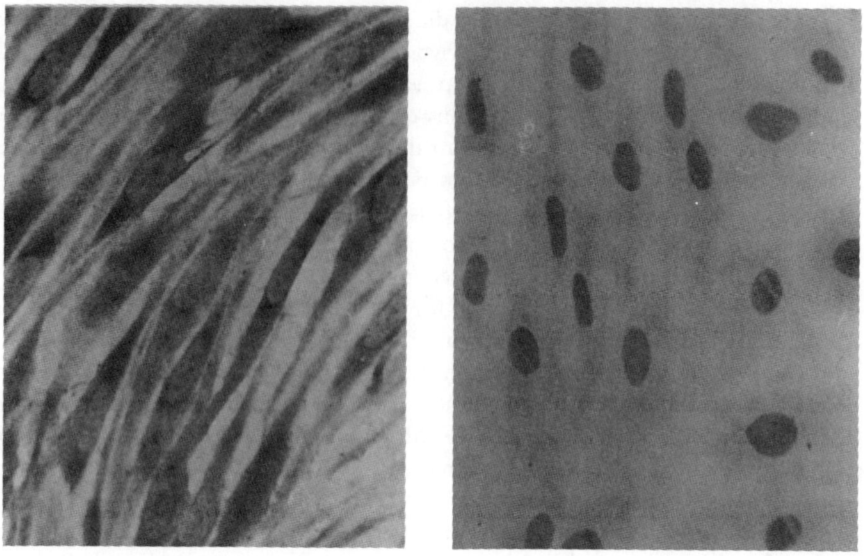

Figure 6. Immunostaining of primary cultured BCEC by the ABC method with the use of hematoxylin counterstain. Left panel: BCEC immunostained with monoclonal antibody MRK16 at 10 µg/mL (×560). Right panel: BCEC immunostained with normal mouse serum as the negative control (×560). (Reproduced with permission from reference 69. Copyright 1992 Pergamon.)

of renal proximal tubules, and apical surface of columnar epithelial cells. Accordingly, we have assumed that *P*-glycoprotein may act as an active efflux pump to reduce the BBB permeability of these lipophilic drugs.

As shown in Figure 6, the primary cultured BCEC were stained with anti-*P*-glycoprotein antibody MRK16 indicating the expression of *P*-glycoprotein in these primary cultured cells (*69*). Moreover, Figure 7 shows the immunoelectron micrograph of the primary cultured BCEC stained with MRK16, demonstrating that *P*-glycoprotein is localized at the apical membrane of the cells. As the character of apical membrane of the primary cultured BCEC was suggested to correspond to the character of the luminal membrane of BCEC (*40*), the expression of *P*-glycoprotein in the BCEC could be attributed to the expression at the luminal membrane of in vivo BCEC. This in vitro evidence is in good agreement with the reported results obtained in the normal human brain (*70*). To characterize the *P*-glycoprotein function of the primary cultured BCEC, intracellular uptake and efflux studies were performed. Figure 8 shows the effect of 10 µM verapamil, a typical MDR-reversing agent, on the time course of the residual amount of VCR. The descending curve represents the release of VCR from the cells, demonstrating that preloading of verapamil caused a significant reduction of the efflux rate of VCR (Figure 8). These effects of verapamil are attributed to specific inhibition of active efflux by *P*-glycoprotein at the apical membrane of the primary cultured BCEC. Table IV shows the effect of metabolic inhibitors on the steady-state cell-to-medium (C/M) ratio of VCR. The C/M ratio of VCR was significantly increased in the presence of sodium azide or 2,4-DNP in the absence of D-glucose (Table IV). Moreover, by replacing D-glucose with 3-O-methylglucose (3-OMG), an unmetabolizable sugar analogue, the C/M ratio of VCR was increased in the presence of 2,4-DNP (Table IV). However, the addition of 2,4-DNP in the presence of D-glucose failed to increase the steady-state uptake of VCR; this result could be attributed to the significant utilization of intracellular glucose to generate sufficient ATP in the BCEC.

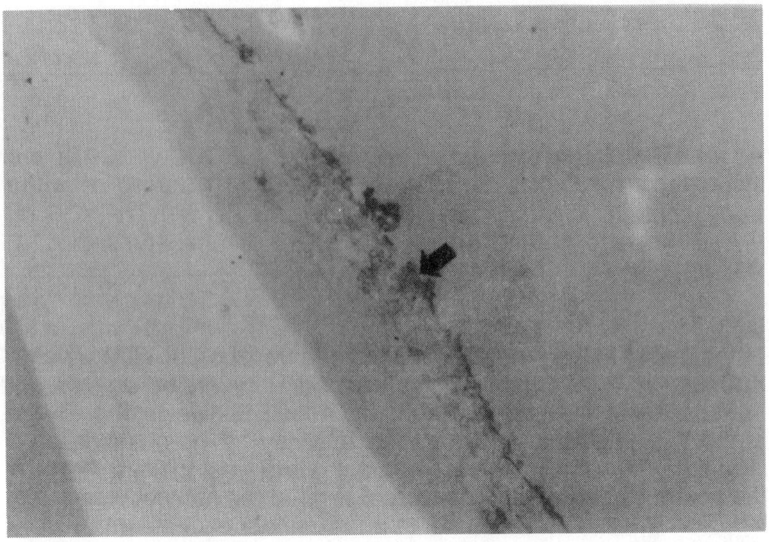

Figure 7. Electron micrograph of primary cultured BCEC immunodetected by the ABC method using MRK16 (10 μg/mL). The arrow indicates the apical side of the monolayer of primary cultured BCEC (×25,000). (Reproduced with permission from reference 69. Copyright 1992 Pergamon.)

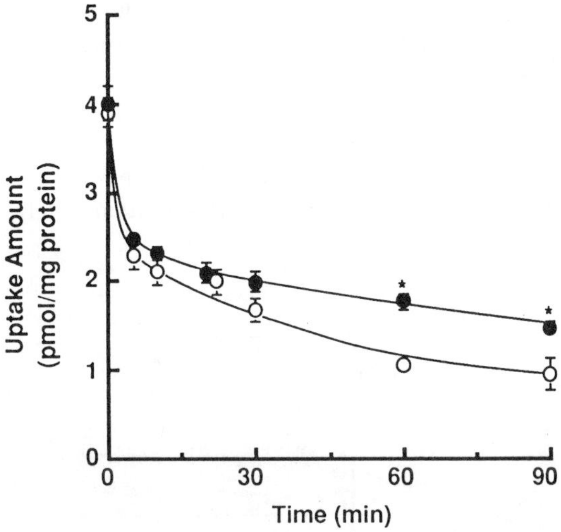

Figure 8. Effect of verapamil on the efflux of VCR from primary cultured BCEC. Cells were incubated with [^{3}H]VCR (190 nM) at 37 °C for 60 min, and the efflux was measured in the presence (●) or absence of 10 μM verapamil (○) as the control. Each point represents the mean ± SE of four experiments. Key: (*) P < 0.05 vs. control in Student t test. (Reproduced with permission from reference 69. Copyright 1992 Pergamon.)

Table IV. Effect of Metabolic Inhibitors on the Uptake of VCR by Primary Cultured BCEC

Metabolic Inhibitor	*Sugar*	*Percent of Control*
Control	D-Glucose	100
2,4-DNP	—[a]	145 ± 5[b]
Sodium azide	—[a]	139 ± 11[b]
2,4-DNP	3-OMG	165 ± 6[b]
—[a]	—[a]	96 ± 7
2,4-DNP	D-Glucose	102 ± 4

NOTE: Cells were preincubated for 30 min at 37 °C with each metabolic inhibitor in the presence or absence of sugar, and then the uptake of [^{3}H]VCR (190 nM) was measured at 37 °C for 60 min in the presence or absence of metabolic inhibitor 2,4-DNP (1 mM), sodium azide (10 mM), or sugar (10 mM). Each value represents the percentage of control value (mean ± SE of four experiments).

[a] —, None used.

[b] $P < 0.05$ vs. control by Student *t* test.

SOURCE: Reproduced with permission from reference 69.

These enhancement effects of metabolic inhibitors on the steady-state C/M ratio of VCR also support the contention that an energy-dependent efflux system is functioning in the BCEC, which is consistent with the function of *P*-glycoprotein in MDR tumor cells (*71*). Interestingly, the steady-state C/M ratio of VCR was significantly increased by MRK16, a monoclonal antibody to *P*-glycoprotein (Figure 9). Because the epitope region of MRK16 is the extracellular domain of *P*-glycoprotein, this evidence strongly supports the contention that this specific active efflux of VCR is attributable to the function of *P*-glycoprotein at the apical membrane of primary cultured BCEC. During the course of these studies, similar results were reported with a mouse brain capillary endothelial cell line (*72*).

In contrast to typical peptides, CsA, a cyclic endecapeptide, has a significantly high octanol–water partition coefficient (log $P = 3.0$ at pH 7.4) because of its unique structure that consists of *N*-methyl aliphatic amino acids that have intramolecular hydrogen bondings and that lack terminal carboxyl and amino groups. Interestingly, CsA distributes into several

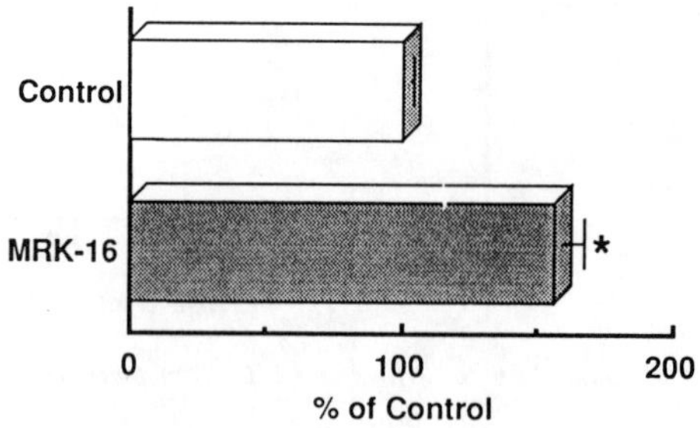

Figure 9. Effect of MRK16 on the uptake of VCR by primary cultured BCEC. Cells were preincubated at 4 °C for 30 min with MRK16 (10 µg/mL) or without antibody (control), and incubated at 37 °C for 60 min with [^{3}H]VCR (190 nM). Each bar represents the mean ± SE of four experiments. Key: (*) P *$<$ 0.05 vs. control in Student* t *test. (Reproduced with permission from reference 69. Copyright 1992 Pergamon.)*

organs after oral administration and is effective as an immunosuppressor, whereas the BBB permeability of CsA is strictly limited (*63*). CsA competitively inhibits the binding of Vinca alkaloid to *P*-glycoprotein of MDR tumor cells (*66*), similar to results with VCR (*69*), so it is possible to assume that CsA is pumped out by the function of *P*-glycoprotein at the BBB. The steady-state C/M ratio of CsA (Table V) was increased threefold in the presence of verapamil (*73*). Moreover, the steady-state C/M ratio of CsA was significantly increased in the presence of several kinds of MDR-reversing agents, such as quinidine, chlorpromazine, testosterone, progesterone, and VCR (Table V). Similar to the study with VCR, the energy requirement of the efflux pump was also examined. Addition of 2,4-DNP and replacement of 2-deoxyglucose by D-glucose in the uptake medium significantly decreased the steady-state C/M ratio of CsA, supporting the idea that CsA is also pumped out by the energy-dependent efflux system in the BCEC (Table V). Moreover, the C/M ratio of CsA was significantly increased by MRK16, whereas no effect of mouse IgG was observed (Table V). These results demonstrate that *P*-glycoprotein acts as an active efflux pump for CsA at the apical membrane of the BCEC (*73*). It was reported, using Western blotting technique, that the in vivo expression of *P*-glycoprotein at the brain capillary endothelial membrane was significantly higher than that of in vitro primary cultured BCEC (*61*). Accordingly, the efflux rate of *P*-glycoprotein at the in vivo BBB would be significantly greater than that observed in in vitro cultured BCEC. Intensive studies of these in vivo functions of *P*-glycoprotein are in progress. Because significant differences in carrier protein expression between in vivo and in vitro conditions are also known to be present in the case of BCEC, some kind of trophic factors may be released from astrocytes and pericytes to regulate the expression of membrane proteins in the BCEC. Although we can not completely

Table V. Effect of Various Agents on the Uptake of CsA by Primary Cultured BCEC

Agents	*Concentration*	*Relative Uptake (% of control)*
Control	[a]	100
Verapamil	500 μM	326.3 ± 12.77[b]
Quinidine	500 μM	241.0 ± 2.08[b]
Chlorpromazine	100 μM	138.1 ± 4.93[b]
Testosterone	100 μM	227.2 ± 9.23[b]
Progesterone	100 μM	152.4 ± 14.68[b]
VCR	100 μM	226.1 ± 12.77[b]
Vinblastine	100 μM	165.5 ± 7.22[b]
2,4-DNP	1 mM	399.0 ± 22.78[b]
+ 2-Deoxyglucose	10 mM	[c]
MRK16	10 μg/mL[d]	344.0 ± 13.46[b]
Mouse IgG	10 μg/mL[d]	125.3 ± 21.56[b]

NOTE: After cells were pretreated at 37 °C for 30 min with each compound, the uptake of [^{3}H]CsA (90 nM) was measured at 37 °C for 60 min. Each value represents the percentage of control value (mean ± SE of three to six experiments).
[a] None.
[b] $P < 0.05$ vs. control by Student *t* test.
[c] No data given.
[d] Pretreated at 4 °C for 30 min, and then the uptake was measured at 37 °C for 60 min.
SOURCE: Reproduced with permission from reference 73. Copyright 1993 Pergamon.

rule out the molecular threshold hypothesis at the present time, this evidence strongly supports that *P*-glycoprotein plays a dominant role in restricting the apparent permeability of CsA and MDR-reversing agents (*73*). Only limited information is available on the structural requirement of *P*-glycoprotein needed to transport drugs from intracellular fluid to the luminal side of brain capillaries. Chemical modification to decrease the affinity to *P*-glycoprotein would be an interesting approach to increase the BBB permeability of CNS-acting drugs, including peptide drugs with extremely low permeability at the BBB.

Conclusions

Undoubtedly, the most difficult organ to deliver physiologically active peptide drugs to is the brain, because of its strong defense system attributed to the BBB. In this chapter, we show a strong possibility that oligopeptide drug delivery utilizing a specific transport system will become one of the most successful ways to overcome this difficulty. By chemical modification of physiologically active peptides that are very unstable and that have limited BBB transport (i.e., fragmentation of active region, increasing enzymatic stability and transport activity through the BBB), we could develop several neuropharmaceutical peptides that will be effective after systemic administration. To provide a much more rational strategy to design CNS-acting peptide drugs, further intensive studies on the structure–transport relationship of the BBB transport system are necessary. Our successful results with E-2078 and ebiratide will provide new insights into oligopeptide drug delivery to the brain. Introduction of basic amino acids, addition of amino groups, and blocking of negatively charged moieties to the fundamental structure of native peptides would be the key to such modification. Several novel experimental techniques developed for BBB transport studies and techniques of molecular biology will make remarkable contributions to this new research field.

Acknowledgments

We greatly appreciate the valuable discussions of Drs. Y. Sugiyama and H. Suzuki, University of Tokyo.

References

1. Owman, C. In *Peptidergic Mechanisms in the Cerebral Circulation;* Evinsson, L.; McCulloch, J., Eds.; Ellis Horwood Publishers: London, 1987; pp 191–213.
2. Hock, F. J.; Gerhards, H. J.; Wiemer, G.; Usinger, P.; Geiger, R. *Peptides* **1988,** *9,* 575–581.
3. Nakazawa, T.; Furuya, Y.; Kaneko, T.; Yamatsu, K.; Yoshino, H.; Tachibana, S. *J. Pharmacol. Exp. Ther.* **1990,** *252,* 1247–1254.
4. Triguero, D.; Buciak, J. L.; Pardridge, W. M. *J. Pharmacol. Exp. Ther.* **1991,** *258,* 186–1192.
5. Kumagai, A. K.; Eisenber, J. B.; Pardridge, W. M. *J. Biol. Chem.* **1987,** *262,* 15214–15219.

6. Friden, P. M.; Walus, L. R.; Watson, P.; Doctrow, S. R.; Kozarich, J. W.; Backman, C.; Bergman, H.; Hoffer, B.; Bloom, F.; Granholm, A. C. *Science (Washington, DC)* **1993,** *259,* 373–377.
7. Culver, K. W.; Ram, Z.; Wallbridge, S.; Ishii, H.; Oldfield, E. H.; Blaese, M. *Science (Washington, DC)* **1992,** *256,* 1550–1552.
8. Goldstein, G. W.; Betz, A. L. *Sci. Am.* **1986,** *255,* 74–83.
9. Pardridge, W. M. *Physiol. Rev.* **1983,** *63,* 1481–1535.
10. Pardridge, W. M. *Endocrinol. Rev.* **1986,** 7, 314–330.
11. Fishman, J. B.; Rubin, J. B.; Handrahan, J. V.; Connor, J. R.; Fine, R. E. *J. Neurosci. Res.* **1987,** *18,* 299–304.
12. Ziegler, K.; Lins, W.; Frimmer, M. *Biochim. Biophys. Acta* **1991,** *1061,* 287–296.
13. Spector, R.; Johanson C. E. *Sci. Am.* **1989,** *261,* 48–53.
14. Cornford, E. M.; Oldendorf, W. H. *Biochim. Biophys. Acta* **1975,** *394,* 211–219.
15. Klecker, R. W., Jr.; Collins, J. M.; Yarchoan, R.; Jenkins, J. F.; Broder, S.; Myers, C. E. *Clin. Pharmacol. Ther.* **1987,** *41,* 407–412.
16. Terasaki, T.; Pardridge, W. M. *J. Infect. Dis.* **1988,** *158,* 630–632.
17. Spector, R.; Berlinger, W. G. *J. Neurochem.* **1982,** *39,* 837–841.
18. Suzuki, H.; Sawada, Y.; Sugiyama, Y.; Iga, T.; Hanano, M. *J. Pharmacol. Exp. Ther.* **1987,** *243,* 1147–1152.
19. Banks, W. A.; Kastin, A. J. *Am. J. Physiol.* **1990,** *259,* E1–E10.
20. Banks, W. A.; Kastin, A. J. Barrera, C. M. *Pharm. Res.* **1991,** *8,* 1345–1350.
21. Banks, W. A.; Kastin, A. J. *Brain Res. Bull.* **1985,** *15,* 287–292.
22. Banks, W. A.; Kastin, A. J.; Coy, D. H. *Brain Res.* **1984,** *301,* 201–207.
23. van Bree, J. B. M. M.; de Boer, A. G.; Verhoef, J. C.; Danhof, M.; Breimer, D. D. *J. Pharmacol. Exp. Ther.* **1989,** *249,* 901–905.
24. Durham, D. A.; Banks, W. A.; Kastin, A. J. *Neuroendocrinology* **1991,** *53,* 447–452.
25. Kannan, R.; Kuhlenkamp, J. F.; Jeandidier, E.; Trinh, H.; Ookhtens, M.; Kaplowitz, N. *J. Clin. Invest.* **1990,** *85,* 2009–2013.
26. Pardridge, W. M.; Triguero, D.; Buciak, J.; Yang. J. *J. Pharmacol. Exp. Ther.* **1990,** *255,* 893–899.
27. Yoshino, H.; Nakazawa, T.; Arakawa, Y.; Kaneko, T.; Tsuchiya, Y.; Matsunaga, M.; Araki, S.; Ikeda, M.; Yamatsu, K.; Tachibana, S. *J. Med. Chem.* **1990,** *33,* 206–212.
28. Terasaki, T.; Hirai, K.; Sato, H.; Kang, Y.; Tsuji, A. *J. Pharmacol. Exp. Ther.* **1989,** *251,* 351–357.
29. Terasaki, T.; Deguchi, Y.; Sato, H.; Hirai, K.; Tsuji, A. *Pharm. Res.* **1991,** *8,* 815–820.
30. Knutson, V. P.; Ronnett, G. V.; Lane, M. D. *J. Biol. Chem.* **1983,** *258,* 12139–12142.
31. Pardridge, W. M.; Triguero, D.; Buciak, J. *J. Pharmacol. Exp. Ther.* **1989,** *251,* 821–826.
32. Raub, T.; Audus, K. L. *J. Cell Sci.* **1990,** 97, 127–138.
33. Griffiths, G.; Simons, K. *Science (Washington, DC)* **1986,** *234,* 438–443.
34. van Deurs, B.; Sandvig, K.; Petersen, O. O.; Olsnes, S.; Simons, K.; Griffiths, G. *J. Cell Biol.* **1988,** *106,* 253–267.
35. Reichner, J. S.; Whiteheart, S. W.; Hart, G. W. *J. Biol. Chem.* **1988,** *263,* 16316–16326.
36. Brandli, A. W.; Simons, K. *EMBO J.* **1989,** *8,* 3207–3213.
37. Ungerstedt, U. In *Measurement of Neurotransmitter Release In Vivo;* Marsden, C. A., Ed.; John Wiley & Sons: New York, 1984; pp 81–105.
38. Shimura, T.; Tabata, S.; Terasaki, T.; Deguchi, Y.; Tsuji, A. *J. Pharm. Pharmacol.* **1992,** *44,* 583–588.
39. Triguero, D.; Buciak, J.; Pardridge, W. M. *J. Neurochem.* **1990,** *54,* 1882–1888.

40. Audus, K. L.; Bartel, R. L.; Hidalgo, I. J.; Borchardt, R. T. *Pharm. Res.* **1990,** *7,* 435–451.

41. Terasaki, T.; Takakuwa S.; Moritani, S.; Tsuji, A. *J. Pharmacol. Exp. Ther.* **1991,** *258,* 932–937.

42. Terasaki, T.; Takakuwa, S.; Saheki, A.; Moritani, S.; Shimura, T.; Tabata, S.; Tsuji, A. *Pharm. Res.* **1992,** *9,* 529–534.

43. Applezweig, M. H.; Braudry, F. D. *Psychol. Rep.* **1955,** *1,* 417–420.

44. de Wied, D. *Am. J. Physiol.* **1964,** *207,* 255–259.

45. Levine, S.; Jones, L. E. *J. Comp. Physiol. Psychol.* **1965,** *59,* 357–360.

46. Matsumoto, T.; Oshima, K.; Miyamoto, A.; Sakurai, M.; Gotoh, M.; Hayashi, S. *Neurochem. Res.* **1989,** *14,* 778.

47. Wiemer, G.; Gerhards, H. J.; Hock, F. J.; Usinger, P.; von Rechenberg, W.; Geiger, R. *Peptides* **1988,** *9,* 1081–1087.

48. Shimura, T.; Tabata, S.; Ohnishi, T.; Terasaki, T.; Tsuji, A. *J. Pharmacol. Exp. Ther.* **1991,** *258,* 459–465.

49. Haigler, H. T.; Maxfield, F. R.; Willingham, M. C.; Patan, I. *J. Biol. Chem.* **1980,** *255,* 1239–1241.

50. Kang, Y.; Terasaki, T.; Tsuji, A. *J. Pharmacobio-Dyn.* **1990,** *13,* 10–19.

51. Benveniste, H.; Diemer, N. H. *Acta Neuropathol. (Berl.)* **1987,** *74,* 234–238.

52. Lonnroth, P.; Jansson, P.-A.; Smith, U. *Am. J. Physiol.* **1987,** *253,* E228–E231.

53. Lindefors, N.; Amberg, G.; Ungerstedt, U. *J. Pharmacol. Methods* **1989,** *22,* 141–156.

54. Amberg, G.; Lindefors, N. *J. Pharmacol. Methods* **1989,** *22,* 157–183.

55. Benveniste, H. *J. Neurochem.* **1989,** *52,* 1667–1679.

56. Bungay, P. M.; Morrison, P. F.; Dedrick, R. L. *Life Sci.* **1990,** *46,* 105–119.

57. Stahle, L.; Segersvard, S.; Ungerstedt, U. *J. Pharmacol. Methods* **1991,** *25,* 41–52.

58. Deguchi, Y.; Terasaki, T.; Kawasaki, S.; Tsuji, A. *J. Pharmacobio-Dyn.* **1991,** *14,* 483–492.

59. Deguchi, Y.; Terasaki, T.; Yamada, H.; Tsuji, A. *J. Pharmacobio-Dyn.* **1991,** *15,* 79–89.

60. Terasaki, T.; Deguchi, Y.; Kasama, Y.; Pardridge, W. M.; Tsuji, A. *Int. J. Pharm.* **1992,** *81,* 143–152.

61. Pardrigde, W. M. *Peptide Drug Delivery to the Brain;* Raven: New York, 1991.

62. Cornford, E. M. *Mol. Physiol.* **1985,** *7,* 219–260.

63. Cefalu, W. T.; Pardridge, W. M. *J. Neurochem.* **1985,** *45,* 1954–1956.

64. Levin, V. A. *J. Med. Chem.* **1980,** *23,* 682–684.

65. Trauble, H. *J. Membr. Biol.* **1971,** *4,* 193–208.

66. Tamai, I.; Safa, A. R. *J. Biol. Chem.* **1990,** *265,* 16509–16513.

67. Thiebaut, F.; Tsuruo, T.; Hamada, H.; Gottesman, M. M.; Pastan, I.; Willingham, M. C. *Proc. Natl. Acad. Sci. U.S.A.* **1987,** *84,* 7735–7738.

68. Cordon-Cardo, C.; O'Brien, J. P.; Boccia, J.; Casals, D.; Bertino, J. R.; Melamed, M. R. *J. Histochem. Cytochem.* **1990,** *38,* 1277–1287.

69. Tsuji, A.; Terasaki, T.; Takabatake, Y.; Tenda, Y.; Tamai, I.; Yamashima, T.; Moritano, S.; Tsuruo, T.; Yamashita, J. *Life Sci.* **1992,** *51,* 1427–1437.

70. Sugawara, I.; Hamada, H.; Tsuruo, T.; Mori, S. *Jpn. J. Cancer Res.* **1990,** *81,* 727–730.

71. Juranka, P. F.; Zastawny, R. L.; Ling, V. *FASEB J.* **1989,** *3,* 2583–2593.

72. Tatsuta, T.; Naito, M.; Oh-hara, T.; Sugawara, I.; Tsuruo, T. *J. Biol. Chem.* **1992,** *267,* 20383–20391.

73. Tsuji, A.; Tamai, I.; Sakata, A.; Tenda, Y.; Terasaki, T. *Biochem. Pharmacol.* **1993,** *46,* 1096–1099.

RECEIVED for review September 9, 1993. ACCEPTED revised manuscript December 27, 1993.

CHAPTER 14

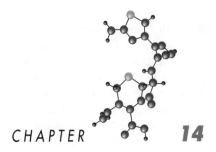

Molecular Packaging

Peptide Delivery to the Central Nervous System by Sequential Metabolism

Nicholas Bodor
and Laszlo Prokai

Intracellular or transcellular transport (i.e., directly through the endothelial cell membrane) is the principal route into and out of the central nervous system (CNS) (*1*) because of the blood–brain barrier (BBB). As a result, the BBB exhibits a low permeability to hydrophilic substances, such as small ions and polar compounds, that do not have specific transport mechanisms. Various enzymes are also present in the BBB (*2*), and metabolically unstable substances may be rapidly degraded before they can reach the brain tissue.

Transport of peptide molecules across the BBB cannot be ruled out, but it is unlikely that endogenous peptides pass the BBB in physiologically significant amounts. Within the circumventricular organs, the peptide molecules actually reach the cellular elements of the tissue. However, no evidence of penetration to deeper layers exists. Most of the naturally occurring neuropeptides are hydrophilic and thus, do not cross the BBB in the absence of a specific transport system in the BBB. Carrier-mediated transport of several dipeptides and tripeptides occurs in the brain (*3*). Certain larger peptides (IGF-I, IGF-II, insulin, transferrin, and cationized albumin) have receptors on the BBB. These receptors are identified on the luminal surface of the brain capillaries and are believed to act as transcytosis systems because they are expected to be present on both luminal and antiluminal borders (*4*). Recent studies (*5*) indicated that Leu-enkephalin is also taken up intact at the luminal side of the BBB. However, these studies do not rule out the possibility that Leu-enkaphalin may be metabolized during the next steps in one of the compartments in parallel. These compartments may involve cytosolic endothelial space, luminal surface of the BBB, glial-end foot layer in apposition with the antiluminal side of the capillary endothelium, and enkephalinergic synaptic regions juxtaposed to the brain microvessels (*6*). The crucial characteristic of the active (carrier-mediated) transport systems is, however, their saturability.

The controversy as to whether peptides are able to cross the BBB also originated in their metabolic instability. BBB enzymes recognize and rapidly degrade most naturally occurring neuropeptides (*2, 7*). Metabolically stable peptide analogues may exhibit increased resistance to certain peptidases, but overall protection against the variety of neuropeptide-degrading enzymes is generally not possible until the peptide nature of the molecule is changed.

Various strategies are available for directing centrally active peptides into the brain (*4*). The strategies can be grouped into three categories: invasive procedures, physiological-based strategies, and pharmacological-based strategies. Invasive strategies include implantation of an intraventricular catheter, followed by pharmaceutical infusion into the ventricular compartment. Because of the deep convolutions of the human brain surface, most regions of the brain are 0.5–1 cm away from the ependymal surface or from the cortical surface of the cerebrospinal fluid (CSF) compartment. Taking an average diffusion coefficient (2×10^{-6} cm^2 s^{-1}) for a peptide of 2000–5000 molecular weight, the effective diffusion distance of the substance is substantially <1 mm, if the degradation rate is rapid ($t_{1/2} < 10$ min). Therefore, intrathecal administration of peptides would be expected to allow for delivery of the compound only to the surface of the brain (*8*).

Another invasive method relies on the effect of intracarotid infusion of high concentration (>1 M) of osmotically active substances, such as mannitol or arabinose. Their high local concentration causes shrinkages of the brain capillary endothelial cells, resulting in a transient opening of the tight junctions (*9*) that may facilitate the transport of molecules that otherwise cannot cross the BBB. The considerable toxic effects of the procedure should be taken into account; these toxic effects can lead to inflammation, encephalitis, and so on, and to the incidence of seizures (as frequently as in 20% of the applications). However, an indiscriminate delivery occurs. In conclusion, invasive procedures are only justified for some life-threatening conditions, but these surgical routes are not acceptable for less dramatic illnesses.

The physiological-based strategies involve the formation of chimeric peptides (*10*). It has been assumed that specific peptide receptor transcytosis systems exist in the BBB for various longer peptides. This scheme relies on covalently coupling (e.g., via disulfide bond) the peptide, such as β-endorphin (which is not normally transported through the BBB), to insulin, immunoglobulin F-II (IGF-II), or transferrin that would undergo receptor-mediated transcytosis (transport vectors). These so-called chimeric peptides may then be transported through the BBB by way of the receptors for the transportable peptide or protein (i.e., insulin, IGF-II, or transferrin). After gaining access to brain interstitial space, the active peptide (β-endorphin) is released and it may then interact with the corresponding receptor to initiate pharmacological action in the brain. Claims of brain delivery of β-endorphin coupled to insulin (*11*) have occurred, but the active form of the peptide could not be detected in the CNS. It appears, however, that cleavage of the active neuropeptide from the chimeric peptide is necessary, and that the coupling chemistry is needed to be perfected to allow postdelivery release and pharmacological action. It has also been claimed that cationized albumins and cationized liposomes carry peptides to the brain (*7*), but no pharmacological effect has been reported. Recently, antibodies to brain capillary endothelial cells also were suggested as brain-specific transport vectors (*7*). However, some serious drawbacks may arise from the physiologically based approach. The poor stoichiometry of the neuropeptide to the carrier molecule limits the mass transport of the target peptide (a 500 to 4000 Da peptide is loaded onto a 5000 to >100,000 Da carrier molecule). In addition, the carrier- or receptor-mediated cellular transport has physiologically limited transported capacity (saturability) that also prevents pharmacologically significant amounts from entering the brain. Also, these carriers are not brain specific, as indicated by observations of uptake by nonneural cells or by cells outside the CNS (*12*). Finally, release of the active peptide from the conjugates has not been documented.

Hydrophobic substances, because of their high lipid solubility, can generally diffuse freely across the BBB. The objective of the pharmacological-based strategies is to turn the water-soluble substance into lipid-soluble ones. Encapsulating the peptide in liposomes, which were shown to be taken up by cells lining the reticuloendothelial system of liver and spleen, has resulted in no measurable transport across the BBB (*13*). Peptide latentiation (by forming cyclic derivatives or diketopiperazines), which increases lipid solubility, has been applicable only to small peptides such as the C-terminal dipeptide of the thyrotropin releasing hormone (TRH) (*14*). However, these modifications on the peptide structure may not be successful for larger molecules. Cyclosporin, a cyclic endecapeptide of fungal origin, which has no free carboxy and amino termini, no charged groups arising from acidic or basic amino acid residues, and which has four amide nitrogens methylated, is one of the most lipid-soluble natural compounds. However, transport of cyclosporin across the BBB is paradoxically low (*15*). This, however, can be explained by the observed peptide degradation (*16*). Again, the reason for this cleavage is the recognition of the substance as a peptide ("enzymatic BBB"). A recent report (*17*) on the penetration of cyclosporin to the brain also pointed out that this lipid-soluble peptide is transported in the blood in combination with erythrocytes, leukocytes, and plasma proteins, all of which account for >90% of the concentration of the drug in the blood. Accordingly, the concentration of the free drug in the plasma is greatly limited. Because the carriers for cyclosporin are unable to cross the BBB, the drug has apparently no vehicle to transport it into the cerebral compartment. The exception is the choroid plexus, where no such restriction to the contact of the cyclosporin carriers with cells exists. Generally, lipid-soluble compounds that are able to cross the BBB can maintain active concentrations in the CNS only if their blood concentrations are maintained at adequately high levels.

Another factor that may restrict BBB transport of large lipophilic molecules is size exclusion (*18*). The cyclic peptides cyclosporin and ocreotide, which are both relatively rigid molecules due to their specific structure and the intramolecular hydrogen bonding, are transported across the BBB at rates that are at least two orders less than predicted by their partition coefficients, even after correction for the molecular weight. The controversy surrounding the transportability of cyclosporin to the brain, as discussed in the previous paragraph, suggests that the steric hindrance strictly preventing drugs with molecular weight >1000 Da from penetrating the BBB is unfounded. The size exclusion related to the transport through lipid membranes must be associated primarily with the molecular volume determined by the actual geometric size of the molecule, the overall conformation, and the heteroatom content. For flexible molecules, the movement across the membranes should actually be assisted by the thermal fluctuation of the membrane lipid (*19*), even if the molecular weight exceeds 1000 Da.

Our approach is an enzyme-based strategy, and is distinct from a simple pharmacological-based approach in which a lipophilic peptide prodrug is applied (*20, 21*). Although the acquired lipophilicity of these prodrugs may assure penetration to the BBB (and to other membranes), this is not the sole factor in the transportability of peptides in the CNS. The enzymatic degradation, and the consequent attenuation or loss of biological activity, should also be prevented during the passage of the substance from the general circulation to the brain tissue. In the strategy we call "molecular packaging", the peptide unit in the delivery system appears as a perturbation on the bulky molecule dominated by lipophilic modifying groups that direct BBB penetration, and they also prevent it from recognition by peptidases. A specific functional group attached to the target peptide should also provide retention in the CNS, and controlled release of the biologically active substance is achieved by predictable sequential metabolism (*22*).

Molecular Packaging: A Chemical Delivery System Approach

A chemical delivery system (CDS) is defined as a biologically inert molecule that requires several steps in its conversion to the active drug and that enhances drug delivery to a particular organ or site. [The difference from prodrugs (*23, 24*) usually require a single activating step and, thus, generally lack site-specificity.] In designing a CDS for the CNS, the unique architecture of the BBB was exploited. As with a prodrug, a CDS should be sufficiently lipophilic to allow for brain uptake. Subsequent to this step, the molecule should undergo an enzymatic or other conversion to promote retention within the CNS but, at the same time, to accelerate peripheral elimination of the entity. Finally, the intermediate should be further metabolized and the active compound should be released in a sustained manner. One system that possesses these attributes is summarized in Scheme I.

The molecule (D) can be modified to provide increased lipophilicity through biolabile functional groups (F_1,...,F_n), which are susceptible to easy removal. The (dihydropyridine-type) targetor ($-T_D$), a specific functional group on the molecule, also enhances BBB pene-

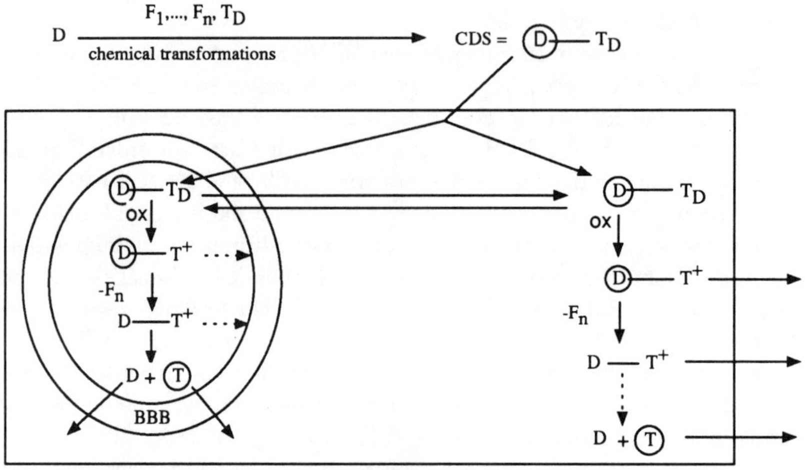

Scheme I. Brain targeting by the CDS approach.

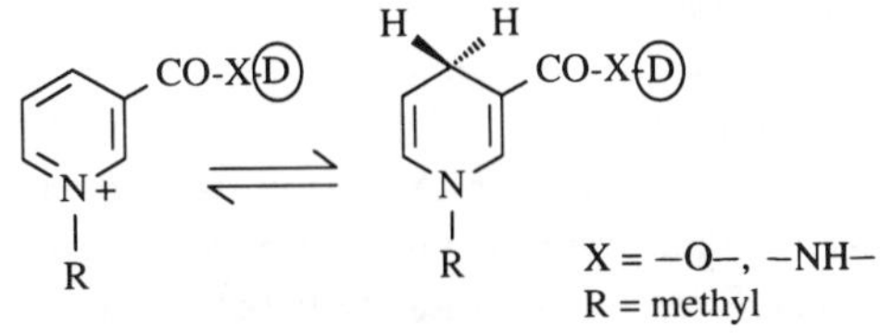

Scheme II. The trigonellinate ⇌ 1,4-dihydrotrigonellinate redox system.

tration and, most importantly, can be converted by enzymatic oxidation to a water-soluble, lipid-insoluble (quaternary pyridinium) salt ($-T^+$). Many moieties may serve such functions, but the trigonellinate ⇌ 1,4-dihydrotrigonellinate redox pairs (Scheme II) have proved to be the most useful. In this approach, a hydroxy, amino, or carboxylic acid containing drug is esterified, amidated, or otherwise covalently linked to nicotinic acid or nicotinic acid derivative. This compound is then quanternized to generate the 1-methylnicotinate salt or trigonellinate and chemically reduced to give the 1,4-dihydrotrigonellinate targetor. This dihydro moiety also enhances the lipophilicity of the drug to which it is attached. Following systemic administration, the CDS can partition into several body compartments because of its enhanced lipophilicity; some compartments are inaccessible to the unmanipulated compound. At this point, the CDS is simply working as a lipoidal prodrug. The targetor moiety is, however, specifically designed to undergo an enzymatically mediated oxidation that converts the membrane-permeable dihydrotrigonellinate to a hydrophilic, membrane-impermeable trigonellinate salt. This conversion occurs ubiquitously. The mechanism of this oxidation was extensively examined, and it was suggested to be analogous to the oxidation of NAD(P)H, a coenzyme associate with numerous oxidoreductases and cellular respiration (*25*). The now polar, oxidized targetor–drug conjugate is trapped behind the lipoidal BBB and, in essence, remains "locked-in" in the CNS. Any of the oxidized salt that is present in the periphery will be rapidly lost because it is now polar and is an excellent candidate for elimination by the kidney and bile. The conjugate that is trapped behind the BBB can then undergo biotransformation to give the active species in a slow and sustained manner. By the design of the system, concentrations of the active drug are very low in the periphery, thus reducing systemic, dose-related toxicities. In addition, the drug in the CNS is present mostly in the form of an inactive conjugate, thus reducing the potential for any central toxicities. This approach should allow for a more potent compound because a larger portion of the administered dose is delivered to its site of action. The redox targetor has proved to be widely applicable for brain targeting of a variety of substances (*26–28*), and its attachment alone results in brain-specific delivery for small molecules such as dopamine (*29*). For peptides, the NH_2-terminus of the molecule is available for coupling to the targetor to the molecule. However, the attachment of 1,4-dihydrotrigonellyl to the NH_2-terminus alone will not furnish sufficient increase in lipid solubility to the peptide and will only protect against aminopeptidases such as aminopeptidase M (EC 3.4.11.2). The unmodified COOH-terminal portion of the molecule will be susceptible to cleavage by numerous exo- and endopeptidases present in the BBB (*2*).

A bulky and lipophilic moiety (L) attached to the COOH-terminus of the peptide through an ester bond may substantially increase the lipid solubility and may also hinder this part of the molecule from being recognized by peptide-degrading enzymes. Cholesteryl esters of amino acids and dipeptides were prepared, and they were sufficiently stable chemically to be considered as a suitable protecting function (*30*). This part of the molecule is, however, labile toward esterase and lipase, which permits the molecule's removal after delivery.

The final step of the drug delivery scheme is the release of the biologically active peptide from the targetor–peptide conjugate. Attachment of the targetor directly to the N-terminal residue may not yield the target peptide because of the low amidase activity of the brain tissue. Therefore, a spacer function (S) separates the peptide sequence (P) to be delivered from the targetor part of the CDS with an additional amino acid residue or with a pair of residues. This portion of the molecule should be selected based on the peptidolytic activity prevalent at the site of the action, so that release of the desired peptide in this reaction is favored over degradation induced by other peptidases.

Altogether, the molecular architecture will conform to the desired attribute in that the peptide part (P) will appear only as a perturbation on a bulky, lipophilic molecule dominated by the lipophilic steroidal portion (L) and the targetor (T); this conformation is reflected by our reference to "molecular packaging". To date, delivery systems based on this concept were developed for two important classes of neuropeptides; the synthetic analogues of the opioid pentapeptide leucine enkephalin and of TRH.

Brain Targeting of Leucine Enkephalin Analogues

The endogenous opioids are part of an extensive neuronal network, involving multiple ligands and multiple receptors, in both the CNS and peripheral nervous system. The endorphins, enkephalins, and dynorphins are opioid peptides derived from the glycoprotein precursors proopiomelanocortin (POMC), proenkephalin, and prodynorphin, respectively. These substances exhibit complex pharmacology and exert diverse physiological effects in mammals. Studies have implicated endogenous opioid peptides in the regulation of seizure threshold (*31–34*), motor behavior (*35*), feeding and drinking (*36*), mental disorders (*34, 37, 38*), immune responses (*39, 40*), and regulation of gastrointestinal (*34, 41, 42*), cardiovascular (*43–45*), neuroendocrine (*46–49*), and cognitive (*34, 50*) functions. Analgesia is, however, the opioid peptides best-recognized central effect. This particular activity of the opioid peptides mostly was demonstrated only after central (intracerebroventricular) administration, but systemically administered compounds failed to exert any CNS effect. This failure is only partly due to the metabolic instability of the endogeneous opioid peptides, because most metabolically stable peptide analogues also lack analgesic activity following intravenous administration (*51*).

The controversy whether enkephalins can cross the BBB may have originated in their metabolic instability, both in plasma and brain tissue. Therefore, the "brain uptake index" (BUI) technique in this case is not sufficiently sensitive and specific to show the uptake of trace amounts, which involves a single intracarotid injection in the rat with analysis after 15 s. Although passage of naturally occurring opioid peptides from blood into the brain has been suggested from BUI studies, more recent data indicate a significant hydrolysis of Leu-enkephalin (*52*). The presence of high "arylamidase" activity in the BBB is why radiolabeling the tyrosine (Y) for the BUI technique yields misleading results, because this residue is usually cleaved first from the peptide molecule (*53*), is taken up rapidly by carrier-mediated transport, and shows up in the brain. Thus, the relationship between the analgesic doses after peripheral versus central administration was unfavorable, suggesting that the entry of these analogues into the brain is quite sluggish, and very low CSF–plasma ratios were determined (*54*). It should be recognized that generally the capillary-bound BBB aminopeptidase is very high (*55–57*) and possibly other "enkephalinases" are also present (*58*).

We have designed the "prototype" of the system for brain delivery of the analogue for Leu-enkephalin (designated Tyr-Gly-Gly-Phe-Leu or YGGFL by using the three- or one-letter abbreviations of the amino acid residues present in the peptide, respectively) based on several requirements. The target peptide is especially sensitive to cleavage and, thus, to deactivation by endopeptidases at the Tyr^1 end and at the Gly^3-Phe^4 position. Whereas the cleavage of the terminal Tyr is adequately hindered by incorporation of a D-amino acid, like D-Ala, at the second position, the architecture of the entire molecule should give protection of potential cleavage sites located closer to the carboxy-terminus. This is provided by the bulky, lipophilic steroidal moiety **(L)** as an ester function. Brain targeting is achieved by placing the 1,4-dihydrotrigonellinate targetor **(T)** at the N-terminus through an amide-type covalent bond. The design should also account for the release of the parent peptide from the CDS molecule. Esterase and lipase enzymes are capable of removing the cholesteryl part after (or during) penetration through the BBB. However, the amidase activity of the brain, which would remove the trigonellinate from the N-terminus, is expected to be low. Because of the involvement of alanine–aminopeptidase in enkephalinergic transmission in brain (*59*), we selected L-Ala as a "spacer" **(S)** to facilitate the cleavage of the enkephalin analogue from the expected locked-in molecule, T^+-AYAGFL [T^+ refers to trigonellyl (the oxidized form of the targetor) and the peptide sequence is given by the one-letter abbreviation of the amino acid residues (A is alanine, Y is tyrosine, G is glycine, F is phenylalanine, and L is leucine)]. Elaborate molecular modeling and calculations showed that, indeed, the peptide part is manifested just as a "perturbation" on the molecular structure.

Stability studies of the CDS (**1**) and its predicted biotransformation products have been performed in phosphate buffer, whole blood, and 20% (w/w) brain homogenate. The 1,4-dihydropyridine compound was very unstable, as expected, in biological fluid [half-life ($t_{1/2}$) is 1.28 min in rat blood and 4.03 min in rat brain homogenate]. The quaternary intermediates [T^+-AYAGFL cholesterate (**2**), and T^+-AYAGFL (**3**)] are stable in the pH range 4–9 for at least 6 h. However, **2** has a $t_{1/2}$ of 2.24 min in whole blood and 37.84 min in brain homogenate, whereas the $t_{1/2}$ of **3** is 101.7 and 31.26 min in blood and brain homogenate, respectively. These results are indicative of possible biotransformation ability in vivo. A receptor binding study (competition with [^{3}H]diprenorphine, a μ-receptor agonist) confirmed that **2** has very low affinity (the 50% inhibitory concentration, IC_{50}, is 5.6×10^{-6} M), whereas **3** has slight affinity ($IC_{50} = 2.0 \times 10^{-7}$ M) to opioid receptors compared with the parent peptide, [D-Ala^2]-Leu-enkephalin ($IC_{50} = 4.0 \times 10^{-8}$ M). Therefore, it can be concluded that only a "true chemical delivery" of the enkephalin analogue would result in profound biological response.

Preliminary in vivo distribution studies were performed with rats as experimental models. Because of the possible pitfalls of the BUI method, we developed analytical procedures that provide the necessary molecular specificity for detection of the species involved in the biotransformation processes. Electrospray ionization (ESI) and fast atom bombardment (FAB) mass spectrometry proved to be specific and sensitive methods for the analysis of thermally labile quaternary trigonelline compounds (*60*).

The CDS (**1**) was dissolved in ethanol/50% (w/w) hydroxypropyl-β-cyclodextrin (1:1) solution. After receiving intravenous (iv) injections of 5- or 20-mg/kg doses through the tail vein, rats were sacrificed by decapitation at appropriate time points (15 min, and 1, 2, and 4 h). Brain and blood samples were collected immediately. The brain samples were subsequently homogenized in methanol–dimethyl sulfoxide–acetic acid (94:5:1, v/v) solution. After centrifuging for 15 min at 12,000 rpm, the supernatant was removed and the solvent was evaporated under reduced pressure. Column chromatography (Supelclean

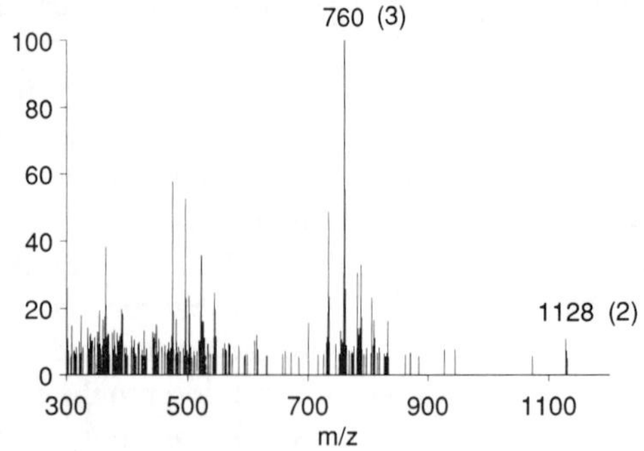

Figure 1. ESI mass spectrum of the brain sample 15 min after the administration of the [D-Ala^2]-enkephalin CDS (1). (Reproduced with permission from reference 22. Copyright 1992 American Association for the Advancement of Science.)

LC-18 cartridges, elution of poorly retained compounds with deionized water first) was performed, and sample was collected by elution with 50% methanol in 3% aqueous acetic acid solution. The effluent was directly analyzed by ESI mass spectrometry. The recorded ESI mass spectra (Figure 1) clearly showed the presence of an *m/z* 760 ion that is attributable to the presence of **3**, which was obviously absent from the analytical sample obtained from the brain of the animal treated with the vehicle solution only. We estimated that amount of the "locked-in" **3** to be 500–700 pmol/g of tissue 15 min after iv administration by comparing the absolute intensity of *m/z* 760 with that of a sample obtained from the brain of an undosed animal to which a known amount of **3** had been added. In tissue collected 1, 2, and 4 h after systemic administration, **2** could no longer be identified and the quantity of **3** was proportionally (with $t_{1/2}$ values of ~40–60 min) decreased with time. However, these compounds were not detectable in the blood withdrawn from the animals and pretreated in essentially the same manner as tissue samples to register ESI mass spectra.

One may speculate about the reasons for the low amount of **2** (*m/z* 1128), which should be the immediate precursor of **3**. Compound **3** has a large lipophilic substituent and, thus, its recovery from the tissue by the simple workup procedure may be low. Additionally, the technique may discriminate so that compounds with a bulky and lipophilic moiety may show low sensitivity. However, the design principles relied on the facile removal of the protective ester function by enzymatic hydrolysis, and thus a significant amount of **2** was not expected to be present.

Quantitative estimation of the studies (*22*) were validated by FAB mass spectrometry. The crude extract was passed through a column containing Dowex-SCX cation-exchange resin to remove inorganic contaminants interfering with the analysis, then concentrated under a stream of nitrogen gas. Ten microliters of the resultant sample was transferred, while gradually evaporating the remaining solvent by gentle heating with a stream of hot air, onto the FAB probe tip. A liquid matrix (3-nitrobenzyl alcohol) was added to dissolve the residue, and the FAB mass spectra were recorded using a 6-keV xenon beam directed onto the surface of the target. Quantification relied on a specific internal standard that was chemically identical to the compound to be quantified. The trideuteromethyl (C^2H_3) analogue of **3** was synthesized and added to the sample during homogenation in a known quantity. This internal standard has a distinctive *m/z* 763 ion due to the incorporation of the heavy isotopes, but no other differences occurred in physicochemical properties from **3**, which is detected at *m/z* 760. Hence, comparison of the signal intensity of the analyte versus internal standard gives accurate quantitative data. To increase reproducibility and detection limit, mass spectra were collected by multichannel signal averaging over a narrow mass range (*m/z* 750–770). The high concentration (~nmol/g tissue) and the sustained level (in vivo $t_{1/2}$ of ~40–60 min) of the locked-in species that contains the intact peptide segment are indications that the concept of molecular packaging is very efficient in overcoming the obstacles imposed by the BBB toward the enkephalin analogue studied. We compared spectra obtained after administration of the parent enkephalin and its cholesteryl ester in equimolar doses. No detectable amount of compounds with the peptide part intact was found. Indications that fragments of the cholesteryl-protected peptides penetrated the BBB occurred, although the peptides disappeared rapidly from the brain. This finding is, however, in accordance with the aminopeptidase, and possibly not with the endopeptidase, activity of the BBB that is able to sequentially remove the N-terminal amino acids of the molecule. Clearly, the targetor function is necessary to furnish stability and "lock-in" by virtue of the ionic trigonellinate moiety, because this moiety is formed by enzymatic oxidation within the brain. Additionally, the cholesteryl ester seems to be cleaved very rapidly after or during the transport through the BBB.

Pharmacological experiments related to the analgesic effect of the CDS (**1**) administration were performed in parallel with the distribution studies. The latency of the tail-flick response was measured for groups of animals injected with the vehicle solution, 10 mg/kg of [D-Ala^2]-Leu-enkephalin, and 5 and 20 mg/kg of CDS (**1**). The iv administration of **1** indeed showed a tendency of increased analgesic activity compared with the parent peptide. The dose dependence of **1** also corroborated this effect. However, the magnitude of the response remains much lower than would be expected from the amount of the locked-in quaternary peptide conjugate (**3**). Because the amount of **3** after delivery decreases profoundly, which cannot simply be attributed to the elimination from the CNS, cleavage of **3** must be involved.

We have obtained conclusive data from an in vitro experiment on the cleavage sites involved, and it appears that the expected Ala-Tyr^1 peptidolysis is not predominant. In the in vitro experiment, 1 mL of brain homogenate (20%, w/w, in pH 7.4 Tris buffer) spiked with 30 nmol of the targetor–peptide conjugate (**3**) was used. After incubation at 37 °C for 0.5 and 2.0 h, 250-μL aliquots were removed, and sample solutions for ESI mass spectrometry were obtained according to the procedure used for the in vivo experiment. The neutral endopeptidase EC 3.4.24.11 ("enkephalinase") is, as with the parent peptide, the major degrading enzyme resulting in an inactive T$^+$-AYAG species (*m/z* 500) due to the cleavage at the Gly^3-Phe^4 bond (*see* mass spectra in Figure 2). Carboxypeptidase is

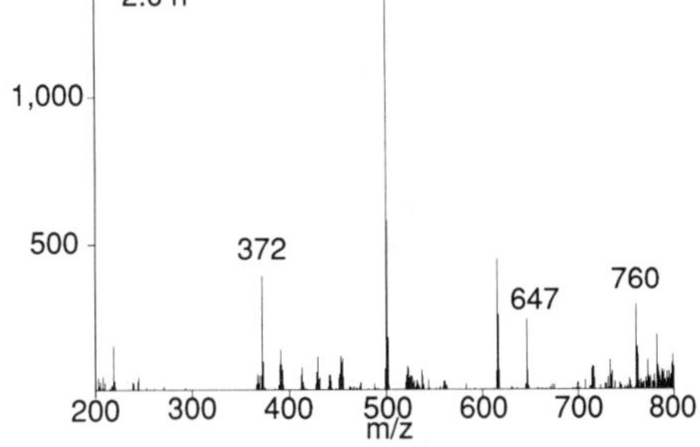

Figure 2. ESI mass spectrum showing the cleavage of the locked-in T$^+$-Ala-Tyr-[D-Ala]-Gly-Phe-Leu (m/z 760) in vitro *(brain homogenate; incubation at 37 °C for 2 h). (Reproduced with permission from reference 22. Copyright 1992 American Association for the Advancement of Science.)*

another notable deactivating enzyme that removes the C-terminal (L)-Leu[5] (T⁺-AYAGF at *m/z* 647). The substrate for the dipeptidyl peptidase (EC 3.4.14.5) is the primary cleavage product T⁺-AYAG, yielding T⁺-AY (*m/z* 372 in the ESI mass spectrum). We have been unable to identify [D-Ala[2]]-Leu-enkephalin among the cleavage products of **3**, although a small amount may be released from the targetor–peptide conjugate in the CNS in vivo, resulting in the slight analgesia that was measured. The molecular mechanism of brain targeting is summarized in Scheme III.

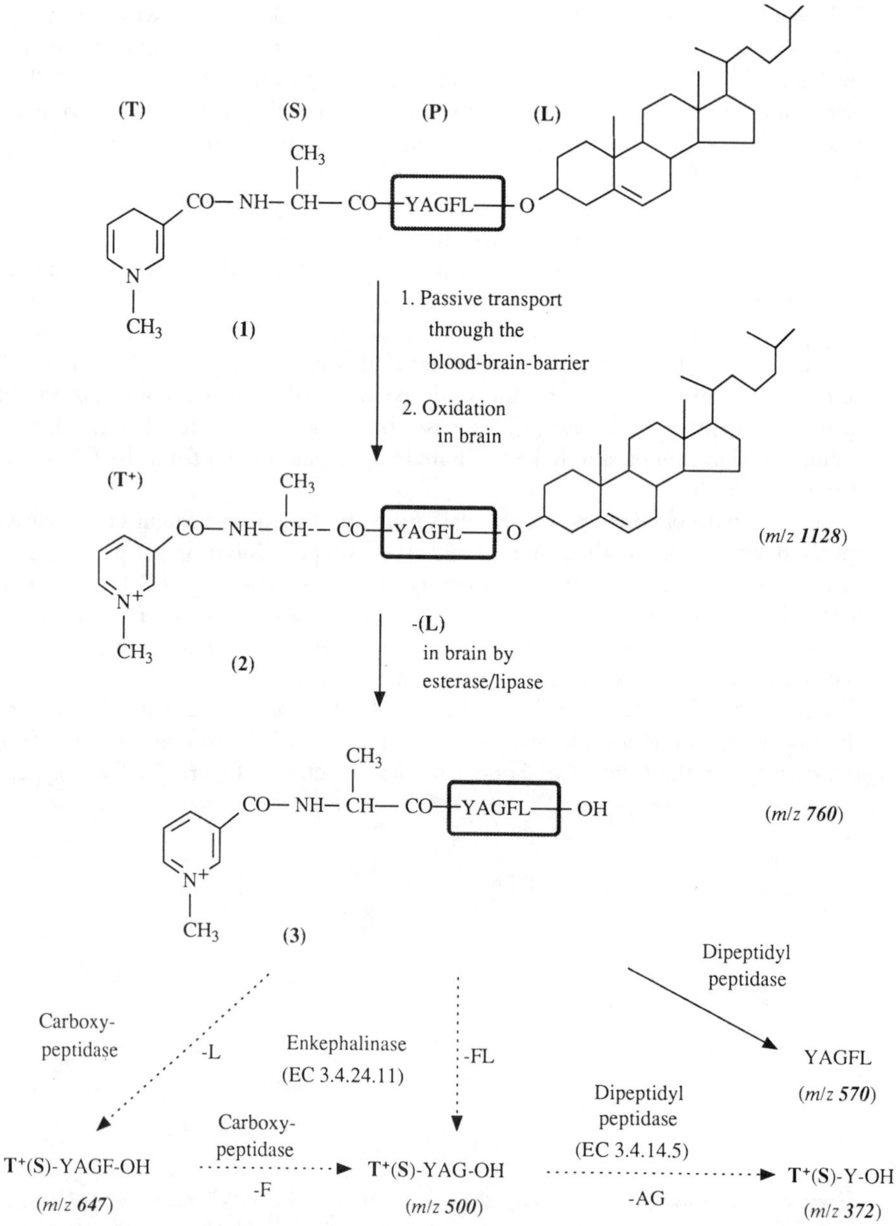

Scheme III. Brain delivery of [D-Ala[2]]-Leu-enkephalin by sequential metabolism (22).

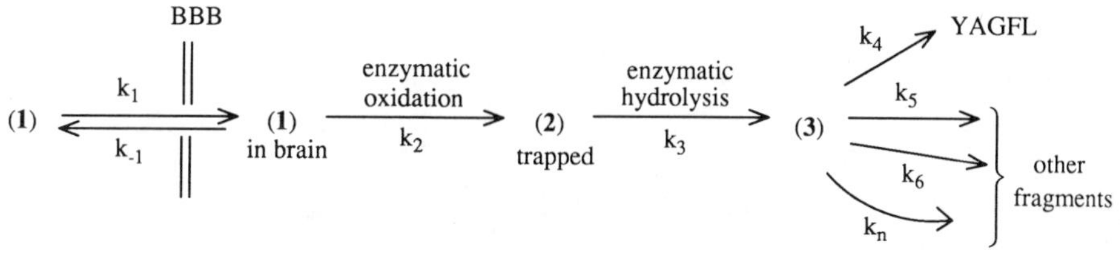

Scheme IV. Rate processes affecting the delivery of the enkephalin analogue.

However, the concentration of **3** cannot be used as the only measure of the efficiency for peptide delivery. Although this compound supports the mechanism proposed, it is an intermediate in a process governed by sequential metabolism and the lock-in mechanism. The concentrations of **1** and **2** in brain were not measured because of difficulties involved in their extraction from biological tissue. Many large drugs delivered by the redox system showed a large concentration of the pyridinium ion derived from the CDS molecule by a single oxidation process, and the concentration of subsequent metabolites may reflect a steady state regulated by the balance between the formation from a precursor and the rate of further metabolic transformations. In the context of the rates of the individual processes shown schematically in Scheme IV, the concentration of **3** is determined by rate constants k_1, k_2, k_3, as well as k_4–k_n. As determined from the receptor binding studies, **3** can be regarded as inactive, and only the release of the enkephalin analogue from this precursor molecule results in a significant pharmacological response. Considering that the precursor of the biologically active peptide is susceptible to deactivation by diverse neuropeptide-degrading enzymes in the CNS, appropriate measures should be implemented to control the metabolism of **3**.

The Leu^5 residue in the structure is important for the pharmacological activity. Thus, resistance to cleavage at this particular site is important. Accordingly, the metabolically more stable [D-Ala^2]-[D-Leu^5]-enkephalin has been applied as a lead compound of the CDS. This modification to the structure also enhances the resistance to enkephalinase. In the in vitro experiment (20% brain homogenate), >90% of the targetor–[D-Ala^2]-[D-Leu^5]-enkephalin conjugate (*m/z* 760) with a single (L)-alanyl spacer was intact after 4 h of incubation at 37 °C (Figure 3). Also, release of the enkephalin analogue was indicated by the detection of the protonated molecular ion of the peptide at *m/z* 570, but the peptidase deactivation characteristic to the targetor–peptide conjugate of the (L)-Leu^5 analogue was

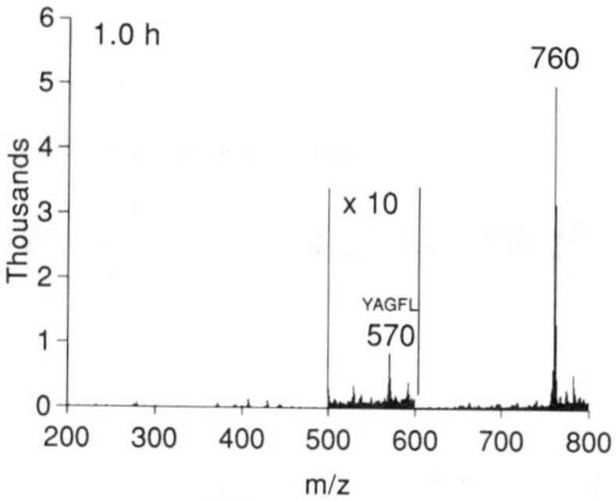

Figure 3. ESI mass spectrum showing the cleavage of the T^+-Ala-Tyr-[D-Ala]-Gly-Phe-[D-Leu] (m/z 760) in vitro (brain homogenate; incubation at 37 °C for 2 h).

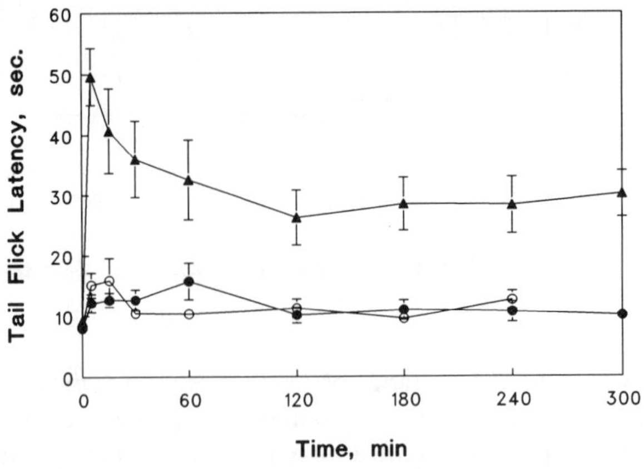

Figure 4. The CNS effect of the iv injection (8.8 μmol per kg of body weight) of [D-Ala²]-[D-Leu⁵]-enkephalin CDS (▲) compared with that of the parent peptide (●) and of the vehicle solution (○) as a control. The measure of analgesia was tail-flick latency in Sprague–Dawley rats. (Reproduced with permission from reference 22. Copyright 1992 American Association for the Advancement of Science.)

absent or very significantly reduced. The slow release at the site of action and subsequent slow degradation by specific peptidases may provide sustained release and a steady concentration level of the biologically active peptide in the CNS.

For the CDS for the [D-Ala^2]-[D-Leu^5]-enkephalin, detailed and time-resolved pharmacological studies (*22*) revealed a long-lasting and statistically significant increase in the tail-flick latency, a measure of the spinal cord-mediated analgesia, of rats after systemic administration. In contrast, the unmanipulated enkephalin analogue and the partially conjugated (either with the targetor or with the cholesterol moiety) peptides showed no effect compared with the group of animals injected with the vehicle solution alone (Figure 4). Considering the confidence intervals for the measured time points indicated on the curves, a statistically very significant effect (~20 s) and sustained increase was obtained even 5 h after iv administration. The prolonged activity should not be misinterpreted. It is due to the specific and most important value of the CDS approach; that is, the "lock-in" mechanism. The type of curve presented in Figure 4 is distinct from those characteristic to simple delivery methods, including direct (intracerebroventricular) injection to the brain. We have reported (*61*) that in other cases, such as with the estradiol redox delivery system, one can achieve a therapeutic, sustained, and steady-state concentration in the brain and subsequent pharmacological activity for weeks after one single dose.

Brain Targeting of a Centrally Active TRH Analogue

TRH was originally isolated from the hypothalamus (*62*), and the highest brain concentration of this tripeptide is found in the median eminence (*63*). The extrahypothalamic distribution of TRH and its receptors indicate that TRH plays roles in nervous system physiology other than its well-known role as the primary neurotrophic hormone for TSH secretion

(*62, 63*). TRH has been shown to exert a variety of extrahypothalamic effects in animals. The most interesting and best-documented effect of this neuropeptide is its analeptic action. High doses of TRH (3–100 mg/kg) administered peripherally (*64–67*) and lower doses administered into specific brain regions (*68, 69*) have been shown to reduce pentobarbital-induced sleeping time by 50% or more in rats, rabbits, and monkeys. A similar effect of TRH was observed for ethanol-induced sleeping (*70*). The analeptic effects of TRH appear to be mediated by a cholinergic mechanism, as indicated by the finding that the TRH effect on barbiturate-induced sleeping is antagonized by the muscarinic receptor blockers scopolamine and atropine (*64–67*). Further support for a cholinergic involvement in the analeptic effects of TRH comes from studies of the central loci of the TRH effect; for example, the septum was 5–10 times more sensitive (median effective dose of 9.5 nM/rat) than the next most sensitive brain regions in inducing an analeptic response (*68*).

In normal animals, TRH has little effect on cholinergic neuronal activity (*71*). By contrast, in animals in which cholinergic activity is depressed by barbiturates, TRH enhances activity. High affinity choline uptake (*72*) is reduced by pentobarbital, and TRH prevents this effect in the hippocampus, cortex, and midbrain (*73*). Additionally, cortical high-affinity choline uptake is enhanced in pentobarbital-treated rats in which TRH was injected into the nucleus basalis (*69*), and the pentobarbital-induced decline in hippocampal choline uptake is blocked by septal infusion of TRH (*74*). Similarly, MK-721, a potent TRH analogue (*75*), antagonizes the barbiturate-induced decrease in high-affinity choline uptake in the hippocampus and cortex when administered intracerebroventricularly (*76*). TRH also enhances [^{3}H]ACh synthesis from [^{3}H]choline in the parietal cortex of unanesthetized rats (*77*) and induces ACh release from the cerebral cortical surface in anesthetized rabbits (*78*). Finally, TRH was reported to potentiate the excitatory action of ACh on cortical neurons (*79, 80*), but this observation has not been confirmed (*81, 82*).

A report by Horita et al. (*83*) argues for a potential cytoprotective role of TRH in rats. These authors observed that the reduction in hippocampal computer-averaged transients (CAT) activity and high-affinity choline uptake induced by ibotenic acid injection into the septum could be reversed by the acute administration of the TRH analogue MK-771. These results indicate that residual cholinergic innervation (~60% of normal) of the hippocampus can respond to TRH administration. In the same study the rate of learning in a 12-arm radial maze was enhanced in septal-lesioned rats treated daily with TRH (*83*).

Two reports (*84, 85*) of positive effects of TRH on memory in patients with probable Alzheimer's disease were published. Mellow et al. (*84*) used an acute dosing paradigm and observed a significant but modest effect of TRH when behavioral tests were conducted within 90 min of iv dosing. They observed no effect of TRH on attention, episodic memory, or visual memory, but a significant 34% improvement in semantic memory. Using a chronic daily ascending dose paradigm (2 to 12 mg, iv) over 5 days in choline-treated patients, Lampe et al. (*85*) observed significant (TRH versus placebo) improvement in 4 of 11 cognitive tests after 5 days of TRH therapy. Additionally, patients treated with TRH showed significantly improved scores versus their baseline scores in 5 of 11 tests. These data are in contrast to two reports showing no effect of TRH on cognition (*86, 87*), both of which used much smaller doses of TRH. Finally, Molchan et al. (*88*) evaluated the effects of TRH (0.5 mg/kg, iv) versus placebo in normal subjects whose cognitive function was reduced by scopolamine pretreatment; they observed that at 45 min after TRH injection, the neuropeptide markedly attenuated the effects of scopolamine in 6 of 12 cognitive tests.

Another cholinergic effect of TRH is seen in amyotrophic lateral sclerosis. As is suspected with Alzheimer's disease, amyotrophic lateral sclerosis is a CNS disorder whose clinical manifestations are a progressive, degenerative loss of a subpopulation of central cholinergic neurons. The stimulatory effects of TRH on motorneurons and consequent electromyographic (EMG) activitation are well established (*78*).

Bearing in mind that TRH has an extremely short $t_{1/2}$, of ~5 min (*89*), and does not effectively penetrate the BBB (*90, 91*), these few clinical evaluations of the efficacy of the tripeptide following peripheral administration are encouraging. The availability of stable analogues of TRH that are active in the brain and that are molecularly packaged for brain-enhanced delivery and release of the active compound would allow for critical tests of the potential for TRH therapy in Alzheimer's patients.

Poor access of TRH to the CNS is attributable to the low distribution coefficient ($K_D < 0.005$ between 1-octanol and water) preventing transport through the BBB (*92*). Thus, large doses (3–100 mg) must be administered to produce neuropharmacological effects. Numerous attempts were made to modify the structure of TRH to obtain metabolically stable analogues with improved potency and selectivity by dissociating the endocrine and CNS function. Modifications to the *p*Glu and Pro rings [e.g., their replacement with heterocyclic amino acids such as α-aminoadipic acid (*93*), orotyl (*94*), 1-oxo-tetrahydroisoquinolinyl (*95*), and so on, residues at the N-terminus, and with thiazolidine-4-carboxyl, pipecolic acid (*93*) or 3,3′-dimethylprolineamide (*96*) at the C-terminus] impart resistance toward enzymes that metabolize the parent structure. Complete dissociation of the CNS effect from thyrotropin-releasing activity may be provided by incorporating aliphatic amino acids in place of His^2 (*97*). These analogues improve metabolic stability and increase the lipid solubility of TRH, but their peptide character often prevents substantial delivery to the brain. Also, even lipophilic molecules obtained by these modifications may rapidly leave the CNS as soon as their concentrations in the blood decrease, so that sustained brain concentrations may not be accomplished.

The packaging strategy applied to the enkephalin analogues may also be adapted to other neuropeptides with free NH_2^- and COOH-termini. In TRH, no free hydroxy or amino groups are present. However, TRH is derived from posttranslational processing of a precursor polyprotein (*98*). The deduced sequence of the 123 amino acid TRH precursor contains three copies of the sequence Lys-Arg-Gln-His-Pro-Gly-Lys–Arg-Arg, and a fourth incomplete sequence (*99*). The TRH progenitor sequences (Gln-His-Pro-Gly) are flanked by dibasic residues that are typical sites of processing by carboxypeptidase B-like enzymes in polyproteins (*100, 101*). The C-terminal glycine functions as an amide donor for amidation of the proline (*102*), an enzymatic activity-designated peptide glycine α-amidating monooxygenase (PAM) that requires Cu^{2+}, ascorbic acid, and molecular oxygen (*103*). In the presence of extracts from the secretory granules of bovine pituitary, conversion of *p*Glu-His-Pro-Gly to TRH was previously demonstrated (*104*). Glutamine (Gln), and not glutamic acid (Glu), is the precursor of the N-terminal pyroglutamyl residue as was also shown for other peptides, including gastrin 17 (*105*) and luteinizing hormone releasing hormone (LHRH) (*106*). Cyclization of the N-terminal glutamine is known to be catalyzed by the specific enzyme glutaminyl cyclase (*107*). Histidine (His) is not essential for the CNS activity. In fact, the respective [Leu^2], [Nle^2], and [Nva^2] analogues have 2.5 to 10 times greater CNS effect in the inhibition of catalepsy compared with TRH (*97*).

We aimed at the "packaging" of a TRH analogue based on a Gln-Leu-Pro-Gly (QLPG) progenitor sequence. It was crucial to the proposed strategy that the locked-in T^+(S)-QLPG peptide precursor could rapidly be converted to the required prolinamide C-terminus. Therefore, we synthesized the hypothetical locked-in precursor T^+-AQLPG and studied its bioactivation or degradation in vitro (Figure 5). We noted that processing to the prolinamide C-terminus by PAM in brain tissue occurs faster than any other enzymatic cleavage reaction examined. Thus, the processing to the desired TRH analogue is only dependent on the release of Gln-Leu-Pro-NH_2, which is similar to the release of the

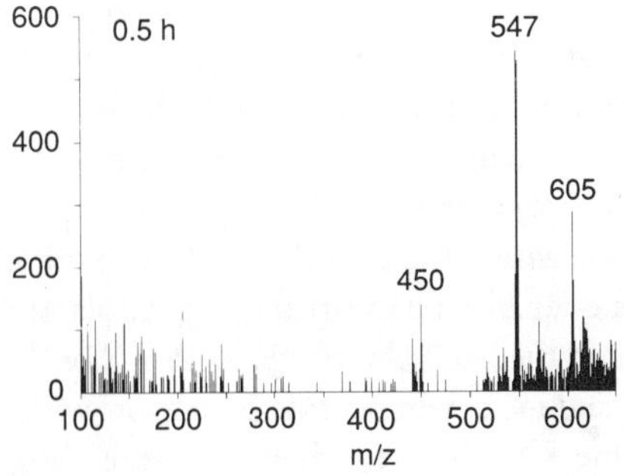

Figure 5. ESI mass spectrum showing the cleavage of the T^+-Ala-Gln-Leu-Pro-Gly (m/z 605) in vitro (brain homogenate; incubation at 37 °C for 0.5 h).

enkephalin analogue from the locked-in precursor obtained by the removal of the C-terminal cholesteryl (dipeptidyl peptidase cleavage). The CNS delivery of a pharmacologically significant amount of THR analogue is evidenced by the profound decrease in the barbiturate-induced sleeping time (the measure of the activational effect on cholinergic neurons) in mice (Figure 6). At equimolar (30 µmol/kg) dose, the iv administration of the TRH analogue showed only a marginal decrease (limited BBB penetration), whereas the CDS with (L)-Ala spacer resulted in ~30% reduction in the sleeping time of the experimental animals.

Although the enhanced CNS delivery of the TRH analogue significantly improved the pharmacological response, further refinement of the CDS may be justified based on the noted endopeptidase cleavage (possibly by enkephalinase, E.C. 3.4.24.11, which may attack internal Phe, Leu, Tyr, and Trp residues) (*7*) that yields a T^+-AQL fragment (*m/z* 450) in the in vitro experiments that is related to the processing of the locked-in precursor T^+-AQLPG. When considering the replacement of one of the residues with the corresponding D-amino acid in the peptide ([Leu^2]-TRH) part of the structure to impart metabolic stability, one also has to weigh the possibility that changes in the biological activity of the neuropeptide may occur. Indeed, these modifications result in a marked decrease in the central anticataleptic activity of this tripeptide (Table I) (*97*). Therefore, one may not consider the D-analogues for molecular packaging of [Leu^2]-TRH to improve the pharmacological effect on CNS delivery, even if the stability of the D-analogues toward neuropeptide degrading enzymes is improved by these modifications to the peptide part (P) of the structure.

Scheme IV highlighted the fact that facile release of the desired peptide from a

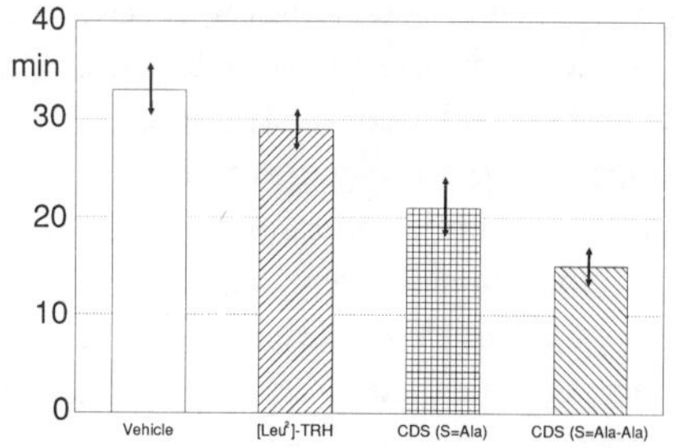

Figure 6. Effect of the TRH analogue pGlu-Leu-Pro-NH_2 and its CDSs on the methohexital-induced sleeping time in mice after iv injection (30-µmol/kg of body weight dose).

locked-in precursor is determined by the balance between the rate of release in this reaction versus the overall rate of other enzymatic degradations leading to the loss of biological activity. This balance may also be influenced by modifying the spacer (S) function of the CDS molecule. It was demonstrated (*108*) that prolyl endopeptidase (EC 3.4.21.26) may play a role in neuropeptide metabolism (*109*). This peptidase cleaves –Pro–X bonds (X is an amino acid residue within the peptide bond) with the exception of the Pro–Pro bond. The proline (Pro) residue can also be substituted by the alanine (Ala) residue (*110*). However, *N*-blocked peptides with the general formula Z-Pro-X or Z-Ala-X are not attacked. Therefore, the brain-targeted and locked-in peptide precursors with a single Ala spacer are not metabolized by this enzyme. On the other hand, by extending the spacer part of the CDS with another amino acid residue, a Z-X′-Ala-X bond is likely to be cleaved by prolyl endopeptidase, where Z represents the trigonellyl targetor (T^+). Considering the implication to the underlying synthetic procedure involving a repeated deprotection–coupling sequence, we have selected Ala as the additional X′ residue for convenience. With the modified CDS for *p*Glu-Leu-Pro-NH_2, the CNS delivery by sequential metabolism is shown in Scheme V. The pharmacological consequence of this modification at the spacer portion of the CDS results in a >50% decrease of the barbiturate-induced sleeping time in mice (Figure 6) compared with the control group. This result is indeed an improvement over the CDS in which a single Ala separates the N-terminal residue (Gln) of the biologically active peptide from the targetor (T) function (*111*).

Conclusions and Future Directions

Molecular packaging of peptides is a rational drug design approach by which the brain delivery of these important biomolecules in a pharmacologically significant amount was documented for the first time. These studies (*22, 111*) include detailed mass spectrometric analyses providing high levels of molecular specificity, as well as observation of relevant pharmacological responses. The strategy underscores the importance of controlling transport and metabolism in efforts to direct peptide-based drugs to the site of action. The design should be based on a thorough understanding of the processes involved in the formation and metabolism (inactivation) of the target neuropeptide in the CNS. We may only realize the promise of biologically active peptides to become a future generation of high-efficiency neuropharmaceuticals by overcoming the obstacle represented by the BBB. The specific feature of the CDS approach is that targeting is achieved through three fundamental steps: crossing the BBB, retention in the brain or lock-in, and release of the biologically active peptide. The overall strategy is based on structural, physicochemical, and enzymatic aspects of the BBB with respect to the molecules transported through this biological membrane, as well as "designed-in" sequential metabolic processes.

To date, this novel method has been applied to enkephalin analogues and to an important TRH analogue. Future directions, beyond the further refinement of the existing delivery system to increase therapeutic efficacy, include the application of the strategy to a variety of other neuropeptides classes. Specific conclusions need to be developed to determine how the size and complexity of the peptide can influence CNS delivery by sequential metabolism to formulate the general design principles.

Table I. CNS and Hormonal Activities of TRH and [Leu²]-TRH

Compound	*Anticataleptic Effect (ED_{50}, mg/kg iv)*	*TSH Releasing Effect (%)*
pGlu-His-Pro-NH_2 (TRH)	113.0	100
pGlu-Leu-Pro-NH_2	40.0	2
D-pGlu-Leu-Pro-NH_2	70.0	0
pGlu-D-Leu-Pro-NH_2	>80	0
pGlu-Leu-D-Pro-NH_2	>80	0

SOURCE: Adapted from reference 97.

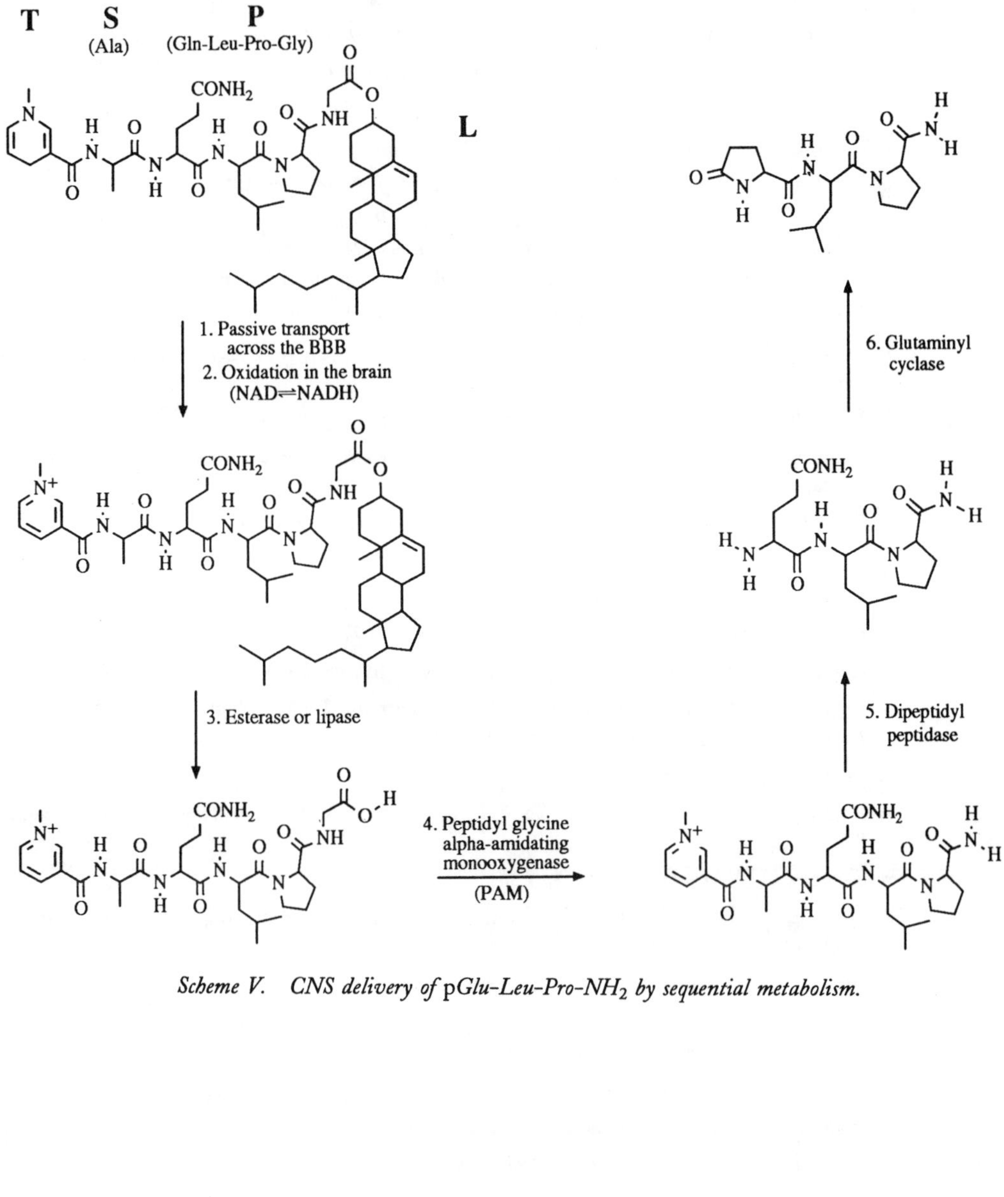

Scheme V. CNS delivery of pGlu-Leu-Pro-NH_2 *by sequential metabolism.*

Acknowledgments

This research was supported by the National Institute for Aging (Grant No. PO1 AG10485). The invaluable contribution of W. M. Wu, H. Farag, S. Jonnalagadda, and J. Simpkins, as well as of X. Ouyang (Graduate Research Assistant supported by Genentech) is kindly acknowledged.

References

1. Brightman, M. W.; Reese, T. S. *J. Cell Biol.* **1969,** *40,* 648.
2. Brownlees, J.; Williams, C. H. *J. Neurochem.* **1993,** *60,* 793.
3. Yamaguchi, T.; Yamaguchi, M.; Lajtha A. *J. Neurol. Sci.* **1970,** *10,* 323.
4. Pardridge, W. *Endocrine Rev.* **1986,** *7,* 314.
5. Zlokovic, B.; Makic, B.; Djuricic, B.; Davson, H. *J. Neurochem.* **1989,** *53,* 1333.
6. Zlokovic, B.; Begley, D.; Segal, M.; Dawson, H.; Rakic, L.; Lipovac, M.; Mitrovic, D. M. In *Peptide and Amino Acid Transport Mechanisms in the Central Nervous System;* Rakic, L.; Begley, D. J.; Davson, H.; Zlokovic, B., Eds.; MacMillan: London, 1988; p 3.
7. Pardridge, W. M. *Peptide Drug Delivery to the Brain;* Raven: New York, 1991.
8. Poplack, D. G.; Blayer, A. W.; Horowitz, M. E. In *Neurobiology of Cerebrospinal Fluid;* Plenum: New York, 1981; p 561.
9. Neuwelt, E.; Rapoport, S. *Fed. Proc.* **1984,** *43,* 214.
10. Pardridge, W. M. In *Directed Drug Delivery: A Multidisciplinary Problem;* Borchardt, R. T.; Repta, A. J.; Stella, V. J., Eds.; Humana: Clifton, NJ, 1985; p 83.
11. Pardridge, W. M.; Triguero, D.; Buciak, J. L. *Endocrinology* **1990,** *126,* 977.
12. Ito, F.; Ito, S.; Shimizu, N. *Mol. Cell Endocrinol.* **1984,** *36,* 165.
13. Patel, H. M., *Biochem. Soc. Trans.* **1984,** *12,* 333.
14. Hoffman, P. L.; Walter, R.; Bulat, M. *Brain Res.* **1977,** *122,* 87.
15. Cefalu, W. T.; Pardridge, W. M. *J. Neurochem.* **1985,** *45,* 1954.
16. Begley, D. J.; Squires, L. K.; Zlokovic, B. W.; Mitrovic, D. M.; Hughes, C. C. W.; Revest, P. A.; Greenwood, J. *J. Neurochem.* **1990,** *55,* 1222.
17. Begley, D. J. *Prop. Brain Res.* **1992,** *91,* 163.
18. Pardridge, W. M.; Triguero, D.; Yang, J.; Cancilla, P. A. *J. Pharmacol. Exp. Ther.* **1990,** *253,* 884.
19. Träuble, H. *J. Membr. Biol.* **1971,** *4,* 193.
20. Tsuzuki, N.; Hama, T.; Hibi, T.; Konishi, R.; Futaki, S.; Kitagawa, K. *Biochem. Pharmacol.* **1991,** *41,* R5.
21. Bundgaard, H.; Moss, J. *Pharm. Res.* **1990,** *7,* 885.
22. Bodor, N.; Prokai, L.; Wu, W.-M.; Farag, H.; Jonnalagadda, S.; Kawamura, M.; Simpkins, J. *Science (Washington, DC)* **1992,** *257,* 1698.
23. Bundgaard, H. *Design of Prodrugs;* Elsevier Science: Amsterdam, The Netherlands, 1985.
24. Bodor, N.; Kaminski, J. *Annu. Reports Med. Chem.* **1987,** *22,* 303.
25. Hoek, J.; Rydstrom, J. *Biochem. J.* **1988,** *254,* 1.

26. Bodor, N.; Brewster, M. *Pharm. Ther.* **1983,** *19,* 337.

27. Bodor, N.; Simpkins, J. *Science (Washington, DC)* **1983,** *221,* 65.

28. Bodor, N. *Ann. N.Y. Acad. Sci.* **1987,** *507,* 289.

29. Bodor, N.; Brewster, M. E. In *Handbook of Experimental Pharmacology;* Springer-Verlag: Berlin, Germany, 1991; Vol. 100, Chapter 7, p 231.

30. Shashoua, V. E.; Jacob, J. N.; Ridge, R.; Campbell, A.; Baldessarini, R. J. *J. Med. Chem.* **1984,** *27,* 659.

31. Frenk H. *Brain Res. Rev.* **1983,** *6,* 197.

32. Tortella, F. C.; Cowan, A.; Adler, M. W. *Life Sci.* **1981,** *29,* 1039.

33. Tortella, F. C.; Long, J. B.; Holaday, J. W. *Brain Res.* **1985,** *332,* 174.

34. Olson, G. A.; Olson, R. D.; Kastin, A. J. *Peptides* **1985,** *6,* 769.

35. Sandyk, R. *Life Sci.* **1985,** *37,* 1655.

36. Baile, C. A.; McLaughlin, C. L.; Della-Fera, M. A. *Physiol. Rev.* **1986,** *66,* 172.

37. Frederickson, R. C. A.; Geary, L. E. *Prog. Neurobiol.* **1982,** *19,* 19.

38. Schmauss, C.; Emrich, H. M. *Biol. Psychiatr.* **1985,** *20,* 1211.

39. Plotnikoff, N. P.; Murgo, J.; Miller, G. C.; Cordes, C. N.; Faith, R. E. *Fed. Proc.* **1985,** *44,* 118.

40. Wybran, J. *Fed. Proc.* **1985,** *44,* 92.

41. Porreca, F.; Burks, T. F. *J. Pharmacol. Exp. Ther.* **1983,** *227,* 22.

42. Schick, R.; Schudziarra, V. *Clin. Physiol. Biochem.* **1985,** *3,* 43.

43. Holaday, J. W. *Annu. Rev. Pharmacol. Toxicol.* **1983,** *23,* 541.

44. Bernton, E. W.; Long, J. B.; Holaday, J. W. *Fed. Proc.* **1985,** *44,* 290.

45. Johnson, M. W.; Mitch, W. E.; Wilcox, C. S. *Prog. Cardiovasc. Dis.* **1985,** *27,* 435.

46. Grossman, A.; Rees, L. H. *Br. Med. Bull.* **1983,** *39,* 83.

47. Bicknell, R. J. *J. Endocrinol.* **1985,** *107,* 437.

48. Millan, M. J.; Herz, A. *Int. Rev. Neurobiol.* **1985,** *26,* 1.

49. Yen, S. S.; Quigley, M. E.; Reid, R. L.; Ropert, J. F.; Cetel, N. S. *Am. J. Obstet. Gynecol.* **1985,** *152,* 485.

50. Izquierdo, I.; Netto, C. A. *Ann. N.Y. Acad. Sci.* **1985,** *444,* 162.

51. Frederickson, R. C. A. In *Analgesics: Neurochemical, Behavioral and Clinical Perspectives;* Kuhar, M.; Pasternak, G., Eds.; Raven: New York, 1984; p 9.

52. Zlokovic, B. V.; Begley, D. J.; Chain-Eliash, D. J. *Brain Res.* **1985,** *336,* 125.

53. Hambrook, J. M.; Morgan, B. A.; Ranee, M. J.; Smith, C. F. C. *Nature (London)* **1976,** *262,* 782.

54. Hill, R. C.; Marbach, P.; Scherrer, D. *Br. J. Pharmacol.* **1981,** *72,* 571P.

55. Pardridge, W. M.; Mietus, L. J. *Endocrinology* **1981,** *109,* 1138.

56. Pardridge, W. M.; Eisenberg, J.; Yamada, T. *J. Neurochem.* **1985,** *44,* 1178.

57. Hersch, L.; Aboukhair, N.; Watson, S. *Peptides* **1987,** *8,* 523.

58. McKelvy, J. F. In *Brain Peptides;* Krieger, D. T.; Brownstein, M. J.; Martin, J. B., Eds.; Wiley Interscience: New York, 1983; p 117.

59. Schwartz, J.; Giros, B.; Gros, C.; Llorens-Cortes, C.; Arrang, J.; Garbarg, M.; Pollard, H. *Cephalagia* **1987,** *7 (Suppl. 6),* 32.

60. Prokai, L.; Hsu, B.-H.; Farag, H.; Bodor, N. *Anal. Chem.* **1989,** *61,* 1723.

61. Anderson, W. R.; Simpkins, J. W.; Brewster, M. E.; Bodor, N. *Pharmacol. Biochem. Behav.* **1987,** *27,* 265.

62. Boler, R.; Enzman, F.; Folkers, K.; Bowers, C.; Schally, A. *Biochem. Biophys. Res. Commun.* **1969,** *37,* 705.

63. Brownstein, M.; Palkovits, M.; Saavedra, R.; Bassiri, R.; Utiger, R. *Science (Washington, DC)* **1974,** *185,* 267.

64. Breese, G.; Cott, J.; Cooper, B.; Prange, A.; Lippton, M.; Plotnikoff, N. *J. Pharmacol. Exp. Ther.* **1975,** *193,* 11.

65. Horita, A.; Carino, M.; Smith, J. *Pharmacol. Biochem. Behav.* **1976,** *5,* 111.

66. Miyamota, M.; Nagai, Y.; Norumi, S.; Saji, Y.; Nagawa, Y. *Pharmacol. Biochem. Behav.* **1982,** *17,* 797.

67. Kraemer, G.; Mueller, R.; Breese, G.; Prange, A., Jr.; Lewis, J.; Morrison, H.; McKinney, W., Jr. *Pharmacol. Biochem. Behav.* **1976,** *4,* 709.

68. Kalivas, P.; Horita, A. *J. Pharmacol. Exp. Ther.* **1980,** *212,* 203.

69. Horita, A.; Carino, M.; Lai, H. *Fed. Proc.* **1986,** *45,* 795.

70. Cott, J.; Breese, G.; Cooper, B.; Barlow, I.; Prange, A. *J. Pharmacol. Exp. Ther.* **1976,** *196,* 594.

71. Yarbrough, G. *Prog. Neurobiol.* **1979,** *12,* 291.

72. Atweh, S.; Simon, J.; Kuhar, M. *Life Sci.* **1975,** *17,* 535.

73. Schmidt, D. *Commun. Psychopharm.* **1977,** *1,* 469.

74. Brunello, N.; Cheney, D. *J. Pharmacol. Exp. Ther.* **1981,** *219,* 489.

75. Porter, C.; Lotti, V.; DeFelice, M. *Life Sci.* **1978,** *21,* 811.

76. Santori, E.; Schmidt, D. *Regul. Pept.* **1980,** *1,* 69.

77. Malthe-Sorenssen, D.; Wood, P.; Cheney, D.; Costa, E. *J. Neurochem.* **1978,** *31,* 685.

78. Yarbrough, G. *Life Sci.* **1983,** *33,* 111.

79. Yarbrough, G. *Science (Washington, DC)* **1976,** *263,* 523.

80. Braitman, D.; Auker, C.; Carpenter, D. *Brain Res.* **1980,** *194,* 244.

81. Pittman, Q.; Blumen, H.; Renaud, L. *Can. Fed. Biol. Sci.* **1978,** *21,* 2.

82. Winokur, A.; Beckman, A. *Brain Res.* **1978,** *150,* 205.

83. Horita, A.; Carino, M.; Zabawska, J.; Lai, H. *Peptides* **1989,** *10,* 121.

84. Mellow, A.; Sunderland, T.; Cohen, R.; Lawlor, B.; Hill, J.; Newhouse, P.; Cohen, M.; Murphy, D. *Psychopharmacology* **1989,** *98,* 403.

85. Lampe, T.; Norris, J.; Risse, S.; Owens-Williams, E.; Keenan, T. In *Alzheimer's Disease: Basic Mechanisms, Diagnosis and Therapeutic Strategies;* Iqbal, K.; McLachlan, D. R. C.; Winbald, B.; Wisniewski, H. M., Eds.; John Wiley & Sons: New York, 1991; p 643.

86. Peabody, C.; DeBlois, T.; Tinklenberg, J. *Am. J. Psychiatr.* **1986,** *143,* 262.

87. Sunderland, T.; Mellow, A.; Gross, M.; Cohen, R.; Tariot, P.; Newhouse, P.; Murphy, D. *Am. J. Psychiatr.* **1986,** *143,* 1318.

88. Molchan, S.; Mellow, A.; Lawlor, B.; Weingartner, H.; Cohen, M.; Sunderland, T. *Psychopharmacology* **1990,** *100,* 84.

89. Bassiri, R.; Utiger, R. *J. Clin. Invest.* **1973,** *52,* 1616.

90. Jackson, I. In *Neurobiology of Cerebrospinal Fluid,* J. H., Wood, Ed.; Plenum: New York, 1980: p. 625.

91. Metcalf, G. *Brain Res. Rev.* **1982,** *4,* 389.

92. Meisenberg, G.; Simmons, W. *Life Sci.* **1983,** *32,* 2611.

93. Nutt, R.; Holly, F.; Homnick, C.; Hirschmann, R.; Veber, D. *J. Med. Chem.* **1981,** *24,* 692.

94. Friedrichs, E.; Schwertner, E.; Herrling, S.; Gunzler, W-A.; Ottig, F.; Flohe, L. In *Structure*

and Activity of Natural Peptides; Voelter, W.; Weitzel, G., Eds.; Walter de Gruyter: Berlin, German, 1981; p 461.

95. Maeda, H.; Suzuki, M.; Sugano, H.; Yamamura, M.; Ishida, R. *Chem. Pharm. Bull.* **1988,** *36,* 190.
96. Metcalf, G.; Dettmar, P.; Fortne, D.; Lynn, A.; Tulloch, I. *Regul. Pept.* **1982,** *3,* 193.
97. Szirtes, T.; Kisfaludy, L.; Palosi, E.; Szporny, L. *J. Med. Chem.* **1984,** *27,* 741.
98. Jackson, I. *Ann. N.Y. Acad. Sci.* **1989,** *553,* 71.
99. Richter, K.; Kawashima, E.; Egger, R.; Kriel, G. *EMBO J.* **1984,** *3,* 617.
100. Docherty, K.; Steiner, D. *Annu. Rev. Physiol.* **1982,** *44,* 625.
101. Gainer, H.; Russell, J.; Loh, Y. *Neuroendocrinology* **1985,** *40,* 171.
102. Bradbury, A.; Finnie, M.; Smyth, D. *Nature (London)* **1982,** *298,* 686.
103. Eipper, B.; Myers, A.; Mains, R. *Endocrinology* **1985,** *116,* 2497.
104. Husain, I.; Tait, S. *FEBS Lett.* **1983,** *152,* 272.
105. Yoo, O.; Powell, C.; Argarwal, K. *Proc. Natl. Acad. Sci. U.S.A.* **1982,** *79,* 1049.
106. Seeburg, P.; Adelman, J. *Nature (London)* **1984,** *311,* 666.
107. Fischer, W.; Spiess, J. *Proc. Natl. Acad. Sci. U.S.A.* **1987,** *84,* 3628.
108. Koida, M.; Walter, R. *J. Biol. Chem.* **1976,** *251,* 7593.
109. Wilk, S. *Life Sci.* **1983,** *33,* 2149.
110. Yoshimoto, T.; Fischl, M.; Orlowski, R. C.; Walter, R. *J. Biol. Chem.* **1978,** *253,* 3708.
111. Prokai, L.; Ouyang, X.; Wu, W.-M.; Bodor, N. *J. Am. Chem. Soc.* **1994,** *116,* 2643.

RECEIVED for review July 12, 1993. ACCEPTED revised manuscript October 14, 1993.

[illegible]
[illegible] Pharm. Bull. [illegible]
[illegible] Chem. [illegible]
[illegible]
[illegible]
[illegible]
[illegible]
[illegible]
104. [illegible]
[illegible]
[illegible]
[illegible]
[illegible]
[illegible]
[illegible]
[illegible]
[illegible]

Received for review [illegible] October 18, 19[illegible]

Peptide Transport in Microorganisms

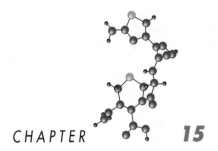

CHAPTER 15

Bacterial Peptide Permeases as a Drug Delivery Target

J. W. Payne

Over the years, several reviews of peptide transport in microorganisms that describe the topic mainly from a physiological viewpoint have been published (*1–7*). The early reviews, largely descriptive accounts of the nutritional role of peptides, include discussions of the complementary aspects of transport and hydrolysis. In these reviews, comparisons are made between the transport of peptides and of amino acids, attention is drawn to the inherently greater efficiency of the process of peptides transport, and the added physiological advantage in possession of both types of transport systems is noted. Today, it may be concluded that the ubiquity of peptide transport systems, not simply among microorganisms but throughout Nature, testifies to their exceptional physiological value. Recently, considerable advances have been made in deciphering the molecular basis of peptide transport in microorganisms. It is, therefore, timely to consider this topic from a molecular point of view rather than from a physiological one.

The concept of "designer drugs" appears increasingly in the scientific literature and even attracts media attention occasionally. However, a discussion of the application of fundamental knowledge concerning peptide transport to its exploitation in drug delivery has generally been only a minor part of these earlier reviews. Furthermore, when the broad topic of antimicrobial peptides has been discussed, only a few authors (*3–14*) have considered the possible exploitation of peptide permeases in the design of peptide-based antimicrobial compounds. Therefore, it is opportune here to focus specifically on this aspect and to evaluate the potential of a knowledge-based approach for exploiting bacterial peptide permeases as a drug delivery target. Coverage will be restricted to bacteria; fungal peptide transporters are described by Naider and Becker in Chapter 16. An overview of the functioning of a typical peptide permease in the cytoplasmic membrane (*see* Models and Mechanisms of Peptide Transport) and of the influence of porins and other components of the overlying bacterial cell envelope on accessibility of peptides to such a permease (*see* The Bacterial Cell Envelope as a Permeability Barrier) is followed by detailed consideration of the molecular architecture, functions, and specificities of the best-characterized systems from a range of bacterial species (*see* Peptide Permeases of Enteric Bacteria, Peptide Transport in Other Bacteria, and Regulations and Energetics of Peptide Permeases). Then, examples are discussed of both naturally occurring and synthetic peptide analogues, whose antibacterial activity requires their transport via peptide permeases (*see* Peptide Permease Exploitation by Antibacterial Peptides). These examples testify to the efficacy of this principle of drug targeting. Certain scientific-based criteria that need to be satisfied in the design of effective peptide carrier prodrugs are also discussed (*see* Prospects for Rationally Designed Peptide Prodrugs). Probably peptide mimetics will be designed in the future that combine desirable features of potency, specificity, low toxicity, antibacterial spectrum, oral absorption, and good pharmacokinetics. This approach should deliver drugs, unfortunately deciding whether they will be commercially successful is not equally amenable to scientific method.

Models and Mechanisms of Peptide Transport

Speculations concerning a variety of hypothetical mechanisms by which peptides might be taken up by microorganisms were considered in detail in several earlier reviews (*4, 5, 11*). Subsequent studies have effectively eliminated all but one scheme for peptide transport, that which involves direct transport of intact peptide. Nevertheless, it is worth reconsidering these other schemes briefly in the context of peptide-based drug delivery because if they were to be found to occur in any species, they would compromise the application of peptide permeases for drug targeting.

These alternative mechanisms envisaged a role for specific membrane-bound peptidases in peptide accumulation (*5, 11*). For example, by a group-translocation mechanism, peptides could be bound by a peptidase on the outer face of the cytoplasmic membrane, with translocation through the membrane and vectorial discharge of peptide residues to the intracellular side of the membrane then being coupled to peptide bond cleavage. The original peptide substrate would undergo obligatory cleavage as a concomitant of the transport step, leading to transfer of free amino acids in the case of dipeptides or a free amino acid residue and smaller peptide with tri- and oligopeptides. Mechanistically, such a sequence is similar to that envisaged for the γ-glutamyl cycle that has been postulated as a transport system for amino acids and peptides in various cells and tissues (*15*); however, evidence against involvement of the cycle in such a process in microorganisms has been presented (*16*).

A second, related mechanism also envisages peptides being bound to specific peptidases in the outer surface of the membrane. However, following peptide hydrolysis, the cleaved residues interact with amino acid transporters in their immediate microenvironment, which then translocate the free residues into the cytoplasm; in this mechanism, these amino

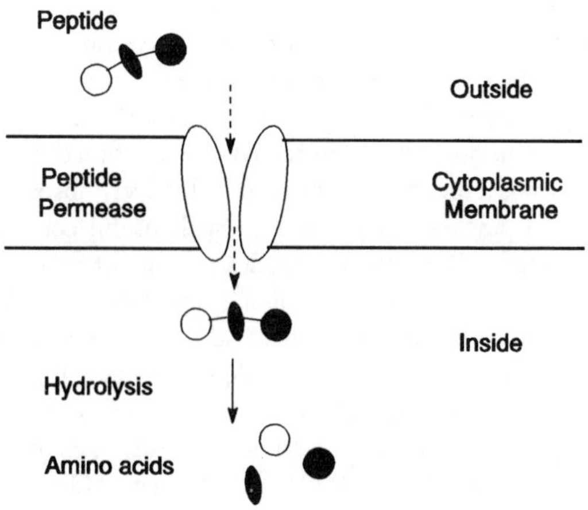

Figure 1. Mechanism of action of a typical peptide permease. Intact peptide accumulation is followed by intracellular peptidase action to liberate the constituent amino acid residues.

acid transporters are accessible only to the products of peptidase action and not to exogenous free amino acids. Specificity for di-, tri-, and oligopeptides could be achieved by coupling to specific di-, tri-, or aminopeptidases, respectively.

Other related variants of these schemes were considered previously (*11*). Fortunately, for the potential exploitation of peptide permeases in drug delivery, no good evidence has ever been obtained for the operation of any of these schemes for uptake of the multitudinous small peptide products of protein hydrolysis. Therefore, given the caveat just made, further consideration of such schemes seems unnecessary.

For present purposes therefore, one can state that there are no relevant features of substrate transport by bacterial peptide permeases that are not best considered through operation of the simplest conceptual model systems. That is, the existence of several complementary, substrate-specific transport systems that are present in the cytoplasmic membrane and that bind peptides directly and translocate them intact into the bacterial cell, where they subsequently may be hydrolyzed (Figure 1). Such a process is simple to conceptualize. However, this process has been shown to be mechanistically complex, involving various protein components that provide substrate specificity, transmembrane pore formation, and energy coupling to the permease. Aspects of these structure–function relationships are considered later (*see* Peptide Permeases of Enteric Bacteria).

The Bacterial Cell Envelope as a Permeability Barrier

Bacteria are classified into two categories on the basis of a simple (i.e., Gram-positive and Gram-negative) staining procedure that depends on differences in cell envelope structure. The cell envelope of Gram-positive bacteria is relatively simpler than that found in Gram-negative bacteria. In all cases, the cell envelope is a multilayered structure, the composition and structure of which can vary quite considerably between different species. In many bacteria, transient alterations in the cell envelope may also occur in response to environmental changes to optimize physiological status of the organism. The cell envelope is a rather rigid structure that confers shape and strength to the bacterial cell and provides a protective barrier against potentially damaging agents (chemical or biological) that may be encountered by a bacterium. This important protective barrier function of the cell envelope presents bacteria with a complementary problem of effectively absorbing nutrients because the cell envelope acts as a permeability barrier to such useful compounds also. Thus, to various extents, the bacterial cell envelope impedes passage of peptides to their transport systems, which are located in the cytoplasmic membrane underlying the cell envelope.

Gram-negative bacteria possess an additional outer membrane (OM), which carries in its outer monolayer a lipopolysaccharide component that provides a barrier to surface-active agents. To allow passage of the wide range of essential nutrients, which are largely polar molecules, the OM possesses a range of diffusion channels formed by proteins known as porins. The characteristics of these OM proteins (Omp) were reviewed (*17, 18*). The main porins in *Escherichia coli* (*E. coli*), OmpC and OmpF, form channels of 1.1- and 1.2-nm diameters, respectively. These porins effectively act as nonspecific, molecular sieves that limit penetration to molecules with a size of ~650 Dal, which for peptides is typically a penta- or hexapeptide (*19*). The important role of OmpF and OmpC in facilitating the passage of peptides across the OM has been shown with appropriate mutant bacterial strains (*20, 21*). The study (*21*) also implicated OmpA in peptide uptake, although its relative importance still has to be clarified (*22*). However, it is important to emphasize that all these porins are general diffusion channels that are not specifically related, either functionally or genetically, to the peptide permeases of the cytoplasmic membrane (*see* Models and Mechanisms of Peptide Transport) (Figure 1). Many other Gram-negative bacteria have been shown to possess analogous porins (*23*). *Pseudomonas aeruginosa* (*Ps. aeruginosa*) is a species of particular interest in this regard because of its high level of intrinsic resistance to many antibacterial compounds. An array of substrate-specific and general porins (Opr) were identified (*18, 24*). However, although several of these porins appear to have larger diameters than analogous Omp in *E. coli,* solute diffusion through them can be 100-fold slower; debate continues as to possible explanations for these observations (*25, 26*). Of particular relevance to peptide penetration is the OprD protein that was shown to function in the uptake of the β-lactam antibiotic imipenem, against which basic peptides containing a C-terminal lysine acted as competitive inhibitors (*27, 28*). However, the OprD protein is best regarded as a diffusion pore for cations rather than having an integrated function with any underlying peptide permease; analogous anion channels (e.g., for phosphates) are known (*18, 24*).

Between the OM and inner (cytoplasmic) membrane is the periplasm, which includes the peptidoglycan, a macromolecular cross-linked structure that is covalently linked to the OM through bonding to OmpA and Braun lipoprotein. The extent to which the peptidoglycan penetrates through the periplasm is still unclear (*29*). No evidence exists that the periplasm acts as a permeability barrier to peptides or similar small molecules. Surprisingly, there is still much uncertainty about the size of the periplasm, although it may occupy about a quarter of the whole cell volume (*30*). Periplasm has the properties of a viscous gel in which proteins move relatively slowly and is isoosmolar with the cytoplasm. Most importantly, the periplasm contains a collection of proteins termed periplasmic binding proteins that are components of many transport systems including peptide permeases (*see* The Oligopeptide Permease).

Gram-positive bacteria differ from Gram-negative organisms in that they lack an OM and a region equivalent to the periplasm. In the former bacteria, thick layers of peptidoglycan and teichoic or teichuronic acid protect the exterior surface of the cytoplasmic membrane, and in some species, further layers of polysaccharide and protein may occur.

These layers do not provide a molecular sieving barrier as found with the OM of Gram-negative bacteria. However, analogues of the freely diffusable periplasmic binding proteins are found in Gram-positive bacteria, although they are attached to the exterior of the cytoplasmic membrane by a lipid anchor (*see* Peptide Transport in Other Bacteria).

Peptide Permeases of Enteric Bacteria

The typical Gram-negative enteric bacteria *E. coli* and *Salmonella typhimurium* (*S. typhimurium*) have been most extensively studied, and the results obtained with these species have provided the basis for our current ideas on the structures and properties of bacterial peptide permeases. It has been proven, as expected, that bacteria inhabiting the peptide-rich lumen of the gut are well adapted to capitalizing on this nutritional resource. Biochemical studies on the substrate specificities for peptide transport, coupled with molecular biological studies on the isolation and characterization of transport mutants, indicate the existence of three peptide permeases. These complementary systems, with somewhat overlapping specificities, are referred to as the oligopeptide, dipeptide, and tripeptide permeases. Related peptide transport systems have been characterized in other bacterial species (*see* Peptide Transport in Other Bacteria).

The Oligopeptide Permease

The presence of the oligopeptide permease (Opp) in *E. coli* was first demonstrated by the isolation of mutants resistant to the toxic peptide triornithine (*31*). In these mutants, transport of not only triornithine had been lost, but also transport of other tripeptides and higher oligomers was impaired, whereas dipeptide transport was only slightly decreased. Using a variety of transport assays, with radioactively labeled peptide substrates and fluorescence-labeling techniques (these various assays, their applications, and their limitations have been reviewed) (*7, 11, 32*), the substrate specificities and kinetic parameters for the Opp have been determined. The Opp has a high affinity for natural peptide substrates with typical affinity for transporter (K_t) values of 0.1–2 μM. The Opp system occurs in high copy number in the bacterial cell and functions efficiently to transport substrates rapidly and with high throughput, with typical maximum transport rate (V_{max}) values of 2–20 nmol min^{-1} (mg $protein)^{-1}$.

Good information on the structural specificities of the system is crucial if it is to be subverted for the uptake of peptide-based drugs. In fact, much useful information on substrate specificity has come from studies using peptide analogues synthesized as putative antibacterial agents. With respect to overall size, oligopeptides, up to and including hexapeptides, are substrates for the Opp. However, in the whole cell, this size restriction is controlled by the OM porins (*19–21*). Thus, by carrying out transport assays in whole cells, it is not possible to establish whether or not the Opp can handle peptides larger than those able to penetrate the porins. In the chemotherapeutic context, the effectiveness of a peptide-based drug may be limited by its penetrability through the OM, but it still remains of some interest to know the actual steric limitations of the Opp itself (this is considered again later).

Further structural requirements have been established using transport assays with chemically modified peptides. Of particular importance is the N-terminal α-amino group; for effective transport, oligopeptides require a positively charged, primary or secondary α-amino group. Modifications that do not satisfy this requirement (e.g., α-*N*-acyl or α-*N*-

dialkyl substitutions) effectively stop transport (*7, 11, 12*). A range of allowed modifications, including various alkyl substituents, have only slight effects on overall transport rates (*33*). In contrast to the crucial role of the N-terminal α-amino group, the C-terminal carboxyl group plays only a minor role in substrate recognition. Thus, the group may be removed completely or modified in various ways (e.g., esterification or amidation) without seriously impairing transport. This observation correlates with the specificity for the amino terminus because a system that can handle peptides of varied length, in which the spatial separation of N- and C-termini is markedly different, is unlikely to have recognition features for both termini. Clearly, this structural feature is important for the design of peptide-based antimicrobial agents. The Opp shows strong stereochemical specificity; peptides composed of all L-residues are transported most effectively. However, oligopeptides containing D-residues at positions at or towards the C-terminus do show a low transport activity; again, this result is compatible with the lesser importance for substrate recognition of this region of the molecule (*11, 34*). For good substrate transport, normal *trans*-α-peptide bonds are favored; however, some modifications can be tolerated, and methylation of the peptide-bond nitrogen still allows significant, if reduced, transport (*11, 35*). This particular observation is potentially important for the design of peptide drugs because this modification dramatically decreases peptidase-dependent hydrolysis, leading to markedly more stable peptide structures. Interestingly, alterations to the peptide backbone itself can be achieved without seriously affecting transport. For example, a range of peptide analogues containing aza or aminoxy residues were well transported (*36–38*); certain of these compounds were also antibacterial and will be considered again later (*see* Synthetic Peptide Carrier Prodrugs). Perhaps of most significance structurally is the extent to which variations in side-chain residues can be tolerated. The system has evolved to handle the enormous structural diversity inherent in the side chains of protein-derived oligopeptides. This versatility is its strength from a normal physiological viewpoint, but also its weakness if potentially deleterious, modified, or synthetic side-chain residues can be transported in peptide form.

In summary, therefore, the results of transport studies using natural and modified peptides showed the Opp to be a system that could be exploited for the targeting and uptake of peptidomimetic antibacterial agents. The Opp belongs to the class of transport systems that is sensitive to cold osmotic shock (*39, 40*). These systems require a periplasmic binding protein for the initial substrate recognition step and are energized directly by adenosine triphosphate (ATP) hydrolysis (*39–43*). Many early studies (*44, 45*) on oligopeptide transport in *E. coli* described the Opp as a binding protein-dependent system, and this was endorsed by the observation that the system in *S. typhimurium* is encoded by an operon of several genes as found for other shock-sensitive systems (*46*). A specific oligopeptide binding protein with an M_r value of ~58 K has been identified in both species (*14, 46, 47*). The protein is usually the most abundant of all periplasmic proteins, typically comprising ~5–8% of the total protein, and is unusual in being about twice the size of all other binding proteins characterized to date (*48*). The Opp binding protein OppA is the product of the first gene (*opp*A) of the *opp* operon (*46, 49, 50*). The genes for *E. coli* and *S. typhimurium* are ~75% homologous and encode proteins of 543 and 542 amino acids, respectively, with signal peptides that facilitate their secretion into the periplasm. The proteins have been purified and their substrate specificities characterized in several laboratories (*7, 14, 47, 51, 52*).

In early studies (*51*) on the substrate binding specificities of OppA from *E. coli*, equilibrium dialysis assays were used. In general, the results were in accord with the conclusions reached from transport assays. However, certain peptides known to be recognized by the Opp (e.g., Lys_3 and Orn_3) showed no binding. The technique of fluorescence emission spectroscopy was also used to study peptide binding, but this also gave confusing results; for example, certain peptides shown to bind to OppA by equilibrium dialysis failed to cause fluorescence shifts. Reasons why these procedures can give rise to anomalous results have been previously discussed (*7*). Assuming that the substrate specificities for overall transport would reside largely if not exclusively in OppA, and thus a convenient binding assay could form the basis for the design and testing of peptide carriers for prodrug delivery, we developed a simple filter-binding assay for OppA (*7, 14*). In this assay, binding of a radioactively labeled peptide can be determined directly as a ligand–OppA complex following its precipitation on a membrane filter; the relative affinities of other ligands can be determined from their abilities to compete for binding with a radioactively labeled substrate. These results supported and extended those obtained in transport assays and overcame the difficulties associated with equilibrium dialysis. The results from these filter binding assays could most simply be interpreted as arising from the interaction of a typical peptide with a single binding site on OppA, rather than with two or more separate binding sites, as had been suggested on the basis of the results of other assays (*51, 53*). The latter results underscore the applicability of the assay for determining the transport potential of novel substrates because the results with well-characterized substrates mirrored the substrate specificities and binding affinities found for their transport via the Opp. In a specific test of this principle (*54*), we found good correlation between binding affinities and antibacterial activities for a range of synthetic oligopeptides (*see* Synthetic Peptide Carrier Prodrugs). A novel finding, different from that observed for uptake in whole cells, was that peptides larger than those able to penetrate the porins (including ones of ~12 or more residues) are able to bind specifically to OppA (*14*).

Another important feature of the interaction of substrates with periplasmic proteins is that ligand binding causes conformational changes that facilitate attachment of the ligand–binding protein complex to its cognate membrane proteins, through which substrate translocation can then take place. Evidence for conformational changes in a range of binding proteins has come from application of various biophysical techniques (*48*). Different conformational forms have also been identified by X-ray crystallography, which typically reveals three conformational forms. To date all binding proteins have been shown to have a similar overall structure; that is, two lobes connected by short polypeptide hinges, with the substrate binding site located in a prominent cleft between the two lobes. In the native state, the proteins exist in an "open conformation". Initial ligand binding involves contact with only one domain (lobe) of the protein, to produce an "open-liganded conformation". Subsequently, binding to both domains occurs to produce a "closed-liganded conformation" in which the substrate is inaccessible to external solvent (*55*). To date, OppA has not been studied in this way, but evidence for substrate-induced conformational shifts has been obtained by isoelectric focusing (IEF). Thus, native OppA has an isoelectric point (pI) of 6.20, and two peptide-bound forms with pI values of 6.26 and 6.55 have been observed (*7, 14*). Conversion of the free to a liganded form of OppA is stoichiometric with peptide substrate. Observation of this pI shift provides a simple assay for substrate binding that complements the filter-binding assay described previously. As such, it offers a useful screening procedure to determine the binding and likely transport properties of peptidomimetic compounds.

It would be very desirable for the design of peptidomimetic compounds able to use Opp not only to have information about structural specificities based on various assays with substrates and analogues but also to obtain information directly about the nature of the active site in OppA. With the availability of such complementary structural details it would be feasible to use computer modeling to optimize and predict substrate docking at the binding site. As a step in this direction we have derivatized OppA from *E. coli* with a radioactively labeled peptide analogue carrying a photolabile azido residue that was shown to bind competitively with Ala_3. After selective proteolysis of the derivatized protein, a radioactively labeled fragment was sequenced to identify putative active site residues. Residues in the gene sequence of oppA corresponding to these putative active-site amino acids of OppA were modified by site-directed mutagenesis, and the mutant proteins were expressed and purified. Transformed cells of *E. coli* containing mutant protein showed alterations in sensitivity to triornithine and ability to transport nutrient peptides, and the purified proteins displayed changed binding characteristics (*56*). These studies are ongoing and should provide insight into the molecular nature of the substrate–protein interaction. However, it remains to be established whether the mutational change is restricted in its effect on substrate binding, or whether the mutation has additional (or primary) deleterious effect on the ability of OppA to undergo the conformational rearrangement required for it to interact with other membrane proteins of the Opp.

The *opp* operon of *S. typhimurium* comprises five genes that are transcribed in the order *opp*A, *opp*B, *opp*C, *opp*D, and *opp*F; all are essential for peptide transport via Opp (*49, 50*). The *opp*B and *opp*C genes encode membrane-bound proteins, the predicted overall topology of which satisfies the two-times-six transmembrane helix paradigm for active transporters (*57*). Similarities with analogous proteins of the prokaryotic maltose and histidine permeases and with functionally similar eukaryotic systems have been noted, and a generalized model for transport has been proposed (*58*). The *opp*D and *opp*F genes encode proteins of very similar amino acid sequence that possess binding sites for ATP (*59*). Studies (*42, 60*) with similar proteins from various permeases indicate that they function in the energization of transport. There is still uncertainty as to the membrane topography of this class of ATP-linked proteins, for which conflicting evidence points to them having either a peripheral, cytoplasmic location or extending through the membrane into the periplasm (*61*). All these membrane proteins presumably exist as a topographically defined complex similar to those described for the maltose and histidine permeases (*61, 62*); however, in these cases, two copies of the ATP-binding components (MalK and HisP) rather than the separate OppD and OppF exist.

The Opp belongs to a group of periplasmic transport systems that have been termed traffic ATPases (*40*). These bacterial systems belong to a wider family of membrane systems including mammalian transporters of considerable medical significance (*40, 43, 63, 64*). The membrane proteins of this super-family share a high level of homology to the conserved components of the bacterial permeases. Typically, the systems have two hydrophobic membrane proteins (analogous to OppB and OppC) and two ATP-binding components (corresponding to OppD and OppF). These four elements may either be present as separate components in a multiprotein complex or as a single fused polypeptide. It remains to be established whether any of the systems present in mammals, plants, and insects require the equivalent of periplasmic binding proteins. Mechanistically, it seems likely that the involvement of binding proteins will be limited to uptake systems needing to bind substrates present at low external concentrations. Nevertheless, the systems characterized in simple bacteria clearly offer excellent models for analogous mammalian systems, such as the multidrug resistance protein (*65*) and the cystic fibrosis transmembrane conductance regulator protein (*66*).

Finally, therefore, one can propose a model based on information from a variety of binding protein-dependent permeases for transport via the Opp. The first specificity-determining step involves peptide interaction with OppA. Through a series of binding stages, substrate-induced conformational changes occur in OppA, causing it to gain increased affinity for the membrane-bound complex, presumably most specifically for OppB and/or OppC. Attachment of the liganded-OppA will then promote a sequence of further conformational changes in the membrane proteins that gives rise to translocation. Details on these proposed changes require further studies. These conformational rearrangements will also lead to the binding and hydrolysis of ATP by OppD and OppF. A translocation pore is created within the transmembrane helical domains of the integral membrane proteins, possibly coordinated with dissociation of substrate via an open-liganded form of OppA. The presence of low affinity, substrate-recognition features vectorially aligned within the transmembrane pore could facilitate forward translocation of substrate; although such a mechanism is hypothetical, some evidence exists that in other permeases, the membrane proteins may possess low-level substrate-recognition properties (*67*).

The Dipeptide Permease

The existence of a specific dipeptide permease (Dpp) in *E. coli* became apparent when mutants defective in the Opp were characterized (*31*). Early studies on this system have been reviewed (*4, 5, 11*).

The substrate specificities for peptide transport by the Dpp are analogous to those for the Opp, but generally much stricter. Thus, specificity towards the N-terminal α-amino group is the same as in the Opp, but loss or derivatization of the C-terminal α-carboxyl group effectively stops uptake through the Dpp. Additionally, the requirement for L-amino acid residues is much more strict. However, dipeptides with modified peptide bonds (*N*-methyl) or extended backbones (aza, aminoxy) are transported. Interestingly, in these latter components, D-residues are better tolerated than normal (*36–38*). Transport is more influenced by the nature of side-chain residues than is observed in the Opp, but a wide range of natural and modified dipeptides is transported, but with varied kinetic parameters (*7, 68, 69*).

The Dpp is also a binding protein-dependent permease, although it has yet to be characterized to the same degree of detail as Opp. The gene *dpp*A, for the dipeptide binding protein DppA, has been sequenced and shown to code for a protein of M_r 56K (*70, 71, 72*); its amino acid sequence shows 27% identity to OppA from *S. typhimurium*. The DppA is an abundant periplasmic protein under typical growth conditions and is produced in slightly lower amount than OppA. The protein has been purified to homogeneity (*14, 71*) and has a pI of 6.1, as determined by IEF; this presumably is the native form, because it can be converted to either of two other conformational forms, with pI values of 5.9 and 6.0, by binding of dipeptide substrates (*14*). The substrate specificities of the Dpp have also been studied by techniques described for the Opp; for example, by assays with purified DppA, with mobility shifts on IEF gels, and with radioactively labeled peptides in a filter-binding assay. Again, these results showed that the structural specificities and relative affinities found for DppA paralleled those found for transport via the Dpp (*14, 69*).

These studies and detailed measurements of transport in appropriate *opp* mutants have somewhat surprisingly revealed that the Dpp can actually recognize tripeptides in addition to dipeptides, but typically at least 100-fold less well (*7, 14, 68*). Furthermore, peptide analogues that may lack peptide bonds but that may retain other important features of charge and spatial separation, such as 5-amino levulinic acid, also show binding to DppA (*54*) and transport (*73*). Thus, these relatively simple in vitro assays with purified proteins can be extremely useful for testing and predicting substrate and analogue uptake via the Dpp.

In *E. coli,* DppA also functions in peptide chemotaxis (*72*), leading to a scenario in which such bacteria may be pictured as taking part in a lemming-like, suicidal dash toward toxic dipeptide-based drugs!

The Tripeptide Permease

The tripeptide permease (Tpp) is the least well characterized of the three peptide permeases. Early studies on peptide uptake indicated the Tpp possessed specificities toward N- and C-termini similar to those reported for the Opp (*74, 75*). Later studies (*7, 68, 76*), in strains carrying better-documented mutations in Opp and Dpp, showed that Tpp can transport a variety of tripeptides and, less effectively, dipeptides, but fails to transport tetra- or higher peptides. Its specificity towards side-chain residues is more restrictive than found with the Opp, with best uptake being seen with peptides composed of hydrophobic residues, especially with N-terminal Met or Val residues (*76, 77*).

Mutations seemingly linked to the *Tpp* locus have been described for both *E. coli* and *S. typhimurium,* although discrepancies exist between the reported results (*76–78*). One such gene locus has been cloned from *S. typhimurium,* although no sequence data are available (*76*). It has yet to be established whether the Tpp belongs to the class of shock-sensitive or shock-resistant permeases, because supportive evidence for it being a shock-sensitive permeases have not been confirmed by identification of a Tpp binding protein, despite considerable efforts in this regard (*79*).

Although the Tpp is generally believed to play a relatively minor role compared with the other two peptide permeases, studies (*80*) on the regulation of its expression indicate that it could become relatively more important in the anaerobic environment of the mammalian gut. In terms of a target for antibacterial drugs, Tpp has surprisingly been shown to be the main route of uptake for the best-characterized synthetic peptidomimetic inhibitor produced to date; that is, the dipeptide analogue alafosfolin (*76*) (*see* Synthetic Peptide Carrier Proteins).

Peptide Transport in Other Bacteria

No other bacterial species have been studied in great detail like *E. coli* and *S. typhimurium.* Nevertheless, a variety of studies indicate the operation of comparable peptide permeases in a wide range of bacteria. In the context of this review, few of the species studied pose clinical problems, with only *Ps. aeruginosa* and certain Gram-positive species, such as *Staphylococcus aureus* (*Staph. aureus*) being likely targets for peptide-based antibacterial agents. Consequently, this discussion will not be extensive, but will only highlight general principles and molecular details that may prove useful in drug exploitation. A more detailed review of uptake in other microorganisms has appeared recently (*7*). The various species studied differ markedly in their cell envelope structures, and to simplify consideration of their peptide transport characteristics, it is convenient to group them according to these broad differences.

Among the Gram-negative organisms, *Ps. aeruginosa* is a clinically significant pathogen that is resistant to many antibacterial compounds widely used against other species; its resistance derives mainly from its atypically complex cell envelope structure. The OM contains a variety of both general and substrate-specific porins (*18, 24, 25*). Of particular relevance to peptide permeation through the OM is the porin OprD, which has been shown to be involved in the uptake of important β-lactams (*27*), such as imipenem. A range of di-, tri-, and tetrapeptides containing positively charged residues, such as lysine, have been shown to enter through this porin and to show competition for passage into the periplasm with imipenem (*27*). In general, penetration through porins in *Ps. aeruginosa* seems to be less effective (more controlled) than in *E. coli,* for example. There is still uncertainty as to the reasons for this because there is evidence that the porin diameters in *Ps. aeruginosa* are actually larger than in *E. coli*; current opinion favors the idea that either the porins are in a closed, gated state most of the time or that there is high friction between solutes and the porin pores (*18, 25*). Whatever the explanation, it seems certain that this feature contributes to the relatively slow rates of peptide transport that have been measured. Thus, systems analogous to the Dpp and Opp of *E. coli* have been described that have comparable affinities (e.g., K_t of ~10–20 μM for di- and tripeptides, respectively), but uptake rates of ~0.5 nmol peptide min^{-1} mg $protein^{-1}$ that are about two orders of magnitude slower than reported for *E. coli* (*7, 81*). Both systems show strict stereospecificity. For the Opp, alanyl peptides up to and including hexaalanine use the system.

Little information is available concerning the genes or the proteins for these systems, but it seems reasonable to expect similarities with the systems found in the enteric bacteria. Indeed antibodies raised against OppA from *E. coli* have been found to cross react with proteins present in the periplasmic fraction of *Ps. aeruginosa* obtained by a modified osmotic shock procedure (*54*). Thus, it is to be expected that the peptide permeases of *Ps. aeruginosa* should be susceptible to exploitation by peptidomimetic drugs in a manner similar to that established for *E. coli.* However, more information is needed concerning their structural requirements and their relationship to porin specificity before extensive progress can be made in designing peptide-based antibacterial agents active against this particularly refractive organism.

Many studies have been performed to determine the utilization of proteins by rumen microorganisms (*7*). In general, the studies have been complicated by the heterogeneous nature of the mixed microbial population in the rumen. However, even more recently when the bacteria primarily involved in protein utilization have been identified and obtained in pure culture, the conclusions are still confused because of the occurrence of secreted and cell-bound proteases and peptidases. However, evidence is accumulating for intact peptide uptake in species such as *Bacteroides ruminicola* and *Streptococcus bovis* (*82*). No particular insights into peptide permease function relevant to the current situation have emerged to date; but, inasmuch as these organisms are strict anaerobes, they may be useful models for other clinically relevant anaerobic organisms.

A similar general conclusion also applies to lactic acid bacteria, which comprise a further group of proteolytic, fastidious organisms that have been extensively studied. Many of these organisms secrete proteases and peptidases and also possess cell wall-bound peptidases. Nevertheless, they also have peptide transport systems able to accumulate intact peptides (*11, 83*). Consequently, peptides generally prove to be nutritionally superior to their constituent free amino acids. Recent studies (*83*) have clarified earlier results and shown that one of the best characterized species, *Lactococcus lactis* (*L. lactis*) possesses both di- and oligopeptide transport systems. The latter system shows broad similarities with the Opp of enteric bacteria. Of particular interest is the finding that the analogous Dpp is energized not by ATP, as in *E. coli,* but by coupling to the proton-motive force (pmf) (*83*). The implications of this observation for the molecular constitution and transport mechanism of the system remain to be explored. Clearly, if the system were to comprise a single protein, analogous to that of the lac permease in *E. coli* that is the paradigm for pmf-dependent systems and is endowed with both substrate recognition features and energy coupling ability, it could have important implications for considerations of peptide transport in a broader biological context. Of particular relevance here are the studies with various Gram-positive species, some of which are certainly candidates for peptide-based drug therapy. However, this possibility may not yet be sufficiently appreciated and, to date, the results are generally only of a preliminary nature. Thus, *Staph. aureus* is certainly susceptible to antibacterial peptide mimetics (*7, 9*). However, only one detailed report on the ability of *Staph. aureus* to transport peptides concluded that its main transport system recognizes both di- and tripeptides and could not readily be equated to either Dpp or Opp of the enteric bacteria (*84*). In contrast, two typical peptide permeases occur (*85*) among the *Enterococci* (formerly *Streptococci*). Thus, in *Enterococcus faecalis,* a predominant dipeptide uptake system is also used by tripeptides, anda relatively minor oligopeptide system exists. The substrate specificities are similar to those of the Dpp and Opp of *E. coli,* respectively. However, rates of peptide transport are up to 10-fold greater in *Enteroccus faecalis,* making it a very attractive target for targeting by peptidomimetic drugs (*85, 86*). Evidence has been presented for a dependence of peptide uptake on a pmf in this organism also (*86*). In addition to these two general peptide permeases, another system exists with specificity apparently restricted to di- and tripeptides with N-terminal glutamyl or aspartyl residues (*87*). If such an observation were to be made in a Gram-negative bacterium, speculation about the possible involvement of an OM porin with specificity for such anionic substrates would occur. However, such a possibility does not arise in this Gram-positive organism. Therefore, the presence of a cytoplasmic membrane permease of limited specificity remains the preferred explanation.

These studies with Gram-positive bacteria indicate strong similarities to the systems found in Gram-negative bacteria. However, Gram-positive bacteria lack a periplasm and therefore were not expected to possess the equivalent of periplasmic binding protein-dependent permeases; surprisingly, however, such is the case. For example, *Streptococcus pneumoniae* (*Strep. pneumoniae*) possesses analogues to periplasmic binding proteins that are lipoproteins anchored to the external surface of the cytoplasmic membrane. Extensive sequence homology occurs between such lipoproteins and analogous Gram-negative periplasmic binding proteins (*88*). In particular, the complete nucleotide sequence of the *ami* operon of *Strep. pneumoniae* shows striking similarity to the *opp* locus of *S. typhimurium* (*89*). The *ami* operon encodes six proteins, with the first gene product AmiA being analogous to OppA; extensive homologies also exist between the membrane protein components of the two species. Analogous results have been reported for peptide transport in *Bacillus subtilis* (*90–93*). Similarly, among the oral streptococci, the species considered of prime importance in the causation of dental caries is *Streptococcus mutans* (*Strep. mutans*), and this also possesses lipoprotein-dependent transport systems (*94*). These organisms would be expected to possess peptide permeases, for peptides are common constituents of their oral environment (*95*). Thus, the model studies with the enteric bacteria should find direct application to these various Gram-positive species, several of which are of potential clinical importance.

Finally, it is noteworthy that the distinctive features of binding proteins from Gram-negative and Gram-positive bacteria seem to have become merged in *Haemophilus influenzae.* This is a Gram-negative bacteria that resembles *Strep. pneumoniae* in many respects and has been reported to possess a lipid-anchored binding protein with marked homology with DppA from *E. coli (96).*

Regulation and Energetics of Peptide Permeases

Appreciation of the possibility that the various peptide permeases present in bacteria may be susceptible to changes in their expression and activity under different environmental and physiological conditions clearly has important implications for their possible exploitation by peptide-based drugs. Thus, a priori, one should not expect to find identical peptide transport characteristics for bacteria grown under typical laboratory conditions and for those found in a clinical situation in a mammalian host. However, when attempts have been made to look for ways in which peptide transport might be regulated, using enteric bacteria as model species, the important general conclusion to emerge is that peptide transport activity is expressed at high activity under almost all tested conditions (*7, 97*). Many of the observations made are understandable in terms of the need to integrate peptide transport with maintenance of intracellular amino acid pools (*97*).

In enteric bacteria, the *opp* operon is transcribed from a promoter upstream of the *opp*A gene. However, OppA, like other binding proteins, is produced at a level ~30–100 fold greater than its cognate membrane proteins; the molecular mechanism that regulates this differential expression is presently unclear (*49*). In general, the presence of amino acids or peptides in growth media does not regulate peptide permeases in *E. coli* (*46, 98*). However, leucine can enhance peptide uptake, an effect mediated by the leucine-responsive regulatory protein Lrp (*78, 80*). The OppA synthesis also appears sensitive to the availability of phosphate, with limiting phosphate causing a decrease in OppA synthesis (*99*). Nutritional limitation by phosphate may be important for growth of bacteria in vivo, making this observation of potential significance for application of peptidomimetic agents. Very similar conclusions have been reached with the Dpp to those just described for the Opp. However, in a particular strain of *E. coli*, JM101, an exceptionally high level of DppA is produced when cells are grown in minimal glucose media. This high level of DppA production is decreased to more normal levels by addition of a mixture of amino acids; the control appears to operate at the level of transcription, but the possible implication of this finding for regulation of the Dpp in general is presently unclear (*71*). In *S. typhimurium*, anaerobiosis has been shown to enhance expression of the Tpp (*80*), although the situation in *E. coli* is less clear. In fact, even for the few effectors found (e.g., Leu, anaerobiosis), varied effects have been noted between these two species. A more detailed consideration of this topic has been published (*7*).

An understanding of the way in which peptide transport is energized is also important when considering the possible application of peptide-based drugs in vivo. In enteric bacteria, binding protein-dependent permeases are energized by ATP (*42*). Thus, peptide uptake via the Dpp and Opp depends on a continuing source of intracellular ATP, but details are scant of exactly how and when ATP is used in the translocation cycle. The stoichiometry of ATP to substrate transport has yet to be established. Other species apparently have peptide permeases that use a proton motive force directly (*7, 83, 86*). This situation is similar to that for peptide transport in the mammalian gut and in higher plants (*5, 100*). The inherent efficiency in translocating several amino acid residues simultaneously in the form of a peptide, compared with their individual transport via amino acid permeases cannot be overemphasized. It is presumably this efficiency that will guarantee the existence of exploitable peptide permeases in all clinically important microorganisms. With peptide uptake, there is also the possibility of additional energy gains from the efflux of certain of the peptide-derived amino acid residues down an electrochemical gradient (*45*). This type of efflux, which derives in large measure from the need to maintain optimum concentrations of individual amino acids in the cytoplasmic pool, has been described in many species varying from *E. coli* through *Staph. aureus* to *L. lactis* (*44, 84, 101*). Ways in which this efflux might produce energy gradients and drive counterflux were discussed previously (*45*), and examples were described in which the energy arising from solute efflux was recycled to energize further substrate uptake (*83, 102*).

Peptide Permease Exploitation by Antibacterial Peptides

The rationale for this chapter derives from the current acceptance that bacterial peptide permeases offer an excellent target for drug delivery. It has taken ~20 years for this view to become firmly established. Nevertheless, further insight is needed into various fundamental aspects before the pharmaceutical industry will be in a position to exploit fully the commercial potential of this designer drug approach.

The experimental basis for current ideas in this area extends back a very long way. The observation that peptides could be nutritionally superior to the equivalent mixtures of constituent amino acids for most microorganisms was made in the last century, but only when knowledge of the mechanism of protein synthesis and the existence of specific transport systems became established did an explanation emerge; the relevance of these early studies have been reviewed (*2, 3, 5*). When in the 1950s and 1960s the first synthetic peptides containing inhibitory amino acid analogues were studied for their antibacterial activities, their modes of action were often interpreted incorrectly (*3, 8*). Subsequently, systematic studies provided evidence for the ubiquity of microbial peptide permeases and led to characterization of their substrate specificities. A developing appreciation of the enormous structural variety inherent in these general substrate specificities raised the possibility that peptides comprising derivatized amino acid residues or unnatural analogues might also act as substrates. It then became quite logical to suggest that if any such derivatized residues or unnatural analogues were intrinsically unable to enter a microorganism in free form, then if they were incorporated into a peptide structure without vitiating the substrate specificities of the peptide permeases, the impermeability problem should be overcome. Thus, the idea was born that if such impermeant moieties were potentially inhibitory to intracellular targets, then they could be delivered into the cytoplasm in the form of peptide carrier prodrugs targeted through peptide permeases (*33*).

Since the concept was proposed, a variety of terms have been proposed to describe peptide prodrug transport; these terms include illicit transport, coined by Ames and coworkers (*103*); warhead delivery, proposed by Ringrose (*8, 9*); and portage transport, suggested by Gilvarg (*10*). The peptide–carrier complexes have been termed smugglins by Payne (*3*). To function as a prodrug, the smugglin must be cleaved intracellularly to release its toxic amino acid mimetic. A crucial enhancement of the toxicity of the warhead derives from the fact that, having been actively transported, it can accumulate to a high intracellular concentration. This accumulation is because the warhead is unable to cross the membrane except as a smugglin component and consequently cannot undergo exodus once it has been released. Figure 2 illustrates this process of peptide prodrug delivery.

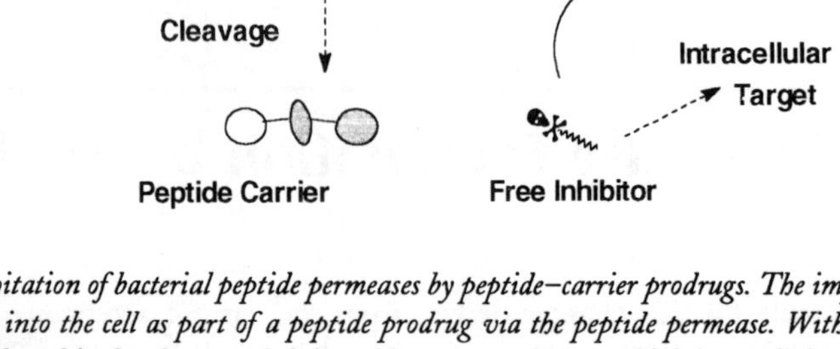

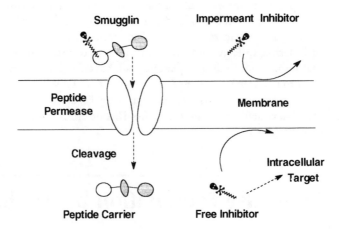

Figure 2. Exploitation of bacterial peptide permeases by peptide–carrier prodrugs. The impermeant inhibitor is transported into the cell as part of a peptide prodrug via the peptide permease. Within the cytoplasm, the inhibitor is released in free form, and, being able to accumulate to a high intracellular concentration, its activity against its target is increased.

The first experimental validation of the concept of illicit transport came independently and simultaneously from two groups, with smugglins containing histidinol phosphate (*103*) and homoserine phosphate (*104*). However, these compounds contained nutritionally active residues rather than inhibitory moieties. Subsequently, however, attention has been focused on antibacterial smugglins. To aid consideration of these varied compounds, it is helpful to consider examples of natural and synthetic smugglins separately.

Natural Antibacterial Peptides

It was somewhat humbling to those working on smugglins and illicit transport to realize, rather belatedly, that Nature had long before recognized the merits of the concept and devised many examples to capitalize on it. A range of natural peptide smugglins has now been recognized, and discovery of additional ones can be expected. In fact, the pace of discovery of new natural smugglins might be anticipated to increase given the recent appreciation that while searching for such compounds, the peptide components commonly incorporated into antibiotic test media must be omitted.

Among the natural smugglins found to date, alanine is the most commonly occurring carrier residue in both dipeptides and oligopeptides. The first natural smugglin to be characterized was the dipeptide bacilysin (bacillin and tetaine) (*see* Chart IA) produced by *Bacillus subtilis* (*105*). The toxic mimetic moiety is anticapsin [L-β-(2,3-epoxycyclohexanono-4) alanine] that is linked to an *N*-terminal alanine carrier residue. Anticapsin itself is not antibacterial because it cannot enter bacteria in its free form; but, when combined in bacilysin, it inhibits both Gram-negative and Gram-positive bacteria (*105, 106*). Bacilysin is transported almost exclusively through dipeptide permeases and after transport is hydrolyzed to yield intracellular anticapsin, which functions as a glutamine analogue to inhibit glucosamine synthetase and thus peptidoglycan synthesis (*107*). Another antibiotic dipeptide is lindenbein (Chart IB), which carries the N-terminal residue N^3-fumaramoyl-L-2,3-diaminopropanoyl-L-alanine (*8*). A range of synthetic analogues has also been prepared that has contributed to an understanding of the contribution of the various components to transport, hydrolysis, and enzyme inhibition (*106*).

Many natural smugglins use phosphorus-containing amino acid mimetics as the toxic warhead. Among these is phaseolotoxin, which is produced by *Pseudomonas syringae* pv. *phaseolicola* as a mixture of components. The structure originally proposed for the main component was (*N*-phosphosulfamyl)-ornithylalanylhomoarginine, although a reinvestigation of its structure has indicated it to be a phosphotriamide (Chart IC) (*108*). The compound inhibits a range of bacteria through action on ornithine carbamoyltransferase after uptake through the oligopeptide permease (*109*). The genes for phaseolotoxin synthesis are clustered and have recently been cloned (*110*). Another interesting example is the tripeptide bialaphos (phosphinothricyl-Ala-Ala) (Chart ID), the toxic moiety of which is L-2-amino-4-methylphosphinobutyric acid, which is a δ-phosphinate analogue of glutamic acid. The free mimetic, but not bialaphos, inhibits isolated glutamine synthetase, whereas only the smugglin shows antibacterial activity (*111–113*). Bialaphos is actually the main component of a mixture of related smugglins produced by *Streptomyces*. Each smugglin contains the mimetic phosphinothricin (PPT) linked to a variety of amino acid residues; for example, PPT-Gly-Ala, PPT-Ala-Ser, PPT-Ala-Gly, PPT-Ala-α-aminobutyric acid, and PPT-Ala-Ala-PPT (*113*). The recurrent observation that natural smugglins are produced as mixtures has potentially important implications for the design of synthetic smugglins and, as such, will be considered again in the section Prospects for Rationally Designed Peptide Prodrugs. Other tripeptide antibiotics related to bialaphos in structure and antibacterial profile are L-(N^5-phosphono)-methionyl-*S*-sulphoximinyl-Ala-Ala (Chart IE) (*111, 112*)

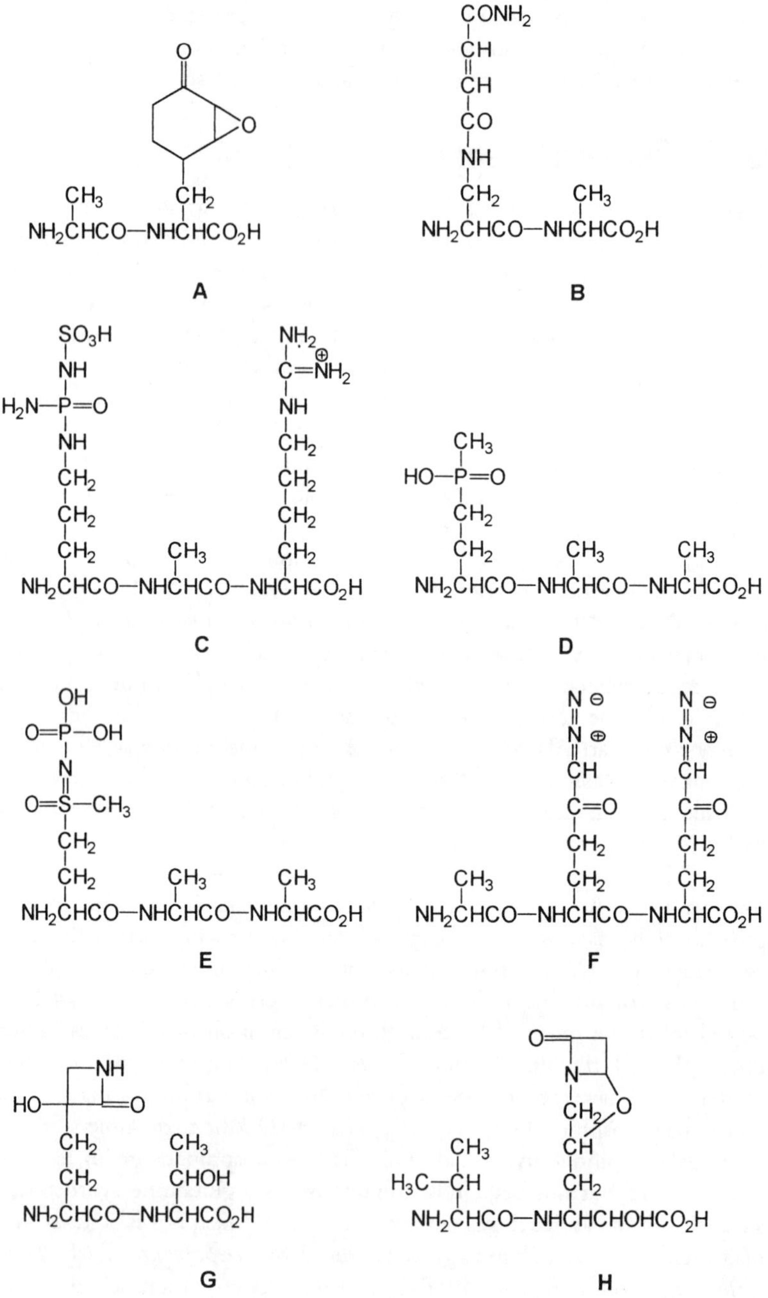

Chart I. Structures of some natural antibacterial smugglins: A, bacilysin; B, lindenbein; C, phaseolotoxin; D, bialaphos; E, (N^5-phosphono)-methionyl-S-sulphoximinyl-Ala-Ala; F, alazopeptin; G, tabtoxin; H, valclavam.

and Gly-Leu-2-amino-2-propenyl phosphonate (*114*). Related tripeptides that include the glutamine analogue 6-diazo-5-oxo-norleucine are alazopeptin (Chart IF) and duazomycin B. Many of these natural smugglins incorporate glutamine analogues that function by interfering with glutamine utilization or synthesis, which means they are likely to show mammalian toxicity and therefore be unsuitable models for clinical compounds.

A novel class of natural smugglins includes those that contain β-lactam rings; for example, the phytotoxin tabtoxin (Chart IG) (*8*). A particularly interesting class is the clavams, which differ from conventional β-lactams in possessing an oxygen atom in place of the sulfur atom at position 1 (*115*). Valclavam (Chart IH) is one such antibiotic that incorporates the β-lactam residue hydroxyethylclavam, the toxicity of which is enhanced by about three orders of magnitude when incorporated into smugglin form. Valclavam is transported by both the Dpp and Opp, a highly desirable structural feature that should be noted when designing synthetic smugglins. Other antimicrobial compounds that possess β-lactam structures and are currently used clinically are orally absorbed through intestinal peptide transporters (as discussed in Chapter 7 by Kramer and co-workers). Thus, there would seem to be several good reasons why compounds of this type merit serious consideration as synthetic smugglins. However, conventional β-lactam drugs are believed not to cross the cytoplasmic membrane, their target being particular enzymes of peptidoglycan synthesis that occur in the cell envelope. This situation raises two considerations. First, the β-lactam moieties contained within smugglins such as valclavam must clearly have different intracellular targets from those of conventional β-lactams. Second, the possibility exists that the mechanism involved in the induction and secretion of β-lactamases active against conventional β-lactams might in part be triggered by limited uptake of β-lactams (or their cleavage products) through peptide permeases.

More examples of natural smugglins were identified based on different carrier residues and varied toxic amino acid mimetics, and their structures and general properties were reviewed (*7–9*). In addition, a number of natural smugglins show selective activity against fungi, amongst which the polyoxins and nikkomycins have been extensively studied and are described by Naider and Becker in Chapter 16.

Synthetic Peptide Carrier Prodrugs

In the last decade, the synthesis and testing of various synthetic smugglins have been described in a range of studies. These studies have contributed significantly to the current understanding of peptide permease function and specificity. They have also led to the discovery of novel antibacterial compounds, some of which have shown clinical potential. However, the full potential of this rational approach to drug design has yet to be realized. The most extensive studies are those carried out within the pharmaceutical industry by the Roche group. These studies involved production of a wide range of phosphonopeptides, the best studied example of which is the dipeptide alafosfalin (alaphosphin, Ro-03-7008 (Chart II A) in which an N-terminal alanyl residue serves to transport the toxic alanine mimetic L-1-aminoethylphosphonic acid [Ala(P)] (*9*). The mimetic itself is not taken up by bacteria and when transported as a smugglin it cannot be effectively effluxed (although its carrier moiety can be), allowing it to become accumulated to concentrations 1000-fold greater than that of the exogenous smugglin.

The alanine mimetic inhibits alanine racemase and consequently prevents peptidoglycan synthesis. Alafosfalin shows broad-spectrum antibacterial activity. Many other prodrug variants were synthesized in which the peptide carrier component was varied; variation in the antibacterial activities of these compounds could be related to the individual substrate specificities of the peptide permeases and to intracellular peptidases of the tested species

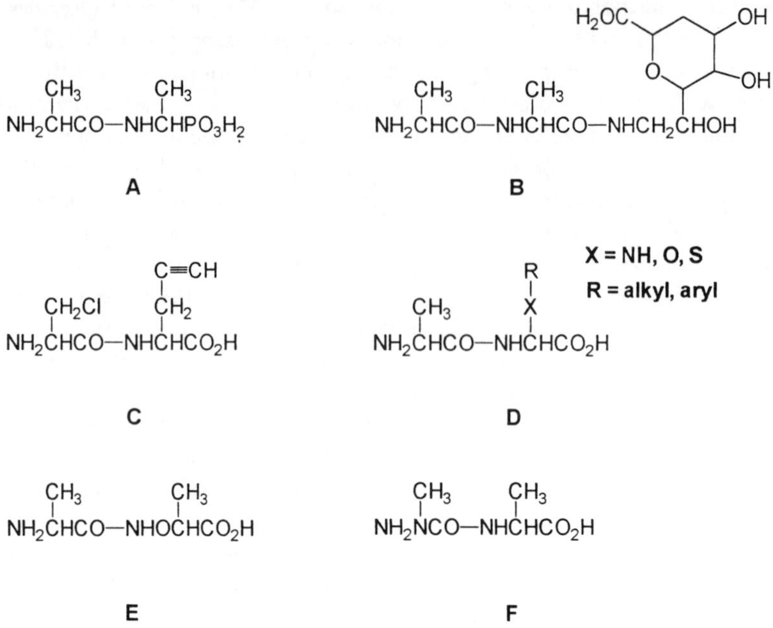

*Chart II. Structures of some synthetic antibacterial smugglins: A, alafosfalin; B, Ala-Ala-α-C-[1,5-anhydro-7-amino-2,7-dideoxy-D-*manno*-heptopyranosyl]-carboxylate; C, β-chloroalanylpropargylglycine; D, various α-glycyl substituted derivatives of Ala-Gly; E, Ala-aminoxyAla; F, α-AzaAla-Ala.*

(*116–119*). In general, oligopeptide forms, such as Ala-Ala-Ala(P), possessed a wider antibacterial spectrum in vitro than their corresponding dipeptide homologues. Thus, these studies showed how iteration between smugglin synthesis and testing could lead to the elaboration of optimized compounds, designed either as wide-spectrum antibacterial agents or more specifically tailored for use against particular species. Ironically, although much of the early work was driven by information relating to the Opp and Dpp of *E. coli* (*120*), alafosfalin is actually transported predominantly through the Tpp, with minor accumulation through the other two permeases (*7, 12, 76, 119*). Resistance to alafosfalin is not a significant problem; when resistant mutants are selected for in vitro, permease-deficient strains are generally found most frequently, although peptidase-defective mutants can also be isolated (*7, 12, 38, 85, 118*). In later studies, the related mimetic aminomethylphosphonic acid was incorporated into a range of smugglin prodrugs (*121*). This mimetic acted as an analogue of both L- and D-alanine and inhibited not only alanine racemase but also UDP-*N*-acetylmuramyl-L-alanine synthetase and D-Ala-D-Ala synthetase. However, smugglins based on this mimetic showed less antibacterial potency than those containing Ala(P), and phosphonate analogues of other amino acids were also generally inactive (*121*).

Alafosfalin and analogues were tested in vivo for oral absorption and pharmacokinetics (*122*). The enhanced activity of oligopeptide analogues in vitro was generally not carried through into the studies in vivo, probably because the analogues were hydrolyzed by host peptidases. When prodrugs were designed specifically to be resistant to such hydrolysis [e.g., Sar-L-Nva-L-Nva-Ala(P)], a wider antibacterial spectrum and potency resulted, although the benefits of stability were countered by lowered intestinal absorption (*121*). Thus, these studies did much to promote the concept of smugglin prodrugs. Although compounds of useful chemotherapeutic potential were identified, problems with long-term toxicity of Ala(P) caused the compounds to be withdrawn from extended clinical trials. Thus, although this class of mimetic was not commercially successful, the studies provided models and morals for future commercial success.

Trying to capitalize on the potential of this mimetic, other researchers synthesized related analogues. A series of phosphonodipeptides containing various C-terminal aminoalkylphosphonic acids showed good antibacterial activities (*123*). Compounds based on Ala-MeAla(P) and LeuMeAla(P) were good inhibitors of alanine racemase and showed significant antibacterial activity (*124*). Other good inhibitors of this enzyme, such as β-chloro-α-aminoethylphosphonic acid have yet to be tested in smugglin form (*125*).

A good example of the application of a rational approach to smugglin design is seen with compounds based on analogues of β-ketodeoxyoctonate (KDO). These mimetics act as inhibitors of Cmp-KDO synthetase and, thus, as prodrugs targeted specifically against lipopolysaccharide synthesis in Gram-negative bacteria. The peptide derivatives were accumulated predominantly through the Opp and showed broad-spectrum antibacterial activity (Chart IIB) (*126–129*).

A range of other mimetics has been incorporated into novel smugglins. Suicide substrates of alanine racemase, based on 3-halovinylglycines, were endowed with enhanced antibacterial activities when incorporated into dipeptides (*130*). A mechanism-based inhibitor of alanine racemase, β-chloroalanine, had its antibacterial activity increased by over three orders of magnitude when linked in dipeptide form (*131*). Similarly, a mechanism-based inhibitor of cystathionine-γ-synthase, L-propargylglycine (2-amino-4-pentynoate), also showed limited antibacterial activity as dipeptide-based prodrugs. However, dipeptides composed of two residues of each of these mimetics displayed enhanced activity, and mixed dipeptides containing one residue of each mimetic showed synergistic antibacterial action (Chart IIC) (*131*). More detailed studies showed that uptake of these smugglins was mainly through the Dpp and that their antibacterial activity paralleled transport activity (*132, 133*).

In an interesting extension of the smugglin concept, mixed dipeptides containing both these mimetics were incorporated into novel "pre–pro-drugs" of the general type of cephalosporin-C_{10}-dipeptide esters (*134*). The idea was to deliver these compounds intact to the vicinity of the Dpp, where the presence of periplasmic β-lactamases would cleave the "pre–pro" version to generate the prodrug. Good antibacterial activity resulting from inhibition of alanine racemase by β-chloroalanine was found with these compounds. The general idea of using peptidases to activate prodrugs in the vicinity of their targets has been discussed with various examples by Ringrose (*9*).

Gilvarg and co-workers (*135*) synthesized a novel class of smugglin in which the toxic moiety formed the side chain of the peptide carrier. For example, the thiol compounds 2-mercaptopyridine or 4-[*N*-(2-mercaptoethyl)-aminopyridine]-2,6-dicarboxylic acid were transported through the Dpp or Opp of *E. coli* as mixed dithiols with a cysteine residue of di- or tripeptides, respectively. The attached compounds underwent intracellular release by dithiol exchange in the cytoplasm.

The loose specificity of peptide permeases with respect to side-chain residues has been exploited in the synthesis of various smugglins that contained nucleophilic moieties, such as aniline, phenol, or thiophenol, attached to the α-carbon of a glycine residue (Chart IID) (*136*). The smugglins are accumulated through the Dpp and Opp and, following intracellular peptidase action, the α-substituted glycine is destabilized and its nucleophilic substituent is released. When present in prodrug form, several of these compounds showed antibacterial activity.

In looking for targets against which smugglins might be designed, the lysine biosynthetic pathway appears attractive. Following up this idea, analogues of diaminopimelic acid have been incorporated into peptide form and have been shown to have antibacterial activity based on interference with peptidoglycan synthesis (*137, 138*).

Most of the synthetic smugglins produced to date have involved linkage of a mimetic to either the side-chain residues or N- or C-termini of the peptide carrier. In contrast, a limited range of studies in which the effects of modifications to the peptide backbone have been investigated were conducted (*36–38*). For example, the normal peptide linkage (–CO–NH–) has been extended to produce aminoxy (–CO–NH–O) (Chart IIE) or hydrazine (–CO–NH–NH–), reversed to produce retro-bonds (–NH–CO–), or the α-carbon was replaced by a nitrogen atom (α-aza) (Chart IIF). Studies (*36–38*) with these compounds showed that many were able to be transported, and this further defined the substrate specificities of the permeases. Most, however, did not possess antibacterial activity, with the exceptions of the hydrazino and aminoxy derivatives. Particularly interesting were the aminoxy compounds; analogues containing D-aminoxy propionic acid (D-OAla) were strongly antibacterial and, in contrast to peptides containing typical D-residues, were well transported.

An interesting collection of peptides containing several different derivatives of 2,3-diaminopropanoic acid (including the N^3-fumaramoyl substituent found in lindebein, *see* Chart IB) possess both antibacterial and antifungal activity (*139*). The mimetics are glutamine analogues that are active against glucosamine-6-phosphate synthase. The uptake of a range of these smugglins occurs through the Opp and Dpp, and a good correlation exists between their affinities for the purified binding proteins and their respective antibacterial activities (*54*).

Finally, as a variant on the idea of synthesizing smugglins containing residues of toxic amino acid mimetics, there are also a number of examples in which recognized antimicrobial agents have been attached to peptide carriers in attempts to enhance their efficacy (*9*). However, in no instance has increased potency ensued; the reasons for this are unknown and no information was obtained on the transport ability of the compounds.

In addition to the examples of antibacterial smugglins just considered, a variety of synthetic peptide carrier prodrugs active against fungi have been developed. These are considered by Becker and Naider in Chapter 16.

Prospects for Rationally Designed Peptide Prodrugs

The ability to deliver an almost limitless resource of synthetic inhibitors against normally inaccessible, and hence protected, intracellular targets by incorporating them into peptide carrier prodrugs is a singularly attractive idea. It seems certain that antibacterial drugs designed to capitalize on this Trojan horse concept will be forthcoming. More work is needed, however, to guide the design of novel synthetic compounds with clinical utility. Certainly, there is no other drug-delivery mechanism for which naturally occurring examples can be found that offers so much potential for rational exploitation and thus commercial reward.

A framework for the design of antibacterial smugglins is now in place. Nevertheless, it is still not possible to suggest the optimal peptide carriers for synthesis of prodrugs active against particular species. Further studies are needed to clarify this situation. In considering some of these features, the discussion will be limited to relevant aspects of bacterial physiology. Many other factors contribute to the efficacy of a peptide prodrug, but such features as oral availability and pharmacokinetics are beyond the scope of the present discussion, although some are considered elsewhere in this volume.

Ideally, for smugglin application against particular pathogenic bacteria, one would wish to know the following: (1) the number of peptide permeases present, (2) the substrate specificities and kinetic parameters of these permeases, (3) their regulation and their expression and activity in vivo, and (4) information on the likely mechanisms of resistance to the peptide prodrugs. Some discussion of these topics has appeared previously (*12, 14*).

Information on the first of these topics is already available for many pathogenic species, and has just been discussed (*see* Peptide Permeases of Enteric Bacteria and Peptide Transport in Other Bacteria). Extension of the techniques used in gaining this information should make it relatively simple to obtain this fundamental information for other model and problem organisms. Information on unknown permease genes may be obtained using clones of well-characterized peptide permeases to probe the genomes of heterologous species, if sufficient homology exists. In a complementary manner, antibodies raised against well-characterized peptide permease proteins can be used to search for cross-reacting proteins in other species. The isolation of peptide permease mutants resistant to defined peptide smugglins should also aid characterization.

The structural and kinetic parameters for any newly identified peptide permeases can, in the first instance, be established by comparison with other well-characterized ones. Procedures for determining transport rates with radioactively labeled substrates or by fluorescence-labeling techniques are in place (*see* The Oligopeptide Permease). Alternatively, substrate specificities can be explored using various ligand-binding assays (*see* The Oligopeptide Permease). A small collection of model peptides possessing relevant structural variations is useful for such studies.

Understanding of the controls and conditions that can influence the expression of peptide permeases in a clinically relevant situation is far from complete (*see* Regulation and Energetics of Peptide Permeases). It is a reasonable assumption that the permeases are well expressed and active under such conditions, and evidence from studies in vivo and clinical trials with alafosfalin and analogues support this assumption (*see* Synthetic Peptide–Carrier Prodrugs). Nevertheless, the need for fundamental studies to clarify regulatory mechanisms, even with such a well-studied bacterium as *E. coli,* still remains.

Successful application of an antibacterial, peptide prodrug will depend not only on its inhibitory activity and antibacterial spectrum, but also on the ease with which bacterial resistance arises. Inhibitory activity, as measured typically in vitro by the index of minimum inhibitory concentration (MIC), will depend largely on the effectiveness of the carrier moiety in delivering the warhead, and on the intrinsic toxicity of this component against its target. In judging the utility of a new agent, perhaps too much attention is currently paid to the MIC component and too little to the matter of resistance. With peptide prodrugs, resistance in principle can arise in several ways. First, the target of the drug moiety may mutate to resistance; this possibility is common to all antibacterial agents and, therefore, need not be considered specifically in this context. Activation of the prodrug, typically through intracellular peptidase activity with peptide-based smugglins, is essential, and resistance could arise through mutational loss of this ability. However, this does not appear to be a particular problem because, in general, bacteria possess an array of peptidases, several of which may function to release the toxic moiety. Nevertheless, it is clearly helpful to acquire information on the enzymes that could serve to activate a prodrug in a target pathogenic species. Finally, resistance may arise from mutational loss of peptide permease activity. However, the presence of several peptide permeases in most species should help to minimize this potential problem. The observation that the Opp in *E. coli* can mutate at an unusually high frequency (*11, 31*) has yet to be shown to be a problem for the application of peptide smugglins in vivo.

It follows from these considerations that in the design and use of synthetic peptide smugglins, one should be cognizant of the desire to achieve high antibacterial activity combined with the ability to counter the development of resistance. In approaching this objective we should be well advised to learn from the ways in which Nature has responded to a similar challenge in various situations. One approach is for the smugglin to carry two or more distinct warheads that are targeted at different enzymes; phaseolotoxin provides a good natural example of this idea (*see* Natural Antibacterial Peptides). Adoption of this strategy could produce compounds with possible synergistic effects between the individual toxic moieties. A few synthetic examples of this approach already exist (*see* Synthetic Peptide Carrier Prodrugs). An alternative strategy is to devise a carrier that allows a particular smugglin to be accumulated through more than one permease. Again, several natural examples exist that successfully capitalize on this principle (*see* Natural Antibacterial Peptides); however, good information on permease specificities is needed to achieve this end with synthetic compounds. As an alternative to this, use of a mixture of smugglins seems to be an attractive approach. Thus, the drug component could be linked to two or more peptide carriers that would ensure uptake through several peptide permeases. This particular ploy is one that Nature has used in many instances, and several peptide-based antibiotics are produced as mixtures with different types and numbers of amino acid residues (*see* Natural Antibacterial Peptides). Relevant examples are phaseolotoxin, bialaphos, nikkomycins, and polyoxins. It seems certain that the production of such mixtures is functionally beneficial (rather than reflecting the chance involvement of biosynthetic enzymes of loose specificity) because they offer advantageous variety in routes of delivery (by various permeases) and prodrug activation (by various peptidases) together with greater inherent resistance to enzymatic inactivation. The use of synthetic smugglins as mixtures in combination therapy seems particularly attractive, although the pharmacokinetics of each component would need to be compatible. As discussed, this approach should guard against development of bacterial resistance and offers potential for synergy between different compounds showing toxicity toward varied intracellular targets. In addition, peptide prodrugs may display enhanced characteristics in combination with other types of compounds. For example, smugglins can show improved antibacterial effects in combination with β-lactams (*129, 131, 140*), other antibiotics (*129, 139, 141*), and even amino acids (*111, 112*).

In the process of deciphering the optimal characteristics for a generic peptide carrier, it is now convenient to combine the results obtained from the application of biochemical and molecular biological techniques with those of computer modeling. Thus, data for "good" and "poor" transport substrates can be conveniently studied and compared using graphical presentations of computer-derived peptide models. Such models should take into account current advances in the structural analysis of small peptides using X-ray crystallography studies on peptide crystals, and other biophysical approaches to the determination of preferred peptide conformations in aqueous solution. In pursuing this integrated approach to the design of peptide prodrugs, it is our expectation that structural templates for ideal peptide carriers will emerge. With such three-dimensional templates available, it is our aim to use techniques analogous to that of homology modeling to identify non-peptide carrier mimetics that match these molds. In this way, proposals for peptidomimetic carriers and drugs should emerge that retain desirable transport and activity functions but, for example, are resistant to hydrolysis by mammalian enzymes. In this search, the rate-limiting element continues to be resources rather than ideas; this particular Holy Grail may be just around the corner, but long may the excitement of such a quest continue.

Acknowledgments

I am grateful to my many collaborators over the years for sharing my appetite for peptides. Thanks are due to current members of my laboratory whose work is quoted here:

Gillian M. Payne, David R. Tyreman, Mark W. Smith, Neil Marshall, and Cordelia M. Schuster.

References

1. Payne, J. W.; Gilvarg, C. *Adv. Enzymol. Relat. Areas Mol. Biol.* **1971,** *35,* 187–244.
2. Payne, J. W. In *Peptide Transport in Protein Nutrition;* Matthews, D. M.; Payne, J. W., Eds.; North Holland: Amsterdam, The Netherlands, 1975; pp 283–364.
3. Payne, J. W. *Adv. Microb. Physiol.* **1976,** *13,* 55–113.
4. Payne, J. W.; Gilvarg, C. In *Bacterial Transport;* Rosen, B. P., Ed.; Marcel Dekker: New York, 1978; pp 325–383.
5. Matthews, D. M.; Payne, J. W. *Curr. Top. Membr. Transp.* **1980,** *14,* 331.
6. *Microorganisms and Nitrogen Sources;* Payne, J. W., Ed.; John Wiley & Sons: Chichester, England, 1980.
7. Payne, J. W.; Smith, M. W. *Adv. Microb. Physiol.* **1994,** *36,* 1–80.
8. Ringrose, P. S. In *Microorganisms and Nitrogen Sources;* Payne, J. W., Ed.; John Wiley & Sons: Chichester, England, 1980; pp 641–692.
9. Ringrose, P. S. In *Scientific Basis of Antimicrobiol Chemotherapy;* Greenwood, D.; O'Grady, F., Eds.; Cambridge University: Cambridge, England, 1985; pp 219–266.
10. Gilvarg, C. In *The Future of Antibiotherapy and Antibiotic Research;* Ninet, L.; Bost, P. E.; Bouanchand, D. H.; Florent, J., Eds.; Academic: Orlando, FL, 1981; pp 351–361.
11. Payne, J. W. In *Microorganisms and Nitrogen Sources;* Payne, J. W., Ed.; John Wiley & Sons: Chichester, England, 1980, pp 211–257.
12. Payne, J. W. *Drugs Exp. Clin. Res.* **1986,** *12,* 585.
13. Thomas, W. A. *Biochem. Soc. Trans.* **1986,** *14,* 383.
14. Tyreman, D. R.; Smith, M. W.; Payne, G. M.; Payne, J. W. In *Molecular Aspects of Chemotherapy;* Shugar, D.; Rode, W.; Borowski, E., Eds.; Springer Verlag: Berlin, Germany, 1992; pp 127–142.
15. Jaspers, C.; Penninckx, M. *FEBS Lett.* **1981,** *132,* 41.
16. Payne, G. M.; Payne, J. W. *Biochem. J.* **1984,** *218,* 147.
17. Benz, R. *Ann. Rev. Microbiol.* **1988,** *42,* 359.
18. Nikado, H. *Mol. Microbiol.* **1992,** *6,* 435.
19. Payne, J. W.; Gilvarg, C. *J. Biol. Chem.* **1968,** *243,* 6291.
20. Andrews, J. C.; Short, S. A. *J. Bacteriol.* **1985,** *161,* 484.
21. Alves, R. A.; Gleaves, J. T.; Payne, J. W. *FEMS Microbiol. Lett.* **1985,** *27,* 333.
22. Sugawara, E.; Nikaido, H. *J. Biol. Chem.* **1992,** *267,* 2507.
23. Jeanteur, D.; Lakey, J. H.; Pattus, F. *Mol. Microbiol.* **1991,** *5,* 2153.
24. Hancock, R. E. W.; Egli, C.; Benz, R.; Siehnel, R. J. *J. Bacteriol.* **1992,** *174,* 471.
25. Nikaido, H.; Nikaido, K.; Harayama, S. *J. Biol. Chem.* **1991,** *266,* 770.
26. Bellido, F.; Martin, N. L.; Siehnel, R. J.; Hancock, R. E. W. *J. Bacteriol.* **1992,** *174,* 5196.
27. Trias, J.; Nikaido, H. *J. Biol. Chem.* **1992,** *265,* 15680.
28. Yoshihira, E.; Nakae, T. *FEBS Lett.* **1992,** *306,* 5.
29. Labischinski, H.; Goodell, E. W.; Goodell, A.; Hochberg, M. L. *J. Bacteriol.* **1991,** *173,* 751.
30. Ferguson, S. J. *Trends Biochem. Sci.* **1990,** *15,* 377.

31. Payne, J. W. *J. Biol. Chem.* **1968,** *243,* 3395.

32. Higgins, C. F.; Gibson, M. M. *Methods Enzymol.* **1986,** *125,* 365.

33. Payne, J. W. *CIBA Found. Symp.* **1972,** *4,* 17.

34. Payne, J. W.; Gilvarg, C. *J. Biol. Chem.* **1968,** *243,* 335.

35. Payne, J. W. *CIBA Found. Symp.* **1977,** *50,* 305.

36. Morley, J. S.; Hennessey, T. D.; Payne, J. W. *Biochem. Soc. Trans.* **1983,** *11,* 798.

37. Morley, J. S.; Payne, J. W.; Hennessey, T. D. *J. Gen. Microbiol.* **1983,** *129,* 3701.

38. Payne, J. W.; Morley, J. S.; Armitage, P.; Payne, G. M. *J. Gen. Microbiol.* **1984,** *130,* 2253.

39. Ames, G. F. L. *Annu. Rev. Biochem.* **1986,** *55,* 397.

40. Ames, G. F. L.; Mimura, C. S.; Shyamala, V. *FEMS Microbiol. Rev.* **1990,** *75,* 429.

41. Ames, G. F. L. *J. Bioenerg. Biomembr.* **1988,** *20,* 1.

42. Ames, G. F. L.; Joshi, A. K. *J. Bacteriol.* **1990,** *172,* 4133.

43. Ames, G. F. L.; Lecar, H. *FASEB J.* **1992,** 2660.

44. Payne, J. W.; Bell, G. *J. Bacteriol.* **1979,** *137,* 447.

45. Payne, J. W. In *Microorganisms and Nitrogen Sources;* Payne, J. W., Ed.; John Wiley & Sons; Chichester, England, 1980; pp 359–381.

46. Higgins, C. F.; Hardie, M. M. *J. Bacteriol.* **1983,** *153,* 1434.

47. Guyer, C. A.; Morgan, D. G.; Osheroff, N.; Staros, J. V. *J. Biol. Chem.* **1985,** *260,* 10812.

48. Furlong, C. E. In *Escherichia coli and Salmonella typhimurium: Cellular and Molecular Biology;* Neidhardt, F. C., Ed.; American Society for Microbiology; Washington, DC, 1987; pp 768–796.

49. Hiles, I. D.; Gallagher, M. P.; Jamieson, D. J.; Higgins, C. F. *J. Mol. Biol.* **1987,** *195,* 125.

50. Hiles, I. D.; Powell, L. M.; Higgins, C. F. *Mol. Gen. Genet.* **1987,** *206,* 101.

51. Guyer, C. A.; Morgan, D. G.; Staros, J. V. *J. Bacteriol.* **1986,** *168,* 775.

52. Kashiwagi, K.; Yamaguchi, Y.; Sakai, Y.; Kobayashi, H.; Igarashi, K. *J. Biol. Chem.* **1990,** *265,* 8387.

53. Goodell, E. W.; Higgins, C. F., *J. Bacteriol.* **1987,** *169,* 3861.

54. Marshall, N. J.; Payne, J. W., School of Biological Sciences, University of Wales, Bangor, Gwynedd LL57 2UW, United Kingdom, unpublished results.

55. Quiocho, F. A. *Curr. Opin. Struc. Biol.* **1992,** *1,* 922.

56. Payne, J. W.; Payne, G. M.; Marshall, N. J.; Klucar, L., School of Biological Sciences, University of Wales, Bangor, Gwynedd LL57 2UW, United Kingdom, unpublished results.

57. Maloney, P. C. *Res. Microbiol.* **1990,** *141,* 374.

58. Kerppola, R. E.; Ames, G. F. L. *J. Biol. Chem.* **1992,** *267,* 2329.

59. Higgins, C. F.; Hiles, I. D.; Whalley, K.; Jamieson, D. J. *EMBO J.* **1985,** *4,* 1033.

60. Higgins, C. F. *Res. Microbiol.* **1990,** *141,* 353.

61. Baichwal, V.; Liu, D.; Ames, G. F. L. *Proc. Natl. Acad. Sci. U.S.A.* **1993,** *90,* 620.

62. Traxler, B.; Beckwith, J. *Proc. Natl. Acad. Sci. U.S.A.* **1992,** *89,* 10852.

63. Higgins, C. F.; Hyde, S. C.; Mimmack, M. M.; Gileadi, U.; Gill, D. R.; Gallagher, M. P. *J. Bioenerg. Biomembr.* **1990,** *22,* 571.

64. Hyde, S. C.; Emsley, P.; Hartshorn, M. J.; Mimmack, M. M.; Gileadi, U.; Pearce, S. R.; Gallagher, M. P.; Gill, D. R.; Hubbard, R. E.; Higgins, C. F. *Nature (London)* **1990,** *345,* 362.

65. Pastan, I.; Gottesman, M. M. *Annu. Rev. Med.* **1991,** *42,* 277.

66. Anderson, M. P.; Gregory, R. J.; Thompson, S.; Souza, D. W.; Paul, S.; Mulligan, R. C.; Smith, A. E.; Welsh, N. J. *Science (Washington, DC)* **1991,** *253,* 202.

67. Speiser, D. M.; Ames, G. F. L. *J. Bacteriol.* **1991,** *173,* 1444.

68. Alves, R. A.; Payne, J. W. *Biochem. Soc. Trans.* **1980,** *8,* 704.

69. Perry, D.; Gilvarg, C. *J. Bacteriol.* **1984,** *160,* 943.

70. Abouhamad, W. N.; Manson, M.; Gibson, M. M.; Higgins, C. F. *Mol. Microbiol.* **1991,** *5,* 1035.

71. Olson, E. R.; Dunyak, D. S.; Jurss, L. N.; Poorman, R. A. *J. Bacteriol.* **1991,** *173,* 234.

72. Manson, M. D.; Blank, V.; Brade, G.; Higgins, C. F. *Nature (London)* **1986,** *321,* 253.

73. Verkamp, E.; Backman, V. M.; Bjornsson, J. M.; Soll, D.; Eggertsson, G. *J. Bacteriol.* **1993,** *175,* 1452.

74. Barak, Z.; Gilvarg, C. *J. Bacteriol.* **1975,** *122,* 1200.

75. Naider, F.; Becker, J. M. *J. Bacteriol.* **1975,** *122,* 1208.

76. Payne, J. W. *Biochem. Soc. Trans.* **1983,** *11,* 794.

77. Gibson, M. M.; Price, M.; Higgins, C. F. *J. Bacteriol.* **1984,** *160,* 122.

78. Andrews, J. C.; Short, S. A. *J. Bacteriol.* **1986,** *165,* 434.

79. Payne, J. W.; Smith, M. W., School of Biological Sciences, University of Wales, Bangor, Gwynedd LL57 2UW, United Kingdom, unpublished results.

80. Jamieson, D. J.; Higgins, C. F. *J. Bacteriol.* **1984,** *160,* 131.

81. Payne, J. W.; Alves, R., School of Biological Sciences, University of Wales, Bangor, Gwynedd LL57 2UW, United Kingdom, unpublished results.

82. Wallace, R. J.; McKain, N. *J. Gen. Microbiol.* **1991,** *137,* 2259.

83. Kunji, E. R. S.; Smid, E. J.; Plapp, R.; Poolman, B.; Konings, W. N. *J. Bacteriol.* **1993,** *175,* 2052.

84. Perry, D. *J. Gen. Microbiol.* **1981,** *124,* 425.

85. Nisbet, T. M.; Payne, J. W. *J. Gen. Microbiol.* **1982,** *128,* 1357.

86. Payne, J. W.; Nisbet, T. M. *J. Appl. Biochem.* **1982,** *3,* 447.

87. Payne, J. W.; Payne, G. M.; Nisbet, T. M. *FEMS Microbiol. Lett.* **1982,** *14,* 123.

88. Gilson, E.; Alloing, G.; Schmidt, T.; Claverys, J. P.; Dudler, R.; Hofnung, M. *EMBO J.* **1988,** *1,* 3971.

89. Alloing, G.; Trombe, M. C.; Claverys, J. P. *Mol. Microbiol.* **1990,** *4,* 633.

90. Perego, M.; Higgins, C. F.; Pearce, S. R.; Gallagher, M. P.; Hoch, J. A. *Mol. Microbiol.* **1991,** *5,* 173.

91. Rudner, D. Z.; LeDeaux, J. R.; Ireton, K.; Grossman, A. D. *J. Bacteriol.* **1991,** *173,* 1388.

92. Mathiopoulos, C.; Mueller, J. P.; Slack, F. J.; Murphy, C. G.; Patankar, A.; Bukusoglu, G.; Sonenshein, A. L. *Mol. Microbiol.* **1991,** *5,* 1903.

93. Slack, F. J.; Mueller, J. P.; Strauch, M. A.; Mathiopoulos, C.; Sonenshein, A. L. *Mol. Microbiol.* **1991,** *5,* 1915.

94. Russell, R. R. B.; Opoku-Aduse, J.; Sutcliffe, I. C.; Tao, L.; Ferretti, J. J. *J. Biol. Chem.* **1992,** *267,* 4631.

95. Gharbia, S. E.; Shah, H. N. *Curr. Microbiol.* **1991,** *22,* 159.

96. Hanson, M. S.; Slaughter, C.; Hansen, E. J. *Infect. Immun.* **1992,** *60,* 2257.

97. Payne, J. W. In *Microorganisms and Nitrogen Sources;* Payne, J. W., Ed.; John Wiley & Sons; Chichester, England, 1980; pp 335–356.

98. Payne, J. W.; Bell, G. *FEMS Microbiol. Lett.* **1977,** *2,* 301.

99. Smith, M. W.; Payne, J. W. *FEMS Microbiol. Lett.* **1992,** *100,* 183.

100. Hardy, D. J.; Payne, J. W. *Plant Physiol. Life Sci. Adv.* **1992,** *11,* 59.

101. Payne, J. W.; Nisbet, T. M. *FEBS Letts.* **1980,** *119,* 73.

102. Brink, B. T.; Otto, R.; Hansen, U. P.; Konings, W. N. *J. Bacteriol.* **1985,** *162,* 383.

103. Ames, B. N.; Ames, G. F. L.; Young, J. D.; Isuchiya, D.; Lecocq, J. *Proc. Natl. Acad. Sci. U.S.A.* **1973,** *70,* 456.

104. Fickel, T. E.; Gilvarg, C. *Nature New Biol.* **1973,** *241,* 161.

105. Kenig, M.; Abraham, E. P. *J. Gen. Microbiol.* **1976,** *94,* 37.

106. Chmara, H.; Milewski, S.; Dzieduszycka, M.; Smulkowski, M.; Sawlewicz, P.; Borowski, E. *Drugs Exp. Clin. Res.* **1982,** *8,* 11.

107. Chmara, H. *J. Gen. Microbiol.* **1985,** *131,* 265.

108. Moore, R. E.; Niemczura, W. P.; Kwok, O. C. H.; Patil, S. S. *Tetrahedron Lett.* **1984,** *25,* 3931.

109. Staskawicz, B. J.; Panopoulos, N. J. *J. Bacteriol.* **1980,** *142,* 474.

110. Peet, R. C.; Lindgreen, P. B.; Willis, D. K.; Panopoulos, N. J. *J. Bacteriol.* **1986,** *166,* 1096.

111. Diddens, H.; Zahner, H.; Kraas, E.; Gohring, W.; Jung, G. *J. Biochem.* **1976,** *66,* 11.

112. Diddens, H.; Dorgerloh, M.; Zahner, H. *J. Antibiot.* **1979,** *32,* 87.

113. Kumada, Y.; Imai, S.; Nagaoka, K. *J. Antibiot.* **1991,** *44,* 1006.

114. Hunt, A. H.; Elzey, T. K. *J. Antibiot.* **1988,** *37,* 802.

115. Rohl, F.; Rabenhorst, J.; Zahner, H. *Arch. Microbiol.* **1987,** *147,* 315.

116. Atherton, F. R.; Hall, M. J.; Hassall, C. H.; Lambert, R. W.; Ringrose, P. S. *Antimicrob. Agents Chemother.* **1979,** *15,* 677.

117. Atherton, F. R.; Hall, M. J.; Hassall, C. H.; Lambert, R. W.; Lloyd, W. J.; Ringrose, P. S. *Antimicrob. Agents Chemother.* **1979,** *15,* 696.

118. Atherton, F. R.; Hall, M. J.; Hassall, C. H.; Holmes, S. W.; Lambert, R. W.; Lloyd, W. J.; Ringrose, P. J. *Antimicrob. Agents Chemother.* **1980,** *18,* 897.

119. Atherton, F. R.; Hall, M. J.; Hassall, C. H.; Lambert, R. W.; Lloyd, W. J.; Lord, A.; Ringrose, P. S.; Westmacott, D. *Antimicrob. Agents Chemother.* **1983,** *24,* 522.

120. Hassall, C. H. *Antimicrob. Agents Chemother.* **1985,** *12,* 193.

121. Atherton, F. R.; Hassall, C. H.; Lambert, R. W. *J. Med. Chem.* **1986,** *29,* 29.

122. Allen, J. G.; Lees, L. J. *Antimicrob. Agents Chemother.* **1980,** *17,* 973.

123. Lejczak, B.; Kafarski, P.; Sztajer, H.; Mastalerz, P. *J. Med. Chem.* **1986,** *29,* 2212.

124. Zboinska, E.; Sztajer, H.; Lejczak, B.; Kafarski, P. *FEMS Microbiol. Lett.* **1990,** *70,* 23.

125. VoQuang, Y.; Carniato, D.; VoQuang, L.; Lacoste, A. M.; Neuzil, E.; Le Goffic, F. *J. Med. Chem.* **1986,** *29,* 148.

126. Hammond, S. M.; Claesson, A.; Jansson, A. M.; Larsson, L. G.; Pring, B. G.; Town, C. M.; Ekstrom, B. *Nature (London)* **1987,** *327,* 730.

127. Goldman, R.; Kohlbrenner, W.; Lartey, P.; Pernet, A. *Nature (London)* **1987,** *329,* 162.

128. Claesson, A.; Jansson, A. M.; Pring, B. G.; Hammond, S. M.; Ekstrom, B. *J. Med. Chem.* **1987,** *30,* 2309.

129. Smith, M. W.; Payne, J. W. *FEMS Microbiol. Lett.* **1990,** *70,* 311.

130. Patchett, A. A.; Taub, D.; Weissberger, B.; Valiant, M. E.; Gadebusch, H.; Thornberry, N. A.; Bull, H. G. *Antimicrob. Agents Chemother.* **1988,** *32,* 319.

131. Cheung, K. S.; Wasserman, S. A.; Dudek, E.; Lerner, S. A.; Johnston, M. *J. Med. Chem.* **1983,** *26,* 1733.

132. Cheung, K. S.; Boisvert, W.; Lerner, S. A.; Johnston, M. *J. Med. Chem.* **1986,** *29,* 2060.

133. Boisvert, W.; Cheung, K. S.; Lerner, S. A.; Johnston, M. *J. Biol. Chem.* **1986,** *261,* 7871.

134. Mobashery, S.; Johnston, M. *Biochemistry* **1987,** *26,* 5878.

135. Boehm, J. C.; Kingsbury, W. D.; Perry, D.; Gilvarg, C. *J. Biol. Chem.* **1983,** *258,* 14850.

136. Kingsbury, W. D.; Boehm, J. C.; Perry, D.; Gilvarg, C. *Proc. Natl. Acad. Sci. U.S.A.* **1984,** *81,* 4573.

137. Berges, D. A.; DeWolf, W. E.; Dunn, G. L.; Grappel, S. F.; Newman, D. J.; Taggart, J. J.; Gilvarg, C. *J. Med. Chem.* **1986,** *29,* 89.

138. Leroux, P.; Blanot, D.; Menginlecreulx, D.; Vanheijenoort, J. *Int. J. Pept. Protein Res.* **1991,** *37,* 103.

139. Milewski, S.; Mignini, F.; Borowski, E. *J. Gen. Microbiol.* **1991,** *137,* 2155.

140. Atherton, F. R.; Hall, M. J.; Hassall, C. H.; Lambert, R. W.; Lloyd, W. J.; Ringrose, P. S.; Westmacott, D. *Antimicrob. Agents Chemother.* **1982,** *22,* 571.

141. Angehrn, P.; Hall, M. J.; Lloyd, W. J.; Westmacott, D. *Antimicrob. Agents Chemother.* **1984,** *25,* 607.

RECEIVED for review September 9, 1993. ACCEPTED revised manuscript December 20, 1993.

CHAPTER 16

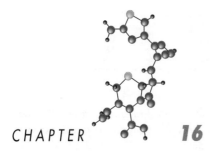

Fungal Peptide Transport as a Drug Delivery System

Jeffrey M. Becker and Fred R. Naider

Since the last review of peptide transport in fungi several years ago (*1*), much of the literature in this area has focused on the pathogenic yeast *Candida albicans.* This development has been influenced and facilitated by the dramatic rise of *C. albicans* infections in patients compromised by acquired immunodeficiency syndrome (*2*), transplantation regimens, and long-term administration of chemotherapeutic drugs (*3*). Recently, exciting progress in the study of peptide transport in fungi and other eukaryotes has been achieved by using the yeast *Saccharomyces cerevisiae* as a paradigm for molecular and genetic analysis; the first reports on the cloning of a eukaryotic peptide transport gene were published recently. In this chapter we will describe, analyze, and evaluate the literature in the area of peptide transport in fungi. Our review will cover some of the earlier studies to set the framework for what are currently the generally accepted notions of fungal peptide transport. We will concentrate, however, on publications from 1987 to the present and will emphasize the possible use of peptide transporters as drug delivery targets.

Before initiating the literature review, we believe some points concerning the fungal kingdom, and especially the relationship of *S. cerevisiae* to *C. albicans,* will be of interest.

Table I. Characteristics Distinguishing *C. albicans* from *S. cerevisiae*

Characteristic	C. albicans	S. cerevisiae
Exhibits a sexual cycle	No	Yes
Generation time, rich medium (min)	35	70
Carbon sources metabolized	Many	Few
Yeast-to-mold transition	Yes	No
Opportunistic pathogen	Yes	No

Fungi are a genetically, physiologically, and ecologically diverse kingdom composed of perhaps 500,000 species. Although known to humans through their physiological activities since antiquity, few fungi have been studied in any detail; about 10% of the half million estimated species have even been identified and classified. Mushrooms, molds, and yeasts are all included within the disparate fungal kingdom. Yeasts are defined as fungi with generally a single-celled thallus and that reproduce usually by budding and occasionally by fission. The literature on peptide transport in fungi is limited to a few reports on *Neurospora crassa* (*4, 5*), a filamentous ascomycete used extensively in genetic and biochemical studies since the 1940s, and more extensive literature on the yeasts *S. cerevisiae* (*6*) and *C. albicans* (*1*). Although both of these yeasts are related as part of the fungal kingdom, they have many distinguishing characteristics (Table I).

S. cerevisiae has been used extensively as a model cell for genetic studies because of the powerful tools developed to manipulate the genotype of this organism (*7–9*). In contrast, *C. albicans* has no known sexual cycle and has only recently yielded to modern analysis by molecular cloning methods (*10*). Most interest in the study of peptide transport in *C. albicans* is due to its opportunistic pathogenicity and its ability to undergo a yeast-to-mold transition, which serves as a model for fungal morphogenesis (*11*). Within the fungi, *C. albicans* is the most prevalent and dominant human pathogen. Moreover, it is currently listed as the third or fourth most widespread opportunistic pathogen among all microbes, including bacteria (*12*). *S. cerevisiae* is distinct from *C. albicans* with regard to pathogenicity. Reports of human infections caused by *S. cervisiae* are extremely rare (*13*). The evolutionary distance between these two yeasts is quite large, perhaps greater than the distance between mouse and human. Therefore, it is not possible to group peptide transports in *S. cervisiae* and *C. albicans* as if it was a singular or unified phenomenon. Indeed, significant differences have been noted in the structural specificities of peptide transport systems in various strains of these fungi (*1*). However, even though we must exert caution in comparing peptide transport between *S. cerevisiae* and *C. albicans, S. cerevisiae* has provided a heterologous host for the cloning and study of *C. albicans* peptide transport genes.

Structural Specificity

Conflicting results on the structural specificity, multiplicity, and regulation of peptide transport in fungi have been reported. The lack of uniformity in comparative studies is due to a variety of factors, including the particular strain used in the investigation, the physiological state of the strain assayed, the method applied to measure transport, and the complexity or multiplicity of the genetic and biochemical components of the transport systems. These opposing results and conclusions make lucid analysis difficult to achieve. However, the structural specificities of peptide transport in fungi appear to have one universal characteristic: Hydrophobic residues are preferred. Nevertheless, full understanding of this structural specificity will be available only after biochemical analysis of the system(s) and genetic dissection, including cloning of the genes encoding each transport component.

Although peptides containing a variety of amino acids can be transported by yeasts, amino acid sequence and composition appear to exert a marked effect on peptide transport. In an early report of peptide transport in *S. cerevisiae* (*14*), tripeptides composed of glycine and methionine were excellent (Met–Met–Met, Gly–Met–Met, Gly–Gly–Met, and Met–Gly–Met), intermediate (Met–Met–Gly and Met–Gly–Gly), or inadequate (Gly–Met–Gly) sources of methionine for a methionine auxotroph, even though these peptides were hydrolyzed to similar extents by intracellular peptidases. However, generalizations can not be made as evidenced by reports from various investigators of wide variations in the uptake of specific peptides depending on the organism tested or the assay used to measure peptide transport. For example, the dipeptide Lys–Lys was unable to support the growth of lysine auxotrophic *S. cerevisiae* strains Z1-2D (*15*) and S288C-24 (*16*), whereas *S. cerevisiae* Σ1278b (*17*) transported this dipeptide well, as measured by disappearance of peptide from an incubation medium. Amino acid composition and sequence dependence in the transport of peptides containing different amino acid residues has been reported for peptide transport in *C. albicans* (*18–23*). For example, Milewski et al. (*24*) showed that Ala–Ala was transported at a fourfold higher rate than Ala–Gly, whereas Ala–Ala–Ala and Leu–Gly–Gly were transported at about the same rate.

In a recent detailed analysis of the uptake of dipeptides containing N^3-(4-methoxyfumaroyl)-L-2,3-diaminopropanoic acid (FMDP), transport rates of *X*-FMDP dipeptides were generally high when *X* was a hydrophobic residue such as Leu or Val, whereas the rates were lower for dipeptides containing Gly or an aromatic residue in this position (*25*). Also, FMDP-*X* dipeptides were transported more slowly than their *X*-FMDP counterparts. Interestingly, Lys–FMDP showed a high rate of transport, which was verified in studies on the affinity (K_m; Michaelis constant) and capacity (V_{max}; limiting velocity) of Lys–FMDP and various other peptide substrates. Despite its relatively favorable transport characteristics, Lys–FMDP was 20-fold less effective as an anticandidal agent in comparison with Nva–FMDP. This difference was due to the slow intracellular cleavage of Lys–FMDP (*25*).

Successful drug delivery using the peptide transport system requires rapid internalization and subsequent release of the warhead delivered by the peptide substrate. High kill rates will only be obtained if the net result of the aforementioned processes is a sufficient intracellular concentration of the toxic moiety. In the study on the FMDP-containing peptides, the velocity of transport of a particular dipeptide was a more important determinant of potency than was intracellular hydrolysis (*25*). The aforementioned attributes of Lys–FMDP transport clearly show that this generalization can not be applied to any given peptide. Indeed, careful scrutiny of the data suggests that the minimum inhibitory concentrations for many FMDP-containing dipeptides also reflect the hydrolysis rates. Thus, peptides should be designed to have high transport rates and to be good substrates for intracellular peptidase. A discussion of peptidase cleavage is beyond the scope of this review. However, we note that fungal peptidases have a different spectrum of activity compared with serum proteases and that additional comparative analysis on microbial versus human peptidases will be an important component of the ultimate use of peptide drugs.

Studies on stereospecificity of the residues of transported peptides have led to the general notion that a D-residue is not tolerated in dipeptide transport. Dipeptides containing D-residues in either the first or second position were not taken up by *C. albicans* (*18, 23, 26, 27*) or *S. cerevisiae* (*28*). Recently, Milewski and co-workers (*25*) reported that

D-Ala–FMDP was transported at a higher rate than Tyr–FMDP. However, the slow uptake of dipeptides containing D-residues, together with their resistance to hydrolysis, probably eliminates their consideration as delivery vehicles. In larger peptides, D-residues are accepted by the transport system depending on their position in the peptide. Nisbet and Payne (*28*) and Marder et al. (*15*) demonstrated that L-Ala–L-Ala–D-Ala was not taken up by *S. cerevisiae*. A similar conclusion was reached by Shallow et al. (*23*) for *C. albicans*. The lack of transport of a tripeptide containing a D-residue in the third position for *C. albicans* (*19*) or *S. cerevisiae* (*29*) was also shown for L-Met–L-Met–D-Met. In contrast, D-Met–L-Met–L-Met was a good transport substrate for *C. albicans* (*19, 20*).

Peptides larger than tripeptides do not appear to be tolerated by the peptide transport system of *S. cerevisiae* with the exception of a report of tetra- and pentamethionine use in strain G1333 (*14*). Nisbet and Payne (*28*) showed that tetra- and pentaalanine were not taken up in agreement with results on Gly- and Leu-containing tetrapeptides in strain Z1-2D (*15*). In contrast, *C. albicans* has been shown by many investigators to transport peptides up to hexapeptides. Recently, Basrai et al. (*30*) showed that tetra- and pentapeptides containing the toxic amino acid oxalysine inhibited growth of *C. albicans* H317. Pentamethionine supported the growth of the methionine auxotroph *C. albicans* WD 18-4 (*19*), tetraalanine was taken up by *C. albicans* 124 (*31*) and *C. albicans* B2620 (*23*), and pentaalanine and hexamethionine but not $(Ala)_3$–Tyr–$(Ala)_3$ were transported by strain 26278 (*22*). Interestingly, peptides of three to six amino acids in length showed a higher rate of uptake by a mycelial form of *C. albicans* than a yeast form (*22*). The size limit to transport is an important factor in drug design strategies. At present, peptides larger than five or six residues are not known to act as carriers or potentially toxic agents against pathogens such as *C. albicans*. However, the presence of endocytotic pathways in this yeast (*32*) opens the possibility that large peptides may be used as carriers or toxic agents.

Since our last review (*1*), the influence of modifications of the peptide N- and C-termini on peptide transport has not been clarified. The prevailing opinion was that some, but not all, strains of *S. cerevisiae* tolerated acylated di- and oligopeptides and that esterification of the α-carboxyl group on the C-terminal residue did not interfere with uptake by the peptide transport system (*6*). In *C. albicans*, a different conclusion may be drawn; α-*N*-Acylated peptides enter some strains, whereas peptide esters do not appear to be transported into *C. albicans* via the peptide transport system (*1*). The fact that azidobenzoyl–Ala–Ala photoaffinity labeled the oligopeptide transport system in *C. albicans* but had virtually no effect on dipeptide transport (25) confirms the conclusion that *N*-α-acylated peptides are recognized by the oligopeptide transport system in *C. albicans*. The data also suggest that the dipeptide transport system of this fungus can not tolerate *N*-α-acylated peptides. The ability to *N*-acylate putative peptide carriers is very important for drug design. *N*-Acylation can stabilize the carrier–toxic agent conjugate to aminopeptidase activity, thereby increasing serum half-lives. At the same time, certain *N*-α-acylated amino acids are not susceptible to enzymatic cleavage. In such a circumstance, construction of a peptide drug with a *N*-α-acylated warhead would be futile, and this must be considered in assembling the peptide conjugate. Definitive conclusions concerning the effect of termini on peptide uptake await isolation of the components of the peptide transport system and biochemical analysis under controlled conditions.

Multiplicity

A number of indications suggested the presence of a single peptide transport system in *S. cerevisiae* before the definitive studies of Island et al. (*33*) established that at least three genes are involved in the major, perhaps only, peptide transport system of this yeast. On the other hand, many investigations have indicated that multiple systems exist in *C. albicans*, although no definitive genetic studies have supported this hypothesis. The evidence for multiplicity in *C. albicans* is based on a number of observations: (1) mutants were isolated that were resistant to certain toxic peptides but were not totally deficient in peptide transport; (2) competition for transport indicated that some peptides were good competitors for uptake of a peptide substrate, whereas other peptides known to be taken up by the cell were poor competitors of that substrate; and (3) uptake of di- and tripeptides was regulated differently from peptides containing four or five amino acids.

For *S. cerevisiae*, strains selected for resistance to various toxic dipeptides lost the ability to take up both di- and oligopeptides (*17, 34, 35*). Furthermore, competition studies indicated that only one peptide transport system operated (*29*), certain dipeptides inhibited the use of a required tripeptide, and tripeptides inhibited dipeptide use (*34*). Finally, Island et al. (*33*) established conclusively, by genetic methodology, that only one peptide transport system, designated *PTR*, exists in *S. cerevisiae*. They identified three genetic components of the system, and it appeared that *PTR* was the only peptide transport system regulated by amino acids (*see* Regulation). In Cloning of Yeast Peptide Transporter, a more detailed description of the cloning and sequencing of one of the genes encoding a component of the system will be presented.

Although a thorough genetic analysis of peptide transport has not yet been carried out in *C. albicans* and one report led to the conclusion of only one system for di- and tripeptide uptake (*18*), convincing evidence points to a multiplicity of systems in this pathogenic yeast. Mutants isolated as resistant to nikkomycin, known to be transported by a route shared by peptides, were deficient in uptake of dipeptides but had the same (*21, 31*) or increased (*36*) capacity for transport of tripeptides. Milewski et al. (*22*) isolated mutants resistant to a toxic tetrapeptide in which the capacity to take up tetrapeptides was lost. In this mutant, tripeptide uptake was reduced, whereas dipeptide transport was practically unaffected. The terms dipeptide and oligopeptide systems for *C. albicans* have been used by Payne, Shallow, and co-workers (*23, 37, 38*) on the basis of mutant isolation and competition for transport assays. Yadan et al. (*36*) demonstrated the existence of multiple transport systems by a kinetic analysis of di- and tripeptide uptake, and McCarthy et al. (*26*) showed multiple systems by transport competition with novel chromophoric substrates to measure peptide transport. Photoaffinity labeling resulted in a significant decrease in uptake of tripeptides and tetrapeptides in *C. albicans*, but virtually no effect was seen on dipeptide transport (*25*). Finally, it was shown that growth inhibition by various peptides containing toxic amino acids is regulated differentially by amino acids and the source of nitrogen for growth (*30*). Oxalysine-containing di- and tripeptide toxicity was increased by micromolar amounts of various amino acids, but no effect on toxicity of oxalysine-containing tetra- or pentapeptides was noted.

Regulation

In both *S. cerevisiae* and *C. albicans*, peptide transport appears to be controlled by several environmental factors. Although cells seem to have the ability to grow on peptides under most conditions, the level of transport is repressed by growth on rich sources of nitrogen, such as ammonium sulfate. In addition, micromolar amounts of amino acids have been found to up-regulate transport. An understanding of the regulation by the nitrogen source seems obvious if one imagines that the cell requires lower levels of nitrogen-containing compounds when growing on a rich nitrogen source, whereas growth on a poor nitrogen source would force the cell to turn on all systems from which nitrogen can be extracted from the environment. The activation of peptide transport by small amounts of amino acids may be explained by similar reasoning. The cell senses a need for peptides from its milieu as a source of amino acids because the levels of amino acids in the environment are not sufficient to satisfy its growth requirement. Whatever the reason for the regulatory phenomenon, regulation at the level of gene transcription was observed (*39*) (*see* Cloning of Yeast Peptide Transporter).

Specific examples of regulation of peptide transport were demonstrated in both *S. cerevisiae* and *C. albicans*. Peptide use was repressed by growth of *S. cerevisiae* in the presence of ammonium ions as a source of nitrogen in comparison to growth on organic nitrogen sources, such as allantoin, isoleucine, or proline (*6, 28, 40*). In *C. albicans*, Yadan et al. (*36*) showed that the apparent K_m for uptake of trimethionine was much lower in rich medium than in minimal medium. The toxicity of polyoxin, a peptidyl-nucleoside transported into *C. albicans* via a peptide transport system, is dependent on the specific nitrogen source in the medium (*41, 42*). Payne et al. (*38*) demonstrated that addition of peptide mixtures to culture medium induced a fivefold increase in the transport of dipeptides and oligopeptides.

Regulation of dipeptide transport by micromolar amounts of certain amino acids was shown in *S. cerevisiae* by Island et al. (*43*). Incubation of cells with 150 μM tryptophan led to a 25-fold increase in the initial rate of Leu–Leu uptake. This effect was not seen in the presence of the protein synthesis inhibitor cycloheximide. With allantoin as the nitrogen source, growth inhibition by toxic peptides was increased in the presence of leucine, tryptophan, lysine, isoleucine, threonine, phenylalanine, valine, histidine, alanine, serine, tyrosine, and asparagine, whereas cysteine, glycine, aspartate, proline, glutamine, glutamate, and arginine did not increase toxicity. The presence of leucine, tryptophan, or lysine resulted in the greatest increase in peptide toxicity in comparison with the other amino acids. Working with *C. albicans* and adding micromolar amounts of amino acids to the growth medium, Basrai et al. (*30*) showed increases both in toxicity to di- and tripeptides and in uptake of a radioactive dipeptide. In comparison to *S. cerevisiae*, arginine, lysine, histidine, glutamine, aspartate, and serine up-regulated peptide transport in *C. albicans*, but alanine, asparagine, cysteine, glutamate, glycine, isoleucine, leucine, phenylalanine, threonine, tyrosine, valine, methionine, proline, and tryptophan did not increase di- and tripeptide toxicity in *C. albicans*. Furthermore, no increase of tetra- or pentapeptide toxicity was noticed, leading the authors to conclude that multiple peptide transport systems in *C. albicans* were differentially regulated by amino acids. The different array of amino acids that regulate peptide transport in *S. cerevisiae* compared with *C. albicans* may reveal some basic metabolic differences between these yeasts or may reflect alternative adaptive mechanisms resulting from evolution in their different ecological niches.

Demonstration of Illicit Transport in *C. albicans*

In 1973, Ames et al. (*44*) and Fickel and Gilvarg (*45*) demonstrated that peptides could be used to deliver toxic agents into bacterial cells. This phenomenon of using peptides as carriers of molecules that otherwise could not enter the cellular cytoplasm was called illicit transport, portage transport, or warhead delivery (*46, 47*). Shortly thereafter, exploitation of such a drug conveying scheme in fungi was proposed and demonstrated (*19, 48, 49*). The use of peptides for drug delivery can be expanded to encompass additional perspectives. The peptide can be used to target a toxic moiety to a pathogen based on a differential structural specificity from the host. Also, peptides or peptide-related compounds, such as the peptidyl-nucleoside group represented by polyoxins and nikkomycins, that are toxic themselves and taken up via peptide transport systems can be used without the necessity of release of a toxic moiety from the conjugate.

The establishment of illicit transport in *C. albicans*, the possibility of using the structural specificity of the transport system for targeted delivery, and the observation of toxicity of polyoxin all served as catalysts for developing peptides as antifungal drugs. Another factor favoring peptides for drug delivery is the apparent multiplicity of transport systems in *C. albicans*. If more than one route for drug entry exists, then the possibility is exceedingly low of the organism becoming drug resistant by a mutation in the uptake mechanism. Even so, demonstrations of the utility of such agents in animal model systems are not numerous. Becker et al. (*50*) demonstrated that nikkomycin could prolong the survival of *C. albicans*-infected mice, although once drug treatment was terminated, the animals eventually died from the *C. albicans* infection.

In the following sections, we discuss two aspects of peptide-based antifungal drugs: combinations of amino acids or peptides with either impermeable or normally permeable moieties to form a peptide–drug conjugate, and peptides that are themselves toxic. At present, the paucity of knowledge concerning peptide transport in human cells has precluded the design of drugs selectively targeted to a fungal pathogen.

Peptide–Drug Conjugates

More than 40 years ago, Dunn and Dittmer (*51*) showed that dipeptides were a source of toxic amino acids. However, the peptides they investigated were not recognized initially as carriers of toxic moieties into the target fungal cell. Kenig and Abraham (*52*) showed that the dipeptide antibiotic bacilysin was active against *C. albicans* and suggested on the basis of antagonism of toxicity by peptides that peptide transport may play a role in the inhibitory action of the drug.

To validate the uptake of toxic peptide conjugates via peptide transport systems, investigators used various experimental protocols, such as competition of toxic peptides by nontoxic peptides and peptide transport mutants in which toxic peptides were no longer active. In the first systematic attempts to demonstrate peptide smuggling in *C. albicans*, 5-fluorocytosine (*53*) and 5-fluoroorotic acid (*49*) were conjugated to peptides. On the basis of competition assays and activity of these conjugates, the conclusion was drawn that the fluorinated pyrimidines were carried into the yeast cell via a peptide transport system. Peptides containing 5-fluorouracil attached as an α-substituted glycine were shown to enter *C. albicans* via a peptide transport system (*54*) Glucosamine-6-phosphate synthase inhibitors, such as FMDP, coupled to norvaline, dimethionine, or trimethionine (*25, 55–58*) and N^3-(iodoacetyl)-L-2,3-diaminopropanoic acid (*56*) attached to amino acids were demonstrated to be toxic and absorbed through a peptide transport system. Interestingly, FMDP inhibits cell wall biosynthesis in *C. albicans*, and a conjugate of FMDP with norvaline was effective at microgram-per-milliliter amounts against *C. albicans, Cryptococcus neoformans* and *Aspergillus sp.* and was somewhat effective in a mouse model of candidiasis (*58*).

A number of studies have been carried out with bacilysin, also called tetaine (*27, 31, 35, 37, 52, 59, 60*). This dipeptide enters *C. albicans* by a peptide transport system, as established by competition experiments and by the isolation of peptide transport mutants. Once having entered the cell, bacilysin is cleaved by cellular peptidases to release an epoxy amino acid (anticapsin) that appears to irreversibly inactivate glucosamine-6-phosphate synthetase in a manner similar to that of FMDP (*24*). One consequence of the inhibition of this enzyme is inhibition of the biosynthesis of chitin and mannoprotein, a finding that supports the observation that the mycelial form of *C. albicans,* which is richer in chitin than the yeast form, is more sensitive to this antifungal compound (*60*).

A large number of studies have demonstrated inhibition of fungal growth by toxic amino acids when these analogues are contained within a peptide. The toxic amino acids used were fluorophenylalanine (*21, 35, 42, 61*), ethionine (*34*), oxalysine (*30*), and unusual D-amino acids (*62*). Such peptides were especially useful in the isolation of peptide transport mutants of *S. cerevisiae* for subsequent genetic and gene-cloning studies (*33*).

Peptide Transport in Drug Delivery

In the 1960s, a group of compounds known as polyoxins were shown to be potent competitive inhibitors of the cell-wall enzyme chitin synthase found in a number of plant pathogenic fungi (*63–66*). Additional chitin synthase inhibitors were found as natural products from *Streptomyces* broths and were called nikkomycins or neopolyoxins (*67*). Both polyoxins and nikkomycins are peptidyl-nucleosides. Although their target enzyme is found in all fungal cells examined, polyoxins and nikkomycins were used only as inhibitors of phytopathogenic fungi, even though studies showed the compounds were effective inhibitors of chitin synthase from *S. cerevisiae* (*68*). The presence of chitin synthase in *C. albicans* was demonstrated convincingly by Braun and Calderone (*69, 70*). Becker et al. (*41*) demonstrated that polyoxin D at millimolar concentrations caused marked morphological alterations of *C. albicans* and inhibited the growth of *C. neoformans,* a pathogen of humans. These effects were reversed by addition of peptides to the growth medium. Furthermore, chitin synthase in cell extracts of *C. albicans* was inhibited by polyoxin D. On the basis of these and other studies (*1*), we concluded that this class of chitin synthase inhibitors were taken into *C. albicans* by a peptide transport system.

Hector and Pappagianis (*71*) showed that polyoxin D was active against the human fungal pathogen *Coccidioides immitis.* Other studies have demonstrated in vitro activity of nikkomycins against diverse fungi such as *C. immitis, Blastomyces dermatitidis, C. albicans,* and *C. neoformans,* with only moderate susceptibility of *Rhizopus arrhizus* and *Aspergillus fumigatus,* and resistance of *Candida tropicalis* (*72, 73*). Unfortunately, none of these studies has directly demonstrated entrance of polyoxin or nikkomycin via a peptide transport system. Speculation remains as to whether sensitivity to these peptidyl-nucleosides demonstrates the existence of peptide transport systems in these human pathogenic fungi.

A number of other investigators presented data confirming and extending the conclusion that polyoxins and nikkomycins are taken up by a peptide transport mechanism in *C. albicans* (*21, 31, 35, 36, 37, 42*). Although both polyoxins and nikkomycins enter *C. albicans* through peptide transport systems, polyoxin D, in contrast to nikkomycin Z, has a very poor affinity for transport (*21, 36, 42*). Apparently, the affinity or capacity of the peptide transport system for polyoxins is not great enough to accumulate amounts that effectively inhibit cellular growth. We reasoned that the low affinity of polyoxin D for the peptide transport system in *C. albicans* was a consequence of its unusual amino acid composition. Therefore, we synthesized analogues of this antibiotic containing amino acids with hydrophobic side chains presumed to be optimal substrates for the peptide transport system. Several of these were stable inside the yeast cell, were excellent inhibitors of chitin synthetase

(K_1 = 10^{-6} M), and competed with dileucine uptake (*74*). Many additional synthetic alterations of the polyoxin or nikkomycin nucleus have been attempted to further improve efficacy against *C. albicans*. Di- and tripeptidyl polyoxin and nikkomycin prodrugs (*75, 76*), polyoxin L analogues with derivitized *amine*-terminus and containing α-amino fatty acids (*77, 78, 79*), and analogues resistant to cellular peptidases (*78, 80*) were effective chitin synthase inhibitors and, in some cases, competed with peptide transport; however, none was an effective anti-candidal agent. Another strategy was attempted by Zahner's group by biosynthesizing various nikkomycins with a mutagenized *Streptomyces* strain fed with various precursors of the antibiotic (*81, 82*). Many derivatives of both the nucleoside and the peptide moieties were obtained in this way. Again, none of these compounds represented drugs with improved antifungal activity.

During the course of these studies that were designed to obtain new and improved peptidyl-nucleosides with better antifungal activity, a series of papers from Cabib, Robbins, and co-workers described the use of genetics and biochemistry to identify multiple chitin synthases in fungi, including at least three chitin synthase enzymes in *C. albicans* (*83*). The multiplicity of targets for the polyoxins and nikkomycins obscure an interpretation of the effects of different analogues on cells. Except for the study by Cabib (*84*), which showed that nikkomycins and polyoxin D differentially inhibited chitin synthases 1 and 2 in *S. cerevisiae*, little is known about the effects of different analogues on the various chitin synthases. Obviously, a more detailed study on the interaction of various inhibitors with individual enzymes is needed.

Animal studies showed that nikkomycin can prolong the survival of mice infected by *C. albicans*, but the animals died when administration of the drug ceased (*50*). The fungistatic nature of nikkomycin protection against *C. albicans* was confirmed in a subsequent study using a mouse model (*85*). The fungistatic action of nikkomycin might reflect the serum concentrations obtained in these animals, so routes to increase potency are urgently needed. Toward this end, efforts to use nikkomycin in combination with other antifungal drugs have been successful in several studies (*24, 85, 86*). Furthermore, if the flux of the polyoxins or nikkomycins into *C. albicans* can be increased, perhaps the preparation of an efficacious antifungal chitin synthase inhibitor delivered by the peptide transport system is feasible. Studies are still on-going in our lab to obtain this goal. Some indirect support for this approach comes from studies demonstrating nikkomycin as highly efficacious in a mouse model of *C. immitis* (*87*). In the same study infections caused by *Blastomyces dermitidis* and *Histoplasma capsulatum* in mice were also eliminated by treatment with nikkomycin. However, to fully understand the structural features of peptidyl-nucleosides that determine their uptake by these fungi, the cellular components of the fungal peptide transport system must be defined.

Cloning Components of the Yeast Peptide Transport System

The isolation and characterization of peptide transport mutants of *S. cerevisiae* (*33*) opened a facile route to the cloning of peptide transport genes of this yeast, and as subsequently described, unlocked a method to clone other fungal and eukaryotic peptide transporters. By using conditions previously found to induce specific components of the *Saccharomyces* peptide permease, 24 spontaneous mutants and 31 *N*-methyl-*N*-nitrosoguanidine-induced mutants were isolated that were deficient in peptide uptake. Genetic analysis allowed these mutants to be separated into complementation groups called *PTR1*, *PTR2*, and *PTR3*. *PTR1* and *PTR2* were totally peptide-transport deficient, as defined by sensitivity to toxic peptides and direct uptake of radioactive peptide substrates. This study (*33*) represented the first genetic characterization of peptide permeation in a eukaryotic cell.

The *S. cerevisiae PTR2* gene was isolated by transforming the recessive *S. cerevisiae* mutant strain PB1X-9B with a YCp50-based genomic DNA library (*88*). Positive transformants were assessed on dipeptide media and recovered in *Escherichia coli* strain HB101, and the plasmid was isolated. A number of deletions and frame-shift mutations within the cloned insert were constructed, and a final 3.1-kb fragment was found to restore dipeptide use and toxic peptide sensitivity to *ptr2* mutant strains. The 3.1-kb fragment was sequenced by dideoxynucleotide chain termination methods. Analysis of the *PTR2* sequence revealed a 1803-bp open reading frame preceded by a 883-bp leader sequence containing many elements of known yeast promoters. A hydrophobic protein composed of 601 amino acid residues with a M_r of 68 100 Da and exhibiting 12 probable transmembrane spanning regions was deduced from the nucleotide sequence (Figure 1). A search of the Protein Data Base showed that a nitrate transporter of the plant *Arabidopsis thaliana* had significant sequence similarity to Ptr2p (Figure 2).

The availability of well-characterized peptide transport mutants of *S. cerevisiae* and of a cloned *PTR2* gene allowed us to investigate the genetic basis for peptide transport in

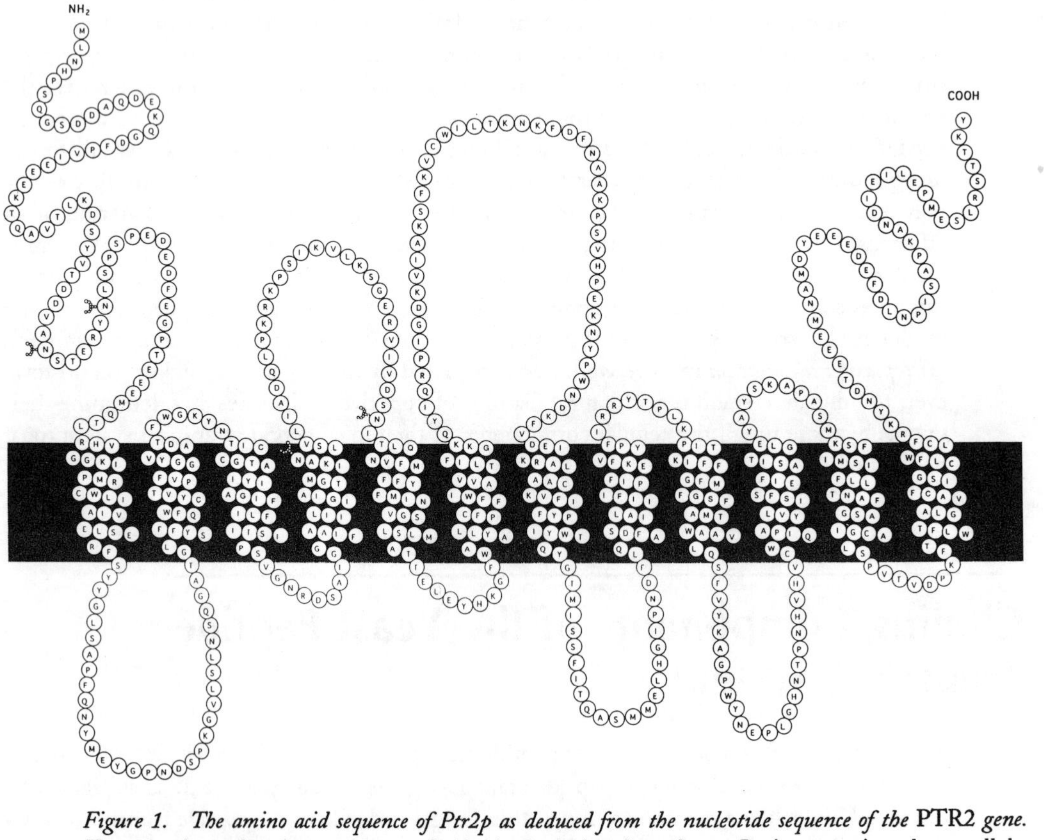

Figure 1. The amino acid sequence of Ptr2p as deduced from the nucleotide sequence of the PTR2 *gene. Transmembrane domains are shown embedded in the darkened membrane. Regions are oriented extracellular or intracellular on the basis of potential extracellular glycosylation sites.*

```
                    .         .         .         .         .
PTR2   23 IEEEKTQAVTLKDSYVTDDVANSTERYNLSPSPEDEDFEGPTEEEMQTLR 72
          :.|.|.:.: |...                      ||:|...:  .|
CHL1    3 LPETKSDDILLDAW......................DFQGRPADRSKTGG 30

                    .         .         .         .         .
       73 HVGGKIPMRCWLIAIVELSERFSYYGLSAPFQNYMEYGPNDSPKGVLSLN 122
           .::          :| .:|  ||:. .|::. : .|:         .|.: |.
       31 WASAA......MILCIEAVERLTTLGIGVNLVTYL........TGTMHLG 66

                    .         .         .         .         .
      123 SQGATGLSYFFQFWCYVTPVFGGYVADTFWGKYNTICCGTAIYIAGIFIL 172
          ...|..     |   :::  .::||::||||:|:| ||..  .||   .|: ||
       67 NATAANTVTNFLGTSFMLCLLGGFIADTFLGRYLTIAIFAAIQATGVSIL 116

                    .         .         .         .         .
      173 FITSI.............PSVGNRDSAI..GGFIAAIILIGIATGMIKAN 207
           :..|             .| .:..|:|  ..:. |:.|.:::|| :||.
      117 TLSTIIPGLRPPRCNPTTSSHCEQASGIQLTVLYLALYLTALGTGGVKAS 166

                    .         .         .         .         .
      208 LSVLIADQLPKRKPSIKVLKSGERVIVDSNITLQNVFMFFYFMINVGSLS 257
          :|.: .||:.. .|. :           |.:|.  .|  |:| ||||  |
      167 VSGFGSDQFDETEPKER...........SKMTY..FFNRFFFCINVG..S 201

                    .         .         .         .         .
      258 LMATTELEYHKG........FWAAYLLPFCFFWIAVVTLIFGKKQYIQRP 299
          |:|.| | | .:           .|: :.:.: : ..|  :  |..|..|
      202 LLAVTVLVYVQDDVGRKWGYGICAFAIVLALSVFLAGTNRYRFKKLIGSP 251

                    .         .         .         .         .
      300 IGDKVIAKSFKVCWILTKNKF...........DFNAAKPSVHPEKNYPWN 338
          :.:  :|  : ..|   | .:           |: ||..|:......| .
      252 MTQ..VAAVIVAAWRNRKLELPADPSYLYDVDDIIAAEGSMKGKQKLPHT 299

                    .         .         .         .         .
      339 DKF..............................VDEIKRALAACKVFIFY 358
          :.|                              |:|:|. :   .:: :
      300 EQFRSLDKAAIRDQEAGVTSNVFNKWTLSTLTDVEEVKQIVRMLPIWATC 349

                    .         .         .         .         .
      359 PIYWTQYGTMISSFITQASMME....LHGIPNDFLQAFDSIALIIFIPIF 404
           ::|| .: :..  :.|.. ::       :|| . :..|   :|:: .:::
      350 ILFWTVHAQLTTLSVAQSETLDRSIGSFEIPPASMAVFYVGGLLLTTAVY 399

                    .         .         .         .         .
      405 EKFVYPFIRRY....TPLKPITKIFFGFMFGSFAMTWAAVLQSFVYKAGP 450
          ::....:.::.     .|:|: :| :|::|||:||. ||:::    .:.:.
      400 DRVAIRLCKKLFNYPHGLRPLQRIGLGLFFGSMAMAVAALVELKRLRTAH 449

                    .         .         .         .         .
      451 WYNEPLGHNTPNHVHVCWQIPAYVLISFSEIFASITGLEYAYSKAPASMK 500
           ....:    .. .: .:: ||.|::::::| :  ...|:: ....| :||
      450 AHGPTV...KTLPLGFYLLIPQYLIVGIGEALIYTGQLDFFLRECPKGMK 496

                    .         .         .         .         .
      501 SFIMSIFLLTNAFGSAIGCALSPVTVDPKFT.....WLFTGLA....... 538
          ::  :::| | |:|  :::.|  ||: .|||     |: .:|.
      497 GMSTGLLLSTLALGFFFSSVL..VTIVEKFTGKAHPWIADDLNKGRLYNF 544

                    .         .         .         .
      539 ...VACFIS.GCLFWLCFRKYNDTEEEMNAMDYEEEDEFDLNPISAPKAN 584
             ||.::. . |::|.|.|:   .|.. |    | : |:| :| | |.::
      545 YWLVAVLVALNFLIFLVFSKWYVVYKEKRLA...EVGIELDDEP.SIPMGH 590
```

Figure 2. Alignment of the amino acid sequences of Ptr2p and the deduced amino acid sequence of the Arabidopsis thaliana *nitrate transport protein (CHL1). The proteins exhibit 26% identity and 52% similarity over the entire sequence.*

C. albicans (*89*). Preliminary hybridization studies showed that no complementarity existed between *PTR2* DNA and DNA in various *Candidal* libraries. However, a *C. albicans PTR2* homologue that complemented the *ptr2* mutation in *S. cerevisiae* PB1X-9B has been cloned using a *C. albicans* library in the yeast YEp24 high-copy shuttle vector. Restriction enzyme analysis of plasmid DNA isolated from seven independent *S. cerevisiae* transformants showed that three different inserts, each with a common 8-kb sequence, were present. The plasmids complemented a *ptr2* mutation in another independently isolated mutant and in a *PTR2* gene-deleted strain, showing that the complementing *C. albicans* DNA was indeed a functional peptide transport homologue. We are currently sequencing the putative *C. albicans* peptide transport gene.

Exploitation of the *S. cerevisiae* peptide transport mutants has led to further progress in the molecular cloning of additional putative peptide transporters. The fact that *S. cerevisiae* has been used often (*90–92*) for the heterologous functional expression of a number of eukaryotic genes supported the rationale for this approach. Plasmids containing DNA from *S. cerevisiae* that complement *ptr1* mutants have been obtained. In collaboration with the laboratory of Gary Stacey of the University of Tennessee, we have cloned and sequenced a gene from *A. thaliana* that appears to be a plant homolog of the *S. cerevisiae PTR2* gene (*93*). We are excited by the apparent preliminary success of the use of yeast mutants to clone peptide transport genes of eukaryotic cells.

Additional Modes of Peptide Transport

In addition to the traditional view of transport of small peptides across the cytoplasmic membrane by a carrier-mediated process from the external environment into the cytoplasm, other modes of transport of peptides exist in the fungal cell. We refer to examples of peptides larger than the normally studied di- and tripeptide transport substrates associated with carrier-mediated peptide transport systems. For example, internalization of the *S. cerevisiae* tridecapeptide mating pheromone, α-factor, by receptor-mediated endocytosis has been documented (*94*). *C. albicans* was shown to manifest pinocytosis, which may enable uptake of large peptides and proteins by this process (*30*). The α-factor is exported from the producing MATα-cell by the secretory system involving membrane vesicles (*95, 96*), and the export process of the a-factor mating pheromone of *S. cerevisiae* has recently been shown to involve a novel secretory pathway, as described in the following paragraph.

The a-factor is a dodecapeptide containing both a thioether-linked farnesyl group and a carboxy methyl ester on the carboxyl-terminal cysteine residue (*97*). This lipopeptide does not traverse the normal pathway of endoplasmic reticulum to Golgi cells and then to secretory vesicles used by signal-sequence tagged proteins, as does its reciprocal pheromone, α-factor. Rather, the secretion of a-factor is mediated by the STE6 protein (*98*), a member of the adenosine-5′-triphosphate binding cassette (ABC) protein superfamily (*99*). Other members of the ABC family include the multidrug resistance (MDR) protein, the cystic fibrosis transmembrane conductance regulator (CFTR), and transporters associated with antigen processing proteins. The a-factor is the only known natural peptide substrate for members of the superfamily, although the synthetic tripeptide *N*-acetyl-leucyl-leucyl-norleucinal is transported by the P170 protein, the mammalian *mdr1* gene product (*100*). This discovery raised the possibility that *MDR*-like genes present in the genome of mammalian cells may be involved in secretion of peptides, as is the case of the *STE6* gene product of *S. cerevisiae*. Functional complementation of the yeast STE6 by a mammalian *mdr* gene was performed (*101, 102*), and STE6-CFTR chimeras expressed in yeast are functional and were used to uncover aspects of the structure and function of CFTR (*103*). We expect that in the future, *S. cerevisiae* will continue to be an excellent tool for the study of different modes of peptide transport found in a variety of eukaryotic cells.

Summary and Conclusions

The implementation of modern molecular techniques to the study of peptide transport in fungi will lead to a better understanding of many aspects of peptide transport that are not well-defined at present. For example, the multiplicity of systems and the structural specificity of each system can be elucidated once individual components can be studied separately after genetic manipulation. The structure of the proteins mediating peptide transport and the regulation of the genes encoding those proteins will soon be analyzed, and the ability to manipulate peptide transport in fungi will be realized. Such knowledge should guide advances in drug design against opportunistic fungi whose incidence is greatly increasing. Furthermore, the use of *S. cerevisiae* for heterologous cloning of peptide transporters from a variety of eukaryotic cells will facilitate advances in our understanding of peptide transport in many biological systems. This information will be invaluable for medicinal chemists seeking to deliver peptide drugs through membrane barriers such as the human intestinal lining. The future of study of peptide transport in fungi should be full of exciting opportunities and discoveries.

Acknowledgments

We are thankful for the support of the American Cancer Society and for the help and inspiration provided by our many wonderful technicians, students, and postdoctoral fellows, including Wayne Lichliter, Robert Marder, David Logan, Ali Gulumoglu, Sam Lewis, Shabbir Khan, Kim Dunsmore, Bijoy Kundu, Lu Hilenski, Stevan Marcus, John Tullock, David Parker, David Ti, Alvin Steiffeld, Ponniah Shenbagamurthi, Ram Khare, Herb Smith, Cynthia Pousman, David Miller, Eduardo Krainer, Charlotte Boney, Nancy Covert, Michael Island, Hong-Long Zhang, Jack Perry, Munira Basrai, and Henry-York Steiner.

References

1. Naider, F.; Becker, J. M. In *Current Topics in Medical Mycology;* McGinnis, M. M. Ed.; Springer–Verlag: New York, 1987; Vol. II, pp 170–198.
2. Meyer, R. D.; Holmberg, K. In *Diagnosis and Therapy of Systemic Fungal Infections;* Holmberg, K.; Meyer, R., Eds.; Raven: New York, 1989; pp 79–100.
3. Paya, C. V. *Clin. Infect. Dis.* **1993,** *16,* 677–688.
4. Wolfinbarger, L., Jr.; Marzluf, G. A. *J. Bacteriol.* **1974,** *119,* 371–378.
5. Wolfinbarger, L., Jr.; Marzluf, G. A. *Arch. Biochem. Biophys.* **1975,** *171,* 637–644.
6. Becker, J. M.; Naider, F. In *Microorganisms and Nitrogen Sources;* Payne, J. W., Ed.; John Wiley & Sons. New York, 1980; pp 257–279.
7. Botstein, D.; Fink, G. R. *Science (Washington, DC)* **1988,** *240,* 1439–1443.
8. Hartwell, L. H. *Cancer* **1992,** *69,* 2615–2621.
9. Herskowitz, I. *Nature (London)* **1985,** *316,* 878.
10. Kirsch, D. R.; Kelly, R.; Kurtz, M. B. *The Genetics of Candida;* CRC. Boca Raton, FL, 1990.

11. Odds, F. C. In *Candida and Candidosis: A Review and Bibliography*, 2nd ed.; Odds, F. C., Ed.; Bailliere Tindall: London, 1988; pp 3–5.

12. Edwards, J. E., Jr.; Filler, S. G. *Clin. Infect. Dis.* **1992,** *14(Suppl),* S106–113.

13. Sobel, J. D.; Vazquez, J.; Lynch, M.; Meriwether, C.; Zervos, M. J. *Clin. Infect. Dis.* **1993,** *16,* 93–99.

14. Naider, F.; Becker, J. M.; Katzir-Katchalski, E. *J. Biol. Chem.* **1974,** *249,* 9–20.

15. Marder, R.; Becker, J. M.; Naider, F. *J. Bacteriol.* **1977,** *131,* 906–916.

16. Becker, J. M.; Naider, F.; Katchalski, E. *Biochim. Biophys. Acta* **1973,** *291,* 388–397.

17. Nisbet, T. M.; Payne, J. W. *FEMS Microbiol. Lett.* **1979,** *6,* 193–196.

18. Davies, M. B. *J. Gen. Microbiol.* **1980,** *121,* 181–186.

19. Lichliter, W. D.; Naider, F.; Becker, J. M. *Antimicrob. Agents Chemother.* **1976,** *10,* 483–490.

20. Logan, D. A.; Becker, J. M.; Naider, F. *J. Gen. Microbiol.* **1979,** 114, 179–186.

21. McCarthy, P. J.; Newman, D. J.; Nisbet, L. J.; Kingsbury, W. D. *Antimicrob. Agents Chemother.* **1985,** *28,* 494–499.

22. Milewski, S.; Andruszkiewicz, R.; Borowski E. *FEMS Microbiol. Lett.* **1988,** *50b,* 573–578.

23. Shallow, D. A.; Barrett-Bee, K. J.; Payne, J. W. *FEMS Microbiol. Lett.* **1991,** *79,* 9–14.

24. Milewski, S.; Mignini, F.; Borowski, E. *J. Gen. Microbiol.* **1991,** *137,* 2155–2161.

25. Milewski, S.; Andruszkiewicz, R.; Kasprzak, L.; Mazerski, J.; Mignini, F.; Borowski, E. *Antimicrob. Agents Chemother.* **1991,** *35,* 36–43.

26. McCarthy, P. J.; Nisbet, L. J.; Boehm, J. C.; Kingsbury, W. D. *J. Bacteriol.* **1985,** *162,* 1024–1029.

27. Milewski, S.; Chmara, H.; Borowski, E. *Arch. Microbiol.* **1983,** *135,* 130–136.

28. Nisbet, T. M. Payne, J. W. *J. Gen. Microbiol.* **1979,** *115,* 127–133.

29. Becker, J. M.; Naider, F. *Arch. Biophys. Biochem.* **1977,** *178,* 245–255.

30. Basrai, M.; Zhang, H.-L; Miller, D.; Naider, F.; Becker, J. M. *J. Gen. Microbiol.* **1992,** *138,* 2353–2362.

31. McCarthy, P. J.; Troke, P. F.; Gull, K. *J. Gen. Microbiol.* **1985,** *131,* 775–780.

32. Basrai, M. A.; Naider, F.; Becker, J. M. *J. Gen. Microbiol.* **1990,** *136,* 1059–1065.

33. Island, M. D.; Perry, J. R.; Naider, F.; Becker, J. M. *Curr. Genet.* **1991,** *20,* 457–463.

34. Marder, R.; Rose, B.; Becker, J. M.; Naider, F. *J. Bacteriol.* **1978,** *136,* 1174–1177.

35. Moneton, P.; Sarthou, P.; Le Goffic F. *J. Gen. Microbiol.* **1986,** *132,* 2147–2153.

36. Yadan, J. C.; Gonneau, M.; Sarthou, P.; Le Goffic, F. *J. Bacteriol.* **1984,** *160,* 884–888.

37. Payne, J. W.; Shallow, D. A. *FEMS Microbiol. Lett.* **1985,** *28,* 55–60.

38. Payne, J. W.; Barrett-Bee K. J.; Shallow, D. A. *FEMS Microbiol. Lett.* **1991,** *79,* 15–20.

39. Steiner, H. Y.; Barnes, O. B.; Naider, F.; Becker, J. M., University of Tennessee, Knoxville, TN, unpublished results.

40. Moneton, P.; Sarthou, P.; Le Goffic, F. *FEMS Microbiol. Lett.* **1986,** *36,* 95–98.

41. Becker, J. M.; Covert, N. L.; Shenbagamurthi, P.; Naider, F. *Antimicrob. Agents Chemother.* **1983,** *23,* 926–929.

42. Mehta, R. J.; Kingsbury, W. D.; Valenta, J.; Actor, P. *Antimicrob. Agents Chemother.* **1984,** *25,* 373–374.

43. Island, M. D.; Naider, F.; Becker, J. M. *J. Bacteriol.* **1987,** *169,* 2132–2136.

44. Ames, B. N.; Ames, G. F.; Young, J. D.; Isuchiya, D.; Lecocq. J. *Proc Natl. Acad. Sci. U.S.A.* **1973,** *70,* 456–458.

45. Fickel, T. E.; Gilvarg, C. *Nature (London)* **1973,** *241,* 161–163.

46. Gilvarg, C. In *The Future of Antibiotherapy and Antibiotic Research;* Ninet, L.; Bost, P. E.; Bouanchaud, D. H.; Florent, J., Eds.; Academic: Orlando, FL, 1981; pp 351–361.

47. Ringrose, P. S. In *Symposium of the Society of General Microbiology;* Greenwood, D.; O'Grady, F., Eds.; Cambridge University: London, 1985; pp 219–266.

48. Becker, J. M.; Steinfeld, A.; Naider, F. *Proceedings of the Fourth International Conference on the Mycoses;* Scientific Publication No. 356; Pan American Health Organization: Brasilia, Brazil, 1977; pp 303–308.

49. Ti, J.-S.; Steinfeld, A. S.; Naider, F.; Culumoglu, A.; Lewis, S. V.; Becker, J. M. *J. Med. Chem.* **1980,** *23,* 913–918.

50. Becker, J. M.; Marcus, S.; Tullock, J.; Miller, D.; Krainer, E.; Naider, F. *J. Infect. Dis.* **1988,** *157,* 212–214.

51. Dunn, F. W.; Dittmer, K. *J. Biol. Chem.* **1951,** *188,* 263–272.

52. Kenig, M.; Abraham, E. P. *J. Gen. Microbiol.* **1976,** *94,* 37–45.

53. Steinfeld, A. S.; Naider, F.; Becker, J. M. *J. Med. Chem.* **1979,** *22,* 1104–1109.

54. Kingsbury, W. D.; Boehm, J. C.; Mehta, R. J.; Grappel, S. F.; Gilvarg, C. *J. Med. Chem.* **1984,** *184,* 1447–1451.

55. Andruszkiewicz, R.; Chmara, H.; Milewski, S.; Borowski, E. *J. Med. Chem.* **1987,** *30,* 1715–1719.

56. Andruszkiewicz, R.; Milewski, S.; Zieniawa, T.; Borowski, E. *J. Med. Chem.* **1990,** *33,* 132–135.

57. Andruszkiewicz, R.; Chmara, H.; Milewski, S.; Zieniawa, T.; Borowski, E. *J. Med. Chem.* **1990,** *33,* 2755–2759.

58. Milewski, S.; Chmara, H.; Andruszkiewicz, R.; Borowski, E.; Zaremba, M.; Borowski, J. *Drugs Exp. Clin. Res.* **1988,** *14,* 461–465.

59. Chmara, H.; Smulkowski, M.; Borowski, E. *Drugs Exp. Clin. Res.* **1980,** *6,* 7–14.

60. Milewski, S.; Chmara, H.; Borowski, E. *Arch. Microbiol.* **1986,** *145,* 234–240.

61. Kingsbury, W. D.; Boehm, J. C.; Mehta, R. J.; Grappel, S. F. *J. Med. Chem.* **1983,** *26,* 1725–1729.

62. Meyer-Glauner, W.; Bernard, E.; Armstrong, D.; Merrifield, B. *Zbl. Bakt. Hyg. Orig. A.* **1982,** *252,* 274–278.

63. Hori, M.; Kakiki, K.; Suzuki, S.; Misato, T. *Agric. Biol. Chem.* **1971,** *35,* 1280–1291.

64. Isono, K.; Nagatsu, J.; Kobinata, K.; Sasaki, K.; Suzuki, S. *Agric. Biol. Chem.* **1965,** *31,* 190–199.

65. Isono, K.; Asahi, K.; Suzuki, S. *J. Am. Chem. Soc.* **1969,** *91,* 7490–7505.

66. Suzuki, S.; Isono K.; Nagatsu, J.; Mizutani, T.; Kawashima, Y.; Mizuno, T. *J. Antibiot.* **1965,** *18,* 131.

67. Dahn, U.; Hagenmaier H.; Hohne, H.; Konig, W. A.; Wolf, G.; Zahner, H. *Arch. Microbiol.* **1976,** *107,* 143–160.

68. Bowers, B.; Levin, G.; Cabib, E. *J. Bacteriol.* **1974,** *119,* 564–575.

69. Braun, P. C.; Calderone, R. A. *J. Bacteriol.* **1978,** *135,* 1472–1477.

70. Braun, P. C.; Calderone, R. A. *J. Bacteriol.* **1979,** *14,* 666–670.

71. Hector, R. F.; Pappagianis, D. *J. Bacteriol.* **1983,** *154,* 488–498.

72. Hector, R. F.; Zimmer, B. L.; Pappagianis, D. In *Recent Progress in Antifungal Chemotherapy;* Yamaguchi, H., Ed.; Marcel Dekker: New York, 1991; pp 341–353.

73. Perfect, J. R.; Wright K. A.; Hector, R. F. In *Recent Progress in Antifungal Chemotherapy;* Yamaguchi, H., Ed.; Marcel Dekker: New York, 1991; pp 369–379.

74. Smith, H. A.; Shenbagamurthi, P.; Naider, F.; Kundu, B.; Becker, J. M. *Antimicrob. Agents Chemother.* **1986,** *29,* 33–39.

75. Naider, F.; Shenbagamurthi, P.; Steinfeld, A. S.; Smith, H. A.; Boney, C.; Becker, J. M. *Antimicrob. Agents Chemother.* **1983,** *24,* 787–796.

76. Krainer, E.; Becker, J. M.; Naider, F. *J. Med. Chem.* **1991,** *34,* 174–180.

77. Shenbagamurthi, P.; Smith, H. A.; Becker, J. M.; Steinfeld, A.; Naider, F. *J. Med. Chem.* **1983,** *26,* 1518–1522.

78. Shenbagamurthi, P.; Smith, H. A.; Becker, J. M.; Naider, F. *J. Med. Chem.* **1986,** *29,* 802–809.

79. Khare, R. K.; Becker, J. M. *J. Med. Chem.* **1988,** *83,* 650–656.

80. Emmer, G.; Ryder, N. S.; Grassberger, M. A. *J. Med. Chem.* **1984,** *27,* 278–281.

81. Decker, H.; Zahner, H.; Heitsch, H.; Konig, W. A.; Fiedler, H.-P. *J. Gen. Microbiol.* **1991,** *137,* 1805–1813.

82. Delzer, J.; Fiedler, H.-P.; Mueller, H.; Zahner, H.; Rathmann, R.; Ernst, R.; Konig, W. A. *J. Antibiot.* **1984,** *37,* 80–82.

83. Bulawa, C. *Annu. Rev. Microbiol.* **1993,** *47,* 505–534.

84. Cabib, E. *Antimicrob. Agents Chemother.* **1991,** *35,* 170–173.

85. Hector, R. F.; Schaller, K. *Antimicrob. Agents Chemother.* **1992,** *36,* 1284–1289.

86. Hector, R. F.; Braun, P. C. *Antimicrob. Agents Chemother.* **1986,** *29,* 389–394.

87. Hector, R. F.; Zimmer, B. L.; Pappagianis, D. *Antimicrob. Agents Chemother.* **1990,** *34,* 587–593.

88. Perry, J. R.; Basrai, M. A.; Steiner, H.-Y.; Naider, F.; Becker, J. M. *Mol. Cell. Biol.* **1994,** *14,* 104–115.

89. Basrai, M., University of Tennessee, Knoxville, TN, unpublished results.

90. Colicelli, J.; Birchmeier, C.; Michaeli, T.; O'Neill, K.; Riggs, M.; Wigler, M. *Proc. Natl. Acad. Sci. U.S.A.* **1990,** *86,* 3599–3603.

91. Dougherty, K. M.; Brandriss, M. C.; Valle, D. *J. Biol. Chem.* **1992,** *267,* 871–875.

92. Lew, D. J.; Dulic, V.; Reed, S. I. *Cell* **1991,** *66,* 1197–1206.

93. Steiner, H. Y.; Song, W.; Zhang, L; Naider, F.; Becker, J. M.; Stacey, G. *Plant Cell* **1994,** *6,* 1289–1299.

94. Riezman, H.; Chvatchko, Y.; V. Dulic. *Trends Biochem. Sci.* **1986,** *11,* 325–328.

95. Fuller, R. S.; Brake, A. J.; Thorner, J. *Science (Washington, DC)* **1989,** *246,* 482–486.

96. Sprague, G. F., Jr.; Thorner, J. In *The Molecular Biology of the Yeast Saccharomyces,* 2nd ed.; Broach, J. R.; Pringle, J. R.; Jones, E. W.; Eds.; Cold Spring Harbor Laboratory: Cold Spring Harbor, New York, 1992; pp 657–744.

97. Anderegg, R. J.; Betz, R.; Carr, S. A.; Crabb, J. W.; Duntze, W. *J. Biol. Chem.* **1988,** *263,* 18236–18240.

98. Michaelis, S. *Sem. Cell Biol.* **1993,** *4,* 17–27.

99. Higgins, C. *Annu. Rev. Cell Biol.* **1992,** *8,* 67–113.

100. Sharma, R. C.; Inoue, S.; Roitelman, J.; Schimke, R. T.; Simoni, R. D. *J. Biol. Chem.* **1992,** *267,* 5731–5734.

101. Kuchler, K.; Thorner, J. *Proc. Natl. Acad. Sci. U.S.A.* **1992,** *89,* 2302–2306.

102. Raymond, M.; Gros, P.; Whiteway, M.; Thomas, D. Y. *Science (Washington, DC)* **1992,** *256,* 232–234.

103. Teem, J. L.; Berger, H. A.; Ostedgaard, L. S.; Rich, D. P.; Tsui, L.-C.; Welsh, M. J. *Cell* **1993,** *73,* 335–346.

RECEIVED for review September 9, 1993. ACCEPTED revised manuscript November 23, 1993.

Peptide Metabolism

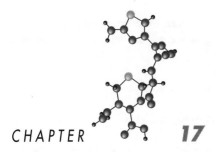

CHAPTER 17

Peptidomimetic Design and Chemical Approaches to Peptide Metabolism

T. K. Sawyer

Peptide-Based Drug Design Principles

Contemporary peptide-based drug design (*1–14*) is an interdisciplinary scientific endeavor composed of synthetic chemistry interfaced with biophysical chemistry (e.g., NMR spectroscopy and X-ray crystallography), computational chemistry (e.g., molecular modeling), biochemistry, and pharmacology.

Opportunities in Peptide-Based Therapeutics

The development of peptide-based therapeutics has been achieved with noteworthy success; examples of these therapeutics (Table I) include biologically or synthetically derived insulin, calcitonin, adrenocorticotropin, oxytocin, gonadotropin-releasing hormone, somatotropin-releasing factor, and somatostatin (*2, 15–24*). Examples of human diseases or disorders for which peptide-based drug (i.e., agonist or antagonist) therapy exists include osteoporosis (calcitonin), diabetes (insulin), prostrate cancer and endometriosis (gonadotropin-releasing

Table I. Some Peptide, Peptidomimetic, and Peptidoligand Drugs or Lead Compounds

Compound	*Known or Proposed Therapeutic Application (Protein, Peptide, or Nonpeptide Origin)*
Recombinant Peptides	
Atrial natriuretic factor	Potential use in prophylaxsis and treatment of acute renal failure
Epidermal growth factor	Potential use in skin grafting, eye surgery, and treatment of burns or ulcers
Somatotropin	Treatment of growth defects
Insulin	Treatment of type-I diabetes
Glucagon-like insulinotropic peptide	Potential treatment of insulin-insensitive diabetes
Transforming growth factor-β	Potential treatment of wound healing and burns
Nerve growth factor	Potential treatment of neural plasticity defects possibly related to Alzheimer's disease
Interleukin-1 (α–β)	Potential use in cancer therapy and inflammation
Hirudin	Potential use in prevention or treatment of venous blood clots
Tissue plasminogen activator	Potential use in treatment of blood clots in heart attacks
Factor VII	Potential use as a blood clotting factor in major forms of hemophiliacs
Erythropoietin	Potential uses in treatment of anemia in kidney dialysis patients, AIDS, and cancer
Platelet-derived growth factor	Potential uses in promoting growth of fibroblasts and keratinocytes and formation of new blood vessels
α-Interferon	Potential use as an immune stimulant for cancer therapy
β-Interferon	Potential use as immune stimulant for treatment of viral diseases and multiple sclerosis
γ-Interferon	Potential use as an immune stimulant for treatment of infectious diseases, cancer, and rheumatoid arthritis
Synthetic Peptides	
Aspartame	Artificial sweetener (Asp-Phe-OMe; a dipeptidyl glucose mimetic)
Sandostatin	Symptomatic treatment of acromegly and carcinoid syndrome (Hexapeptide agonist analogue of somatostatin)
Desmopressin	Treatment of severe diabetes insipidus (Nonapeptide agonist analogue of vasopressin)
Buserelin	Treatment of prostrate cancer and endometriosis (Nonapeptide agonist analogue of gonadotropin-releasing hormone)
A-75889	Potential use in treatment of prostrate cancer and endometriosis (Nonapeptide antagonist analogue of gonadotropin-releasing hormone)
HOE-140	Potential treatment of pain, inflammation, rhinitis, and asthma (Decapeptide antagonist analogue of BK)
Ebiratide	Potential application for CNS disorders; cognition treatment (Hexapeptide agonist analogue of adrenocorticotropin)
Melanotan	Potential use as a stimulant of skin pigmentation (Tridecapeptide agonist analogue of α-melanotropin)
BQ-123	Potential treatment of hypertension, restenosis, and related disorders (Cyclic pentapeptidyl, natural product antagonist of endothelin)
Metkephamid	Potential analgesic (Pentapeptide agonist analogue of ENK)
Pentigetide	Potential antiallergic agent (Pentapeptide antagonist fragment derivative of IgE)

Table I. *Continued*

Compound	*Known or Proposed Therapeutic Application (Protein, Peptide, or Nonpeptide Origin)*
MDL-28050	Potential use as anticoagulant agent (Hirudin-based decapeptide antagonist of thrombin)
L365209	Potential use as uterine relaxant for prevention of premature labor (Pseudo-hexapeptidyl, natural product antagonist of oxytocin)
RX-7736	Potential use for CNS disorders; cognition enhancement (Tripeptide agonist analogue of TRH)
GHRP-6	Potential growth hormone secretagogue lead (Hexapeptide agonist; native ligand unknown)
Calcitonin	Treatment of hypercalcemia, Paget's disease, osteoporosis, and pain affiliated with bone cancer (Synthetic derivative of natural peptide)
Peptidomimetics	
Captopril, enalapril	Antihypertensive drugs (Teprotide/Ang-I related peptidomimetic inhibitors of ACE)
A-72517	Potential antihypertensive drug (Angiotensinogen-based peptidomimetic inhibitor of renin)
Ro-318539	Potential anti-HIV/AIDS drug (*gag-pol* substrate-based peptidomimetic inhibitors of HIV protease)
SC-47643	Potential use for prevention of thrombus formation (Fibrinogen-based peptidomimetic antagonist at GPIIa–IIIb receptor)
Ro-24,9975	Potential use for CNS disorders; cognitive enhancement (TRH-based peptidomimetic agonist at TRH receptor)
CI-988	Potential anxiolytic drug (CCK-based peptidomimetic antagonist at CCK_B–gastrin receptor)
FK-888	Potential analgesic/antiinflammatory drugs (Substance P-based peptidomimetic antagonists at NK_1 receptor)
SCH-34826	Potential antihypertensive drug; inhibitor of atrial natriuretic factor (ANP) degradation (Peptidomimetic inhibitor of NEP)
CT-0543	Potential antimetastatic agent (Peptidomimetic inhibitor of gelatinase)
MD-805	Potential antithrombolytic drug (Peptidomimetic inhibitor of thrombin)
Peptidoligands	
Morphine	Analgesic drug (Nonpeptidyl natural product agonist at μ-opioid receptor)
CP-96,345	Potential analgesic and antiinflammatory drug (Nonpeptide antagonist at NK_1 receptor)
DuP-753	Potential antihypertensive drug (Nonpeptide antagonist at AT_1-type AII receptor)
OPC-21268	Potential use as a diuretic agent (Nonpeptide antagonist at V_1 vasopressin receptor)
L-366509	Potential use for prevention of premature labor (Nonpeptide antagonist of oxytocin)
Devazepide, SC-50998	Potential use for treatment of pancreatitis (Nonpeptide antagonists at CCK_A receptor)
CI-988, LY-262,691	Potential use for treatment of anxiety (Nonpeptide antagonists at CCK_B–gastrin receptor)
L-692429	Potential use as a GH secretagogue (Nonpeptide agonist at GHRP-6 receptor)
Ampicillin, amoxycillin	Antibacterial drugs (Nonpeptide penicillin-based inhibitors of bacterial cell wall peptidoglycan biosynthesis)

NOTE: *See:* Table VI and Figures 5 and 6 for chemical structures of some of the compounds.
SOURCE: Adapted from references 2, 3, 15–17, 24, 30, and 39.

hormone), acromegaly and ulcers (somatostatin), diuresis and hypertension (vasopressin), hypoglycemia (glucagon), and hypothyroidism (thyrotropin-releasing hormone [TRH]). The success of the aforementioned natural peptide hormones or chemically modified analogues thereof, which are currently marketed as prescription medications, inspired significant basic and pharmaceutical research dedicated to discovery, design, and development of other peptides, pseudopeptides (*19, 20*), peptidomimetics (*21–30*), and screening-derived nonpeptides (*26–30*) (Table I). Such compounds may prove useful as therapeutic or diagnostic agents for other life-threatening or disabling afflictions, including immunological dysfunction (e.g., acquired immunodeficiency syndrome), cancer, neural dysfunction (e.g., Alzheimer's disease, anxiety, or panic), and cardiovascular dysfunction.

Given the known pharmacokinetic and metabolism properties of peptides (*31–39*), considerable potential exists for chemical transformation of peptides to synthetic derivatives (e.g., pseudopeptides and peptidomimetics) by appropriate integration of molecular design principles that exploit lipophilicity, resistance to peptidase degradation, minimized molecular size, and prodrug derivatization. This chapter is dedicated to summarizing past and current research studies related to peptidomimetic molecular design and chemistry approaches to peptide metabolism.

Traditional Approaches in Peptide Analogue Lead Finding

In retrospect, a significant emphasis to accelerate peptide-based drug design focused on receptor or enzyme binding to identify high-affinity and selective leads, which may eventually be evaluated in vivo to provide some initial pharmacological data to indicate the therapeutic potential of such compounds. Pioneering work by a number of academic, government, and industrial research groups (*1–14*) led to a fundamental approach (Scheme I) to modify peptides by systematic fragment analysis, side-chain substitution, backbone replacement,

Peptide

Molecular Design Strategy

Primary Structure Analysis
* Contiguous Binding/Active Site
* Discontiguous Binding/Active Site

Sidechain Substructure Analysis
* Conservation of Cα–Cβ
* Chirality Inversion

Backbone Substructure Analysis
* Cis/Trans Isomerization
* 3-D Flexibility; H-Bonding

Secondary Structure/Topography Analysis
* Local Conformational Constraint
* Local Topographical Constraint

Pharmacophore/Model Analysis
* Identification of Key Functionalities

Synthetic Chemistry Strategy

Peptide Sequence Reduction
* N-/C-Terminal Truncation
* Internal Deletion

Amino Acid Substitutions
* Ala Scan
* D-Amino Acid Scan

Amide Replacements
* N-Alkylation
* Amide Surrogates

Peptide Sidechain/Backbone Modification
* Steric or Cyclic Amino Acid/Dipeptide Replacement
* Cyclic Peptide; Secondary Structure-Mimetics

Peptidomimetic Template Genesis
* 3-D Structure-Based, Template Functionalization

Peptide Analogue, Pseudopeptide or Peptidomimetic

Scheme I. Strategies for peptide-targeted molecular design and peptidomimetic chemistry.

stereochemical modification, and conformational constraints of the native molecule. Indeed, this approach has proven useful to identify prototypic leads for pharmacological evaluation in vivo as well as to advance potential peptidyl, pseudopeptidyl, and peptidomimetic (scaffold- and/or template-based) drug candidates. Nevertheless, peptide-targeted lead finding generally does not take into consideration predictively or experimentally determined metabolism, oral bioavailability, clearance, or other in vivo properties that are known to contribute to the pharmacokinetic and pharmacodynamic profile of a drug candidate until much later in the preclinical research timetable.

Emerging Strategies in Peptidomimetic Drug Discovery

In contrast to the aforementioned traditional strategy for advancing peptide-based drug design, a trend is developing to integrate aspects of metabolism, absorption, clearance, and related pharmacokinetic or pharmacodynamic studies much earlier within the discovery and design process. Such efforts may expedite the identification of chemical modifications that enhance the in vivo pharmacological efficacy of a particular lead or provide discrimination within a series of potential leads. In this regard, two reviews, by Powell (*34*) and Fauchere and Thurieau (*39*), provide excellent summaries of the metabolism of peptides in vitro and, to a lesser extent, in vivo by peptidases. Furthermore, Spatola and Darlak (*19*) and Spatola (*20*) established the concept of partial synthetic transformation of peptides to yield pseudopeptides, which incorporate amide bond replacements (Ψ[xxx] as a term refers to the replacement of the CONH moiety by typically nonhydrolyzable surrogates). Applications of pseudopeptidyl analogues of receptor- or enzyme-targeted peptides include stabilization to peptidase degradation, transition state mimicry, and antagonist discovery. Pseudopeptidyl modifications themselves may also provide synthons to construct peptidomimetics that have no natural dipeptide substructure.

Another sophisticated approach of peptide-targeted design, which is still emerging in both concept and experimental analysis, is related to exploiting the three-dimensional (3-D) structural (conformation and topography) and molecular dynamic properties of a native or synthetic peptide (*4, 22, 24*). This approach may lead to conformationally constrained peptide analogues and peptidomimetics depending on the degree of peptide backbone substructure (or scaffold) transformation. Such compounds may have strong stability against enzymatic degradation. A recent twist (*40*) to such research studies is the design of N-substituted Gly-oligopeptides, which are metabolically stable and provide prototypic leads for a variety of selected receptor and/or enzyme targets.

In addition to the aforementioned two approaches, which are ultimately related to peptide scaffold transformation, an increasing number of nonpeptidyl compounds that were discovered by high-volume (mass) biological screening are being discovered by analysis of natural product sources or synthetic chemical files. Typically, these compounds (*23–26*) are antagonists at peptide receptors or enzyme inhibitors, and such a screening approach has resulted in the successful development of highly potent and selective receptor- or enzyme-targeted therapeutic candidates. Insofar as nomenclature is concerned, the term "nonpeptide" will be used to describe such mass screening-derived compounds that competitively displace the naturally occurring peptide from its target receptor or enzyme. The genesis of combinatorial synthesis of primarily peptide chemical entities is intimately related to the approach of mass screening (*41–57*), and this relationship greatly extended the scope and opportunity for peptide lead discovery. Specifically, synthetic peptide combinatorial libraries (SPCLs) may incorporate novel building blocks of an amino acid or dipeptide intermediate, and the molecular design of the SPCL may address issues like peptide metabolism, 3-D structural properties, molecular recognition, and biological mechanisms, in an expeditious manner. The SPCL technology is, indeed, a powerful tool to investigate peptide interaction with proteins (e.g., receptors, enzymes, or antibodies), nucleic acids, or carbohydrate targets.

Peptide Chemical Structure and Synthetic Modification

The vast number of naturally occurring peptides were discovered over the past 50 years and provide the basis for the emerging basic and pharmaceutical research targeted on peptides. The structural determinations of, for example, peptide hormones, neurotransmitters, growth factors, and cytokines (Table II) illustrated the tremendous scope of the molecular size and complexity of peptides.

Primary Structure, Conformation, and Topography

Peptides are structurally diverse by virtue of the number of unique combinations of amino acid building blocks that constitute their molecular framework. The precise linear arrangement, (i.e., amino → carboxy [N → C] directionality) of amino acids in a peptide is referred to as its primary structure (Table I). The 3-D folding of the peptide onto itself through covalent (e.g., disulfide) or noncovalent (e.g., hydrogen or ionic) bonding determines the secondary structure, or conformation, of the peptide. Also contributing to such secondary structure is the tendency of peptides to adopt spatial orientations of both the backbone and side-chain functionalities to bring hydrophobic amino acids (e.g., Trp, Tyr, Phe, Leu, Ile, Val, and Met) into proximity to form a hydrophobic surface. Similarly, hydrophilic amino acids (e.g., Lys, Arg, Glu, Asp, Thr, Ser, Gln, Asn and His) residues may be distributed within such 3-D structures to form a hydrophilic surface that may provide a high degree of solvation by water. However, topography, or the "relative and co-operative 3-D arrangement of side-chain groups" (*4*), may not necessarily correlate with conformation. Similar topography may be achieved by different conformations, and different topographies may exist as the result of a similar conformation.

The 3-D substructure of peptides may be further described in terms of torsion angles between the backbone amine nitrogen (N^{α}), backbone carbonyl carbon (C'), backbone hydrocarbon (C^{α}), and side chain hydrocarbon functionalization (e.g., C^{β}, C^{γ}, C^{δ}, and C^{ϵ} of Lys) depending on the amino acid sequence (Chart I). The torsion angle nomenclature is as follows: Ψ, N^{α}-C^{α}-C'-N^{α}; ω, C^{α}-C'-N^{α}-C^{α}; ϕ, C'-N^{α}-C^{α}-C'; χ_1, N^{α}-C^{α}-C^{β}-R; χ_2, C^{α}-C^{β}-C^{γ}-R; and so forth (where ω is amide bond torsion angle and R is any atom except hydrogen). A Ramachandran plot of Ψ versus ϕ for peptides possessing intrinsic secondary structure further illustrate that particular combinations of torsion angles for a helical, reverse-turn, or extended conformation do exist in a predominant fashion. With respect to ω, the trans geometry is more energetically favored for most typical dipeptide substructures; however, when the C-terminal partner is Pro or other N-alkylated or cyclic amino acids, the cis geometry is quite likely (Chart I). In summary, local 3-D structural flexibility is directly related to covalent or noncovalent bonding interactions within the amino acid sequence of a particular peptide, and even modest chemical modifications, such as N^{α}-methylation, C^{α}-methylation, or C^{β}-methylation, can have significant consequences. Furthermore, these backbone or side-chain modifications may afford stability of the parent peptide to peptidases

Table II. Primary Structures of Some Biologically Active Peptides

Peptide	*Primary Structure*
TRH	<Glu^{1}-His-Pro-NH_2
FMRF-amide	Phe^{1}-Met-Arg-Phe-NH_2
ENK (Met)	Tyr^{1}-Gly-Gly-Phe-Met
CCK-8	Asp^{1}-Tyr[SO_3H]-Met-Gly-Trp-Met-Asp-Phe-NH_2
Angiotensin II	Asp^{1}-Arg-Val-Tyr-Ile-His-Pro-Phe
GnRH	<Glu^{1}-His-Trp-Ser-Tyr-Gly-Leu-Arg-Pro-Gly-NH_2
Oxytocin	Cys^{1}-Tyr-Ile-Gln-Asn-Cys-Pro-Leu-Gly-NH_2
Vasopressin	Cys^{1}-Tyr-Phe-Gln-Asn-Cys-Pro-Arg-Gly-NH_2
Teprotide	<Glu-Trp-Pro-Arg-Pro-Gln-Ile-Pro-Pro
BK	Lys^{1}-Arg-Pro-Gly-Phe-Ser-Pro-Phe-Arg
SP	Arg^{1}-Pro-Lys-Pro-Gln-Gln-Phe-Phe-Gly-Leu-Met-NH_2
Cyclosporin A	cyclo(MeLeu-MeLeu-Me-Thr[*4R*-4(*E*-2-butenyl)-4-methyl]-Abu-Sar-MeLeu-Leu-MeLeu-Ala-D-Ala-MeLeu)
α-Melanotropin	Ac-Ser^{1}-Tyr-Ser-Met-Glu-His-Phe-Arg-Trp-Gly-Lys-Pro-Val-NH_2
Neurotensin	<Glu^{1}-Leu-Tyr-Glu-Asn-Lys-Pro-Arg-Arg-Pro-Tyr-Ile-Leu
Somatostatin	Ala^{1}-Gly-Cys-Lys-Asn-Phe-Phe-Trp-Lys-Thr-Phe-Thr-Ser-Cys
Endothelin	Cys^{1}-Ser-Cys-Ser-Ser-Leu-Met-Asp-Lys-Glu-Cys-Val-Tyr-Phe-Cys-His-Leu-Asp-Ile-Ile-Trp^{21}
VIP	His^{1}-Ser-Asp-Ala-Val-Phe-Thr-Asp-Asn-Tyr-Thr-Arg-Leu-Arg-Lys-Gln-Met-Ala-Val-Lys^{21}-Tyr-Leu-Asn-Ser-Ile-Leu-Asn-NH_2
Glucagon	His^{1}-Ser-Gln-Gly-Thr-Phe-Thr-Ser-Asp-Tyr-Ser-Lys-Tyr-Leu-Asp-Ser-Arg-Arg-Ala-Gln-Asp^{21}-Phe-Val-Gln-Trp-Leu-Met-Asp-Thr
Galanin	Gly^{1}-Trp-Thr-Leu-Asn-Ser-Ala-Gly-Tyr-Leu-Leu-Gly-Pro-His-Ala-Ile-Asp-Asn-His-Arg-Ser^{21}-Phe-His-Asp-Lys-Tyr-Gly-Leu-Ala-NH_2
ACTH	Ser^{1}-Tyr-Ser-Met-Glu-His-Phe-Arg-Trp-Gly-Lys-Pro-Val-Gly-Lys-Lys-Arg-Arg-Pro-Val-Lys^{21}-Val-Tyr-Pro-Asn-Gly-Ala-Glu-Asp-Glu-Ser-Ala-Glu-Ala-Phe-Pro-Leu-Glu-Phe
Neuropeptide-Y	Tyr^{1}-Pro-Ser-Lys-Pro-Asp-Asn-Pro-Gly-Glu-Asp-Ala-Pro-Ala-Glu-Asp-Leu-Ala-Arg-Tyr-Tyr^{21}-Ser-Ala-Leu-Arg-His-Tyr-Ile-Asn-Leu-Met-Thr-Arg-Gln-Arg-Tyr-NH_2
Corticotropin-releasing factor	Ser^{1}-Gln-Glu-Pro-Pro-Ile-Ser-Leu-Asp-Leu-Thr-Phe-His-Leu-Leu-Arg-Glu-Val-Leu-Glu-Met^{21}-Thr-Lys-Ala-Asp-Gln-Leu-Ala-Gln-Gln-Ala-His-Ser-Asn-Arg-Lys-Leu-Leu-Asp-Ile-Ala^{41}-NH_2
GHRP	Tyr^{1}-Ala-Asp-Ala-Ile-Phe-Thr-Asn-Ser-Tyr-Arg-Lys-Val-Leu-Gly-Gln-Leu-Ser-Ala-Arg-Lys^{21}-Leu-Leu-Gln-Asp-Ile-Met-Ser-Arg-Gln-Gln-Gly-Glu-Ser-Asn-Gln-Glu-Arg-Gly-Ala^{41}-Arg-Ala-Arg-Leu-NH_2
Insulin	Gly^{1}-Ile-Val-Glu-Gln-Cys-Cys-Thr-Ser-Ile-Cys-Ser-Leu-Tyr-Gln-Leu-Glu-Asn-Tyr-Cys-Asn (A-Chain, above; B-Chain, below) Phe^{1}-Val-Asn-Gln-His-Leu-Cys-Gly-Ser-His-Leu-Val-Glu-Ala-Leu-Tyr-Leu-Val-Cys-Gly-Glu^{21}-Arg-Gly-Phe-Phe-Tyr-Thr-Pro-Lys-Thr
Parathyroid hormone	Ser^{1}-Val-Ser-Glu-Ile-Gln-Leu-Met-His-Asn-Leu-Gly-Lys-His-Leu-Asn-Ser-Met-Glu-Arg^{21}-Val^{22}-Glu-Trp-Leu-Arg-Lys-Lys-Leu-Gln-Asp-Val-His-Asn-Phe-Val-Ala-Leu-Gly-Ala-Pro-Leu^{41}-Ala-Pro-Arg-Asp-Ala-Gly-Ser-Gln-Arg-Pro-Arg-Lys-Lys-Glu-Asp-Asn-Val-Leu-Val-Glu^{61}-Ser-His-Glu-Lys-Ser-Ser-Leu-Gly-Glu-Ala-Asp-Lys-Ala-Asp-Val-Asp-Val-Leu-Thr-Lys^{81}-Ala-Lys-Ser-Gln

Backbone/Side-Chain Dihedral Angles of Phe in a Peptide

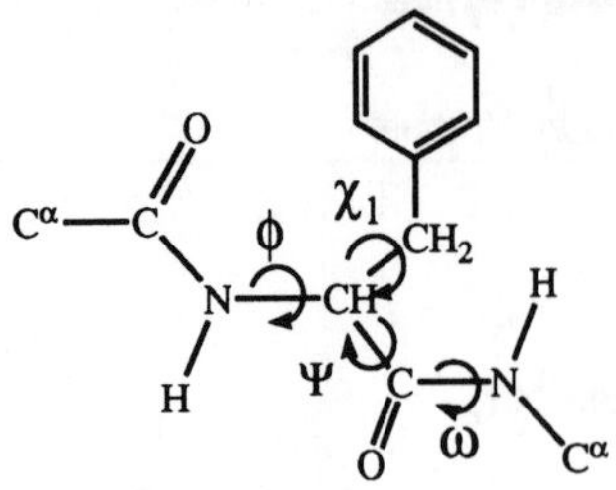

Cis and Trans Peptide Bond Isomers of Phe-Pro

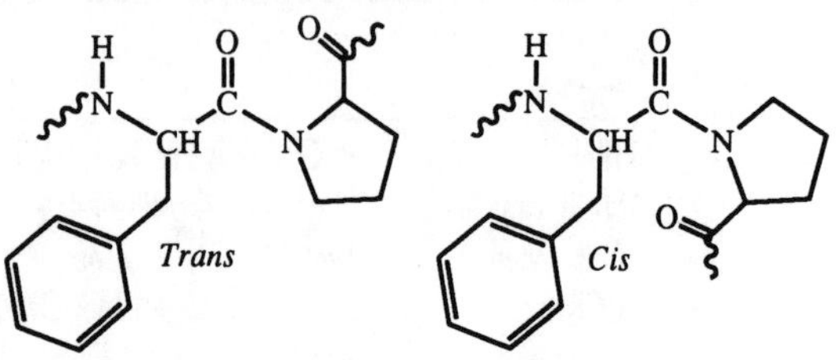

Possible Backbone/Side-Chain Cα–Cβ Spatial Projections of Phe

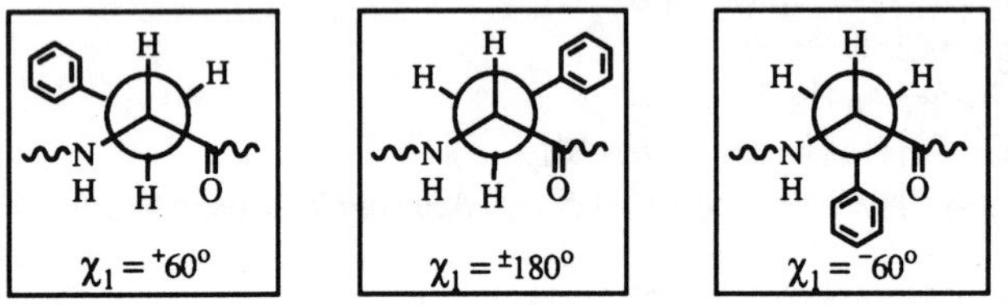

Intramolecular H-Bonding in Peptide Secondary Structure

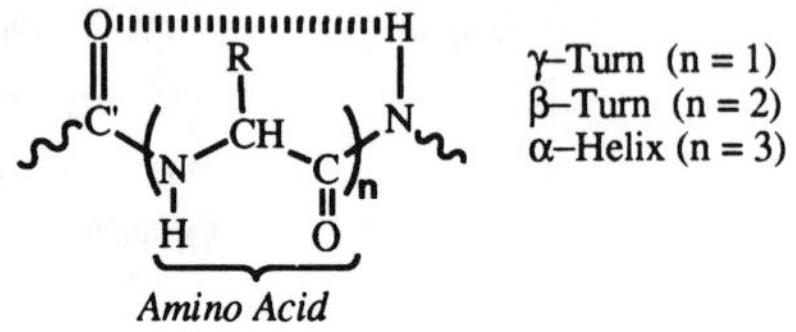

Conformationally-Restricted Phe Analogues

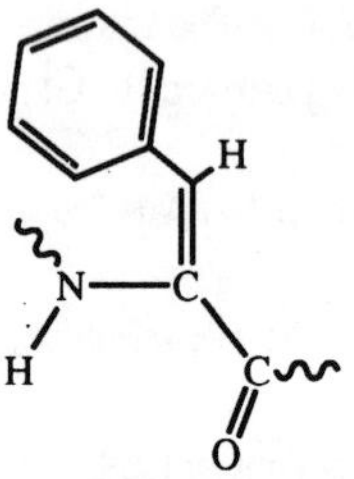

χ_1 constrained
Z-configuration
(loss of chirality)

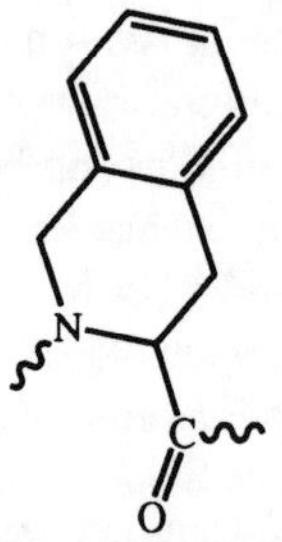

Φ, Ψ, and χ_1 constrained
cis-trans (amide) isomerism

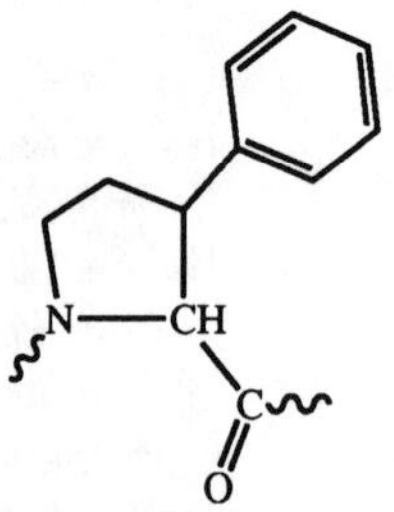

Φ, Ψ, and χ_1 constrained
cis-trans (amide) isomerism

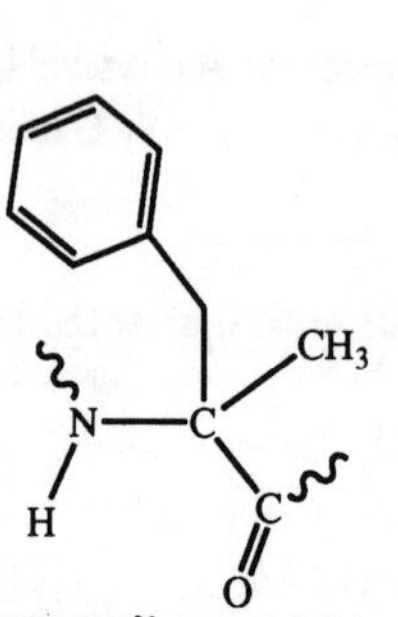

Φ, Ψ, and χ_1 constrained
α, α–disubstitution

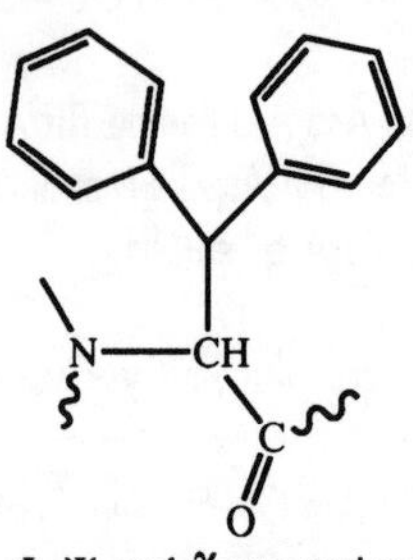

Φ, Ψ, and χ_1 constrained
β, β–disubstitution

Chart I. Three-dimensional structural properties of peptides.

and may provide conceptual impetus for yet more sophisticated molecular design and peptidomimetic chemistry efforts (*4–8*).

Synthetic Modification of Peptide Backbone or Side-Chain Substructures

A plethora of sophisticated synthetic chemistry approaches have entered into the arena of peptide-based molecular design, including well-established applications of unusual amino acids and dipeptide surrogates and other types of chemical modifications. Specifically, the following amide bond (CONH) replacements were reported (*see* Chart II for chemical structures): aminomethylene or $\Psi[CH_2NH]$, **1** (*58–62*); ketomethylene or $\Psi[COCH_2]$, **2** (*63–66*); ethylene or $\Psi[CH_2CH_2]$, **3** (*67–69*); olefin or $\Psi[CH{=}CH]$, **4** (*70–74*); ether or $\Psi[CH_2O]$, **5** (*75–78*); thioether or $\Psi[CH_2S]$, **6** (*79–81*); tetrazole or $\Psi[CN_4]$, **7** (*82–84*); thiazole or Ψ[thz], **8** (*85*); retroamide or Ψ[NHCO], **9** (*86, 87*); thioamide or Ψ[CSNH], **10** (*88–91*); ester or $\Psi[CO_2]$, **11** (*92, 93*); hydroxymethylene or Ψ[CH(OH)], **12** (*94–96*); hydroxyethylene or $\Psi[CH(OH)CH_2]$ and $\Psi[CH_2CH(OH)]$, **13** (*97–100*) and **14** (*101*), respectively; dihydroxyethylene or Ψ[CH(OH)CH(OH)], **15** (*102–105*); hydroxyethylamine or $\Psi[CH(OH)CH_2N]$, **16** (*106, 107*). Such amide substitutions provide insight into the conformational and H-bonding characteristics (intra- or intermolecular) at the particular site of chemical modification, and as relative to the parent compound, these backbone replacements can impart stability toward peptidase cleavage. The discovery of yet other peptidase-resistant or nonhydrolyzable CONH isosteres continues to be investigated.

Such backbone modifications may have particular promise in the design of peptidase inhibitors as the result of transition-state mimicry of the hypothetical tetrahedral intermediate aminol ($\Psi[C(OH)_2NH]$), which might exist within the mechanistic pathway of substrate cleavage. However, the more synthetically simple N-substituted amide (*108–112*) modifications, including those of cyclic amino acids (*113–119*), provide an excellent starting point to explore conformational (e.g., Chart I) and H-bonding properties and to impart stability toward peptidase cleavage. Furthermore, the concept of N-substituted amides to exploit the aforementioned chemical and biological properties was extended to a scaffold approach in which an N-substituted oligo-Gly is systematically modified at the amide nitrogen by functionalization identical to that of the C^{α}-side chain, which results in an achiral scaffold (**17**, Chart II) that can be synthetically elaborated in a broad-scope manner (*40*). Finally, the C^{α}-carbon itself provides a backbone site that can be modified by hydrogen substitution (*120–126*), C–H → C–R, and dehydration (*127–129*), C^{α}–C^{β} → C=C (*E*- or *Z*-geometry), as exemplified in Chart I. Furthermore, the C^{β}-carbon may be substituted (*130–132*) singularly or doubly (Chart I), and the scope of such chemistry extends to so-called chimeric amino acids (*22*). Most importantly, the aforementioned backbone (N^{α}, C^{α}, or C′) and side-chain (C^{α} or C^{B}) substructure modifications can impart stability against or inhibition of peptidases (*133*).

Progress in the Design of Novel Peptidomimetic Scaffolds

In addition to the aforementioned backbone and side-chain modifications that were investigated in peptidyl and peptidomimetic compounds, a noteworthy effort has emerged in both the molecular design and peptidomimetic chemistry of secondary structure replacements (*25, 134*). These secondary structure mimetics may be advanced as the result of preliminary structure–conformation studies based on substitution of D-amino acids, C^{α}-Me-amino acids, or dehydroamino acids within the peptide. Such modifications may induce or stabilize inherent turn conformations (β-turn, γ-turn, β-sheet, or α-helix) that exemplify the role of the backbone amide groups in intramolecular H-bonding (*7*). Furthermore, given the premise that such turn conformations were identified as common biological recognition motifs (*135*), the design of secondary structure mimetics is particularly intriguing. As illustrated in Chart II, a variety of secondary structure mimetics were designed (*25, 136–145*) and tested over the past few years after being incorporated in peptides of varying size.

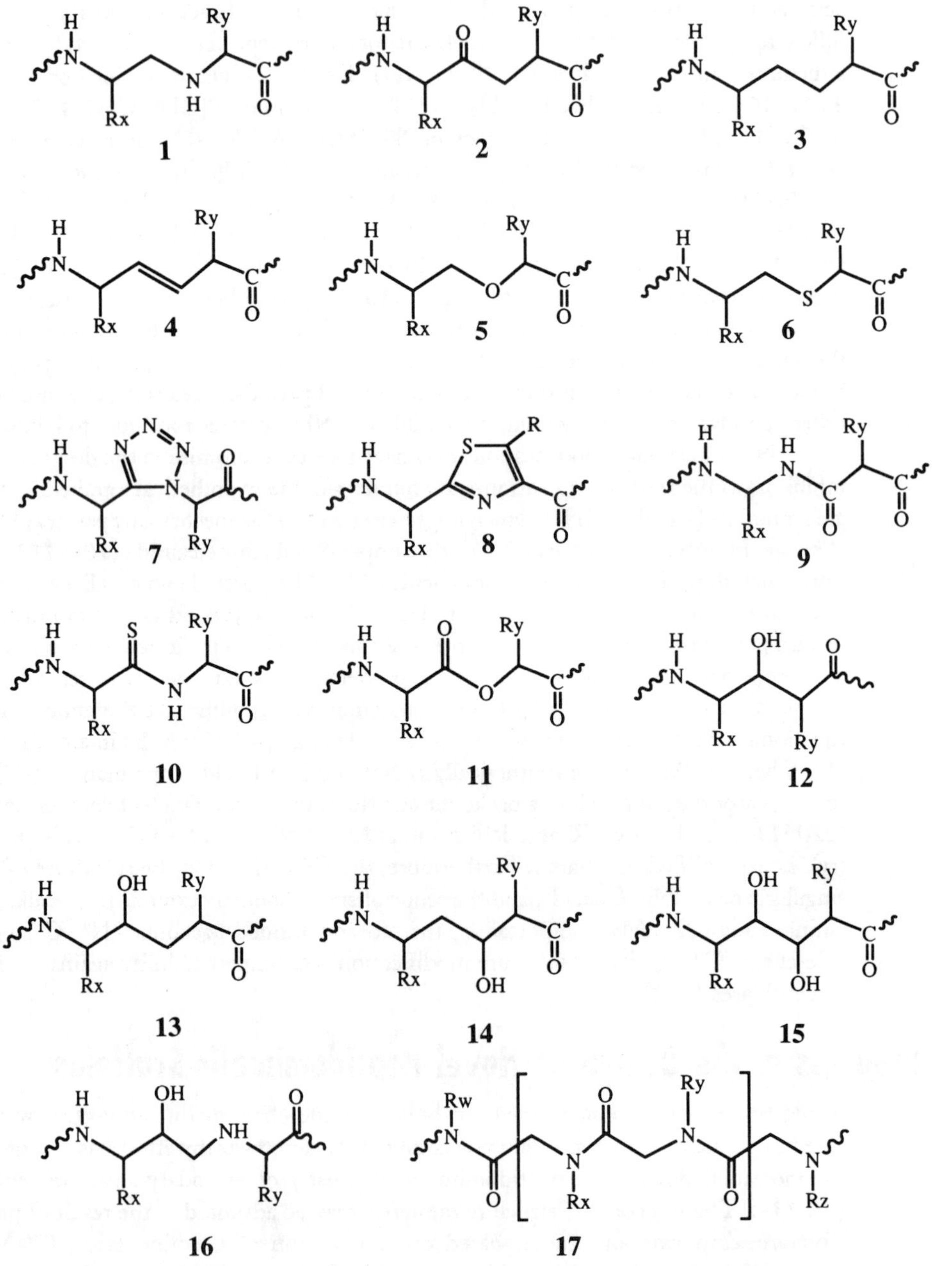

Chart II. Backbone and side-chain modifications imparting stability to peptidases.

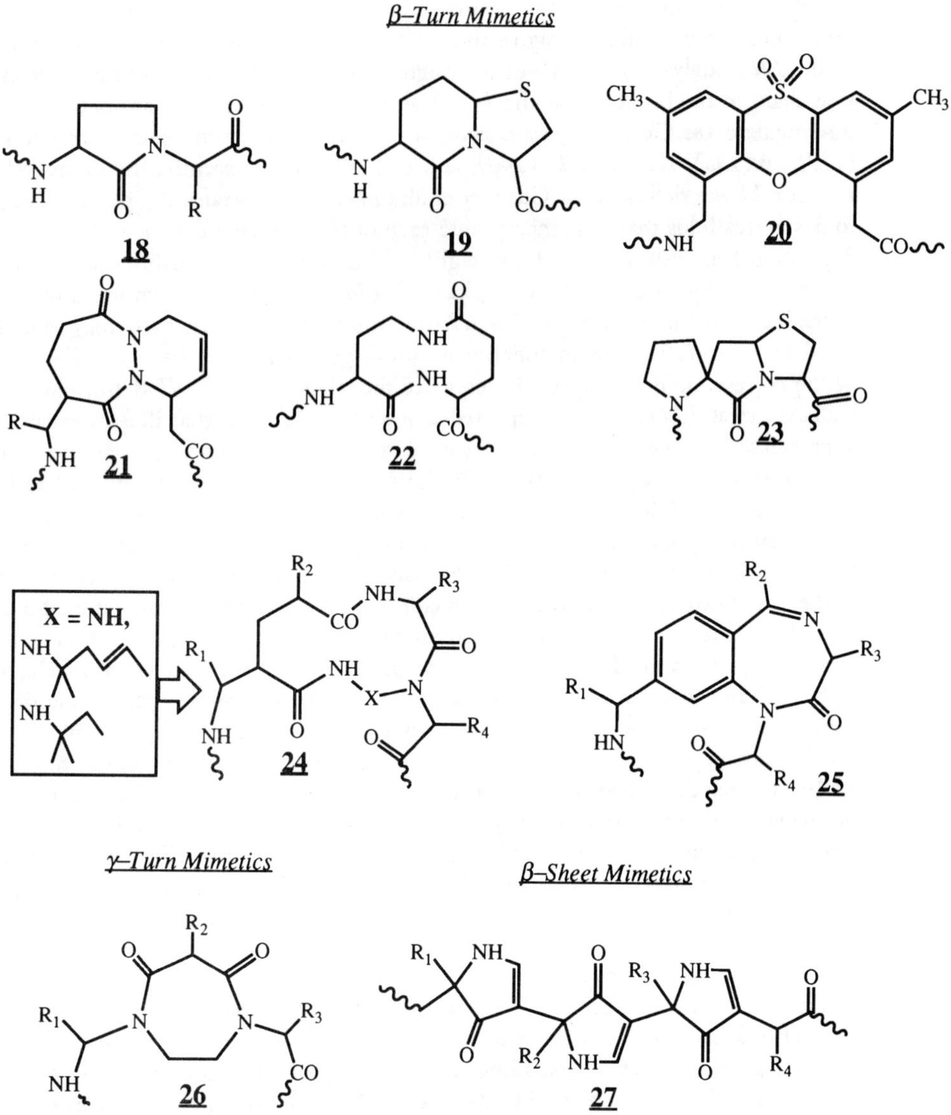

Chart II. Continued.

Specifically, the β-turn presents a challenging secondary structure because its backbone conformation is highly variable compared with those of the α-helix and β-sheet. The topographical properties of the side-chain substructure need to be exploited further by synthetic strategies as well as by studies addressing molecular recognition. The conventional description of β-turns is that they are secondary structural motifs of a tetrapeptide sequence in which the separation of the first and fourth Cα atoms is ≤7 Å. β-Turns occur in a nonhelical region and possess a 10-membered intramolecularly H-bonded ring (Chart I). Furthermore, the classification of β-turns has been based on the geometry of the peptide backbone (i.e., the ϕ and Ψ torsion angles at the second $[i+1]$ and third $[i+2]$ amino acids). Such β-turns include Types I, I′, II, II′, III, III′, IV, V, V′, VI, and VII. Despite the complexity of definitions, some noteworthy efforts (*136–144*) have advanced the discovery of β-turn mimetics based on scaffold design (Chart III). Specifically, the pioneering work of Freidinger et al. (*136*) catalyzed interest in the design of β-turn mimetics by using constraint of the side chain to the backbone at the *i* + 1 and *i* + 2 sites through a monocyclic dipeptide intermediate **18.** Over the past decade, a variety of cyclic templates were developed: **19** (*137*), **20** (*138*), **21** (*139*), **22** (*140*), and **23** (*141*). Most recently, the monocyclic β-turn mimetic **24** was described by Gardner et al. (*142*) and illustrates the potential opportunity to design scaffolds that may incorporate each of the side chains (*i*, *i* + 1, *i* + 2, and *i* + 3 positions) as well as five of the eight NH or C=O functionalities within the parent tetrapeptide sequence. Similarly, Ripka et al. (*143, 144*) have shown the intriguing utility of the benzodiazepine template **25** as a β-turn mimetic scaffold that also may be multisubstituted to simulate side-chain functionalization, particularly at the *i* and *i* + 3 positions of the corresponding tetrapeptide sequence modeled in type I–VI β-turn conformations. Callahan et al. (*145*) recently reported a γ-turn mimetic **26** that illustrates an innovative approach to incorporate a retroamide surrogate between the *i* and *i* + 1 amino acid residues with an ethylene bridge between the N′ (nitrogen replacing the carbonyl C′) and N atoms of the *i* and *i* + 2 positions; this template allows the possibility for all three side chains of the parent tripeptide sequence. Finally, the design of a β-sheet mimetic **27** by Smith et al. (*146*) provides an attractive template to constrain the backbone of a peptide to simulate a β-sheet type extended conformation. In conclusion, such efforts in the rational transformation of predicted or experimentally determined secondary structures are quite valuable and provide impetus to the design or application of β-turn, γ-turn, β-sheet, or α-helix mimetics.

More extensive, or radical structural transformation of a peptide, resulting in analogues with essentially no naturally occurring amino acids or dipeptide substructure, has also become popular (*22–30*) because of the premise that such compounds might be more well-suited for drug development in terms of chemical or biological properties (e.g., low molecular mass, metabolic stability, and bioavailability by oral administration). Such designed peptidomimetics (including peptoids and nonpeptides; Chart IV) have historical precedence in the discovery of orally effective and sustained-acting inhibitors of angiotensin-converting enzyme (ACE) [e.g., captopril, **28** (*147*) and enalapril, **29** (*148*)]. These ACE inhibitors may be viewed as modified dipeptides that possess functional groups exhibiting molecular complementarity to the target metallopeptidase active site, including Zn^{2+} coordinative and ancillary hydrophobic binding interactions. In addition to ACE inhibitors, significant efforts have been advanced to discover and develop yet other receptor- or enzyme-targeted peptidomimetics, including the following: TRH agonists, **30** (*24*); fibrinogen (GPIIb and IIIa) antagonists, **31** (*149*) and **32** (*150*); endothelin antagonists, **33** (*151*); cholecystokinin$_B$ (CCK_B)–gastrin antagonists, **34** (*152*); somatostatin agonists and antagonists, **35** (*153, 154*); substance-P (NK_1) antagonists, **36** (*155*); neurokinin-A (NK_2) antagonists, **37** (*156*); Ras farnesyl protein transferase inhibitors, **38** (*157*) and **39** (*158*); human immunodeficiency virus (HIV) protease inhibitors, **40** (*159*), **41** (*160*), and **42** (*161*); renin inhibitors **43** (*162*), **44** (*163*), and **45** (*164*); thrombin inhibitors, **46** (*165*); elastase inhibitors, **47** (*166*); gelatinase inhibitors, **48** (*167*); endopeptidase 24.15 inhibitors, **49** (*168–170*); and neutral endopeptidase (NEP) 24.11 inhibitors, **50** (*171–173*) and **51** (*174–176*).

In synchrony with the aforementioned peptide-targeted design strategy, efforts have been focused on identifying novel lead compounds by screening a variety of natural product, chemical file, or synthetic combinatorial libraries. Philosophically, this approach also has precedence (*177*) from the standpoint that morphine (**52**) is a nonpeptidyl natural product agonist at opioid receptors for which endogenous enkephalin and endorphin peptides act. As just mentioned, the compounds discovered by this receptor- or enzyme-targeted screening strategy may be collectively referred to as nonpeptides. Representative leads (Chart V) that

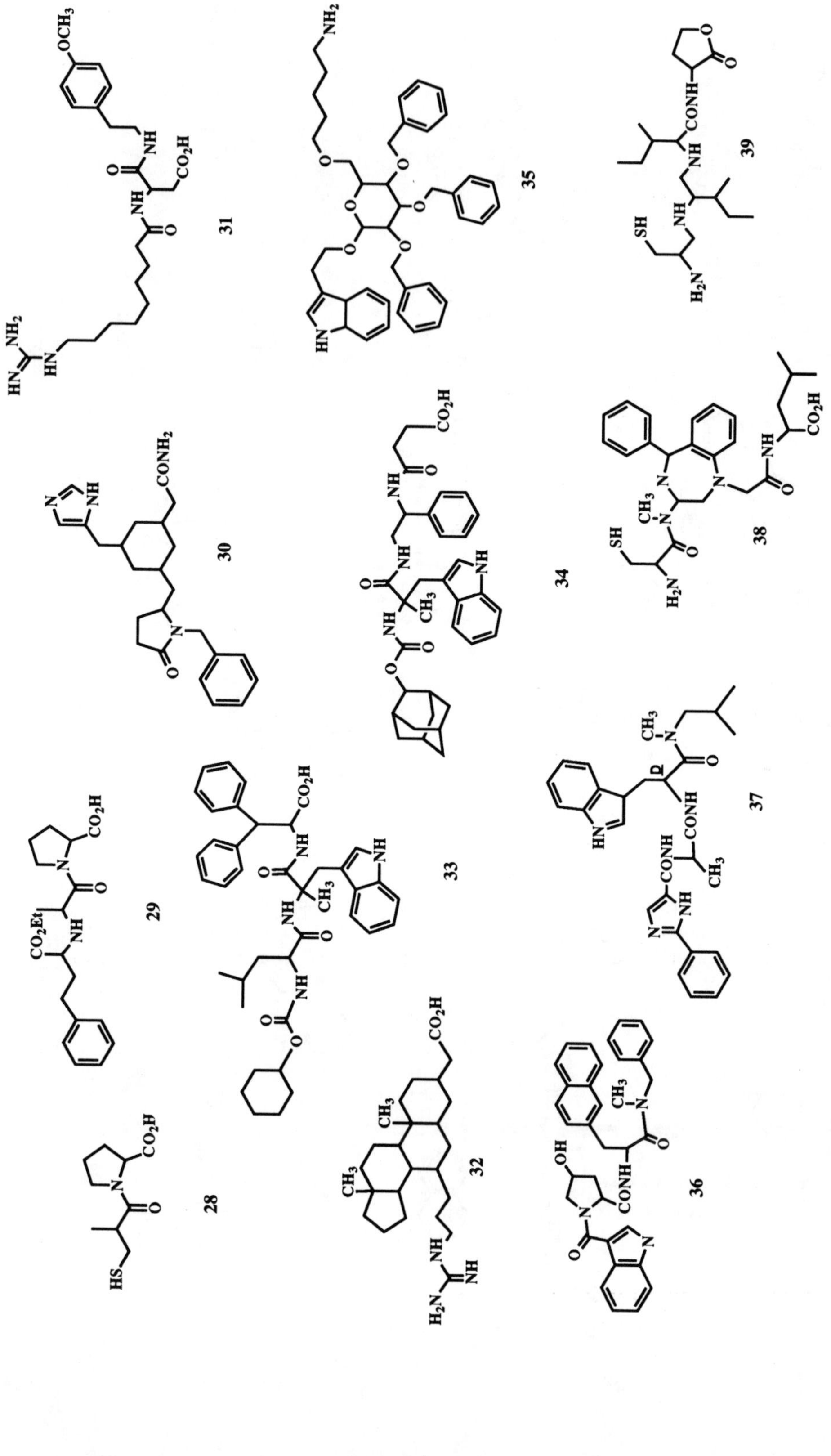

Chart III. Examples of peptidomimetics designed from peptide secondary structures.

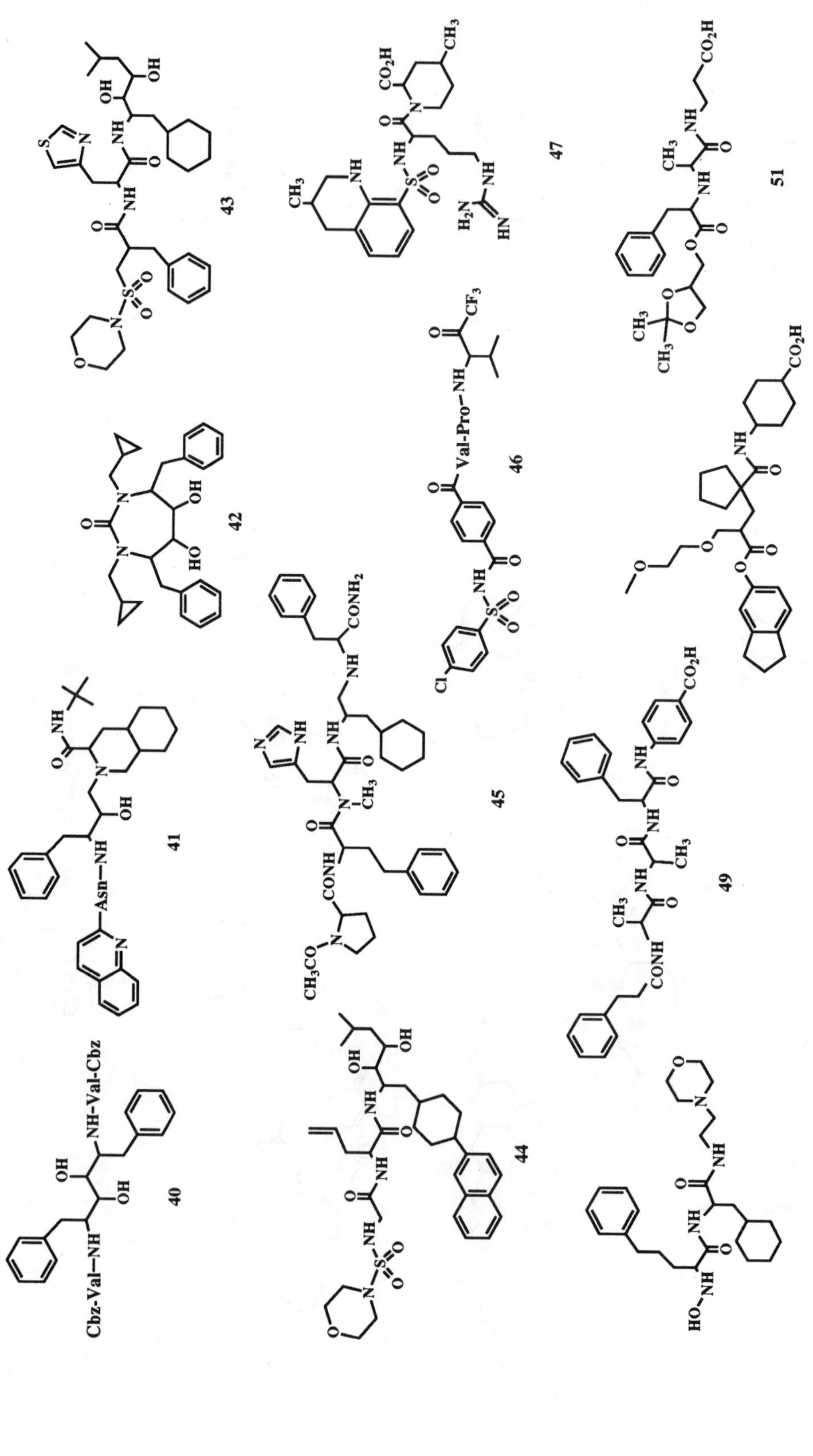

Chart IV. Peptidomimetics developed from peptide leads or structure-based design.

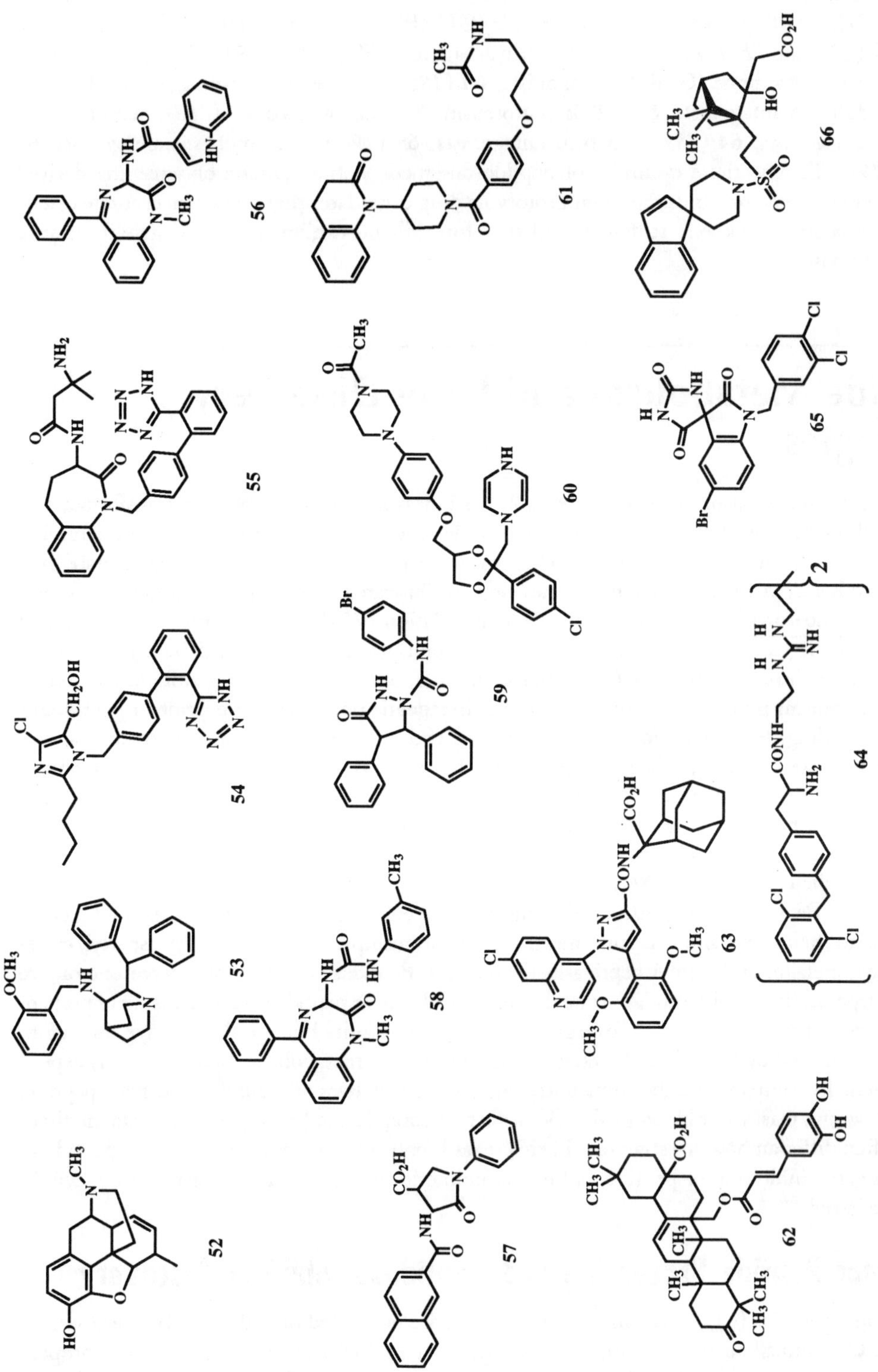

Chart V. Nonpeptides developed from natural products or chemical-file screening.

have contributed significantly to the increasing momentum of this approach include the following: substance P (NK_1) antagonists, **53** (*178*); angiotensin$_1$ (AT_1) antagonists, **54** (*179*); growth hormone-releasing peptide (GHRP) agonists, **55** (*180*); CCK_A antagonists, **56** (*181*) and **57** (*182*); CCK_B and gastrin antagonists, **58** (*183*) and **59** (*184*); gonadotropin-releasing hormone (GnRH) antagonists, **60** (*185*); vasopressin V_1 antagonists, **61** (*186*); endothelin antagonists, **62** (*187*); neurotensin (NT) antagonists, **63** (*188*); neuropeptide-Y antagonists, **64** (*189*); oxytocin antagonists, **65** (*190*); and bombesin antagonists, **66** (*191*). Finally, these examples of peptide-designed peptidomimetic or screening-derived nonpeptide compounds represent prototypic drug candidates that do not incorporate extensive peptide backbone structure, and therefore, should exhibit metabolic stability toward peptidases.

Peptide Metabolism and Drug Discovery Strategies

Peptide metabolism (*31–39, 59, 192, 193*) includes many possible chemical transformations, and the most well-studied are backbone amide cleavage by peptidases. However, natural or synthetic peptides that possess stability toward peptidase cleavage may yet be metabolized by other detoxification systems in the body, as illustrated by the metabolism of cyclosporin by mono- or dihydroxylation and N-demethylation (*194*). Furthermore, achievement of metabolic stability is no guarantee of either oral activity or sustained biological activity because absorption barriers (e.g., intestinal, nasal, and buccal) and hepatobiliary excretion mechanisms may severely compromise the therapeutic potential of a peptide or peptidomimetic drug, thereby subjecting it to acute or chronic intravenous administration requirements. Nevertheless, strategies to modify the oral absorption and duration of action of peptidyl or peptidomimetic drug candidates are being advanced; these strategies include structure-based prodrug approaches (*195, 196*), structure-based enhancement of drug-transport (active or passive) activities (*197–199*), and structure-based modification of molecular lipophilicity properties (*200*).

Peptidases (*201, 202*), including exo- and endopeptidases, may be classified according to their active site structure–mechanism relationships, as exemplified by aspartyl, cysteinyl, metallo, and serinyl peptidases (Table III). Peptidase cleavage may serve one of two purposes; that is, either to generate the biologically active peptide (i.e., processing of propeptides or prepropeptides) or to degrade the peptide into its biologically inactive fragments. As reviewed by Powell (*34*), determination of peptide metabolic stability in vivo is experimentally complex, and the comparative analysis of half-lives of natural or synthetic peptides in serum, plasma, or blood (Table IV) is further complicated because experimental methods often differ among investigators. In this regard, only a few cases of well-studied peptidase-targeted inhibition or peptide and peptidomimetic structure–stability relationships will be reviewed.

Precursor Peptide Processing and Peptidase Inhibitor Discovery

Some peptides (e.g., oxytocin and vasopressin) are not coded directly by DNA. Rather, N- or C-terminally extended amino acid sequences are initially biosynthesized. These propeptides may then be packaged within secretory vesicles along with peptidases to provide site-

Table III. Peptidase Classification and Examples

Aspartic	*Cysteinyl*	*Metallo*	*Serinyl*
Pepsin	Cathepsin-B	Peptidyl dipeptidase-A	Thrombin
Renin	Cathepsin-H	Endopeptidase 24.11	Trypsin
Cathepsin-D	Cathepsin-L	Endopeptidase 24.15	Chymotrypsin-A
Cathepsin-E	Cathepsin-S	Aminopeptidase-M	Elastase
HIV Protease	Cathepsin-M	Carboxypeptidase-A	Kallikrein
	Cathepsin-N	Stromolysin	Cathepsin-A
	Cathepsin-T	Gelatinase-A	Cathepsin-G
	Calpains	Gelatinase-B	Cathepsin-R
	Papain	Collagenase	Tissue plasminogen activator (TPA)
	Proline endopeptidase		
	Interleukin-converting enzyme		

SOURCE: Adapted from references 30 and 201.

specific processing of the precursor peptide to yield the native peptide. Other peptides (e.g., angiotensin and bradykinin [BK]) are also derived by the processing of plasma propeptides. The inhibition of the conversion of such biologically inactive precursor peptides to their active molecular forms is considered a logical drug discovery approach, and a number of processing peptidase targets have been identified (*203, 204*), such as ACE, renin, HIV protease, and thrombin. The precedence of ACE-targeted inhibitors in both peptidomimetic drug design and development has markedly inspired this area of research, and current discovery efforts targeting renin, HIV protease, and thrombin exemplify such strategies to design potent and selective inhibitors of key endogenous precursor peptide processing peptidases (e.g., **40–46;** Chart IV). Several examples of pseudopeptidyl and peptidomimetic inhibitors of these enzymes are detailed in "Design of Metabolically Stable Peptide Analogues and Peptidomimetics".

Table IV. Peptide Stabilities in vitro in Serum or Plasma

Peptide	*Half-Life (min)*
Atrial naturetic factor (99–126)	>60 (rat plasma)
CCK	17 (rat plasma); 50 (human plasma)
ENK-Met	0.4 (rat plasma); 5 (human monkey)
GHRH	90 (rat serum); 435 (human serum)
GHRP	13 (pig plasma)
Melanotropin	92 (goat serum)
Somatostatin	3 (human plasma)
Substance P	70 (human plasma)
TRH	230 (rat serum); 1300 (human serum)
Vasopressin	920 (rat serum)

SOURCE: Adapted from references 34 and 39.

Peptide Degradation and Peptidase Inhibitor Discovery

In contrast to precursor peptide processing roles, peptidases may also serve to inactivate peptides by cleaving the molecules at specific internal peptide bonds. Exopeptidases, such as carboxypeptidases and aminopeptidases, cleave off the C- or N-terminal amino acids, respectively. Endopeptidase cleavage is generally quite specific, and many exo- and endopeptidases (Table III) have been well-characterized (*204*) in terms of mechanistic properties and substrate specificity. As an alternative to the discovery of peptidomimetics or nonpeptides that may exhibit functional properties similar to those of the native peptide agonist (taking into consideration the possibility of receptor subtype selectivity), some research has targeted inhibiting the key peptidases that otherwise degrade the endogenous peptide. Conceptually, such an approach might lead to sustained or enhanced pharmacological activity of the endogenous peptide as the result of degradation-sparing, the short plasma half-lives of most peptide hormones, neurotransmitters, growth factors, and cytokines may be extended if key peptidases (i.e., rate-limiting enzymes) are effectively inhibited. A liability of this approach, however, is that the key peptidase identified may have multiple roles in endogenous peptide processing or degradation. Relative to a number of biologically active peptides that have been investigated (Table V), a single cleavage site may account for complete inactivation of the natural or synthetic peptide. Thus, knowledge of biodegradation mechanisms is considered an important research objective to facilitate the design of either specific inhibitors of key degrading peptidases (e.g., elastase, gelatinase, endopeptidase 24.15, and NEP) or peptides and/or peptidomimetics that are enzymatically stable to such peptidases. Relative to the aforementioned degrading peptidases, significant efforts to design potent and enzyme-selective peptidomimetic inhibitors (e.g., **47–51**; Chart IV) are being advanced. A noteworthy example of in vivo efficacy of an inhibitor of a degrading peptidase was reported (*168–170*) for endopeptidase 24.15. This peptidase is apparently critical to the cleavage-induced inactivation of the vasodilator peptide bradykinin at its Phe^5–Ser^6 site, and the resultant in vivo bradykinin degradation-sparing effect of the inhibitor may be correlated to the hypotensive properties observed after its administration. A series of peptidomimetic inhibitors of NEP 24.11 are also being advanced (*171–176*) as related to the known (*205, 206*) degrading effects of NEP on a number of peptides, including atrial natriuretic factor, enkephalin (ENK), CCK, substance P, NT, angiotensin, and oxytocin. NEP inhibitors may have therapeutic applications in the realm of hypertension or heart failure (sparing of endogenous atrial natriuretic factor) or analgesia (sparing of endogenous enkephalin or administered peptide).

Design of Metabolically Stable Peptide Analogues and Peptidomimetics

Powell (*34*) noted that measurements of peptide half-lives in serum or plasma (Table IV) as well as determinations of structure–cleavage relationships of such compounds (Table V) should be carefully interpreted in relationship to in vivo studies. Specifically, stability studies lack correlation between animal and human serum or plasma, although peptide degradation occurs more expeditiously in rat plasma compared with human plasma. Ideally, any interpretation of in vitro stability measurements and in vivo studies should be restricted, if possible, to the same species. Furthermore, a plethora of factors can contribute markedly to the complexity of interpretation of such data, and these factors include intersubject variation of in vitro peptide half-lives, circulating peptidase levels, and effects of chemical reagents added to the incubation medium. Nevertheless, the stability and structure–cleavage relationships

Table V. Peptide Processing or Degradation by Exo- and Endopeptidases

Peptide or Precursor Peptide	*Peptidase*	*Principal Cleavage Site*
Atrial naturetic peptide	Endo 24.11 (NEP)	Cys^7-Phe^8; Ser^{25}-Phe^{26}
Angiotensinogen	Renin	Leu^{10}-Val^{11}
Angiotensin-I	ACE	Phe^8-His^9
Angiotensin II	NEP	Tyr^4-Ile^5
	Aminopeptidase[a]	Asp^1-Arg^2
	Carboxypeptidase[a]	Pro^7-Phe^8
Bradykinin	NEP	Pro^7-Phe^8
	ACE	Pro^7-Phe^8
	Peptidyl dipeptidase[a]	Pro^7-Phe^8
	Endo 24.15	Phe^5-Ser^6
CCK-8	NEP	Asp^7-Phe^8; Gly^4-Trp^5
	Metalloendo 24.15	Met^3-Gly^4
	Serinyl-endopeptidase[a]	Met^3-Gly^4; Met^6-Asp^7
Gastrin	NEP	Gly^{13}-Trp^{14}; Asp^{16}-Phe^{17}
ENK	NEP	Gly^3-Phe^4
	Aminopeptidase[a]	Tyr^1-Gly^2
GmRH	NEP	Gly^6-Leu^7; His^2-Trp^3
	Endo 24.15	Tyr^5-Gly^6; Gly^6-Leu^7
Insulin	Insulinase 22.11	Tyr^{16}-Leu^{17} (B-chain)
Melanotropin	Endopeptidase[a]	Phe^7-Arg^8
Neurotensin	NEP	Pro^{10}-Tyr^{11}; Tyr^{11}-Ile^{12}
	Cysteinyl endopeptidase	Arg^8-Arg^9
Oxytocin	Post-Pro endopeptidase	Pro^7-Leu^8
	Aminopeptidase[a]	Cys^1-Tyr^2
	NEP	Pro^7-Leu^8
Somatostatin	NEP	Phe^6-Phe^7; Thr^{10}-Phe^{11}; Asn^5-Phe^6
Substance P	NEP	Gln^6-Phe^7; Phe^7-Phe^8; Gly^9-Leu^{10}
	Post-Pro endopeptidase	Pro^2-Lys^3; Pro^4-Gln^5
TRH	Pro endopeptidase	Pro^3-amide
Vasopressin	Trypsin-like peptidase	Arg^8-Gly^9
	Aminopeptidase[a]	Cys^1-Tyr^2
VIP	NEP	Ser^{25}-Ile^{26}; Thr^7-Asp^8

[a] Not specifically defined.
SOURCE: Adapted from references 31 and 39.

of peptides can greatly impact the design of their chemically modified peptide analogue and peptidomimetics. The incorporation of unusual amino acids, amide bond replacements, secondary structure mimetics, and global or local conformational constraints each can enhance the stability property of the peptide analogue to peptidase degradation. Some representative synthetic peptides or pseudopeptides that exhibit pronounced efficacy in vivo due to enhanced metabolic stability (e.g., long-lived serum half-life or resistance to cleavage by specific peptidases) include the following: the somatostatin analogue octreotide (Sandostatin; *207*); the vasopressin analogue desmopressin (*208, 209*); the BK antagonist HOE-140 (*210, 211*); the α-melanotropin (MSH) analogue Melanotan-I (*212*); the endothelin (ET) antagonist BQ-123 (*213, 214*); the adenocorticotropin (ACTH) analogue ebiratide (*215*) active in the central nervous system (CNS); the CCK analogue Ro-23-7014 (*216*); the thrombin antagonist, MDL-28050 (*217*); the TRH analogue RX-77368 (*218, 219*); the NT analogue XL-597 (*220*); the ENK analogue metkephamid acetate (*221*); the growth hormone secretagogue GHRP-6 (*222–224*); the GnRH agonist buserelin acetate (*225, 226*); the GnRH antagonist A-75889 (*227*); and the vasoactive intestinal peptide (VIP) analogue Ro-25-1553 (*228*). Such peptide analogues (*see* Tables I and VI for known or proposed therapeutic activities and chemical structures, respectively) provide prototypic leads

Table VI. Stable Peptide Analogues Designed from Natural Foreign Peptide Leads

Peptide	*Primary Structure*
Sandostatin	Phe-Cys-Phe-D-Trp-Lys-Thr-Cys-Thr-ol (Thr-ol, threoninol)
Desmopressin	Mpa-Tyr-Phe-Cln-Asn-Cys-Pro-D-Arg-Gly-NH_2 (Mpa, 3-mercaptopropionic acid)
HOE-140	D-Arg-Lys-Arg-Hyp-Thi-Ser-D-Tic-Oic-Arg (Hyp, 4-hydroxyproline; Thi, thienylalanine; Oic, octahydroindole-2-carboxylic acid)
Melanotan	Ac-Ser-Tyr-Ser-Nle-Glu-His-D-Phe-Arg-Trp-Gly-Lys-Pro-Val-NH_2 (Nle, norleucine)
FR-139317	Chp(U)-D-Trp(Nin-Me)-D-Pyr (Chp(U), cycloheptylureido; Pyr, 2-[2-pyridyl]-alanine)
Ebiratide	Met(O_2)-Glu-His-Phe-D-Lys-Trp-NH-$(CH_2)_4$-NH_2 (Met[O_2], methionine sulfone)
MDL-28050	Suc-Tyr-Glu-Pro-Pro-Glu-Glu-Tyr-Ala-Cha-Gln (Suc, succinyl; Cha, cyclohexylalanine)
L-365,209	cyclo(Pro-D-Phe-Ile-D-Dhp-Dhp-D-MePhe) (Dhp, dehydropiperazyl)
RX-77368	<Glu-His-Dmp-NH_2 (Dmp = β,β-dimethyl-Pro)
Metkephamide	Tyr-D-Ala-Gly-Phe-MeMet-NH_2
DPDPE	Tyr-D-Pen-Gly-Phe-D-Pen-NH_2 (Pen, penicillamine)
GHRP-6	His-D-Trp-Ala-Trp-D-Phe-Lys-NH_2
Buserelin	<Glu-His-Trp-Ser-Tyr-D-Ser(^{t}Bu)-Leu-Arg-Pro-NH-Et
A-75998	Ac-D-Nal-D-Phe(*p*Cl)-D-Pal-Ser-MeTyr-D-Lys(Nic)-Leu-Lys(Isp)-Pro-D-Ala-NH_2 (Nal, 3-[2-naphthyl]Ala; Phe(pCl), *p*-chloro-Phe; Pal, [3-pyridyl]Ala; Nic, nicotinyl; Isp, isopropyl)
Ro-25-1553	His-Ser-Asp-Ala-Val-Phe-Thr-Glu-Asn-Tyr-Thr-Lys-Leu-Arg-Lys-Gln-Nle-Ala-Ala-Lys-Lys-Tyr-Leu-Asn-Asp-Leu-Lys-Lys-Gly-Gly-Thr-NH_2
Ro-23-7014	Ac-Tyr[SO_3H]-Met-Gly-Trp-Thr[SO_3H]-MePhe-NH_2

NOTE: See text for additional literature citations.
SOURCE: Adapted from reference 39.

for further design of peptidomimetic derivatives or, in some cases, drug candidates for clinical development.

Receptor Agonists

More than 25 years ago, structure–activity studies of ACTH led to the discovery of [D-Ser^1, Lys^{17}]ACTH$_{1-18}$-NH$_2$, an ACTH analogue exhibiting superior in vivo peripheral (steroidogenesis) activity and resistance to degradation relative to the native fragment (*229*). The CNS-active ACTH derivative Met(O$_2$)-Glu-His-Phe-D-Lys-Trp-NH-(CH$_2$)$_8$-NH$_2$ is resistant toward degradation (*215*), and the C-terminal aminoalkylamide moiety contributes to the transport properties (blood–brain barrier) following nasal administration (*230*).

Structure–activity development of CCK agonists has been primarily focused on the C-terminal octapeptide (CCK-8; Table II) or heptapeptide (CCK-7) fragments of the natural peptide. With respect to CCK-8 metabolism studies (Table V), substitution of Gly-4 by D-Ala or D-Trp confers stability at Met–Gly to cleavage by rat brain membrane preparations (*231*). The pseudopeptide Boc-[Nle$^3\Psi$(NHCO)Gly^4, MeNle^6]CCK-7 is stable to NEP and rat brain membrane preparations (*232*). Several studies (*216, 233, 234*) of the comparative in vitro and in vivo structure–activity relationships of CCK analogues with backbone or side-chain modifications led to the discovery of CCK-8 or CCK-7 analogues exhibiting pronounced anorexigenic (satiety) activities. Interestingly, even the *N*-acetylated CCK-8 derivative Ac-[Nle^{3,6}]CCK-8 possess sustained and potent in vivo anorexigenic activity (*234*). The discoveries (*109, 235*) of tetrapeptide (CCK-4-based) agonists at the CCK$_A$ receptor provided new promise for the design of CCK agonist peptidomimetics.

The degradation of ENK by aminopeptidase, carboxypeptidase, and NEP (Table V) is markedly compromised by synthetic modifications such as reported (*221*) for [D-Ala^2,MeMet^5]ENK-NH$_2$, a sustained-acting and orally active agonist. Furthermore, the Gly-2, Phe-4, and Met/Leu-5 positions have provided key substructural sites for a plethora of backbone, stereochemical, conformational (cyclization-related), and secondary structure mimetic approaches, as reviewed by Hruby et al. (*4*) and Schiller (*12, 236*). δ-Selective agonists such as [D-Pen^2, D-Pen^5]ENK (*237*), δ-selective antagonists such as Tyr-TicΨ[CH$_2$NH]]Phe-Phe (*238*), and μ-selective antagonists such as D-Phe-Cys-Tyr-D-Trp-Orn-Thr-Pen-Thr-NH$_2$ (*239*), have been discovered and are stabilized against peptidases by virtue of backbone or side-chain modifications.

Relative to MSH, which is degraded rapidly by serum and the endopeptidases trypsin and chymotrypsin, the synthetic derivatives [Nle^4, D-Phe^7]MSH, [Cys^4,Cys^{10}]MSH and Ac-[Nle^4, Asp^5, D-Phe^7, Lys^{10}]MSH$_{4-10}$-NH$_2$ each exhibit pronounced stabilities (*212, 240, 241*). Furthermore, these MSH peptides are highly potent (*242, 243*), and the identification of both intrinsic and sustained activity in vitro of the central D-Phe-Arg-Trp sequence suggests the possibility for the design of MSH agonist peptidomimetics (*244*). Discoveries of MSH antagonists such as His-D-Arg-Ala-Trp-D-Phe-Lys-NH$_2$ (*245*) and Ac-Nle-Asp-Trp-D-Phe-Nle-Trp-Lys-NH$_2$ (*246*) may advance MSH antagonist peptidomimetics.

Structure–activity studies of NT (*247, 248*) have shown that the NT$_{8-13}$ fragment exhibits essentially identical biological effects as the native peptide, and the development of NT agonists that are active in vivo was achieved (*249–253*). Specifically, (*254*) Arg^8–Arg^9 of NT was shown (*254*) to be susceptible to cleavage by endopeptidase 24.15 (Table V). The same dipeptide component of NT$_{8-13}$ is apparently labile toward degradation or related transport properties (blood–brain barrier) unless modified by Ψ[CH$_2$NH] modification or N-terminal acylation (*251–253*). More elaborate transformation of the Pro^{10}–Tyr^{11} (e.g., XL-597, Table VI) and Ile^{12}–Leu^{13} (e.g., Ψ[C(O)CH$_2$]) substructures to nonpeptidic or pseudopeptidic groups was disclosed (*255*).

Analyses of somatostatin structure–activity relationships and conformation led to the discovery of potent, cyclic somatostatin fragment analogues, including: cyclo(Aha-Cys-Phe-D-Trp-Lys-Thr-Cys) (*256*); D-Phe-Cys-Phe-D-Trp-Lys-Thr-Cys-Thr-ol (*207*); and cyclo(MeAla-Tyr-D-Trp-Lys-Val-Phe) (*257*). The stability of these somatostatin agonist derivatives toward degradation by peptidases and serum may be correlated with their sustained in vivo activities and oral bioavailability. The native peptide is short-lived in plasma (Table IV) and is quite susceptible to cleavage by NEP (Table V) at sites within or proximate to the Phe-Trp-Lys-Thr somatostatin$_{7-10}$ sequence, which is a critical substructural component of the aforementioned analogues.

Receptor Antagonists

The BK antagonist HOE-140 (Table VI) exhibits sustained activity in vivo following intravenous administration (*210, 211*). This peptide analogue is substituted at multiple sites that may enhance stability toward exo- or endopeptidases (aminopeptidase or endopeptidases 24.15 and 24.11, respectively; Table V). The first reported (*258*) competitive antagonist of BK, [D-Phe^7]BK, suggested that modifications at the Pro-7 site might partially stabilize the peptide against degradation.

Intense research has focused on the discovery of potent antagonists for specific peripheral or central therapeutic indications at the CCK$_B$-gastrin receptor. The design and development of potent CCK–gastrin-based pepidomimetics that contain a Cα-Me-D-Trp as a key substructural element were reported (*259*) and such compounds have shown potent antagonist properties at CCK$_B$–gastrin receptors and potential for the development of orally active drugs that may have therapeutic promise for the treatment of specific CNS disorders such as panic or anxiety.

The ET antagonists have been advanced from both ET-based approaches as well as natural product or chemical file screening (*260*), and such compounds may be therapeutically useful in cardiovascular diseases (e.g., hypertension and restenosis). In particular, both peptide and peptidomimetic lead discoveries have resulted in the identification of highly potent ET receptor (ET$_A$ and ET$_B$) antagonists, including: [Dpr^1, Asp^{15}]ET (*261*), where Dpr refers to 1,3-diaminopropionic acid; the C-terminal ET$_{16-21}$ analogue PD-142893 (*262*) or Ac-D-Dip-Leu-Asp-Ile-Ile-Trp; the cyclic pentapeptide BQ-123 (*213, 214*) or cyclo(D-Val-Leu-D-Trp-D-Asp-Pro); and the tripeptide FR-139317 (*263*) or cycloheptyl-ureido-Leu-D-Trp(N^{in}-Me)-D-Pyr (Pyr refers to 2-[2-pyridyl]-alanine). The latter two peptides would be expected to be stable toward peptidase degradation, albeit this stability has not been reported. A recent report (*151*) has shown that C$^\alpha$ and C$^\beta$ modifications of the C-terminal dipeptide substructure of FR-139317 can significantly modify its ET$_A$ and ET$_B$ receptor selectivity, as exemplified by cyclohexyl-ureido-Leu-D-Trp(C$^\alpha$-Me)-D-Dip (**33**, Chart IV).

The integrin receptor GPIIb/IIIa recognizes the tripeptide sequence Arg-Gly-Asp (RGD), which is common to many integrin receptor peptide and protein ligands, including fibrinogen (Fg), vitronectin, fibronectin, von Willebrand factor, osteopontin, thrombospondin, and the collegens (*264*). Recently, the discovery of cyclic peptides that incorporate the RGD sequence have led to the identification of potent antiplatelet antithrombotics (*265*). Given the proposed pharmacophore of the RGD, a variety of novel peptidomimetic derivatives Chart IV were discovered (*149, 150*), and these compounds provide leads in the development of metabolically stable RGD antagonists.

The discovery of GnRH agonists or antagonists has been targeted on the development of therapeutic agents for endocrine diseases (e.g., prostrate cancer, endometriosis, and precocious puberty) as well as nonsteroidal contraception (*266*). As opposed to receptor-desensitization or down-regulation action of chronically administered GnRH agonists (buserelin; Table VI), the application of GnRH antagonists has also been highly desired because such compounds may elicit less undesired side effects. Historically, the first GnRH antagonist was found more than two decades ago (*267*), and over the past 10 years many potent and sustained-acting GnRH antagonist analogues were advanced (*268–271*). A recently described (*271*) GnRH antagonist A-75998 (Table VI) has pronounced stability toward peptidases known to degrade GnRH. Specifically, this decapeptide analogue was resistant to cleavage by chymotrypsin and intestinal degradation using a rat jejunal sac model.

Substance-P (SP) is degraded at multiple sites by NEPs and post-proline cleaving enzyme (Table V). The search for potent and metabolically-stable SP antagonists first led to the finding (*272*) of spantide, [D-Arg^1, D-Trp^{7,9}, Leu^{11}]SP, an analogue that incorporates D-amino acids at several key sites susceptible to degradation. Most recently, the discovery of tripeptide (*273*) and peptidomimetic (*159*) antagonists exemplify prototypic leads toward the design of peptidomimetic derivatives. A potent nonpeptide antagonist of SP has also been reported (*178*).

Enzyme Inhibitors

The development of inhibitors of ACE is considered a significant contribution as well as an inspiration to contemporary peptidomimetic design. In retrospect, ACE inhibitor research has integrated studies on biochemical mechanisms and pharmacophore modeling approaches to develop a number of orally bioavailable, metabolically stable, and sustained-acting drugs (*274–276*). Captopril (*147*) and the prodrug enalapril (*148*) exemplify clinically effective and successful ACE inhibitors (*see* Chart IV) that have revolutionized the treatment of hypertension and congestive heart failure.

The development of orally bioavailable inhibitors of renin facilitated the systematic transformation of the minimal substrate sequence of angiotensinogen (ANG), His-Pro-Phe-His-Leu-Val-Ile-His (ANG_{6-13}), into potent pseudopeptidyl or peptidomimetic inhibitors (*277*). In retrospect, the discoveries (*278–281*) of "transition state" (or collected substrate) bioisosteres of the Leu^{10}–Val^{11} cleavage site (e.g., statine or LeuΨ[CH(OH)]Gly, LeuΨ[CH(OH)CH_2]Val, LeuΨ[CH_2NH]Val) significantly advanced the identification of nanomolar potent pseudopeptidyl inhibitors of renin. Studies to simplify the ANG-based template and take into account the degradation of renin inhibitors have merged to focus primarily on peptidomimetic derivatives of ANG_{7-11}. For example, stabilization of Pro-Phe-His (ANG_{7-9}) against chymotryptic-like peptidases was achieved by the use of Nα^{α}-Me-His-9, Tyr(OMe)-8, Pro(C$^{\alpha}$-Me)-7, D-Pro-7 substitutions (*277, 282–290*). A plethora of known pseudopeptidyl and peptidomimetic inhibitors of renin encompass many different Leu^{10}–Val^{11} (P_1–P_1' site) replacements (*133, 277*), and specific examples in Figure 5 illustrate different design strategies focused on the ANG_{7-11} template. Representative ANG_{7-11}-based, peptidomimetic inhibitors of renin that were advanced (*277, 282–290*) as clinical candidates include A-64662, CGP-38,560A, KRI-1230, A-72517, CI-992, and Ro-42-5892. Most importantly, given the results of pharmacokinetic and bioavailability studies (*162*) performed on A-72517 (**43;** Chart IV), the design of orally active peptidomimetics in the 500–700 molecular mass range may be possible.

The HIV protease HIV Pr inhibitor research strategies have been focused on the design of peptidomimetic and, more recently, nonpeptide derivatives that are selective for the target enzyme and are capable of penetrating into infected cells to block maturation of budding HIV virions (*291–293*). The first reported (*294*) peptidomimetic inhibitor of HIV Pr exhibiting such properties was U-81749E [Tba-ChaΨ[CH(OH)CH_2]Val-Ile-Amp; Tba: ^{t}butylacetyl and Cha: 2-cyclohexylalanine]. Representative peptidomimetic inhibitors which have been advanced as clinical candidates or preclinical leads include Ro-31-8959 (*159*), A-75925 (*160*), and XL-263 (*161*) (Chart IV), corresponding to **40**, **41**, and **42**, respectively. As exemplified by Ro-31-8959, the development of such HIV Pr inhibitors has included the successful attainment of selectivity toward the target enzyme (relative to other aspartyl proteases), cellular activity, and stability toward peptidases. As exemplified by both A-75925 and XL-263, the design of such HIV Pr inhibitors has been structure based in terms of the C_2 symmetry of the target enzyme. The integration of X-ray crystallography in HIV Pr inhibitor drug design research has been exceptionally timely (*295*).

Peptidomimetic inhibitors (*157, 158*) of Ras farnesyl protein transferase (Ras FPT) have advanced convincing data to suggest the potential applications of such compounds to the therapeutic treatment of Ras-related carcinogenesis (*296–298*). Specifically, Ras FPT inhibitors such as L-731,734 **38** (*157*) and BZA-2B **39** (*158*), are modified derivatives (Chart IV) of the tetrapeptide Cys-Val-Ile-Met, which is derived from the C-terminus of p21, a natural protein substrate for Ras FPT (*296–298*). Such compounds are predicted to exhibit increased resistance to peptidase degradation compared with their parent tetrapeptides because of pseudopeptide bond substitutions (e.g., $\Psi[CH_2NH]$) or dipeptidyl replacements (e.g., Val-Ile by 3-methylamino-1-carboxymethyl-5-phenyl-benzodiazepin-2-one).

Summary and Conclusions

A significant achievement in the design of peptide analogues and peptidomimetics has been made over the past two decades in terms of both basic research and pharmaceutical development. The design of metabolically stable, potent, and selective compounds that are structurally based on native (or foreign) peptide leads has been successful in numerous cases, including the discovery of receptor agonists, receptor antagonists, and enzyme inhibitors. Although a correlation of bioavailability and metabolic stability must exist in terms of peptide or peptidomimetic derivatives exhibiting sustained in vivo activities, the design of orally active compounds is not well-understood. Both structure–stability and structure–bioavailability studies are of tremendous importance to peptide-based drug design. Integration of such research earlier in the drug discovery process may greatly impact strategies, which often focus primarily only on structure–conformation and structure–activity studies, and advance prototypic peptide or peptidomimetic leads to the drug candidate stage. Perhaps, future directions in peptide-based drug design will attest to such interdisciplinary strategies within the generic scope of chemical–biological property relationships using well-established biochemical and pharmacological methodologies.

References

1. *The Peptides: Analysis, Synthesis, Biology;* Gross, E.; Meienhofer, J.; Udenfriend, S., Eds.; Academic: Orlando, FL, 1979–1987; Vols. I–IX.
2. Ward, D. J. In *Peptide Pharmaceuticals—Approaches to the Design of Novel Drugs;* Open University: Buckingham, England, 1991.
3. Fauchere, J.-L. In *Advances in Drug Research;* Testa, B., Ed.; Academic: London, 1986; Vol. 15, pp 29–69.
4. Hruby, V. J.; Al-Obeidi, F.; Kazmierski, W. *Biochem. J.* **1990,** *268,* 249–262.
5. Hruby, V. J.; Pettit, B. M. In *Computer-Aided Drug Design, Method and Application;* Perun, T. J.; Propst, C. L., Eds.; Marcel Dekker: New York, 1989; pp 405–461.

6. DeGrado, W. F. *Adv. Prot. Chem.* **1988,** *39,* 51–124.

7. Toniolo, C. *Int. J. Pept. Prot. Res.* **1990,** *35,* 287–300.

8. Rose, G. D.; Gierasch, L. M.; Smith, J. A. *Adv. Prot. Chem.* **1985,** *37,* 1–109.

9. Goodman, M.; Ro, S. In *Medicinal Chemistry and Drug Design; Vol. I. Principles of Drug Discovery,* 5th ed.; Burger, A., Ed.; Wiley: New York, 1994.

10. Kessler, H.; Haupt, A.; Will, M. In *Computer-Aided Drug Design, Method and Application;* Perun, T. J.; Propst, C. L., Eds.; Marcel Dekker: New York, 1989; 461.

11. Hruby, V. J. *LIfe Sci.* **1982,** *31,* 189–199.

12. Schiller, P. W. In *Medicinal Chemistry for the 21st Century;* Wermuth, C. G., Ed.; Blackwell: London, 1992; pp 215–232.

13. Huffman, W. F. In *Medicinal Chemistry for the 21st Century;* Wermuth, C. G., Ed.; Blackwell: London, 1992; pp 247–257.

14. Kemp, D. S. In *Medicinal Chemistry for the 21st Century;* Wermuth, C. G., Ed.; Blackwell: London, 1992; pp 259–277.

15. Eberle, A. N. *Chimia* **1991,** *45,* 145–153.

16. *Polypeptide and Protein Drugs—Production, Characterization and Formulation;* Hider, R. C.; Barlow, D., Eds.; Ellis Horwood: Chichester, England, 1991.

17. Sawyer, T. K.; Cody, W. L.; Leonard, D. M.; Hadley, M. E. In *Encyclopedia of Molecular Biology;* Meyers, R. A., Ed.; VCH: New York, 1994.

18. Spatola, A. F. In *Chemistry and Biochemistry of Amino Acids, Peptides and Proteins;* Weinstein, B., Ed.; Marcel Dekker: New York, 1983; Vol. 7, pp 267–357.

19. Spatola, A. F.; Darlak, K. *Tetrahedron* **1988,** *44,* 821–833.

20. Spatola, A. F. In *Methods in Neurosciences;* Conn, P. M., Ed.; Academic: Orlando, FL, 1993; pp 19–42.

21. Plattner, J. J.; Norbeck, D. W. In *Drug Discovery Technologies;* Clark, C. R.; Moos, W. H., Eds.; Ellis Horwood: Chichester, England, 1990; pp 120–126.

22. Marshall, G. R. *Tetrahedron* **1993,** *49,* 3547–3558.

23. Farmer, P. S. In *Drug Design;* Ariens, E. J., Ed.; Academic: Orlando, FL, 1980; Vol. X, pp 119–143.

24. Olson, G. L.; Bolin, D. R.; Bonner, M. P.; Bos, M.; Cook, C. M.; Fry, D. C.; Graves, B. J.; Hatada, M.; Hill, D. E.; Kahn, M.; Madison, V. S.; Rusiecki, V. K.; Sarubu, R.; Sepinwall, J.; Vincent, G. P.; Voss, M. E. *J. Med. Chem.* **1993,** *36,* 3039–3049.

25. Kahn, M. (Guest Ed.) *Tetrahedron Symp.–in print* **1993,** *50.*

26. Horwell, D. C. (Guest Ed.) *Bioorg. Med. Chem. Lett.* **1993,** *3(5).*

27. Morgan, B. A.; Gainor, J. A. *Annu. Rep. Med. Chem.* **1989,** *24,* 243–252.

28. Freidinger, R. M. *Trends Pharmacol. Sci.* **1989,** *10,* 270–274.

29. Rees, D. C. *Annu. Rep. Med. Chem.* **1993,** *28,* 59–69.

30. Wiley, R. A.; Rich, D. H. *Med. Res. Rev.* **1993,** *13,* 327–384.

31. *Protein Pharmacokinetics and Metabolism;* Ferrariolo, R. L.; Mohler, M. A.; Gloff, C. A.; Eds.; Pharmaceutical Biotechnology Series; Plenum: New York, 1992.

32. *Stability and Characterization of Protein and Peptide Drugs: Case Histories;* Wang, Y. J.; Pearlman, R., Eds.; Pharmaceutical Biotechnology Series; Plenum: New York, 1993.

33. Humphrey, M. J.; Ringrose, P. S. *Drug Metab. Rev.* **1986,** *17,* 283–310.

34. Powell, M. F. *Annu. Rep. Med. Chem.* **1993,** *28,* 285–294.

35. Erickson, R. Chapter 2 in this volume.

36. Krishnamoorthy, R.; Mitra, A. Chapter 3 in this volume.

37. van Nispen, J.; Pinder, R. *Annu. Rep. Med. Chem.* **1986,** *21,* 51–62.

38. van Nispen, J.; Pinder, R. *Annu. Rep. Med. Chem.* **1987,** *22,* 51–62.

39. Fauchere, J. L.; Thurieau, C. *Adv. Drug Res.* **1992,** *23,* 128–159.

40. Simon, R. J.; Kania, R. S.; Zuckermann, R. N.; Huebner, V. D.; Jewell, D. A.; Banville, S.; Ng, S.; Wang, L.; Rosenberg, S.; Marlowe, C. K.; Spellmeyer, D. C.; Tan, R.; Frankel, A. D.; Santi, D. V.; Cohen, F. E.; Bartlett, P. A. *Proc. Natl. Acad. Sci. U.S.A.* **1992,** *89,* 9367–9371.

41. Dower, W. J.; Fodor, S. P. A. *Annu. Rep. Med. Chem.* **1991,** *26,* 271–280.

42. Jung, G.; Beck-Sickinger, A. G. *Angew. Chem. Int. Ed. Engl.* **1992,** *31,* 367–386.

43. Pavia, M. R.; Sawyer, T. K.; Moos, W. H. *Bioorg. Med. Chem. Lett.* **1993,** *3,* 387–396.

44. Andrews, P.; Leonard, D.; Cody, W. L.; Sawyer, T. K. In *Peptide Synthesis and Protocols;* Dunn, B. M.; Pennington, M., Eds.; Humana: Clifton, NJ, 1994.

45. Pavia, M. R.; Sawyer, T. K.; Moos, W. H. *Bioorg. Med. Chem. Lett* **1993,** *3,* 387–396.

46. Geysen, H. M.; Meleon, R. H.; Barteling, S. J. *Proc. Natl. Acad. Sci. U.S.A.* **1984,** *81,* 3998–4002.

47. Geysen, H. M. U.S. Patent 4 708 871, 1987.

48. Maeji, N. J.; Bray, A. M.; Valerio, R. M.; Seldon, M. A.; Wang, J.-X.; Geysen, H. M. *Pept. Res.* **1991,** *4,* 142–146.

49. Houghten, R. A. *Proc. Natl. Acad. Sci. U.S.A.* **1985,** *82,* 5131–5135.

50. Houghten, R. A. U.S. Patent 4 631 211, 1986.

51. Beck-Sickinger, A. G.; Duerr, H.; Jung, G. *Pept. Res.* **1991,** *4,* 88–92.

52. Furka, A.; Sebestyen, F.; Asgedom, M.; Dibo, G. *Int. J. Pept. Prot. Res.* **1991,** *37,* 487–493.

53. Zuckermann, R. N.; Huebner, V.; Santi, D. V.; Siani, M. A. Int. Patent WO01/17823, 1991.

54. Lam, K. S.; Salmoln, S. E.; Hersh, E. M.; Hruby, V. J.; Kazmierski, W. M.; Knapp, R. J. *Nature (London)* **1991,** *254,* 82–84.

55. Lam, K. S.; Salmon, S. E.; Hruby, V. J.; Hersh, E. M.; Al-Obeidi, F. Int. Patent WO92/991, 1991.

56. Houghten, R. A.; Cuervo, J. H.; Pinilla, C.; Appel, J. R., Jr.; Blondelle, S. Int. Patent WO92/9300, 1992.

57. Owens, R. A.; Geselichen, P. D.; Houchins, B. J.; DiMarchi, R. D. *Biochem. Biophys. Res. Commun.* **1991,** *181,* 402–408.

58. Szelke, M.; Leckie, B.; Hallet, A.; Jones, D. M.; Sueiras, J.; Atrash, B.; Lever, A. F. *Nature (London)* **1982,** *299,* 555–557.

59. Sasaki, Y.; Murphy, W. A.; Heiman, M. L.; Lance, V. A.; Coy, D. H. *J. Med. Chem.* **1987,** *30,* 1162–1166.

60. Rodriguez, M.; Bali, J.-P.; Magous, R.; Castro, B.; Martinez, J. *Int. J. Peptide Prot. Res.* **1986,** *27,* 293–299.

61. Coy, D. H.; Hocart, S. J.; Sasaki, Y. *Tetrahedron* **1988,** *44,* 835–841.

62. Ho, P. T.; Chang, D.; Zhong, J. W. X.; Musso, G. F. *Pept. Res.* **1993,** *6,* 10–12.

63. Almquist, R. G.; Chao, W. R.; Ellis, M. E.; Johnson, H. L. *J. Med. Chem.* **1980,** *23,* 1392–1398.

64. Ewenson, A.; Cohen-Suissa, R.; Levian-Teitelbaum, D.; Selinger, Z.; Chorev, M.; Gilon, C. *Int. J. Pept. Prot. Res.* **1988,** *31,* 269–280.

65. Mendre, C.; Rodriguez, M.; Lignon, M.-F.; Galas, M. C.; Gueudet, C.; Worma, P.; Martinez, J. *Eur. J. Pharmacol.* **1990** *186,* 213–222.

66. Kaltenbronn, J. S.; Hudspeth, J. P.; Lunney, E. A.; Michniewicz, B. M.; Nicolaides, E. D.; Repine, J. T.; Roark, W. H.; Stier, M. A.; Tinney, F. T.; Woo, P. D. W.; Essenberg, A. D. *J. Med. Chem.* **1990,** *33,* 838–845.

67. Rodriguez, M.; Aumelas, A.; Martinez, J. *Tetrahedron Lett.* **1990,** *31,* 5153–5156.

68. Tourwe, D.; Couder, J.; Ceusters, M.; Meert, D.; Burks, T. F.; Kramer, T. H.; Davis, P.; Knapp, R.; Yamamura, H. I.; Leysen, J. E.; Van Binst, G. *Int. J. Pept. Prot. Res.* **1992,** *39,* 131–136.

69. Kawasaki, K.; Maeda, M. *Biochem. Biophys. Res. Commun.* **1980,** *106,* 113–116.

70. Cox, M. T.; Gormley, J. J.; Hayward, C. F.; Petter, N. N. *J. Chem. Soc. Chem. Commun.* **1980,** 800–802.

71. Whitesell, J. K.; Lawrence, R. M. *Chirality* **1989,** *1,* 89–91.

72. deGaieta, L. S. L.; Czarniecki, M.; Spaltenstein, A. *J. Org. Chem.* **1989,** *54,* 4004–4005.

73. Fujii, N.; Habashita, J.; Shigemori, N.; Otaka, A.; Ibuka, T. *Tetrahedron Lett.* **1991,** *32,* 4969–4972.

74. Kempf, D. J.; Wang, C.; Spanton, S. G. *Int. J. Pept. Prot. Res.* **1991,** *38,* 237–241.

75. Nicolaides, E. D.; Tinney, F. J.; Kaltenbronn, J. S.; Repine, J. T.; DeJohn, D. A.; Lunney, E. A.; Roark, W. H.; Marriott, J. G.; Davis, R. E.; Voigtman, R. E. *J. Med. Chem.* **1986,** *29,* 959–971.

76. TenBrink, R. E. *J. Org. Chem.* **1987,** *52,* 418–422.

77. Breten, P.; Monsigny, M.; Mayer, R. *Int. J. Pept. Prot. Res.* **1990,** *35,* 346–351.

78. Fincham, C. I.; Higginbottom, M.; Hill, D. R.; Horwell, D. C.; O'Toole, J. C.; Ratcliffe, G.; Rees, D. C.; Roberts, E. *J. Med. Chem.* **1992,** *35,* 1472–1484.

79. Spatola, A. F.; Edwards, J. V. *Biopolymers* **1986,** *25,* 229–244.

80. Benovitz, D. E.; Spatola, A. F. *Peptides* **1985,** *6,* 257–261.

81. Anwer, M. K.; Sherman, D. B.; Spatola, A. F. *Int. J. Pept. Prot. Res.* **1990,** *36,* 392–399.

82. Zabrocki, J.; Dunbar, J. B., Jr.; Marshall, K. W.; Toth, M. V.; Marshall, G. R. *J. Org. Chem.* **1992,** *57,* 202–209.

83. Zabrocki, J.; Smith, G. D.; Dunbar, J. B., Jr.; Iijima, H.; Marshall, G. R. *J. Am. Chem. Soc.* **1988,** *110,* 5875–5880.

84. Boteju, L. W.; Hruby, V. J. *Tetrahedron Lett.* **1993,** *34,* 1757–1760.

85. Gordon, T.; Hansen, P.; Morgan, B.; Singh, J.; Baizman, E.; Ward, S. *Bioorg. Med. Chem. Lett.* **1993,** *3,* 915–920.

86. Chorev, M.; Shavitz, R.; Goodman, M.; Minick, S.; Guillemin, R. *Science (Washington, DC)* **1979,** *204,* 1210–1212.

87. Chorev, M.; Goodman, M. *Acc. Chem. Res.* **1993,** *26,* 266–273.

88. Bartlett, P. A.; Spear, K. L.; Jacobsen, N. E. *Biochemistry* **1982,** *21,* 1608–1611.

89. Clausen, K.; Spatola, A. F.; Lemieux, C.; Schiller, P.; Lawesson, S. O. *Biochem. Biophys. Res. Commun.* **1984,** *120,* 305.

90. LaJoie, G.; Lepine, F.; LeMarie, S.; Jolicoeur, F.; Aube, C.; Turcotte, A.; Belleau, B. *Int. J. Pept. Prot. Res.* **1984,** *24,* 316–324.

91. Majer, Zs.; Zewdu, M.; Hollosi, M.; Seprodi, J.; Vadasz, Zs.; Teplan, T. *Biochem. Biophys. Res. Commun.* **1988,** *150,* 1017.

92. Gesellchen, P. D.; Frederickson, R. C.; Tafur, S.; Smiley, D. In *Peptides: Synthesis, Structure and Function;* Rich, D. H.; Gross, E., Eds.; Pierce Chemical: Rockford, IL, 1981; pp 621–624.

93. Roy, J.; Gazis, D.; Shakman, R.; Schwartz, I. L. *Int. J. Pept. Prot. Res.* **1982,** *20,* 35–42.

94. Morishima, H.; Takita, T.; Umezawa, H. *J. Antibiot.* **1973,** *26,* 115.

95. Rich, D. H.; Sun, E. T.; Boparai, A. S. *J. Org. Chem.* **1978,** *43,* 3624.

96. Woo, P. W. K. *Tetrahedron Lett.* **1985,** *26,* 2973–2976.

97. Evans, B.; Rittle, K.; Homnick, C.; Springer, J.; Hirshfield, J.; Veber, D. *J. Org. Chem.* **1985,** *50,* 4615–4625.

98. Wuts, P. G. M.; Putt, S. R.; Ritter, A. R. *J. Org. Chem.* **1988,** *53,* 4503–4508.
99. Ghosh, A. K.; McKee, S. P.; Thompson, W. J. *J. Org. Chem.* **1991,** *56,* 6500–6503.
100. Jones, D. M.; Nilsson, B.; Szelke, M. *J. Org. Chem.* **1993,** *58,* 2286–2290.
101. Chen, H.-G.; Sawyer, T. K.; Wuts, P. G. M. *Acta Pharmacol. Sinica* **1994,** *15,* 33–35.
102. Thaisrivongs, S.; Pals, D. T.; Kroll, L. T.; Turner, S. R.; Han, F.-S. *J. Med. Chem.* **1987,** *30,* 976–982.
103. Atsuumi, S.; Nakano, M.; Koike, Y.; Tanaka, S.; Funabashi, H.; Hahimoto, J.; Morishima, H. *Chem. Pharm. Bull.* **1990,** *38,* 3460–3462.
104. Rosenberg, S. H.; Boyd, S. A.; Mantei, R. A. *Tetrahedron Lett.* **1991,** *32,* 6507–6508.
105. Baker, W. R.; Condon, S. L. *J. Org. Chem.* **1993,** *58,* 3277–3284.
106. Rich, D. H.; Green, J.; Toth, M. V.; Marshall, G. R.; Kent, S. B. H. *J. Med. Chem.* **1990,** *33,* 1285–1288.
107. Ryono, D. E.; Free, C. A.; Neubeck, R.; Samaniego, S. G.; Godfrey, J. D.; Petrillo, E. W., Jr. In *Peptides, Structure and Function. Proceedings of the Ninth American Peptide Symposium;* Deber, C. M.; Hruby, V. J.; Kopple, K. D., Eds.; Pierce Chemical: Rockford, IL, 1985; p 739.
108. Hruby, V. J.; Fang, S.; Knapp, R.; Kazmierski, W.; Lui, G. K.; Yamamura, H. I. In *Peptides: Chemistry, Structure and Biology;* Rivier, J. E.; Marshall, G. R., Eds.; ESCOM: Leiden, The Netherlands, 1990; pp 53–55.
109. Shiosaki, K.; Lin, C. W.; Kopecka, H.; Craig, R. A.; Bianchi, B. R.; Miller, T. R.; Witte, D. G.; Stashko, M.; Nazdan, A. M. *J. Med. Chem.* **1992,** *35,* 2007–2114.
110. Yamazaki, T.; Ro, S.; Goodman, M.; Chung, N. N.; Schiller, P. W. *J. Med. Chem.* **1993,** *36,* 708–719.
111. Aubry, A.; Marraud, M. *Biopolymers* **1989,** *28,* 109–122.
112. Kazmierski, W.; Wire, W. S.; Liu, G. K.; Knapp, R. J.; Shook, J. E.; Burks, T. F.; Yamamura, H. I.; Hruby, V. J. *J. Med. Chem.* **1988,** *31,* 2170–2177.
113. Samanen, J.; Cash, T.; Narindray, D.; Brandeis, E.; Adams, W. T.; Weideman, H.; Yellin, T.; Regoli, D. *J. Med. Chem.* **1991,** *34,* 3036–3043.
114. Tiley, J. W.; Danho, W.; Madison, V.; Fry, D.; Swistok, J.; Makofske, R.; Michalewsky, J.; Swartz, A.; Weatherford, S.; Triscari, J.; Nelson, D. *J. Med. Chem.* **1992,** *35,* 4229–4252.
115. Bardi, R.; Riazzesi, A. M.; Toniolo, C.; Sukumar, M.; Balaram, P. *Biopolymers* **1986,** *25,* 1635–1644.
116. Valle, G.; Kazmierski, W. M.; Crisma, M.; Bonora, G. M.; Toniolo, C.; Hruby, V. J. *Int. J. Pept. Prot. Res.* **1992,** *40,* 222–232.
117. Holladay, M. W.; Nazdan, A. M. *J. Org. Chem.* **1991,** *56,* 3900–3905.
118. Schiller, P. W.; Weltrowska, G.; Nguyen, T. M. D.; Lemieux, C.; Chung, N. N.; Marsden, B. J.; Wilkes, B. C. *J. Med. Chem.* **1991,** *34,* 3125–3132.
119. Deeks, T.; Crooks, P. A.; Waigh, R. D. *J. Med. Chem.* **1983,** *26,* 762–765.
120. Toniolo, C.; Benedetti, E. *Trends Biochem. Sci.* **1991,** *16,* 350–353.
121. Nagaraj, R.; Balaram, P. *FEBS Lett.* **1978,** *96:* 273–276.
122. Samanen, J.; Cash, T.; Narindary, D.; Brandeis, E.; Adams, W., Jr.; Weideman, H.; Yellin T. *J. Med. Chem.* **1991,** *34,* 3036–3043.
123. London, R. E.; Stewart, J. M.; Cann, J. R. *Biochem. Pharmacol.* **1990,** *40,* 41–48.
124. Toniolo, C.; Crisma, M.; Valle, G.; Bonora, G. M.; Polinelli, S.; Becker, E. L.; Freer, R. J.; Sudhanand; Rao, R. B.; Balaram, P.; Sukumar, M. *Peptide Res.* **1989,** *2,* 275–281.
125. Toniolo, C.; Formaggio, F.; Crisma, M.; Valle, G.; Boesten, W. H. J.; Shoemaker, H. E.; Kamphuis, J.; Temussi, P. A.; Becker, E. L.; Precigoux, G. *Tetrahedron* **1993,** *49,* 3641–3653.

126. Schiller, P. W.; Weltrowska, G.; Nguyen, T. M.-D.; Lemieux, C.; Chung, N. N.; Marsden, B. J.; Wilkes, B. C. *J. Med. Chem.* **1991,** *34,* 3125–3132.

127. Shimohigashi, Y.; Stammer, C. H. *Int. J. Pept. Prot. Res.* **1982,** *20,* 199–206.

128. Kaur, P.; Uma, K.; Balaram, P.; Chauhan, V. S. *Int. J. Pept. Prot. Res.* **1989,** *33,* 103–109.

129. Gupta, A.; Chauhan, V. S. *Int. J. Pept. Prot. Res.* **1993,** *41,* 421–426.

130. Meraldi, J.-P.; Hruby, V. J.; Brewster, A. I. R. *Proc. Natl. Acad. Sci. U.S.A.* **1977,** *74,* 1373–1377.

131. Mosberg, H. I.; Hurst, R.; Hruby, V. J.; Gee, K.; Yamamura, H. I.; Galligan, J. J.; Burks, T. F. *Proc. Natl. Acad. Sci. U.S.A.* **1983,** *80,* 5871–5874.

132. Pelton, J. T.; Gulya, K.; Hruby, V. J.; Duckles, S. P.; Yamamura, H. I. *Proc. Natl. Acad. Sci. U.S.A.* **1985,** *82,* 236–239.

133. Rich, D. H. In *Comprehensive Medicinal Chemistry;* Hansch, C.; Sammes, P. G.; Taylor, J. B., Eds.; Pergamon: Oxford England, 1990; Vol. 2, pp 391–441.

134. Ball, J. B.; Alewood, P. F. *J. Mol. Recog.* **1990,** *3,* 55–64.

135. Marshall, G. R. *Curr. Opinion Struct. Biol.* **1992,** *2,* 904–919.

136. Freidinger, R. M.; Veber, D. F.; Perlow, D. S.; Brooks, J. R.; Saperstein, R. *Science (Washington, DC)* **1980,** *210,* 656–658.

137. Nagai, U.; Sato, K. *Tetrahedron Lett.* **1985,** *26,* 647–650.

138. Feigl, M. *J. Am. Chem. Soc.* **1986,** *108,* 181–182.

139. Kahn, M.; Wilke, S.; Chen, B.; Fujita, K. *J. Am. Chem. Soc.* **1988,** *110,* 1638.

140. Kemp, D. S.; Stites, W. E. *Tetrahedron Lett.* **1988,** *29,* 5057–5060.

141. Genin, M. J.; Johnson, R. L. *J. Am. Chem. Soc.* **1992,** *114,* 8316–8318.

142. Kahn, M.; Wilke, S.; Chen, B.; Fujita, K.; Lee, Y. H.; Johnson, M. E. *J. Mol. Recognit.* **1988,** *1(2),* 75–79.

143. Ripka, W. C.; DeLucca, G. V.; Bach, A. C., II; Pottorf, R. S.; Blaney, J. M. *Tetrahedron* **1993,** *49,* 3593–3608.

144. Ripka, W. C.; DeLucca, G. V.; Bach, A. C., II; Pottorf, R. S.; Blaney, J. M. *Tetrahedron* **1993,** *49,* 3609–3628.

145. Callahan, J. F.; Newlander, K. A.; Burgess, J. L.; Eggleston, D. S.; Nichols, A.; Wong, A.; Huffman, W. F. *Tetrahedron* **1993,** *49,* 3479–3488.

146. Smith, A. B.; Keenan, T. P.; Holcomb, R. C.; Sprengeler, P. A.; Guzman, M. C.; Wood, J. L.; Caroll, P. J.; Hirschmann, R. S. *J. Am. Chem. Soc.* **1992,** *114,* 10672–10674.

147. Ondetti, M. A.; Rubin, B.; Cushman, D. W. *Science (Washington, DC)* **1977,** *196,* 441–444.

148. Patchett, A. A.; Harris, E.; Tristam, E. W.; Wyvratt, M. J.; Wu, M. T.; Taub, D.; Peterson, E. R.; Ikeler, T. J.; TenBroeke, J.; Payne, L. G.; Ondeyka, D. L.; Thorsett, E. D.; Greenlee, W. J.; Lohr, N. S.; Hoffsommer, R. D.; Joshua, H.; Ruyle, W. V.; Rothrock, J. W.; Aster, S. D.; Maycock, A. L.; Robinson, F. M.; Hirschmann, R.; Sweet, C. S.; Ulm, E. H.; Grosse, D. M.; Vassil, T. C.; Stone, C. A. *Nature (London)* **1980,** *288,* 280–283.

149. Alig, L.; Edenhofer, A.; Muller, M.; Trzeciak, A.; Weller, T. U.S. Patent 5 039 805, 1991.

150. Hirschmann, R.; Sprengeler, P. A.; Kawasaki, T.; Leahy, J. W.; Shakespeare, W. C.; Smith, A. B., III *J. Am. Chem. Soc.* **1992,** *114,* 9699–9701.

151. Vara Prasad, J. V. N.; Cody, W. L.; Cheng, X.-M.; Doherty, A. M.; DePue, P. L.; Dunbar, J. B., Jr.; Welch, K. M.; Flynn, M. A.; Reynolds, E. E.; Sawyer, T. K. In *Proceedings of the 13th American Peptide Symposium;* Hodges, R.; Smith, J., Eds.; ESCOM: Leiden, The Netherlands, 1994.

152. Horwell, D. C. *Neuropeptides* **1991,** *19 (Suppl.),* 57–64.

153. Nicolaou, K. C.; Salvino, J. M.; Raynor, K.; Pietranico, S.; Reisine, T.; Freidinger, R.; Hirsch-

mann, R. In *Peptides—Chemistry, Structure and Biology;* Rivier, J. E.; Marshall, G. R., Eds., ESCOM: Leiden, The Netherlands, 1990; pp 881–884.

154. Hirschmann, R.; Nicolaou, K. C.; Pietranico, S.; Salvino, J.; Leahy, E. M.; Sprengeler, P. A.; Furst, G.; Smith, A. B., III; Strader, C. D.; Cascieri, M. A.; Candelore, M. R.; Donaldson, C.; Vale, W.; Maechler, L. *J. Am. Chem. Soc.* **1992,** *114,* 9217–9218.

155. Schilling, W.; Bittiger, H.; Brugger, F.; Criscione, L.; Hauser, K.; Ofner, S.; Olpe, H.-R.; Vassout, A.; Veenstra, S. In *Proceedings of the 12th International Symposium on Medicinal Chemistry;* Vienna, Austria, 1993.

156. Smith, P. W.; McElroy, A. B.; Pritchard, J. M.; Deal, M. J.; Ewan, G. B.; Hagen, R. M.; Ireland, S. J.; Ball, D.; Beresford, I.; Sheldrick, R.; Jordan, C. C.; Ward, P. *Bioorg. Med. Chem. Lett.* **1993,** *3,* 931–935.

157. Kohl, N. E.; Mosser, S. D.; deSolms, S. J.; Giuliani, E. A.; Pompliano, D. L.; Graham, S. L.; Smith, R. L.; Scolnick, E. M.; Oliff, A.; Gibbs, J. B. *Science (Washington, DC)* **1993,** *260,* 1934–1936.

158. James, G. L.; Goldstein, J. L.; Brown, M. S.; Rawson, T. E.; Somers, T. C.; McDowell, R. S.; Crowley, C. W.; Lucas, B. K.; Levinson, A. D.; Marsters, J. C., Jr. *Science (Washington, DC)* **1993,** *260,* 1937–1942.

159. Kempf, D. J.; Codacovi, L.; Wang, X. C.; Kohlbrenner, W. E.; Wideburg, N. E.; Saldivcar, A.; Vasavanonda, S.; March, K. C.; Bryant, P.; Sham, H. L.; Green, B.; Betebenner, D. A.; Erickson, J.; Norbeck, D. W. *J. Med. Chem.* **1993,** *36:* 320–330.

160. Roberts, N. A.; Martin, J. A.; Kinchington, D.; Broadhurst, A. V.; Craig, J. C.; Duncan, I. B.; Galpin, S. A.; Handa, B. K.; Kay, J.; Krohn, A.; Lambert, R. W.; Merrett, J. H.; Mills, J. S.; Parkes, K. E. B.; Redshaw, S.; Ritchie, A. J.; Taylor, D. L.; Thomas, G. J.; Machin, P. J. *Science (Washington, DC)* **1990,** *248,* 358–361.

161. Jadhav, P. K.; Lam, P. Y.; Eyermann, C. J.; Hodge, C. N.; Woerner, F. J.; Bacheler, L. T.; Meek, J. L.; Otto, M. J.; Rayner, M. M.; Wong, N. Y.; Chang, C. H.; Weber, P. C.; Jackson, D. A.; Sharpe, T. R.; Erickson-Viitanen, S. *Chemical Abstracts;* 205 National Meeting of the American Chemical Society, Denver, CO; American Chemical Society: Washington, DC, 1993; 107.

162. Rosenberg, S. H.; Spina, K. P.; Condon, S. L.; Polakowski, J.; Yao, Z.; Kovar, P.; Stein, H. H.; Cohen, J.; Barlow, J. L.; Klinghofer, V.; Egan, D. A.; Tricaro, K. A.; Perun, T. J.; Baker, W. R.; Kleinert, H. D. *J. Med. Chem.* **1993,** *36,* 460–467.

163. Plummer, M.; Hamby, J.; Hingorani, G.; Batley, B.; Rapundalo, S. *Bioorg. Med. Chem. Lett.* **1993,** *3:* 2119–2124.

164. Sawyer, T. K.; Maggiora, L. L.; Liu, L.; Staples, D. J.; Bradford, V. S.; Mao, B.; Pals, D. T.; Dunn, B. M.; Poorman, R.; Hinzmann, J.; DeVaux, A. E.; Affholter, J. A.; Smith, C. W. In *Peptides: Chemistry and Biology;* Marshall, G. R.; Rivier, J., Eds.; ESCOM: Leiden, The Netherlands, 1990; pp 855–857.

165. Mellott, M. J.; Connolly, T. M.; York, S. J.; Bush, L. R. *Thromb. Haem.* **1990,** *64,* 526–534.

166. Williams, J. C.; Stein, R. L.; Giles, R. E.; Krell, R. D. *Ann. N. Y. Acad. Sci.* **1991,** *624,* 230–243.

167. Beeley, N. R. A. *CHI Conference on Developing Small Molecule Mimetics;* Philadelphia, PA, 1993.

168. Orlowski, M.; Michaud, C.; Molineaux, C. J. *Biochemistry* **1988,** *27,* 597–602.

169. Lasdun, A.; Reznik, S.; Molineaux, C. J.; Orlowski, M. *J. Pharmacol. Exp. Ther.* **1989,** *251,* 439–447.

170. Genden, E. M.; Molineaux, C. J. *Hypertension* **1991,** *18,* 360–365.

171. Danilewicz, J. C.; Barclay, P. L.; Barnish, I. T.; Brown, D.; Campbell, S. F.; James, K.; Samuels, G. M. R.; Terrett, N. K.; Wythes, M. J. *Biochem. Biophys. Res. Commun.* **1989,** *164,* 58–65.

172. Brown, D.; Barclay, P. L.; Barnish, I. T.; Campbell, S. F.; Danilewicz, J. C.; Ellis, P.; James,

K.; Samuels, G. M. R.; Terrett, N. K.; Wythes, M. J. In *Peptides: Chemistry, Structure and Biology;* Rivier, J.; Marshall, G. R., Eds.; ESCOM: Leiden, The Netherlands, 1990; pp 247–248.

173. O'Connell, J. E.; Jardine A. G.; Davidson, G.; Connell, J. M. C. *J. Hypertension* **1992,** *10,* 271–277.

174. Chipkin, R. E.; Berger, J. G.; Billard, W.; Iorio, L. C.; Chapman, L. C.; Barnett, A. *J. Pharmacol. Exp. Ther.* **1988,** *245,* 829–838.

175. Sybertz, E. J.; Chiu, P. J. S.; Vermalapalli, S.; Watkins, R.; Haslanger, M. F. *Hypertension* **1990,** *15,* 152–161.

176. Burnier, M.; Ganslmayer, M.; Perret, F.; Porchet, R. N.; Kosoglou, T.; Gould, A.; Nussberger, J.; Waeber, B.; Brunner, H. R. *Clin. Pharmacol. Ther.* **1991,** *50,* 181–191.

177. Lord, J. A.; Waterfield, A. A.; Hughes, J.; Kosterlitz, H. W. *Nature (London)* **1977,** *267,* 495–499.

178. Snider, M. R.; Constantine, J. W.; Lowe, J. A.; Longo, K. P.; Lebel, W. S.; Woody, H. A.; Drozda, S. E.; Desaia, M. C.; Vinick, F. J.; Spencer, R. W.; Hess, H.-J. *Science (Washington, DC)* **1991,** *251,* 435–437.

179. Chiu, A. T.; McCall, D. E.; Aldrich, P. E.; Timmermans, P. B. M. W. M. *Biochem. Biophys. Res. Commun.* **1990,** *172,* 1195–1202.

180. Smith, R. G.; Cheng, K.; Schoen, W. R.; Pong, S.-S.; Hickey, G.; Jacks, T.; Butler, B.; Chan, W. W.-S.; Chaung, L.-Y. P.; Judith, F.; Taylor, J.; Wyvratt, M. J.; Fisher, M. H. *Science (Washington, DC)* **1993,** *260,* 1640–1643.

181. Evans, B. E.; Rittle, K. E.; Bock, M. G.; DiPardo, R. M.; Freidinger, R. M.; Whitter, W. L.; Lundell, G. F.; Veber, D. F.; Anderson, P. S.; Chang, R. S. L.; Lotti, V. J.; Cerno, D. J.; Chen, T. B.; Kling, P. J.; Kunkel, K. A.; Springer, J. P.; Hirshfield, J. J. *J. Med. Chem.* **1988,** *31,* 2235–2246.

182. Flynn, D. L.; Villamil, C. I.; Becker, D. P.; Gullikson, G. W.; Moummi, C.; Yang, D.-C. *Bioorg. Med. Chem. Lett.* **1992,** *2,* 1251–1256.

183. Bock, M. G.; DiPardo, R. M.; Evans, B. E.; Rittle, K. E.; Whitter, W. L.; Veber, D. F.; Anderson, P. S.; Freidinger, R. M. *J. Med. Chem.* **1989,** *32,* 13–16.

184. Howbert, J. J.; Kobb, K. L.; Britton, T. C.; Mason, N. R.; Bruns, R. F. *Bioorg. Med. Chem. Lett.* **1993,** *3,* 875–880.

185. De, B.; Plattner, J. J.; Bush, E. N.; Jae, H.-S.; Diaz, G.; Johnson, E. S.; Perun, T. J. *J. Med. Chem.* **1989,** *32,* 2038–2041.

186. Yamamura, Y.; Ogawa, H.; Chihara, T.; Kondo, K.; Onogawa, T.; Nakamura, S.; Mori, T.; Tominaga, M.; Yabuuchi, Y. *Science (Washington, DC)* **1991,** *252,* 572–574.

187. Fujimoto, M.; Mihara, S.; Nakajima, S.; Ueda, M.; Nakamura, M.; Sakkurai, K. *FEBS Lett.* **1992,** *305,* 41–44.

188. Boigegrain, R.; Gully, D.; Jeanjean, F.; Molimard, J.-C. Eur. Patent 477,049, 1992.

189. Doughty, M. B.; Chu, S. S.; Misse, G. A.; Tessel, R. *Bioorg. Med. Chem. Lett.* **1992,** *2,* 1497–1502.

190. Evans, B. E.; Leighton, J. J.; Rittle, K. E.; Gilbert, K. F.; Lundell, G. F.; Gould, N. P.; Hobbs, D. W.; DiPardo, R. M.; Veber, D. F.; Pettitbone, D. J.; Clineschmidt, B. V.; Anderson, P. S.; Freidinger, R. M. *J. Med. Chem.* **1992,** *35,* 3919–3927.

191. Valentine, J. J.; Nakanishi, S.; Hageman, D. L.; Snider, R. M.; Spencer, R. W.; Vinick, F. J. *Bioorg. Med. Chem. Lett.* **1992,** *2,* 333–338.

192. Lee, V. H. L.; Traver, R. D.; Taub, M. E. In *Peptide and Protein Drug Delivery;* Lee, V. H. L., Ed.; Marcel Dekker: New York, 1993; pp 303–358.

193. Brownlees, J.; Williams, C. H. *J. Neurochem.* **1992,** *60,* 793–803.

194. Maurer, G. *Transplant. Proc.* **1985,** *17 (Suppl. 1),* 19–26.

195. Stewart, B. H.; Taylor, M. D. Chapter 9 in this volume.

196. Moss, J. Chapter 18 in this volume.

197. Tsuji, A. Chapter 5 in this volume.

198. Amidon, G. Chapter 6 in this volume.

199. Kramer, W.; Girbig, F.; Gutjahr, U.; Kowalewski, S. Chapter 7 in this volume.

200. Nestor, J., Jr. Chapter 19 in this volume.

201. Barrett, A. J.; Salveson G. In *Proteinase Inhibitors;* Elsevier: Amsterdam, The Netherlands, 1986.

202. Barrett, A. J.; Rawlings, N. D. *Biochem. Soc. Trans.* **1991,** *19,* 707–715.

203. Navia, M. A.; Murcko, M. A. *Curr. Opin. Struct. Biol.* **1992,** *2,* 202–210.

204. Scharpe, S.; DeMeester, I.; Hendriks, D.; Vanhoof, G.; van Sande, M.; Vriend, G. *Biochimie* **1991,** *73,* 121–126.

205. Lecomte, J.-M.; Costentin, J.; Vlaiculescu, A.; Chaillet, P.; Marcais-Collado, H.; Llorens-Cortes, C.; Leboyer, M.; Schwartz, J.-C. *J. Pharmacol. Exp. Ther.* **1986,** *273,* 937–944.

206. Sybertz, E. J. *Cardiovasc. Drug Rev.* **1990,** *8,* 71.

207. Bauer, W.; Briner, U.; Doepfner, W.; Haller, R.; Huguenin, R.; Marbach, P.; Petcher, T. J.; Pless, J. *Life Sci.* **1982,** *31,* 1133–1140.

208. Vilhardt, H. *Drug Invest.* **1990,** *2 (Suppl. 5),* 2–8.

209. Kahns, A. H.; Burr, A.; Bundgaard, H. *Pharm. Res.* **1993,** *10,* 68–74.

210. Rhaleb, N. E.; Rouissi, N.; Jukic, D.; Regoli, D.; Henke, S.; Breipohl, G.; Knolle, J. *Eur. J. Pharmacol.* **1992,** *210(2),* 115–120.

211. Wirth, K.; Hock, F. J.; Albus, U.; Linz, W.; Alpermann, H. G.; Anagnostopoulos, H.; Henke, S.; Breipohl, G.; Konig, W.; Knolle, J.; Scholkens, B. A. *Br. J. Pharmacol.* **1991,** *102,* 774–777.

212. Sawyer, T. K.; Sanfilippo, P. J.; Hruby, V. J.; Engel, M. H.; Heward, C. B.; Burnett, J. B.; Hadley, M. E. *Proc. Natl. Acad. Sci. U.S.A.* **1980,** *77,* 5754–5758.

213. Ihara, M. K.; Noguchi, T.; Saeki, T.; Fukuroda, S.; Tsuchida, S.; Kimura, S.; Fukami, T.; Ishikawa, K.; Nishikibe, M.; Yano, M. *Life Sci.* **1992,** *50,* 247–255.

214. Ishikawa, K.; Fukumui, T.; Nagase, T.; Fujita, K.; Hayama, T.; Niyama, K.; Mase, T.; Ihara, M.; Yano, M. *J. Med. Chem.* **1992,** *35,* 2139–2142.

215. Wiemer, G.; Gerhards, H. J.; Hock, F. J.; Usinger, P.; vonRechenberg, W.; Geiger, R. *Peptides* **1988,** *9,* 1081–1087.

216. Danho, W.; Makofske, R. C.; Swistok, J.; Michalewsky, J.; Gabriel, T.; Marks, N.; Berg, M. J.; Baird, L.; Geiler, V.; Mackie, G.; Nelson, D.; Triscari, J. *Pept. Res.* **1991,** *4,* 59–65.

217. Krstenansky, J. L.; Broersma, R. J.; Owen, T. J.; Payne, M. H.; Yates, M. T.; Mao, S. J. *Thromb. Haemostas.* **1990,** *63,* 208–214.

218. Morgan, B. A.; Bower, J. D.; Dettmar, P. W.; Metcalf, G.; Schafer, D. J. In *Proceedings of the 6th American Peptide Symposium;* Gross, E.; Meienhofer, J., Eds.; Pierce Chemical: Rockford, IL, 1979; pp 909–912.

219. Metcalf, G. *Brain Res. Rev.* **1982,** *4,* 389–408.

220. Maduskuie, T. P., Jr.; Bleicher, L. S.; Caccioloa, J.; Cheatham, W.; Fevig, J. M.; Johnson, A. L.; McComb, S. A.; Nugiel, D. A.; Schimdt, W. K.; Spellmey, D. A.; Voss, M. E.; Wagner, R. M. *J. Cell. Biochem.* **1993,** *17 (Suppl.),* 232.

221. Frederickson, R. C. A.; Smithwick, E. L.; Shuman, R.; Bemis, K. G. *Science (Washington, DC)* **1981,** *211,* 603–605.

222. Momany, F. A.; Bowers, C. Y.; Reynolds, G. A.; Hong, A.; Newlander, K. *Endocrinology* **1984,** *114,* 1531–1536.

223. Bowers, C. Y.; Momany, F. A.; Reynolds, G. A.; Hong, A. *Endocrinology* **1984,** *114,* 1537–1545.

224. McCormick, G. F.; Millard, W. J.; Badger, T. M.; Bowers, C. Y.; Martin, J. B. *Endocrinology* **1985,** *117,* 97–105.

225. Waxman, J.; Williams, G.; Sandow, J.; Hewitt, G.; Abel, P.; Farah, N.; Fleming, J.; Cox, J.; O'Donoghue, E. P.; Sikora, K. *Am. J. Clin. Oncol.* **1988,** *11 (Suppl. 2),* S152–S155.

226. Sandow, J.; Clayton, R. H. In *Hormone Biochemistry and Pharmacology;* Briggs, M.; Corgin, Eds.; Eden: Montreal, Canada, 1988; Vol. 2, pp 63–106.

227. Haviv, F.; Fitzpatrick, T. D.; Nichols, C. J.; Swenson, R. E.; Mort, N. A.; Bush, E. N.; Diaz, G.; Nguyen, A. T.; Holst, M. R.; Cybulski, V. A.; Leal, J. A.; Bammert, G.; Rhutasel, N. S.; Dodge, P. W.; Johnson, E. S.; Cannon, J. B.; Knittle, J.; Greer, J. *J. Med. Chem.* **1993,** *36,* 928–933.

228. Bolin, D. R.; Cottrell, J. M.; Michalewsky, J.; Garippa, R.; Rinaldi, N.; O'Donnell, M.; Selig, W. In *Proceedings of the Thirteenth American Peptide Symposium;* Hodges, R.; Smith, J., Eds.; ESCOM: Leiden, The Netherlands, 1994.

229. Jeffcoate, W. J.; Phenekos, C.; Ratcliffe, J. G.; Williams, S.; Rees, L.; Besser, G. M. *Clin. Endocrinol.* **1977,** *7,* 1–11.

230. Hayashi, S.; Matsumoto, T.; Oshima, K.; Tanaka, Y.; Yamamoto, T.; Shimura, T. *Jpn. J. Pharmacol.* **1989,** *49,* 18P–19P.

231. Knight, M.; Barone, P.; Tamminga, C. A.; Steardo, L.; Chase, T. N. *Peptides* **1985,** *6,* 631–634.

232. Charpentier, B.; Durieux, C.; Pelaprat, D.; Dor, A.; Reibaud, M.; Blanchard, J. C.; Roques, B. P. *Peptides* **1988,** *9,* 835–841.

233. Rosamund, J. D.; Comstock, J. M.; Thomas, N. J.; Clark, A. M.; Blosser, J. C.; Simmons, R. D.; Gawlak, D. L.; Loss, M. E.; Augello, S. J.; Spatola, A. F.; Benovitz, D. E. In *Peptides, Chemistry and Biology;* Marshall, G. R., Ed.; ESCOM: Leiden, The Netherlands, 1988; pp 610–612.

234. Moran, T. H.; Sawyer, T. K.; Seeb, D. H.; Ameglio, P. J.; McHugh, P. R. *J. Clin. Nutr.* **1992,** *55,* 286S–290S.

235. Shiosaki, K.; Lin, C. W.; Kopecka, H.; Tufano, M. D.; Bianchi, B. R.; Miller, T. R.; Witte, D. G.; Nazdan, A. M. *J. Med. Chem.* **1991,** *34,* 2837.

236. Schiller, P. W. In *Progress in Medicinal Chemistry;* Ellis, G. P.; West, G. B., Eds.; Elsevier: Amsterdam, The Netherlands, 1991; Vol. 28, pp 301–340.

237. Mosberg, H. I.; Hurst, R.; Hruby, V. J.; Gee, K.; Yamamura, H. I.; Galligan, J. J.; Burks, T. F. *Proc. Natl. Acad. Sci. U.S.A.* **1983,** *80,* 5871–5874.

238. Schiller, P. W.; Weltrowska, G.; Nguyen, T. M.-D.; Wilkes, B. C.; Chung, N. N.; Lemieux, C. *J. Med. Chem.* **1993,** *36,* 3182–3187.

239. Kazmierski, W.; Wire, W. S.; Lui, G. K.; Knapp, R. J.; Shook, J. E.; Burks, T. F.; Yamamura, H. I.; Hruby, V. J. *J. Med. Chem.* **1988,** *31,* 2170–2177.

240. Castrucci, A. M.; Hadley, M. E.; Sawyer, T. K.; Hruby, V. J. *Comp. Biochem. Physiol.* **1984,** *78B,* 519–524.

241. Al-Obeidi, F.; Hadley, M. E.; Pettitt, B. M.; Hruby, V. J. *J. Am. Chem. Soc.* **1989,** *111,* 3413–3416.

242. *The Melanotropins:* Hadley, M. E., Ed.; CRC: Boca Raton, FL, 1988; Vols. I-III.

243. Hruby, V. J.; Sharma, S. D.; Toth, K.; Jaw, J. Y.; Al-Obeidi, F.; Sawyer, T. K.; Hadley, M. E. In *Melanotropic Peptides, Proceedings of the New York Academy of Sciences;* Eberle, A. N.; Vaudry, J., Eds.; New York Academy of Sciences: New York, 1993; pp 51–63.

244. Sawyer, T. K.; Staples, D. J.; Al-Obeidi, F. A.; Hruby, V. J.; Castrucci, A.; Hadley, M. E. In *Melanotropic Peptides, Proceedings of the New York Academy of Sciences;* Eberle, A. N.; Vaudry, J., Eds.; New York Academy of Sciences: New York, 1993; pp 597–599.

245. Sawyer, T. K.; Staples, D. J.; Castrucci, A. M.; Hadley, M. E. *Pept. Res.* **1989,** *2,* 1–7.

246. Al-Obeidi, F. A.; Hruby, V. J.; Hadley, M. E.; Sawyer, T. K.; de Castrucci, A. M. L. *Int. J. Pept. Prot. Res.* **1990,** *35,* 228–234.

247. Kitabgi, P.; Checler, F.; Mazella, J.; Vincent, J.-P. *Rev. Clin. Basic Pharmacol.* **1985,** *5,* 397–486.

248. Henry, J. A.; Horwell, D. C.; Meecham, K. G.; Rees, D. C. *Bioorg. Med. Chem. Lett.* **1993,** *3,* 949–952.

249. Eisai Company Ltd., Japanese Patent 01–316399, 1989.

250. Tokumura, T.; Tanaka, T.; Sasaki, A.; Tsuchiya, Y.; Abe, K.; Machida, R. *Chem. Pharm. Bull.* **1990,** *38,* 3094–3098.

251. Cody, W. L.; He, J. X.; Heyl, D. L.; Thieme-Sefler, A. M.; Wustrow, D. J.; Sawyer, T. K.; Akunne, H.; Pugsley, T. A.; Corbin, A. E.; Davis, M. D. In *Peptides 1992;* Schneider, C. H.; Eberle, A. N., Eds.; ESCOM: Leiden, The Netherlands, 1993; pp 677–678.

252. Couder, J.; Tourwe, D.; Van Binst, G.; Schuurkens, J.; Leysen, J. E. *Int. J. Pept. Prot. Res.* **1993,** *41,* 181–184.

253. Lugrin, D.; Vecchini, F.; Doulut, S.; Rodriguez, M.; Martinez, J.; Kitabgi, P. *Eur. J. Pharmacol.* **1991,** *205,* 191–198.

254. Checler, F.; Vincent, J. P.; Kitabgi, P. *J. Pharmacol. Exp. Ther.* **1983,** *227,* 743–748.

255. Johansen, J. L.; Thegersen, H.; Madsen, K.; Suzdak, P.; Hansen, K. T.; Uels, I. U.; Garcia-Lopez, M. T.; Gomez-Monterrey, I.; Gonzalez-Munix, R.; Herranz, R. *Proceedings of the 13th American Peptide Symposium;* ESCOM, Leiden, The Netherlands, 1993; p 216.

256. Veber, D. F.; Freidinger, R. M.; Perlow, D. S.; Paleveda, W. J., Jr.; Holly, F. W.; Strachan, R. G.; Nutt, R. F.; Arison, B. H.; Homnick, C.; Randall, W. C.; Glitzer, M. S.; Saperstein, R.; Hirschmann, R. *Nature (London)* **1981,** *292,* 55–58.

257. Veber, D. F.; Saperstein, R.; Nutt, R. F.; Freidinger, R. M.; Brady, S. F.; Curley, P.; Perlow, D. S.; Paleveda, W. J.; Colton, C. D.; Zacchei, A. G.; Tocco, D. J.; Hoff, D. R.; Vandlen, R. L.; Gerich, J. E.; Hall, L.; Mandarino, L.; Cordes, E. H.; Anderson, P. S.; Hirschmann, R. *Life Sci.* **1984,** *34,* 1371–1378.

258. Vavrek, R. J.; Stewart, J. M. *Peptides* **1985,** *6,* 161–164.

259. Horwell, D. C.; Hughes, J.; Hunter, J. C.; Pritchard, M. C.; Richardson, R. S.; Roberts, E.; Woodruff, G. N. *J. Med. Chem.* **1991,** *34,* 404–414.

260. Doherty, A. M. *J. Med. Chem.* **1992,** *35,* 1493–1508.

261. Spinella, M. J.; Malik, A. B.; Everitt, J.; Andersen, T. T. *Proc. Natl. Acad. Sci. U.S.A.* **1991,** *88,* 7443–7446.

262. Cody, W. L.; Doherty, A. M.; He, J. X.; DePue, P. L.; Rapundalo, S. T.; Hingorani, G. A.; Major, T. C.; Panek, R. L.; Dudley, D. T.; Haleen, S. J.; LaDouceur, D.; Hill, K. E.; Flynn, M. A.; Reynolds, E. E. *J. Med. Chem.* **1992,** *35,* 3301–3303.

263. Nirei, H.; Hamada, K.; Shoubo, M.; Sogabe, K.; Notsu, Y.; Ono, T. *Life Sci.* **1993,** *52,* 1869–1874.

264. Ruoslahti, E.; Pierschbacher, M. D. *Science (Washington, DC)* **1987,** *238,* 491–497.

265. Barker, P. L.; Bullens, S.; Bunting, S.; Burdick, D. J.; Chan, K. S.; Deisher, T.; Eigenbrot, C.; Gadek, T. R.; Gantzos, R.; Lipari, M. T.; Muir, C. D.; Napier, M. A.; Pitti, R. M.; Padua, A.; Quan, C.; Stanley, M.; Struble, M.; Tom, J. Y. K.; Burnier, J. P. *J. Med. Chem.* **1992,** *35,* 2040–2048.

266. Karten, J. J.; Rivier, J. E. *Endocrine Rev.* **1986,** *7,* 44.

267. Vale, W.; Grant, G.; Rivier, J.; Monahan, M.; Amoss, M.; Blackwell, R.; Burgus, R.; Guillemin, R. *Science (Washington, DC)* **1972,** *176,* 933–934.

268. Nestor, J. J., Jr.; Tahilramani, R.; Ho, T. L.; McRae, G. I.; Vickery, B. H. *J. Med. Chem.* **1984,** *27,* 1170–1174.

269. Hocart, S. J.; Nekola, M. V.; Coy, D. H. *J. Med. Chem.* **1987,** *30,* 739–743.

270. Ljungqvist, A.; Feng, D.-M.; Hook, W.; Shen, Z.-X.; Bowers, C.; Folkers, K. *Proc. Natl. Acad. Sci. U.S.A.* **1988,** *85,* 8236–8240.

271. Haviv, F.; Fitzpatrick, T. D.; Nichols, C. J.; Swenson, R. E.; Mort, N. A.; Bush, E. N.; Diaz, G.; Nguyen, A. T.; Holst, M. R.; Cybulski, V. A.; Leal, J. A.; Bammert, G.; Rhutasel, N. S.; Dodge, P. W.; Johnson, E. S.; Cannon, J. B.; Knittle, J.; Greer, J. *J. Med. Chem.* **1993,** *36,* 928–933.

272. Folkers, K.; Hakanson, R.; Horig, J.; Jie-Cheng, X.; Leander, S. *Br. J. Pharmacol.* **1984,** *83,* 449–456.

273. Hagiwara, D.; Miyake, H.; Morimota, H.; Murai, M.; Fujii, T.; Matsuo, M. *J. Med. Chem.* **1992,** *35,* 2015–2025.

274. Petrillo, E. W.; Ondetti, M. A. *Med. Res. Rev.* **1982,** *2,* 1–41.

275. Patchett, A. A.; Cordes, E. H. *Adv. Enzymol.* **1985,** *57,* 1–84.

276. Ehlers, M. R. W.; Riordan, J. F. *Biochemistry* **1989,** *28,* 5311–5318.

277. Greenlee, W. J. *Pharm. Res.* **1990,** *10,* 173–236.

278. Szelke, M.; Leckie, B.; Hallett, A.; Jonees, D. M.; Seiras, J.; Atrash, B.; Lever, A. F. *Nature (London)* **1982,** *299,* 555–557.

279. Boger, J.; Lohr, N. S.; Ulm, E. H.; Poe, M.; Blaine, E. H.; Fanelli, G. M.; Lin, T.-Y.; Payne, L. S.; Schorn, T. W.; LaMont, B. I.; Vassil, T. C.; Stailito, I. I.; Veber, D. F.; Rich, D. H.; Boparai, A. S. *Nature (London)* **1983,** *303,* 81–83.

280. Szelke, M.; Jones, D. M.; Atrash, B.; Hallett, A.; Leckie, B. J. In *Peptides: Structure and Function;* (Hruby, V. J.; Rich, D. H., Eds.; Pierce Chemical: Rockford, IL, 1983; p 579.

281. Sawyer, T. K.; Hester, J. B.; Schostarez, H. J.; Thaisrivongs, S.; Bundy, G. L.; Liu, L.; Bradford, S.; DeVaux, A. E.; Staples, D. J.; Maggiora, L. L.; TenBrink, R. E.; Kinner, J. H.; Smith, C. W.; Pals, D. T.; Couch, S. J.; Hinzmann, J. S.; Poorman, R. A.; Einspahr, H. M.; Finzel, B. C.; Watenpaugh, K. D.; Mao, B.; Epps, D. E.; Kezdy, F. J.; Heinrikson, R. L. In *Advances in Experimental Medicine and Biology;* Dunn, B. M., Ed.; Plenum: New York, 1992; pp 307–323.

282. Thaisrivongs, J.; Pals, D. T.; Harris, D.; Kati, W. M.; Turner, S. R.; DeGraaf, G. L.; Harris, D. W.; Johnson, G. A. *J. Med. Chem.* **1986,** *29,* 2088–2093.

283. Boger, J.; Bennet, C. D.; Payne, L. S.; Ulm, E. H.; Blaine, E. H.; Homnick, C. F.; Schorn, T. W.; LaMont, B. I.; Veber, D. F. *Regul. Pept. Suppl.* **1985,** *4,* 8–13.

284. Sawyer, T. K.; Maggiora, L. L.; Liu, L.; Staples, D. J.; Bradford, V. S.; Mao, B.; Pals, D. T.; Dunn, B. M.; Poorman, R.; Hinzmann, J.; DeVaux, A. E.; Affholter, J. A.; Smith, C. W. In *Peptides: Chemistry and Biology;* Marshall, G. R.; Rivier, J., Eds.; ESCOM: Leiden, The Netherlands, 1990; pp 46–48.

285. Kleinert, H. D.; Martin, D.; Chekal, M. A.; Kadam, J.; Luly, J. R.; Plattner, J. J.; Perun, T. J.; Luther, R. R. *Hypertension* **1988,** *11,* 613–619.

286. Fischli, W.; Clozel, J.-P.; El Amrani, K.; Wostl, W.; Neidhart, W.; Stadler, H.; Branca, Q. *Hypertension* **1991,** *18,* 22–31.

287. Wood, J. M.; Gulati, N.; Forgiarini, P.; Fuhrer, W.; Hobfauer, K. G. *Hypertension* **1985,** *7,* 797–803.

288. Kleinert, H. D.; Rosenberg, S. H.; Baker, W. R.; Stein, H. H.; Klinghofer, V.; Barlow, J.; Spina, K.; Polakowski, J.; Lovar, P.; Cohen, J.; Denissen, J. *Science (Washington, DC)* **1992,** *257,* 1940–1943.

289. Miyazaki, M.; Toda, N.; Etoh, Y.; Kubota, T.; Kinji, I. *Jpn. J. Pharmacol.* **1986,** *40,* 70P.

290. Patt, W. C.; Hamilton, H. W.; Ryan, M. J.; Painchaud, C. A.; Tylor, M. D.; Rapundalo, S. T.; Batley, B. L.; Connolly, C. J. C.; Taylor, D. G., Jr. *J. Med. Chem.* **1992,** *2,* 10–15.

291. Tomasselli, A. G.; Howe, W. J.; Sawyer, T. K.; Wlodawer, A.; Heinrikson, R. L. *Chem. Today* **1991,** *5,* 6–27.

292. Huff, J. R. *J. Med. Chem.* **1991,** *34,* 2305–2314.

293. Norbeck, D. W.; Kempf, D. J. *Annu. Rep. Med. Chem.* **1991,** *26,* 141–150.

294. McQuade, T. J.; Tomasselli, A. G.; Liu, L.; Karacostas, V.; Moss, B.; Sawyer, T. K.; Heinrikson, R. L.; Tarpley, W. G. *Science (Washington, DC)* **1990,** *247,* 454–456.

295. Wlodawer, A.; Erickson, J. W. *Annu. Rev. Biochem.* **1993,** *62,* 543–585.

296. Reiss, Y.; Goldstein, J. L.; Seabra, M. C.; Casey, P. J.; Brown, M. S. *Cell* **1990,** *62,* 81.

297. Goldstein, J. L.; Brown, M. S.; Stradley, S. J.; Reiss, Y.; Gierasch, L. M. *J. Biol. Chem.* **1991,** *266,* 15575–15578.

298. Reiss, Y.; Brown, M. S.; Goldstein, J. L. *J. Biol. Chem.* **1992,** *267,* 6403–6408.

RECEIVED for review September 9, 1993. ACCEPTED revised manuscript January 24, 1994.

CHAPTER 18

Peptide Prodrugs Designed to Limit Metabolism

Judi Møss

Rationale and Applications of the Prodrug Concept for Peptides

Prodrug design comprises an area of drug research that is concerned with the optimization of drug delivery. A prodrug is a pharmacologically inactive derivative of a parent drug molecule that requires spontaneous or enzymatic transformation within the body to release the active drug and that has improved delivery properties over the parent drug molecule (*1*). In general, prodrugs are designed to overcome pharmaceutically and pharmacokinetically based problems associated with the parent drug molecule that would otherwise limit the clinical usefulness of the drug. Figure 1 illustrates the prodrug approach. The usefulness of a drug molecule is limited by its suboptimal physicochemical properties (e.g., it may show poor biomembrane permeability). By attachment of a promoiety to the molecule or by otherwise modifying the compound, a prodrug is formed that overcomes the barrier for the usefulness of the drug. Once past the barrier, the prodrug is reverted to the parent compound

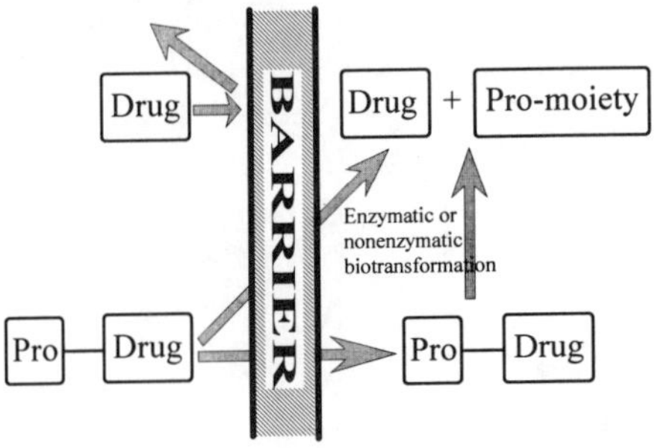

Figure 1. Schematic illustration of the prodrug approach.

by a postbarrier enzymatic or nonenzymatic process. Prodrug formation can thus be considered as conferring a transient chemical cover to alter or eliminate undesirable properties of the parent drug molecule.

A number of barriers may limit the clinical usefulness of a drug; for example, a physical barrier (i.e., poor biomembrane penetration characteristics) or an enzymatic barrier. Several drugs are efficiently absorbed from the gastrointestinal tract but show limited systemic bioavailability due to presystemic (first-pass) metabolism or inactivation before reaching the systemic circulation. This metabolism can occur in the intestinal lumen, at the brush border of the intestinal cells, in the mucosal cells lining the gastrointestinal tract, or in the liver. The prodrug approach can be very useful to reduce first-pass metabolism. The traditional approach has been to mask the metabolizable moiety (i.e., to derivatize a functional group required for an enzymatic attack), but another strategy is to derivatize to lower the affinity of the substrate for the enzymes responsible for the degradation.

Peptides are becoming an important new class of drugs, but their application as clinically useful compounds is seriously hampered because of substantial delivery problems. Peptides generally show poor biomembrane penetration characteristics and rapid enzymatic degradation at the various mucosal absorption sites as well as in the systemic circulation (*2, 3*). Several peptides also suffer from systemic transport problems because they do not readily penetrate cell membranes to reach the receptor biophase or cross the blood–brain barrier (*4*).

A potentially useful approach to solve these delivery problems may be derivatization of the bioactive peptides to produce prodrugs or to transport forms possessing, with respect to delivery and metabolic stability, enhanced physicochemical properties in comparison with the parent compounds. *N*-Alkylation of peptide bonds usually makes these bonds resistant to enzymatic attack (*5–9*). However, because *N*-methyl and similar alkyl derivatives are generally not bioreversible, the approach of simple *N*-alkylation implies the design of a new peptide (the analogue approach). In contrast, prodrug derivatization is a bioreversible approach, the derivatives being converted quantitatively to the parent peptides through a spontaneous or an enzymatic reaction. Thus, such derivatization may protect small peptides against degradation by enzymes present at the mucosal barrier as well as render hydrophilic peptides more lipophilic and therefore facilitate their absorption. However, for derivatization to be a useful approach, the derivatives should be capable of releasing the parent peptide spontaneously or enzymatically in the blood following the derivatives absorption (*10*). A basic requisite for the application of the prodrug approach is the ready availability of chemical derivative types applicable to peptides and satisfying the prodrug requirements, the most prominent of these being reconversion to the parent peptidic drug in vivo (*11, 12*).

In recent years several types of bioreversible derivatives for the functional groups or chemical entities (such as carboxyl, hydroxyl, thiol, amino, and amido groups) occurring in amino acids and peptides have been explored (*11, 13–16*). The purpose of this chapter is to review and discuss various chemical approaches to obtain such prodrug forms. The chapter is arranged by enzyme reaction type, and a short description of the properties of a specific peptidase is followed by examples of prodrug attempts to circumvent the particular hydrolytic reaction. Some of the prodrug approaches are used with several of the enzymes. A more extensive coverage of various types of bioreversible derivatives is published elsewhere (*11, 13–16*).

Protection of Peptides against Carboxypeptidases

Carboxypeptidase A (CPA) is a pancreatic proteolytic enzyme whose primary function is that of a C-terminal exopeptidase. The enzyme catalyzes the hydrolysis of almost any peptide having a terminal free carboxyl group and a C-terminal residue of the L-configuration (*17, 18*). Sensitive peptide substrates for CPA are acylated dipeptides of the type represented in Figure 2, in which the arrow indicates the point of cleavage. The rate of hydrolysis is usually enhanced if the terminal residue is aromatic or branched aliphatic (*17–19*).

Carboxypeptidase digestion can lead to the metabolic degradation of peptides from the C-terminal end of the molecule. One way to eliminate carboxypeptidase-mediated degradation is to mask or remove the C-terminal carboxylic acid and another way is to alter the substrate characteristics by derivatizing a functional group near the cleavable bond.

5-Oxazolidinones as Prodrugs for the α-Amido Carboxy Moiety

Because of masking of the carboxylic acid function in acylated amino acids, the 5-oxazolidinones afford protection of the amide bond against proteolytic cleavage by enzymes requiring a free C-terminal carboxyl group (*20*). The same protection possibly could be achieved by simple C-terminal esterification. However, by involving the α-amido group in the derivatization, the peptide bond might be changed so it is no longer a substrate for proteolytic enzymes.

5-Oxazolidinones can serve as potential prodrug forms for the α-amido carboxy moiety occuring in most peptides. 5-Oxazolidinones are formed by condensing the α-amido

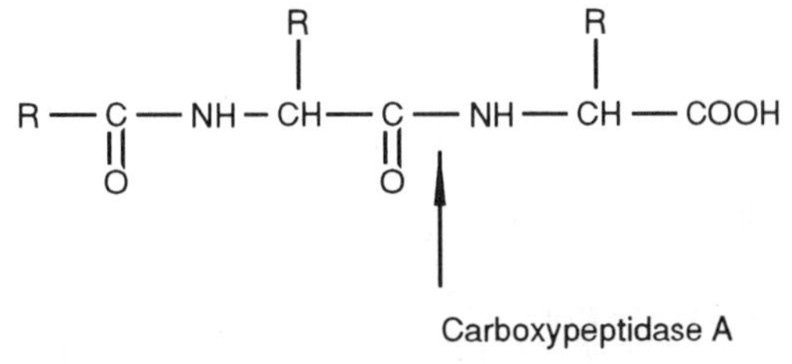

Figure 2. Point of action of CPA.

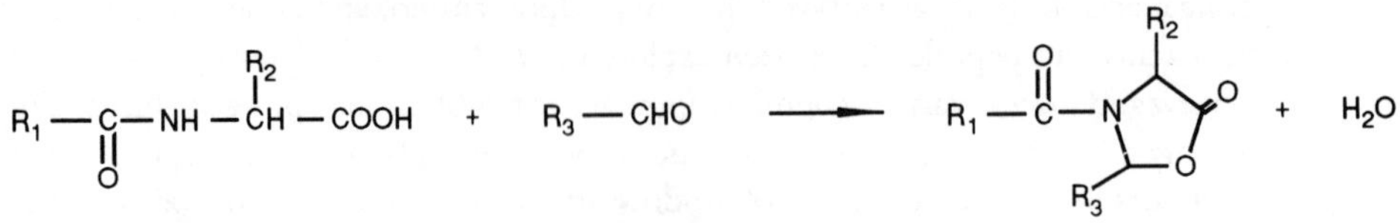

Scheme I. (Reproduced with permission from reference 25. Copyright 1991.)

carboxy group in a peptide with an aldehyde (Scheme I) (*20*). In accordance with the behavior of linear *N*-acyloxyalkyl derivatives of secondary amides or carbamates (*21*), the degradation appears to take place via a two-step reaction, as shown in Scheme II; that is, initial hydrolytic opening of the lactone ring is followed by a spontaneous decomposition of the intermediary *N*-hydroxyalkyl derivative to give the parent *N*-acyl amino acid and aldehyde. The derivatives are hydrolyzed quantitatively in aqueous solution and in plasma solutions at rates dependent on the 2- and 4-substituents in the oxazolidinone as well as on the stability of the initial product of hydrolysis (i.e., the *N*-hydroxyalkyl derivative). The stability of the latter can be controlled by selection of appropriate aldehydes or ketones for the oxazolidinone formation. The rate-determining step in the liberation of the parent *N*-acylated amino acid from the 5-oxazolidinones depends on the relative stabilities of the latter and the intermediate *N*-hydroxyalkyl derivative.

The 5-oxazolidinones prepared from *N*-benzyloxycarbonyl (*Z*)-glycine, (*Z*)-alanine, (*Z*)-valine, and (*Z*)-phenylalanine, studied by Buur and Bundgaard (20), were quite unstable

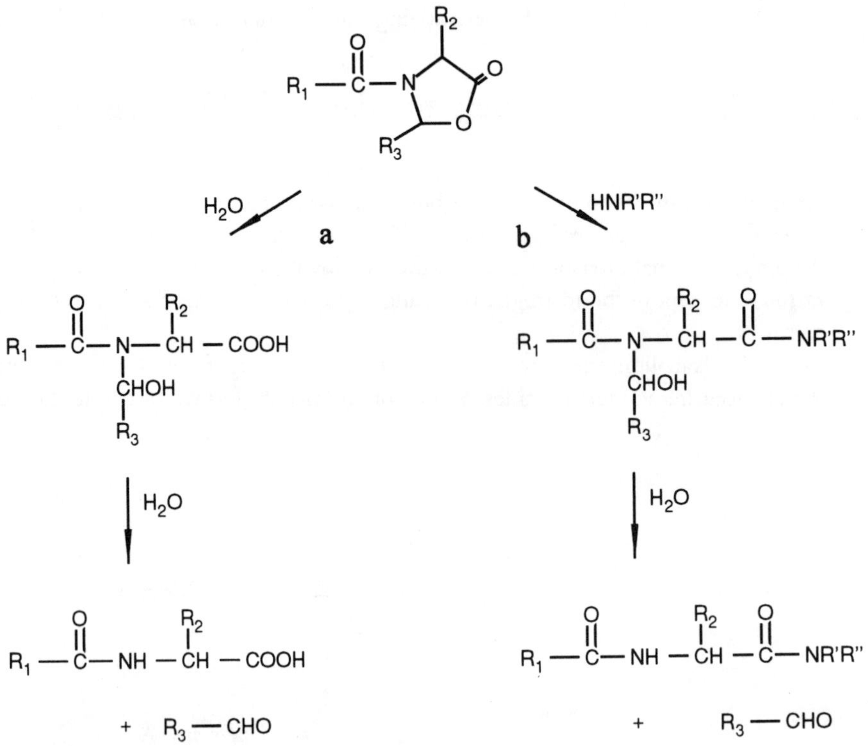

Scheme II. (Reproduced with permission from reference 25. Copyright 1991.)

in aqueous solution. Furthermore, because the slowest step in the overall degradation was the decomposition of the intermediate *N*-hydroxymethyl derivative, the latter derivatives are considered as prodrug forms with the potential to afford protection of the amide bond against proteolytic cleavage.

N-α-Hydroxyalkylation of the Peptide Bond

The strategy of this prodrug approach is to create an *N*-α-hydroxyalkyl derivative of the peptide bond because such derivatives of primary and cyclic amides are known to be spontaneously converted to the parent amide and the corresponding aldehyde in aqueous solution, with a conversion rate dependent on the nature of the alkyl group and the acidity of the amide (*11, 21–24*). A major obstacle to this approach has been the difficulty in performing *N*-α-hydroxyalkylation of secondary amides such as peptide bonds (*24*), but this difficulty can be overcome by making 5-oxazolidinones (*20*) as discussed in the preceding section. The lactone ring in *N*-acyl 5-oxazolidinones is highly reactive and is easily opened by hydrolysis (Scheme IIa) or aminolysis (Scheme IIb), resulting in the intermediate formation of an *N*-α-hydroxyalkyl derivative for both reaction pathways (*20*). A number of such *N*-α-hydroxyalkyl derivatives of various amino acids and peptides were prepared, and their stability in aqueous solution was determined (*25*). In aqueous solution at pH 1–11, all the derivatives are quantitatively converted to their parent *N*-acylated amino acid or to peptide and aldehyde but at vastly different rates. At pH 7.4 and 37 °C, the half-lives ($t_{1/2}$) of decomposition ranged from 4 min to 2050 h (Table I). Similar rates are seen in human plasma solutions, indicating the absence of enzymatic catalysis of the decomposition as for other *N*-α-hydroxyalkyl derivatives (*24, 26*). The structural factors influencing the stability include both steric and polar effects within the acyl and *N*-α-hydroxyalkyl moiety as well as within the amino acid or amine attached to the *N*-α-hydroxyalkylated *N*-acyl amino acid. The stability of a given peptide can be controlled by selection of appropriate aldehydes for the initial 5-oxazolidinone formation. Thus, by using acetaldehyde, benzaldehyde, or chloraldehyde (i.e., the R_2-moiety) instead of formaldehyde, a higher rate of prodrug conversion is achieved (Table I).

To determine if *N*-α-hydroxyalkylation of the second peptide bond in a dipeptide would make the terminal bond stable toward CPA, the rates of decomposition of the derivatives were determined in the presence of CPA at a concentration of 50 U/mL. Under these conditions, the parent *N*-benzyloxycarbonyl (*Z*) derivatives of the dipeptides Gly-L-Leu and Gly-L-Ala were hydrolyzed by CPA, with $t_{1/2}$ of 6 min and 8.7 h, respectively. The *N*-hydroxymethylated compounds [i.e., Z-Gly(CH_2OH)-Leu and Z-Gly(CH_2OH)-Ala] were totally resistant to cleavage by the enzyme as revealed by their similar rates of decomposition (i.e., $t_{1/2}$ = 980 and 137 h, respectively) in the presence or absence of the enzyme at pH 7.4 and 37 °C. These results convincingly demonstrate that *N*-hydroxymethylation of the *N*-terminal peptide bond in the *N*-acylated dipeptides Z-Gly-L-Leu and Z-Gly-L-Ala completely protects the C-terminal peptide bond against CPA. Thus, the presence and integrity of the second peptide bond appear to be important for the enzymatic reactivity. This is in agreement with the findings that the rate of the CPA-catalyzed hydrolysis of *N*-acyl dipeptides is greatly decreased if *N*-methylglycine (sarcosine) or β-alanine is the second amino acid (*27, 28*).

Direct protection of the susceptible C-terminal peptide bond of an *N*-acylated dipeptide (or a higher homologue) by *N*-α-hydroxyalkylation of this bond should also be possible by transforming an *N*-acyl dipeptide into the corresponding 5-oxazolidinone through condensation with an aldehyde and through subsequent hydrolysis. This has been found to be true by *N*-hydroxymethylation of the C-terminal peptide bond in the *N*-acylated dipeptide

Table I. Half-Lives of Conversion of Various *N*-α-Hydroxyalkyl Derivatives to the Parent *N*-Acylated Amino Acids or Peptides in Aqueous Solution at 37 °C

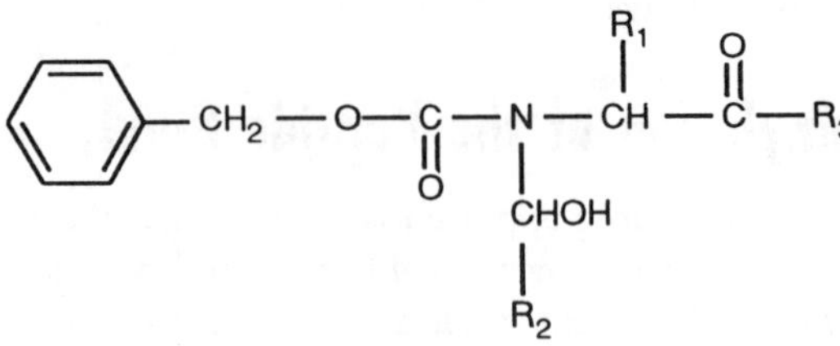

R_1	R_2	R_3	$t_{1/2}$ (h)
H	H	OH	2050
CH_3	H	OH	199
$CH(CH_3)_2$	H	OH	169
$CH_2C_6H_5$	H	OH	391
H	CH_3	OH	3.6
H	C_6H_5	OH	0.08
H	CCl_3	OH	5.7
H	H	$NHCH_2C_6H_5$	10.8
H	H	$NHCH_2CH_2C_6H_5$	22.3
H	H	$N(C_2H_5)$	17.5
H	H	Gly	15.3
H	H	Ala	137
H	H	Leu	980
H	H	Val	910
H	H	Gly-Gly	14.1

SOURCE: Adapted with permission from reference 25. Copyright 1991.

Z-Ala-L-Ala. This location derivatization does also completely protect the C-terminal peptide bond against CPA (Bundgaard, H.; Rasmussen, G. J., Royal Danish School of Pharmacy, Denmark, unpublished data).

Thus, it has been shown (*29*) that *N*-α-hydroxyalkylation of a peptide bond may be a useful prodrug approach to protect the bond or an adjacent peptide bond against specific proteolytic cleavage, as illustrated with CPA. Further derivatization of the hydroxyl group in the *N*-α-hydroxyalkyl derivatives (e.g., by esterification to afford *N*-α-acyloxyalkyl derivatives) from which the parent peptide may be released in a two-step reaction did not result in a stabilization of the hydroxymethyl derivatives toward decomposition in acidic and neutral aqueous solution (*29*).

Protection of Peptides against Aminopeptidases

Aminopeptidases have a broad substrate specificity. By definition, an N-terminal exopeptidase hydrolyzes peptide bonds at or near the amino-terminus but not at other bonds in the peptide chain. Generally, this type of enzyme requires a free α-amino (or α-imino) group

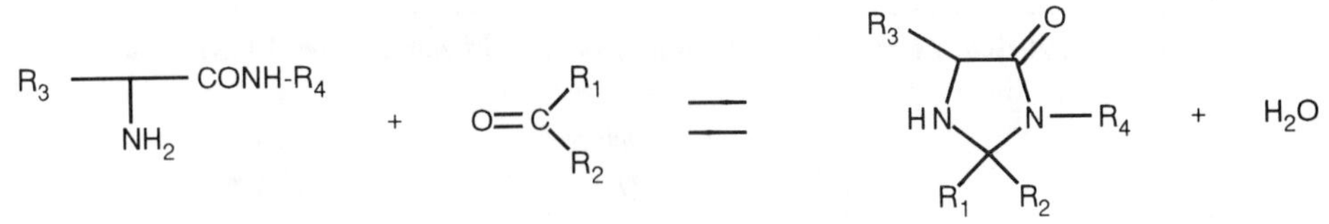

Scheme III.

and releases free amino acids from the N-terminus of the peptide substrate (*30*). Aminopeptidase-catalyzed hydrolysis of several peptides, such as enkephalins, is an important enzymatic barrier for the absorption of peptides (*3*). If it is known that the peptide in question is being degraded by aminopeptidases, the first logical step is to introduce an N-terminal modification that will prevent the enzyme from interacting with the peptide.

4-Imidazolidinones as Prodrug Derivatives for the α-Aminoamide Moiety

The metabolic inactivation by aminopeptidases can be circumvented by transforming the peptides into 4-imidazolidinone prodrug derivatives. Such derivatives are readily formed by condensing peptides that have a free N-terminal amino group with aldehydes or ketones (Scheme III) (*31, 32*). The usefulness of these derivatives as prodrug forms is due to their ready hydrolysis in aqueous solution with release of the parent peptide, with the rate of hydrolysis being dependent on the pH and the structure of the R_1, R_2, R_3, and R_4 substituents (Scheme III); that is, the structure of the parent peptide and the carbonyl component (*32–34*). The mechanism of degradation is similar to that of various N-Mannich bases of amides (Scheme IV) (*35, 36*). In fact, 4-imidazolidinones can be considered to be cyclic N-Mannich bases in which the amide and amino functions are placed in the same molecule.

It has been shown for various 4-imidazolidinones derived from different dipeptides, tripeptides, and pentapeptides (enkephalins), that the derivatives are completely inert toward

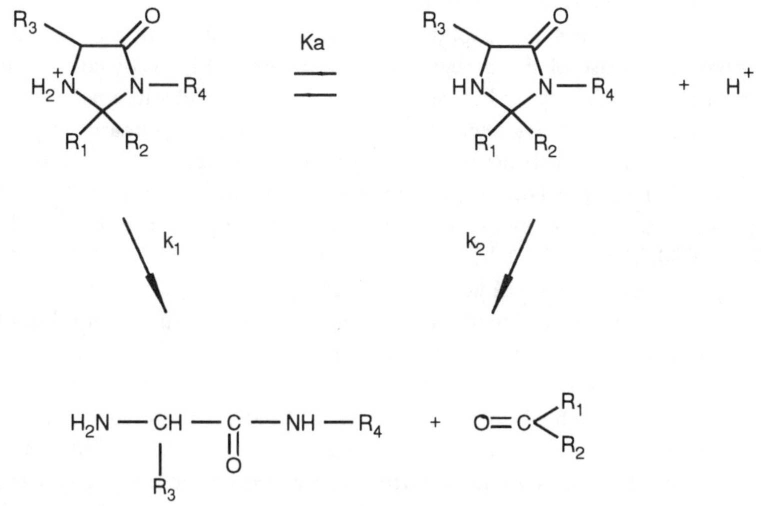

Scheme IV. (Reproduced with permission from reference 34. Copyright 1991.)

Table II. Half-Lives of Hydrolysis of Various Di- and Tripeptides and 4-Imidazolidinones

Compound	Buffer, pH 7.4 (h)	Aminopeptidase (20 U mL^{-1}, pH 7.4)	80% Human Plasma	10% Rabbit Intestinal Homogenate
Phe-Ala		1.7 min	42 min	4.6 min
4-Im-Ph-Ala	15.2	14.4 h	16.3 h	13.9 h
Ala-Phe		1.9 min	18 min	2.0 min
4-Im-Ala-Phe	6.9	5.9 h	7.0 h	6.4 h
Gly-Phe		480 min	210 min	2.8 min
4-Im-Gly-Phe	2.0	1.9 h	2.0 h	2.0 hr
Gly-Phe-Phe		55 min	3.9 min	1.6 min
4-Im-Gly-Phe-Phe	0.3	0.3 h	0.3 h	0.3 h
Ala-Phe-Gly		0.9 min	2.7 min	1.1 min
4-Im-Ala-Phe-Gly	0.4	0.4 h	0.3 h	0.4 h
Tyr-Gly-Gly		54 min	14 min	40 min
4-Im-Tyr-Gly-Gly	11.2	13.2 h	18.6 h	8.4 h

NOTE: Half-lives determined in 0.05 M phosphate buffer solution (pH 7.4) without or with aminopeptidase, in 80% human plasma, and in 10% rabbit intestinal homogenate at 37 °C. The half-lives refer to the disappearance of either parent peptide (e.g., Phe-Ala; enzymatic cleavage of the dipeptide) or the derivatized peptide (e.g., 4-Im-Phe-Ala; conversion of the derivative to the parent peptide according to Scheme IV).
SOURCE: Adapted with permission from reference 33. Copyright 1991.

aminopeptidases capable of degrading the parent peptides [i.e., purified leucine aminopeptidase as well as human plasma and rabbit intestinal homogenates (Table II) (*33, 34*)]. This observed resistance is in accordance with the general substrate specificities of aminopeptidases and other exopeptidases cleaving peptides at their N-terminals. These specificities include a free N-terminal amino group, preferably a primary amino group and an unmodified peptide (*30*). In the 4-imidazolidinone structure, the primary amino group of the parent peptide was transformed into a secondary amino group and, most importantly, the vulnerable N-terminal peptide bond was alkylated.

In designing 4-imidazolidinone prodrug derivatives of a given peptide, the rate of spontaneous release of the parent peptide at neutral pH can be controlled by selection of an appropriate aldehyde or ketone (i.e., the R_1 and R_2 substituents). Using enkephalins as model compounds, it has been shown that the reactivity primarily depends on the steric properties of these substituents, the rate of hydrolysis being increased with increased steric effects (*34*). Thus, the 4-imidazolidinone derivatives of Leu-enkephalin (Scheme V) formed with acetaldehyde, acetone, or cyclopentanone decompose at pH 7.4 and 37 °C, with $t_{1/2}$ values of 30, 10.9, and 3.1 h, respectively (*34*).

In considering 4-imidazolidinones as a potential prodrug type for peptides, the large decrease obtained in basicity of the reaction N-terminal amino group should be appreciated. The 4-imidazolidinones just mentioned are much weaker bases (pK_a = 3–3.5) than the parent dipeptides, and such depression of amino protonation brings about an increase in the lipophilicity of the N-terminal amino acid part at physiological pH as confirmed by partition experiments in octanol–aqueous buffer systems (*32, 33*). The increased lipophilicity attained, which obviously will be further influenced by the lipophilicity of the substituents of the carbonyl component, may be of value in situations where delivery problems of peptide drugs are due to low lipophilicity.

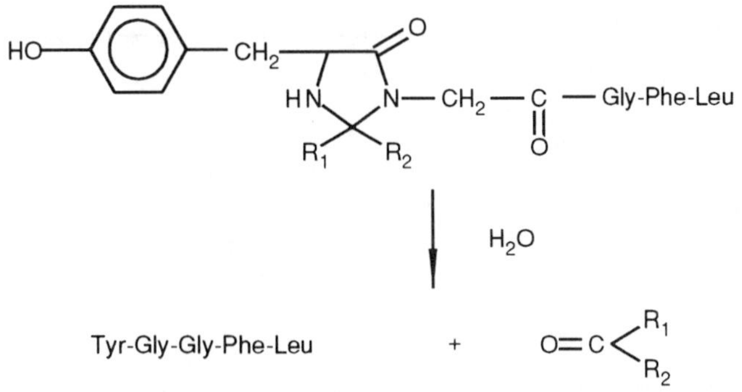

Scheme V. (Reproduced with permission from reference 34. Copyright 1991.)

Protection of Peptides against α-Chymotrypsin

α-Chymotrypsin is an endopeptidase (serine protease) that catalyzes the hydrolysis of peptide bonds in which the reactive carbonyl group belongs to the L-amino acids tryptophan, tyrosine, phenylalanine and, to a lesser extent, leucine and methionine (*37*). It also catalyzes the hydrolysis of various simple esters and amides (*38*). Several peptide drugs are rapidly degraded by α-chymotrypsin, such as adrenocorticotropin [ACTH-(1–10)] (*39*), oxytocin (*40*), [D-Ala^1]-peptide T amide (*41*), and various renin inhibitors (*42*). Stabilization of such peptides against α-chymotrypsin-catalyzed degradation is important to make oral absorption feasible. The normal concentration of α-chymotrypsin in the human gut including the stool is ~0.5 mg mL^{-1} or 2×10^{-5} M (*43, 44*).

N-α-Hydroxyalkylation of a Peptide Bond

N-α-Hydroxyalkylation of a peptide bond not only protects peptides against degradation by carboxypeptidase but also may protect peptides against hydrolysis by other proteolytic enzymes, such as α-chymotrypsin and trypsin, because these enzymes usually also require the free NH-moiety of the peptide bond (*45*). Kahns et al. (*46*) reported that *N*-hydroxymethylation of the N-terminal peptide bond in *N*-acetyl-L-phenylalanine amides protects the C-terminal amide bond against cleavage by α-chymotrypsin. The *N*-hydroxymethyl derivatives are readily bioreversible, undergoing spontaneous hydrolysis at physiological pH, and are thus promising prodrugs to use to overcome the enzymatic barrier to absorption of various peptides.

Stabilization of Peptide Amides

The C-terminal primary amide group, occuring in many bioactive peptides, is generally important for the biological activity and stability of these peptides (*47, 48*). The C-terminal peptide amide bond is cleavable by α-chymotrypsin, and in an attempt to protect these peptides against enzymatic degradation, this bond has been derivatized in various ways to evaluate whether chemical modification would stabilize the amide toward α-chymotrypsin-catalyzed cleavage (*49*). The derivatives studied were previously investigated as potential prodrug forms of amides (i.e., N-Mannich bases, *N*-hydroxymethyl derivatives, glyoxylic acid adducts, and *N*-acyl and *N*-phthalidyl derivatives; the structures are given in Scheme

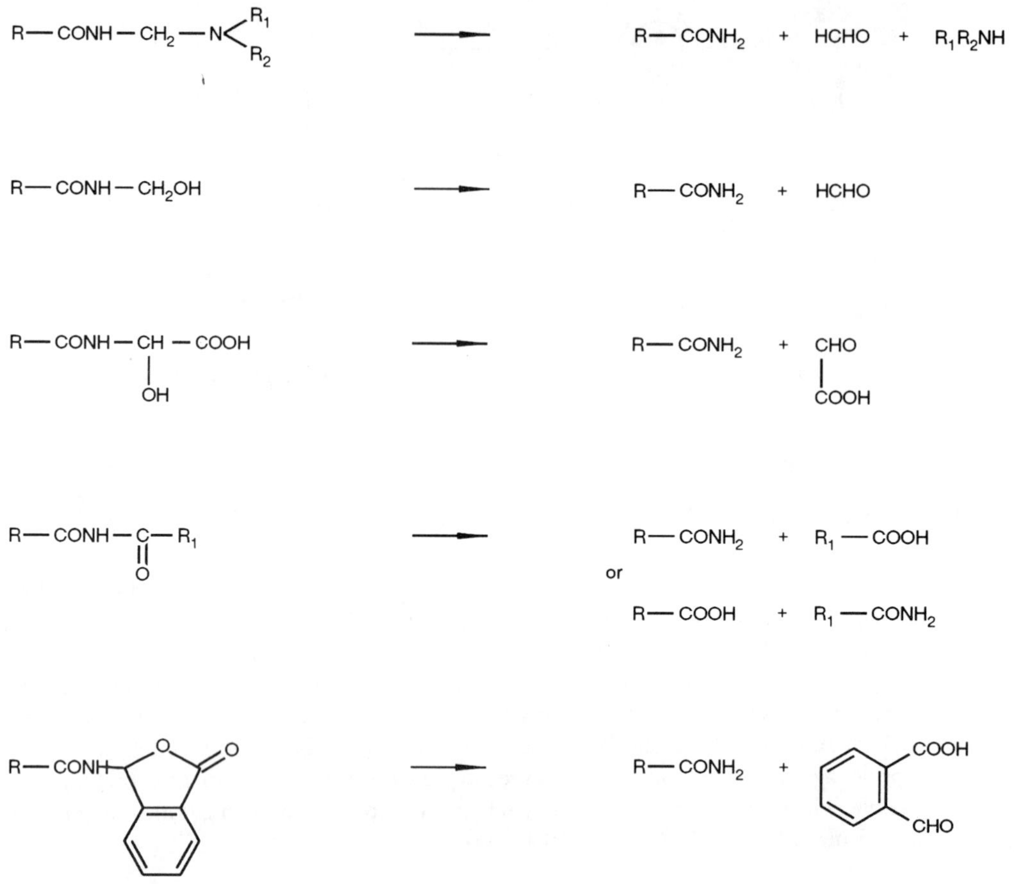

Scheme VI.

VI). N-Mannich bases (*N*-aminomethylated amides) and *N*-hydroxymethyl derivatives (*11*) are spontaneously hydrolyzed to the parent amide in aqueous solution at rates dependent on pH, whereas the hydrolysis of α-hydroxyglycine derivatives (glyoxylic acid adducts) (*50, 51*), *N*-acylated amides (*52*), and *N*-phthalidyl derivatives (*53*) is also catalyzed by plasma enzymes. These conversions proceed quantitatively except for the *N*-acylated amides, where a competing reaction may be hydrolysis of the "wrong" imide moiety (Scheme VI) (*52*).

Condensation of the C-terminal amide group with glyoxylic acid to produce peptidyl-α-hydroxyglycine derivatives was found to be particularly effective in protecting against α-chymotrypsin (*49*). Thus, whereas *N*-Ac-L-Phe-NH_2 was readily hydrolyzed by α-chymotrypsin, its glyoxylic acid adduct proved to be completely resistant to degradation by the enzyme (*49*), (Table III). The decomposition of the derivative proceeds with the quantitative formation of *N*-Ac-Phe-NH_2, which subsequently is hydrolyzed by the enzyme to *N*-Ac-Phe-OH (Scheme VII). The degree of protection obtained may depend on the structure and enzymatic susceptibility of the parent amide.

In addition to protecting the terminal, derivatized amide moiety from cleavage by α-chymotrypsin, α-hydroxyglycine derivatization can also protect the internal underivatized peptide bond to a significant extent as has been shown with such derivatives of Z-Phe-Gly-NH_2, Z-Phe-Ala-NH_2, and Z-Tyr-Leu-NH_2 (Table III) (*49*). The usefulness of this

Table III. Half-Lives of Hydrolysis of Different *N*-Protected Amino Acid and Dipeptide Amides and α-Hydroxyglycine Derivatives

Compound	*Buffer, pH 7.40 (h)*	*Buffer (pH 7.40) with α-Chymotrypsin (0.5 mg mL^{-1})*	*80% Human Plasma (h)*
N-Ac-Phe-NH_2		3.8 h	
N-Ac-Phe-CH(OH)-COOH	8.1	6.7 h	2.9
N-Z-Phe-Ala-NH_2		0.9 min	
N-Z-Phe-Ala-CH(OH)-COOH	13.3	20 min	1.8
N-Z-Phe-Gly-NH_2		11 min	
N-Z-Phe-Gly-CH(OH)-COOH	7.7	1.3 h	1.5
N-Z-Tyr-Leu-NH_2		0.2 min	
N-Z-Tyr-Leu-CH(OH)-COOH	10.0	15 min	1.6

NOTE: The $t_{1/2}$ values were determined in human plasma, in the presence of α-chymotrypsin, and in buffer solution at 37 °C. The $t_{1/2}$ refers to the disappearance of either parent peptide (i.e., enzymatic cleavage of the peptide amide) or the derivatized peptide (i.e., conversion of the derivative to the parent peptide amide, which subsequently is hydrolyzed according to Scheme VII).

SOURCE: Reproduced with permission from reference 49. Copyright 1991.

approach stems from the ready conversion of the glyoxylic acid adduct to the parent peptides under physiological conditions of pH and temperature. Thus, at pH 7.4 and 37 °C, the $t_{1/2}$ values of the conversion of various peptidyl-α-hydroxyglycine derivatives ranged from 7 to 13 h (*50*). However, in several cases, this conversion is markedly accelerated in the presence of human plasma, most probably due to an α-hydroxyglycine amidating dealkylase enzyme (*50*).

N-Aminomethylation (i.e., N-Mannich base formation) also protects the peptide amides; for instance, the morpholine derivatives of *N*-Ac-Phe-NH_2 and *N*-Ac-Trp-NH_2 are stabilized by factors of 3 and 24, respectively (*49*). Using *N*-Ac-L-PheNH_2 as a model, neither *N*-acetylation, *N*-hydroxymethylation, nor *N*-phthalidylation is seen to afford any protection of cleavage of the terminal amide bond. In fact, such derivatization makes the amide bond more susceptible to α-chymotrypsin-catalyzed hydrolysis (*49*).

Stabilization of Tyrosyl Peptides

In 1963, Peterson et al. (*54*) reported that the facile α-chymotrypsin-catalyzed hydrolysis of *N*-acetyl-L-tyrosine methyl ester is diminished by several orders of magnitude when the tyrosine phenol group is replaced by a methoxy, ethoxy, or isoproxy group. Later, Kundu et al. (*55*) reported that the chymotryptic hydrolysis of *N*-acylated tyrosine esters could also be diminished by esterification of the phenolic group with acetic, propionic, or pivalic acid, the stabilizing effect being ascribed to steric factors hindering the interaction of the compounds with the enzyme. In 1988, Plattner et al. (*42*) showed that whereas the aromatic side chains of phenylalanine and tyrosine fit into the hydrophobic binding pocket of α-chymotrypsin, the *O*-methyl group in *O*-methoxytyrosine collides with the wall in the pocket. An ester group may act similarly, but esterification of the phenolic group, in contrast to alkylation, is bioreversible by the action of nonspecific esterases in plasma or liver.

Using *N*-Ac-Tyr-NH_2 as a α-chymotrypsin-sensitive model compound for tyrosyl peptides containing a C-terminal amide group, some means of esterification of the tyrosin phenol group were explored (*56*). The usefulness of this approach depends greatly on the

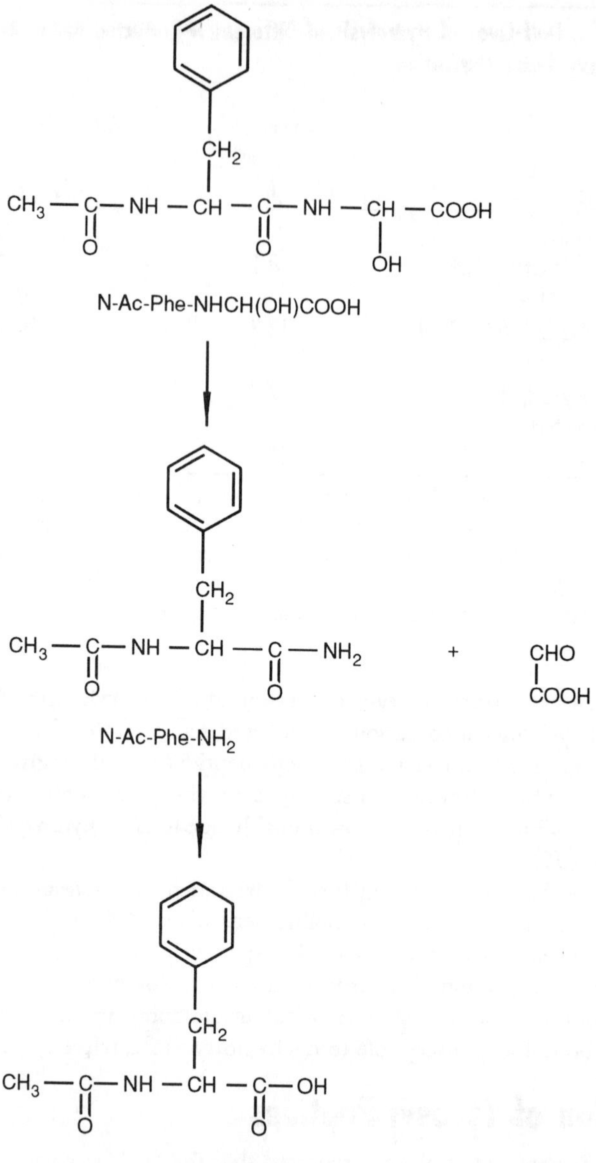

Scheme VII.

type of esterification performed because some esters are themselves readily attacked by the enzyme (Table IV). Inspection of the rate data in Table IV reveals a widely different pattern of reactivity of the OH-modified tyrosine peptide amides. It is readily apparent that esterification of the tyrosine phenol group can afford a large degree of protection against α-chymotrypsin-catalyzed hydrolysis of the C-terminal amide bond in the model compound *N*-Ac-L-Tyr-NH_2, but this is quite dependent on the chain length of the ester. The degradation of the ester derivatives proceeds via an initial and quantitative formation of *N*-Ac-L-Tyr-NH_2 which is subsequently hydrolyzed in a more rapid step to *N*-Ac-L-Tyr (Scheme VIII). Acetylation of the tyrosine phenolic group affords complete protection of the tyrosyl

Table IV. Half-Lives of Hydrolysis of Various Ester Derivatives of *N*-Acetyl-L-Tyrosine Amide in 0.1 M

Compound	*Buffer, pH 7.4 (h)*	*α-Chymotrypsin (0.50 mg mL⁻¹)*
N-Ac-Tyr-NH_2		72 min
Esters thereof:		
Acetate	40	4.8 h
Propionate	76	64 min
Isobutyrate	137	40 min
Pivalate	905	4.4 h
Valerate	136	41 s
Hexanoate	134	25 s
Benzoate	215	4.9 min
Methylcarbonate	50	2.7 h
Butylcarbonate	76	2.3 min
Glutarate	2.4	2.6 h
Ethylcarbamate	32	36 h

NOTE: Half-lives determined in 0.1 M phosphate buffer solution with or without α-chymotrypsin at 37 °C.
SOURCE: Reproduced with permission from reference 56. Copyright 1991.

amide bond against α-chymotrypsin. Although the *O*-acetylation makes the tyrosyl amide bond resistant to the enzyme, the esterolytic activity of α-chymotrypsin is responsible for the lower stability of the derivatives in the enzyme solutions as compared with pure buffer solutions. This cleavage of the protecting ester group becomes more predominant as the alkyl chain of the esters is enlarged. Besides being influenced by the chain length of the acyl group, the α-chymotrypsin-catalyzed hydrolysis of the protecting ester moiety is dependent on steric factors. Thus, the *O*-pivaloyl ester derivative is almost as stable as the *O*-acetyl ester (*56*). Because the protecting ester moiety itself may be a substrate for α-chymotrypsin, this prodrug concept is limited to certain types of esters. Such esters include aliphatic carboxylic or carbonate esters with short or branched side chains, a glutarate ester, and carbamate esters. Most of these esters are readily bioreversible, the conversion to the parent peptide being catalyzed by plasma or liver esterases (Scheme IX) (*56*).

Prodrugs of Desmopressin

Desmopressin [1-(3-mercaptopropanoic acid)-8-D-arginine vasopressin; dDAVP] (*see* structure 1) is a synthetic analogue of the antidiuretic hormone vasopressin used in the treatment of central diabetes insipidus and nocturnal enuresis and as a hemostatic agent (*57*). The drug is usually administered orally or intranasally. However, the bioavailability of the peptide is only 2–10% for the nasal route (*58–60*) and <1% for the oral route (*59*). The poor bioavailability of the peptide is due largely to its low lipophilicity (*61, 62*), but degradation of the peptide by enzymes in the gut lumen and intestinal and nasal mucosal tissues may also play a role in the low bioavailability (*63–65*). Desmopressin is degraded by α-chymotrypsin with a $t_{1/2}$ of 19 min (Table V). The mechanism of degradation has not been reported but is most likely to occur at the Tyr–Phe bond and the Phe–Gln bond due to the substrate specificity of α-chymotrypsin (*37, 49*).

Esterification of the tyrosine phenolic group in desmopressin to give more stable and lipophilic prodrug forms may be a useful approach to enhance its bioavailability (*66*). The rates of hydrolysis of the desmopressin esters are comparable to those of similar esters

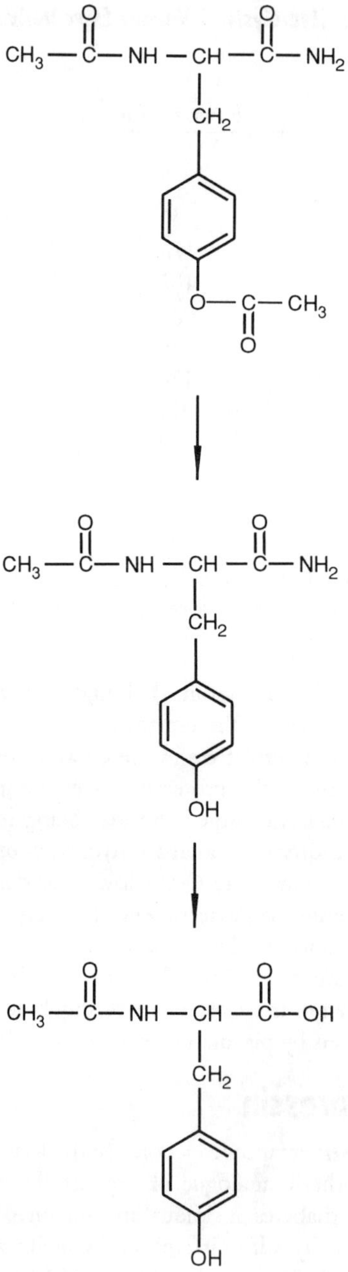

Scheme VIII. (Reproduced with permission from reference 49. Copyright 1991.)

of *N*-Ac-L-Tyr-NH_2, indicating no noticeable influence of the rest of the desmopressin molecule on the tyrosine ester stability. As illustrated in Figure 3 for the propionate ester, plasma enzymes quantitatively catalyze the ester hydrolysis. As is the case for chemical hydrolysis, the sterically hindered esters (i.e., pivalate and 2-ethyl-hexanoate) are the most stable compounds toward plasma-catalyzed hydrolysis.

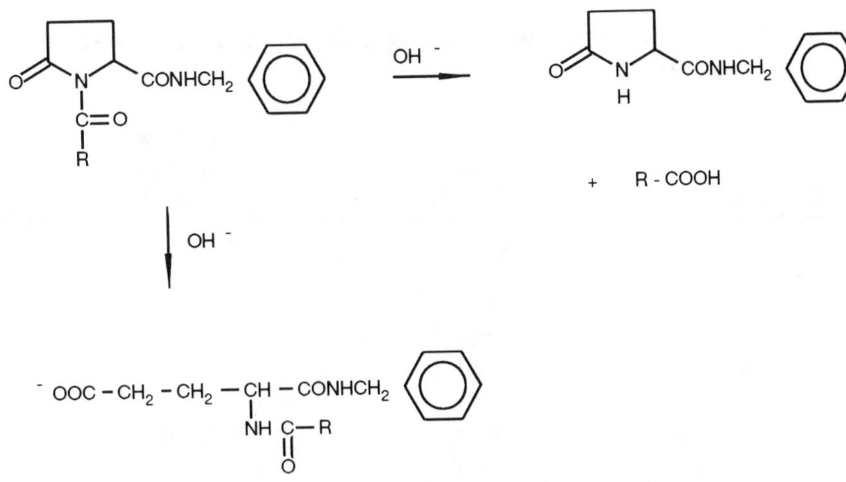

Scheme IX.

Apart from the pivalate ester, the desmopressin esters are more readily degraded by α-chymotrypsin than the parent peptide (Table V). The rapid degradation of the hexanoate and the octanoate ester, however, did not result from increased cleavage of the peptide bond but from cleavage of the tyrosine ester bond. Thus, incubation of hexanoate and octanoate esters with α-chymotrypsin resulted in quantitative release of desmopressin, which then degraded more slowly. The rate data obtained for the desmopressin esters qualitatively agree with the study on various esters of *N*-α-acylated tyrosine amides (*56*) in which aliphatic carboxylic acid and carbonate esters with a short or branched side chain were shown to be the most resistant esters toward hydrolysis by α-chymotrypsin.

Derivatization of the peptide produces prodrugs that are markedly more lipophilic than the parent desmopressin. This property may render the prodrug forms more capable of penetrating the biological membranes than the parent peptide. In fact, recent investigations (*66*) showed that all prodrug esters studied, except the isobutyl carbonate ester, showed increased permeability relative to the parent peptide across Caco-2 cell monolayers. However, no apparent correlation exists between the permeability and the lipophilicity of these

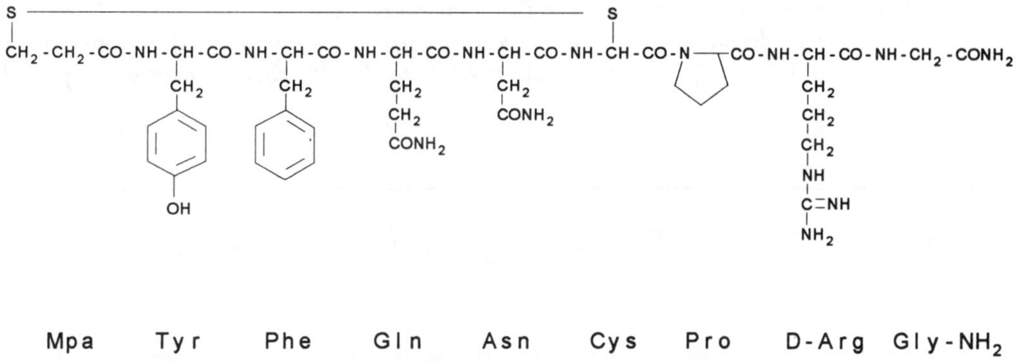

Structure 1. Desmopressin.

Table V. Half-Lives of Hydrolysis of Desmopressin and Various Ester Derivatives of the Tyrosyl Residue

Compound	Buffer, pH 7.40 (h)	Buffer (pH 7.40) with α-Chymotrypsin (0.50 mg mL^{-1})	80% Human Plasma
Desmopressin	Stable	19 min	>100 h
Esters thereof:			
Propionate	66 h	10 min	34 min
Pivalate	784 h	40 min	5.9 h
Hexanoate	190 h	6 s	19 min
Octanoate	165 h	<2 s	7 min
2-Ethyl-hexanoate	4460 h	2.0/15 min[a]	14.1 h
Isobutylcarbonate	115 h	3 min	8.2 min

NOTE: Half-lives determined in buffer solution, in the presence of α-chymotrypsin, and in 80% human plasma at 37 °C.
[a] Different reactivities of the stereoisomeric forms of the ester.
SOURCE: Reproduced with permission from reference 66. Copyright 1993.

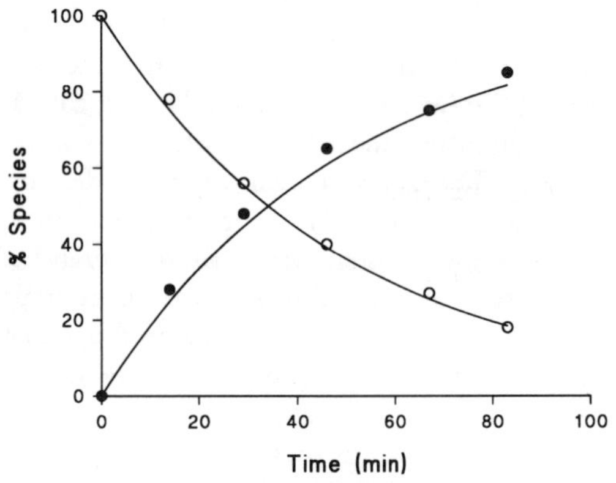

Figure 3. Plots showing the time courses of the degradation of O*-propionyl desmopressin (○) and formation of desmopressin (●) in 80% human plasma at 37 °C.*

derivatives. In contrast, transport of the desmopressin esters (*O*-pivaloyl- and *O*-octanoyl-) could not be detected across human skin in vitro (*67*) or across oral mucosa from pigs (Kahns, A. H.; Bundgaard, H., Royal Danish School of Pharmacy, Denmark, unpublished data).

Protection of Peptides against Pyroglutamyl Aminopeptidase

An N-terminal pyroglutamyl residue occurs in several peptides, such as thyrotropin-releasing hormone (TRH), luteinizing-hormone-releasing hormone (LHRH), neurotensin, bombesin, fibrinopeptides, and gastrin. The specific cleavage of this residue is affected by pyroglutamyl aminopeptidase (PAPase I; also called L-pyroglutamyl-peptide hydrolase, EC 3.4.11.8) (*68–73*) or, in the case of the degradation of TRH, by a TRH-specific pyroglutamyl aminopeptidase (PAPase II) as well (*74–78*). PAPase I is a cysteine protease that occurs in many different tissues, such as liver, kidney, and brain, but not in the blood (*70, 79*). It cleaves almost all *p*Glu-peptide bonds, including that in TRH (*69, 70, 77*). In contrast, PAPase II (EC 3.4.19), which is found predominantly in brain (*80–82*), and the TRH-specific serum enzyme [also termed thyroliberinase (*83*)] exhibit a high degree of substrate specificity (*74, 77, 81, 84, 85*). These enzymes exhibit similar physical and chemical characteristics and probably are identical or derived from the same gene (*74*). The enzyme catalyzes the hydrolysis of TRH but does not hydrolyze other *p*Glu-containing peptides (*74*).

Derivatives of the Pyroglutamyl Group

Using L-pyroglutamyl benzylamide (structure 2, compound **I**) as a model compound for a *p*Glu-peptide, a number of potential prodrug derivatives of the pyroglutamyl group were studied to protect the *p*Glu-peptide bond against enzymatic cleavage by PAPase I (*86–88*). The derivatives studied include various *N*-acyl derivatives, N-Mannich bases, and *N*-α-hydroxyalkyl derivatives derived from glyoxylic acid and esters thereof. The stability of these compounds was determined at 37 °C in aqueous buffer solutions, in human plasma, and in the presence of PAPase I (from calf liver). Whereas the parent pyroglutamyl benzylamide was rapidly hydrolyzed to pyroglutamic acid in the presence of PAPase ($t_{1/2}$ = 10 min), none of the *N*-substituted derivatives was attacked by the enzyme under identical reaction conditions. This is apparent from the observation that the rates of hydrolysis of the derivatives in the presence of the enzyme were quite similar to those occurring in buffer solutions (Table VI). The derivatives tested comprise varying chemical structures and they all were totally resistant to cleavage by the enzyme, so the results might indicate that any replacement of the hydrogen in the cyclic amide moiety of the pyroglutamyl group would lead to derivatives not tolerated by PAPase I. To be a useful prodrug approach, the derivatization performed should be readily bioreversible in addition to affording protection against the specific cleavage by PAPase. Most of the derivatives fulfill this requirement, the exceptions being

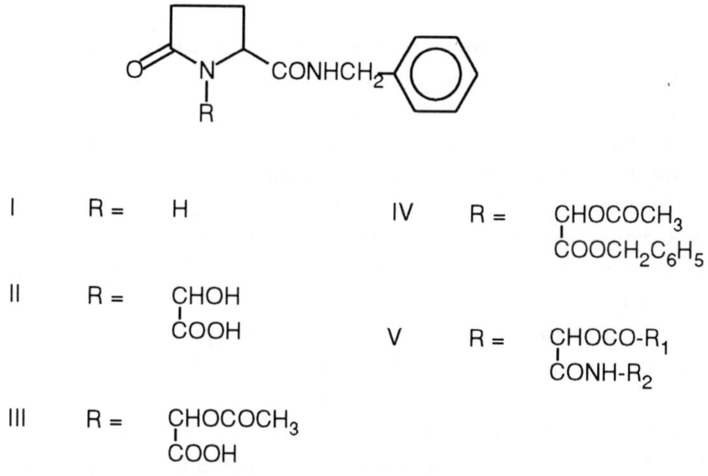

Structure 2. Derivatives of L-pyroglutamyl benzylamide. (Reproduced with permission from reference 56. Copyright 1991.)

Table VI. Rate Data for the Hydrolysis of Various Derivatives of L-Pyroglutamyl Benzylamide

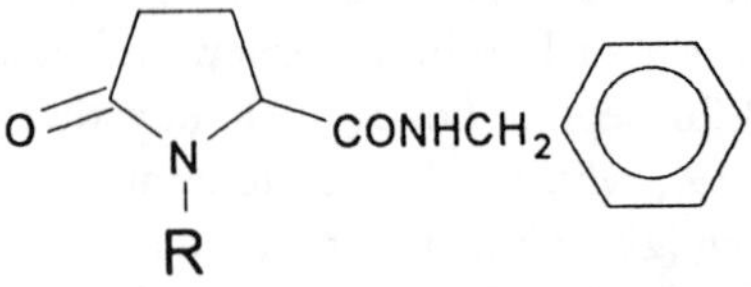

	$t_{1/2}$		
-R	*Buffer, pH 7.4*	*PAPase I[a]*	*Human Plasma*
-H	Stable	10 min	Stable
$-COCH_3$ (**1**)	18.0 h	19.7 h	14.0 h
$-COCH_2CH_3$ (**2**)	28.9 h	28.1 h	4.5 h
$-COCH_2CH_2CH_3$ (**3**)	39.3 h	37.2 h	7.8 h
$-COCH_2Cl$ (**4**)	14 min	15 min	8 min
$-COC_6H_5$ (**5**)	21.5 h	20.1 h	4.3 h
$-CH_2NHCH_3$ (**6**)	41 min	44 min	43 min
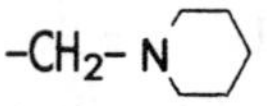 (**7**)	2 min	2 min	2 min
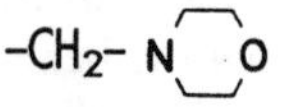 (**8**)	4.7 h	4.8 h	5.0 h
-CHOH COOH	51 min	62 min	32 min
-CHOH $COOCH_2C_6H_5$	6 min	6 min	[b]
$-CHOCOCH_3$ COOH	133 h	143 h	121 h
$-CHOCOCH_3$ $COOCH_2C_6H_5$	64 min	61 min	12 min

NOTE: Determined at 37 °C (R = H).

[a] Half-lives for the degradation in buffer solution (pH 7.4) containing PAPase I at a concentration fo 0.01 U mL^{-1}.

[b] Not determined.

SOURCE: Adapted from data in references 86–88.

the *N*-acyl derivatives (compounds **1–5** in Table VI) which are only partly (55–75%) converted to the parent compound in human plasma (*86*). The *N*-propionyl derivative (compound **2**, Table VI), however, was quantitatively cleaved to pyroglutamyl benzylamide in plasma by a plasma enzyme-catalyzed process. The degradation routes for the *N*-acyl derivatives involve hydrolytic ring opening and hydrolysis of the *N*-acyl moiety as depicted in Scheme X. At pH 7.4 these reactions occur simultaneously and to about the same extent. Plasma enzymes predominantly catalyze the cleavage of the external *N*-acyl moiety and for the propionyl derivative, a quantitative conversion to the parent pyroglutamyl benzylamide was observed (*86*).

The hydrolysis of the N-Mannich bases (compounds **6–8**, Table VI) proceeds spontaneously in aqueous solution, with no influence of plasma on the rate of hydrolysis. The compounds are cleaved quantitatively to the parent pyroglutamyl benzylamide, formaldehyde, and the corresponding amine. The stability of N-Mannich bases is strongly influenced by the steric effect and basicity of the amine component (*11, 35*), and it is therefore readily feasible to select N-Mannich bases of pyroglutamyl compounds with widely different chemical stabilities and, hence, rates of conversion at physiological pH.

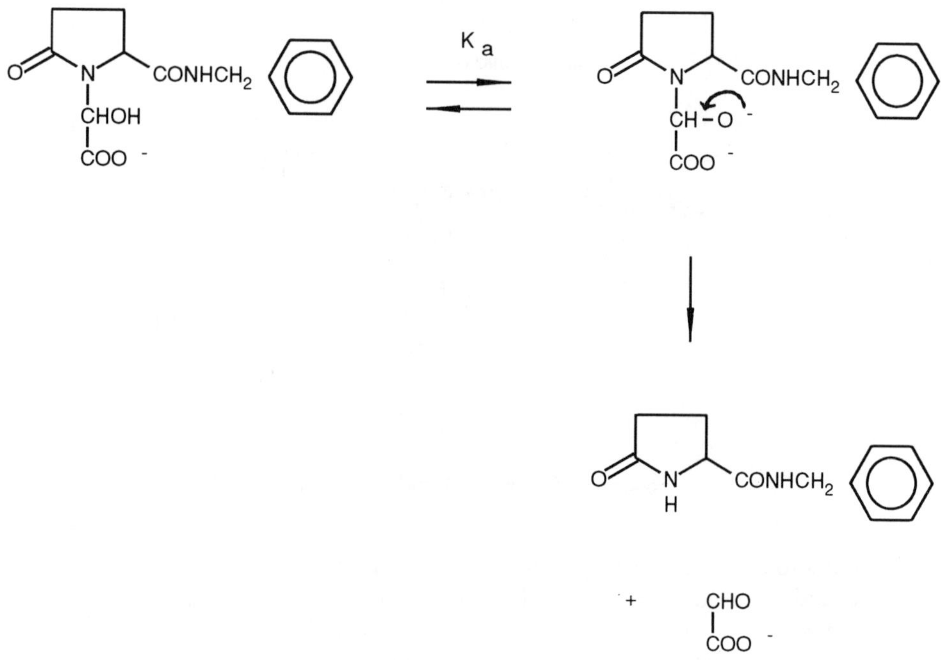

Scheme X.

The *N*-α-hydroxyalkyl derivative (structure 2, compound **II**), formed by reacting pyroglutamyl benzylamide with glyoxylic acid, is also spontaneously hydrolyzed in aqueous solution with a $t_{1/2}$ of hydrolysis of 51 min at pH 7.4 and 37 °C (*87*). Blocking the hydroxyl group by esterification [e.g., by acetylation (structure 2, compound **III**)], greatly stabilizes the compound. The decomposition of the hydroxyalkyl derivative (structure 2, compound **II**) in neutral and alkaline solutions most likely involves a stepwise pathway with anionic *N*-α-hydroxyalkyl amide as an intermediate undergoing rate-determining N–C bond cleavage (Scheme X) (*51, 87*). By blocking the hydroxyl group, the decomposition along the pathway shown can no longer take place. Therefore, the degradation of compound **III** (structure 2) requires a hydrolytic cleavage of its ester group prior to undergoing this decomposition. The hydrolysis of this ester moiety is not significantly catalyzed by plasma. This may be due to the negative charge of the compound at pH 7.4, because several other esters containing an ionized carboxylate group are also only slowly hydrolyzed in the presence of plasma (*89*). The relatively poor stability of the diester (structure 2, compound **IV**) in aqueous solution was shown to be due to a facile hydrolysis of its benzyl ester moiety resulting in the formation of compound **III.** As previously suggested (*87*), the glyoxylic acid derivative, in which a high solution stability is combined with a high lability in vivo, may be a derivative in which an amide is used in conjunction with *O*-acylation (structure 2, compound **V**). The studies (*87*) with the pyroglutamyl peptide and its derivatives demonstrate clearly the utility of the prodrug approach to protect a peptide against specific enzymatic cleavage.

Prodrug Derivatives of TRH

TRH is the hypothalamic peptide that regulates the synthesis and secretion of thyrotropin from the anterior pituitary gland. Since its discovery in 1969, TRH has been shown to have not only a variety of endocrine- and central nervous system (CNS)- related biological activity,

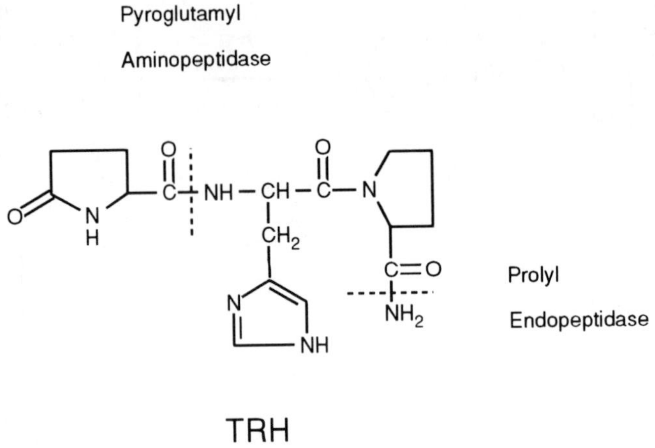

Figure 4. Enzymatic degradation pathways of TRH.

but also potential as a drug in the management of various neurologic and neuropsychiatric disorders including depression, brain injury, acute spinal trauma, Alzheimer's disease, and schizophrenia (*90–97*). The clinical utilization of the neuropharmacological properties of TRH is, however, greatly hampered by its rapid metabolism and clearance as well as by its poor access to the CNS (*91, 94, 96, 98*). Following parenteral administration in humans, TRH has a plasma $t_{1/2}$ of only 6–8 min (*99–101*) mainly because of rapid enzymatic degradation of the peptide in the blood, in particular by the so-called TRH-specific serum enzyme PAPase II (*74–77, 102*). This enzyme catalyzes the hydrolysis of TRH at the *p*Glu–His bond (Figure 4), yielding pyroglutamic acid and His-Pro-NH_2. The lipophilicity of TRH is very low (*88*) and this low lipophilicity may be a primary reason for the limited ability of the peptide to penetrate the blood–brain barrier (*103, 104*).

Because the degradation of TRH occurs mainly at the *p*Glu–His bond, derivatization of the peptide in the proximity of this bond will provide enzymatic resistance because the molecule no longer fits into the enzyme. Such derivatization may also improve the lipophilicity and penetration characteristics and thereby solve some delivery problems of the peptide. In 1990, the prodrug approach was applied to TRH by derivatization of the imidazole group of the histidine residue (*105*). By reacting TRH with various chloroformates, a number of *N*-alkoxycarbonyl derivatives of TRH were obtained. The compounds are hydrolyzed quantitatively to TRH in aqueous solution (Scheme XI), the $t_{1/2}$ values ranging from 9 to 37 h at pH 7.4 and 37 °C (Table VII).

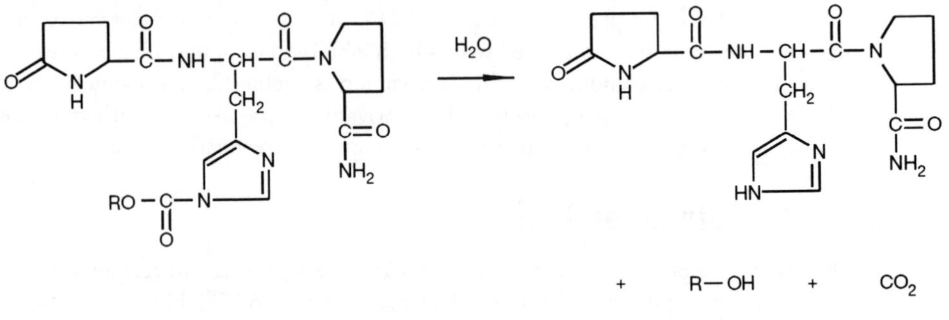

Scheme XI.

Table VII. Rate Data for the Hydrolysis and Lipophilicity of TRH and Its Various *N*-Alkoxycarbonyl Prodrug Derivatives of the Imidazole Group

	$t_{1/2}$ (37 °C)			
Compound	*Buffer, pH 7.4 (h)*	*Human Plasma*	*PAPase I*[a] *(min)*	*log P*[b]
TRH		9.4 min[c]	4	−2.46
R				
CH_3	9.3	2.8 h	12	−1.88
C_2H_5	14.2	3.8 h	23	−1.30
i-C_3H_7	35.8	6.6 h	29	−0.80
C_4H_9	19.0	4.3 h	11	−0.47
C_6H_{13}	17.9	1.2 h	8	0.71
C_8H_{17}	17.5	0.4 h	24	1.88
Cyclo-C_6H_{11}	36.8	6.4 h	22	0.60

[a] Half-lives of degradation in buffer solution containing calf liver PAPase (0.01 U mL^{-1}).
[b] Partition coefficients between octanol and phosphate buffer (pH 7.40, 21 °C).
[c] Half-life of hydrolysis at a concentration $<5 \times 10^{-6}$ M.
SOURCE: Adapted from data in reference 105.

The *N*-alkoxycarbonyl derivatives are fairly rapidly hydrolyzed in the presence of PAPase I, but they are more stable than TRH itself (Table VII). The enzymatic reaction taking place is cleavage of the pyroglutamyl bond in both TRH and the derivatives. The relatively low degree of protection against PAPase I achieved by modification of the imidazole group of TRH is not unexpected because, as noted before, the specificity of this enzyme encompasses almost all pyroglutamyl-containing peptides. As previously mentioned, the degradation of TRH in human plasma or blood is entirely due to hydrolytic cleavage of its *p*Glu–His bond by virtue of the TRH-specific serum enzyme (PAPase II). This reaction displays classical Michaelis–Menten kinetics, the Michaelis constant (K_m) and maximum velocity (V_{max}) values being 1.9×10^{-5} M and 1.4×10^{-6} M min^{-1}, respectively (*102*). At high TRH concentrations (greater than K_m), the rate of hydrolysis of the peptide followed zero-order kinetics, with a rate constant of 1.4 μmol min^{-1}. In contrast, at a low substrate concentration (less than K_m), the enzymatic reaction is first-order, with a rate constant of 0.074 min^{-1}, which corresponds to a $t_{1/2}$ of 9.4 min (*102*). The stability of the *N*-alkoxycarbonyl derivatives in human plasma solutions at 37 °C differed greatly from that of TRH. The kinetics of degradation of the derivatives followed first-order kinetics at substrate concentrations even up to 10^{-3} M, and the derivatives showed a markedly higher stability than TRH when the comparison is made at conditions where the degradation of TRH also occurs according to first-order kinetics. As can be seen from the rate data in Table VII, the $t_{1/2}$ values of degradation range from 0.4 to 6.6 h, whereas TRH degrades with a $t_{1/2}$ of 9.4 min. Comparison of the plasma and buffer rate data for the derivatives reveals a significant plasma-catalyzed hydrolysis.

For the consideration of the derivatives as prodrug forms of TRH, a crucial point is the derivatives degradation pathway in the presence of plasma. The plasma-catalyzed degradation observed may be due to a catalyzed hydrolysis of the acyl bond at the imidazole moiety with liberation of TRH or it may be due to other degradation processes, such as hydrolysis of the *p*Glu–His bond. Ideally, the compounds should be hydrolyzed in plasma at the imidazole-protecting moiety and should release TRH in quantitative amounts. Kinetic analysis of concentration–time profiles for TRH demonstrated that for all derivatives studied, the sole or predominant (i.e., >95%) reaction taking place in human plasma is hydrolysis at the imidazole carbamate moiety to yield TRH. These experiments (*105*) demonstrate that *N*-acylation of the imidazole moiety of TRH to give *N*-alkoxycarbonyl derivatives is a promising prodrug approach to make the *p*Glu–His bond resistant to cleavage by the TRH-specific PAPase serum enzyme. The derivatives are totally resistant to cleavage by this enzyme, but are readily bioreversible as the parent TRH is formed quantitatively from the derivatives by spontaneous or plasma esterase-catalyzed hydrolysis.

In addition to being potentially useful to prolong the duration of action of TRH in vivo, the *N*-alkoxycarbonyl prodrug derivatives possess greatly increased lipophilicity relative to TRH as assessed by octanol–buffer partition coefficients (Table VII). This property may render the prodrug forms more capable of penetrating the blood–brain barrier or various other biomembranes than the parent peptide. In fact, recent experiments (*106*) showed that it may be possible to achieve transdermal delivery of TRH by using a lipophilic TRH prodrug such as *N*-octyloxycarbonyl-TRH (R = C_8H_{17}).

The feasibility of utilizing the TRH prodrugs to improve the oral delivery characteristics of TRH has also been evaluated (*107*). The bioavailability of TRH after oral administration is only ~1–2% (*100, 108*) probably because of the very low lipophilicity of TRH (Table VII). Enzymatic degradation of the peptide at the absorption site may also contribute to the poor bioavailability, although Yokohama et al. (*109*) reported that TRH is stable in the presence of gastrointestinal proteolytic enzymes and in rat intestinal homogenates. In vitro penetration studies (*107*) using the modified Ussing chamber showed that the *N*-alkoxycarbonyl derivative do not improve the penetration of TRH across various intestinal segments of the rat or albino rabbit. This lack of penetration may be due to a very facile degradation of the compounds affected by prolyl endopeptidase (*110*). This enzyme cleaves the C-terminal prolineamide residue both in the prodrugs and in TRH (Scheme XI) (*107*). In the TRH cleavage, the acid TRH-OH (*p*Glu-His-Pro) is formed.

Although the higher lipophilicities of the prodrugs should lead to enhanced permeability, the greater susceptibility of the prodrugs to undergo enzymatic degradation than TRH apparently more than offsets the improved lipophilicity characteristics. Therefore, prodrugs suitable for improving the oral absorption of TRH should not only possess a certain lipophilicity but also should be resistant to the prolyl endopeptidase enzyme. This resistance is not fulfilled by the *N*-alkoxycarbonyl derivatives. More suitable prodrug derivatives may be compounds in which the susceptible prolineamide moiety of TRH is derivatized, eventually in combination with the modification described at the imidazole function.

Conclusions

Many different strategies were employed to overcome the enzymatic and penetration barriers associated with the delivery of peptide and protein drugs, such as synthesis of analogues, use of protease inhibitors and absorption enhancers and the development of new dosage forms (*3, 111–113*). Among these various approaches, the prodrug technology has been mostly overlooked in the past, except for the more traditional drugs of the peptide- or amino acid-type, like amino-containing β-lactam antibiotics and L-DOPA. The prodrug approach may, however, be a highly useful means of improving the absorption of peptides. With this bioreversible derivatization technique, it is readily feasible to obtain derivatives with increased lipophilicity and, in some cases, with metabolic stability. The prodrug technique may find its greatest use for peptide drugs containing not more than 7–10 amino acids, and a large number of future peptide drugs will certainly fall in this range. For example, the various renin inhibitors presently under development (*114*) belong in this category. The use of prodrugs for improving the delivery characteristics of larger molecules such as proteins should, however, not be overlooked. In this regard, the increasing attention being paid to acyl enzymes as prodrugs should be mentioned (*115*). Whereas knowledge of the design of prodrug derivatives for various functional groups occurring in peptides and proteins has increased greatly in recent years, more work is needed to enable rational use of this knowledge in the design of prodrugs capable of protecting a peptide bond against enzymatic cleavage and hence peptide inactivation. Thus, several peptides are rapidly cleaved by the gastrointestinal enzymes α-chymotrypsin, CPA, and trypsin. The most efficient way to protect the vulnerable peptide bonds against proteolytic cleavage by the prodrug approach may be derivatizing the peptide bond itself. However, the possibility of obtaining the same results by bioreversible modification of the peptide elsewhere in the molecule should also be considered. Thus, it can be imagined that a pentapeptide with a C-terminal amide group (which is often seen for endogenous peptide hormones), cleavable by the pancreatic enzymes at, for example, the bond between the second and third amino acid, can be stabilized against these enzymes just by modifying the C-terminal amide function in a bioreversible manner. Although great knowledge exists about the substrate specificities of the various proteolytic enzymes (*116, 117*) more work is needed to take advantage of this feature of the enzymes in making proteolytic enzyme-resistant prodrugs of peptides.

References

1. Albert, A. *Nature (London)* **1958,** *182,* 421–423.
2. Humphrey, M. J.; Ringrose, P. S. *Drug Metab. Rev.* **1986,** *17,* 283–310.
3. Lee, V. H. L.; Yamamoto, A. *Adv. Drug Del. Rev.* **1990,** *4,* 171–207.
4. Meisenberg, G.; Simmons, W. H. *Life Sci.* **1983,** *32,* 2611–2623.
5. Farmer, P. S. In *Drug Design;* Ariëns, E. J., Ed.; Academic: London, 1980; Vol. X, pp 119–143.
6. Sandberg, B. E. B.; Lee, C.-M.; Hanley, M. R.; Iversen, L. L. *Eur. J. Biochem.* **1981,** *114,* 329–337.
7. Thaisrivongs, S.; Pals, D. T.; Harris, D. W.; Kati, W. M.; Turner, S. R. *J. Med. Chem.* **1986,** *29,* 2088–2093.
8. Veber, D. F.; Freidinger, R. M. *Trends Neurosci.* **1985,** *8,* 392–396.
9. Fauchère, J.-L. *Adv. Drug Res.* **1986,** *15,* 29–69.
10. Bundgaard, H. In *Delivery Systems for Peptide Drugs;* Davis, S. S.; Illum L.; Tomlinson, E., Eds.; Plenum: New York, 1986; pp 49–68.
11. Bundgaard, H. In *Design of Prodrugs;* Bundgaard, H., Ed.; Elsevier: Amsterdam, The Netherlands, 1985; pp 1–91.
12. Bundgaard, H. In *Bioreversible Carriers in Drug Design. Theory and Application;* Roche, E. B., Ed.; Pergamon: New York, 1987; pp 13–94.
13. Bundgaard, H. *Adv. Drug Del. Rev.* **1989,** *3,* 39–65.
14. Bundgaard, H. In *A Textbook of Drug Design and Development;* Krogsgaard-Larsen, P.; Bundgaard, H., Eds.; Harwood Academic: London, 1991; pp 113–191.
15. Bundgaard, H. *Adv. Drug Del. Rev.* **1992,** *8,* 1–38.

16. Bundgaard, H. *J. Controlled Rel.* **1992,** *21,* 63–72.

17. Smith, E. L. *Adv. Enzymol.* **1951,** *12,* 191–257.

18. Hartsuck, J. A.; Lipscomb, W. N. In *The Enzymes;* Boyer, P. D., Ed.; Academic: Orlando, FL, 1971; pp 1–56.

19. Stahmann, M. A.; Fruton, J. S.; Bergmann, M. *J. Biol. Chem.* **1946,** *164,* 753–760.

20. Buur, A.; Bundgaard, H. *Int. J. Pharm.* **1988,** *46,* 159–167.

21. Bundgaard, H.; Nielsen, N. M. *Acta Pharm. Suec.* **1988,** *24,* 233–246.

22. Johansen, M.; Bundgaard, H. *Arch. Pharm. Chem. Sci. Ed.* **1979,** *7,* 175–192.

23. Bundgaard, H.; Johansen, M. *Int. J. Pharm.* **1980,** *5,* 67–77.

24. Bundgaard, H.; Johansen, M. *Int. J. Pharm.* **1984,** *22,* 45–56.

25. Bundgaard, H.; Rasmussen, G. J. *Pharm. Res.* **1991,** *8,* 313–322.

26. Møss, J.; Bundgaard, H. *Int. J. Pharm.* **1989,** *52,* 255–263.

27. Hanson, H. T.; Smith, E. L. *J. Biol. Chem.* **1948,** *175,* 833–848.

28. Snoke, J. E.; Neurath, H. *J. Biol. Chem.* **1949,** *181,* 789–902.

29. Bundgaard, H.; Rasmussen, G. J. *Pharm. Res.* **1991,** *8,* 1238–1242.

30. Delange, R. T.; Smith, E. L. In *The Enzymes;* Boyer, P. D., Ed.; Academic: Orlando, FL, 1971; Vol. III, pp 81–118.

31. Hardy, P. M.; Samworth, D. J. *J. Chem. Soc. Perkin Trans. 1* **1977,** 1954–1960.

32. Klixbüll, U.; Bundgaard, H. *Int. J. Pharm.* **1984,** *20,* 273–284.

33. Rasmussen, G. J.; Bundgaard, H. *Int. J. Pharm.* **1991,** *71,* 45–53.

34. Rasmussen, G. J.; Bundgaard, H. *Int. J. Pharm.* **1991,** *76,* 113–122.

35. Bundgaard, H.; Johansen, M. *J. Pharm. Sci.* **1980,** *69,* 44–46.

36. Bundgaard, H.; Johansen, M. *Arch. Pharm. Chem. Sci. Ed.* **1980,** *8,* 29–52.

37. Hess, G. P. In *The Enzymes;* Boyer, P. D., Ed.; Academic: Orlando, FL, 1971; Vol. III, pp 213–248.

38. Bender, M. L.; Kilheffer, J. V. *Crit. Rev. Biochem.* **1973,** *1,* 149–199.

39. Wiedhaup, K. In *Topics in Pharmaceutical Sciences;* Breimer, D. D.; Speiser, D., Eds.; Elsevier: Amsterdam, The Netherlands, 1981; pp 307–324.

40. Barth, T.; Pliska, V.; Rychlik, I. *Coll. Czech. Chem. Comm.* **1967,** *2,* 1058–1063.

41. Kahns, A. H.; Bundgaard, H. *Int. J. Pharm.* **1991,** *77,* 65–70.

42. Plattner, J. J.; Marcotte, P. A.; Kleinert, H. D.; Stein, H. H.; Greer, J.; Bolis, G.; Fung, A. K. L.; Bopp, B. A.; Luly, J. R.; Sham, H. L.; Kempf, D. J.; Rosenberg, S. H.; Dellaria, J. F.; De, B.; Merits, I.; Perun, T. J. *J. Med. Chem.* **1988,** *31,* 2277–2288.

43. Dockter, G.; Hoppe-Seyler, F.; Appel, W.; Sitzmann, F.-C. *Clin. Biochem.* **1986,** *19,* 329–332.

44. Goldberg, D. M.; Cambell, R.; Roy, A. D. *Biochim. Biophys. Acta* **1968,** *167,* 613–615.

45. Blow, D. M. In *The Enzymes;* Boyer, P. D., Ed.; Academic: Orlando, FL, 1971; Vol. III, pp 185–212.

46. Kahns, A. H.; Friis, G. J.; Bundgaard, H. *Biomed. Chem. Letts.* **1993,** *3,* 809–812.

47. Bradbury, A. F.; Smyth, D. G. *Biosci. Rep.* **1987,** *7,* 907–916.

48. Eipper, B. A.; Mains, R. E. *Ann. Rev. Physiol.* **1988,** *50,* 333–344.

49. Kahns, A. H.; Bundgaard, H. *Pharm. Res.* **1991,** *8,* 1533–1538.

50. Bundgaard, H.; Kahns, A. H. *Peptides* **1991,** *12,* 745–748.

51. Bundgaard, H.; Buur, A. *Int. J. Pharm.* **1987,** *37,* 185–194.

52. Kahns, A. H.; Bundgaard, H. *Int. J. Pharm.* **1991,** *71,* 31–43.

53. Bundgaard, H.; Buur, A.; Hansen, K. T.; Larsen, J. D.; Møss, J.; Olsen, L. *Int. J. Pharm.* **1988**, *45,* 47–57.

54. Peterson, R. L.; Hubele, K. W.; Niemann, C. *Biochemistry* **1963**, *2,* 942–946.

55. Kundu, N.; Roy, S.; Maenza, F. *Eur. J. Biochem.* **1972**, *28,* 311–315.

56. Kahns, A. H.; Bundgaard, H. *Int. J. Pharm.* **1991**, *76,* 99–112.

57. Vilhardt, H. *Drug Invest.* **1990**, *2 (Suppl. 5),* 2–8.

58. Harris, A. S.; Nielsson, I. M.; Wagner, Z. G.; Alkner, U. *J. Pharm. Sci.* **1986**, *75,* 1085–1088.

59. Vilhardt, H.; Lundin, S. *Gen. Pharmacol.* **1986**, *17,* 481–483.

60. Köhler, M.; Harris, A. *Eur. J. Clin. Pharmacol.* **1988**, *35,* 281–285.

61. Lundin, S.; Artursson, P. *Int. J. Pharm.* **1990**, *64,* 181–186.

62. Lundin, S.; Pantzar, N.; Broeders, A.; Ohlin, M.; Weström, B. R. *Pharm. Res.* **1991**, *8,* 1274–1280.

63. Saffran, M.; Bedra, C.; Kumar, G. S.; Neckers, D. C. *J. Pharm. Sci.* **1988**, *77,* 33–38.

64. Lundin, S.; Bengtsson, H.-I.; Folkesson, H. G.; Weström, B. R. *Pharmacol. Toxicol.* **1989**, *65,* 92–95.

65. Morimoto, K.; Yamaguchi, H.; Iwakura, Y.; Miyazaki, M.; Nakatani, E.; Iwamoto, T.; Okashi, Y.; Nakai, Y. *Pharm. Res.* **1991**, *8,* 1175–1179.

66. Kahns, A. H.; Buur, A.; Bundgaard, H. *Pharm. Res.* **1993**, *10,* 68–74.

67. Kahns, A. H.; Møss, J.; Bundgaard, H. *Acta Pharm. Nord.* **1992**, *4,* 187–188.

68. Doolittle, R. F.; Armentrout, R. W. *Biochemistry* **1968**, *7,* 516–521.

69. Orlowski, M.; Meister, A. In *The Enzymes;* Boyer, P. D., Ed.; Academic: Orlando, FL, 1971; Vol. IV, pp 123–151.

70. Abraham, G. N.; Podell, D. N. *Mol. Cell. Biochem.* **1981**, *38,* 181–190.

71. Fujiwara, K.; Kobaysashi, R.; Tsuru, D. *Biochim. Biophys. Acta* **1979**, *570,* 140–148.

72. Browne, P.; O'Cuinn, G. *Eur. J. Biochem.* **1983**, *137,* 75–87.

73. Griffiths, E. C.; McDermott, J. R. *Mol. Cell. Endocrinol.* **1983**, *33,* 1–25.

74. Bauer, K. *Biochimie* **1988**, *70,* 69–74.

75. Wilk, S.; Suen, C.-S.; Wilk, E. K. *Neuropeptides* **1988**, *12,* 43–47.

76. Bauer, K.; Nowak, P. *Eur. J. Biochem.* **1979**, *99,* 239–246.

77. Wilk, S. *Ann. N. Y. Acad. Sci.* **1989**, *553,* 252–264.

78. Møss, J.; Bundgaard, H. *Pharm. Res.* **1990**, *7,* 751–755.

79. Szewczuk, A.; Kwiatkowska, J. *Eur. J. Biochem.* **1970**, *15,* 92–96.

80. Friedman, T. C.; Wilk, S. *J. Neurochem.* **1986**, *46,* 1231–1239.

81. Wilk, S.; Wilk, E. K. *Ann. N. Y. Acad. Sci.* **1989**, *553,* 556–558.

82. Wilk, S. *Life Sci.* **1986**, *39,* 1487–1492.

83. Bauer, K.; Nowak, P.; Kleinkauf, H. *Eur. J. Biochem.* **1981**, *118,* 173–176.

84. Emerson, C. H. *Methods Enzymol.* **1989**, *168,* 365–371.

85. Lanzara, R.; Liebman, M.; Wilk, S. *Ann. N. Y. Acad. Sci.* **1989**, *553,* 559–562.

86. Bundgaard, H.; Møss, J. *J. Pharm. Sci.* **1989**, *78,* 122–126.

87. Møss, J.; Bundgaard H. *Int. J. Pharm.* **1989**, *52,* 255–263.

88. Bundgaard, H.; Møss, J. *Trans. Biochem. Soc.* **1989**, *17,* 947–949.

89. Nielsen, N. M.; Bundgaard, H. *Int. J. Pharm.* **1987**, *39,* 75–85.

90. Jackson, I. M. D. *N. Engl. J. Med.* **1982**, *306,* 145–155.

91. Metcalf, G. *Brain. Res.* **1982**, *4,* 389–408.

92. Griffiths, E. C. *Psychoneuroendocrinology* **1985,** *10,* 225–235.
93. Griffiths, E. C. *Nature (London)* **1986,** *322,* 212–213.
94. Griffiths, E. C. *Clin. Sci.* **1987,** *73,* 449–457.
95. Horita, A.; Carino, M. A.; Lai, H. *Annu. Rev. Pharmacol. Toxicol.* **1986,** *26,* 311–332.
96. Loosen, P. T. *Progr. Neuropsychopharmacol. Biol. Psychiatry* **1988,** *12,* S87–S117.
97. Metcalf, G.; Jackson, I. M. *Ann. N.Y. Acad. Sci.* **1989,** *553,* 1–631.
98. Hichens, M. *Drug Metab. Rev.* **1983,** *14,* 77–98.
99. Bassiri, R. M.; Utiger, R. D. *J. Clin. Invest.* **1973,** *52,* 1616–1619.
100. Duntas, L.; Keck, F. S.; Pfeiffer, E. F. *Dtsch. Med. Wochenschr.* **1988,** *113,* 1354–1357.
101. Iversen, E. J. *Endocrinology* **1988,** *118,* 511–516.
102. Møss, J.; Bundgaard, H. *Pharm. Res.* **1990,** *7,* 751–755.
103. Nagai, Y.; Yokohama, S.; Nagawa, Y.; Hirooka, Y.; Nihei, N. *J. Pharm. Dyn.* **1980,** *3,* 500–506.
104. Banks, W. A.; Kastin, A. J. *Brain Res. Bull.* **1985,** *15,* 287–292.
105. Bundgaard, H.; Møss, J. *Pharm. Res.* **1990,** *7,* 885–892.
106. Møss, J.; Bundgaard, H. *Int. J. Pharm.* **1990,** *66,* 39–45.
107. Møss, J.; Buur, A.; Bundgaard, H. *Int. J. Pharm.* **1990,** *66,* 39–45.
108. Yokohama, S.; Yamashita, K.; Toguche, H.; Takeuchi, J.; Kitamori, N. *J. Pharm. Dyn.* **1984,** *7,* 101–111.
109. Yokohama, S.; Yoshioka, T.; Yamashita, K.; Kitamori, N. *J. Pharm. Dyn.* **1984,** *7,* 445–451.
110. Wilk, S. *Life Sci.* **1983,** *33,* 2149–2157.
111. Lee, V. H. L. *CRC Crit. Rev. Ther. Drug Carrier Syst.* **1988,** *5,* 69–97.
112. Banga, A. K.; Chien, Y. W. *Int. J. Pharm.* **1988,** *48,* 15–50.
113. Verhoef, J. C.; Boddé, H. E.; de Boer, A. G.; Bouwstra, J. A.; Junginger, H. E.; Merkus, F. W. H. M.; Breimer, D. D. *Eur. J. Drug Metab. Pharmacokinet.* **1990,** *15,* 83–93.
114. Greenlee, W. J. *Med. Res. Rev.* **1990,** *10,* 173–236.
115. Markwardt, F. *Pharmazie* **1989,** *44,* 521–526.
116. *The Enzymes;* Boyer, P. D., Ed.; Academic: Orlando, FL, 1971; Vol. III.
117. Banerjee, P. K.; Amidon, G. L. In *Design of Prodrugs;* Bundgaard, H., Ed.; Elsevier: Amsterdam, The Netherlands, 1985; pp 93–133.

RECEIVED for review July 12, 1993. ACCEPTED revised manuscript April 29, 1994.

CHAPTER 19

Improved Duration of Action of Peptide Drugs

John J. Nestor, Jr.

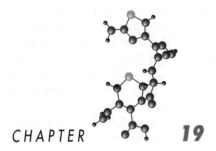

Although peptides play critical roles throughout the body in endocrine, paracrine, and autocrine functions, they are very underrepresented as drugs. Despite the very high potency and specificity of peptide hormones, as pharmaceutical agents they suffer from the shortcomings of a short duration of action and a very low oral bioavailability. Much work over the past 20 years has focused on attempts to overcome these critical flaws (*1*). Approaches to improved oral bioavailability have centered on novel formulations (*2*) or active uptake systems, such as peptide linkage to vitamin B_{12} (*3*), and to date have been only partially successful. Attempts to affect the duration of action of peptides have usually been directed at stabilization of the peptide bond (*4, 5*). Approaches by myself and co-workers to peptide pharmaceuticals with improved duration of action have employed not only protection from proteolysis, but also have incorporated the novel concept of using the body as its own depot for the compound (whole-body depot effect). In this chapter, I will review several methods that have been used to achieve a prolonged duration of action, and I will focus on our results using the depoting concept for several compounds that have reached clinical trials.

Factors Affecting Potency

The characteristics of a suitable pharmaceutical candidate include high potency, low toxicity, high oral bioavailability, and prolonged duration of action. Analogues of polypeptide hormones typically fulfill the first two criteria admirably but fail to be absorbed and show rapid clearance. The high potency typically reflects very tight binding due to multiple interactions within receptors of the "seven transmembrane" class (*6*). Such tight binding, particularly for antagonists, frequently is a consequence of a very slow receptor off-rate (*7*), which can contribute to a prolonged apparent duration of action. The ability of a polypeptide to be absorbed after oral administration appears to be a function of size and polarity (*8*). Most peptides are too polar to pass through the cell membranes of the gut lining, but paracellular transport through openings between the cells (average pore radius, 8 Å) appears to take place (*9*). Small, polar polypeptides (up to tripeptide) have significant oral bioavailabilities. Tripeptide thyrotropin releasing hormone has up to 10% bioavailability (*10*), whereas the oral bioavailability of longer polypeptides drops off to <1%. Hydrophobic polypeptides of substantial size may retain high bioavailability, however. For example, the cyclic undecapeptide cyclosporin (*11*) has an oral bioavailability of >30% in an optimized lipoidal formulation. This finding may relate to a different route of absorption (perhaps transmembrane passage for very hydrophobic peptides) and is probably better characterized as being inversely related to the number of exposed H-bonds (*8*) rather than to global hydrophobicity. Although some absorption enhancers increase paracellular transport (e.g., surfactants and ethylenediaminetetraacetic acid), apparently by doubling pore size (*9*), the repeated defeat of these physiological barriers in the gut is felt to be potentially dangerous (*12*). This area remains very controversial (*13*), with frequent reports of novel formulations or transport systems that result in dramatic increases in absorption (*2*). The reproducibility of such reports has been a problem in the past. Interspecies and interindividual variation can be a confounding factor in trying to evaluate the results of such experiments; however, these limits seem likely to hold for the near future for peptides that are not actively transported. The problem of low oral bioavailability of peptides appears to be with us for some time to come.

Although peptides can be administered successfully by nasal (*14*), parenteral, or controlled-release formulations (*15, 16*), their very rapid clearance has remained a challenge. The amide bond is susceptible to cleavage by many proteases in the body, and many hormones do not survive even one pass through the major sites of proteolysis, such as the lungs or liver (*17*). As a result, typical half-life ($t_{1/2}$) values for small peptides are in the seconds-to-minutes time frame. Protection from proteolysis, therefore, has been the predominant focus for research directed at the design of polypeptide drugs (*4, 5*). The use of amide bond replacements has been successful occasionally (*18–20*), but the products frequently suffer from decreased potency due to the loss of critical binding interactions (perhaps H-bonding) or altered ligand conformation. Although prevention of proteolysis must certainly play a role in the design of a successful peptide pharmaceutical, attention must be paid to other routes of clearance.

A frequently overlooked fact is that small, polar molecules are rapidly cleared by glomerular (kidney) filtration. For example, a molecule like inulin (M_r ca. 5000) is not metabolized in the body but is still cleared rapidly (*21*) through the kidney, and it has a $t_{1/2}$ of ca. 1 h. This fact should not be surprising because with each pass through the glomerulus, ca. 25% of the plasma volume (and its low molecular weight components) is filtered into the urine (*21*). Even a completely nonmetabolized compound like inulin would be unacceptable as a drug for most applications because it would need to be administered many times per day. Studies with gonadotropin-releasing hormone (GnRH, also denoted LHRH) analogues also have shown no clear-cut relationship between resistance to proteolysis and duration of action (*22*). Interestingly, GnRH is cleared much less rapidly in renally impaired subjects than in subjects with liver dysfunction or normal subjects; these findings also point to the kidney as an important site for clearance (*23*). Similarly, most of the metabolically stable, third-generation cephalosporins still have $t_{1/2}$ values of only 2 h or less (*24*). The plasma $t_{1/2}$ values for these agents roughly correlate with their degree of binding to protein. Thus, a polypeptide that was fully protected from proteolysis still might be expected to show an unacceptably short $t_{1/2}$ if it spent the majority of its time free in the plasma. For this reason, I focused early on the need to get the polypeptide out of the plasma to protect it from clearance by glomerular filtration. This concept was the genesis of our use of the whole body depot concept, which will be discussed later. Clearly, a combination of approaches that are directed at protection from both proteolysis and glomerular filtration is necessary for the design of successful polypeptide pharmaceuticals.

Protection from Proteolysis

Multiple replacements have been used to avoid the susceptibility of the amide bond to cleavage; for example, D-amino acid, *N*-alkyl amino acid (*25*), retro-inverso amide (*26, 27*), *C*-α-alkyl- amino acid (*28*), α-aza- amino acids (*29*), *N*-hydroxy- amino acid (*30*), sterically demanding amino acids (*31*), thioamide linkage (*32, 33*), and amide bond replacements gathered under the "pseudopeptide" linkage rubric (*18*). An approach used in nature, N- or C-terminal blockade by acylation or amidation, has occasionally been fruitful [e.g., N-terminal pyroGlu (*34*)]. The search for amide bond replacements has resulted in great diversity (List I), but no clear rules can predict where the replacements will be successful.

List I. Approaches to the Blockade of Proteolysis

1. N-terminal acylation, pGlu formation
2. C-terminal amidation
3. N-alkyl amino acid substitution
4. D-amino acid substitution
5. Amide bond reduction
6. Amide backbone modifications ("pseudopeptide linkage")
7. Retro-inverso substitution
8. Rigidification, cyclization
9. Unnatural amino acid substitution
10. Concomitant use of enzyme inhibitors

Probably the most successful substitution has been D-amino acid incorporation. In fact, the scanning of new peptide structures by their incorporation at each position has become routine (*35, 36*). D-Amino acid incorporation is expected to protect the resulting amide bond from proteolysis, but incorporation also has frequently resulted in the development of antagonism or increased binding affinity. Substitution for a Gly residue has frequently been effective, and the ability of a D-amino acid to stabilize a type-II β-turn may be a factor here (*37*). Important examples are in the GnRH and enkephalin area, where D-amino acid substitution led to significant increases in resistance to proteolysis and potency. [D-Ala^6]GnRH (*37*) and [D-Ala^2]Met enkephalinamide (*38*) have five and >20 times the potency of the parent molecules, respectively. An interesting attribute that is little noted is that D-amino acid substitution frequently results in a more hydrophobic analogue (*39*), a result that may contribute to depoting of the molecule (discussed later). This substitution can result in a profoundly different conformation for this region of the molecule, however.

One of the more useful substitutions is the reduced amide bond (*40*). In several analogue series [e.g., tetragastrin (*41*) and bombesin (*42*)], its incorporation into a peptide has resulted in a potent antagonist. The development of simple procedures for synthesis also has made the reduced amide bond an attractive substitution (*43*). Two important features of this substitution are its flexibility and the positive charge that it brings to an analogue. A net positive charge of a polypeptide may affect the way it is handled by the negatively charged surface of a cell [e.g., the tendency for concentration near the surface in accord with Gouy–Chapman theory (*44*)].

The use of unnatural amino acid residues (*45*) has been a favorite approach, not only because the residues frequently are difficult enzyme substrates, but also because of the ability to use their unusual physical properties to tune the global properties of the peptide (*46*). Thus, amino acid residues with moderate (*47*) to extreme hydrophobic character (*48, 49*) have seen increasing use (discussed later). Co-workers and I have made extensive use of *N*-alkylated Arg (*50*) analogues as a way to modulate phospholipid membrane affinity. The incorporation of α-aza- amino acids has been a useful modification at selected positions in several peptides (*29*). In each of these cases, the residues are resistant to metabolism by the commonly encountered enzymes (*48, 51, 52*).

An important approach for future development may be in the direction of cyclization or rigidification [*see* review (*53*)]. Hypothetically, rigid analogues that fit the appropriate receptor may not be flexible enough to conform to the active site of a catabolic enzyme and therefore may be resistant to degradation. For example, an enkephalin analogue cyclized between position 2 and the C-terminus (D-α,γ-diaminobutyric acid at position 2) showed high receptor affinity and was not degraded by rat-brain membranes (*54*). Such an analogue might also benefit in potency because it pays less of an entropic price to adopt the biologically active conformation in the receptor binding site. In addition, such analogues may provide structural information useful for peptidomimetic design (*53*). Conformationally restricted analogues have been shown to offer an approach to the generation of receptor subtype-specific compounds (*55*).

These approaches have been developed during the past 15 years, and several have been very fruitful. Their applications remain largely empirical exercises, however, and the search for compounds with a sufficient duration of action can be a long and difficult process. For example, the approach to renin inhibitors (*56*) and later to human immunodeficient virus protease inhibitors (*57*) with sufficient duration of action (to say nothing about sufficient oral bioavailability!) has occupied sizeable synthesis groups at many pharmaceutical companies for the past decade.

The Drug Depot Concept

Small organic molecules, the classical targets of medicinal chemists, also have problems with rapid clearance. They are frequently metabolized by oxidative processes (e.g., cytochrome *P*-450 mediated) and by conjugation with glucuronic acid, but glomerular filtration is recognized as being an important route for clearance of very water soluble molecules (*58*). Optimum hydrophobicity frequently is correlated with the highest in vivo potency (*59*). This correlation may be due to depoting in the body as well as to the more usually recognized effects on receptor or enzyme binding (*58*). Such considerations had not been applied to peptide analogue design when we began our studies in the GnRH area a number of years ago, and then are still rarely cited.

Focus only on protection from proteolysis was not sufficient because, as just noted, small (<25000 Da), polar, very water soluble molecules that spend most of their time free in the plasma would be cleared rapidly by filtration into the urine. The approach needed, therefore, was to find a way to cause the polypeptide analogue to use biological structures to form a depot in the body; that is, a "whole-body depot" effect. In this way the peptide is protected from both proteolysis and clearance by glomerular filtration, but can it partition back to the plasma later to maintain a biologically active concentration. The initial approach toward removing the peptide from the plasma was to use very hydrophobic, unnatural amino acids to reduce the solubility of the analogue in the plasma (*48*). I visualized the peptide as partitioning into hydrophobic compartments of the body (e.g., lipids, hydrophobic carrier proteins, and cell membranes) for later release. This approach has been highly successful in yielding very long acting peptides. Our first and most carefully analyzed application of this approach was in the GnRH field [*see* review (*60*)]. I will use this experience to illustrate the results and confirm the mechanism for this strategy before moving on to recent extensions of the approach.

Hydrophobic Approaches to Peptide Depoting

Initial studies on GnRH (*p*Glu-His-Trp-Ser-Tyr-Gly-Leu-Arg-Pro-Gly-NH_2) showed that substitution of a D-amino acid in position 6 (*37*) resulted in a three- to fivefold increase in potency. Further work involved the incorporation of the D-form of essentially all of the proteinogenic amino acids (*61, 62*) and indicated that a trend toward increased potency with increased hydrophobicity existed in the data. Our first analogues incorporated very hydrophobic, unnatural amino acids (Table I) for the reasons just cited and were immediately successful (*48*). Thus, [3-(2-naphthyl)-D-Ala^6]GnRH (nafarelin, trade name Synarel for endometriosis) had an in vivo potency 190-fold that of the native hormone and subsequently (*63*) was shown to have a $t_{1/2}$ of 3–5 h (Figure 1) in humans, depending on dose and mode of administration (in contrast to 5–30 min for GnRH). Another analogue of interest was [3-(2,4,6-trimethylphenyl)-D-Ala^6]GnRH, which had a potency ca. 200 times that of GnRH (*48*). These analogues remain among the most potent GnRH agonists (*46*).

We studied the retention behavior of our compounds on a reversed-phase column (k' values used as a measure of the global hydrophobicity of the compound) and found a roughly parabolic correlation between potency and hydrophobicity. After the completion of our work, a parabolic quantitative structure–activity relationship (QSAR) derived for GnRH agonists was published (*64*). The QSAR predicted that more hydrophobic analogues than those published at that time would be more potent. Our analogue series was hypothesis-directed rather than QSAR-directed but began at a hydrophobicity region just beyond that previously published and ended with some extremely hydrophobic compounds (e.g., [3-dicyclohexylmethyl)-D-Ala^6]GnRH with 90-fold GnRH potency). Not all of our biological data fell strictly on a parabolic line, and we assumed steric considerations were a confounding factor for some analogues with β-branched amino acids such as Nal(1) in position 6. However, data were in general agreement with the notion of an optimum hydrophobicity. In effect, our analogue series filled in precisely the region of hydrophobicity that was independently predicted to contain more potent analogues.

Table I. Selected GnRH Analogues—Estrus Suppression Activity

Analogue[a]	*k′ (HPLC)*[b]	*Potency*[c]
[D-Tmo^6]GnRH	0.59	140
[D-Trp^6,Pro^{9}NHEt]GnRH	1.29	100 (standard)
[D-Cha^6]GnRH	1.45	60
[D-Pfp^6]GnRH	1.73	110
[D-Nal(1)^6]GnRH	1.91	50
[D-Nal(2)^6]GnRH	1.95	200
[D-Mtf^6]GnRH	2.05	100
[D-Ptf^6]GnRH	2.14	120
[D-Tmp^6]GnRH	2.18	200
[D-Dca^6]GnRH	26.27	90

[a] Abbreviations: Tmo, 3-[2-(3,4,5-trimethoxyphenyl)]-Ala; Cha, 3-(cyclohexyl)-Ala; Pfp, 3-(pentafluorophenyl)-Ala; Nal(1), 3-(1-naphthyl)-Ala; Nal(2), 3-(2-naphthyl)-Ala; Mtf, 3-(*m*-trifluoromethylphenyl)-Ala; Ptf, 3-(*p*-trifluoromethylphenyl)-Ala; Tmp, 3-(2,4,6-trimethylphenyl)-Ala; Dca, 3-(dicyclohexylmethyl)-Ala.
[b] k' = (retention volume − void volume)/void volume; results are from reversed-phase high performance liquid chromatography (HPLC) (C_{18} column).
[c] Calculated relative to that of GnRH in a 2-week rat estrus-suppression assay (twice daily sc injection). The potency of the standard was 100-fold that of GnRH in this series of direct comparisons.

The reasons for the high potency of nafarelin were examined in some detail. The receptor binding affinity (*65*) of nafarelin was significantly higher than that of GnRH (0.2×10^{-9} M versus 1.4×10^{-9} M), and its potency for gonadotropin release for in vitro cellular assays was also greater. However, an additional four-fold increase in apparent potency was seen in whole animal assays. Furthermore, when maximal efficacy for receptor activation was evaluated by quantitation of protein kinase C activation (*66*) (measured by relocalization to the cell membrane) in cells in vitro, nafarelin showed 1.4 times the maximal efficacy of GnRH. The current uses of agonistic GnRH analogues rely on hyperstimulation, and thereby, desensitization (*67*) and uncoupling of the GnRH receptor from its signal transduction mechanisms. Therefore, this ability to cause increased maximal efficacy is an advantage. The in vitro data appear to relate to a slower rate of turning the receptor off (*7*) for this hydrophobic molecule, but the in vivo data also must be a function of the increased duration of action.

When [^{3}H]nafarelin was studied by equilibrium dialysis (*68*) against plasma, it was found to be ca. 80% bound to the serum albumin (SA) fraction. This binding was not saturable over several logs of concentration and was the same for plasma from males and females of several species (Figure 2). In contrast, GnRH is 18–20% bound under similar conditions. Thus, one can visualize nafarelin as circulating in a protected form, noncovalently bound to the well-known hydrophobic carrier patch on SA, and then being slowly released back to the plasma for receptor binding. In this form, nafarelin is protected from both proteolysis and glomerular filtration.

It is important to note that, while nafarelin has been stabilized against proteolysis at one site, cleavage at other sites still takes place in vivo. The principle metabolites seen in the rhesus monkey appear to arise from proteolysis at the 4–5 bond (*69*). The 1–2 and 9–10 bonds also are expected to be cleaved (*22*).

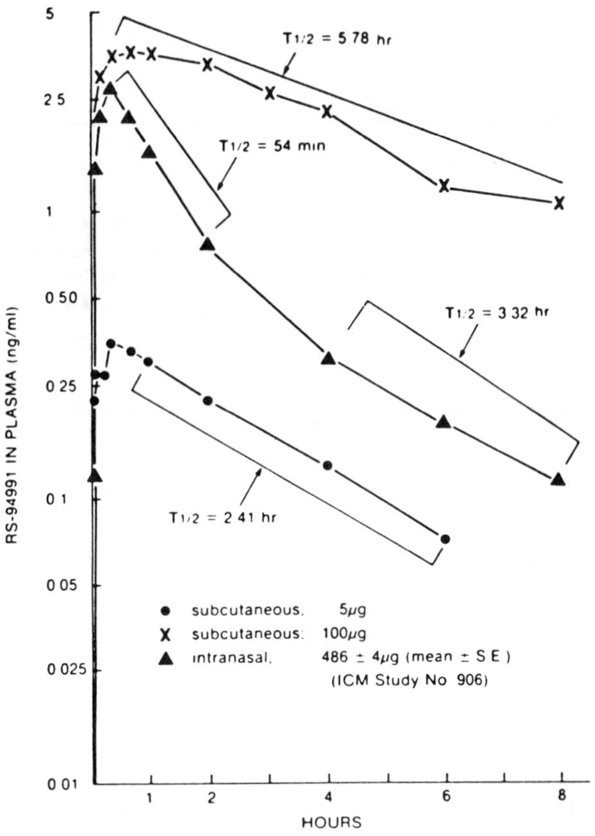

Figure 1. Concentrations of [D-Nal(2)⁶]GnRH (RS-94991) in the plasma of female volunteers after a single sc or intranasal dose. (Reproduced with permission from reference 112. Copyright 1986 Royal Society of Chemistry.)

Subsequent studies by Felgner et al. (*70*) showed that nafarelin also has a substantial binding affinity for liposomes with a composition similar to that of cell membranes, and this finding may point to another depot site. A hydrophobic agent capable of transferring fluorescent energy (dansyl phosphatidylethanolamine: absorption at 340 nm, emission at 520 nm) was used in these studies in the membrane as a reporter for bound peptide (naphthyl and Trp side-chain absorption at ca. 280 nm and emission at ca. 340 nm). Energy transfer only occurs when the distance between the donor and acceptor is <4 Å; thus, a direct measure of the amount of peptide bound is provided. Saturable binding (K_D of 5×10^{-6} M) was measured for nafarelin on these naked liposomes. Thus, binding to cell membranes in the body also appears to be an important mechanism for the formation of depots of nafarelin.

These data illustrate the power of the approach and confirm the hypothesized mechanism of action. Nafarelin and a close analogue [(D-Nal(2)^6,aza-Gly^{10}]GnRH (potency is 230 times that of GnRH) remain the most potent agonistic analogues of GnRH yet reported in whole animal assays (*46*), although they are not the most potent receptor binders. The demonstrated affinity for SA and phospholipid membranes suggests that the peptide can depot on such structures as a way to avoid both proteolysis and renal clearance. This concept

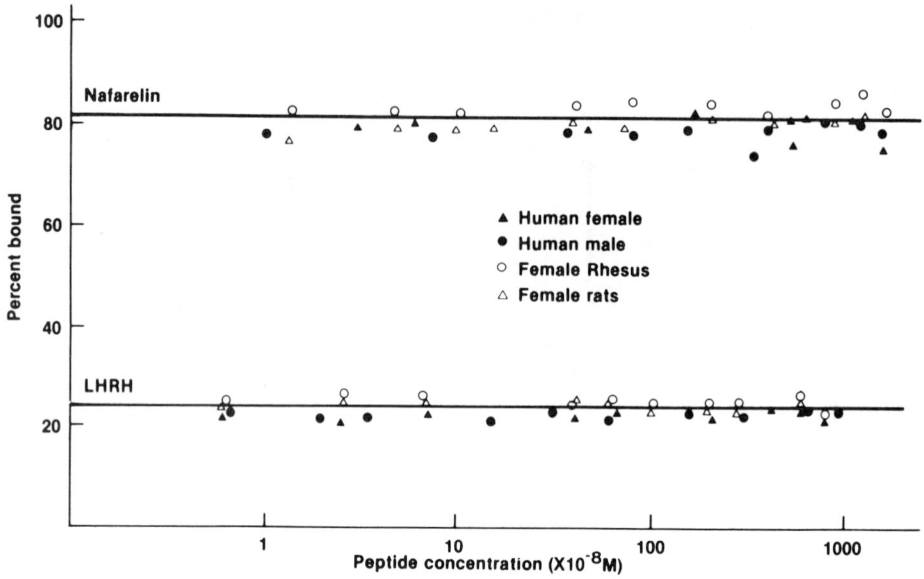

Figure 2. Equilibrium dialysis study of the binding of [^{3}H]nafarelin and [^{3}H]GnRH to plasma components from human, rhesus monkey, and rat. (Reproduced with permission from reference 68. Copyright 1985 Academic.)

offers a new way to analyze peptide structure–activity relationships and may be an important factor in the final optimization of peptide drug candidates.

Hydrophilic Approaches to Depoting

Alternatives to hydrophobic interactions exist for the formation of depots in the body. I have focused on electrostatic interactions as a second route for binding peptides in the tissues rather than allowing them to circulate freely. This approach was developed from hypothesized interactions between Arg residues and the phosphate head groups on cell membrane phospholipids (*71*).

The study of GnRH antagonists demonstrated that the most potent analogues would be exceptionally hydrophobic molecules. Although the agonistic analogue plays only a relatively transitory role in stimulating, and thereby desensitizing (*67*), the receptor, the challenge for a GnRH antagonist is substantially greater because it must be present on the receptor continuously to block access for GnRH. As little as 10% receptor occupancy by GnRH causes luteinizing hormone (LH) release. A very long duration of action, therefore, should be required for GnRH antagonists. Based on the concept outlined earlier, one might expect that very hydrophobic analogues would fit this requirement. In fact, the most potent analogues at one point had to be dissolved in corn oil for administration, and because this formulation can form a depot, it also may have contributed to the apparent duration of action.

To generate a molecule with increased water solubility for ease of formulation (*72*), charged residues were incorporated into position 6 of [*N*-Ac-D-*p*Cl-Phe^1,D-*p*Cl-Phe^2, D-

Trp^3,D-X^6,D-Ala^{10}]GnRH. The D-Phe^6 analogue was one of the most potent analogues to that date and fit the picture of a very hydrophobic, long-acting compound (50% effective dose [ED_{50}] < 10 μg/rat; administered at noon during proestrus). Incorporation of positively charged residues in position 6 greatly increased the water solubility but was expected to decrease the duration of action according to the foregoing discussion. Surprisingly, the Arg^6 compound was both highly potent and very long acting (ED_{50} = 1.2 μg/rat when administered at noon during proestrus and 5 μg/rat when administered 24 h earlier) for blockade of ovulation (*73*). This less than fivefold ratio of the ED_{50} for doses >24 h before the preovulatory GnRH surge compared with that for doses given 1.5 h before the event contrasts to a 40-fold dose ratio at the same time points for [*N*-Ac-Pro^1,D-*p*F-Phe^2, D-$Nal(2)^{3,6}$]GnRH, which is one of the most potent members of the hydrophobic class of compounds (*73*).

I hypothesized that the D-Arg-containing compound forms an effective depot in vivo as well, and ascribed it to an additional type of interaction, which can be visualized between a negatively charged phospholipid head group and a positively charged Arg-guanidine function. Our hypothesis for the role of this position 6 substitution did not involve specific receptor interactions, and therefore, the length or bulk of the side chain was not considered to be critical. We therefore generated a new class of amino acids (*50*), the N^g,$N^{g\prime}$-dialkylhomoarginines [*h*$Arg(R_2)$], designed to further stabilize this hypothesized interaction (Figure 3) by the additional interaction of alkyl groups with choline chains or with the hydrophobic membrane interior. Alkyl groups from methyl to palmityl were investigated. These amino acids were prepared conveniently from a suitably protected Lys derivative (e.g., *Z*-D-Lys-*O*-Bzl) by reaction with the corresponding *N,N′*-dialkylcarbodiimide. More recently, we adopted a synthesis that relies on the selective reaction of the *N*-epsilon amino function of unprotected Lys with *N,N′*-dialkyl-*S*-Me-isothiourea to directly produce the *h*$Arg(Et_2)$ residue directly in good yield (*74*).

We evaluated the effectiveness of this type of substitution by incorporation of several members into a series of GnRH antagonists. These analogues were tested in an assay (*75*) designed to measure their duration of action (Figure 4), all of these *h*$Arg(R_2)$ modifications (including the minimal Me_2 substitution) yielded analogues that appeared to be longer acting than the corresponding Arg-containing parent structure (Figure 4A). The behavior

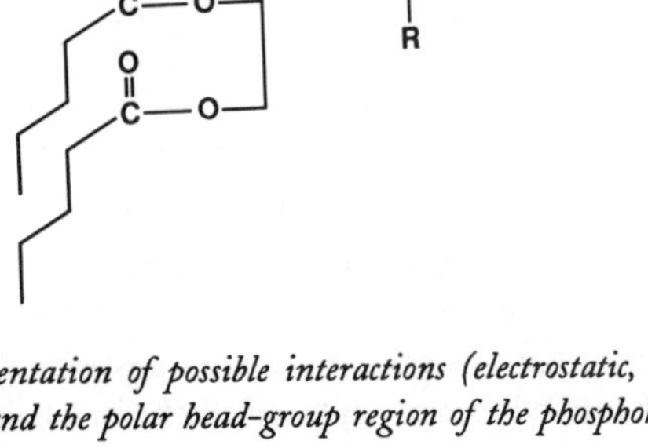

Figure 3. Schematic representation of possible interactions (electrostatic, H-bonding, and hydrophobic) between h*$Arg(R_2)$ residues and the polar head-group region of the phospholipid cell membrane.*

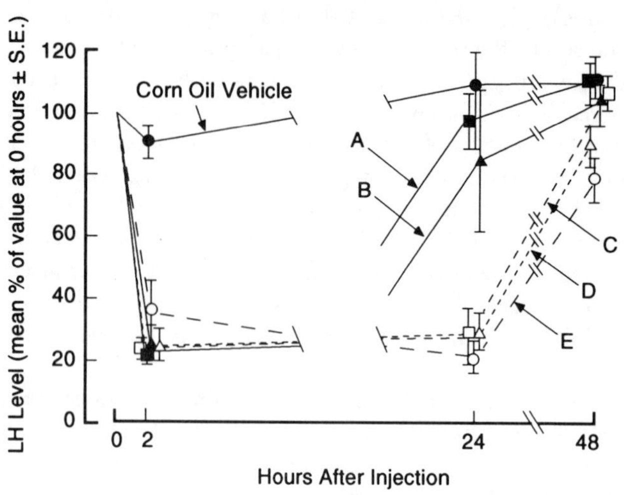

Figure 4. LH levels in castrated male rats following sc injection of GnRH antagonists at 50 μg/kg. ***A,*** [N-*Ac*-D-pC*l*-*Phe*^{1},D-pC*l*-*Phe*^{2},D-*Trp*^{3},D-*Arg*^{6},D-*Ala*^{10}]*GnRH;* ***B,*** [N-*Ac*-D-*Nal(2)*^{1},D-pF-*Phe*^{2},D-*Trp*^{3},D-*Arg*^{6}]*GnRH;* ***C,*** [N-*Ac*-D-*Nal(2)*^{1},D-pF-*Phe*^{2},D-*Trp*^{3},D-h*Arg(Et*$_2$*)*^{6}]*GnRH;* ***D,*** [N-*Ac*-D-*Nal(2)*^{1},D-pF-*Phe*^{2},D-*Trp*^{3},D-h*Arg(Me*$_2$*)*^{6}]*GnRH; and* ***E,*** [N-*Ac*-D-*Nal(2)*^{1},D-pC*l*-*Phe*^{2},D-*Trp*^{3}, D-h*Arg(Et*$_2$*)*^{6},D-*Ala*^{10}]*GnRH. (Reproduced with permission from reference 75. Copyright 1983 Pierce Chemical.)*

of [*N*-Ac-D-Nal(2)^{1},D-*p*Cl-Phe^{2},D-Trp^{3},D-*h*Arg(Et$_2$)^{6},D-Ala^{10}]GnRH (detirelix) in this assay is unusual. Detirelix did not achieve complete LH suppression by 2 h (Figure 4E), but maintained a profound suppression through 24 h. At the 48-h point, it is the only analogue that is still significantly different ($p < 0.05$) from control. The Arg(Et$_2$) residue was the most effective of this short series, and detirelix was chosen for further study.

Detirelix proved to be highly potent and extremely long acting in animal studies (*76*) and in humans (*77, 78*). A single subcutaneous (sc) injection of detirelix at 1 mg/kg to male dogs caused a precipitous fall in testosterone (T) levels, to the range seen in castrated animals, that remained for 5 days (*76*). When detirelix was tested in a 9-week study using once daily sc injection of 100 μg/kg in male dogs, rapid suppression of T levels and the anticipated cessation of spermatogenesis was seen. When detirelix administration was eliminated, however, T levels did not return to pretreatment levels for 2 weeks (*76*). Pharmacokinetic studies in monkeys yielded $t_{1/2}$ values of 1.5 and 4 h for 0.04 mg/kg doses given by the intravenous (iv) and sc routes, respectively (Table II; *79*). When a larger dose (1 mg/kg) was given sc, the $t_{1/2}$ was a remarkable 32 h.

Clinical trials (*78*) with detirelix also indicated a very prolonged duration of action. Injection of 1-, 5-, and 20-mg doses sc to normal female volunteers resulted in $t_{1/2}$ values $\gg$24 h and blood levels of detirelix detectable for >72 h for the lowest dose (Figure 5). This type of $t_{1/2}$ for a peptide drug is very impressive and was virtually unprecedented at the time. More recently, studies with the very hydrophobic analogue antide ([*N*-Ac-D-Nal(2)^{1},D-*p*Cl-Phe^{2},D-Pal(3)^{3},Lys(Nic)^{5},D-Lys(Nic)^{6},Lys(iPr)^{8},D-Ala^{10}]GnRH) have also shown an exceptionally long duration of action in humans ($t_{1/2}$ = 6.5 days); the concept of depoting by protein binding has been invoked to explain this observation (*80*). Whether this compound actually precipitated from solution in the body and whether a useful dosing regimen can be found for this compound are questions that remain to be answered.

It is clear that two types of depoting can be involved in the pharmacokinetic behavior

Table II. Pharmacokinetic Data for GnRH Analogues

Species	Pharmacokinetic Parameter	GnRH	Nafarelin	Detirelix	Ganirelix
Rat	CLs (mL min^{-1} kg^{-1})	57.6	12	3.3	2.5
	Vdβ (L/kg)	0.56	0.58	0.45	0.29
	PB (%)	22	80	—	82
	$t_{1/2}$ (h)	0.11	0.56	1.6	1.4
Monkey	CLs (mL min^{-1} kg^{-1})	31.4 ± 5.8	2.7 ± 0.3	1.3 ± 0.3	0.8 ± 0.2
	Vdβ (L/kg)	1.53 ± 0.37	0.47 ± 0.02	0.86 ± 0.32	0.32 ± 0.04
	PB (%)	22	80	—	84
	$t_{1/2}$ (h)	0.55 ± 0.08	1.8 ± 0.1	7.1 ± 1.5	5.1 ± 0.8
Human	C_{max} (ng/mL)	—	14[a]	23[b]	37[b]
	$t_{1/2}$ (h)	—	3	29	15

NOTE: GnRH is gonodotropin-releasing hormone, CLs is clearance rate, Vdβ is volume of distribution, PB is plasma binding, $t_{1/2}$ is half-life, and C_{max} is maximal drug concentration.

[a] Dose was 0.4 mg, sc.

[b] Dose was 1 mg, sc.

of a compound like detirelix. Interaction between this positively charged molecule and negatively charged phospholipid headgroups, as originally visualized, takes place. In addition, a physical depot appears to be formed at the site of injection. Detirelix forms a copious white precipitate in the presence of anionic materials, including plasma components (e.g., SA) and phospholipid vesicles. A mixed precipitate containing the analogue and components from the extracellular fluid form at the injection site when large doses are administered sc. The administration of smaller doses iv allows the compound to rapidly distribute, and the resulting $t_{1/2}$ reflects the whole-body depot effect referred to previously.

A significant negative effect of the use of peptides with a high positive charge, such as these antagonists, is the peptide's ability to cause mast-cell degranulation, which results in histamine release, edema (*81*), hypotension (*82*), and a local inflammatory response.

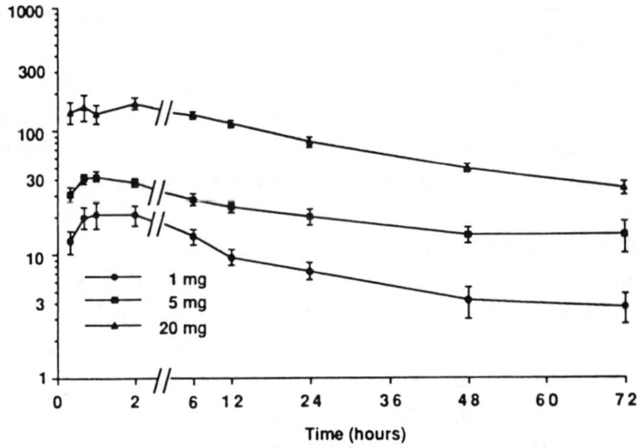

Figure 5. Concentrations of detirelix in the serum of female volunteers after single sc injections of the doses shown. (Reproduced with permission from reference 78. Copyright 1992 The Endocrine Society.)

Although this problem was initially greeted with surprise, it is in accord with the known ability of neuropeptides and positively charged small molecules to cause mast-cell degranulation (*83*). Study of the structure–activity relationship for degranulation of rat peritoneal mast cells in vitro indicated a general increase in potency for this toxic side effect as the hydrophobicity or positive charge of the molecule was increased (*84*). Both types of modification are also expected to increase the membrane affinity of such molecules, an attribute that we expected to enhance the $t_{1/2}$ of the peptide analogues due to depoting.

Despite these apparently conflicting considerations of membrane affinity for potency and toxicity, we and others have found analogues that can show high potency and reduced histamine release. We noted that the Arg^8-Pro^9 sequence in GnRH fits a pattern for potent histamine releasers (*85*), and incorporation of hArg(R_2) residues in position 8 results in a compound with reduced histamine releasing potency, ganirelix ([*N*-Ac-D-Nal(2)^1,D-*p*Cl-Phe^2,D-Pal(3)^3,D-hArg(Et_2)^6,*h*Arg(Et_2)^8,D-Ala^{10}]GnRH; *74*). Other workers have found that the incorporation of an N^{ω}-substituted basic residue in position 8 (usually an N^{ϵ}-*i*Pr-Lys) leads to decreased histamine-releasing potency (*86*), despite the increased basicity of such substitutions. Such a modification also increases the hydrophobicity of the residue (expected to increase histamine release), but I hypothesized that shielding of the positive charge by the alkyl groups may be responsible for the overall beneficial effect (*71*).

An important goal during the design phase for the series that led to ganirelix was the achievement of an early, high peak level of compound in the circulation. We noted (*77*) that when detirelix was tested in the clinical trial it had a similar potency to the "4-F antagonist", although we found (*50*) a higher potency in the rat (50% inhibitory concentration [IC_{50}] of 0.6 μg and 2.4 μg for detirelix and the 4-F antagonist, respectively). We hypothesized that the differences in behavior might be related to the physical properties of the molecules. We had expected that the increased hydrophobicity of detirelix coupled with its increased "stickiness" on phospholipid membranes would cause it to have a slow influx and in turn result in a blunted peak (lower maximum plasma concentration, C_{max}) and a slower elimination phase (longer $t_{1/2}$). A general rule of thumb is that animals metabolize drugs at rates that are inversely proportional to their body weights (*87*). Thus, the rat is known to metabolize peptides far more rapidly than the human; the $t_{1/2}$ for detirelix is 1.6 h in the rat (*79*) and 30 h in humans (*78*). We reasoned that a very slow clearance would be more important for potency in rat than in humans and that a compound with a very slow influx and very slow clearance (such as detirelix) would be favored in the rat. In contrast, the human system would be less demanding metabolically, and a compound with a rapid influx, and therefore higher C_{max}, would give a higher percentage of receptors loaded. Unloading of antagonists from receptors is typically very slow; therefore, this higher initial loading might be reflected in higher potency in the human. An additional advantage of a faster influx and distribution phase would be that daily administration would be appropriate. The detirelix $t_{1/2}$ of $\gg$24 h suggested administration at 2- or 3-day intervals, which is an impractical dosage scheme for patients to remember. We therefore sought a more soluble compound.

Ganirelix has also been the subject of clinical trials. In a recent phase I trial in postmenopausal women (*88*), the $t_{1/2}$ (Figure 6) varied from 15 h (1 mg, sc) to 27 h (6 mg, sc). The time (T_{max}) to maximal drug concentration (C_{max}) was on the order of 1–2 h, in accord with our desire for rapid influx and therefore higher maximal blood levels. Thus, the C_{max} seen for the 6-mg dose of ganirelix is 152 ng/mL compared with 45 ng/mL for a 5-mg dose of detirelix (Table II) (*78*).

Although the $t_{1/2}$ and T_{max} data reported for detirelix and ganirelix in humans are similar ($t_{1/2}$ of 30 and 15–27 h, respectively; T_{max} of 1–2 h for both), but pharmacokinetic curves (Figures 5 and 6) and volume of distribution (*Vd*β) from animal studies (Table II) suggest that the desired objective of a more rapid influx was achieved. For compounds with

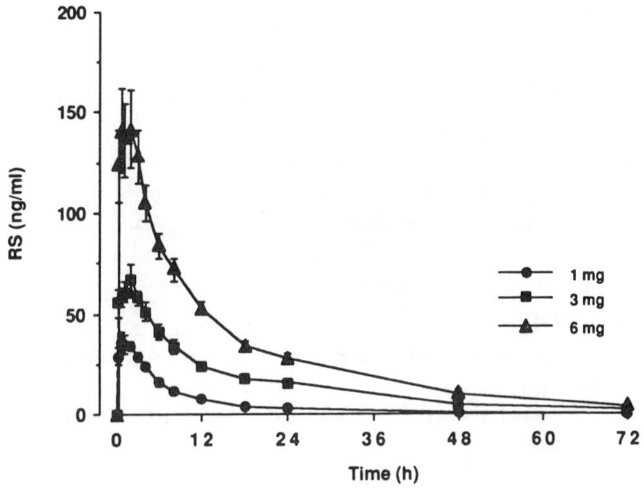

Figure 6. Concentrations of ganirelix (RS-26306) in the serum of female volunteers after single sc injections of the doses shown. (Reproduced with permission from reference 88. Copyright 1992 The Endocrine Society.)

such a slow elimination, accurate estimation and comparison of $t_{1/2}$ values are difficult. Comparison of the uptake and early distribution phases shows a higher, sharper peak for ganirelix (Figure 6) and a more rapid falloff as the compound distributes to the tissues. The curves show that substantially higher drug levels are present at 72 h for detirelix (Figure 5) than for ganirelix, despite a higher C_{max} for ganirelix. In addition, the $Vd\beta$ value in monkeys for detirelix is more than twice that for ganirelix (*89*; Table II). This parameter is a measure of how widely dispersed a compound is in the tissues and is usually larger for a more tissue bound compound than for one that is more localized to the extracellular fluid phase. For comparison, the $Vd\beta$ value for nafarelin was 0.5 L/kg (*90*). These data confirm that ganirelix is less tissue bound than detirelix and suggest a more rapid distribution and elimination phase.

The strategy of optimization of a GnRH antagonist for humans through pharmacokinetic considerations appears to have been successful. Although ganirelix is only twice as potent as detirelix in the rat antiovulatory assay (*74*; ED_{50} of 0.3 versus 0.6 μg/rat), it was at least fivefold more potent in humans (*88*). Thus, a single, 1-mg of dose of ganirelix to postmenopausal women was capable of profound suppression of LH secretion for >24 h. This finding suggests that ganirelix is the most potent GnRH antagonist that has been tested in humans to date. On the basis of the plasma clearance curve, one would expect some compound accumulation in the body ("dose stacking") on repeated dosing, and the effective dose on repeated administration would be expected to be lower. Because of its very high potency and much reduced propensity to cause histamine release, ganirelix is considered to be "potentially the first clinically promising agent for the treatment of gonadotropin-dependent disorders" (*88*).

Hydrophilic Depoting of GnRH Agonists

To extend the scope of our experience with the new *h*Arg(R_2) class of amino acids, we incorporated them into a series of agonistic GnRH analogues. As just described, a clear-cut trend existed toward increased potency with the incorporation of more hydrophobic

Table III. Biological Activity in vivo of Selected Hydrophilic GnRH Agonists

Compound	*k′ (HPLC)[a]*	*Potency[b]*
[D-*h*Arg(Et$_2$)^6]GnRH	0.5	40
[D-*h*Arg(CH$_2$CH$_2$)^6,Pro^9-NHEt]GnRH	0.6	130
[D-*h*Arg(Me,Bu)^6]GnRH	0.7	40
[D-*h*Arg(Et$_2$)^6,Pro^9-NHEt]GnRH	0.8	150
[D-*h*Arg(Me,Bu)^6,Pro^9-NHEt]GnRH	1.4	80
[D-Nal(2)^6]GnRH[c] (standard)	1.8	200
[D-*h*Arg(hexyl$_2$)^6]GnRH	15.3	50

[a] Measure of relative hydrophobicity; k' = (retention volume–void volume)/void volume; C_{18} packing.
[b] Calculated relative to that of GnRH in a 2-week rat estrus-suppression assay (twice daily sc injection).
[c] Reference 48.

amino acids into position 6, and our work pushed the testing of this hypothesis to the extreme limits of hydrophobicity (*48*). With our new insight into an alternative type of hydrophilic depoting (*50, 71*), we sought to prove the generality of the hypothesis by incorporation into the agonistic GnRH analogue area. In reexamining the early data (*61*), we found that [D-Arg^6]GnRH and [D-Arg^6, Pro^9-NHEt]GnRH were made and exhibited increased potency in vitro relative to GnRH (4- and 17-fold, respectively). However, this observation was not followed-up.

We incorporated members of our new series of *h*Arg(R$_2$) analogues into position 6, a position previously thought to prefer very hydrophobic residues. We used alkyl groups from ethyl to hexyl for R. In several cases we used unsymmetrically substituted *h*Arg analogues in an attempt to reach the hydrophobic region of the membrane while retaining an electrostatic interaction with the PO_4 group [e.g., *h*Arg(Me,hexyl)]. The resulting analogues (*91*) were exceptionally potent (Table III) and approached the potency of the hydrophobic series previously prepared (*48*). A very interesting analogue prepared subsequently ([D-*h*-Arg(CH$_2$CF$_3$)$_2^6$,Pro^9-NHEt]GnRH) exhibited an in vivo potency (190 times the GnRH potency) within the range of the most potent analogues (*46*) ever reported in the GnRH agonist area. This analogue was originally prepared for ^{19}F NMR studies, but the *h*Arg(CH$_2$CF$_3$)$_2$ residue is protonated at physiological pH (pK_a = 9.5) and appears to function well in vivo. Whether the increased potency relative to *h*Arg(Et$_2$) is due to the increased hydrophobicity of the CH_2CF_3 substituent or to its unique blend of hydrophobicity and polarity (partial negative charge on F) is not clear. Thus, position 6 of GnRH can optimally accept either a very hydrophobic amino acid side chain like naphthyl or a polar side chain like the guanidine function on Arg or *h*Arg(R$_2$). This remarkable observation is best explained by our hypothesis that the primary role for the side-chain functional group in position 6 in these analogues deals with pharmacokinetic effects, not specific receptor interactions.

Receptor Binding Considerations

We have focused on an enhanced interaction with the phospholipid head groups of the cell membrane for this class of positively charged amino acids and have outlined the reasons why this interaction may have an important effect on the pharmacokinetics of analogues containing them. Such an interaction may be envisioned as having two other benefits as well. As mentioned earlier, the Gouy–Chapman equation (*44*) predicts that positively charged molecules would form an increasing concentration gradient toward a negatively charged surface such as the cell membrane. This phenomenon could be envisioned as facilitating membrane-bound peptide receptor loading due to the increased apparent ligand concentration. The hypothesis by Schwyzer et al. (*92*) that flexible small-peptide ligands adopt their "biologically active" conformation on or in the cell membrane suggests a second benefit from mechanisms for increased membrane affinity such as the one we devised. Schwyzer et al. suggested that initial interaction between a positively charged region of the ligand is followed by penetration of the membrane by a less polar region of the ligand in an α-helical conformation. Membrane binding by positively charged or neutral peptides has been implicated in the mechanism of receptor subtype selectivity, e.g., μ-, κ-, δ-opioid receptor selectivity (*93*). This type of receptor-loading mechanism would fit nicely with the older concept (*94*) of ligand-receptor targeting by "address" (positively charged) and "message" (less polar, α-helical region) sequences within a peptide hormone. Also, the mechanism would provide an explanation of the source of the energy to pay the large, unfavorable entropic price of rigidification of a peptide from its random solution conformation (*95*). While there has been experimental confirmation of the ability of small peptide hormones to adopt this organized, membrane-bound structure (*96*), this mechanism has not been proven to be a step on the way to receptor binding. Our present picture of the structure of the "seven transmembrane" class of receptors (*6*) does not show a direct route from the membrane surface or interior to the presumed binding site in the lumen of the barrel formed by the transmembrane helices. Nonetheless, prolonged pharmacokinetic profiles and concentration due to the Gouy–Chapman fixed charge potential (*44*) offer the possibility of important benefits in drug design.

Additional Applications of the Peptide Depoting Approach

To further broaden the applicability of this approach, I applied it to several other peptide drug discovery projects, including that of ACE inhibition (*see* reference *97* for review). This ACE inhibition seemed to be a productive area to explore because duration of action appeared to be an unsolved problem with the molecules then extant (captopril, enalapril, and lisinopril) and a clear-cut structure–activity relationship for the P_1 and P_1' side chains was lacking (*98*). Also, lisinopril and enalapril are excreted predominantly in the urine without metabolism of the active species, yet still have relatively short $t_{1/2}$ values (*99*).

ACE is a Zn-containing metalloendoprotease (peptidyldipeptide hydrolase, EC 3.415.1) that converts angiotensin I to the potent vasopressor substance angiotensin II. Significantly, the hypotensive hormone bradykinin is inactivated by this enzyme as well, and potentiation of the hypotensive effects of bradykinin (apparently NO mediated) may play a substantial role in the therapeutic effects seen (*100*). Inhibition of ACE has become front-line therapy for hypertension and other cardiovascular diseases.

Early on, two major classes (*97*) of ACE inhibitor were recognized: captopril ana-

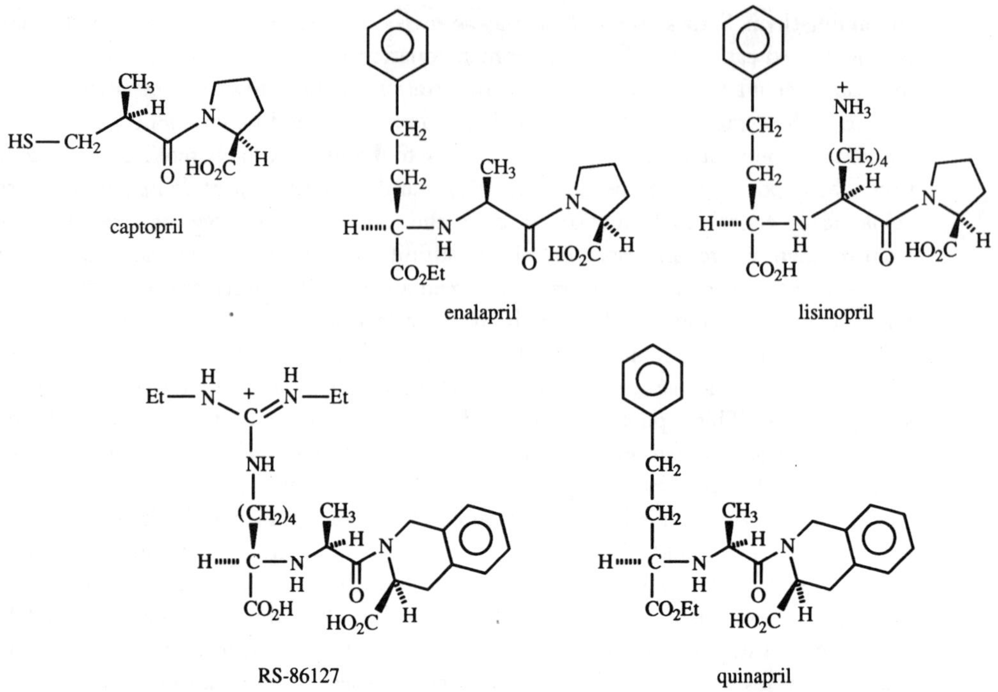

Chart I. Chemical structures of RS-86127 and ACE inhibitor standards.

logues (*101*) and enalapril analogues (*102;* Chart I). In the enalapril series, most compounds were tested as the ethyl ester prodrug of the active diacid species (i.e., enalaprilate) to prolong the duration of action and enhance oral bioavailability. I was intrigued by the observation that a homophenylalanine (*h*Phe) residue could be accepted in the putative P_1 position, a position in which Pro would be expected to bind for the substrate. This substitution is found in the majority of the later enalapril analogues. This residue appeared to be important for potency, but I questioned the specificity of its interaction with the enzyme. I noted that a Lys side chain in the P_1' site could replace the CH_3 of Ala and lead to lisinopril, a potent molecule (*102*), but again questioned the specificity of the interaction. I hypothesized that in both cases the side chains were having relatively nonspecific, hydrophobic interactions with the enzyme [e.g., with the $(CH_2)_4$ of Lys for lisinopril] and that they were also contributing to hydrophobic and hydrophilic depoting of the molecules. I therefore sought to test these hypotheses by specifically enhancing these contributions. As a stringent test of this concept, I replaced the aromatic side chain (*h*Phe) of enalaprilate with positively charged *h*Arg(R_2) residues. I thus sought to replace what I considered a hydrophobic depoting residue with a hydrophilic depoting residue. I also increased the global hydrophobicity of the molecule by exchanging the C-terminal Pro residue with the more hydrophobic tetrahydroisoquinoline carboxylic acid residue. These analogues were initially made and tested as the active forms on the enzyme, the diacid analogues. The most potent analogue of the series was RS-86127 (Chart I).

Whereas RS-86127 was a potent molecule in vitro, its most interesting properties were demonstrated in vivo (*103*). The IC_{50} for RS-86127 was essentially identical to that of enalaprilate and its closest structural comparator quinaprilate (reported after this work

Table IV. Inhibitory Activity against Human ACE

Compound	IC_{50} (nM)[a]	$t_{1/2}$ (h)[b]	Duration of Hypotensive Effect (dose)[c]
Captopril	23	1.7	8 h (75 mg/kg; po)
Enalaprilate[d]	1.2	11	8 h (12.5 mg/kg; po)
Lisinopril	1.2	12.6	NT[e]
Quinaprilate[d]	2.8	1	NT[e]
RS-86127	1.4	NT[a]	>30 h (12.5 mg/kg; po)

[a] 50% inhibitory concentration (IC_{50}) was determined against human ACE.
[b] Values determined in humans, using the prodrug form for enalapril and quinapril.
[c] See data in Figure 7; number of hours of blood pressure decrease of >20 mm Hg.
[d] Compound is a dicarboxylic acid metabolite, active species.
[e] NT is not tested.

was completed), but the value was substantially lower than that of captopril (Table IV). When tested in the spontaneously hypertensive rat (12.5 mg/kg, orally) versus the same dose of enalapril and a larger dose of captopril (75 mg/kg), an unusual pattern developed. The literature standards were active as expected on day one (hypotensive effects for 1–8 h), whereas no effect was seen for RS-86127 in most experiments. However, when a second dose was given on day two, the standards behaved as before, but RS-86127 caused a profound hypotensive response lasting for >24 h (Figure 7). Thus, the efficacy of the active species against the enzyme is not a predictor of the duration of action in vivo for this analogue. RS-86127 was tested against enalapril, a longer-acting prodrug form of the active species enalaprilate, so the results are even more impressive. The closest structural type to RS-86127 is quinaprilate, the active diacid form of the prodrug quinapril (*104*) (Chart I). The

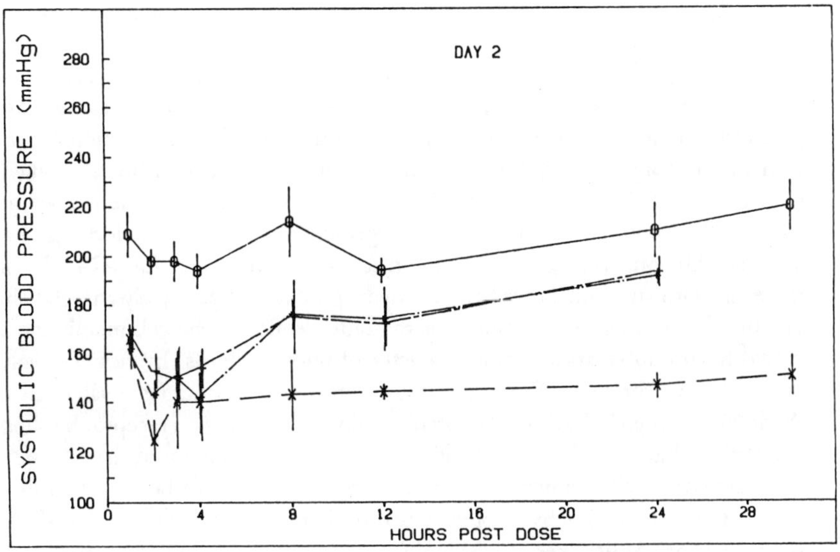

*Figure 7. Duration of antihypertensive effect in conscious spontaneously hypertensive rats for RS-86127 and ACE inhibitor standards. Compounds were administered orally at time zero. Key: +, captopril (75 mg/kg); *, enalapril (12.5 mg/kg); X, RS-86127 (12.5 mg/kg); and O, vehicle.*

$t_{1/2}$ of quinapril is much shorter than that of enalapril (1 versus 11 h) in humans (*105*), and the corresponding data in the rat are expected to be much shorter. Thus, even though we have no data directly comparing RS-81627 with quinapril, its impressive duration of effect in comparison with that of enalapril demonstrates the remarkable improvement caused by replacement of the *h*Phe residue with *h*Arg(Et_2). In later studies, the Ca salt of RS-86127 was evaluated with more reproducible and even longer lasting hypotensive effects at the same 12.5-mg/kg dose (>72 h; end of study).

Although the precise mechanisms responsible for this remarkable duration of action have not been proven, we attribute it to the membrane affinity that we have designed into the molecule. We envision a combination of electrostatic (phosphate head group polar region) and hydrophobic (membrane interior) interactions as being responsible for this affinity. This combination would result in a depoting of the molecule on tissues throughout the body as initially visualized. One would also expect that RS-86127 might be depoting on tissues in the gut and result in delayed uptake and a further delaying of clearance from the body.

Summary and Conclusions

The thrust of this chapter was to illustrate that the design of long-acting peptide pharmaceuticals can not rely solely on the blockade of metabolism. The approach that we focused on couples a degree of protection from proteolysis with physical properties designed to assure depoting of the compound in the body. This whole-body depot approach protects the peptide from both proteolysis and the normally rapid clearance by glomerular filtration. Thus, although ganirelix and detirelix are still subject to proteolysis at the Ser^4-Tyr^5 site, as is GnRH itself, the drugs have $t_{1/2}$ values in the 15–48-h range.

During the past decade, we have modified our analogue design considerations from purely hydrophobic depoting to a more balanced approach. Whereas hydrophobic compounds like [D-Nal(2)^6]GnRH and similar analogues were able to achieve substantial increases in serum $t_{1/2}$ values (up to 3–5 h), solubility eventually became a problem. When positively charged residues were coupled with the hydrophobic character of members of the GnRH antagonist series, some relatively soluble, but extraordinarily long-lived peptides ($t_{1/2}$ in the 15–>30-h range) resulted. In the ACE inhibitor series, we again sought a balanced hydrophobic and positively charged character for the molecule and again achieved a remarkably long-acting compound (antihypertensive effects for >24 h by oral dosing). A molecule with such a balance between hydrophobic and positively charged character can be visualized as having good affinity for the outer leaflet of the cell membrane because of the mixed hydrophobic and anionic character of phospholipids. In fact, we evaluated some of our analogues for such affinity and found affinity constants (K_a) values in the 10^6 range. As discussed, such behavior also may have beneficial effects on receptor binding and pharmacokinetic behavior. The considerations described are not often used to interpret peptide structure–activity relationships, but we have found them to be very fruitful.

Overuse of the hydrophobic approach to increasing $t_{1/2}$ is possible, however. In several recent studies, very hydrophobic peptides (renin inhibitors and somatostatin analogues) were rapidly cleared by the liver. Remarkably, active uptake of these peptides by hepatocytes appeared to use the same pathway as that used to transport bile salts (*106*). Therefore, the use of a balanced hydrophobic and hydrophilic approach to depoting may be of increased importance. A new type of partition coefficient, that with membrane rather than octanol, may be the critical measurement in the future.

The examples presented in this chapter and by others working in the peptide area show that successful peptide pharmaceuticals can be designed with a long duration of action. If injection or controlled-release from an erodable depot is acceptable, quite useful results can be obtained (e.g., GnRH agonists for endometriosis, prostatic cancer, and precocious puberty). A number of peptides can achieve therapeutic levels and a sufficient $t_{1/2}$ by intranasal application (GnRH agonists, calcitonin, and vasopressin analogues).

The fundamental and still unsolved barrier to the more widespread use of peptide pharmaceuticals is that of poor oral bioavailability. The protective barrier of the gut lining excludes polar molecules above a size of about 8 $Å^2$ cross-section unless an active transport system is being used. Studies aimed at understanding the structural requirements for uptake by the dipeptide transport system of the gut are under way and offer the promise of drugs or prodrugs designed to be taken up by this route. The ACE inhibitor enalapril, for example, is actively absorbed via this mechanism (*107*).

Because breaching this barrier in a nondestructive manner is so difficult, the best current approach to reproducible, high oral bioavailability appears to be peptidomimetic design or high-throughput natural-product screening for a nonpeptide lead (*see* reference 108 for review). The advent of "nonpeptide library" approaches to generate structural diversity offers perhaps the most exciting possibilities for lead generation (*109, 110*). Unfortunately, many examples of the nonpeptide library screening approach are successful, but few clear examples of de novo peptidomimetic design have been reported (however, *see* reference 111). This search remains a challenge for the near future, but in the interim, the list of successful peptide pharmaceuticals continues to lengthen.

Acknowledgments

I thank my colleagues for their efforts in carrying out the chemical and biological work described in this chapter. Without the diligent and creative work of Teresa L. Ho and Brian H. Vickery, these studies would not have been possible. Georgia I. McRae, David Waterbury, Arthur Strosberg, and Philip Felgner made substantial contributions to the biological and biophysical characterization of these molecules. I am grateful to Mary Smith for her preparation of this manuscript.

References

1. Dutta, A. S. In *Advances in Drug Research:* Testa, B., Ed.; Academic: Orlando, FL, 1991; Vol. 21, pp 147–286.
2. Davis, S. S. *Scrip Mag.* **1992,** *May,* 34–38.
3. Smith, P. L.; Wall, D. A.; Gochoco, C. H.; Wilson, G. *Adv. Drug Deliv. Rev.* **1992,** *8,* 253–290.
4. Veber, D. F.; Freidinger, R. M. *Trends Neurosci.* **1985,** *8,* 392–396.
5. Fauchère, J.-L.; Thurieau, C. In *Advances in Drug Research;* Academic: Orlando, FL, 1992; Vol. 23, pp 127–159.

6. Savarese, T. M.; Fraser, C. M. *Biochem. J.* **1992,** *283,* 1–19.

7. Heber, D.; Dodson, R.; Swerdloff, R. S.; Channabasavaiah, K.; Stewart, J. M. *Science (Washington, DC)* **1982,** *216,* 420–421.

8. Conradi, R. A.; Hilgers, A. R.; Ho, N. F. H.; Burton, P. S. *Pharm. Res.* **1991,** *12,* 1453–1460.

9. Tomita, M.; Shiga, M.; Hayashi, M.; Awazu, S. *Pharm. Res.* **1988,** *5,* 341–346.

10. Haigler, E. D., Jr.; Hershman, J. M.; Pittman, J. A., Jr. *J. Clin. Endocrinol. Metab.* **1972,** *35,* 631–635.

11. Wood, A. J.; Maurer, G.; Niederberger, W.; Beveridge, T. *Transplant Proc.* **1983,** *15 (Suppl 1),* 2409–2412.

12. Davis, S. S. *Trends Pharmacol. Sci.* **1990,** *11,* 353–355.

13. Edgington, S. M. *Bio/Technology* **1991,** *9,* 1327–1331.

14. Vickery, B. H. In *LHRH and Its Analogs: Contraceptive and Therapeutic Applications;* Vickery, B. H.; Nestor, J. J., Jr., Eds.; MTP: Lancaster, England, 1987; Part 2, pp 547–556.

15. Sanders, L. M.; Kell, B. A.; McRae, G. I.; Whitehead, G. W. *J. Pharm. Sci.* **1986,** *75,* 356–360.

16. Langer, R. *Science (Washington, DC)* **1990,** *249,* 1527–1533.

17. Bennett, H. P. J.; McMartin, C. *Pharmacol. Rev.* **1979,** *30,* 247–292.

18. Spatola, A. F. In *Chemistry and Biochemistry of Amino Acids, Peptides and Proteins;* Weinstein, B., Ed.; Marcel Dekker: New York, 1983; Vol. 7, Chapter 5.

19. Fauchère, J. L. In *Advances in Drug Research;* Testa, B., Ed.; Academic: Orlando, FL, 1986; Vol. 15, pp 29–69.

20. Finchan, C. I.; Higginbottom, M.; Hill, D. R.; Horwell, D. C.; O'Toole, J. C.; Ratcliffe, G. S.; Rees, D. C.; Roberts, E. *J. Med. Chem.* **1992,** *35,* 1472–1484.

21. Dworkin, L. D.; Brenner, B. M. In *The Kidney: Physiology and Pathophysiology,* 2nd ed.; Seldin, D. W.; Giebisch, G., Eds.; Raven: New York, 1992; Chapter 29.

22. Horsthemke, B.; Knisatschek, H.; Rivier, J.; Sandow, J.; Bauer, K. *Biochem. Biophys. Res. Commun.* **1981,** *100,* 753–759.

23. Pimstone, B.; Epstein, S.; Hamilton, S. M.; LeRoith, D.; Hendricks, S. *J. Clin. Endocrinol. Metab.* **1977,** *44,* 356–360.

24. Barriere, S. L.; Flaherty, J. F. *Clin. Pharm.* **1984,** *3,* 351–373.

25. Cash, W. D.; Mahaffey, L. M.; Buck, A. S.; Nettleton, D. E., Jr.; Romas, C.; duVigneaud, V. *J. Med. Pharm. Chem.* **1962,** *5,* 413–423.

26. Nakagawa, S. H.; Yang, F.; Kato, T.; Fluoret, G.; Hecter, O. *Int. J. Pept. Prot. Res.* **1976,** *8,* 465–479.

27. Chorev, M.; Shavitz, R.; Goodman, M.; Minick, S.; Guillemin, R. *Science (Washington, DC)* **1979,** *204,* 1210–1212.

28. Hardy, P. M.; Lingham, I. N. *Int. J. Pept. Prot. Res.* **1983,** *21,* 392–405.

29. Dutta, A. S.; Furr, B. J. A.; Giles, M. B.; Valcaccia, B.; Walpole, A. L. *Biochem. Biophys. Res. Commun.* **1978,** *81,* 382–390.

30. Polónski, T.; Chimiak, A. In *Peptides: Structure and Biological Function;* Gross, E.; Meienhofer, J., Eds.; Pierce Chemical: Rockford, IL, 1979; pp 265–268.

31. Metcalf, G.; Dettmar, P. W.; Lynn, A. G.; Brewster, D.; Havler, M. E. *Regul. Pept.* **1981,** *2(5),* 277–284.

32. Jones, W. C.; Nestor, J. J., Jr.; du Vigneaud, V. *J. Am. Chem. Soc.* **1973,** *95,* 5677–5679.

33. Clausen, K.; Anwer, M. K.; Bettag, A. L.; Benovitz, D. E.; Edwards, J. V.; Lawesson, S.-O.; Spatola, A. F.; Winkler, D.; Browne, B.; Rowell, P.; Schiller, P.; Lemieux, C. In *Peptides: Structure and Function;* Hruby, V. J.; Rich, D. H., Eds.; Pierce Chemical: Rockford, IL, 1983; pp 307–310.

34. Sandberg, B. E. B.; Lee, C-M.; Hanley, M. R.; Iversen, L. L. *Eur. J. Biochem.* **1981,** *114,* 329–337.

35. Rivier, J.; Ling, N.; Monahan, M.; Rivier, C.; Brown, M.; Vale, W. In *Peptides: Chemistry, Structure, and Biology;* Walter, R.; Meienhofer, J., Eds.; Ann Arbor Science: Ann Arbor, MI, 1975; pp 863–870.

36. Tam, J. P.; Lin, Y.-Z.; Wu, C.-R.; Shen, Z. Y.; Galantino, M.; Liu, W.; Ke, X.-H. In *Peptides: Chemistry, Structure, Biology;* Rivier, J.; Marshall, G. R., Eds.; ESCOM: Leiden, The Netherlands, 1990; pp 75–77.

37. Monahan, M.; Amoss, M.; Anderson, H. A.; Vale, W.; Guillemin, R. *Biochemistry* **1973,** *12,* 4616–4620.

38. Pert, C. B.; Pert, A.; Chang, J.; Fong, B. T. W. *Science (Washington, DC)* **1976,** *194,* 330–332.

39. Pietrzyk, D. J.; Smith, R. L.; Cahill, W. R., Jr. *J. Liq. Chromatogr.* **1983,** *6,* 1645–1671.

40. Atherton, E.; Law, H. D.; Moore, S.; Elliott, D. F.; Wade, R. *J. Chem. Soc C* **1971,** 3393–3396.

41. Martinez, J.; Bali, J. P.; Rodriguez, M.; Castro, B.; Magous, R.; Laur, J.; Ligon, M. F. *J. Med. Chem.* **1985,** *28,* 1874–1879.

42. Coy, D. H.; Heinz-Erian, P.; Jiang, N.-Y.; Sasaki, Y.; Taylor, J.; Moreau, M.-P.; Wolfrey, W. T.; Gardner, J. D.; Jensen, R. T. *J. Biol. Chem.* **1988,** *263,* 5056–5660.

43. Sasaki, Y.; Coy, D. H. *Peptides* **1987,** *8,* 119–121.

44. Sargent, D. F.; Bean, J. W.; Kosterlitz, H. W.; Schwyzer, R. *Biochemistry* **1988,** *27,* 4974–4977.

45. Roberts, D. C.; Vellaccio, F. In *The Peptides;* Gross, E.; Meienhofer, J., Eds.; Academic: Orlando, FL, 1983; Vol. 5, Chapter 6.

46. Ho, T. L.; Nestor, J. J., Jr.; McRae, G. I.; Vickery, B. H. *Int. J. Pept. Prot. Res.* **1984,** *24,* 79–84.

47. Do, K. Q.; Thanei, M.; Caviezel, M.; Schwyzer, R. *Helv. Chim. Acata* **1980,** *62,* 956–964.

48. Nestor, J. J., Jr.; Ho, T. L.; Simpson, R. A.; Horner, B. L.; Jones, G. H.; McRae, G. I.; Vickery, B. H. *J. Med. Chem.* **1982,** *25,* 795.

49. Hsieh, K.-h.; LaHann, T. R.; Speth, R. C. *J. Med. Chem.* **1989,** *32,* 898–903.

50. Nestor, J. J., Jr.; Tahilramani, R.; Ho, T. L.; McRae, G. I.; Vickery, B. H. *J. Med. Chem.* **1988,** *31,* 65–72.

51. Bosshard, H. R.; Berger, A. *Helv. Chim. Acta* **1973,** *56,* 1838–1845.

52. Benoiton, N. L.; Seely, J. H. *Can. J. Biochem.* **1970,** *48,* 1122–1131.

53. Hruby, V. J. *Life Sci.* **1982,** *31,* 189–199.

54. DiMaio, J.; Schiller, P. W. *Proc. Natl. Acad. Sci. U.S.A.* **1980,** *77,* 7162–7166.

55. Mosberg, H. I.; Hurst, R.; Hruby, V. J.; Galligan, J. J.; Burks, T. F.; Gee, K.; Yamamura, H. I. *Biochem. Biophys. Res. Commun.* **1982,** *106,* 506–512.

56. Kleinert, H. D.; Baker, W. R.; Stein, H. H. *Adv. Pharmacol.* **1991,** *22,* 207–250.

57. Martin, J. A. *Antiviral Res.* **1992,** *17,* 265–278.

58. Ariëns, E. J. In *Drug Design;* Ariëns, E. J., Ed.; Academic: Orlando, FL, 1971; Vol. 2, Chapter 1.

59. Hansch, C.; Steward, A. R.; Anderson, S. M.; Bentley, D. *J. Med. Chem.* **1967,** *11,* 1–11.

60. Karten, M. J.; Rivier, J. E. *Endocrine Rev.* **1986.** *7,* 44–66.

61. Rivier, J.; Ling, N.; Monahan, M.; Rivier, C.; Brown, M.; Vale, W. In *Peptides: Chemistry, Structure, and Biology;* Walter, R.; Meienhofer, J., Eds.; Ann Arbor Science: Ann Arbor, MI, 1975; pp 863–870.

62. Coy, D. H.; Vilchez-Martinez, J. A.; Coy, E. J.; Schally, A. V. *J. Med. Chem.* **1976,** *19,* 423–425.

63. Nestor, J. J., Jr.; Ho, T. L.; Tahilramani, R.; McRae, G. I.; Vickery, B. H. In *LHRH and Its*

Analogs; Labrie, F.; Belanger, A.; Dupont, A., Eds.; Elsevier Science Publishers B. V.: Amsterdam, The Netherlands, 1984; pp 24–35.

64. Nadasdi, L.; Medzihradszky, K. *Biochem. Biophys. Res. Commun.* **1981,** *99,* 451–457.
65. Clayton, R. N.; Catt, K. J. *Endocrine Rev.* **1981,** *2,* 186–209.
66. Strulovici, B.; Tahilramani, R.; Nestor, J. J., Jr. *Biochemistry* **1987,** *26,* 6005–6111.
67. McArdle, C. A.; Conn, P. M. In *LHRH and Its Analogs: Contraceptive and Therapeutic Applications;* Vickery, B. H.; Nestor, J. J., Jr.; Hafez, E. S. E., Eds.; MTP: Lancaster, England, 1984; Part 2, pp 77–100.
68. Chan, R. L.; Chaplin, M. D. *Biochem. Biophys. Res. Commun.* **1985,** *127,* 673–679.
69. Chan, R. L.; Chaplin, M.D. *Drug Metab. Dispos.* **1985,** *13,* 566–571.
70. Felgher, A.; Felgner, P. L.; Louie, P.; Malone, E.; Nestor, J. J., Jr., Syntex Research, Palo Alto, CA, unpublished results.
71. Nestor, J. J., Jr. In *LHRH and Its Analogs: Contraceptive and Therapeutic Applications;* Vickery, B. H.; Nestor, J. J., Jr., Eds.; MTP: Lancaster England, 1987; Part 2, pp 3–16.
72. Coy, D. H.; Horvath, A.; Nekola, M. V.; Coy, E. J.; Erchegyi, J.; Schally, A. V. *Endocrinology (Baltimore)* **1982,** *110,* 1445–1447.
73. Nestor, J. J., Jr.; Tahilramani, R.; Ho, T. L.; McRae, G. I.; Vickery, B. H. *J. Med. Chem.* **1984,** *27,* 1170–1174.
74. Nestor, J. J., Jr.; Tahilramani, R.; Ho, T. L.; Goodpasture, J. C.; Vickery, B. H.; Ferrandon, P. *J. Med. Chem.* **1992,** *35,* 3942–3948.
75. Nestor, J. J. Jr.; Tahilramani, R.; Ho, T. L.; McRae, G. I.; Vickery, B. H.; Bremner, W. J. In *Peptides: Structure and Function;* Hruby, V. J.; Rich, D. H., Eds.; Pierce Chemical: Rockford, IL, 1983; pp 861–864.
76. Vickery, B. H.; McRae, G. I.; Goodpasture, J. C. In *LHRH and Its Analogs: Contraceptive and Therapeutic Applications;* Vickery, B. H.; Nestor, J. J., Jr., Eds.; MTP: Lancaster, England, 1987; Part 2, pp 517–543.
77. Pavlou, S. N.; Wakefield, G. B.; Kovacs, W. J. In *LHRH and Its Analogs: Contraceptive and Therapeutic Applications;* Vickery, B. H.; Nestor, J. J., Jr., Eds.; MTP: Lancaster, England, 1987; Part 2, pp 245–256.
78. Andreyko, J. L.; Monroe, S. E.; Marshall, L. A.; Fluker, M. R.; Nerenberg, C. A.; Jaffee, R. B. *J. Clin. Endocrinol. Metab.* **1992,** *74,* 399–405.
79. Chan, R. L.; Ho, W. L.; Webb, A. S.; LaFargue, J.; Nerenberg, C. A. *Pharm Res.* **1988,** *5,* 335–340.
80. Danforth, D. R.; Gordon K.; Leal, J. A.; Williams, R. F.; Hodgen, G. D. *J. Clin. Endocrinol. Metab.* **1990,** *70,* 554–556.
81. Schmidt, F.; Sundaram, K.; Thau, R. B.; Bardin, C. W. *Contraception* **1984,** *29,* 283–289.
82. Lee, C.-H.; Van Antwerp, D.; Hedley, L.; Nestor, J. J., Jr.; Vickery, B. H. *Life Sci.* **1989,** *45,* 697–702.
83. Forman, J.; Jordan, C. *Agents Actions* **1983,** *13,* 105–116.
84. Karten, M. J.; Hook, W. A.; Siraganian, R. P.; Coy, D. H.; Folkers, K.; Rivier, J. E.; Roeske, R. W. In *LHRH and Its Analogs: Contraceptive and Therapeutic Applications,* Part 2; Vickery, B. H.; Nestor, J. J., Jr.; Eds.; MTP: Lancaster, England, 1987; pp 179–190.
85. Sydbom, A.; Terenius, L. *Agents Actions* **1985,** *16,* 269–272.
86. Roeske, R. W.; Chaturvedi, N. C.; Hrinyo-Pavlina, T.; Kowalczuk, M. In *LHRH and Its Analogs: Contraceptive and Therapeutic Applications;* Vickery, B. H.; Nestor, J. J., Jr., Eds.; MTP: Lancaster, England, 1987; Part 2, pp 17–24.
87. Clark, B.; Smith, D. A. In *Progress in Drug Metabolism;* Gibson, G. G., Ed.; Wiley Interscience: Chichester, England, 1989; Vol. 11, pp 175–215.

88. Rabinovici, J.; Rothman, P.; Monroe, S. E.; Nerenberg, C.; Jaffee, R. B. *J. Clin. Endocrinol. Metab.* **1992,** *75,* 1220–1225.

89. Chan, R. L.; Hsieh, S. C.; Haroldsen, P. E.; Ho, W.; Nestor, J. J., Jr. *Drug Metab. Dispos.* **1991,** *19,* 858–864.

90. Chu, N. I.; Chan, R. L.; Hama, K. M.; Chaplin, M. D. *Drug Metab. Dispos.* **1985,** *13,* 560–565.

91. Nestor, J. J., Jr.; Tahilramani, R.; Ho, T. L.; McRae, G. I.; Vickery, B. H. In *Peptides: Structure and Function;* Deber, C. M.; Hruby, V. J.; Kopple, K. D., Eds.; Pierce Chemical: Rockford, IL, 1985; pp 557–560.

92. Schwyzer, R.; Gremlich, H.-U.; Gysin, B.; Sargent, D. F. In *Peptides: Structure and Function;* Hruby, V. J.; Rich, D. H., Eds.; Pierce Chemical: Rockford, IL, 1983; pp 657–664.

93. Schwyzer, R. *Biochemistry* **1986,** *25,* 6335–6342.

94. Schwyzer, R. *Ann. N. Y. Acad. Sci.* **1977,** *297,* 3–26.

95. Schwyzer, R. In *Peptides 1986;* Theodoropoulos, D., Ed.; de Gruyter: New York, 1987; pp 7–23.

96. Erne, D.; Sargent, D. F.; Schwyzer, R. *Biochemistry* **1985,** *24,* 4261–4263.

97. Petrillo, E. W., Jr.; Ordetti, M. D. *Med. Res. Rev.* **1982,** *2,* 1–41.

98. Hirschman, R. In *Peptides: Structure and Function;* Hruby, V. J.; Rich, D. H., Eds.; Pierce Chemical: Rockford, IL, 1983; pp 1–32.

99. Vertes, J.; Haynie, R. *Am. J. Cardiol.* **1992,** *69,* 8c–16c.

100. Wiemer, G.; Schölkens, B. A.; Becker, R. H. A.; Busse, R. *Hypertension* **1991,** *18,* 558–563.

101. Ondetti, M. A.; Rubin, B.; Cushman, D. W. *Science (Washington, DC)* **1977,** *196,* 441–44.

102. Patchett, A. A.; Harris, E.; Tristram, E. W.; Wyvratt, M. J.; Wu, M. T.; Taub, D.; Peterson, E. R.; Ikeler, T. J.; ten Broeke, J.; Payne, L. G.; Ondeyka, D. L.; Thorsett, E. D.; Greenlee, W. J.; Lohr, N. S.; Hoffsommer, R. D.; Joshua, H.; Ruyle, W. V.; Rothrock, J. W.; Aster, S. D.; Maycock, A. L.; Robinson, F. M.; Hirschmann, R.; Sweet, C. S.; Ulm, E. H.; Gross, D. M.; Vassil, T. C.; Stone, C. A. *Nature (London)* **1980,** *288,* 280–283.

103. Hedley, L.; Van Antwerp, D.; Lee, C.-H.; Strosberg, A. M. *Pharmacologist* **1987,** *29,* 478.

104. Klutchko, S.; Blankley, C. J.; Fleming, R. W.; Hinkley, J. M.; Werner, A. E.; Nordin, I.; Holmes, A.; Hoefle, M. L.; Cohen, D. M.; Essenburg, A. D.; Kaplan, H. R. *J. Med. Chem.* **1986,** *29,* 1953–1961.

105. Frishman, W. H. *Am. J. Cardiol.* **1992,** *69,* 17c–25c.

106. Bertrams, A. A.; Ziegler, K. *Biochim. Biophys. Acta* **1991,** *1091,* 337–348.

107. Yuasa, H.; Amidon, G. L.; Fleisher, D. *Pharm. Res.* **1993,** *10,* 400–404.

108. Wiley, R. A.; Rich, D. H. *Med. Res. Rev.* **1993,** *13,* 327–384.

109. Bunir, B. A.; Ellman, J. A. *J. Am. Chem. Soc.* **1992,** *114,* 10997–10998.

110. DeWitt, S. H.; Kiely, J. S.; Stankovic, C. J.; Schroeder, M. C.; Cody, D. M. R.; Pavia, M. R. *Proc. Natl. Acad. Sci. U.S.A.* **1993,** *90,* 6909–6913.

111. Lam, P. Y. S.; Jadhav, P. K.; Eyermann, C. J.; Hodge, C. N.; Ru, Y.; Bacheler, L. T.; Meek, J. L.; Otto, M. J.; Rayner, M. M.; Wong, Y. N.; Chang, C.-H.; Weber, P. C.; Jackson, D. A.; Sharpe, T. R.; Erickson-Viitanen, S. *Science (Washington, DC)* **1994,** *263,* 380–384.

112. Nestor, J. J., Jr. In *Third SCI–RSC Medicinal Chemistry Symposium;* Lambert, R. W., Ed.; Royal Society of Chemistry: London, 1986; pp 362–384.

RECEIVED for review July 12, 1993. ACCEPTED revised manuscript February 16, 1994.

Methods and Systems for Evaluating Transport and Metabolism of Peptide-Based Drugs

CHAPTER 20

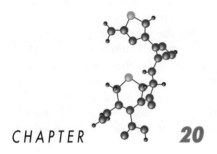

Cell Culture Models for Examining Peptide Absorption

Donald W. Miller,
Akira Kato,
Ka-yun Ng,
Elsbeth G. Chikhale,
and Ronald T. Borchardt*

The potential application of peptides and proteins as therapeutic agents represents an area of growing interest to pharmaceutical scientists. The potency and specificity of peptides are desired advantages, but their low bioavailability and decreased absorption across cellular barriers pose significant problems for peptide therapeutics. Therefore, the successful design and development of peptide therapeutics must take into consideration the physicochemical properties of peptides that influence absorption and metabolism as well as the cellular mechanisms controlling macromolecule transport and clearance.

Until recently, information concerning the transport and clearance of a therapeutic agent relied almost exclusively on pharmacokinetic profiles from in vivo and perfused-organ experiments. However, advances in cell culture technology have provided an additional tool

*Corresponding author

for the study of membrane absorption, transport, and clearance of therapeutic agents (*1*). Cell culture models are particularly advantageous for examining the cellular mechanisms controlling transport and metabolism of compounds and identifing ways in which these systems can be used or circumvented to facilitate absorption. Cell culture models also provide a time- and cost-efficient way to determine the influence of chemical or molecular modifications on the transport and metabolism of potential therapeutic agents.

Many different cellular barriers are of interest for the absorption and transport of peptides. However, our discussion will focus on the relevant cell culture models available to examine peptide absorption and metabolism in the intestine, the blood–brain barrier (BBB), and the liver. Special emphasis will be given to the considerations necessary for studying peptide absorption in cell culture systems and the pharmaceutical applications of the various cell culture models in the study of peptide transport and clearance. The correlation between data obtained with cell culture methods and other in vitro and in vivo techniques will also be examined.

Selection and Validation of Cell Culture Models

The utility of cell culture models in examining peptide absorption and metabolism depends on the extent to which the models reproduce the in vivo situation. Ideally, the culture models selected would retain all the morphological features and biochemical characteristics present in the in vivo cellular barrier. The highly differentiated and multicellular nature of most biological barriers involved in peptide transport and metabolism makes complete reproduction of such barriers with cell culture models difficult. However, as the following section attempts to highlight, current cell culture models of the gastrointestinal barrier, BBB, and liver have many morphological and biochemical similarities to their respective in vivo systems.

Morphological Similarities of Culture Models

The mucosal surface of the small intestine comprises various cell types at different stages of differentiation, but the predominent cell type found in the intestinal barrier is the epithelial-derived enterocyte. The prominent morphological features of enterocytes include tight junctional complexes and the presence of microvilli (*2, 3*). Development of primary cultures of enterocytes as a model of the gastrointestinal barrier has proven difficult because of the nonproliferative nature of the isolated mucosal cells (*2, 3*). By using the Caco-2 cell line derived from human colon carcinoma cells, the problem of nonproliferation observed with primary cell cultures is minimized. Furthermore, upon reaching confluency, cultured Caco-2 cells become columnar with tight junctional complexes and microvilli at the mucosal surface (*4, 5*). Although some differences exist (*5, 6*), Caco-2 cells display many of the morphological characteristics that are important in producing the barrier properties of enterocytes in the small intestine and, in this regard, are an acceptable model for studying peptide absorption in the gastrointestinal tract.

The clearance and subsequent metabolism of circulating peptides in the liver limits the bioavailability of most peptides. Therefore, successful delivery of peptides will have to address their hepatic clearance from the bloodstream. In this regard, both nonparenchymal cells (liver endothelia cells) and parenchymal cells (hepatocytes) play an important role. The hepatic absorption of peptides from the blood and their subsequent secretion into the bile requires uptake across the sinusoidal membrane of the hepatocyte followed by secretion across the canalicular membrane of the hepatocyte. To accomplish this, the hepatocytes develop a polarized and functionally differentiated membrane. Both primary hepatocytes and hepatoma cell lines have been shown to re-establish this morphological and functional membrane polarization with cell–cell contacts and bile canaliculi developing over time (*7*). Ultrastructural studies of primary cultured hepatocytes demonstrate that within 24 h in culture, hepatocytes become flattened and polygonal, and biliary membrane polarity is established (*8, 9*).

Although membrane polarity can be achieved, conventional culturing techniques for hepatocytes are limited in their ability to maintain an established epithelial-like polarity. Because the intended applications of hepatocyte cultures are limited to uptake and absorption, as opposed to both absorption and transport in the culture models of the intestinal barrier and BBB, the lack of sustained membrane polarity does not in itself limit the usefulness of the model. However, modifications in the culturing process have improved the time period over which functional and morphological membrane polarity can be maintained (*10–12*). Recent studies of primary cultured hepatocytes embedded in the extracellular matrix report a more complete and sustained polarization of the hepatocyte membrane (*13, 14*).

In contrast to the multiple cell types that compose the intestinal mucosal barrier and the hepatic clearance system, brain microvessel endothelial cells (BMEC) are the primary cells forming the BBB. The barrier properties of the BMEC are attributable, in large part, to the tight junctions and the virtual absence of fenestrations and pinocytic vesicles (*15*). Primary cultured BMEC form similar tight junction complexes with no detectable fenestrations (*16, 17*). In addition, primary cultures of BMEC have little pinocytic activity and abundant mitochondrial organelles (*16, 18*). Both primary and passaged BMEC have been used as cell culture models of the BBB, but most of the models use primary bovine BMEC. Recently, primary BMEC from rat and human sources have been used (*17, 19*).

Biochemical Characteristics of Cell Culture Models

The ability to express specific carriers, receptors, and enzymes in a polarized manner is a characteristic common to enterocytes, BMEC, and hepatocytes, and is essential for the barrier and clearance functions of these cells. Caco-2 cells have been shown to express specific transporter systems for various nutrients and essential macromolecules (*see* Table I). In addition, marker enzymes, such as alkaline phosphatase, γ-glutamyltranspeptidase (GTP), and angiotensin-converting enzyme (ACE), are expressed in Caco-2 cells in a polarized manner (*4, 5*) (*see* Table II). Compared with HT-29 cells, and alternative cell culture model of the intestinal barrier, Caco-2 cells display a greater number of peptidases on the cell surface (*20*) and may therefore be a more appropriate model for evaluating peptide metabolism in the intestinal tract.

Cultured BMEC also retain many biochemical characteristics of endothelial cells in general and BBB endothelial cells in particular (*see* Tables I and II). Biochemical studies with primary cultured bovine BMEC demonstrated a >98% positive staining for factor VIII antigen, a specific endothelial cell marker (*18*). Furthermore, bovine BMEC monolayers also express low-density lipoprotein receptors on the apical cell surface, which is characteristic of endothelial cells (*18*). The similar binding characteristics of various lectins in the in vivo BBB and primary cultured BMEC (*21*) suggest that the glycoproteins found in the BBB are retained in the culture model as well. The presence of transport carriers for various amino acids (*22, 23*), glucose (*24, 25*), biotin (*26*), choline (*27, 28*), and transferrin (*29*) has also been reported in primary cultured bovine BMEC (Table I). Given the nature of

Table I. Representative Receptor and Transport Systems Expressed in the Various Cell Culture Models

	Caco-2		*BMEC*		*Hepatocytes*	
Receptor and Transporter	*Exp.*	*Ref.*	*Exp.*	*Ref.*	*Exp.*	*Ref.*
Amino acid	+	60	+	22, 23	+	152, 153
Angiotensin II			+	154	+	48, 155
Asialoglycoprotein					+	34, 36
Atrial natruiretic factor			+	103		
Bile acid (taurocholate)	+	73			+	47, 108, 114
Biotin	+, −	74, 156	+	26	−	157
Cobalamine (Vitamin B_{12})	+	57, 89, 90				
Choline			+	27, 28		
Dipeptide	+	82, 83				
Epidermal growth factor	+	87			+	117
Ferritin					+	158
Folate	+	159				
Glucose (Na-dependent)	+	56, 160	+	24, 25	+	161
Immunoglobulin A					+	162, 163
Insulin			+	101, 102	+	121
Insulin-like growth factor I			+	104		
Low-density lipoprotein	+	164	+	18	+	165
Methyldopa	+	166				
Monocarboxylic acid			+	167		
Nucleotide			+	28		
P-glycoprotein	+	168	+	169	+	170
Peptide YY	−	3				
Phosphate	+	171				
Spermidine	−	156				
Transferrin	+	88	+	17, 29	+	119, 120
Tissue plasminogen activator					+	46, 116
Vasoactive intestinal peptide (VIP)	+	172			+	155
Vasopressin			+	93	+	48

NOTES: BMEC is brain microvessel endothelia cells. Exp is the expression, either functional or biochemical, of the various receptor and transport systems listed. (+) indicates studies showing expression and (−) indicates studies that were unable to show expression.

the BBB, such transport systems would be essential for delivering nutrients and necessary macromolecules to the brain. Cultured BMEC also express many enzymes, including γ-GTP (*16, 30, 31*), monoamine oxidase (*30*), and catechol-*O*-methyl transferase (*30*), which are characteristic of the BBB (Table II). Furthermore, the presence of various intracellular peptidases (*32*) as well as membrane-bound enzymes such as ACE on the surface (*17, 30*) indicates that cultured BMEC may also be suitable for the study of peptide metabolism in the BBB.

Cultured hepatocytes also express transport carriers for several nutrients, macromolecules, and amino acids (Table I). In addition, primary cultured hepatocytes express glycoproteins that are specific for the liver and may have important functions in the hepatic clearance

Table II. Representative Enzyme Systems That Are Either Characteristic of the Various Cell Types or Involved in Peptide Metabolism

Cell Type	Enzyme Class	Enzyme Example	References
Caco-2	Hydrolases	Sucrase-isomaltase	6, 173
		Lactase-phlorizin hydrolase	173
		Alkaline phosphatase	4, 5
	Peptidases	Aminopeptidase N	20
		Dipeptidylpeptidase IV	6, 20, 173
		Angiotensin-converting enzyme	173
		Aminopeptidase P	173
		Aminopeptidase W	173
		Endopeptidase-24,11	173
		γ-Glutamyl transpeptidase	173
		Membrane dipeptidase	173
	Other enzymes	Ornithine decarboxylase	174
		Diamine oxidase	174
		Phenol sulfotransferase	175
		UDP-glucuronyl transferase	176
		Acyl CoA: cholesterol acyltransferase	177, 178
		3-Hydroxy-3-methylglutaryl CoA reductase	178
		17-β-hydroxysteroid dehydrogenase	179
BMEC	Hydrolases	Acid hydrolase	180
		Alkaline phosphatase	16, 30
	Peptidases	Aminopeptidases	32
		γ-Glutamyl transpeptidase	16, 30, 31
		Angiotensin-converting enzyme	17, 30
	Other enzymes	Catechol-*O*-methyltransferase	30
		Monoamine oxidase	30
		Phenol sulfotransferase	30, 181–183
Hepatocytes	Phase I	Cytochrome *P*-450	10, 37–40
	Phase II	Glutathione *S*-transferase	10, 39, 40
		Cystathionase	49
		3-Hydroxy-3-methylglutaryl CoA reductase	42, 43
		Acyl CoA: cholesterol acyltransferase	41
	Other enzymes	γ-Glutamyl transpeptidase	184
		Peptidases	36, 44, 45

NOTE: BMEC is brain microvessel endothelia cells.

of glycosylated peptides and other macromolecules (*33–36*). Because the liver serves an enormous metabolic function, the models for examining hepatic absorption and metabolism should also retain as many enzymatic properties as possible. As shown in Table II, primary cultured hepatocytes express enzymes from both phase I (cytochrome *P*-450 enzymes) and phase II (glutathione *S*-transferase) reactions (*10, 37–40*). Furthermore, enzymes involved in cholesterol metabolism were detected in cultured hepatocytes (*41–43*). Metabolism of peptides and proteins in cultured hepatocytes was examined by using various peptidase inhibitors (*36, 44–46*). The inhibition of peptide and protein degradation following treatment with peptidase inhibitors indicates that the culture model has the capacity for peptide degradation and could be useful in evaluating peptide metabolism in the liver.

Although cultured hepatocytes express many of the transporter and enzyme systems present in the intact liver, the level of expression is often lower in the culture model (*37, 39, 40, 47–49*). In most cases, the differences are a result of a reduced number of binding sites (B_{max}) in cultured hepatocytes, with the affinity of the carrier or enzyme system (K_d) remaining similar to that found in the intact liver (*47, 48*). In several cases, either a delayed loss of expression or an induced expression of transporter and enzyme systems was observed following alterations in the culture conditions (*10–14, 39, 40*). Recent studies using primary hepatocytes that are embedded in or sandwiched between the extracellular matrix report significant improvements in enzyme induction and expression compared with the results from the more conventional primary hepatocyte monolayers (*13, 50*).

Additional Considerations in Selecting Cell Culture Models

In establishing in vitro cell culture models, reproducibility of the model and accessibility of the cells are important considerations. Although primary cell culture models have a more heterogeneous population of cells and may more closely resemble the in vivo system than cells that have been passaged or established cells lines, interculture variation is more typical with primary cells. Increased reproducibility can be obtained by optimizing the isolation and culturing procedures for the various cell types (*51*). The use of primary cultures also requires access to fresh tissue or organs, which are not always readily available. This problem may be compounded when human organs, such as brain or liver, are desired for the primary cell culture model. In contrast to primary cell cultures, many human-derived malignant cells (including Caco-2 and hepatoma cells) are maintained as established cell lines and are obtainable from the American Type Culture Collection (ATCC). Whereas cells from established cell lines are in general more homogeneous, care must be taken to ensure that passaging the cells does not cause substantial changes in the barrier characteristics (*3*).

For the barriers discussed in the present chapter, few choices in cell culture models are currently available. Although both primary cells and established cell lines are used for examining hepatocyte function, the present culture models for the BBB and gastrointestinal tract rely on primary cells and immortalized cell lines, respectively. However, as more cell culture models become available and improvements are made on existing culture systems, such considerations may become important in selecting the appropriate cell culture model to be used.

Selection and Validation of Microporous Filter Support and Transport Apparatus

Development of suitable cell culture models for evaluating peptide absorption and metabolism requires selection not only of the appropriate cell type but also the appropriate surface for culturing the cells and the transport apparatus for measuring absorption. In this section, the criteria used in selecting the cell support and transport apparatus for cell culture models will be considered.

Types and Considerations of Filter Supports for Transport Studies

The uptake and absorption of compounds can be examined using cells cultured on solid supports, but actual transport studies require culturing of the cells onto permeable membrane or filter supports. Fortunately, microporous filter supports of various chemical composition, area, pore size, and design are now commercially available for cell culture studies. Ideally, the microporous filter support should be sufficiently translucent for direct viewing and monitoring of cell growth by ordinary light microscopy (*52*). In addition, the barrier properties of the cell culture system should be provided by the cell monolayer, not the microporous filter or extracellular matrix on which the cells are grown. In general, the greater the pore size and density, the less the potential for the filter to act as a significant barrier to adsorption. However, practical limitations affect how large the pore size can be before the filter loses its ability to support the cell monolayer. Furthermore, the potential for the cells to migrate through the filter and begin to grow on the underside of the membrane is increased with the larger diameter pore sizes (*53*). This growth results in a poorly defined monolayer and should be avoided whenever possible.

The physicochemical properties of the solute may cause the filter insert and extracellular matrix to act as barriers to peptide transport or absorption. Nonspecific binding of the solute with the filter insert or extracellular matrix has been reported (*54*) and may pose significant problems with some peptides. Often the addition of proteins, such as albumin, to the assay buffer or medium will reduce the amount of nonspecific binding to the filter insert or extracellular matrix. Alternatively, the filter insert and extracellular matrix may act as a molecular sieve by restricting the passage of larger macromolecules. The result of either nonspecific binding or molecular sieving with the filter insert and extracellular matrix is an underestimate of the amount of transport occurring across the cell monolayer. To ensure the selection of the appropriate filter insert and extracellular matrix for examining absorption of peptides in cell culture systems, the passage of the peptide must be examined across the microporous filter and extracellular matrix alone. Information from these studies can then be used to make the appropriate corrections for determining the permeability of peptides across cell monolayers.

Diffusion Apparatus Used for Examining Peptide Absorption in Cell Culture Models

The type of diffusion apparatus that is commonly employed for studying solute transport across cell monolayers is the system diagramed in Figure 1A that uses a cell culture insert. Variations from this basic design involve the type of filter used and configuration of the diffusion apparatus. In general, this system allows easy and independent access to both the apical and basolateral plasma membrane domains. However, transport studies with this diffusion apparatus are usually performed under stagnant conditions (*55–60*) and result in the formation of a large and potentially nonphysiological unstirred water layer (UWL). The effects of the UWL on permeability are greatest for lipophilic compounds, but peptide absorption and transport may also be influenced. Furthermore, potential modifications or conjugations of peptides to lipophilic carriers may result in a peptide with significant lipophilicity. Under these conditions, the contribution of the UWL may be an important factor in determining peptide permeability.

Although the contribution of the UWL can be reduced by agitating the insert system, the design limits the degree of agitation. Therefore, in situations where the UWL is contributing to the barrier properties of the cell culture system, a side-by-side diffusion apparatus

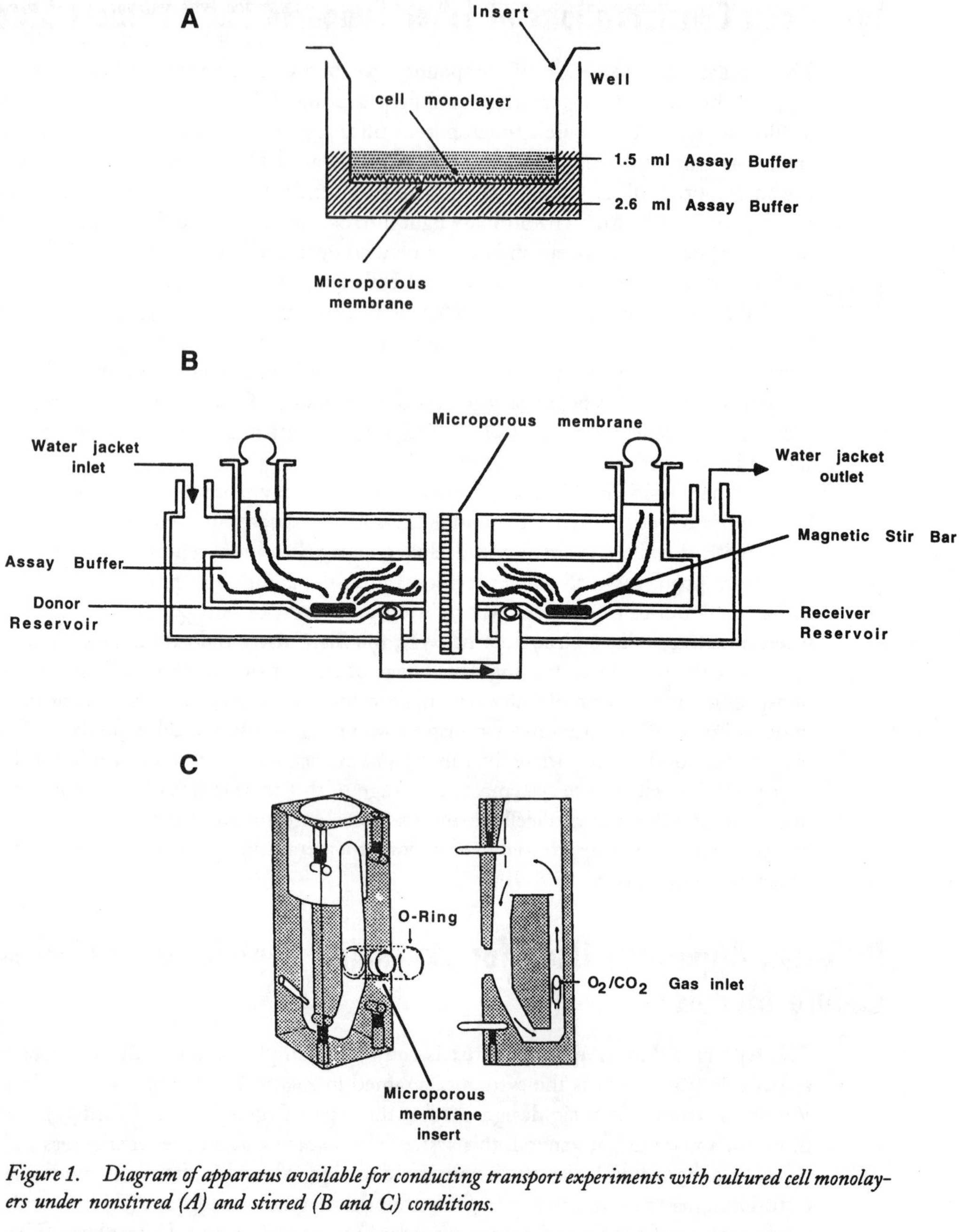

Figure 1. Diagram of apparatus available for conducting transport experiments with cultured cell monolayers under nonstirred (A) and stirred (B and C) conditions.

(Figure 1B) may be more appropriate. The mechanical stirring of the apical and basolateral compartments in the side-by-side system reduces the UWL. However, damage to the cell monolayer can occur during the process of mounting the membrane in the diffusion apparatus (*61*). Recently, a new side-by-side diffusion cell that allows the study of transport across cell monolayers grown on microporous membranes was developed (Figure 1C). The new diffusion apparatus provides a well-stirred environment for transport studies and minimizes the chance of inflicting edge damage to the cell monolayers during mounting or during the course of the transport experiment. The use of this device was validated for the Caco-2 (*62*) and the BMEC cell culture systems (*63*). Overall, this apparatus appears to be superior to the previous diffusion apparatus for studying lipophilic compounds where UWL effects are present.

Influence and Importance of Culture and Assay Conditions

The successful use of in vitro cell culture systems to study solute transport across biological barriers will depend on how closely the in vitro system mimics the in vivo biological barrier. Replication of the in vivo barrier will depend on the culture conditions under which the cells are grown and the assay conditions used for determining absorption or metabolism. Detailed descriptions of the methods and conditions required for Caco-2 (*5, 64*), BMEC (*65*), and hepatocyte (*13, 66*) cell culture were reported previously. The following section will highlight some of the culture and assay conditions that may influence absorption and metabolism of peptides and proteins in the various cell culture models.

Culture Conditions

Most cells in culture generate some form of extracellular matrix, but the amount and composition vary considerably. The extracellular matrix is involved in many different functions, from cell adhesion and proliferation to gene expression and cellular differentiation (*see* references 67–69 for review). With the exception of Caco-2 cells, the culture models presently being discussed require the addition of an extracellular matrix for proper cell monolayer development. The actual extracellular matrix varies; collagen type I (*70*) or collagen type I plus fibronectin (*65, 71*) are used in BMEC. The importance of selecting the appropriate extracellular matrix is best illustrated in cultured hepatocytes, where the extracellular matrix used influences both enzyme expression and induction as well as morphological development of the hepatocyte monolayer (*11*).

Other factors, such as cell seeding density, stage of cellular differentiation, number of times the cells have been passaged, and phenotypic stability, can influence the properties of the cell monolayers and, hence, the cellular absorption or metabolism process. For example, the permeability of confluent BMEC monolayers to the marker molecule, sucrose, decreased proportionately with increasing cell seeding density (*53*). The explanation for this is currently unknown, but increased cell density at seeding may produce a confluent monolayer with a different differentiation state than that observed in confluent monolayers seeded at a lower density. The influence that differentiation can have on peptide absorption is best illustrated for carrier- or receptor-mediated transport systems. Studies examining the carrier-mediated transport of phenylalanine (*62, 72*), taurocholate (*73*), and biotin (*74*) in Caco-2 cells indicate that transport is dependent on the number of days the cells are cultured. A similar time dependency was reported for the expression of the bile acid carrier in primary cultured hepatocytes (*47*). Changes in receptor and carrier activity and expression as a function of cell passage is an additional consideration. In Caco-2 cells, uptake of both phenylalanine and biotin was dependent on the number of times the cells had been passaged (*62, 72, 74*). The decreased transport of these nutrients by their respective transporters suggests that a loss of phenotypic stability occured in the cultured cells during passaging.

Whereas phenotypic changes are possible with any cultured cell (*75*), loss of phenotypic stability is even more pronounced in nontransformed cell lines, such as primary cultured bovine BMEC (*76*). Thus, intitial experiments to determine the optimal seeding density, culturing time, and passaging number should be performed before conducting the transport experiments.

Culture Media

The obvious role of culture medium is to supply the cultured cells with nutrients and growth factors. For this purpose, the culture medium is usually supplemented with serum, which provides hormones and specific growth factors required for cell proliferation and differentiation. The choice of which culture medium to use for growing cells is usually empirical. However, culture media and various media supplements can be formulated in such a way as to enhance particular metabolic or morphological characteristics in the cell culture system. In primary cultured bovine BMEC, heparin is routinely added to the culture medium to stimulate endothelial cell growth and inhibit the growth of smooth muscle cells that may be present in the primary bovine BMEC cultures (*65*). The influence of culture media in cell monolayer development and proliferation is most apparent in cultured hepatocytes where an amino acid-rich medium in the absence of serum enhanced and prolonged the proliferation of primary cultured hepatocytes (*12*). Furthermore, recent studies suggested that coculturing or the addition of conditioned media from other cell types can alter the morphological and biochemical properties of both primary cultured BMEC and hepatocytes (*10, 53, 70, 71*). These studies demonstrate the importance of culture medium in influencing the absorption and metabolic characteristics of the various culture models.

Transport Buffers

To design an in vitro transcellular transport experiment, care must be taken to provide conditions that mimic, as closely as possible, the environment that exists at the in vivo biological barrier of interest. For example, the addition of a soluble cAMP analog and a phosphodiesterase inhibitor to the culture medium and transport buffer produces a low-permeability BMEC monolayer that more closely resembles the permeability observed in the in vivo BBB (*71*). For carrier-mediated transport events, all aspects of the composition of the transport buffers, such as the concentration of the solute, temperature, pH, presence or absence of proteins that might bind the solute, and presence or absence of competing solutes, can influence the transport properties of the cell monolayers. Therefore, identification of each component of the transport buffer is important to eliminate any false interpretation.

Analysis of Peptide Transport and Metabolism in Cell Culture Models

An important advantage in using cell culture models to examine peptide absorption and metabolism is the ability to precisely monitor the location of the peptide and its metabolites throughout the absorption process. This monitoring can be done by examining the accumulation of peptides and metabolites in the apical and basolateral extracellular compartments as well as in the cell monolayer. Exposure of the cells to acid wash or mild trypsin solution allows the cell-associated fraction to be further separated into extra- and intracellular components. For most peptides, samples that require little preparation before analysis can be obtained from the basolateral or apical compartments. Determination of cell-associated peptides and metabolites is often more complicated because of the presence of endogenous macromolecules and may require further sample preparation before analysis. However, the increased sample preparation does not prevent the analysis of cell-associated peptide and metabolite accumulation from providing useful information for identifying the cellular mechanisms responsible for peptide transport and metabolism.

Sample Detection and Separation

The evaluation of peptide absorption and metabolism relies on the ability to accurately detect such events. The physicochemical properties of peptides and the potential for reduced expression of enzymes and proteins that influence transport and metabolism in some cell culture models require an analytical system with sensitive detection capabilities. The most commonly used methods for measuring peptides and their metabolites are liquid chromatography (LC) and radiolabel detection. Detection of peptides by LC can be accomplished by measuring their UV absorption. The use of LC coupled with fluorescence detection gives improved sensitivity over UV detection and is an option for peptides with fluorescent functional groups such as tryptophan in the sequence. Absorption and metabolism can also be analyzed using radiolabeled peptides. The use of radiolabeled peptides offers a relatively easy and sensitive method for determining peptides transport in cell culture models. Evaluation of peptide metabolism can also be done either qualitatively using radiolabeled peptides (trichloroacetic acid precipitation) or quantitatively (LC analysis of radiolabeled fragments). The aforementioned analytical procedures are the most commonly used, and a more detailed discussion of the techniques available for peptide separation and detection can be found in recent reveiwes (*77–79*).

Pharmaceutical Applications of Cell Culture Systems for the Study of Peptide Transport and Metabolism

Two general pathways exist for the absorption of peptides across cellular barriers: the paracellular route (between the cell) and the transcellular route (through the cell). In most cases, the molecular size of peptides excludes significant intestinal, brain, or liver absorption through the paracellular route. Therefore, the main route for peptide passage across the cellular barriers is through transcellular processes. The transcellular absorption of peptides can be further subdivided into passive diffusion or nonpassive transport. In the following section, the use of the various cell culture models to examine peptide absorption through both passive diffusion and nonpassive transport mechanisms will be discussed.

Cell Culture Models for Examining Intestinal Absorption of Peptides

The passive diffusion of most peptides across the intestinal mucosa is characteristically low and is subject to enzymatic degradation. Similar findings (*80, 81*) were reported using the Caco-2 culture model, and only a small number of peptides underwent low-level, nonsaturable, bidirectional passive diffusion across the cell monolayer. The reduced amount of passive diffusion of peptides across Caco-2 cells provides researchers with a model system for examining the physicochemical properties that influence passive diffusion of peptides across the intestinal mucosa. Using a series of D-phenylalanine-containing peptides with various degrees of amide methylation, Conradi et al. (*82, 83*) examined the importance of hydrogen-bonding potential in passive diffusion across Caco-2 cell monolayers. The results of these studies indicated that the hydrogen-bonding potentials of the model peptides were inversely related to their corresponding permeabilities across Caco-2 cell monolayers. Such a correlation was especially evident when other factors, such as molecular size or lipophilicity, were kept constant (*82, 83*) and indicates the importance of hydrogen-bonding in the design of orally active peptide therapeutics.

The intestinal absorption of many di- and tripeptides, including the cephalosporin antibiotics, is facilitated by a dipeptide transporter. The presence of functional dipeptide transporter systems in the Caco-2 culture model was examined by using the enzymatically stable cephalosporins (*59, 84*). Cellular uptake of cephalexin was pH dependent, and maximum uptake occured at pH 6.0. Uptake was also saturable under physiological conditions (*59*). Furthermore, the uptake of cephalexin in Caco-2 monolayers was inhibited competitively by various dipeptides (*59*). Recently, Inui et al. (*84*) demonstrated that specific transport systems were involved in the trancellular transport of cephradine across Caco-2 cell monolayers. In these studies, the presence of both a proton-dependent transport system on the apical plasma membrane and a second unidentified carrier on the basolateral membrane was suggested to be involved in cephradine transport. Recent studies by Smith et al. (*85*) present evidence that suggests the γ-GTP located on the apical plasma membrane may be involved in dipeptide transport. However, the extent to which γ-GTP contributes to the transport of dipeptides in vivo remains to be determined. Further characterization of the dipeptide transporter in Caco-2 cells should provide useful information concerning the possible intestinal transport of dipeptides and the structural requirements necessary for transport by this particular carrier.

Active intestinal absorption of peptides by either fluid-phase or receptor-mediated endocytosis represents another route for transporting peptides from the gastrointestinal tract into the blood stream. The use of Caco-2 cells to assess endocytic transport through the intestine is a potential application for which the culture model is well-suited. Intestinal absorption of proteins by fluid-phase endocytosis was evaluated (*5, 86*) in Caco-2 cell cultures by using horseradish peroxidase (HRP). The results of these studies indicated that HRP was internalized and transported in a temperature-dependent, bidirectional manner. However, the HRP undergoing fluid-phase transcytosis was subject to substantial degradation during the process and was significantly greater in the basolateral-to-apical direction (*5*). Similar findings were reported for the receptor-mediated transcytosis of epidermal growth factor (EGF) in Caco-2 cells (*87, 88*). The receptor-mediated endocytosis of transferrin has also been examined using Caco-2 cells. Although the processing of transferrin appears to be different than that observed for EGF, both receptor binding and transferrin uptake were polarized to the basolateral side of the Caco-2 membrane (*88*). Unless ways to enhance the capacity and directionality of both fluid-phase and receptor-mediated endocytosis are found, the utility of the aforementioned systems for enhancing the intestinal absorption of peptides and proteins will be minimal. The use of the Caco-2 culture model to examine the cellular events controlling the rate, directionality, and processing of macromolecules in the intestine could begin to address such issues.

In contrast to both EGF and transferin, the receptor-mediated binding and uptake of cobalamine (vitamin B_{12}) occurs exclusively on the apical side of the enterocytes. Intestinal absorption occurs by receptor-mediated endocytosis of a complex formed by the binding of cobalamine to an intrinsic factor complex. Following internalization, the cobalamine is released from the intrinsic factor and secreted into the bloodstream bound to the glycoprotein transcobalamine II. Studies using Caco-2 cell monolayers demonstrated the presence of an apical receptor that recognizes the cobalamine intrinsic-factor complex (*89, 90*). Furthermore, proteins similar to transcobalamine II were expressed in Caco-2 cell monolayers and secreted into the basolateral side of the cell monolayer (*57, 90*). The apical uptake of cobalamine and its subsequent transcytosis bound to proteins similar to transcobalamine II in Caco-2 cells are similar to the processes that occur in vivo and indicate that the cell culture system may be useful in evaluating the cobalamine transport system as a means to enhance the intestinal absorption of peptides.

In addition to the morphological characteristics of intestinal epithelia that limit the passage of peptides and proteins, such cells also present an enzymatic barrier to peptide absorption. Augustijns et al. (*91*) studied the enzymatic degradation of delta-sleep-inducing peptide (DSIP). In these studies, the major metabolites of DSIP were Trp-Ala and Trp residues. Furthermore, the appearance of metabolites was more prevelant in the apical buffer solution than in the basolateral buffer solution, a finding that indicates the presence of apically polarized enzyme systems for the metabolism of DSIP in Caco-2 cells (*91*). Although metabolism of DSIP was significantly reduced by the addition of bestatin (aminopeptidase inhibitor), diprotin A (dipeptidylpeptidase IV inhibitor), and captopril (ACE inhibitor), no increase in the permeability of DSIP or its analogues in the presence of the various enzyme inhibitors was observed (*91*).

Cell Culture Models for Examining Peptide Absorption in the BBB

Peptides that are known to cross the in vitro and in vivo BBB via a passive diffusion pathway include DSIP (*92*) and the apical-to-basolateral passage of arginine vasopressin (AVP) (*93, 94*). Currently, work is ongoing in our laboratory to study the effect of the conformational structure of DSIP on its in vitro BBB permeability. We hope information from this study will help explain the ability of DSIP to penentrate the BBB and thus provide a basis for future design of peptide molecules with increased BBB permeability.

Passive diffusion involves the partitioning of the solute from an aqueous environment into the lipophilic plasma membrane. Therefore, the generally hydrophilic nature of peptides presents a major obstacle to passive diffusion across BMEC of the BBB. In an attempt to increase the peptide diffusion through the BBB, the functional groups of peptide molecules have been modified to produce more lipophilic compounds. For example, substitution of D-alanine at position 4 of DSIP for a glycine residue increases the lipophilicity of the analogue and results in increased transport across primary cultured bovine BMEC compared with DSIP (*91*). Similar findings were also reported by Weber and colleagues (*95*) in that modifications of Met-enkephalin to produce more lipophilic peptide analogues also resulted in an increased absorption of the peptides in primary bovine BMEC monolayers. The influence of hydrogen-bonding potential on peptide permeability was also examined by using primary cultured bovine BMEC monolayers (*96*). Similar to the studies of Conradi et al. (*82, 83*) in the Caco-2 cell culture model, the passive diffusion of a series of model peptides across BMEC monolayers was inversely proportional to the hydrogen-bonding potential of the peptides (*96*). Together, these studies suggest that both lipophilicity and hydrogen-bond potential play important roles in determining the permeability of peptides across BMEC. In addition, these studies demonstrate the utility of the BMEC culture model in evaluating the structural properties that influence peptide absorption in the BBB.

Studies evaluating carrier-mediated transport of peptides across cultured BMEC are somewhat limited. A saturable, selective carrier system was identified for AVP in cultured bovine BMEC (*93*). However, carrier-mediated transport of AVP was in the basolateral-to-apical direction, which is unsuited for use in the delivery of peptides to the brain. Carrier systems for the apical-to-basolateral transport of both glucose (*24, 25*) and biotin (*26*) were reported in cultured BMEC. The extent to which such carriers could be used for the transport of peptides across the BBB remains to be determined.

Transport of peptides by both fluid-phase and absorptive endocytosis represents additional pathways for the delivery of peptides to the brain. Although fluid-phase endocytosis was reported to be responsible for the linear, noncompetitive, temperature-dependent passage of albumin across cultured bovine BMEC, the permeability was relatively limited (*97*). In contrast, cationization or glycosylation of albumin resulted in a significant increase in the permeability of the protein in cultured BMEC (*97*). The increased permeability observed was attributable to adsorptive endocytosis occurring through the electrostatic interactions of the protein at the cell surface (*97*). Recent studies (*98*) indicate that ebiratide, a cationic adrenocortocotropic hormone (ACTH) analogue, is transported across primary cultured bovine BMEC through an adsorptive endocytosis pathway. Such findings have been applied in the design of a chimeric peptide consisting of β-endorphin and cationized albumin. The resulting chimeric peptide had a significantly higher permeability across the BBB than β-endoprhin alone (*99, 100*). These studies suggest that cationization or chimeric peptide formation may be a viable method for the enhancement of peptide and protein transport across the BBB.

The expression of several specific receptors on endothelial cells forming the BBB indicates that receptor-mediated transport of selected peptides may be possible. In this regard, specific receptor sites for insulin (*101, 102*), atrial natriuretic factor (ANF) (*103*), insulin growth factor (IGF) (*104*), and transferrin (*29*) were identified in cultured bovine BMEC. In most cases, the binding of the specific ligands to these receptors triggered the rapid internalization of the receptor–ligand complexes with a modest amount of the peptides undergoing transcytosis (*29, 102*). Chemical modifications of peptides and proteins undergoing receptor-mediated endocytosis have received less attention, probably because binding of ligand to receptor is such a highly specific event that even slight modifications in the chemical structure of the ligand may abolish its interaction with the receptor. However, recent studies (*105, 106*) attempted to use antibodies to receptor proteins as possible carriers for drugs across the BBB.

Cell Culture Models for Examining Hepatic Absorption of Peptides

Similar to the epithelial intestinal barrier and the endothelial BBB, the tight junctions on the sinusoidal and cannicular membranes of hepatocytes limit the passive paracellular clearance of peptides (*107*). Therefore, passive diffusion of peptides must occur through interactions with the plasma membranes of the hepatocytes. The size and hydrophilic nature of most peptides prevent their clearance through passive intracellular diffusion, but some small cyclical peptides may be cleared in this manner. One such example is the cyclic undecapeptide cyclosporin A, a potent immunosuppressant used in organ transplantation and cancer therapy. Although interactions of cyclosporin A with membrane proteins associated with the bile-acid carrier are likely, given the inhibition of taurochloric acid uptake observed in primary cultured hepatocytes exposed to cyclosporin A (*108*), the peptide itself does not appear to be a substrate for the bile-acid carrier. These results support the studies of Ziegler et al. (*109*) in isolated hepatocytes, which suggested that cyclosporin A was cleared by the liver through nonspecific hydrophobic interactions with the hepatic membrane.

An undesirable side effect of cyclosporin A is the potential hepatic toxicity associated with the peptide. Studies by Boelsterli et al. (*110*) examined the hepatic uptake of cyclosporin

A and various analogues and their relationship to cyclosporin A-induced hepatic damage in primary hepatocyte cultures. The results of these studies indicated that hepatocyte injury was directly related to the accumulation of cyclosporin A in cultured hepatocytes, because the analogues of cyclosporin A that displayed the highest absorption rates were also the most toxic to the hepatocytes. Although attempts to correlate the lipophilicity of the cyclosporin analogues with their clearance in cultured hepatocytes were favorable for four of the five analogues, other physicochemical parameters, such as size and hydrogen-bonding potential, may also be of importance. These results illustrate the utility of cultured hepatocytes in studies examining the hepatic absorption and toxity of peptides and the influence that peptide structure modifications have on both parameters. Given the clearance and accumulation of circulating peptides by the liver, determination of hepatic toxicity and the development of strategies to reduce it may be important considerations in the design of peptide therapeutics.

Although some smaller lipophilic peptides may undergo passive clearance by hepatocytes, the majority of peptides are absorbed through active carrier- or receptor-mediated processes (*see* Table I for partial listing of carriers and receptors in cultured hepatocytes). One such important carrier present on hepatocytes is the bile-acid transporter, which is responsible for the hepatic transport of bile acids from the portal bloodstream to the bile. The transport of bile acids in hepatocytes is an energy-dependent and, in most cases, a sodium ion-dependent process (*107*). In addition to transporting bile acids, the bile-acid transporter also appears to be involved in the clearance of various foreign substrates, including peptides, from the bloodstream (*107*). The use of cultured hepatocytes to examine peptide absorption by the bile-acid transporter is somewhat limited because of the time-dependent loss of bile-acid transporter in cultured hepatocytes (*47, 108*). However, the studies of Petzinger and Frimmer (*111*), which demonstrated that the uptake of the cyclic heptapeptide phalloidin and the bile-acid cholate in cultured hepatocytes was subject to a similar time-dependent reduction in absorption, are at least consistent with the possible clearance of cyclic peptides by the bile acid-transporter. Recent studies by Muenter and colleagues (*112*) also reported a loss of carrier-mediated uptake of phalloidin in cultured hepatocytes. In these studies, the carrier-mediated absorption of phalloidin could be partially protected by prior exposure of the monolayers to the peptide.

Cultured hepatocytes appear to express many of the glycoproteins found in the intact liver (*33–36*). These glycoproteins present on the plasma surface of hepatocytes can also contribute to the removal of peptides from the bloodstream by receptor-mediated endocytosis and subsequent degradation. Because the glycoproteins recognize specific carbohydrate moieties on the peptide or protein, a potential reduction in clearance may be obtained through alterations in peptide design. A possible example of this is found in tissue plasminogen activator (tPA), a thrombolytic agent that has promising potential in the treatment of fibrin emboli. A major disadvantage in using tPA is the rapid clearance of plasma tPA by the hepatic system (*113*). Examination of the clearance of tPA in primary cultured hepatocytes indicated a receptor-mediated mechanism, because uptake of tracer tPA was inhibited by the addition of excess unlabeled tPA (*46*). Furthermore, the tPA accumulated in the hepatocytes appeared to be targeted for degradation because various endocytotic inhibitors prevented the uptake and degradation of the protein (*46*). Additional studies revealed that the recognition site for receptor-mediated endocytosis of tPA in cultured hepatocytes is a glycoprotein present on hepatocytes that recognizes a carbohydrate side chain of tPA (*114*). Endocytosis and degradation of tPA were inhibited in the primary cultured hepatocytes by galactose, ovalbumin, and ethylenediaminetetraacetic acid (EDTA), a finding that suggested that receptor-mediated internalization and metabolism of tPA was occurring through galactose binding proteins on the hepatocytes. Of possible therapeutic importance was the demonstration that nonglycosylated derivatives of tPA that retained the desired biological activity were not as readily absorbed in cultured hepatocytes (*114*). Other studies examining tPA clearance in cultured human hepatoma cells (*115*) and in primary hepatocytes (*116*) reported receptor-mediated uptake and metabolism of the protein. However, the uptake and degradation of tPA in these studies was not dependent on the glycosalation pattern of the protein nor was it inhibited by ovalbumin (*115, 116*).

The presence of several receptors for various biological peptides in hepatocytes (*see* Table I) suggests that other receptors, in addition to the asialoglycoprotein receptors, may also be involved in hepatic peptide absorption. Recent studies by Kato et al. (*117*) indicated the presence of both high- and low-affinity binding sites for EGF on the plasma membrane of primary cultured hepatocytes. Because EGF binding to the high-affinity sites appears to be responsible for the biological effects of the peptide (*118*), the low-affinity site present in hepatocytes may function as a clearance receptor for the peptide. The increased number of low-affinity sites compared with high-affinity sites in cultured hepatocytes (*117*) is consistent with the clearance role of such receptors. The presence of hepatic tPA receptors may also represent a pathway for the removal of tPA-related proteins from the circulation (*115, 116*). Furthermore, receptors for transferrin (*119, 120*), insulin (*121*), angiotensin II (*48*), and vasopressin (*48*) may have similar clearance functions in cultured hepatocytes. As more peptides are identified, the number of peptide receptors present on hepatocytes will likely increase. The possibility of designing biologically active peptides with less affinity for the hepatocyte receptors may be a logical approach to increasing the bioavailability of peptide therapeutics.

Comparison of Peptide Transport and Metabolism in Cell Culture Models to In Situ and In Vivo Models

Successful prediction of peptide absorption and metabolism is an important part of peptide-based drug design and development. As discussed in this chapter, the use of cell culture models for examining peptide transport and metabolism provides pharmaceutical scientists with an additional tool for the design of peptide therapeutics. However, the utility of cell culture models in studying peptide absorption and metabolism will depend on the extent to which information obtained from the cell culture models can be applied to the in vivo situation. Several studies have shown a relatively good correlation between the in vivo absorption of small molecules and their permeability in the various culture models (*76, 122*), but studies comparing peptide transport and metabolism in cell culture systems to that occurring in vivo are rather limited. The aim of this section is to compare the available data regarding the permeability and metabolism of various peptides in the cell culture models to absorption with in situ and in vivo preparations.

Caco-2 Cell Culture Model

In general, absorption of most peptides across the intestinal barrier is limited due to the physicochemical properties of the peptides and the characteristics of the cells forming the intestinal barrier. As previously described, the passive diffusion of a series of model peptides

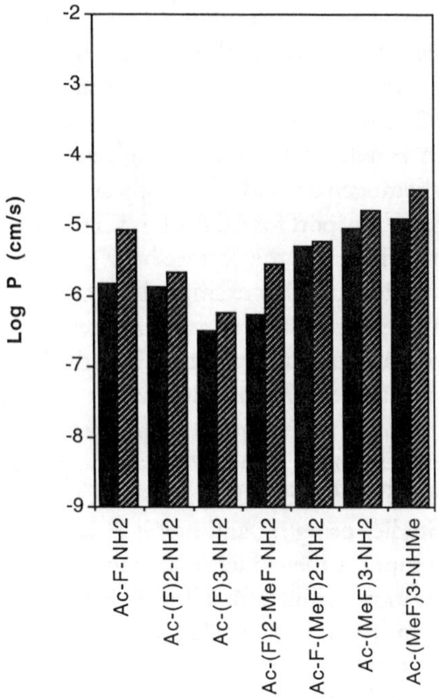

Figure 2. Comparison of the apparent permeability coefficients for selected peptides as determined from cultured Caco-2 cells (hatched bars) and from in situ studies (solid bars).

through Caco-2 monolayers was shown to be dependent on the hydrogen-bonding potential rather than lipophilicity or molecular size (*82, 83*). From these studies, hydrogen-bonding potential was concluded to be an important factor in determining peptide absorption in Caco-2 cells. In this regard, the Caco-2 cell model accurately predicted the permeability of the model peptides both in situ (perfused rat ileum system) (*123*) (*see* Figure 2) and in vivo (anesthetized rats) (*124*). The better reproducibility in determining the permeability coefficients for the model peptides in the Caco-2 cell model compared with the perfused ileum suggests that the Caco-2 model may be a more sensitive method for evaluating the influence that such chemical modifications may have on peptide permeability (*82, 83*). Although in situ and cell culture systems were comparable ($r^2 = 0.883$), the in vivo absorption of the smaller model peptides (Ac-F-NH_2 and Ac-F2-NH_2) was greater than that observed with the Caco-2 cell model (*124*). The differences observed with the lower molecular weight peptides in the two systems may be related to the more complete formation of tight junctions in the Caco-2 cell model compared with the in vivo rat model, and these junctions may result in less paracellular diffusion of the smaller peptides in Caco-2 monolayers. The increased transepithelial electrical resistance found in Caco-2 cell monolayers relative to that found in normal intestinal mucosa is consistent with this explanation (*5*).

Similar results were reported regarding the absorption of 1-deamino-8-D-arginine-vasopressin (dDAVP), an enzymatically stable analogue of the hormone AVP, in the Caco-2 culture model and in vivo preparations. Results from both Caco-2 cell culture (*80*) and everted rat intestinal sac models (*125*) support the passive diffusion of dDAVP in the gastrointestinal tract, most likely through the paracellular route. The observed low absorption rate of dDAVP in the Caco-2 model is comparable to the low in vivo absorption of dDAVP in colonic epithelium (*126*) but smaller than the absorption rate observed in vitro using everted rat intestinal sacs (*125*). This comparison indicates that the permeability of the Caco-2 monolayers is more similar to that of colonic than small-intestinal epithelium. The aforementioned studies using dDAVP also highlight the appropriateness of the Caco-2 culture model in predicting peptide metabolism in vivo. Analysis of dDAVP metabolism in cultured Caco-2 cells indicated that the peptide was enzymatically stable. This result is in agreement with metabolism studies using intestinal segments (*125*), but studies using intestinal mucosa homogenates indicated substantial degradation of dDAVP (*127*). Because the proposed route of transport for dDAVP is via the paracellular pathway, the peptide would not be expected to encounter the intracellular cytosolic enzymes present in the intestinal homogenate preparation. As this example illustrates, the ability of the Caco-2 cell culture model to mimic the enzymatic systems encountered in vivo is an important consideration in selecting this preparation for evaluating intestinal peptide metabolism.

Opinions about the transport mechanism of thyrotropin releasing hormone (TRH), a tripeptide hormone, are still divided. Transport of TRH across Caco-2 monolayers was reported by some investigators to be solely via the paracellular route (*81, 128*), whereas one study reported that 25% of the TRH transport across Caco-2 cells was via carrier-mediated transport (*129*). Studies using rat and rabbit proximal small-intestinal brush-border-membrane vesicles also support the paracellular route (*130*). However, other in situ and in vitro everted sac investigations of intestinal TRH transport suggest that TRH is transported by an active mechanism dependent on Na^+ (*131*).

The extent to which the Caco-2 culture model will reliably predict the in vivo absorption of peptides by active transport mechanisms remains to be elucidated. However, studies examining the active transport system for various dipeptides on the apical plasma membrane of Caco-2 cell monolayers show remarkable similarities to dipeptide absorption in vivo (*59, 84, 132*). The apparent Michaelis constant (K_m) values for cephradine, cephalexin, and cefaclor were 8.3, 7.5, and 7.6 mM, respectively, in the Caco-2 model (*59, 84, 132*) compared with K_m values of 5.0, 2.2–19, and 3.0–16.1 mM, respectively, observed in other in vitro and in situ models (*133–138*). The transport of the cephalosporins was temperature and pH dependent and competitively inhibited by various dipeptides in all the experimental preparations examined. Furthermore, the hydrogen-ion coupled dipeptide transport system reported on the basolateral membrane of Caco-2 cells (*139*) was similar to those previously described in rabbit enterocyte basolateral membrane vesicles (*140*). Taking all these observations into account, we can conclude that Caco-2 cell monolayers provide a useful model for studying properties of the human H^+/dipeptide cotransport system in the intestinal mucosa.

BMEC Culture Model

The absorption of two centrally active peptides, AVP and DSIP, was examined by using both primary cultured bovine BMEC and in situ/in vivo preparations. Studies using primary cultured bovine BMEC indicated that both AVP and DSIP undergo low-level, passive diffusion across the cell monolayers, and permeability coefficients were $2.4–5.1 \times 10^{-3}$ and 1.2×10^{-4} cm/min, respectively, for AVP and DSIP (*92–94, 141*). The permeability coefficients for AVP and DSIP were somewhat lower for in vivo studies (2.9×10^{-5} and 9.3×10^{-6} cm/min, respectively), but the BBB passage of both peptides appeared to be via passive diffusion mechanisms (*142, 143*). However, carrier-mediated transport systems for the blood-to-brain transport of both peptides have been observed using the in situ perfused guinea pig brain preparation (*144, 145*). The presence of carrier-mediated transport of AVP and DSIP in the in situ preparation may be attributable to differences in the species or peptide concentrations used in the in situ studies.

The metabolism of DSIP was examined in primary cultured BMEC (*92*), whole rat preparations (*146*), and mouse brain homogenates (*147*). In both the BMEC and mouse brain homogenates, the major metabolite of DSIP was tryptophan, a finding that suggested metabolism through aminopeptidases in both experimental preparations (*92, 147*). However, the actual rate of DSIP degradation in the BMEC culture model (*92*) was significantly slower than the degradation rate observed following incubation of the peptide with either the whole rat or mouse brain homogenate preparations (*146, 147*). The differences in metabolic rate in cultured BMEC and other preparations may be due to the limited penetration of DSIP into intracellular compartments expressing aminopeptidase activity in brain endothelial cells. Alternatively, whole-brain homogenates may have a higher aminopeptidase activity than the brain microvascular endothelial cells. Support for this latter theory is the observation that aminopeptidase activity in the brain is not located predominantly in the BBB (*32*).

Primary cultured BMEC have also been used to examine the physicochemical properties that influence the permeability of peptides in the BBB. Studies determining the importance of hydrogen-bonding potential in the permeability of peptides in cultured BMEC indicated an inverse relationship between hydrogen-bonding potential and peptide permeability. Whereas the actual permeability coefficients were higher in magnitude in the BMEC culture model, a similar relationship between hydrogen-bonding potential and peptide permeability was observed by using an in situ brain perfusion method (Figure 3; *96*). Primary cultured BMEC have also been used to accurately predict the absorption of a series of opioid peptide analogues. In these studies, the permeabilities of several [Met5]enkephalin analogues were examined in both cultured bovine BMEC and intact conscious rats (*95*). The rank order of peptide permeability in the bovine BMEC cell culture model was similar to the accumulation of peptide in the brain following intravenous administration (*95*). In this particular study, modifications that increased lipophilicity of the peptides resulted in increased permeabilities in both the cell culture model and the in vivo preparation. Taken together, these findings support the concept that physicochemical properties that influence diffusion in the cell culture model have similar effects on BBB diffusion in vivo.

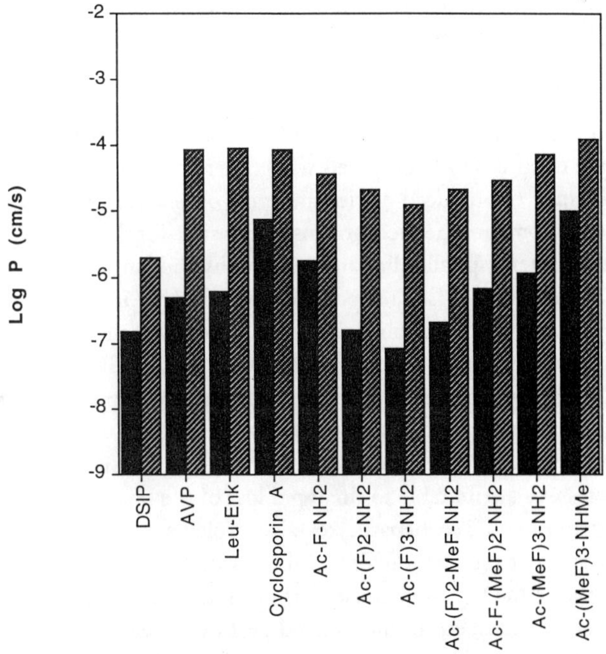

Figure 3. Comparison of the apparent permeability coefficients for selected peptides as determined from BMEC cultures (hatched bars) and from in situ/in vivo studies (solid bars).

Cultured BMEC have also proven to be useful in evaluating the physicochemical parameters involved in the active absorption of macromolecules in the BBB. Results from primary cultured bovine BMEC (*97*), isolated brain capillaries (*99*), and perfused brain studies (*99*) all support the low-level transport of albumin by fluid-phase endocytosis. Cationization or glycosylation of native albumin significantly enhanced the absorption of the protein in cultured bovine BMEC by a saturable and time- and temperature-dependent absorptive endocytotic mechanism (*97*). These findings in the cell culture model are in agreement with earlier studies showing increased absorption of cationized albumin compared with native albumin in isolated brain capillaries and in situ perfusion models (*100*). A similar route of absorption has been proposed to explain the BBB permeability of a cationic neuromodulating ACTH analogue, ebiratide. Analysis of ebiratide transport in cultured bovine BMEC (*98*), isolated bovine brain capillaries (*148*), and in vivo rat preparations (*149*) indicated the saturable uptake of the peptide could be inhibited by endocytotic inhibitors and cationic agents such as polylysine. The dissociation constant for ebiratide uptake in primary cultured bovine BMEC (K_d = 15.9 μM) was of similar magnitude to that obtained in isolated bovine brain capillaries (K_d = 62.1 μM) and indicated the utility of the cell culture system in evaluating the physicochemical properties influencing active absorption of peptides and proteins in the BBB.

Cultured Hepatocyte Model

In comparison to Caco-2 and BMEC culture models, very few experiments have evaluated hepatic peptide absorption in both cell culture and in vivo preparations. This deficiency is attributable, in part, to the reduced activity of uptake and metabolism systems for many compounds in cultured hepatocytes compared with either isolated hepatocytes or perfused liver models (*47*). The explanation for the discrepancy in carrier-mediated uptake of compounds in the experimental systems is not known. However, the reduction in maximum uptake (V_{max}) despite similar K_m values in the culture and perfused liver model suggests that fewer carrier sites are available for mediating transport in cultured hepatocytes. Such reductions could be due to a decrease in the effective surface area available for uptake in confluent monolayers or a decrease of the number of active transport sites expressed in hepatocyte cultures (*47*) compared with other in vitro or in situ preparations.

Despite the reduced carrier transport observed in cultured hepatocytes, studies have attempted to compare the mechanisms responsible for the uptake of the hepatotoxic peptides phalloidin, dimethylphalloidin, and tetramethylrhodamylphalloidin with those occurring in isolated hepatocyes and isolated perfused livers (*111, 150, 151*). Studies using freshly isolated hepatocytes suggested that phalloidin and dimethylphalloidin were taken up predominantly by the bile-acid transport system of liver parenchymal cells and, to a lesser extent, by endocytosis (*107*). However, studies examining the uptake of phalloidin and tetramethylrhodaminylphalloidin in primary cultured hepatocytes indicated uptake by an exclusively endocytotic pathway (*150, 151*). The lack of carrier-mediated transport of the phallotoxins in cultured hepatocytes was attributable to the rapid loss of the bile-acid carrier (*111*). The loss of bile-acid carriers in cultured hepatocytes also explains the lack of toxicity associated with the administration of the phallotoxins in the culture model compared with in vivo preparations (*151*). Because the bile-acid carrier has been implicated in the hepatic absorption of several small cyclic peptides, either the isolated perfused liver or freshly isolated hepatocytes might provide a more suitable experimental model than cultured hepatocytes for examining peptide uptake by carrier-mediated processes.

Summary and Conclusions

The application of cell culture models to evaluate the absorption and metabolism of peptides has tremendous potential in the effective design of peptide and protein therapeutics. The ability to examine the cellular mechanisms involved in peptide transport and metabolism in a controlled system is clearly an advantage of cell culture models. The ability to rapidly assess the influence of chemical modifications is also a desirable feature. However, the acceptance of cell culture models depends on the extent to which the information obtained with the cell culture system is applicable to the in vivo situation. In the present discussion, the current cell culture models for studying intestinal (Caco-2 cells), BBB (BMEC), and hepatic (primary cultured hepatocytes) absorption and metabolism were described. All three culture models possess many of the morphological and biochemical characteristic found in their respective in vivo systems. With the possible exception of cultured hepatocytes, available evidence suggests the culture models discussed represent additional tools for evaluating peptide absorption and transport. Given the current culturing techniques and the continual advances in the culturing process, the use of cell culture models to mechanistically determine and predict peptide absorption and metabolism in the various biological barriers will have a growing impact on the design and development of peptide drugs.

Acknowledgments

Donald W. Miller is a Scientific Education Partnership postdoctoral fellow funded by the Marion Merrell Dow Foundation. Ka-yun Ng is the recipient of a Pharmaceutical Manufacturers Association postdoctoral fellowship.

References

1. Wilson, G.; Davis, S. S.; Illum, L. *Pharmaceutical Applications of Cell and Tissue Culture;* Plenum: New York, 1990.
2. Neutra, M.; Louvard, D. *Functional Epithelial Cells in Culture;* Liss: New York, 1989; pp 363–398.
3. Zweibaum, A.; Laburthe, M.; Grasset, E.; Louvard, D. *Handbook of Physiology: The Gastrointestinal System IV;* American Physiological Society: Bethesda, MD; 1991, Chapter 7, pp 223–255.
4. Pinto, M.; Robine-Leon, S.; Appay, M.-D.; Kedinger, M.; Triadou, N.; Duddsalx, E.; Lacroix, B.; Simom-Assmann, P.; Haffen, K.; Fogh, J.; Zweibaum, A. *Biol. Cell.* **1983,** *47,* 323–330.
5. Hidalgo, I. J.; Raub, T. J.; Borchardt, R. R. *Gastroenterology* **1989,** *96,* 736–749.
6. Ruseet, M.; Trugnan, G.; Brun, J.-L.; Zweibaum, A. *FEBS Lett.* **1986,** *208,* 34–38.
7. Maurice, M.; Rogier, E.; Cassio, D.; Feldmann, G. *J. Cell Sci.* **1988,** *90,* 79–92.
8. Chapman, G. S.; Jones, A. L.; Meyer, U. A. *J. Cell Biol.* **1973,** *59,* 735–747.
9. Wanson, J.-C.; Drochmans, P.; Mosselmans, R.; Ronveaux, M.-F. *J. Cell Biol.* **1977,** *74,* 858–877.
10. Guguen-Guillouzo, C.; Gripon, P.; Vandenberghe, Y.; Lamballe, F.; Ratanasavanh, D.; Guillouzo, A. *Xenobiotica* **1988,** *18,* 773–783.

11. Schuetz, E. G.; Li, D.; Omiecinski, C. J.; Muller-Eberhard, U.; Kleinman, H. K.; Elswick, B.; Guzelian, P. S. *J. Cell. Physiol.* **1988,** *134,* 309–323.

12. Mitaka, T.; Sattler, G. L.; Pitot, H. C. *J. Cell. Physiol.* **1991,** *147,* 495–504.

13. Musat, A. I.; Sattler, C. A.; Sattler, G. L.; Pitot, H. C. *Hepatology* **1993,** *18,* 198–205.

14. LeCluyse, E. L.; Audus, K. L.; Hochman, G. H. *Am. J. Physiol.* **1994,** *266,* C1764–C1774.

15. Risau, W.; Wolburg, H. *Trends Neurosci.* **1990,** *13,* 174–178.

16. Audus, K. L.; Borchardt, R. T. *Pharm. Res.* **1987,** *3,* 81–87.

17. Abbot, N. J.; Hughes, C. C. W.; Revest, P. A.; Greenwood, J. *J. Cell Sci.* **1992,** *103,* 23–37.

18. Guillot, F. L.; Audus, K. L.; Raub, T. J. *Microvasc. Res.* **1990,** *39,* 1–14.

19. Dorovini-Zis, K.; Prameya, R.; Bowman, P. D. *Lab. Invest.* **1991,** *64,* 425–436.

20. Howell, S.; Kenny, A. J.; Turneer, A. J. *Biochem. J.* **1992,** *284,* 595–601.

21. Raub, T. J.; Audus, A. J. *J. Cell Sci.* **1990,** *97,* 127–138.

22. Audus, K. L.; Borchardt, R. T. *J. Neurochem.* **1986,** *47,* 484–488.

23. Cancilla, P. A.; DeBault, L. E. *J. Neuropathol. Exp. Neurol.* **1983,** *42,* 191–199.

24. Takakura, Y.; Kuentzel, S. L.; Raub, T. J.; Davis, A.; Baldwin, S. A.; Borchardt, R. T. *Biochim. Biophys. Acta* **1991,** *1070,* 1–10.

25. Takakura, Y.; Trammel, A. M.; Kuentzel, S. L.; Raub, T. J.; Davis, A.; Baldwin, S. A.; Borchardt, R. T. *Biochim. Biophys. Acta* **1991,** *1070,* 11–19.

26. Shi, F.; Bailey, C.; Malick, A. W.; Audus, K. L. *Pharm. Res.* **1993,** *10,* 282–288.

27. van Bree, J. B. M. M.; Audus, K. L.; Borchardt, R. T. *Pharm. Res.* **1988,** *5,* 369–371.

28. Trammel, A. M.; Borchardt, R. T. *Pharm. Res.* **1987,** *4,* S-41.

29. Raub, T. J.; Newton, C. R. *J. Cell. Physiol.* **1991,** *149,* 141–151.

30. Baranczyk-Kuzma, A.; Audus, K. L.; Borchardt, R. T. *J. Neurochem.* **1986,** *46,* 1956–1960.

31. Mischeck, U.; Meyer, J.; Galla-H.-J. *Cell Tissue Res.* **1989,** *256,* 221–226.

32. Baranczyk-Kuzma, A.; Audus, K. L. *J. Cereb. Blood Flow Metab.* **1987,** *7,* 801–805.

33. Dini, L. *Cell. Mol. Biol.* **1991,** *37,* 167–171.

34. Dini, L.; Lentini, A.; Mantile, G.; Massimi, M.; Devirgiliis, L. C. *Biol. Cell.* **1992,** *74,* 217–224.

35. Kato, H.; Takahashi, S. I.; Takenaka, A.; Funabiki, R.; Noguchi, T.; Naito, H. *Int. J. Biochem.* **1989,** *21,* 483–495.

36. Falasca, L.; Lentini, A.; Dini, L. *Cell. Mol. Biol.* **1992,** *38,* 621–627.

37. Wortelboer, H. M.; DeKruif, C. A.; Van Iersel, A. A. J.; Falke, H. E.; Noordhoek, J.; Blaauboer, B. J. *Biochem. Pharmacol.* **1990,** *40,* 2525–2534.

38. Sinclair, J. F.; McCaffrey, J.; Sinclair, P. R.; Bement, W. J.; Lambrecht, L. K.; Wood, S. G.; Smith, E. L.; Schenkman, J. B.; Guzelian, P. S.; Park, S. S.; Gelboin, H. V. *Arch. Biochem. Biophys.* **1991,** *284,* 360–365.

39. McMillan, J. M.; Shaddock, J. G.; Casciano, D. A.; Arlotto, M. P.; Leakey, J. E. A. *Mutat. Res.* **1991,** *249,* 81–92.

40. Rogiers, V.; Vandenberghe, Y.; Callaerts, A.; Verleye, G.; Cornet, M.; Mertens, K.; Sonck, W.; Vercruysse, A. *Biochem. Pharmacol.* **1990,** *40,* 1701–1706.

41. Mayorek, N.; Grinstein, I.; Bar-Tana, J. *Eur. J. Biochem.* **1989,** *182,* 395–400.

42. Stange, E. F.; Fleig, W. E.; Schneider, A.; Nother-Fleig, M.; Alavi, M.; Preclik, G.; Ditschuneit, H. *Atherosclerosis* **1982,** *41,* 67–80.

43. Forte, T. M. *Annu. Rev. Physiol.* **1984,** *46,* 403–415.

44. Tanaka, K.; Ikegaki, N.; Ichihara, A. *Arch. Biochem. Biophys.* **1981,** *208,* 296–304.

45. Neff, N. T.; DeMartino, G. M.; Goldberg, A. L. *J. Cell. Physiol.* **1979,** *101,* 439–458.

46. Einarsson, M.; Smedsrod, B.; Pertoft, H. *Thromb. Haemostasis* **1988,** *59,* 474–479.

47. Follmann, W.; Petzinger, E.; Kinne, R. K. H. *Am. J. Physiol.* **1990,** *258,* C700–C712.
48. Bouscarel, B.; Augert, G.; Taylor, S. J.; Exton, J. H. *Biochim. Biophys. Acta* **1990,** *1055,* 265–272.
49. Meredith, M. J. *Cell Biol. Toxicol.* **1987,** *3,* 361–377.
50. Sidhu, J. S.; Farin, F. M.; Omiecinski, C. J. *Arch. Biochem. Biophys.* **1993,** *301,* 103–113.
51. Vinters, H. V.; Reave, S.; Costello, P.; Girvin, J. P.; Moore, S. A. *Cell Tissue Res.* **1987,** *249,* 657–667.
52. Pitt, A. M.; Gabriels, J. E.; Badmington, F.; McDowell, T.; Gonzales, L.; Waugh, M. E. *Biotechniques* **1987,** *5,* 162–171.
53. Raub, T.; Kuentzel, S. L.; Sawada, G. A. *Exp. Cell Res.* **1992,** *199,* 330–340.
54. Rim, S.; Audus, K. L.; Borchardt, R. T. *Int. J. Pharm.* **1986,** *32,* 79–84.
55. Hilgers, A. R.; Conradi, R. A.; Burton, P. S. *Pharm. Res.* **1990,** *7,* 902–910.
56. Blais, A.; Bissonnette, P.; Berteloot, A. *J. Membr. Biol.* **1987,** *99,* 113–125.
57. Dix, C. J.; Hassan, I. F.; Obray, H. Y.; Shah, R.; Wilson, G. *Gastroenterology* **1990,** *98,* 1272–1279.
58. Cho, M. J.; Thompson, D. P.; Cramer, C. T.; Vidmar, T. J.; Scieszka, J. F. *Pharm. Res.* **1989,** *6,* 71–77.
59. Dantzig, A.; Bergin, L. *Biochim. Biophys. Acta* **1990,** *1027,* 211–217.
60. Hidalgo, I. J.; Borchardt, R. T. *Biochim. Biophys. Acta* **1990,** *1028,* 25–30.
61. Bonsdorff, C. H. V.; Fuller, S. D.; Simons, K. *EMBO J.* **1985,** *4,* 2781–2792.
62. Hidalgo, I. J.; Hillgren, K. M.; Grass, G. M.; Borchardt, R. T. *In Vitro Cell. Dev. Biol.* **1992,** *28A,* 578–580.
63. Ng, K.; Grass, G. M.; Lane, H.; Borchardt, R. T. *In Vitro Cell. Dev. Biol.* **1993,** *29A,* 627–629.
64. Wilson, G. *Eur. J. Drug Metab. Pharmacokinet.* **1990,** *15,* 159–163.
65. Miller, D. W.; Audus, K. L.; Borchardt, R. T. *J. Tiss. Cult. Methods* **1992,** *14,* 217–224.
66. Guillouzo, A. In *Research in Isolated and Cultured Hepatocytes;* Guillouzo, A.; Guguen-Guillouzo, C., Eds.; John Libbey Eurotex: London, 1986; pp 313–332.
67. Donjacour, A. A.; Cunha, G. R. *Cancer Treat. Res.* **1991,** *53,* 335–364.
68. Shimizu, Y.; Shaw, S. *Fed. Am. Soc. Exp. Biol. J.* **1991,** *5,* 2292–2299.
69. Ingber, D. E.; Folkman, J. *Cell* **1989,** *58,* 803–805.
70. Dehouck, M. P.; Meresse, S.; Dehouck, B.; Fruchart, J. C.; Cecchelli, R. *J. Controlled Release* **1992,** *21,* 81–92.
71. Rubin, L. L.; Hall, D. E.; Porter, S.; Barbu, K.; Cannon, C.; Horner, H. C.; Janatpour, M.; Liaw, C. W.; Manning, K.; Morales, J.; Tanner, L. I.; Tomaselli, K. J.; Bard, F. *J. Cell Biol.* **1991,** *115,* 1725–1735.
72. Hu, M.; Borchardt, R. T. *Biochim. Biophys. Acta* **1992,** *1135,* 233–244.
73. Hidalgo, I. J.; Borchardt, R. T. *Biochim. Biophys. Acta* **1990,** *1035,* 97–103.
74. Ng, K.; Borchardt, R. T. *Life Sci.* **1993,** *53,* 1121–1127.
75. Harris, M. *Dev. Biol.* **1989,** *6,* 79–95.
76. Pardridge, W. M.; Triguero, D.; Yang, J.; Cancilla, P. A. *J. Pharmacol. Exp. Ther.* **1990,** *253,* 884–891.
77. Pearson, J. D.; McCroskey, M. C.; DeWald, D. B. *J. Chromatogr.* **1987,** *418,* 245–276.
78. Grego, B.; Hearn, M. T. W. *J. Chromatogr.* **1984,** *336,* 25–40.
79. Causon, R. C.; McDowall, R. D. *J. Controlled Release* **1992,** *21,* 37–48.
80. Lundin, S.; Artursson, P. *Int. J. Pharm.* **1990,** *64,* 181–186.
81. Lundin, S.; Moss, J.; Bundgaard, H.; Artursson, P. *Int. J. Pharm.* **1991,** *76,* R1–R4.
82. Conradi, R. A.; Hilgers, A. R.; Ho, N. F. H.; Burton, P. S. *Pharm. Res.* **1991,** *8,* 1453–1460.

83. Conradi, R. A.; Hilgers, A. R.; Ho, N. F. H.; Burton, P. S. *Pharm. Res.* **1992,** *9,* 435–439.

84. Inui, K.; Yamamoto, M.; Saito, H. *J. Pharmacol. Exp. Med.* **1992,** *261,* 195–201.

85. Smith, T. K.; Gibson, C. L.; Howlin, B. J.; Pratt, J. M. *Biochem. Biophys. Res. Commun.* **1991,** *178,* 1028–1035.

86. Heyman, M.; Crain-Denoyelle, A.-M.; Nath, S. K.; Desjeux, J.-F. *J. Cell. Physiol.* **1990,** *143,* 391–395.

87. Hidalgo, I. J.; Kato, A.; Borchardt, R. T. *Biochem. Biophys. Res. Commun.* **1989,** *160,* 317–324.

88. Hughson, E. J.; Hopkins, C. *J. Cell Biol.* **1990,** *110,* 337–348.

89. Dix, C. J.; Obray, H. Y.; Hassan, I. F.; Wilson, G. *Biochem. Soc. Trans.* **1987,** *15,* 439–440.

90. Ramanajam, K. S.; Seetharam, S.; Ramasamy, M.; Seetharam, B. *Am. J. Physiol.* **1991,** *260,* G416–G422.

91. Augustijns, P. F.; Ng, K.; Borchardt, R. T., University of Kansas, Lawrence, KS, unpublished results.

92. Raeissi, S.; Audus, K. L. *J. Pharm. Pharmacol.* **1989,** *41,* 848–852.

93. Reardon, P. M.; Audus, K. L. *S.T.P. Pharma.* **1993,** *3,* 63–68.

94. van Bree, J. B. M. M.; de Boer, A. G.; Verhoef, C. J.; Danhof, M.; Briemer, D. D. *J. Pharmacol. Exp. Ther.* **1989,** *249,* 901–905.

95. Weber, S. J.; Abbruscato, T. J.; Brownson, E. A.; Lipkowski, A. W.; Polt, R.; Misicka, A.; Haaseth, R. C.; Bartosz, H.; Hruby, V. J.; Davis, T. P. *J. Pharmacol. Exp. Ther.* **1993,** *266,* 1649–1655.

96. Chikhale, E. G.; Ng, K.; Burton, P. S.; Borchardt, R. T. *Pharm. Res.* **1993,** *11,* 412–419.

97. Smith, K. R.; Borchardt, R. T. *Pharm. Res.* **1989,** *6,* 466–473.

98. Terasaki, T.; Takaakuwa, S.; Saheki, A.; Moritani, S.; Shimura, T.; Tabata, S.; Tsuji, A. *Pharm. Res.* **1992,** *9,* 529–534.

99. Kumagai, A. K.; Eisenberg, J. B.; Pardridge, W. M. *J. Biol. Chem.* **1987,** *262,* 15214–15219.

100. Pardridge, W. M.; Triguero, D.; Buciak, J.; Yang, J. *J. Pharmacol. Exp. Ther.* **1990,** *255,* 893–899.

101. Keller, B. T.; Smith, K. R.; Borchardt, R. T. *Pharm. Weekbl. Sci. Ed.* **1988,** *10,* 38–39.

102. Miller, D. W.; Keller, B. T.; Borchardt, R. T. *J. Cell Physiol.*, in press.

103. Smith, K. R.; Kato, A.; Borchardt, R. T. *Biochem. Biophys. Res. Commun.* **1988,** *157,* 308–314.

104. Rosenfeld, R. G.; Pham, H.; Keller, B.; Borchardt, R. T. *Biochem. Biophys. Res. Commun.* **1987,** *149,* 159–166.

105. Friden, P. M.; Walus, L. R.; Musso, G. F.; Taylor, M. A.; Malfroy, B.; Starzyk, R. M. *Proc. Natl. Acad. Sci. U.S.A.* **1991,** *88,* 4771–4775.

106. Friden, P.; Walus, L. R.; Watson, P.; Doctrow, S. R.; Kozarich, J. W.; Backman, C.; Bergman, H.; Holter, B.; Bloom, F.; Granholm, A. C. *Science (Washington, DC)* **1993,** *259,* 373–377.

107. Frimmer, M.; Ziegler, K. *Biochim. Biophys. Acta* **1988,** *947,* 75–99.

108. Kukongviriyapan, V.; Stacey, N. H. *J. Pharmacol. Exp. Ther.* **1988,** *247,* 685–689.

109. Ziegler, K.; Polzin, G.; Frimmer, M. *Biochim. Biophys. Acta* **1988,** *938,* 44–50.

110. Boelsterli, U. A.; Brouillard, B. J.-F.; Donatsch, P. *Toxicol. Appl. Pharmacol.* **1988,** *96,* 212–221.

111. Petzinger, E.; Frimmer, M. *Biochim. Biophys. Acta* **1988,** *937,* 135–144.

112. Muenter, K.; Mayer, D.; Faulstich, H. *Toxicol. In Vitro* **1991,** *4,* 201–206.

113. Emeis, J. J.; Van den Hoogen, C. M.; Jense, D. *Thromb. Haemostasis* **1985,** *54,* 661–664.

114. Smedsrod, B.; Einarsson, M.; Pertoft, H. *Thromb. Haemostasis* **1988,** *59,* 480–484.

115. Owensby, D. A.; Sobel, B. E.; Schwartz, A. L. *J. Biol. Chem.* **1988,** *263,* 10587–10594.

116. Seydel, W.; Stang, E.; Roos, N.; Krause, J. *Arzneim.-Forsch./Drug Res.* **1991,** *41,* 182–186.

117. Kato, S.; Kohno, H.; Ohkubo, Y. *Res. Commun. Chem. Pathol. Pharmacol.* **1991,** *73,* 145–152.

118. Bellot, F.; Molenaar, W.; Kris, K.; Mirakhur, B.; Verlaan, D.; Ullrich, A.; Schlessinger, J.; Felder, J. *J. Cell Biol.* **1990,** *110,* 491–502.

119. Trinder, D.; Batey, R. G.; Morgan, E. H.; Baker, E. *Am. J. Physiol.* **1990,** *259,* G611–G617.

120. Trinder, D.; Morgan, E.; Baker, E. *Hepatology* **1986,** *6,* 852–858.

121. Kashiwagi, A.; Harano, Y.; Kosugi, K.; Nakano, T.; Hidaka, H.; Shigeta, Y. *J. Biochem.* **1985,** *97,* 679–684.

122. Artursson, P.; Karlsson, J. *Biochem. Biophys. Res. Commun.* **1991,** *175,* 880–885.

123. Kim, D. C.; Burton, P. S.; Borchardt, R. T. *Pharm. Res.* **1993,** *10,* 1710–1714.

124. Karls, M. S.; Rush, B. D.; Wilkinson, K. F.; Vidmar, T. J.; Burton, P. S.; Ruwart, M. J. *Pharm. Res.* **1991,** *8,* 1477–1481.

125. Vilhardt, H.; Lundin, S. *Acta Physiol. Scand.* **1986,** *126,* 601–607.

126. Lundin, S.; Vilhardt, H. *Acta Endocrinol.* **1986,** *112,* 457–460.

127. Lundin, S.; Bengtsson, H. I.; Folkesson, H. G.; Westrom, B. R. *Pharmacol. Toxicol.* **1989,** *65,* 92–95.

128. Twaites, D. T.; Simmons, N. L.; Hirst, B. H. *Pharm. Res.* **1993,** *10,* 667–673.

129. Nicklin, P. L.; Irwin, W. J. *J. Pharm. Pharmacol.* **1991,** *43,* 103P.

130. Thwaites, D. T.; Hirst, B. H.; Simmons, N. L. *Pharm. Res.* **1993,** *10,* 674–681.

131. Yokohama, S.; Yoshioka, T.; Yamashita, K.; Kitamori, N. *J. Pharm. Dyn.* **1984,** *7,* 445–451.

132. Dantzig, A. H.; Tabas, L. B.; Bergin, B. *Biochim. Biophys. Acta* **1992,** *1112,* 167–173.

133. Inui, K.; Okano, T.; Maegawa, H.; Kato, M.; Takano, M.; Hori, R. *J. Pharmacol. Exp. Ther.* **1988,** *247,* 235–241.

134. Inui, K.; Okano, T.; Takano, M.; Kitazawa, S.; Hori, R. *Biochem. Pharmacol.* **1983,** *32,* 621–626.

135. Nakashima, E.; Tsuji, A.; Mizuo, H.; Yamana, T. *Biochem. Pharmacol.* **1984,** *33,* 3345–3352.

136. Quay, J. F. *Physiologist* **1975,** *18,* 359.

137. Yoshikawa, T.; Muranushi, N.; Yoshida, M.; Oguma, T.; Hirano, K.; Yamada, H. *Pharm. Res.* **1989,** *6,* 302–307.

138. Sinko, P. J.; Amidon, G. L. *Pharm. Res.* **1988,** *5,* 645–650.

139. Thwaites, D. T.; Brown, C. D. A.; Hirst, B. H.; Simmons, N. L. *J. Biol. Chem.* **1993,** *268,* 7640–7642.

140. Dyer, J.; Beechey, R. B.; Gorvel, J. P.; Smith, R. T.; Wootton, R.; Shirazi-Beechey, S. P. *Biochemistry* **1990,** *269,* 565–571.

141. Guillot, F. L.; Raub, T. J.; Audus, K. L. *J. Cell Biol.* **1987,** *105,* 312a.

142. Banks, W. A.; Kastin, A. J. *Pharm. Res.* **1991,** *8,* 1345–1350.

143. Banks, W. A.; Kastin, A. J. *Life Sci.* **1987,** *41,* 1319–1338.

144. Zlokovic, B. V.; Hyman, S.; McComb, J. G.; Lipovac, M. N.; Tang, G.; Davson, H. *Biochim. Biophys. Acta* **1990,** *1025,* 191–198.

145. Zlokovic, B. V.; Susic, V. T.; Davson, H.; Berley, D. J.; Jankov, R. M.; Mitrovic, D. M.; Lipovac, M. N. *Peptides* **1989,** *10,* 249–254.

146. Marks, N.; Stern, F.; Kastin, A. J.; Coy, D. H. *Brain Res. Bull.* **1977,** *2,* 491–493.

147. Huang, J. T.; Lajtha, A. *Res. Commun. Chem. Pathol. Pharmacol.* **1978,** *19,* 191–199.

148. Shimura, T.; Tabata, S.; Ohnishi, T.; Terasaki, T.; Tsuji, A. *J. Pharm. Exp. Ther.* **1991,** *258,* 459–465.

149. Shimura, T.; Tabata, S.; Hayashi, S. *Peptides* **1991,** *12,* 509–512.

150. Faulstich, H.; Trischmann, H.; Mayer, D. *Exp. Cell Res.* **1983,** *144,* 73–82.

151. Mayer, D.; Faulstich, H. *Biol. Cell* **1983,** *48,* 143–150.
152. Dorio, R. J.; Hoek, J. B.; Rubin E. *J. Biol. Chem.* **1984,** *259,* 11430–11435.
153. Takada, A.; Bannai, S. *J. Biol. Chem.* **1984,** *259,* 2441–2445.
154. Guillot, F. L.; Audus, K. L. *J. Cerebr. Blood Flow Metab.* **1990,** *10,* 827–834.
155. Fishman, J. B.; Dickey, B. F.; Bucher, N. L.; Fine, R. E. *J. Biol. Chem.* **1985,** *260,* 12641–12646.
156. Cogburn, J. N.; Donovan, M. G.; Schasteen, C. S. *Pharm. Res.* **1991,** *8,* 210–216.
157. Weiner, D.; Wolf, B. *Biochem. Med. Metab. Biol.* **1990,** *44,* 271–281.
158. Osterloh, K.; Aisen, P. *Biochim. Biophys. Acta* **1989,** *1011,* 40–45.
159. Mason, J. B.; Shoda, R.; Haskell, M.; Selhub, J.; Rosenberg, I. H. *Biochim. Biophys. Acta* **1990,** *1024,* 331–335.
160. Riley, S. A.; Warhurst, G.; Crowe, P. T.; Turnberg, L. A. *Biochim. Biophys. Acta* **1991,** *1066,* 175–182.
161. Rhoades, D. B.; Takano, M.; Gattoni-Celli, S.; Chen, C.-C.; Isselbacher, K. J. *Proc. Natl. Acad. Sci. U.S.A.* **1988,** *85,* 9042–9046.
162. Gebhardt, R.; Robenek, H. *J. Histochem. Cytochem.* **1987,** *35,* 301–309.
163. Gregoire, C. D.; Zhang, L.; Daniels, C. K. *Gastroenterology* **1992,** *103,* 296–301.
164. Fabricant, M.; Broitman, S. A. *Cancer Res.* **1990,** *50,* 632–636.
165. Babaev, V. R.; Kosykh, V. A.; Tsibulsky, V. P.; Ivanov, V. O.; Repin, V. S.; Smirnov, V. N. *Hepatology* **1989,** *10,* 56–60.
166. Hu, M.; Borchardt, R. T. *Pharm. Res.* **1990,** *7,* 1313–1319.
167. Shah, M. V.; Borchardt, R. T. *Pharm. Res.* **1989,** *6,* s-77.
168. Burton, P. S.; Conradi, R. A.; Hilgers, A. R.; Ho, N. F. *Biochem. Biophys. Res. Commun.* **1993,** *190,* 760–766.
169. Tsuji, A.; Terasaki, T.; Takabatake, Y.; Tenda, Y.; Tamai, I.; Yamashima, T.; Moritani, S.; Tsuruo, T.; Yamashita, J. *Life Sci.* **1992,** *51,* 1427–1437.
170. Fardel, O.; Ratanasavanh, D.; Loyer, P.; Ketterer, B.; Guillouzo, A. *Eur. J. Biochem.* **1992,** *205,* 847–852.
171. Mohrmann, I.; Mohrmann, M.; Biber, J.; Murer, H. *Am. J. Physiol.* **1986,** *250,* G323–330.
172. Grasset, E.; Bernabeu, J.; Pinto, M. *Am. J. Physiol.* **1985,** *248,* C410–418.
173. Hauri, H.-P.; Sterchi, E. E.; Bienz, D.; Fransen, J. A. M.; Marxer, A. *J. Cell Biol.* **1985,** *101,* 838–851.
174. D'agostino, L.; Danielle, B.; Pignata, S.; Gentile, R.; Tangliaferri, P.; Contegiacomo, A.; Silvestro, G. *Gastroenterology* **1989,** *97,* 888–894.
175. Baranczyk-Kuzuma, A.; Garren, J. A.; Hidalgo, I. J.; Borchardt, R. T. *Life Sci.* **1991,** *49,* 1197–1206.
176. Bjorge, S.; Hamelehle, K. L.; Homan, R.; Rose, S. E.; Turluck, D. A.; Wright, D. S. *Pharm. Res.* **1991,** *8,* 1441–1443.
177. Kam, N. T. P.; Albright, E.; Mathur, S.; Field, F. J. *J. Lipid Res.* **1989,** *30,* 371–377.
178. Field, F. J.; Albright, E.; Mathur, S. *Lipids* **1991,** *26,* 1–8.
179. Buur, A.; Mork, N. *Pharm. Res.* **1992,** *9,* 1290–1294.
180. Baranczyk-Kuzma, A.; Raub, T. J.; Audus, K. L. *J. Cereb. Blood Flow Metab.* **1989,** *9,* 280–289.
181. Baranczyk-Kuzma, A.; Audus, K. L.; Borchardt, R. T. *Neurochem. Res.* **1989,** *14,* 689–691.
182. Scriba, G. K. E.; Borchardt, R. T. *J. Neurochem.* **1989,** *53,* 610–615.
183. Scriba, G. K. E.; Borchardt, R. T. *Brain Res.* **1989,** *501,* 175–178.
184. Stenius, U.; Rubin, K.; Gullberg, D.; Hogberg, J. *Carcinogenesis* **1990,** *11,* 69–73.

RECEIVED for review July 12, 1993. ACCEPTED revised manuscript October 13, 1993.

CHAPTER 21

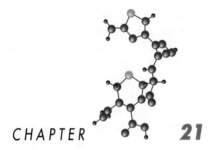

Gastrointestinal Transport of Peptides

Experimental Systems

David Fleisher

Absorption Rate Limits and Choice of Experimental System

Experimental techniques to study the intestinal absorption of peptide drugs are described in this chapter. To date, oral delivery of this class of agents has been limited by low systemic availability or substantial variability in plasma levels of drug. The two major barriers to oral delivery of peptide drugs are gastrointestinal (GI) metabolism and low intestinal membrane permeation. To outline the experimental methods available for the study of these barriers and evaluate potential methods to overcome them, a substantial amount of literature, including absorption studies of nonpeptide drugs, is drawn upon. Descriptions and illustrations of experimental systems and tools to study intestinal absorption are outlined from whole animal pharmacokinetic studies to mechanistic studies of drug transport and metabolism at the level of the mucosal membrane.

Choosing the appropriate experimental system to investigate the potential for oral delivery of a peptide, peptidomimetic drug, or prodrug candidate is based on a projection of the rate-limiting step for intestinal absorption. This projection is, in turn, gleaned from knowledge of the drug's physical–chemical properties. If the candidate possesses poor aqueous solubility, absorption will be controlled by dissolution rate. On a quantitative basis, this condition should occur when the dimensionless dose number is $\geqq 1$. This dimensionless dose number is defined as the ratio of the oral dose to drug aqueous solubility divided by the orally administered fluid volume (*1*). Under these circumstances, gastric emptying, GI motility, and biliary and GI fluid secretion will play critical roles in controlling the rate and (possibly) extent of absorption, and an in vivo animal model is usually most appropriate to assess absorption. Fed versus fasted-state differences are often pronounced when dissolution rate controls absorption rate.

When the candidate possesses high aqueous solubility, membrane partitioning or aqueous desolvation limitations (*2*) may dictate that absorption is controlled by membrane permeability. This rate limit is best studied with in situ or in vitro systems that isolate the membrane permeation step. If dissolution and membrane permeability are not rate-limiting (i.e., a water-soluble drug traverses the intestinal epithelial barrier predominantly via a paracellular or carrier-mediated pathway), gastric emptying may regulate the rate of absorption. Under these conditions, in situ and in vitro experiments should provide mechanistic information on the effects of intestinal contents on drug absorption. These effects include the influence of solvent drag and enhancer expansion on paracellular transport and the influence of drug concentration, potential inhibitors, and ion gradients on carrier-mediated transport. In situ and in vitro treatment differences should also be studied in vivo because the coupling of gastric emptying variability with membrane transport parameters may significantly affect rate and extent of absorption of drugs undergoing paracellular or carrier-mediated membrane transport.

Finally, the digestive function of the GI tract and its initial delivery of absorbed solutes to the liver may limit the driving force for transport to the systemic circulation following oral administration of a peptide drug. Dimensionless numbers quantitating these "reaction" rate limits to absorption were defined by Sinko et al. (*3*) for intestinal clearance and by Roberts and Rowland (*4*) for hepatic clearance. In vitro studies may indicate that a peptide drug candidate is enzymatically stable or provide the kinetic details for particular metabolic or reaction rate limits, but in situ experiments must be performed to determine drug appearance kinetics in the mesenteric or systemic circulation. Furthermore, more than one animal model should be tested because species differences are observed more often with biochemical transformation than drug transport.

In Vivo Animal Models

Evaluation of the rate and extent of absorption from plasma level data following oral administration is afforded by pharmacokinetic analysis to obtain an absorption rate constant, k_a, and the area under the curve (AUC) of the plasma level of drug versus time (t), respectively. Equations for estimating k_a or mean absorption time by compartmental and noncompartmental analysis, with oral data alone or oral and intravenous data, were provided by Wagner (*5*). However, large ranges of variability in absorption rate parameters arise when either gastric emptying, or dissolution dosage form release, or membrane permeability do not singularly control the rate of absorption. Additional problems complicate compartmental analysis of k_a (even when absorption is slow) if k_a does not dominate distribution and elimination effects on drug plasma levels (*6*).

For peptide drugs, assessment of the extent of absorption may also prove difficult because plasma assays for intact drug peptide levels are difficult to develop. Measurements of radiolabeled peptide drugs in plasma are of special concern for this class of drugs because of the potential for these measurements to include a labeled metabolite. Biological activity assays are most commonly used to determine rate and extent of absorption (*7*) of peptide drugs, and such measurements may be confounded by biological controls on activity other than by peptide drug absorption. Given these absorption assessment limitations, supplementary experimental input both from GI measurements in vivo and more isolated studies in situ and in vitro is desirable to address the potential for and significance of peptide drug absorption and plasma level variability.

Transit and Dissolution Rate Limits

The polarity of the amino acid building blocks of peptides suggests that most small peptide drugs should not have dissolution rate limits for absorption, but some larger peptides and chemically modified small peptides may possess substantial hydrophobicity and poor aqueous solubility. The cyclic endekapeptide cyclosporine A has very low aqueous solubility, which certainly contributes to its low and highly variable oral bioavailability (*8*). In the smaller peptide category, a number of potent renin inhibitors have been developed by isostere modification of a natural peptide, which has resulted in very low aqueous solubility (*9*). Their solubility is typically a strong function of pH (*10*), but poor solubility at intestinal pH results in low and highly variable oral bioavailability of these smaller peptides.

When dose and dissolution number calculations (*1*) suggest that pharmaceutical formulation will not improve oral availability, in vivo experiments must be performed to assess absorption. The expense and effort required for a clinical study may dictate that preliminary studies in animals are appropriate. If species-dependent metabolism is not an issue for the peptide drug candidate, the dog will often serve as a good animal model. Gastric emptying, GI pH, and overall GI transit time comparisons between dog and human have shown that the dog is a good in vivo animal model to study drug absorption and to evaluate particular oral dosage forms (*11*). When the dog can be used as an animal model, additional tools and techniques are available so that gastric emptying and GI pH, motility, and transit can be simultaneously monitored while drug blood levels are determined as a function of time after oral drug administration.

Coupling Plasma Level Versus Time Data with GI Measurements

Gastric emptying time can be determined by radiotelemetry, and the timing of this event with respect to the time of oral administration can have a significant effect on the rate and extent of absorption of a poorly water-soluble drug (*12*). Oral drug coadministration of a miniature pH-reading (Heidelberg) capsule converts GI hydrogen ion concentration to a radio signal that is transmitted to an antenna belt around the waist of the subject or animal. A jump in the pH readout from 1 to 6–7 signifies that the Heidelberg capsule has emptied into the small intestine (Figure 1). In the fasted state, a migrating motor complex (MMC) is responsible for clearing the stomach of its contents, and this event will occur randomly (within a 2-h monitoring period in dogs and humans) with respect to the time of oral drug administration. Noncaloric liquids (such as water and saline) display very different fasted-state gastric emptying patterns depending on the volume of administration (*13*). Low volumes (<100 mL) tend to empty in a zero-order fashion and are therefore more affected by fasted-state MMCs. High volumes (>200 mL) empty in a first-order manner (exponentially as a function of the initial volume load), providing rapid gastric clearance of liquid

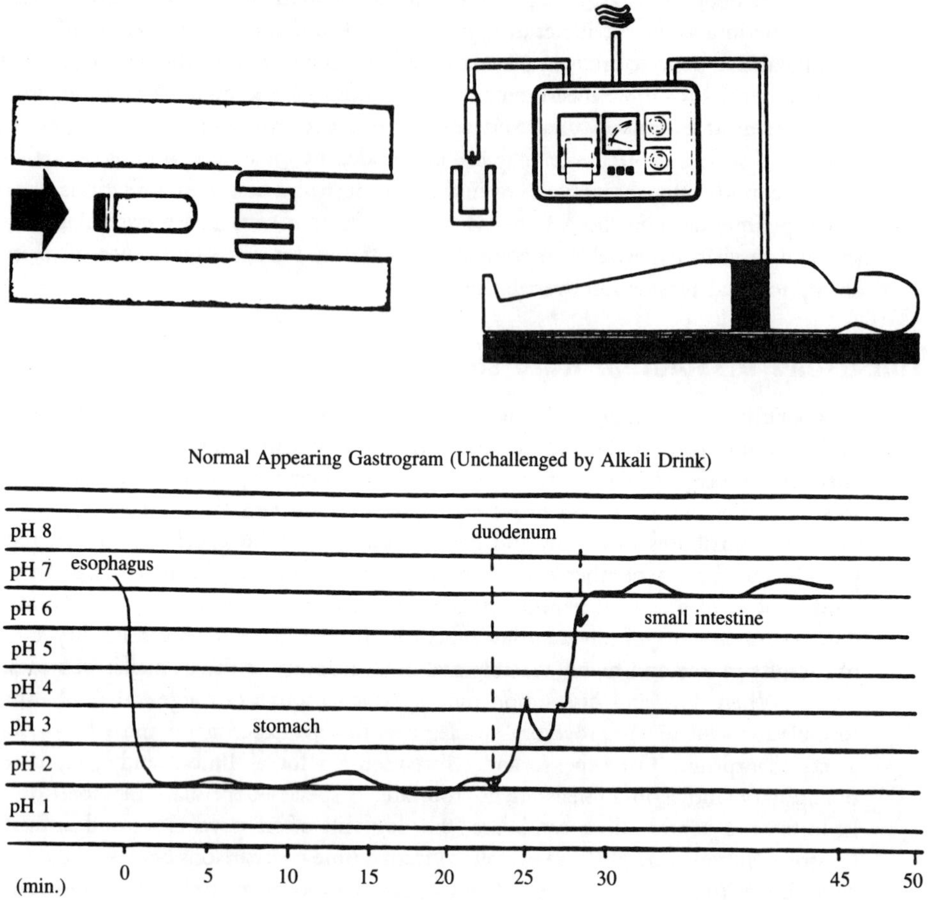

Figure 1. The GI transit of a Heidelberg capsule (which converts hydrogen ion concentration to a radio signal) permits determination of gastric emptying time.

and therefore less MMC influence. In this regard, drug hydrophobicity and resultant disintegration are critical factors in determining if gastric drug contents will enter the intestine in coordination with the liquid emptying pattern.

Meyer et al. (*14*) showed that solid particles of radius >0.5 mm will not empty from the stomach with the liquid phase until an MMC dumps the stomach contents. The consequences of this property for hydrophobic drugs is tremendous variability in the emptying of nondisintegrated and undissolved drug particles from the stomach to the intestine in the fasted state. Therefore, the fasted-state gastric emptying pattern of an orally administered drug will be highly dependent both on the volume of noncaloric fluid administered as well as the relative timing of oral dose and random MMC. These events dictate variable intestinal drug dissolution with respect to GI transit time, which results in highly variable absorption rates for poorly water-soluble drugs.

Fed-State Interactions

Following a short period (<10 min) of volume-controlled emptying, fed-state gastric emptying of liquid contents is dictated by feedback control from upper intestinal nutrient sensors. As a result, nutrient liquids empty in a zero-order fashion at a set rate (*x* cal/min). (Physiologically, this behavior serves to promote efficient nutrient absorption in the fed state.) This rate (x) will depend on both the source of calories and the animal species. For example, a gastric glucose load will empty at 2 kcal/min in humans (*15*). The length of time before fed-state gastric emptying patterns return to fasted-state random MMCs is, therefore, a function of the gastric caloric load (kilocalories per milliliter) as well as the total volume intake.

In vivo radiotelemetry experiments show that random MMCs do not dump the stomach until after the caloric load has completely emptied [as long as 6–8 h depending on caloric character (*12*)]. Because the particle size cutoff for solid drug is similar under fasted- and fed-state conditions, larger undissolved particles will be held up in the stomach, and this condition ensures more complete dissolution and a more uniform delivery of small particles and dissolved drug in the fed than in the fasted state. If the drug does not physically interact with nutrients and is chemically and biochemically stable under fed-state conditions, it is not surprising that the absorption of poorly water-soluble drugs would be less variable in the fed state (particularly if the meal has a reasonable lipid content) than in the fasted state. Furthermore, if drug dissolution is incomplete over the entire GI length in the fasted state, longer gastric residence time in the fed state may ensure more complete dissolution, and a greater extent of absorption is possible. This result has been observed for drugs like griseofulvin, for which a large dose (500 mg) coupled with very poor aqueous solubility results in a larger fraction of dose absorbed under simulated fed-state as opposed to fasted-state conditions (*16*). A more pronounced increase with a high-fat meal compared with a high-protein meal suggests that lipid-bile salt mixed micelles add to the enhanced residence time effects of the fed state on drug absorption.

Residence Time Considerations

For both poorly soluble and enzymatically or chemically labile peptide drugs, variability in GI residence time is a critical issue for drug absorption variability. Gastric and lower intestinal (colonic) residence times of lumenal contents are highly variable, but these regions do not typically display high capacities for drug absorption. However, these regions may contribute to variable drug absorption for poorly soluble or controlled-release drug dosage forms because the relationship between dissolution time and regional residence time determines the magnitude of the concentration driving force for drug absorption. Any factors contributing to extended residence times of "stable" drugs in these regions would be projected to magnify this driving force and may contribute to an increase in the extent of absorption. Some evidence suggests that drug toxicity may result from substantial increases in the extent of absorption as the result of abnormal colonic residence times. This condition may be particularly important for lipophilic drugs that have high membrane permeability because continued absorption in the colon is supplied by continued drug dissolution or release (*17*) for as long as the drug resides in the colon. The residence time in the upper intestine (duodenum, jejunum, and ileum), where drug absorption in the GI tract is maximum, does not vary as much as gastric and colonic residence time. This property is primarily due to the fact that fasted-state intestinal motility patterns show cyclic bursts of activity (phase III) in continuum with longer periods of minimal activity (phase I), whereas the fed-state provides for sustained periods of intermediate activity (phase II-like). However, differences in meal content produce significant transit differences in the jejunum (*18*).

Species differences in upper intestinal residence time may dictate significant variability in the extent of drug absorption as a function of dosage form. Overall GI residence times are similar in dogs and humans, but the length of the small intestine in dogs is about half that of humans, and upper intestinal residence times in humans have been reported to be up to 3 times greater than in dogs (*19*). These measurements were made by measuring lumenal concentration changes in nonabsorbable markers sampled through swallowed intestinal tubes as well as by gamma scintigraphy. If the rate of absorption or the rate of delivery (via dosage-form release or dissolution) is slow, large differences in the extent of absorption might be anticipated between dogs and humans. When permeability is rate-limiting, in situ and in vitro models represent more appropriate systems to study absorption variability. Intestinal drug metabolism and the subsequent availability of the parent compound for absorption may also be a strong function of upper intestinal residence time. Species differences in the extent of drug absorption are more pronounced for controlled-release dosage forms than for immediate-release dosage forms (*20*).

Dosage Form and Drug Delivery Factors

In addition to in vivo monitoring of gastric emptying time, radiotelemetry has proven useful in monitoring drug release from dosage forms in the small intestine (*21*). Some information can be deciphered from GI transit because the noise in the radio signal indicates whether the Heidelberg capsule is in the small or large intestine, but the location of the dosage form in the GI tract is more accurately evaluated by gamma scintigraphy (*22*). Dosage form position may be of special importance for the oral delivery of peptide drugs because their site-specific release in the ileum or colon may circumvent upper intestinal metabolism and provide a high driving force for absorption in the lower intestine (*23*). The use of instrumentation to simultaneously measure GI motility and pH in vivo yield valuable information on the variability of profiles of plasma drug and metabolite concentration versus time as a function of GI physiology (Figure 2). Such measurements make use of thin-fiber, liquid-pressure and hydrogen-ion transducers that are easily swallowed and positioned in the GI tract by radiographic monitoring (*24*). Such measurements can provide valuable information to predict the magnitude of variability in absorption lag time, t_{max}, and absorption rate for poorly soluble peptide drugs. In this regard, GI transit variability has been implicated as an important contribution to the variable bioavailability of oral cyclosporin A (*25*).

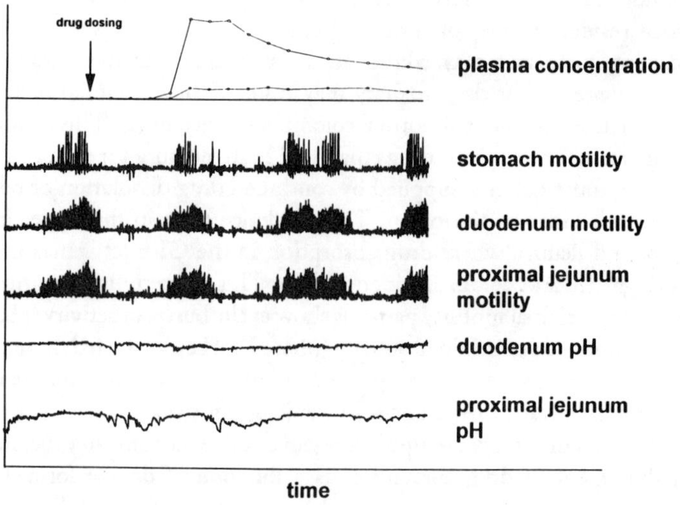

Figure 2. Simultaneous monitoring of drug and metabolite plasma levels, intestinal motility, and pH.

Site-Specific Investigations

In vivo studies with dogs also were used to determine the effect of site-specific delivery on drug plasma levels. Vascular access ports (Figure 3) were employed as noninvasive delivery ports into dog duodenum, jejunum, ileum, and colon. This in vivo system has proven valuable for comparing oral and intestinal site-specific delivery (unpublished data from author's laboratory). This comparison is especially important for small peptide drugs both for designing dosage forms to avoid site-specific metabolism and for determining small peptide concentration dependence and nutrient competition for transport by the mucosal peptide carrier. Radiotelemetry, gamma scintigraphy, pH–fiber-motility transducers, and lumenal access port technologies each provide information on GI physiology changes with time. When this information is correlated with peptide drug plasma level versus time profiles, the range of variability in rate and extent of absorption parameters can be projected in a given population of experimental subjects or patients on the basis of variable GI physiology. This projection may be particularly important for peptide drug candidates for oral delivery because their potency may override low bioavailability limitations if consistent oral delivery can be documented for a given dosage form. These techniques for monitoring variable GI physiology are noninvasive, and telemetry, scintigraphy, and swallowed-transducers can be used in human subjects following drug absorption studies in an appropriate animal model.

Significant information has been obtained from more invasive experimental procedures in dogs on the influence of variable GI physiology on absorption variability. Dogs with surgically implanted GI fistulae and Thiry-Vella loops have been used to study fed-state controls with regard to drug absorption (*13–15, 19, 26*) and to monitor contents in the GI lumen and the relationship of contents to dosage form transit and drug absorption (*13*). These dog models have proven valuable in relating variable GI physiology to drug absorption, but their invasive nature and animal survival times limit extensive scientific use. Recent advances in gut peptide research have provided agonists and antagonists to explore GI physiology effects on absorption in a noninvasive manner. In this regard, a cholecystokinin A receptor antagonist reversed a fed-state effect on the absorption of a poorly soluble drug in a dog model (*27*).

In Situ Animal Models

Permeability and Intestinal Metabolism Rate Limits

For peptide drugs possessing high water solubility, dissolution is not rate limiting to intestinal absorption. For many of these drugs, the intestinal mucosa is the rate-limiting barrier to absorption, and the relative resistance to transport can be assessed by evaluating the effective intestinal permeability. In fact, the permeability (or conductivity for an inorganic ion) is defined as the reciprocal of the barrier resistance (units of time per length). The permeability has units of length per time (i.e., centimeters per second) and can be regarded as a transport velocity. Just as electrical conductivity is defined as the ratio of current flow to the driving force electrical potential (gradient) difference by Ohm's law, permeability (P) is defined as the ratio of drug flow or flux (J) to the concentration gradient across a given barrier through a rearrangement of Fick's first law. [A more complex representation (*28*) is required when the diffusional driving force is in parallel with fluid convection or carrier-mediated processes.] If drug is immediately removed from the receiving compartment (i.e.,

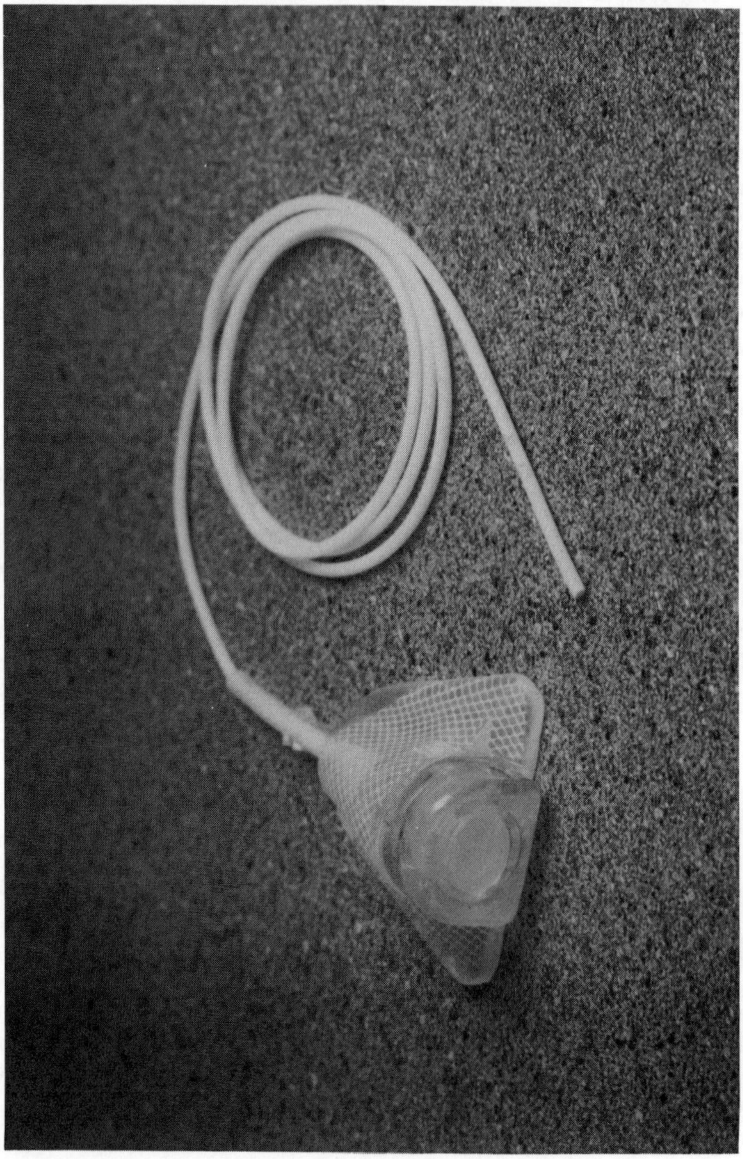

Figure 3. Surgically implanted intestinal access ports allow for site-specific drug administration through a rubber port above the skin of the animal's back.

villus blood carries absorbed drug to the systemic circulation), the concentration gradient driving force is simply the drug concentration (C) in the intestinal lumen, and permeability is defined as $P = J/C$, where drug or mass flux, J, must have units of mass per area-time and the area of interest is the exposed intestinal surface area. A high permeability usually characterizes poorly water-soluble drugs because they frequently possess high membrane–water partition coefficients. Such a measurement can be made from in situ intestinal perfusions of drug solution concentrations below their aqueous solubility. However, a simple in vitro measurement of organic phase–water partition coefficient (K_p) should suffice to predict whether or not membrane transport will be rate-limiting to drug absorption in vivo (*2*). For example, the rate of absorption of a drug with log $K_p = 1$ ($K_p = 10$) and a low hydrogen-bonding index will not be limited by membrane permeation (*28*) and variable membrane permeability would not be expected to contribute significantly to absorption variability in vivo.

Because the intestinal barrier is not a simple aqueous–lipid interface, a highly water-soluble drug (low partition coefficient) may also yield a high membrane permeability. Carrier-mediated transport and transport through paracellular pathways may result in high permeabilities even when the partition coefficient is low or the hydrogen-bonding index is high. Carrier-mediated transport can be evaluated by conducting permeability studies as a function of drug concentration (*28*) because such pathways will be capacity limited and permeability should drop with increasing drug concentration (Figure 4). Paracellular transport is limited by molecular size and should not be significant (under normal fasted-state conditions) for drugs with molecular weights of >250 Da. The relative importance of paracellular transport for low molecular weight, water-soluble drugs can be tested by stimulating intestinal water absorption (*29*) because convective transport by solvent drag should increase the permeability for this class of drugs (Figure 5). When carrier-mediated or paracellular transport results in high permeability for water-soluble drugs, gastric emptying may control the absorption rate. This behavior is observed for the water-soluble analgesic drug acetaminophen (APAP) (Figure 5), which has a molecular weight of 151 Da (*29, 30*). When dissolution rate or gastric emptying rate controls the rate of absorption, in vivo experiments are required to fully evaluate absorption variability. In situ permeability measurements of water-soluble drugs subject to carrier-mediated or paracellular transport can be used to investigate conditions under which these pathways may contribute to absorption variability.

In situ permeability experiments are of greatest value for predicting in vivo absorption limitations for water-soluble drugs. Measurement of a low permeability (consistent with low K_p and independent of drug concentration and solvent drag) indicates that absorption will limit systemic availability from oral dosage. If so, chemical modification and development of an alternate drug candidate with better permeability should be considered before any extensive in vivo studies are attempted to assess systemic availability. A number of experimental in situ techniques are available to evaluate intestinal membrane transport rate limits. The rat is the experimental animal most commonly used in such studies, but in situ absorption experiments in guinea pigs and rabbits have also been performed. Some species differences may exist in the magnitude of drug permeability between animals, but rank-order drug permeability correlations for different drugs should be expected between animal species.

The form of the resulting permeability data depends on the type of flow pattern used in the in situ system. Various pump setups (*31*) are designed to perfuse drug solution through intestinal segments by oscillating, recirculating, and single-pass flow schemes (Figures 6a, 6b, and 6c, respectively). Alternatively, intestinal segments can be filled with drug solution and tied off in the stagnant isolated loop system (Figure 6d). The oscillating setup serves to enhance fluid mixing in the intestinal segment because motility is typically absent

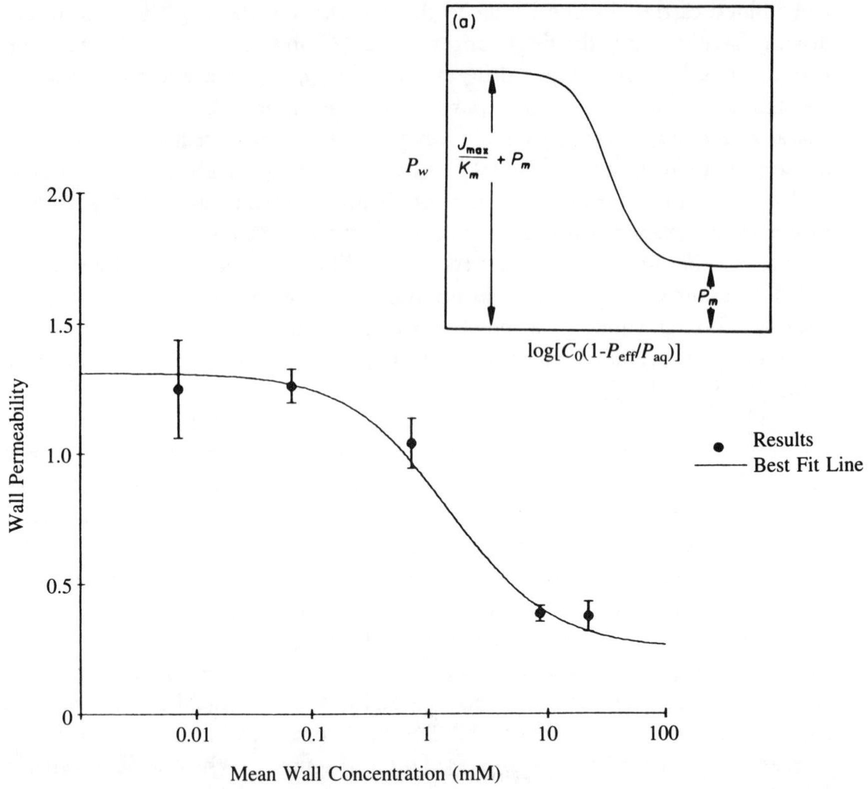

Figure 4. Intestinal permeability of a carrier-mediated drug. Theoretical inset (a) shows that high carrier permeability (J_{max}/K_m) decreases as perfusion drug concentration increases, demonstrating saturation of the carrier and dominance of passive permeability (P_m) at high concentration. Experimental permeability vs. aqueous-resistance (R_{aq})-corrected drug concentration data (mean + SEM, n = 4 rats).

in anesthetized animals (*32*). Drug concentration, C, decreases with time, t, because of continued absorption in the oscillating, recirculating, and closed loop systems, and hence outlet- or time-sampled drug concentration data, $C(t)$, provide for the calculation of non-steady-state permeabilities. The single-pass perfusion system supplies a constant-input drug concentration, C_0, maintaining initial conditions over the time course of the experiment. Following a short (depending on flow rate and absorption rate) nonsteady-state period over which the intestinal segment filling and drug absorption processes equilibrate, a time-independent output concentration, C_m, will be achieved, and will permit the calculation of steady-state permeabilities (Figure 6).

The relationship between an intestinal permeability obtained from an isolated in situ experiment and a pharmacokinetic absorption rate constant is straightforward. An absorption rate constant represents a first-order kinetic rate constant (k_a) characterizing the rate of drug disappearance ($-k_a$) from the intestinal compartment or appearance ($+k_a$) in the mesenteric blood and has units of reciprocal time (i.e., reciprocal minutes):

$$C(t) = C_0 \exp(\pm k_a t) \quad \text{or} \quad \ln(C/C_0) = \pm k_a t$$

Permeability is the product of k_a and the volume (V)-to-membrane surface area (A) ratio

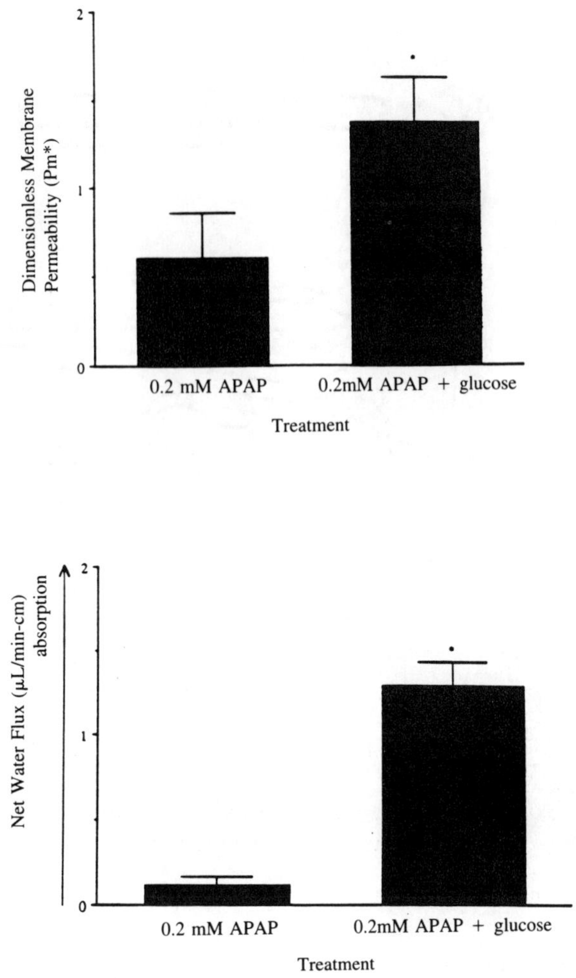

Figure 5. Acetaminophen (APAP) permeability correlates with jejunal water absorption induced by glucose in rat perfusion, suggesting solvent drag of drug through paracellular pathways.

of the perfused intestinal segment ($k_a V/A$) and therefore has units of length per time. Because the intestinal radius is regarded as constant, the only geometric measurement that is required for this calculation is the length of the perfused segment. Alternatively, if a value for the intestinal radius is assumed or measured, the product of k_a and V of the perfused segment is the drug clearance, CL (volume per time):

$$CL = k_a V$$

Because the permeability is essentially the reciprocal of the resistance, R, to mass (drug) transport, an effective permeability, P_{eff}, is defined as the reciprocal of the sum of the membrane resistance, R_m, and the aqueous resistance, R_{aq}, to drug transport in the intestinal lumen up to the intestinal membrane:

$$P_{eff} = 1/R_{eff} \quad \text{where} \quad R_{eff} = R_m + R_{aq}$$

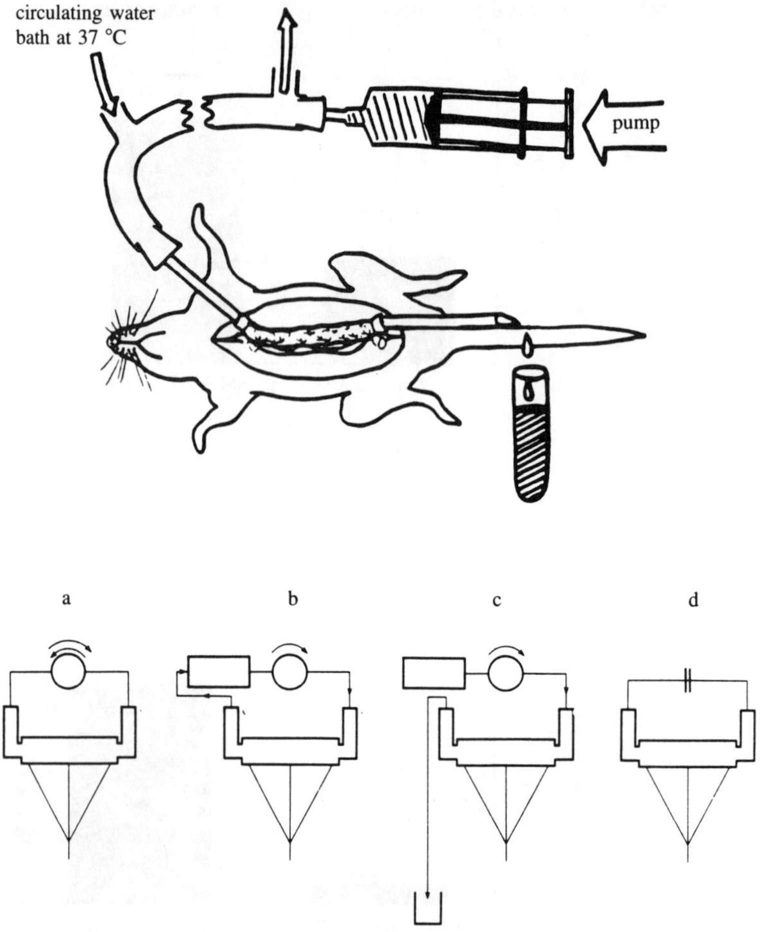

Figure 6. In situ intestinal perfusion system by (a) oscillating, (b) recirculating, (c) single-pass, and (d) stagnant (ligated loop) flow patterns to measure rate of drug disappearance from the intestine (permeability) and appearance in mesenteric blood.

By necessity, R_{aq} is a function of perfusion flow rate. Hydrodynamically, the faster the flow rate, the lower the R_{aq} (boundary layer resistance) to drug transport. In fact, if drug membrane permeability is low (based on its low partition coefficient and membrane permeation), then R_m will dominate R_{aq} regardless of flow conditions. In this case, effective resistance approximates membrane resistance ($R_{eff} \approx R_m$), and likewise effective permeability approximates membrane permeability. Therefore, if the drug does not have a high permeability as the result of paracellular or carrier-mediated transport (or as the result of a high partition coefficient), hydrodynamic flow conditions and aqueous permeability are not important considerations (*28*). When drug permeability is high, variability in the data will be strongly influenced by perfusion flow conditions, and the resistance of the aqueous boundary layer will have to be taken into account. Under these conditions, a more elaborate analysis is in order, such as the theoretical work of Kou et al. (*33*). Such analyses should be necessary only for studying conditions under which high permeability might be compromised (i.e., fed-state inhibition of carrier-mediated absorption of a peptide drug) because high permeation ensures that membrane transport is not rate limiting to drug absorption.

An effective permeability is often reported as a dimensionless number (P^*) so that the permeabilities of different drugs under different perfusion conditions can be compared. This comparison requires that permeabilities be normalized for drug aqueous diffusion coefficient (D), the length (L) of the perfused intestinal segment, and the perfusion flow rate (Q). The aqueous diffusion coefficient (obtained in diffusion experiments or estimated on the basis of molecular size) normalizes for drug transport resistance and has units of length squared per time. Because the absorption rate or permeability can be represented as a first-order loss from the intestinal lumen, the logarithmic ratio of C_m to C_0 defines the rate of absorption. The dimensionless effective permeability is calculated by the following equation:

$$P^*_{\text{eff}} = [\ln(C_m/C_0)]/[(-DL/Q)]$$

The experimental normalization factor in the denominator (D has units of length squared per time, and Q has units of volume per time) is unitless, and thus the calculated effective permeability is in dimensionless form.

An important consideration with respect to in situ permeability calculations is that C_m or $C(t)$ data are not only a function of drug absorption but are also influenced by intestinal water transport (absorption–secretion). Inclusion of a nonabsorbable marker–solute, like radiolabeled polyethylene glycol (PEG) 4000, can be used to calculate the magnitude of net water secretion (via marker dilution measurements) or net water absorption (via marker concentration) in the outlet perfusate. The drug concentration ratio C_m/C_0 can be directly corrected for water transport, in either case, by multiplying by marker C_0/C_m. Intestinal water transport data from these in situ experiments can provide additional information when paracellular transport is a potentially significant component of drug membrane permeability. Stimulating intestinal water absorption by including D-glucose in the intestinal perfusate has been used to confirm the importance of the paracellular pathway in the jejunal absorption of low molecular weight drugs (*29*). The extent of water absorption generated is equivalent in anesthetized animals in situ and in unanesthetized animals in vivo (*32*). Such a demonstration is important if in situ studies are to be regarded as representative of GI physiological responses in vivo. An advantage of the in situ experiment is its provision for sampling directly from GI lumen and the mesenteric blood (vs. systemic blood samples in vivo) so that drug absorption can be isolated from in vivo clearance processes. In situ perfusate composition and flow rate must be restricted to ranges that are relevant in vivo if in situ absorption data are to be used to justify in vivo experiments. However, useful mechanistic information may be obtained with special testing perfusion conditions in situ.

Sampling from the intestinal compartment may be particularly important for peptide drugs because the digestive function of the GI tract may result in chemical transformation of this class of drugs. In addition, the appearance of phase II drug metabolites (i.e., glucuronide, sulfate, and glutathione conjugates) in outlet perfusate is not an uncommon occurrence in the course of in situ intestinal perfusion of drug solutions (*29*), and drug absorption calculations should take this information into account. It is fairly easy to determine metabolite perfusate composition with a sensitive assay because outlet perfusate is fairly clean and extraction procedures are not usually required. It may be important to determine to what extent drug transformation is occurring in the intestinal lumen compared with the intestinal mucosa. This determination can be readily accomplished by perfusing drug-free solution and then spiking the exiting perfusate with drug. By comparing drug degradation kinetics in unperfused versus perfused buffer solution, the extent of lumenal metabolism can be ascertained. Lumenal metabolism can occur as the result of gastric and pancreatic enzymes that are adsorbed to the mucosa and that are released during perfusion as well as by metabolic components of sloughed-off epithelial cells. The kinetic parameters can then be readily used for lumenal metabolism to separate drug absorption and transformation components (*3*).

Drug metabolite clearance by intestinal tissue is vectorial; some metabolites appear in the intestinal lumen and some in the mesenteric blood (*34*). This information is difficult to obtain in situ, and drug permeability corrections for intestinal tissue metabolism present a special problem requiring in vitro experiments. If radiolabeled peptide and high-pressure liquid chromatography (HPLC) assay are both available, such information can sometimes be obtained by perfusing inhibitors of intestinal metabolism. In determining the jejunal permeability of a growth hormone releasing hexapeptide (GHRP) from perfusions in rats (*35*), tritium counts in the mesenteric blood correlated with HPLC measurements of disappearance of peptide from the intestinal lumen. Furthermore, GHRP metabolites were not detected in the outlet perfusate by the HPLC assay, and glucose-induced water absorption increased both loss of GHRP from the intestinal lumen and increased tritium counts in the mesenteric blood. In vitro studies indicated that GHRP was not a substrate for aminopeptidases, but GHRP coperfusion with amastatin, an aminopeptidase inhibitor, reduced both GHRP loss from the exiting jejunal perfusate and tritium counts in the mesenteric blood (Figure 7). The data suggest that a tritiated GHRP metabolite is absorbed and that lumenal glucose enhances the absorption of the metabolite but not the intact peptide. This information might have been obtained in an alternative fashion by a biological assay for changes in plasma growth hormone. However, the kinetics of plasma growth hormone release might not necessarily mirror the appearance of growth hormone releasing peptide in the blood, and other biological controls (i.e., surgical stress or anesthesia) could complicate data interpretation.

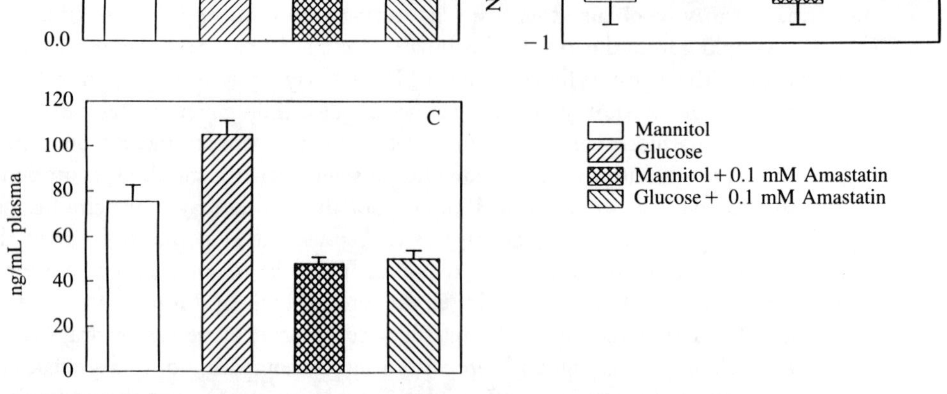

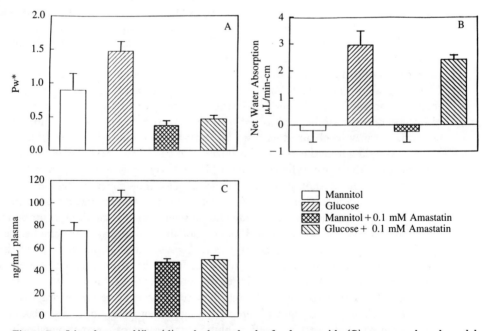

Figure 7. Jejunal permeability (A) and plasma levels of a hexapeptide (C) appear to be enhanced by glucose-induced water absorption (B). However, co-perfusion with an aminopeptidase inhibitor suggests absorption data represents uptake of a hexapeptide metabolite.

Hepatic Clearance Information

In situ sampling of lumenal compartment and mesenteric blood in conjunction with intestinal metabolism studies directly addresses factors influencing drug absorption at the level of the intestinal membrane, and systemic drug levels will also be influenced by hepatic clearance. Hepatic metabolism of orally administered peptide drugs is typically obtained with the bile-duct-cannulated rat model. Surgical installation of jugular vein and bile-duct cannulae allows for exterior sampling from these two compartments subsequent to surgical recovery of the animal. Coupled with total urine and feces collection, rate and extent of drug absorption can be assessed (subsequent to oral administration of radiolabeled drug) from the plasma and elimination compartments, respectively, when compared with intravenous dosing. Radioactivity in each of these compartments can also be analyzed by HPLC, with a radiomatic flow detector permitting an indirect separation of parent drug from biotransformation products. Such information is especially important for peptide drugs because hepatic metabolism often limits systemic availability of the parent drug past the limitations of intestinal absorption. The chapters by LaRusso, Ruwart, and Sugiyama in this text give detailed studies on hepatic clearance of peptide drugs.

Initial studies of oral availability and distribution of a peptide drug in a number of animal species are usually performed by measuring radioactivity in organ as well as elimination compartments subsequent to oral administration of labeled compound. Low counts in compartments other than the intestinal and fecal contents document poor absorption, but high counts in other compartments do not constitute proof that absorption of intact peptide has occurred unless HPLC analysis verifies parent drug distribution. Intestinal metabolism followed by absorption and distribution of a radiolabeled metabolite may often account for high radioactivity in other compartments.

Peptide GI Pharmacology and Transport Issues

A number of peptide drugs with promising therapeutic potential may have pharmacologic activity in the GI tract. Such drugs may alter normal GI physiology and, in turn, contribute to drug absorption variability following oral administration. As an example, somatostatin analogues, for which parenteral dosage forms are available, were studied to assess the potential for an oral dosage form (*36, 37*). Because endogenous somatostatin-28 inhibits the secretion of other gut peptides that mediate the fed-state response (*38*), it is not surprising that oral administration of somatostatin analogues would influence GI physiology and in turn play a role in the absorption variability of these analogues. Following oral administration of octreotide (analogue of Sandoz somatostatin octapeptide) parenteral solution to dogs with a Heidelberg capsule, rapid gastric emptying and accompanying elevations in both gastric and intestinal pH were observed. In situ jejunal perfusions of octreotide in rats increased fasted-state water absorption and depressed glucose-stimulated water absorption (Figure 8) (*35*). Rapid absorption (consistent with gastric emptying time) was evident from octreotide plasma levels in the dog studies, but intestinal water transport did not influence octreotide permeability in the in situ rat perfusions (Figure 8). Furthermore, low permeability values were consistent with low absolute bioavailability in rats, as documented in other studies (*39*). The potential for any peptide or peptidomimetic drug to influence variable intestinal absorption through a local effect on GI pharmacology is worthy of consideration because peptide receptors and endocrine function have been discovered for drugs with other systemic actions (*40, 41*).

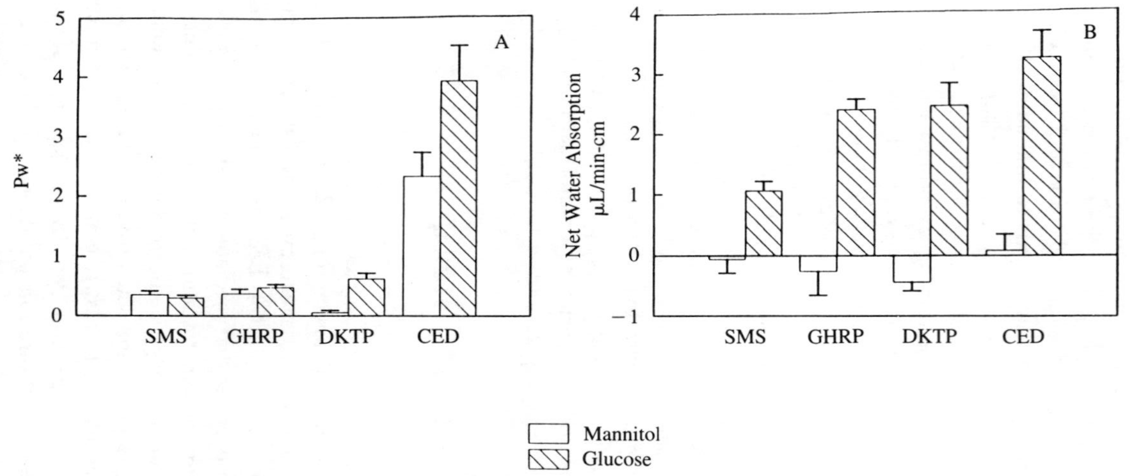

Figure 8. Glucose-induced water absorption (B) enhanced the jejunal permeability (P^*_w) (A) of the dipeptide D-kytorphin (DKTP), and the tripeptide-like cephradine (CED), but not a growth hormone releasing hexapeptide (GHRP). Glucose did not enhance permeability of a cyclic octapeptide somatostatin analogue (SMS), and SMS depressed glucose-induced water absorption.

In Vitro Systems of Study

Isolation of individual transport and metabolism components of peptide drug absorption are often carried out at the tissue, cellular, and membrane levels. As in the physical sciences, such isolation permits a mechanistic focus on component details. However, a number of precautions should be observed when extrapolating in vitro data to in vivo significance. Integration of these details in biological systems may be of limited value if the focus is not on the rate-limiting step. Furthermore, isolation of biological system components may remove controls on transport–metabolism processes normally operative in vivo, and simply establishing system viability does not ensure in vivo relevance. Potential contributions to drug absorption variability as determined in these more isolated studies must be addressed under physiologically realizable conditions. Finally, if distribution and elimination dominate absorption process controls on drug blood levels, isolated studies on absorption variability may constitute a clinically meaningless exercise. When these experimental limitations have been considered, isolated in vitro experiments may provide simple screening tools for projecting rate and extent of absorption as well as mechanistic details accounting for expected ranges of variability.

Tissue Level Studies: Transport–Metabolism and Pharmacology

A number of experimental systems use isolated intestinal tissue to study drug transport and metabolism. A useful procedure is to anesthetize the animal, isolate and remove the tissue, oxygenate the tissue, and then sacrifice the animal because sacrifice followed by organ removal may limit tissue viability. Drug uptake by intestinal tissue is studied in everted intestinal rings or sleeves, whereas mucosal-to-serosal drug flux across intestinal tissue is studied in intestinal sacs or diffusion chambers. Drug metabolism can be compared in homogenized versus intact intestinal tissue.

Everted Ring Uptake

If radiolabeled drug is available, measuring the initial drug uptake rate in everted intestinal rings (*42*) constitutes one of the simplest experimental techniques to screen for absorption rate limitations. Following removal of the intestinal segment of interest, the segment is placed in iced, oxygenated iso-osmotic buffer at pH 7.4. A glass rod is placed into the intestinal lumen and tied to the rod at one end with surgical thread. The intestine is pulled back over the rod from the other end to evert (turn inside out) the segment of interest. The everted segment is then quickly cut into rings ~$^1/_4$ cm long corresponding to a wet weight of 25–50 mg, while oxygenation continues in the iced buffer solution. These rings are quickly transferred to test tubes containing physiologic solutions of radiolabeled drug at body temperature.

In the first ring study, uptake is plotted as a function of time to determine the time frame over which drug uptake is linear (initial uptake rate). Because these small samples have a low capacity for uptake, significant back flux out of the tissue dictates a drug uptake versus time plateau depending on the rate of absorption. All future experiments should be performed at a time point at which uptake is linear. If drug uptake is carrier mediated, uptake versus time plateaus may be seen as early as 1 min. Uptake experiments should not be carried out much past 10 min because viability is quickly lost and the tissue will become leaky. Tissue viability can be checked by including a nonabsorbable marker (like inulin or

PEG 4000) with a different radiolabel. These markers can also be used to subtract drug uptake into the "extracellular" or water-adhering space.

Furthermore, the extent of binding of drug to tissue can be assessed by extrapolating the uptake versus time data back to zero time. Because the binding kinetics should establish equilibrium much more rapidly than tissue uptake, a nonzero intercept provides a measure of the extent of binding. Alternatively, binding can be studied at a single time point in the linear range as a function of incubation solution osmolality. By plotting drug uptake versus the inverse of osmolality, extrapolation to infinite osmolality defines the extent of binding of drug to tissue.

The aqueous resistance to drug transport can be separated from the membrane resistance (inverse tissue uptake) by studying uptake at various shaking rates. For a drug with high uptake, a strong dependence on shaking rate should be observed because aqueous resistance will dominate tissue resistance (*43*). If drug uptake is low, uptake would be expected to be independent of shaking rate because tissue resistance to drug transport will dominate. A plot of uptake versus inverse shaking rate will yield membrane uptake unbiased by aqueous resistance when the data are extrapolated to infinite shaking rate (zero aqueous resistance). The units of uptake are drug mass or counts per time divided by tissue weight. This tissue weight can be obtained by preweighing scintillation vials. Uptake can be converted to a permeability by normalizing for incubation drug concentration if the mucosal surface area for a given ring weight is known (*44*).

Carrier-mediated drug transport can be documented in the everted ring system by studying drug uptake as a function of solution concentration and by performing inhibition experiments. Passive drug uptake limitations as well as membrane metabolism have been verified by this experimental technique. Radiolabel tissue counts are the standard measure of drug uptake in this system, but HPLC identification of metabolites has been obtained via extraction from freeze-dried intestinal rings (*45*). Solid statistical analysis is provided by this simple experimental technique because 40–50 rings can be obtained from the same experimental animal to compare treatment variables. Less exposure of serosal surface to the incubation media can be achieved by using longer everted intestinal sleeves (*46*) in place of rings.

Diffusion Cell Studies

In vitro techniques for evaluating intestinal absorption as uptake by intestinal rings and sleeves are classified as tissue accumulation methods. Methods designed to assess drug transport from the intestinal mucosal to serosal compartments monitor drug flux across the tissue as opposed to drug accumulation in the tissue. These methods include "uptake" (flux) by everted intestinal sacs and drug flux across intestinal tissue mounted in diffusion cells. Because the tissue surface area is defined and fixed by the diffusion cell chambers (Figure 9), the flux is simply the rate of drug appearance in the serosal compartment, and permeability is defined by normalizing the flux by the mucosal drug concentration (*see* In Situ Models). Surface area is less clearly defined in the everted sac experiment, and a length or tissue weight correlation with surface area is also required to normalize permeability calculations.

Everted sacs are similar to everted intestinal sleeves except that they are tied off at either end of the segment. Rather than measuring drug accumulated in the tissue, following incubation of sacs in drug solution, drug appearance in the lumen (serosal compartment) of the everted sac is measured as a function of time. Thus drug flux, J, is actually measured from the mucosal to serosal sides of the tissue (*47*). It is more time-consuming to measure drug transport in these experiments as opposed to drug uptake in intestinal rings and sleeves,

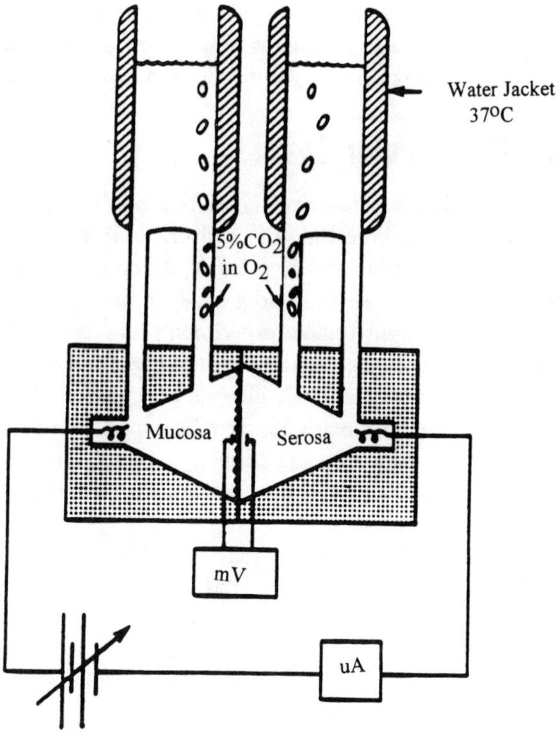

Figure 9. Ussing-chamber–voltage clamp setup to measure mucosal-to-serosal drug flux as well as tissue resistance (conductance).

and viability limitations have been demonstrated by histological examination (*48*). Because of these limitations, this technique is less commonly employed than diffusion cell studies.

Diffusion cell systems have been designed to study drug delivery across a number of epithelial tissues including nasal, buccal, bronchial, corneal, and epidermal (*49*). Mammalian transdermal drug delivery is routinely evaluated in specially designed Franz diffusion cells. However, the first studies of transepithelial transport were performed with amphibian skin for the purpose of studying ion transport in Ussing chambers designed to monitor electrical conductance. The Ussing chamber system can be used to study both drug flux across intestinal epithelia and, because current and voltage measurements can be made in this system (Figure 9), to assess transepithelial resistance. Tissue integrity is confirmed by monitoring transepithelial resistance coupled with leakiness to nonabsorbable markers. Tissue viability can be tested by monitoring electrical current changes in response to carrier-mediated ion transport. Experimental measurement of electrical parameters can be made by electrodes at the mucosal and serosal membrane under "open-circuit" conditions and under "short-circuit" or "voltage-clamped" conditions in which voltage across the tissue is zeroed by passing current from electrodes in the mucosal and serosal solutions (Figure 9). A study of the voltage dependence of drug transport across intestinal tissue permits an assessment of both paracellular (*50*) and carrier-mediated transport (*51*).

Ussing chambers have also been employed to study intestinal metabolism (*52*) and pharmacological (electrical) response (*53*) in conjunction with drug transport. In this regard, a particular strength of this in vitro system is the mucosal and serosal accessibility for both solution additions and sampling for transport, metabolism, and pharmacologic response measurements. Sensitive electrical measurements are easily made, but low transepithelial mass flux (exposed tissue surface area is only ~1 cm^2) may analytically limit measurement of metabolite concentrations or marker dilutions by pharmacological water transport effects.

Interactions at the Cellular Level

Both drug uptake and flux measurements can be made at the level of the intestinal epithelial cell as well as in or across intestinal tissue. Isolated intestinal epithelial cells (enterocytes) can be obtained by perfusing solutions of hyaluronidase and calcium chelators through intestinal segments (*54*) of anesthetized animals in situ. Suspensions of these collected cells can be made in which drug uptake experiments can be conducted. Uptake experiments similar to those discussed for everted intestinal rings can be performed if radiolabeled drug is available by rapid microfiltration (Millipore) of cells from drug incubation solutions (*55*). Both transport and metabolism studies have been performed with isolated cell suspensions because system capacity can be increased by pooling cells from different animals. However, because these cells dedifferentiate (lose mucosal-to-serosal polarity), drug and metabolite directional information is lost.

Epithelial cell polarity may be maintained if cultured epithelial cells are grown in monolayers. This type of growth is not possible for long periods of time because cellular turnover time is fairly short for normal intestinal epithelial cells. Several immortal intestinal cancer cell lines can be grown in monolayers on porous membrane filters and can establish cell polarity after a period of growth. The most popular of these is the Caco-2 cell line originally obtained from human colon cancer tissue (*56*). The transepithelial resistance of this cell line is similar to that of the normal colon (indicative of tight epithelia and a restricted paracellular pathway), and many of the carrier-mediated pathways (e.g., small peptide transport) characteristic of leaky upper intestinal epithelia are present in this cell line (*57*). Because small peptide transport is mediated by a transmucosal proton gradient, an important consideration for choosing an in vitro system might be the capacity to maintain mucosal microclimate pH. Although intestinal tissue goblet cells secrete mucus to slow proton diffusion away from the mucosal membrane, the Caco-2 cell monolayer does not possess mucus-secreting capacity. (When mucus-secreting capacity is required in a cell monolayer study, subclones of a human rectal cell line, HT-29 cells, may be used.) Filter-backed Caco-2 cell monolayers can be mounted in place of intestinal tissue in Ussing chambers, and drug flux, metabolism, and pharmacologic response measurements may be obtained. Again, the limited cell surface area afforded in these systems may dictate capacity limits of these measurements. Borchardt's chapter in this text outlines peptide drug transport studies with cell culture models.

Cellular Membrane Investigations

To evaluate drug transport at the level of the mucosal membrane, procedures to uncouple this process from cellular metabolism and transport across other cellular membranes must be used. Both intestinal mucosal brush border membrane vesicles and basolateral membrane vesicles (*58*) can be prepared to isolate peptide drug transport steps at either cellular pole. The relative purity of such preparations is determined by demonstrating significant enrichment of enzymes specific to those membranes over their activity in a crude intestinal homogenate. Enrichment of sucrase or alkaline phosphatase by 16–20-fold is indicative of adequate separation of brush border membranes, and enrichment of Na^+K^+-ATPase factors by only 10–12-fold has been secured for basolateral membrane preparations. These preparations are usually made from scrapings of freshly removed animal intestine or from human biopsy tissue, but mucosal membrane vesicles have been prepared from Caco-2 cells (*59*) to ensure that membrane transport in this carcinoma cell line mimics that of normal intestinal tissue.

Drug uptake kinetic studies similar to those outlined for everted intestinal ring experiments can be performed to demonstrate concentration dependence and competitive inhibition of peptide drug transport. In this regard, saturable uptake of a dipeptide-like cephalosporin as well as uptake inhibition by a dipeptide-like angiotensin converting enzyme (ACE) inhibitor was reported in small intestinal brush border membrane vesicles (*60*). Effects of inhibitors from the same (mucosal) side of the membrane ("cis" effects) can be tested by studying vesicular drug uptake from incubation solutions containing both drug and inhibitor. Effects of inhibitors from the opposite (mucosal–cytosolic) side of the membrane ("trans" effects) can be tested by preloading vesicles with inhibitor prior to vesicle incubation in drug solution. Preloading is carried out by isolating membranes and forming vesicles in solutions containing a potential inhibitor (or stimulator) of drug uptake. Vesicle preloading of one cephalosporin will trans-stimulate the peptide-carrier-mediated uptake of another cephalosporin (*61*).

The primary utility for mucosal membrane vesicle studies is to evaluate carrier-mediated uptake of di- and tripeptide drugs and to determine to what extent nutrient and other drug peptides might compromise their absorption. However, brush border membrane vesicles have also been used to study passive drug uptake as a function of partition and distribution coefficients based on membrane lipid content (*62*). Furthermore, sites of peptide drug degradation can be assessed by studying their stability in membrane vesicle preparations as compared with crude intestinal homogenates or intestinal perfusate, and the enzymes involved can be identified by peptidase and protease inhibitor studies. Separation of vesicular peptide drug uptake from metabolism has been successfully achieved both with brush border peptidase inhibitors and animal strains genetically lacking a membrane-metabolizing enzyme (*63*) to make vesicles. A primary limitation is that the interior vesicular compartment is very small, thereby presenting limited uptake capacity and short initial uptake times and demanding rapid mixing and filtration procedures (*64*). However, nonspecific membrane binding can still be separated from uptake kinetics by studying drug uptake as a function of incubation medium osmolality.

Summary Comments

Given a broad spectrum of techniques to study limitations to peptide drug absorption and the potential for absorption variability, the choice of an experimental system is based on projected transport rate limits in accordance with drug physical–chemical properties. Mechanistic detail on drug transport and metabolism absorption components can be obtained from isolated in vitro experiments, but in situ and in vivo studies define the impact of absorption variability on systemic response. Absorption parameter correlations between isolated mechanistic investigations and human pharmacokinetic studies (*28, 44, 60*) should provide direction to drug discovery–delivery collaborations maximizing peptide drug therapeutic activity and oral delivery potential. In addition, absorption parameters obtained in more isolated studies can be used in models to project the range of absorption variability anticipated in clinical studies (*1, 3, 4, 16*).

References

1. Oh, D. M.; Curl, R. L.; Amidon, G. L. *Pharm. Res.* **1993,** *10,* 264–270.
2. Karls, M. S.; Rush, B. D.; Wilkinson, K. F.; Vidmar, T. J.; Burton, P. S.; Ruwart, M. J. *Pharm. Res.* **1991,** *8,* 1477–1481.
3. Sinko, P. J.; Leesman, G. D.; Amidon, G. L. *Pharm. Res.* **1993,** *10,* 271–275.
4. Roberts, M. S.; Rowland, M. *J. Pharm. Sci.* **1985,** *74,* 585–587.
5. Wagner, J. G. *J. Pharm. Sci.* **1983,** *72,* 838–842.
6. Chan, K. K. H.; Gibaldi, M. *J. Pharm. Sci.* **1985,** *74,* 388–393.
7. Walker, R. F.; Codd, E. E.; Barone, F. C.; Nelson, A. H.; Goodwin, T.; Campbell, S. A. *Life Sci.* **1990,** *47,* 29–36.
8. Gupta, S. K.; Benet, L. Z. *Biopharm. Drug Dispos.* **1989,** *10,* 591–596.
9. Bundy, G. L.; Pals, D. T.; Lawson, J. A.; Couch, S. J.; Lipton, M. F.; Mauragis, M. A. *J. Med. Chem.* **1990,** *33,* 2276–2283.
10. Garren, K. W.; Pyter, R. A. *Int. J. Pharm.* **1990,** *66,* 167–172.
11. Dressman, J. B. *Pharm. Res.* **1986,** *3,* 123–131.
12. Fleisher, D.; Lippert, C. L.; Sheth, N.; Reppas, C.; Wlodyga, J. *J. Controlled Release* **1990,** *11,* 41–49.
13. Gupta, P. K.; Robinson, J. R. *Int. J. Pharm.* **1988,** *43,* 45–52.
14. Meyer, J. H.; Dressman, J. B.; Fink, A.; Amidon, G. L. *Gastroenterology* **1985,** *89,* 805–813.
15. Brener, W.; Hendrix, T. R.; McHugh, P. R. *Gastroenterology* **1983,** *85,* 76–82.
16. Dressman, J. B.; Fleisher, D. *J. Pharm. Sci.* **1986,** *75,* 109–116.
17. Weaver, D. F.; Camfield, P.; Fraser, A. *Neurology* **1988,** *38,* 755–758.
18. Schemann, M.; Ehrlein, H. J. *Gastroenterology* **1986,** *90,* 991–1000.
19. Dressman, J. B.; Yamada, K. In *Drugs in Pharmaceutical Science: Biopharmaceutical Equivalence;* Welling, P. G.; Tse, F., Eds.; Marcel Dekker: New York, 1990; Vol. 148, pp 235–265.
20. Cressman, W. A.; Sumner, D. *J. Pharm. Sci.* **1971,** *60,* 132–134.
21. Lui, C. Y.; Oberle, R.; Fleisher, D.; Amidon, G. L. *J. Pharm. Sci.* **1986,** *75,* 469–474.
22. Coupe, A. I.; Davis, S. S.; Wilding, I. R. *Pharm. Res.* **1991,** *8,* 360–364.
23. Brondsted, H.; Kopecek, J. *Pharm. Res.* **1992,** *9,* 1540–1545.
24. Amidon, G. L., University of Michigan, Ann Arbor, MI.
25. Quijano, R. J.; Ohnishi, N.; Umeda, K.; Komada, F.; Iwakawa, S.; Okumura, K. *Drug Metab. Dispos.* **1993,** *21,* 141–143.
26. Bastidas, J. A.; Orandle, M. S.; Zinner, M. J.; Leo, C. J. *Surgery* **1990,** *108,* 376–383.
27. Fleisher, D.; Rana, K.; Miles, C.; Lippert, C. *Pharm. Res.* **1992,** *9,* S-268.
28. Amidon, G. L.; Sinko, P. J.; Fleisher, D. *Pharm. Res.* **1988,** *5,* 651–654.
29. Lu, H. H.; Thomas, J.; Fleisher, D. *J. Pharm. Sci.* **1992,** *81,* 21–25.
30. Clements, J. A.; Heading, R. C.; Nimmo, W. S.; Prescott, L. F. *Clin. Pharmacol. Ther.* **1978,** *24,* 420–431.
31. Schurgers, N.; Bijdendijk, J.; Tukker, J. J.; Crommelin, D. J. A. *J. Pharm. Sci.* **1986,** *75,* 117–119.
32. Lu, H. H.; Thomas, J. D.; Tukker, J. J.; Fleisher, D. *Pharm. Res.* **1992,** *9,* 894–900.
33. Kou, J. H.; Fleisher, D.; Amidon, G. L. *Pharm. Res.* **1991,** *8,* 298–305.
34. Wollenberg, P.; Rummel, W. *Biochem. Pharmacol.* **1984,** *33,* 205–208.

35. Hu, Z.; Tse, E. G. C.; Oh, C. K.; Monkhouse, D.; Fleisher, D. *Life Sci.* **1994,** *54,* 1977–1985.

36. Fuessl, H. S.; Domin, J.; Bloom, S. R. *Clin. Sci.* **1987,** *72,* 255–257.

37. Laszlo, F.; Pavo, I.; Penke, B.; Balint, G. A. *Life Sci.* **1989,** *44,* 1573–1578.

38. Schusdziarra, V.; Harris, V.; Arimura, A.; Unger, R. H. *Endocrinology* **1979,** *104,* 1705–1708.

39. Drewe, J.; Fricker, G.; Vonderscher, J.; Beglinger, C. *Br. J. Pharmacol.* **1993,** *108,* 298–303.

40. Takahashi, K.; Jones, P. M.; Kanse, S. M.; Lam, H. C.; Spokes, R. A.; Ghatei, M. A.; Bloom, S. R. *Gastroenterology* **1990,** *99,* 1660–1667.

41. Brown, M. A.; Smith, P. L. *Regul. Pept.* **1991,** *36,* 1–19.

42. Fleisher, D.; Stewart, B. H.; Amidon, G. L. In *Methods in Enzymology;* Academic: New York, 1985; Vol. 112, pp 360–381.

43. Fleisher, D.; Sheth, N.; Griffin, H.; McFadden, M.; Aspacher, G. *Pharm. Res.* **1989,** *6,* 332–337.

44. Stewart, B. H.; Kugler, A. R.; Thompson, P. R.; Bockbrader, H. N. *Pharm. Res.* **1993,** *10,* 276–281.

45. Amidon, G. L.; Stewart, B. H.; Pogany, S. *J. Controlled Release* **1985,** *2,* 13–26.

46. Karasov, W. H.; Diamond, J. M. *J. Comp. Physiol.* **1983,** *152,* 105–116.

47. Gibaldi, M.; Grundhofer, B. *J. Pharm. Sci.* **1972,** *61,* 116–119.

48. Levine, R. R.; McNary, W. F.; Kornguth, P. J.; LeBlanc, R. *Eur. J. Pharmacol.* **1970,** *9,* 211–219.

49. Rojanasakul, Y.; Wang, L. Y.; Bhat, M.; Glover, D. D.; Malanga, C. J.; Ma, J. K. H. *Pharm. Res.* **1992,** *9,* 1029–1034.

50. Ishizawa, T.; Hayashi, M.; Awazu, S. *J. Pharmacobio-Dyn.* **1991,** *14,* 583–589.

51. Abe, M.; Hoshi, T.; Tojima, A. *J. Physiol.* **1987,** *394,* 481–499.

52. Rogers, S. M.; Back, D. J. *Prog. Pharmacol. Clin. Pharmacol.* **1989,** *7,* 43–53.

53. Smith, P. L.; Montzka, D. P.; McCafferty, G. P.; Wasserman, M. A.; Fondacaro, J. D. *Am. J. Physiol.* **1988,** *255,* G175–183.

54. Velasco, G.; Dominguez, P.; Shears, S. B.; Lazo, P. S. *Biochim. Biophys. Acta* **1986,** *889,* 361–365.

55. Calonge, M. L.; Ilundain, A.; Bolufer, J. *J. Cell. Physiol.* **1989,** *138,* 579–585.

56. Hidalgo, I. J.; Raub, T. J.; Borchardt, R. T. *Gastroenterology* **1989,** *96,* 736–749.

57. Dantzig, A. H.; Tabas, L. B.; Bergin, L. *Biochim. Biophys. Acta* **1992,** *1112,* 167–173.

58. DelCastillo, J. R.; Robinson, J. W. L. *Biochim. Biophys. Acta* **1982,** *688,* 45–56.

59. Souba, W. W.; Copeland, E. M. *Ann. Surg.* **1992,** *215,* 536–544.

60. Yuasa, H.; Amidon, G. L.; Fleisher, D. *Pharm. Res.* **1993,** *10,* 400–404.

61. Yuasa, H.; Amidon, G. L.; Fleisher, D. *Pharm. Res.* **1991,** *8,* S-221.

62. Alcorn, C. J.; Simpson, R. J.; Leahy, D.; Peters, T. J. *Biochem. Pharmacol.* **1991,** *42,* 2259–2264.

63. Tiruppathi, C.; Ganapathy, V.; Leibach, F. H. *J. Biol. Chem.* **1990,** *265,* 2048–2053.

64. Malo, C.; Berteloot, A. *J. Memb. Biol.* **1991,** *122,* 127–141.

RECEIVED for review July 12, 1993. ACCEPTED revised manuscript September 28, 1993.

CHAPTER 22

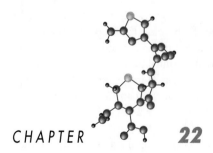

In Vitro Models of Hepatic Uptake

Methods To Determine Kinetic Parameters for Receptor-Mediated Hepatic Uptake

**Yuichi Sugiyama
and Yukio Kato**

Pharmacokinetics of Polypeptides

Recently, many types of polypeptides such as cytokines or lymphokines have been identified. These compounds induce the differentiation and growth of a great variety of cells at very low concentrations (within nano- or picomolar ranges) in vitro and were expected to be developed as therapeutic drugs for an equal variety of diseases. Advancements in gene recombination techniques, which allow the mass production of such compounds, have added to such expectations. However, although these polypeptides show dramatic biological activity in vitro, many stumbling blocks to their administration in vivo exist. One of the most critical issues is efficacy; that is, how efficiently can such compounds be delivered to target organs and display their pharmacological effect after administration into the systemic circulation. In general, polypeptide compounds have very short biological half-lives (*1–3*), which may be one of the reasons they do not always produce a strong enough effect in vivo. For the efficient delivery of polypeptides to their targets, a drug delivery system (DDS) must be developed. Polypeptide pharmacokinetic mechanisms and subsequent expression of biological activity must be clarified before developing a DDS, which is necessary for such polypeptide compounds to be used as medical therapy.

Many variables must be considered in determining the pharmacokinetics of polypeptides, such as binding to plasma protein, permeation through capillaries, degradation by peptidase either in plasma or on the cell surface, specific binding to receptors, uptake and metabolism, and glomerular filtration. The most notable pharmacokinetic feature of polypeptides, compared with that of small molecules, is the contribution of receptors on cell surfaces in liver or kidney to peptide distribution and elimination in the body (*1, 3*). Receptor-mediated endocytosis (RME) is recognized as a general mechanism in the uptake of polypeptides. The RME transport mechanism has been widely analyzed, mainly with in vitro isolated or cultured cell systems (*4–8*).

Traditionally, receptors on the cell surface have been recognized as specific binding sites for the expression of biological effects. The discovery of RME added another dimension to the concept of receptors in which receptors act as specific binding sites for polypeptide transport and clearance. A representative example is the biologically silent receptor (C-receptor) of atrial natriuretic factor (ANF) in kidney (*9*). In perfused rat kidney, a ring-deleted analogue of ANF (C-ANF) almost completely blocked the specific binding of [^{125}I]ANF, but had no biological activity and did not antagonize the effect of ANF. However, C-ANF increased both the plasma level of endogenous ANF and sodium excretion in vivo. The increase in the plasma level of endogenous ANF is believed to come from the inhibition of renal clearance of ANF by C-ANF. These results suggest the existence of biologically silent ANF receptors that are not related to the biological effect expressed by ANF (*9*). Such receptors for transport or clearance are also suggested for other polypeptides; for examples, the transferrin receptor for iron transport (*10*), the galactose receptor for clearance of desialylated serum glycoproteins (*11, 12*), and the interleukin-1 type II receptor (*13*). The concept of "transport receptor" is now well established.

Because the hepatic receptors contribute significantly to the pharmacokinetics of polypeptides, it is necessary to determine the time profiles of not only the polypeptide itself but also of its receptor to clarify polypeptide pharmacokinetics. In this chapter we introduce methodology (in vivo system, liver perfusion system, and isolated or cultured hepatocyte system) to analyze RME processes in hepatocytes based on our studies of epidermal growth factor (EGF) and hepatocyte growth factor (HGF). These methodologies can be applied to the analysis of other polypeptides that are transported into other types of cells by RME. The advantages and shortcomings of each of the methods are presented.

Receptor-Mediated Endocytosis

A schematic diagram for the RME process (*1*) and a kinetic model for RME (*3*) are shown in Figures 1A and 1B, respectively. RME is initiated by the specific binding of a ligand to its receptor on the cell surface. In general, this binding depends on the pH value (*14*) or on metallic cation concentration (*15*). After ligand–receptor binding occurs, coated pits form at the cell surface, and the pits pinch off from the plasma membrane and become coated vesicles (Figure 1A). The process of coated-vesicle formation occurs simultaneously with the loss of the ligand–receptor complex from the cell surface and is called internalization. The coated vesicles fuse with the other intracellular vesicles and become endosomes. Because the pH value in an endosome is low, the ligand dissociates from its receptor in this compartment, and ligands and receptors are then sorted to their own destinations. In hepatocytes, the ligand is finally degraded in a lysosome or transcytosed to the bile duct

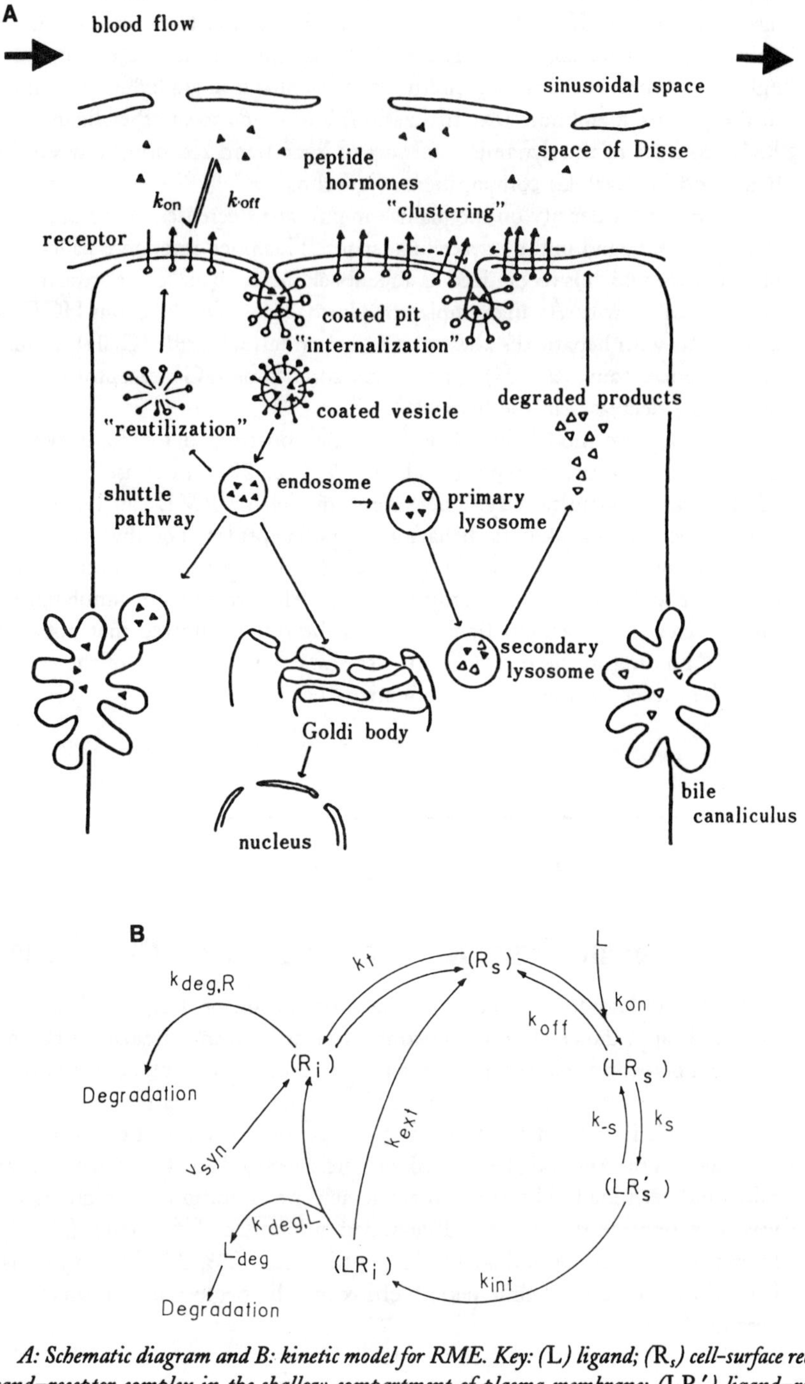

Figure 1. A: Schematic diagram and B: kinetic model for RME. Key: (L) ligand; (R_s) cell-surface receptor; (LR_s) ligand–receptor complex in the shallow compartment of plasma membrane; (LR_s') ligand–receptor complex in the deep compartment of plasma membrane; (LR_i) ligand–receptor complex inside the cell; (R_i) receptor inside the cell. (Reproduced with permission from reference 1. Copyright 1989; and from reference 3. Copyright 1990.)

(*16–18*). The degradation product of the ligand is transported back to the blood capillary or partly into bile (*18, 19*). In 1987, it was reported (*20, 21*) that a portion of internalized ligand, such as EGF, is transported to nuclei and that there is a specific binding site for polypeptide on nuclei; however, the physiological significance of this binding site is still unknown. On the other hand, internalized receptor is degraded in lysosomes (*22*) or returns to the plasma membrane (externalization) and binds to another ligand (*23*). In this way, RME consists of the dynamic transport of ligand and receptor between the plasma membrane and intracellular compartments (recycling).

Receptor density on the plamsa membrane decreases in the presence of excess concentration of ligand in the extracellular space. This phenomenon is called "down regulation" of receptors, which is recognized as a feedback mechanism to suppress the excess expression of biological activity. As an example, the plasma level of endogenous HGF increases remarkably in rats with hepatic diseases such as carbon tetrachloride (CCl_4)–induced hepatitis and partial hepatectomy (*24, 25*). Under such conditions HGF receptor density on the plasma membrane decreases in the liver (*26*).

A kinetic model for RME has been constructed for many types of polypeptides transported into various types of cells (*1, 3, 7, 8, 27*) and is useful for understanding the RME process quantitatively. The kinetic model for RME can be employed to simulate the in vivo pharmacokinetic behaviors of polypeptides. For that purpose, a physiological pharmacokinetic model can be used (*1*). The time profiles of both plasma concentration of EGF and EGF receptor density on liver cell surface were simulated with a simplified physiological pharmacokinetic model, and the results showed that not only the saturation of EGF binding to the receptors but also the down regulation of the cell-surface receptors could be causes for the nonlinear kinetics (*1*).

Analysis with an In Vivo System

Analysis of Plasma Disappearance and Tissue Distribution

Our methodology for estimation of the contribution of drug distribution in each organ to plasma disappearance after administration of drug to the circulation begins with measurement of polypeptide concentrations in plasma and organs. Uptake clearance is then calculated for each organ by the following simple kinetic analysis. After the intravenous administration of ^{125}I-labeled polypeptide of rats, the polypeptide concentration in plasma is determined. After a short time period (1–10 min), rats are sacrificed and tissue concentrations of polypeptide are determined. The trichloroacetic acid precipitation technique is a simple and easy way to determine the intact and degraded amounts of [^{125}I]polypeptide (*28, 29*), but the limitations of this method should be kept in mind (*28, 30*). When the tissue distribution is measured for a short time period, efflux may be neglected and tissue uptake rate is then described by:

$$d\,[X_t]_t/dt = CL_{uptake}\,[Cp]_t \qquad (1)$$

where $[Cp]_t$, $[X_t]_t$, and CL_{uptake} are plasma concentration, the amount of polypeptide associated with tissue at time t, and tissue uptake clearance, respectively. When equation 1 is integrated from time 0 to t, equation 1 gives:

$$[X_t]_t = CL_{\text{uptake}}\,\text{AUC}_{(0-t)} + [X_t]_0 \qquad (2)$$

where $\text{AUC}_{(0-t)}$ and $[X_t]_0$ represent the area under the plasma concentration–time profile from time 0 to t and the amount of polypeptide in tissue at time 0, respectively. Thus, CL_{uptake} can be estimated from the initial slope of a plot of $[X_t]_t$ versus $\text{AUC}_{(0-t)}$ (*29, 31*). This plot is called the "integration plot".

The tissue uptake of EGF (*31*) and HGF (*29*) were analyzed with this technique. In each organ integration plot, a straight light passes through the y-axis intercept, which represents the amount of polypeptide associated with tissue immediately after its administration ($[X_t]_0$). Therefore, dividing the $[X_t]_0$ value by the initial plasma concentration of polypeptide ($[Cp]_0$) yields the initial distribution volume of the tissue ($[Kp]_0$). The $[Cp]_0$ value can be easily obtained from the pharmacokinetic analysis of the plasma disappearance curve of the polypeptide. When the polypeptide is small enough to pass through the endothelial cells of the tissue referred to in equation 2 (i.e., low enough molecular weight), the $[Kp]_0$ value is equal to the volume of tissue extracellular space. When the polypeptide does not easily get into the interstitial space due to relatively larger molecular weight, however, the $[Kp]_0$ value is equal to the volume of tissue capillary space. The $[Kp]_0$ value sometimes becomes larger when the polypeptide binds rapidly to the surface of endothelial cells or epidermal cells.

For polypeptides that do not bind or get into the erythrocyte, the division of CL_{uptake} by the plasma flow rate gives the early phase extraction ratio, which is defined as the fraction of drug extracted by the organ during a single passage. We previously found that the extraction ratio of EGF was 80, 40, 30, and 70% in liver, kidney, small intestine, and spleen, respectively (*3*), and that of HGF was 25% in liver (*29*). In addition, when CL_{uptake} is converted to a per-body-weight value, the contribution of each organ to the disappearance from the systemic circulation can be estimated. The CL_{uptake} value per kilogram of body weight in liver was >60% (*3*) and >90% (*29*) of the sum of CL_{uptake} values for each organ, suggesting that liver is the major clearance organ for EGF and HGF.

Nonlinear kinetics in vivo also can be analyzed with a similar technique (*31*). Because RME is a saturable process, polypeptide pharmacokinetics show a dose dependency. In the analysis of nonlinear kinetics of EGF, a mixture of labeled and various amounts of unlabeled polypeptide is administered, and tissue uptake clearance is determined by similar kinetic analysis (*31*) (Figure 2). The integration plots for EGF are linear until 5 min after administration and suggest that efflux can be neglected until that time. Therefore, CL_{uptake} can be calculated from the plasma concentration–time profile until 3 min and from the amount of EGF in tissue after 3 min (*31*). In this way, CL_{uptake} in liver, kidney, small intestine, stomach, and spleen was found to be dose dependent (*31*) (Figures 2B–2F). Furthermore, the dose dependency of the early-phase plasma half-life (*3*) was parallel only to that of the CL_{uptake} value in liver (Figures 2A–2F), suggesting that saturation of hepatic uptake is a main factor in nonlinear plasma disappearance kinetics.

In practice, the clinical application of polypeptides will be targeted to patients, not to healthy people, for a particular disease. Therefore, the analysis of pharmacokinetics in diseased model animals is important. A basically similar methodology can be applied to such diseased animals. The pharmacokinetics of HGF in CCl_4-intoxicated rats were studied (*32*). After the CCl_4 treatment, a significant decrease and subsequent recovery of CL_{uptake} in liver was observed, whereas no change was seen in kidney, spleen, or lung (*32*). The time–plasma clearance profile after CCl_4 treatment was parallel to that of CL_{uptake} in liver (Figure 3). In addition, >50% of plasma clearance was attributed to CL_{uptake} in liver at any time after CCl_4 treatment. These results suggest that liver is also the major clearance organ in the hepatic disease condition (*32*).

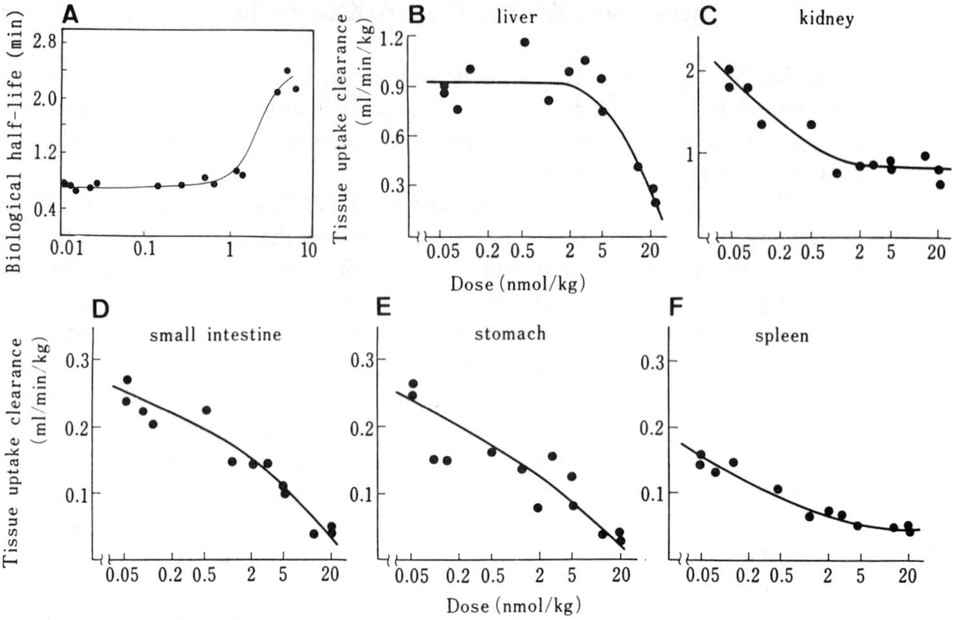

Figure 2. A: Dose dependence of early-phase plasma half-life and B–F: tissue uptake clearance of [^{125}I]EGF. Early-phase plasma half-life was determined from the slope of the plasma concentration–time profile until 3 min after intravenous administration of [^{125}I]EGF plus unlabeled EGF. The tissue uptake clearance was estimated from both the amount of [^{125}I]EGF in tissue at 3 min after the injection and the plasma concentration–time profile until 3 min. (Reproduced with permission from reference 3. Copyright 1990; and from reference 31. Copyright 1988.)

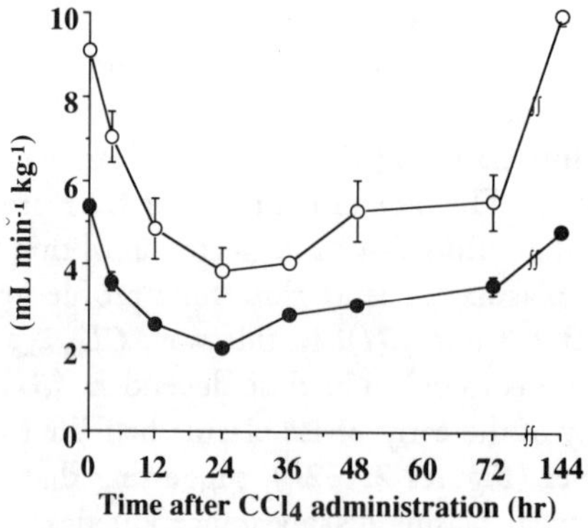

Figure 3. Change in early-phase plasma clearance and hepatic uptake clearance of [^{125}I]HGF after CCl_4 administration. Key: (○) plasma clearance; (●) hepatic uptake clearance. (Reproduced with permission from reference 32. Copyright 1988.)

Analysis of Recovery from Down Regulation of Receptors

A useful technique for analyzing the down regulation and subsequent recovery of polypeptide receptors on liver cell surfaces is first to excise the liver and then isolate the plasma membrane. This is followed by assay of binding of polypeptide to receptor to determine the receptor density by Scatchard analysis (*26*). This technique has the advantage that the change in receptor density on the cell surface can be estimated directly. In this technique, however, it is not easy to convert the density per membrane fraction protein into that per whole liver.

However, the effect of down regulation and the subsequent receptor recovery on the clearance of polypeptide can be estimated relatively easily (*33*) (Figure 4). In this technique a tracer amount of [^{125}I]EGF is administered, and CL_{uptake} is calculated from the plasma concentration–time profile and the amount of [^{125}I]EGF in tissue 3 min following the induction of down regulation of EGF receptors (*33*). The CL_{uptake} value thus obtained can be considered to be proportional to the available receptor density on the cell surface (*33*). Therefore, a change in CL_{uptake} represents the down regulation and recovery of EGF receptors. The half-life of recovery from down regulation was short ($t_{1/2} \approx 22$ min) in liver compared with that in other organs ($t_{1/2} \approx 2$–5 h) (Figure 4).

To analyze down regulation and the subsequent receptor recovery only in liver, a simpler method (*29, 34*), designated as the liver uptake index (LUI) technique, has been developed (Figure 5). In this method, a mixture of a tracer amount of labeled polypeptide and a reference compound that can be distributed only to either extracellular space ([^{14}C]inulin or [^{125}I]bovine serum albumin) or to intra- and extracellular space (^3H_2O or [^{14}C]butanol) is

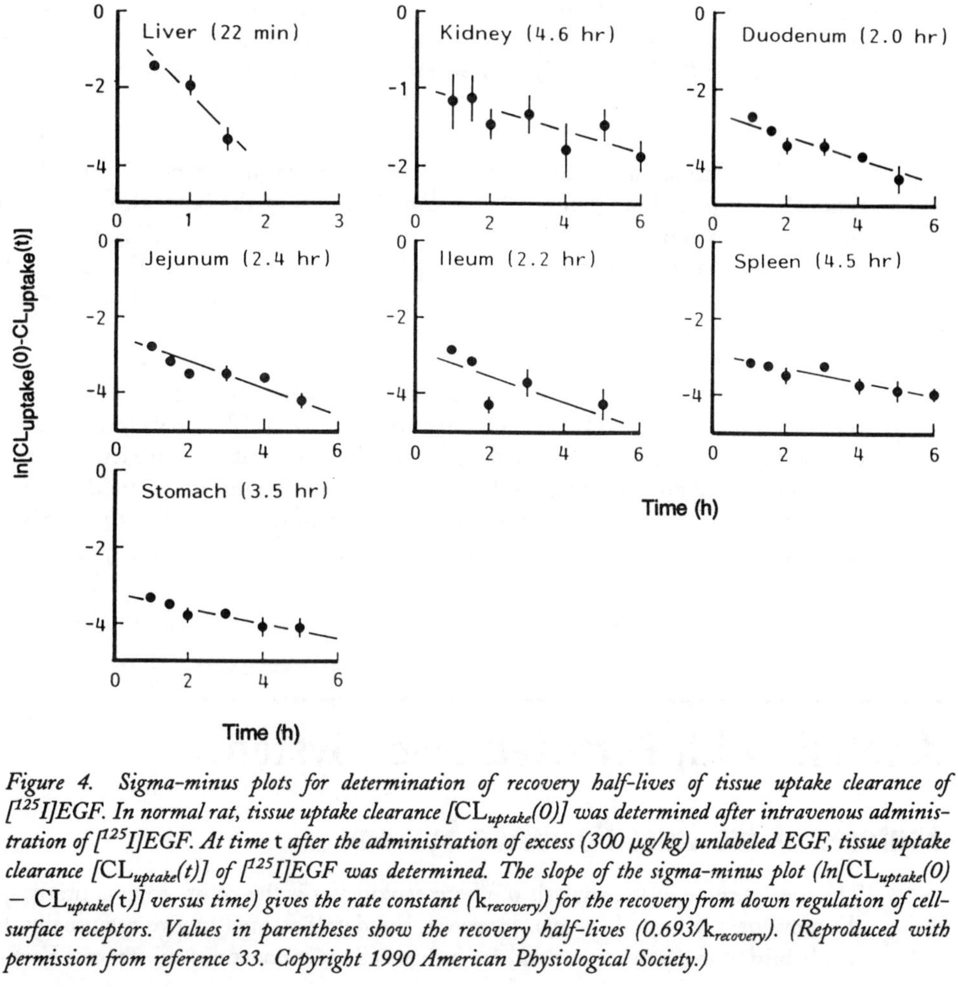

Figure 4. Sigma-minus plots for determination of recovery half-lives of tissue uptake clearance of [^{125}I]EGF. In normal rat, tissue uptake clearance [$CL_{uptake}(0)$] was determined after intravenous administration of [^{125}I]EGF. At time t after the administration of excess (300 µg/kg) unlabeled EGF, tissue uptake clearance [$CL_{uptake}(t)$] of [^{125}I]EGF was determined. The slope of the sigma-minus plot (ln[$CL_{uptake}(0)$ − $CL_{uptake}(t)$] versus time) gives the rate constant ($k_{recovery}$) for the recovery from down regulation of cell-surface receptors. Values in parentheses show the recovery half-lives ($0.693/k_{recovery}$). (Reproduced with permission from reference 33. Copyright 1990 American Physiological Society.)

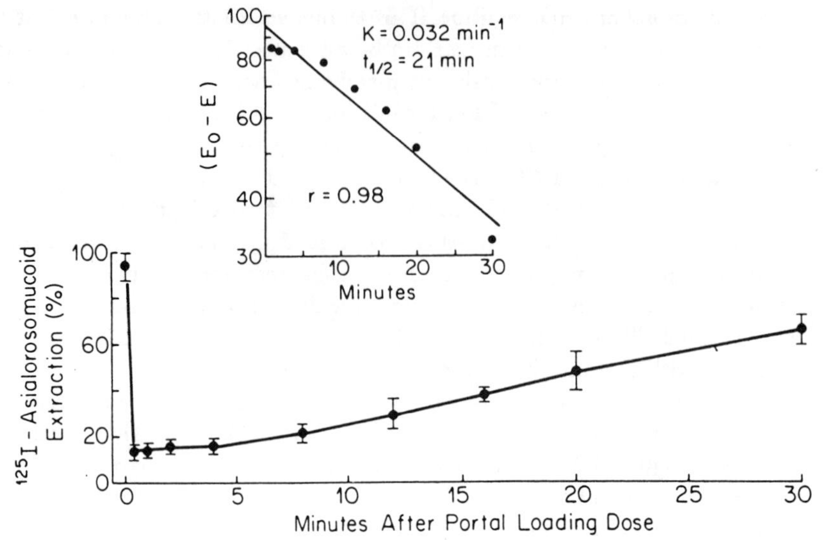

Figure 5. Recovery of the first-pass LUI of [^{125}I]ASOR after the administration of excess loading dose. After the portal vein administration of excess ASOR (400 μg), the liver was excised at 18 s following the second bolus containing [^{125}I]ASOR and tritiated water, as a highly diffusible internal reference. The upper graph is a sigma-minus plot, where E_0 *and* E *represent the extraction ratio of [^{125}I]ASOR after the injection of 0 or 400 μg of ASOR, respectively. (Reproduced with permission from reference 34. Copyright 1983.)*

dissolved in plasma and is administered instantaneously to the portal vein. At specified times (18 s, a period long enough for a complete pass of the bolus through the liver but short enough to prevent recirculation of labeled compound), rats are killed, the livers excised, and the radioactivities counted. As shown in Figure 5 the first-pass LUI, the ratio of hepatic extraction ratio of [^{125}I]asialoorosomucoid (ASOR) to that of highly diffusible internal reference (^3H_2O), decreased after the administration of excess amount of unlabeled ASOR, and then recovered LUI (*34*). The LUI may reflect the available receptor density on the liver cell surfaces, so the decrease of LUI may represent down regulation or occupation of cell-surface galactose receptors. The subsequent increase of LUI may represent the recovery of receptors. The half-life of the recovery from down regulation of galactose receptors was ~21 min (*34*) (Figure 5). This LUI technique is the most simple method to estimate the early-phase hepatic extraction in vivo but has shortcomings, as do the other in vivo systems (*see* next section), and is unsuitable for polypeptides that are minimally extracted by the liver.

Analysis with Perfused Liver Systems

Features of the Liver Perfusion System

The final purpose in the analysis of pharmacokinetics is the ability to quantitatively predict the movement of ligand in vivo. However, the data obtained in vivo are usually the result of a hybrid of many processes and are influenced by endogenous substances. Furthermore, the concentrations of drug, inhibitors, or plasma proteins, the plasma flow rate, or the temperature cannot be set to preferred values in vivo, therefore rendering this system unsuitable for the analysis of a complicated in vivo mechanism. To analyze RME in detail, it is more appropriate to use a perfused liver system and an isolated or cultured cell system. The perfused liver system has the advantage that detailed analysis can be carried out under physiological conditions because the structural architecture of liver as an in vivo system is maintained. The perfused liver system is not without fault however when compared with isolated or cultured cell systems; that is, operation or handling is complicated, and to measure the amount taken up by the liver only one data point can be obtained from a single experiment, except when a special technique like biopsy is used. To overcome these difficulties, we applied the multiple indicator dilution (MID) method and attempted to determine what is happening in the liver without directly measuring the ligand content in the liver.

Multiple Indicator Dilution Method

Features of the MID Method

The MID method was originally developed by Goresky (*35*) to quantify the hepatic transport and handling of small molecules. This method, which is a single-pass perfused liver system, is also an appropriate technique to determine the kinetic parameters representing the relatively rapid interaction between polypeptides and liver cell surfaces. An isolated or cultured hepatocyte system also can be used to obtain such parameters, but the most important advantage of the MID method is that such parameters can be obtained while maintaining the structural architecture of the liver. Furthermore, with this method, multiple experiments can be performed in a single animal. For example, in the analysis of down regulation and the subsequent recovery of polypeptide receptors, receptor density–time profiles can be obtained with a single animal by MID (*see* Analysis of Recovery from Down Regulation of Receptors), whereas many animals are necessary for an in vivo system.

Analysis of the Interaction between Polypeptides and Cell Surface

In the MID method, the injection mixture, which consists of a ligand and extracellular reference compound ([^{14}C]inulin or [^{125}I]bovine serum albumin), is administered as a bolus into portal vein, and hepatic vein effluent is collected in ~0.5–1-s aliquots for ~1 min (*36, 37*). A capillary space marker ([^{51}Cr]erythrocyte) is sometimes included in the injection mixture to determine the sinusoidal volume and a large vessel transit time (*35, 38, 39*). The reference compound distributes only to the extracellular space and passes through the liver without any interaction with the cell surfaces (Figure 6A). On the other hand, the polypeptide ligand, such as [^{125}I]EGF, not only distributes to the extracellular space but also interacts with cell-surface receptors and then is sequestered into the intracellular compartment; the remainder of the polypeptide ligand is detected in the hepatic vein effluent (Figure 6A). Therefore, the concentration–time profile (dilution curve) of polypeptide in the outflow is different from that of the reference compound, depending on the degree of interaction. Therefore, the interaction between polypeptide and cell-surface receptors in liver can be quantitatively estimated by simultaneously analyzing the dilution curves of both compounds based on the distributed model (*36, 39*) (Figure 6B). The apparent rate constant (k_{app}) can be determined from the initial slope of plots of natural logarithms of the ratio of the dose-normalized concentration of reference compound to that of polypeptide in effluent versus time (Figure 6B). When RME is the predominant mechanism for removal of polypeptide from the extracellular space, the k_{app} value thus obtained can be expressed as follows (*36*):

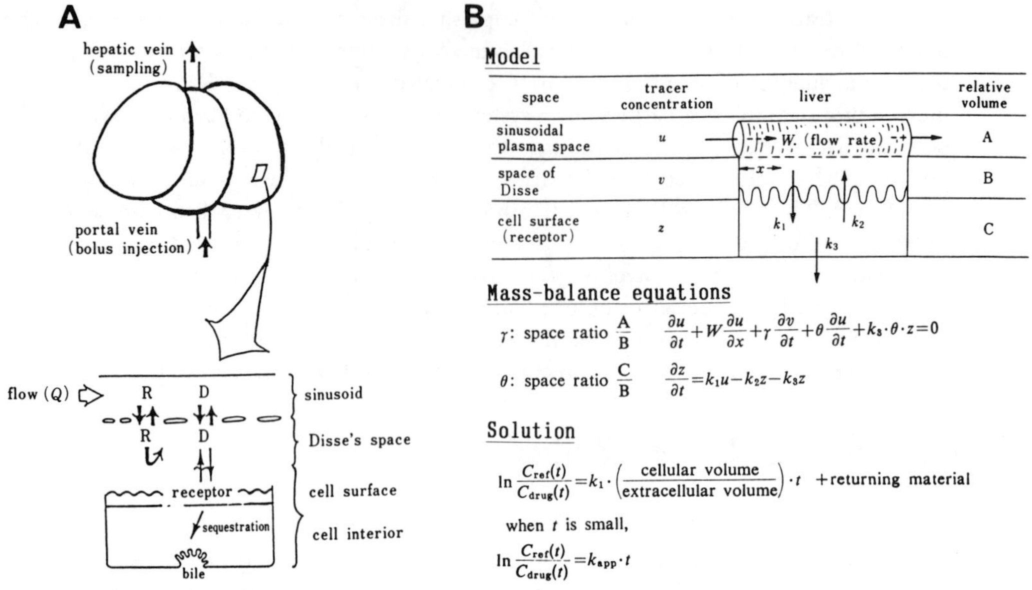

Figure 6. A: Schematic diagram for the MID technique and B: mathematical analysis of the indicator dilution curve based on a distributed model. Key: (D) ligand in question; (R) reference compound that distributes only to extracellular space.

$$k_{app} = k_{on}R_s/V_{d,ref} \tag{3}$$

where k_{on}, R_s, and $V_{d,ref}$ represent the association rate constant, the density of the available receptor on the cell surface, and the extracellular volume, respectively. The $V_{d,ref}$ value can be determined from moment analysis of the dilution curve of the extracellular reference compound (*36*). In addition, k_2 and k_3 (Figure 6B) can be expressed as follows (*36*):

$$k_2 = k_{off} \tag{4}$$

$$k_3 = k_s \tag{5}$$

where k_{off} and k_s represent the dissociation and sequestration rate constants, respectively. The dilution curves of polypeptide at various doses and that of the reference compound were simultaneously fitted to a model equation by a nonlinear least squares method to determine k_{on}, k_{off}, k_s, and R_s separately (*36*). From these analyses, the kinetic parameters representing the association and dissociation between polypeptide and receptor, sequestration of the polypeptide–receptor complex into the intracellular compartment, and receptor density on liver cell surfaces can be determined while maintaining the structural architecture of the liver. The kinetic parameters of EGF and its receptor, determined by the MID method, are listed in Table I (*27, 36*).

We also applied the MID method to characterize the hepatic handling of EGF in D-galactosamine-intoxicated rats. The k_{app} value was decreased in this diseased condition, and this decrease came from the decrease in cell-surface EGF receptor density (*37*).

Analysis of Recovery from Down Regulation of Receptors

Down regulation and the subsequent recovery of EGF receptors were kinetically analyzed with the MID method (*27, 36*). After an injection of excess (20 μg) unlabeled EGF, the recoveries of available cell-surface receptors were "chased" in the MID experiment with a

Table I. Parameters Representing the RME of EGF in the Liver

Parameter	Isolated Hepatocytes 0 °C	Isolated Hepatocytes 37 °C	Perfused Rat Liver 37 °C
k_{on} (min^{-1} nM^{-1})	0.07	0.7–1.1	0.12
k_{off} (min^{-1})	0.15	1.1–1.6	2.1
k_s (min^{-1})	[a]	0.6–1.3	4.1
k_{int} (min^{-1})	[a]	0.09–0.11	0.33
k_t (min^{-1})	[a]	[a]	[a]
$k_{deg,L}$ (min^{-1})	[a]	0.004–0.009	[a]
R_s (t = 0; pmol/g liver)	22	10–16	36
R_i (t = 0; pmol/g liver)	small	[a]	[a]
K_d (nM)	2.1	1.5	18
k_{-s} (min^{-1})	[a]	[a]	0.01
k_{ext} (min^{-1})	[a]	[a]	0.015
k_{bile} (min^{-1})	[a]	[a]	0.0001

[a] Data not available.

SOURCE: Reproduced with permission from reference 24. Copyright 1991.

tracer dose of [^{125}I]EGF (*36*) (Figure 7). Both the hepatic extraction ratio (E_H) and the k_{app} value obtained from the initial slope of the ratio plot decreased to the minimum level just after the injection and then increased (recovery half life ≈ 5–10 min) (*36*) (Figure 7). The k_{app} value obtained from the initial slope based on equation 3 can be considered to be proportional to R_s if it is assumed that both k_{on} and $V_{d,ref}$ remain constant. Therefore, these

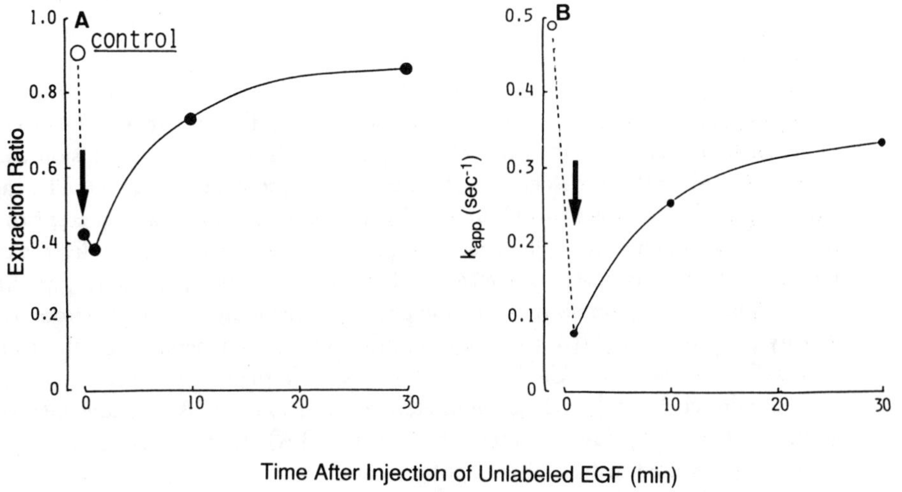

Figure 7. Analysis of the recovery of available EGF receptor on the liver cell surface after injection of an excess amount of unlabeled EGF with the MID method. At certain times after a bolus injection (arrow) of excess amount (20 μg) of EGF into the portal vein, the MID experiment at a tracer dose of [^{125}I]EGF was repeated. A: Extration ratio and B: k_{app} (= $k_{on}R_s/V_{d,ref}$) value of [^{125}I]EGF were obtained by analyzing the dilution curves of [^{125}I]EGF and an extracellular marker, [^{14}C]inulin. (Reproduced with permission from reference 36. Copyright 1988.)

recoveries may be explained by two possible mechanisms: (1) EGF molecules bound to their cell-surface receptors are dissociated during the perfusion and removed by the perfusate flow, resulting in an increase in available receptors; or (2) EGF–receptor complexes are internalized, and free receptors return to the cell surface from the intracellular pool by externalization. Which of these two mechanisms is involved cannot be determined from this experiment. Therefore, to determine the recovery kinetics of internalized receptors, most cell-surface receptors were internalized by perfusing an excess (20 nM) concentration of unlabeled EGF for 20 min followed by a switch to EGF-free perfusate (*27*). Then, the recoveries of available cell-surface receptors were followed by the MID method. Under these conditions, the E_H and k_{app} values decreased to the minimum level just after the perfusate switch and then gradually increased (*27*). A protein synthesis inhibitor, cycloheximide, was perfused throughout this experiment. Therefore, the increase in E_H and k_{app} may represent receptor externalization rather than receptor biosynthesis, and the half-life of such externalization was calculated to be ~30 min (*27*), comparable with the value obtained in an in vivo system (*33*).

The important features of the MID method, compared with an in vivo system, are (1) the effect of endogenous substances can be ignored; (2) repetitive experiments can be performed in a single animal; and (3) the concentrations of drug, inhibitors, or plasma proteins, the plasma flow rate, and the temperature can be controlled.

Analysis of the Internalization Process

Separate Determinations of Surface-Bound and Internalized Polypeptides

Polypeptide is internalized by the cell and then binds to cell-surface receptors. To analyze the internalization process, therefore, the amounts of cell-surface-bound and internalized polypeptide must be determined separately. The binding between many polypeptides and their receptors is known to be pH dependent (*14*). That is, polypeptide dissociates from its receptor in acidic conditions but remains firmly bound in neutral conditions. This phenomenon is applied in the acid-washing technique, a methodology used to detect surface-bound and internalized polypeptide separately (*14, 19*). After the perfusion of ^{125}I-labeled EGF, the perfusate is switched to ice-cold polypeptide-free buffer (pH 7.4) to remove the radioactivity remaining in the extracellular space; this is followed by perfusion with ice-cold acid buffer (pH 3.0) to dissociate surface-bound polypeptide (Figure 8). The radioactivity emerging in the outflow of acid buffer (acid-washable binding) may represent the polypeptide that was bound to the cell-surface receptor. After the acid washing, liver is excised and the radioactivity is counted to determine the internalized polypeptide (Figure 8).

The binding between some polypeptides, including asialoglycoproteins and low-density lipoproteins (LDL), and their specific receptor is dependent on the concentration of metallic cations such as Ca^{2+} or Mn^{2+} (*15, 40*). In the case of such polypeptides, buffer that contains chelators [ethylenediaminetetraacetic acid (EDTA) or ethylene glycol-bis (β-aminoethyl ether)*N,N,N′,N′*-tetraacetic acid (EGTA)] to dissociate surface-bound polypeptide can be used instead of acid buffer.

The time profiles of surface-bound and internalized polypeptide can be determined with a similar technique (*19*) (Figure 9). Note that the rate of increase of internalized polypeptide, which can be estimated by this technique, may be influenced not only by the change in the internalization rate constant but also by other factors such as changes in association and dissociation rate constants between polypeptide and receptors, receptor density on cell surfaces, and ligand concentration. Therefore, to estimate the kinetic parameter

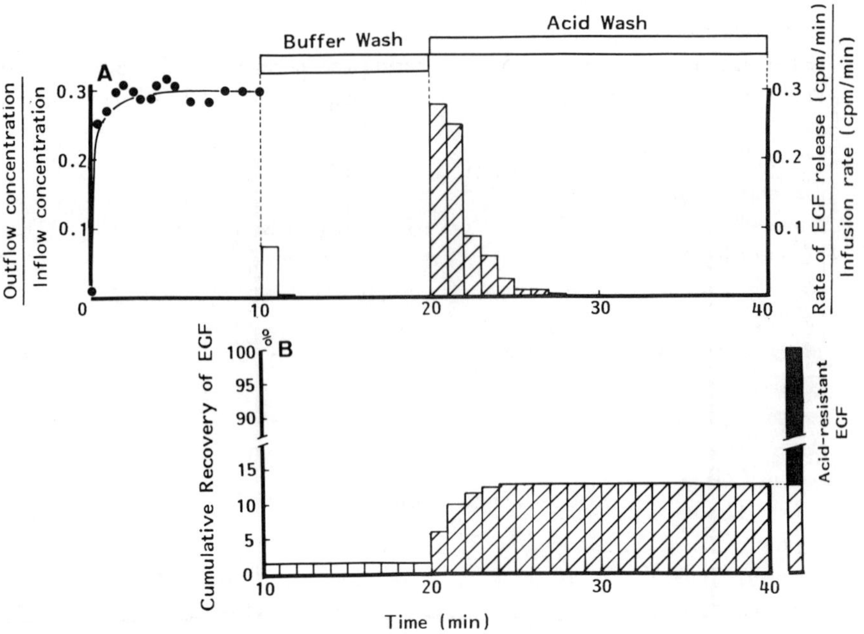

Figure 8. Comparison of radioactivity of [^{125}I]EGF remaining in extracellular space with that from surface-bound [^{125}I]EGF. After infusion of a tracer concentration (0.03 nM) of [^{125}I]EGF for 10 min, the perfusion medium was switched to ice-cold buffer (4 °C) free of EGF. Outflow was immediately collected for 10 min. Perfusion medium was then switched to acid buffer (pH 3.0), and effluent was immediately collected for 20 min. TCA-precipitable radioactivity in each sample and in liver was determined. A: Time profile of outflow-to-inflow concentration ratios (●), and rates of release into effluent by washing with neutral buffer (open bar), followed by acid buffer (hatched bar). B: Cumulative amount of EGF released into effluent by washing with neutral buffer followed by acid buffer. The closed bar represents the radioactivity remaining in the liver after acid washing and corresponds to internalized [^{125}I]EGF. (Reproduced with permission from reference 19. Copyright 1990.)

that represents only the internalization process, such time profiles must be analyzed kinetically (*see* next section).

Kinetic Analysis of the Internalization Process

The internalization rate is given by the following equation at elapsed times before intracellular degradation has occurred:

$$d\,[LR_i]_t/dt = k_{int}\,[LR_s]_t \tag{6}$$

where $[LR_s]_t$ and $[LR_i]_t$ are surface-bound and internalized polypeptide, respectively, and k_{int} represents the internalization rate constant, which is defined as a probability of the ligand–receptor complex being internalized in a unit time. Integration of equation 6 from time 0 to t yields equation 7:

$$[LR_i]_t = k_{int} \int_0^t [LR_s]_t\,dt \tag{7}$$

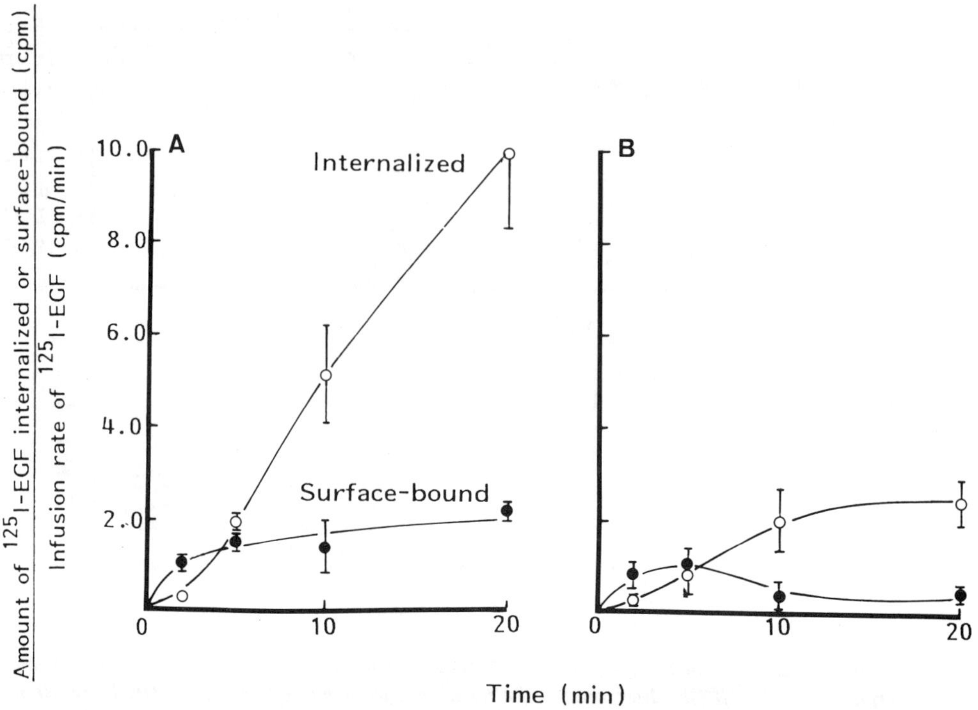

Figure 9. Time course of cell-surface bound (●) and internalized (○) [^{125}I]EGF in perfused liver during perfusion of tracer [^{125}I]EGF (0.03 nM), A and [^{125}I]EGF plus unlabeled EGF (20 nM), B. Infusion of EGF was terminated at indicated times by switching the perfusion medium to ice-cold acid buffer (pH 3.0). Effluent emerging into outflow was collected and used to calculate surface-receptor-bound EGF. After the acid wash, the liver was excised, and radioactivity associated with liver (internalized EGF) was determined. Surface-bound and internalized amounts were normalized by the infusion rate of EGF. (Reproduced with permission from reference 19. Copyright 1990.)

Thus, the k_{int} value can be estimated from the initial slope of a plot of $[LR_i]_t$ versus $\int_0^t [LR_s]_t\, dt$, where $\int_0^t [LR_s]_t\, dt$ is calculated numerically by the trapezoidal rule.

In addition, k_{int} can be estimated more simply as follows. Because the amount of EGF–receptor complex on the cell surfaces reaches steady state promptly, equation 7 can be approximated by the following equation (*41–43*):

$$[LR_i]_t = k_{int}\,[LR_s]_t\, t \tag{8}$$

Therefore, the approximated k_{int} value can be estimated by dividing $[LR_i]_t$ by $[LR_s]_t\, t$. In perfused rat liver, k_{int} for EGF internalization was estimated by the kinetic analysis just described to be 0.33 min^{-1} (*19*) (Table I). This result indicates that EGF bound to cell-surface receptors is internalized with a half-life of ~2 min. It is notable that this k_{int} value is ~0.1 that of the sequestration rate constant (k_s) obtained by the MID method (*see* Analysis of the Interaction between Polypeptides and Cell Surface) (Table I). The difference between k_{int} and k_s values required two compartments for the cell-surface EGF–receptor complex; that is, a "shallow pool" and a "deep pool". The ligand is exchangeable and nonexchangeable with free ligand in perfusate in the shallow and deep pools, respectively. The existence of two such pools was also suggested by the analysis with the isolated hepatocyte system (*27,*

44) (*see* Kinetic Analysis of Internalization and Degradation Process). To observe the deep pool directly, we attempted to measure the dissociation rate of [^{125}I]EGF from cell-surface receptors in the presence of phenylarsine oxide (PAO), which is known to inhibit polypeptide internalization without affecting receptor binding. As expected, a substantial amount of internalized [^{125}I]EGF was not detected in the presence of PAO, and the dissociation rate was very low (half-life ≈ 60 min), compared with the k_{off} value (half-life ≈ 0.3 min) estimated in the MID experiment (Table I). This result also suggests the existence of a deep pool for the EGF–receptor complex on the cell surface. However, it is still unknown what these pools represent. Possibly, they represent two different conformational states or clustering of the complexes to a specific domain on the cell surface.

Temperature-Shift Method

A temperature-shift protocol (*45*) can be used directly to analyze the polypeptide internalization process. Furthermore, this protocol enables us to set up the degree of receptor occupancy by controlling ligand concentration. Polypeptide internalization is almost completely inhibited at low temperature (≈10 °C). This feature is applied in the temperature-shift protocol. First, ^{125}I-labeled polypeptide is perfused at 2–4 °C. After the polypeptide concentration in the outflow attains steady state, the perfusate is switched to ice-cold ligand-free buffer to remove any nonspecific binding. After this process, almost all of the radioactivity still associated with the liver can be considered to represent ligand bound to cell-surface receptors. Next, the perfusate is switched to buffer at 37 °C (temperature-shift) to start the internalization, and the outflow is collected to determine ligand dissociated from cell-surface receptors. After a short period (5–10 min), when no intracellular degradation occurs, the perfusate is switched again to ice-cold buffer both to inhibit further internalization and to wash out nonspecific binding of radioactivity. Finally, the amount of surface-bound and internalized polypeptide is determined by the acid-washing (or chelator-washing) technique (*see* Separate Determinations of Surface-Bound and Internalized Polypeptides).

The amount of internalized polypeptide in such an experiment depends on (1) the amount of surface-bound polypeptide at time 0, when the temperature has just been shifted, (2) the internalization rate constant (k_{int}), and (3) the dissociation rate constant (k_{off}) from cell-surface receptors. Therefore, if the dissociation rate is slow enough and the cell-surface polypeptide at time 0 is determined, the internalization rate constant itself can be estimated directly from the amount of internalized polypeptide detected in this temperature-shift protocol. These protocols also can be used in isolated or cultured hepatocyte systems (*see* Continuous Endocytosis and Single-Round Endocytosis).

Determination of Polypeptide Binding to the Cell-Surface Sites Other Than Receptors

Some polypeptides, such as basic fibroblast growth factor (bFGF) (*46*) and HGF (*47*), have a strong affinity for heparin. These compounds bind not only to their specific receptors, but also to heparin-like substances such as proteoglycans or glycosaminoglycans located on the cell surface (*48*) and in the extracellular matrix (*49*). The physiological meaning of such nonspecific binding is still unknown, although there is some possibility that these heparin-like substances may act as a reservoir for ligand and that they may control the local concentration of polypeptide and subsequent expression of biological activity of the polypeptide.

To determine binding to such heparin-like substances separately from other binding or uptake, heparin-containing buffer is used to remove polypeptide associated with such binding sites (*29, 32, 48*). This heparin-washing technique has been applied in the analysis

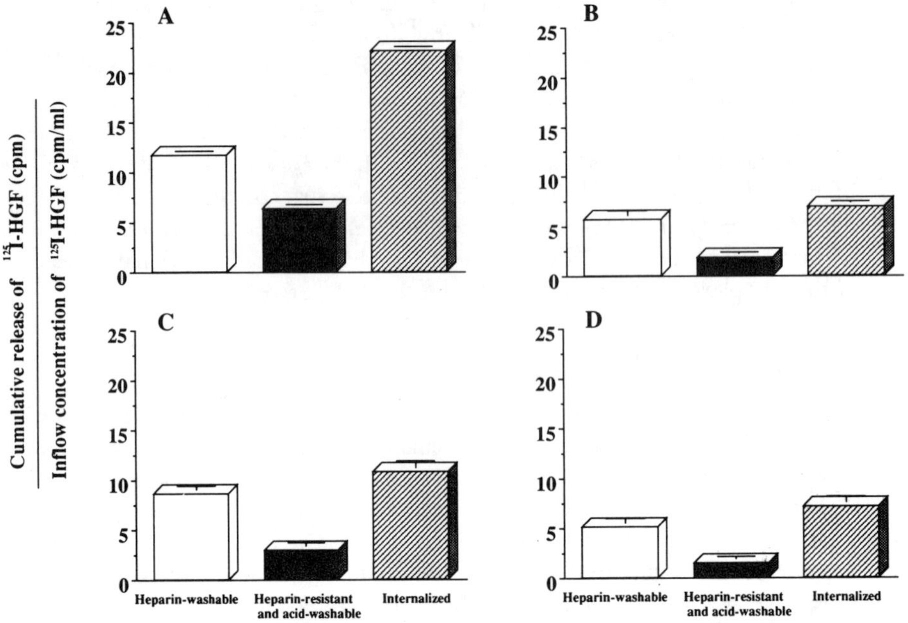

Figure 10. Heparin washable (HW), heparin resistant and acid washable (HRAW), and internalized [^{125}I]HGF in perfused rat liver of normal or CCl_4-intoxicated rat. After the perfusion of a tracer (0.8 pM) amount of [^{125}I]HGF with (B, D) or without (A, C) excess (135 pM) unlabeled HGF for 15 min in normal (A, B) or CCl_4-intoxicated (C, D) rat, the amounts of HW, HRAW, and internalized [^{125}I]HGF were determined. (Reproduced with permission from reference 32. Copyright 1993.)

of the hepatic handling of HGF (Figure 10) (*29, 32*). In this technique, an ice-cold and neutral buffer containing heparin is perfused and the amount of polypeptide in the outflow (heparin-washable binding) is counted just before the perfusion of acid buffer. After [^{125}I]HGF was perfused for 15 min, a significant amount of heparin-washable binding was detected. This binding was followed by the appearance of heparin-resistant but acid-washable binding during the perfusion of acid buffer (*29, 32*). The heparin-resistant but acid-washable binding was decreased greatly in the presence of excess concentration of unlabeled HGF (Figure 10) (*29, 32*). These results suggest that the heparin-washable binding and the heparin-resistant but acid-washable binding mainly represent HGF molecules bound to heparin-like substances and those bound to HGF receptors, respectively (*29*).

We applied the heparin-washing and subsequent acid-washing technique to analyze the hepatic handling of HGF in CCl_4-intoxicated rats (*32*). Use of the perfused liver system in CCl_4-intoxicated rats together with the data (*26*) obtained with isolated liver plasma membrane led to the suggestion that the down regulation of HGF receptors in liver contributes to the decrease in HGF hepatic clearance (*32*) (Figure 10).

The most critical point in this heparin-washing protocol is whether two kinds of binding can be definitely determined. The possibility that ligand bound to its receptor may be dissociated by heparin washing has to be carefully examined. So far, the possibility that such heparin washing may partly remove the cell-surface receptor-bound HGF molecules cannot be excluded. In fact, it was reported that heparin dissociates the complex between LDL and the LDL receptor on the cell surface in vitro, and the heparin-washing technique is used to detect such specific binding of LDL to its receptor (*50*). Whether heparin washing causes dissociation of ligand bound to its receptor probably depends on the experimental conditions during the heparin washing and the kind of ligand and receptor.

For the same purpose, a neutral and hypertonic buffer was used to remove polypeptide bound to heparin-like substances on the surface of cultured cells. This technique uses the characteristics of the binding between polypeptide and heparin (i.e., binding that results from ion–ion interaction) and allows clarification of the existence of two kinds of basic FGF binding sites on baby hamster kidney cells (*51*). Basic FGF bound to a low-affinity binding site with an equilibrium dissociation constant (*Kd*) value of 2 nM can be removed with hypertonic buffer, whereas that bound to a high-affinity binding site with a *Kd* value of 20 pM cannot be removed with hypertonic buffer but can be removed by acid buffer (*51*).

Analysis with Isolated or Cultured Hepatocyte Systems

Features of the Isolated or Cultured Hepatocyte Systems

An isolated hepatocyte suspension system has some advantages compared with an in vivo system; for example, the effect of endogenous substances can be ignored and the concentrations of drug, inhibitors, or plasma proteins, and the temperature can be set to desired values. Also some advantages of the isolated hepatocyte suspension system compared with a perfused liver system are, for example, the direct interaction of ligand and hepatocyte can be estimated without any other structural architecture around the cells, and many data points can be obtained from a single animal. The latter advantage is particularly important in the analysis of time profiles of some ligands taken up by the cells. However, this system also has some faults, for example, the results from this system must be carefully compared with those from an in vivo system because the tissue architecture has been lost, and this system cannot be used for long-term assays because cell viability would not be maintained.

In contrast to a suspension system, a cultured hepatocyte system can be used for long-term assays. The primary culture system of hepatocytes is constructed by culturing isolated hepatocytes in a monolayer on a collagen-coated plastic plate. Because the cultured hepatocytes are in a monolayer, the cells can easily be washed with buffer. Another reported advantage of a cultured hepatocyte system is that some liver-specific functions, which are destroyed by cell isolation, are recovered in primary cultures (*52, 53*). However, some functions, such as the carrier-mediated transport systems and liver-specific enzyme activities, are known to decrease during primary culture (*52, 53*).

The expression of proliferative activity of mitogens for hepatocytes, such as HGF, EGF, and insulin, can also be analyzed in a cultured hepatocyte system. The proliferative ability of cultured hepatocytes is known to depend on cell density (*54, 55*). At high cell density, cell growth is highly suppressed, and the state of the cells resembles that in vivo under normal conditions. At low cell density, cell growth is activated, and the state of the cells may resemble that in vivo in regenerating liver. Therefore, it is interesting to analyze the hepatic handling of polypeptides with a cultured hepatocyte system at various cell densities. For example, down regulation and the subsequent recovery of HGF receptor in hepato-

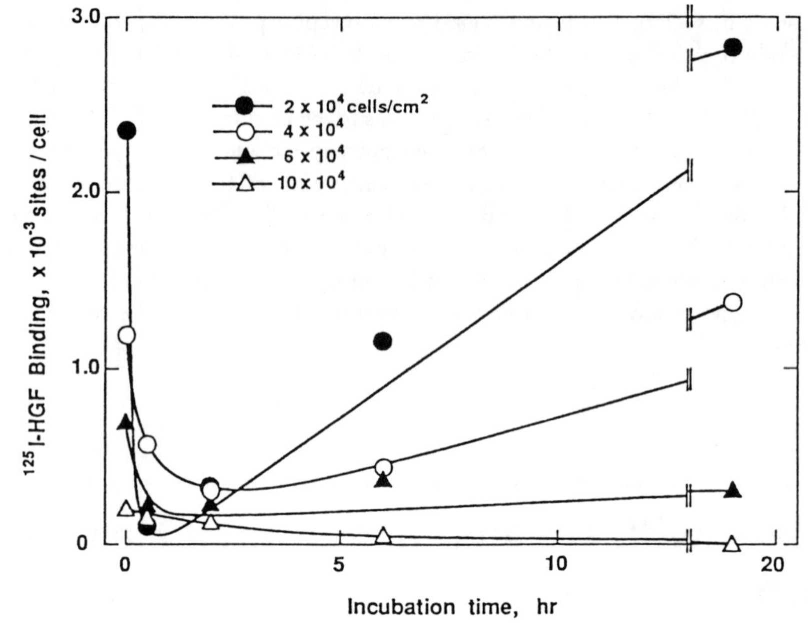

Figure 11. Change in specific binding of [¹²⁵I]HGF to primary cultured rat hepatocytes after the addition of HGF. Hepatocytes cultured at different densities were incubated with 75 pM unlabeled HGF at 37 °C for the indicated time. HGF was removed by an acid-washing technique, and residual HGF receptors on the cell surface were evaluated by the binding assay with [¹²⁵I]HGF. (Reproduced with permission from reference 54. Copyright 1993.)

cytes were analyzed with cultured rat hepatocytes (*54*) (Figure 11). The expression of the HGF receptor on the cell surface was higher at low cell density (Figure 11). In addition, in the presence of HGF in the medium, the rate of both down regulation and recovery of the HGF receptor is higher at low cell density (Figure 11). These results suggest that the HGF receptor is internalized and inserted into the cell surface relatively rapidly in the regeneration state compared with normal conditions.

Whether the events observed in the cultured hepatocyte system actually reflect those in vivo has to be carefully judged. It is advisable to interpret the data from isolated or cultured hepatocytes together with the data from in vivo or perfused liver systems.

Analysis of the Binding between Polypeptide and Receptor

Many data points can be obtained from a single animal by use of an isolated or cultured hepatocyte system, so the concentration dependency of polypeptide binding to its receptor can be easily measured. To analyze only the binding of polypeptide to cell-surface receptors, a binding assay carried out at low temperature (≤10 °C), in which no internalization of polypeptide is observed, would be suitable. The cells are incubated with a mixture of ^{125}I-labeled and various concentrations of unlabeled polypeptides until the binding reaches equilibrium. In an isolated hepatocyte system, a centrifugal filtration technique, in which the incubation mixture is transferred to oil with a gravity of 1.015–1.05 and centrifuged quickly, can be used to determine the amounts of free and bound ligand. In a cultured hepatocyte system, an aliquot of medium is collected and counted to determine the free ligand concentration. After that, the monolayer is washed extensively with ice-cold buffer, is solubilized with detergent or alkali, and is counted to determine the cellular ligand. Scatchard analysis

Table II. Parameter Values for EGF Binding to Hepatocytes Treated with Digitonin or Untreated

Hepatocytes	Treatment[a]	n (pmol/mg protein)	K_d (nM)	α (mL/mg protein)
Control	–	0.087	1.0	0.020
	+	0.086	4.9	0.032
Down regulated	–	0.017	0.9	0.009
	+	0.089	3.4	0.015

NOTE: *n* is the binding capacity; K_d is the dissociation constant; α is the nonspecific binding.
[a] – means untreated; + means treated with digitonin.
SOURCE: Reference 51.

of the data obtained by such a technique reveals the affinity and capacity for ligand binding to receptor. Results of experiments with a similar technique indicate that the affinity of HGF for its receptor (*26*) was very high, with a *Kd* of 20–30 pM, compared with the receptor affinities of EGF (*43, 56*), ASOR (*15*), and vasopressin (*6*), with *Kd* values of 1–6 nM. However, the capacity (B_{max}) of the HGF receptor was reported to be very low (500–600 site per cell) compared with the capacities of EGF, ASOR, and vasopressin (B_{max} values of 1–1.5 × 10^5 site per cell).

It is difficult to analyze the binding of polypeptide to intracellular receptors because a ligand that can easily get access to the intracellular receptor has not been discovered. Some methods to estimate receptor density in the intracellular space were reported, however. For example, cells can be treated with digitonin to increase membrane permeability, thus allowing polypeptide ligand to penetrate the cell (*15, 56*). After this treatment, the binding assay is performed to determine the sum of receptor density on cell surfaces and in the intracellular space. The cell-surface receptor density can be determined by the binding assay with untreated, normal cells, and the intracellular receptor density can be calculated by subtracting this value from the sum. In another technique, the cells are treated with trypsin at low temperature to destroy only cell-surface receptors and are then solubilized with detergent (*57*). The in vitro binding assay with ^{125}I-labeled polypeptide is then carried out, and the receptor density in the intracellular space is determined by Scatchard analysis. Results from experiments with such techniques indicate that most EGF receptors are located on the cell surface (Table II) (*56*) and ~80% of galactose receptor is located intracellularly (*58*) in freshly isolated rat hepatocytes. The method using digitonin treatment indicated that, after the freshly isolated cells were incubated with an excess (100 nM) concentration of unlabeled EGF for 20 min, the cell-surface receptor density decreased with the corresponding increase in intracellular receptor density (Table II) (*56*). This result suggests that EGF receptors internalized in the presence of excess ligand were not degraded within such a short period (i.e., 20 min).

Analysis of Receptor Binding, Internalization, and the Degradation Process

Kinetic Analysis of Internalization and the Degradation Process

Acid- or chelator-washing techniques can be used in both isolated and cultured hepatocyte systems to determine the amounts of surface-bound and internalized polypeptide separately (*see* Separate Determinations of Surface-Bound and Internalized Polypeptides). Alternatively, the cells can be treated with trypsin at low temperature to dissociate ligand bound to the cell-surface receptor (*59*). Finally, the cells and the medium are assayed with the trichloroacetic acid-precipitation or immunoprecipitation method to determine both intact and degraded polypeptides. The degradation products of ^{125}I-labeled polypeptide include, in general, iodine-125 or [^{125}I]tyrosine (*5, 60*), and such products are effluxed to the medium promptly after intracellular degradation. The same kinetic analysis as just described can be used to estimate k_{int} (*see* Kinetic Analysis of the Internalization Process).

To analyze the degradation process, a similar kinetic analysis can be used. The degradation rate is given by equation 9:

$$d\,[L_{deg}]_t/dt = k_{deg,L}\,[LR_i]_t \qquad (9)$$

where $[L_{deg}]_t$ is the degradation product of polypeptide and $k_{deg,L}$ represents the degradation rate constant, which is defined as the probability of internalized ligand degrading per unit time. Integration of equation 9 from time 0 to t yields equation 10:

$$[L_{deg}]_t = k_{deg,L} \int_0^t [LR_i]_t\, dt \qquad (10)$$

Thus, the $k_{deg,L}$ value can be estimated from the slope of an integration plot of $[L_{deg}]_t$ versus $\int_0^t [LR_i]_t\, dt$ (*43, 44*).

We performed a more minute kinetic analysis of RME with isolated rat hepatocytes (*44*). We measured the time profiles of surface-bound and internalized [^{125}I]EGF (Figure 12) and degradation products (Figure 13). Furthermore, the time profile of total cell-associated [^{125}I]EGF was also measured in the presence of varied concentrations of unlabeled EGF in the medium (Figure 14). These experiments were performed for various concentrations of EGF at two temperatures (0 and 37 °C), and the data were fitted simultaneously to the

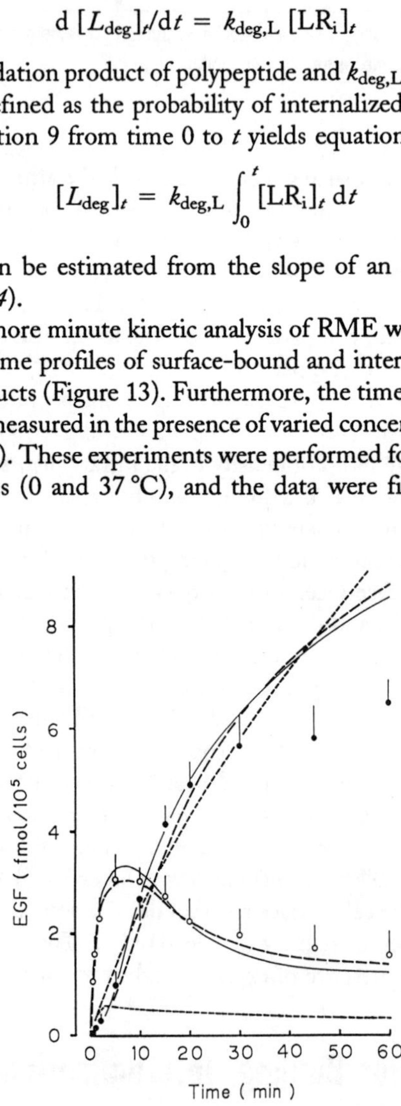

Figure 12. Time profiles of surface-bound and internalized [^{125}I]EGF. Cells were incubated with 0.5 nM [^{125}I]EGF at 37 °C. Lines are fitted curves calculated by nonlinear least squares method according to model 1 or 2 shown in Figure 15. Key: (solid line) fitting to model 1 (no constraint was given to k_e and other parameter values were loosely fixed at ±20% of initial estimates); (dashed line) fitting to model 1 (no constraint was given to k_e and k_s, and other parameters were loosely fixed at ±1 and ±20% of initial estimates, respectively); (dotted line) fitting to model 2 (k_e and other parameters were loosely fixed at ±1 and ±20% of initial estimates, respectively). (Reproduced with permission from reference 44. Copyright 1991.)

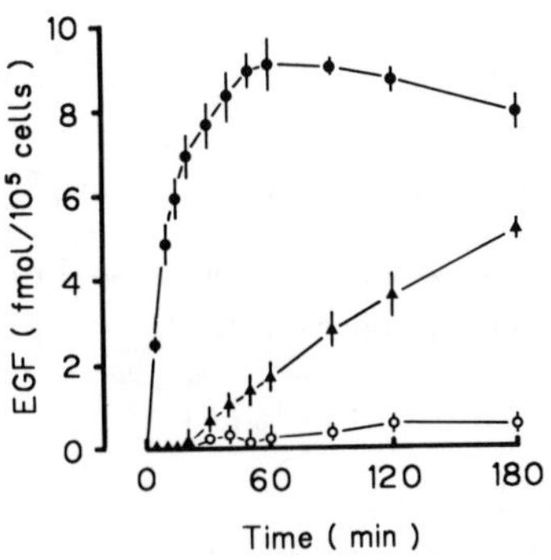

Figure 13. Time profiles of trichloroacetic acid-precipitable and -soluble EGF in isolated rat hepatocytes and medium. Cells were incubated with 0.3 nM [^{125}I]EGF at 37 °C. Key: (●) intact EGF in cell; (▲) degraded EGF in medium; (○) degraded EGF in cell. (Reproduced with permission from reference 44. Copyright 1991.)

mathematical equations based on each model shown in Figure 15. These fittings enabled us to estimate the values for k_{on}, k_{off}, k_s, k_{int}, k_t, and $k_{deg,L}$ simultaneously (Table I), where k_t represents the internalization rate constant of unoccupied receptor (this unoccupied receptor internalization is called "constitutive endocytosis") (*44, 61*). The following results were obtained from this model analysis: (1) the interaction between EGF and receptor at low

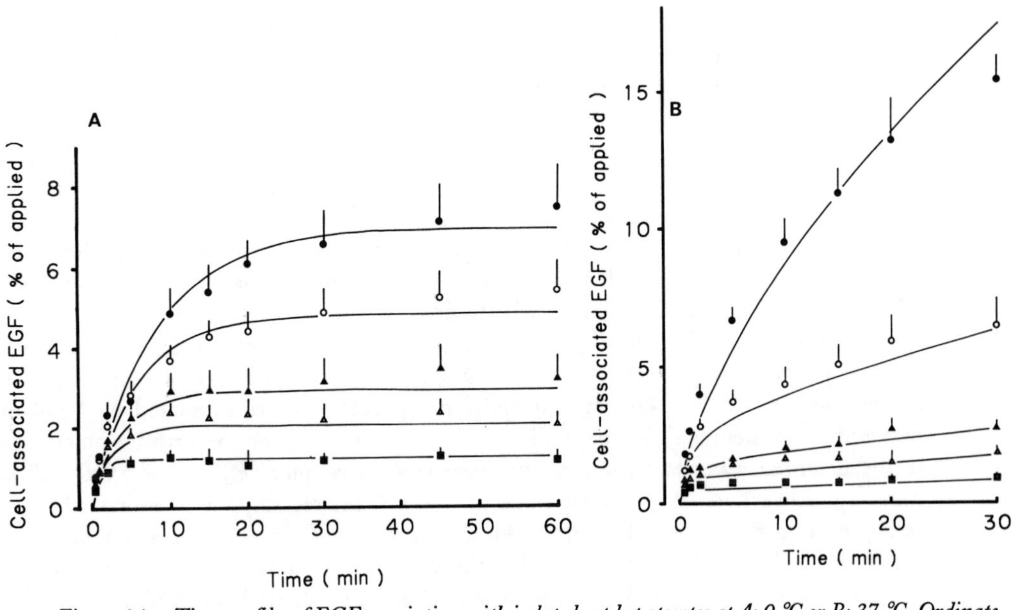

Figure 14. Time profiles of EGF association with isolated rat hepatocytes at A: 0 °C or B: 37 °C. Ordinate shows amount of cell-associated EGF normalized by amount of EGF added to incubation medium. Solid lines in panels A and B are fitted curves determined by nonlinear least square method to model 3 (0 °C) and model 1 (37 °C) respectively, shown in Figure 15. Initial EGF concentrations were (●) 0.3 nM; (○) 1 nM; (▲) 3 nM; (△) 5 nM; and (■) 10 nM. (Reproduced with permission from reference 44. Copyright 1991.)

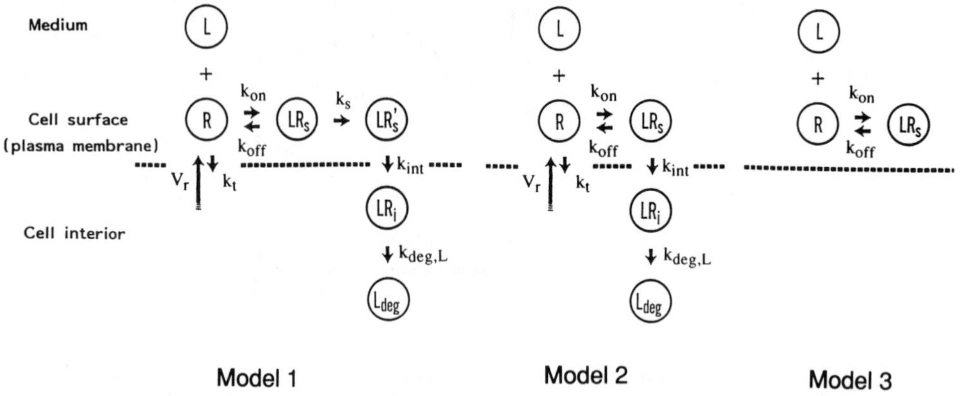

Figure 15. Mathematical models for RME of EGF by isolated rat hepatocytes. Definitions and abbreviations of the parameters and variables are shown in Figure 2 or as follows: (k_{on}) association rate constant; (k_{off}) dissociation rate constant; (k_s) sequestration rate constant of EGF–receptor complex from shallow to deep membranous compartment; (k_{int}) internalization rate constant for EGF–receptor complex; ($k_{deg,L}$) degradation rate constant for internalized EGF; (k_t) internalization rate constant for unoccupied receptor; (V_r) zero-order rate of insertion of receptors into plasma membrane. Model 1: This model includes two pools (LR_s and LR_s') for the EGF–receptor complex on the cell surface. Model 2: This model assumes a single pool (LR_s) for the EGF–receptor complex on the cell surface. Model 3: This model, which describes data obtained on ice, includes neither sequestration nor internalization of the EGF–receptor complex. (Reproduced with permission from reference 44. Copyright 1991.)

temperature can be accounted for by a simple model (Figure 15, Model 3) in which only association and dissociation are considered; (2) two compartments for the cell-surface EGF–receptor complex are required, as shown by the results from the perfused liver system (*see* Kinetic Analysis of the Internalization Process); and (3) k_{int} is much larger than k_t, indicating that the rate of receptor internalization is induced to increase by the binding to EGF (*44*). The third conclusion (i.e., $k_{int} >> k_t$) helped us to understand, in terms of kinetics, the mechanism of down regulation induced by the presence of excess ligand.

These kinetic parameters from isolated hepatocytes were comparable to those obtained in perfused rat liver, except for the k_{on} value (Table I). The k_{on} value in isolated hepatocytes was approximately one order of magnitude larger than that from perfused liver. One of the most feasible explanations for this discrepancy is that the association rate of EGF to the cell-surface receptor may be limited by diffusion in the unstirred water layer that possibly exists in the interstitial space (Disse space) of the liver. Such an unstirred water layer may not exist in a cell suspension system because of the vigorous shaking of the suspension. In fact, it is known that apparent uptake clearance estimated by MID in a perfused liver system has an upper limit value that may correspond to the diffusion clearance in the unstirred water layer; indeed, the association clearance (k_{on} R_s) of EGF (4.3 mL min^{-1} g^{-1} liver) was close to this value (*27, 62*). Most recently, we examined the effect of albumin on the uptake of ligands with a wide range of membrane permeabilities with the MID method (*63*). The finding that the so-called "albumin-mediated transport phenomenon" was observed only for highly membrane-permeable ligands also supports the existence of an unstirred water layer in the extracellular space of perfused rat liver (*63*).

Continuous and Single-Round Endocytosis

When the cells are incubated with ligand at 37 °C, the rate of increase of internalization of polypeptide may be influenced by many factors (*see* Separate Determinations of Surface-Bound and Internalized Polypeptides) because cell-surface binding and the subsequent internalization occur continuously. Such internalization of polypeptide is called "continuous endocytosis". "Single-round endocytosis" of polypeptide can be observed with a different technique in which the cells are incubated with ^{125}I-labeled polypeptide at low temperature to allow receptor binding. These cells are then washed with ice-cold buffer to remove nonspecific binding and are transferred to buffer at 37 °C to start the internalization. This technique corresponds to the temperature-shift protocol in a perfused liver system (*see* Temperature-Shift Method) and has been applied in the following analyses. Although single-round endocytosis of ASOR was not inhibited by adenosine 5′-triphosphate (ATP) depletors, such as NaF and NaN_3, continuous endocytosis was inhibited by these depletors (*64*). These results suggest that the internalization of the galactose receptor is independent of the intracellular ATP level, whereas the externalization of the receptor may require ATP (*64*). Monensin did not affect single-round endocytosis of EGF but did affect continuous endocytosis, suggesting that monensin inhibits receptor externalization without affecting receptor internalization (*61*) (*see* next section).

Inhibitors of Internalization or Degradation Processes

One method to discern whether some polypeptide is taken up by receptor-mediated internalization and then metabolized by lysosomal degradation is to use specific inhibitors of such processes. Some inhibitors of internalization of polypeptide receptor are PAO and dansylcadaverine. In particular, PAO is a well-known inhibitor and is known to inhibit the internalization of EGF receptor (*42, 65*), galactose receptor (*66*), angiotensin-II receptor (*67*), and β-adrenergic receptor (*65*). PAO is thought to inhibit receptor internalization through an interaction with sulfhydryl groups of the receptor molecule. The inhibition by PAO is reversed by bifunctional, but not by monofunctional, sulfhydryl agents (*65*). Although PAO is known to decrease the intracellular ATP level, the inhibition of polypeptide internalization by PAO is not due to a decrease of ATP level (*42, 66*). In addition to such inhibitors, cells can be exposed to hypotonic medium followed by incubation in potassium-free buffer to arrest coated-pit formation (*57*).

The heterogeneity of internalization pathways for some polypeptides was shown by using PAO (*42*) or hypotonic shock (*57*). PAO inhibited EGF internalization almost completely in the presence of a tracer concentration of [^{125}I]EGF, whereas such an effect was minimal when an excess (20 nM) concentration of unlabeled EGF was also present (*42*). These results can be explained if there are two cell-surface binding sites for EGF (high and low affinity) and the internalization via the two sites are PAO sensitive and insensitive, respectively. Similar dual PAO-sensitive and PAO-insensitive pathways were found also for the endocytosis of HGF by liver cells (*29*).

Lysosomotropic agents, such as chloroquine, ammonium chloride, and methylamine, are known as inhibitors of lysosomal degradation of polypeptides. These compounds increase pH in intracellular acidic compartments such as endosomes and lysosomes. As a result, the lysosomotropic agents inhibit both the dissociation of ligand from receptor and lysosomal degradation. Monensin, a proton ionophore, inhibits the proton pump on endosome vesicles and increases pH in endosomes, resulting in inhibition of receptor externalization. Monensin does not affect the internalization process; therefore, monensin inhibits continuous endocytosis of polypeptide without affecting single-round endocytosis (*61*).

Analysis of the Proliferative Activity of Mitogens for Hepatocytes

To develop a polypeptide as a therapeutic cure, it is necessary to clarify not only its pharmacokinetics but also its pharmacodynamics. Some polypeptides, such as HGF and EGF, are mitogens for hepatocytes and may be applied as therapeutic agents for certain types of liver

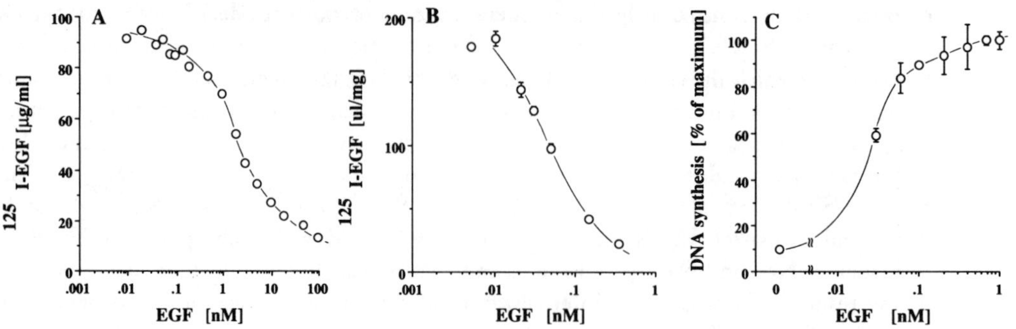

Figure 16. EGF concentration dependence of A, cell-surface binding on ice, or B, cell-surface binding at 37 °C, and C, mitogenic effect in primary cultured rat hepatocytes. A: Cells were incubated with [^{125}I]EGF and unlabeled EGF for 2 h on ice, and total cell-associated radioactivity was determined. B: Cells were incubated with [^{125}I]EGF and unlabeled EGF for 22 h at 37 °C and washed, and cell-surface bound EGF was estimated by the acid-washing technique. C: Twenty-two hours after the addition of EGF to the cells at 37 °C, [^{125}I]deoxyuridine incorporation for 6 h was determined. (Reproduced with permission from reference 43. Copyright 1993.)

diseases. The cultured hepatocyte system is suitable for the analysis of the mitogenic response of hepatocytes to such mitogens. The mitogenic response can be estimated by assessing the incorporation of [^{3}H]thymidine or [^{125}I]deoxyuridine into the trichloroacetic acid-precipitable fraction of the monolayer (*26, 43*).

We studied the hepatic handling and mitogenic effect of EGF with a cultured hepatocyte system. The EGF concentration dependency of DNA synthesis was compared with that of specific binding to cell-surface receptors at 0 °C. The half-effective concentration (EC_{50}) for DNA synthesis was much less (0.02–0.03 nM) than the *Kd* (1–2 nM) for the cell-surface receptor measured at low temperature (*43*) (Figures 16A and 16C). On the other hand, the EGF concentration dependency of specific cell-surface binding at 37 °C and 22 h after the EGF addition, when DNA synthesis starts to increase, was similar to that of DNA synthesis (Figures 16B and 16C), and the apparent dissociation constant (Kd_{app}) for EGF binding to the cell surface at 37 °C was also 0.02–0.03 nM. Kinetic analysis based on an appropriate mathematical model could account for the discrepancy between the two values (0.02–0.03 nM versus 1–2 nM). That is, the Kd_{app} at 37 °C was a lumped constant including not only terms for cell-surface binding (k_{on} and k_{off}), but also those for internalization (k_t and k_{int}), whereas the *Kd* at 0 °C was a constant including only terms for cell-surface binding (*43, 68*). This result suggests that the information about the *Kd* value assessed at 0 °C alone is not enough to determine the appropriate plasma concentration of EGF for clinical use.

References

1. Sugiyama, Y.; Hanano, M. *Pharm. Res.* **1989,** *6,* 194–204.
2. Bennett, H. P. J.; McMartin, C. *Pharmacol. Rev.* **1979,** *30,* 247–292.
3. Sugiyama, Y.; Kim, D. C.; Sato, H.; Yanai, S.; Satoh, H.; Iga, T.; Hanano, M. *J. Controlled Release* **1990,** *13,* 157–174.

4. Bridges, K.; Harfold, J.; Ashwell, G.; Klausner, R. D. *Proc. Natl. Acad. Sci. U.S.A.* **1982,** *79,* 350–354.

5. Shimizu, A.; Kawashima, S. *J. Biol. Chem.* **1989,** *264,* 13632–13638.

6. Fishman, J. B.; Dickey, B. F.; Bucher, N. L.; Fine, R. E. *J. Biol. Chem.* **1985,** *260,* 12641–12646.

7. Willey, H. S.; Cunningham, D. D. *Cell* **1981,** *25,* 433–440.

8. Schwartz, A. L.; Fridovich, S. E.; Lodish, H. F. *J. Biol. Chem.* **1982,** *257,* 4230–4237.

9. Maack, T.; Suzuki, M.; Almeida, F. A.; Nussenzveig, D.; Scarborough, R. M.; McEnroe, G. A.; Lewicki, J. A. *Science (Washington, DC)* **1987,** *238,* 675–678.

10. Ciechanover, A.; Schwartz, A. L.; Varsat, A. D.; Lodish, H. F. *J. Biol. Chem.* **1983,** *258,* 9681–9689.

11. Kawasaki, T.; Ashwell, G. *J. Biol. Chem.* **1976,** *251,* 1296–1302.

12. Kawasaki, T.; Ashwell, G. *J. Biol. Chem.* **1976,** *251,* 5292–5299.

13. Colotta, F.; Re, F.; Muzio, M.; Bertini, R.; Polentarutti, N.; Sironi, M.; Giri, J. G.; Dower, S. K.; Sims, J. E.; Mantovani, A. *Science (Washington, DC)* **1993,** *261,* 472–475.

14. Dunn, W. A.; Hubbard, A. L. *J. Cell Biol.* **1984,** *98,* 2148–2159.

15. Weigel, P. H. *J. Biol. Chem.* **1980,** *255,* 6111–6120.

16. Burwen, S. J.; Barker, M. E.; Goldman, B. I.; Hradek, G. T.; Raper, S. E.; Jones, A. L. *J. Cell. Biol.* **1984,** *99,* 1259–1265.

17. Schiff, J. M.; Fisher, M. M.; Underdown, B. J. *J. Cell. Biol.* **1984,** *98,* 79–89.

18. Renston, R. H.; Jones, A. L.; Christiansen, W. D.; Hradek, G. T.; Underdown, B. J. *Science (Washington, DC)* **1980,** *208,* 1276–1278.

19. Sato, H.; Sugiyama, Y.; Sawada, Y.; Iga, T.; Fuwa, T.; Hanano, M. *Am. J. Physiol.* **1990,** *258,* G682–G689.

20. Burwen, S. J.; Jones, A. L. *Trends Biochem. Sci.* **1987,** *12,* 159–162.

21. Raper, S. E.; Burwen, S. J.; Barker, M. E.; Jones, A. L. *Gastroenterology* **1987,** *92,* 1243–1250.

22. Dunn, W. A.; Connolly, T. P.; Hubbard, A. L. *J. Cell. Biol.* **1986,** *102,* 24–36.

23. Gladhaug, I. P.; Christoffersen, T. *Eur. J. Biochem.* **1987,** *164,* 267–275.

24. Kinoshita, T.; Hirato, S.; Matsumoto, K.; Nakamura, T. *Biochem. Biophys. Res. Commun.* **1991,** *177,* 330–335.

25. Asami, O.; Ihara, I.; Shimidzu, N.; Shimizu, S.; Tomita, Y.; Ichihara, A.; Nakamura, T. *J. Biochem.* **1991,** *109,* 8–13.

26. Higuchi, O.; Nakamura, T. *Biochem. Biophys. Res. Commun.* **1991,** *176,* 599–607.

27. Sugiyama, Y.; Sato, H.; Yanai, S.; Kim, D. C.; Miyauchi, S.; Sawada, Y.; Iga, T.; Hanano, M. In *Topics in Pharmaceutical Sciences;* Breimer, D. D.; Crommelin, D. J. A.; Midha, K. K., Eds.; Elsevier Science Publishers B. V. (Biomedical Division): Amsterdam, The Netherlands, 1989; pp 429–443.

28. Kim, D. C.; Sugiyama, Y.; Fuwa, T.; Sakamoto, S.; Iga, T.; Hanano, M. *Biochem. Pharmacol.* **1989,** *38,* 241–249.

29. Liu, K.; Kato, Y.; Narukawa, M.; Kim, D. C.; Hanano, M.; Higuchi, O.; Nakamura, T.; Sugiyama, Y. *Am. J. Physiol.* **1992,** *263,* G642–G649.

30. Sato, H.; Tsuji, A.; Hirai, K.; Kang, Y. S. *Diabetes* **1990,** *39,* 563–569.

31. Kim, D. C.; Sugiyama, Y.; Sato, H.; Fuwa, T.; Iga, T.; Hanano, M. *J. Pharm. Sci.* **1988,** *77,* 200–207.

32. Liu, K.; Kato, Y.; Yamazaki, M.; Higuchi, O.; Nakamura, T.; Sugiyama, Y. *Hepatology* **1993,** *17,* 651–660.

33. Yanai, S.; Sugiyama, Y.; Iga, T.; Fuwa, T.; Hanano, M. *Am. J. Physiol.* **1990,** *258,* C593–C598.

34. Pardridge, W. M.; Herle, A. J. V.; Naruse, R. T.; Fierer, G.; Costin, A. *J. Biol. Chem.* **1983,** *258,* 990–994.

35. Goresky, C. A. *Am. J. Physiol.* **1964,** *207,* 13–26.

36. Sato, H.; Sugiyama, Y.; Sawada, Y.; Iga, T.; Sakamoto, S.; Fuwa, T.; Hanano, M. *Proc. Natl. Acad. Sci. U.S.A.* **1988,** *85,* 8355–8359.

37. Sato, H.; Sugiyama, Y.; Kim, D. C.; Yanai, S.; Kurita, M.; Fuwa, T.; Iga, T.; Hanano, M. *Biochem. Pharmacol.* **1989,** *38,* 2663–2671.

38. Tsao, C. A.; Sugiyama, Y.; Sawada, Y.; Nagase, S.; Iga, T.; Hanano, M. *J. Pharmacokinet. Biopharm.* **1986,** *14,* 51–64.

39. Zeleznik, A. J.; Roth, J. *J. Clin. Invest.* **1978,** *61,* 1363–1374.

40. Goldstein, J. L.; Brown, M. S.; Anderson, R. G. W. In *International Cell Biology 1976–1977;* Binkley, B. R.; Porter, K. R., Eds.; Rockefeller University: New York, 1977; p 639.

41. Willey, H. S.; Cunningham, D. D. *J. Biol. Chem.* **1982,** *257,* 4222–4229.

42. Kato, Y.; Sato, H.; Ichikawa, M.; Suzuki, H.; Sawada, Y.; Hanano, M.; Fuwa, T.; Sugiyama, Y. *Proc. Natl. Acad. Sci. U.S.A.* **1992,** *89,* 8507–8511.

43. Kato, Y.; Sugiyama, Y. *STP Pharm. Sci.* **1993,** *3,* 75–82.

44. Yanai, S.; Sugiyama, Y.; Kim, D. C.; Iga, T.; Fuwa, T.; Hanano, M. *Am. J. Physiol.* **1991,** *260,* C457–C467.

45. Dunn, W. A.; Wall, D. A.; Hubbard, A. L. *Methods Enzymol.* **1983,** *98,* 225–241.

46. Turnbull, J. E.; Fernig, D. G.; Ke, Y.; Wilkinson, M. C.; Gallagher, J. T. *J. Biol. Chem.* **1992,** *267,* 10337–10341.

47. Nakamura, T.; Teramoto, H.; Ichihara, A. *Proc. Natl. Acad. Sci. U.S.A.* **1986,** *83,* 6489–6493.

48. Zarnegar, R.; DeFrances, M. C.; Oliver, L.; Michalopoulos, G. *Biochem. Biophys. Res. Commun.* **1990,** *173,* 1179–1185.

49. Masumoto, A.; Yamamoto, N. *Biochem. Biophys. Res. Commun.* **1991,** *174,* 90–95.

50. Goldstein, J. L.; Basu, S. K.; Brunschede, G. Y.; Brown, M. S. *Cell* **1976,** *7,* 85–95.

51. Moscatelli, D. *J. Cell. Physiol.* **1987,** *131,* 123–130.

52. Leffert, H. L.; Koch, K. S.; Lad, P. J.; Shapiro, I. P.; Skelly, H.; Hemptinne, B. D. In *The Liver: Biology and Pathology,* 2nd ed.; Arias, I. M.; Jakoby, W. B.; Popper, H.; Schachter, D.; Shafritz, D. A., Eds.; Raven: New York, 1988; pp 833–850.

53. Kukongviriyapan, V.; Stacey, N. H. *J. Cell. Physiol.* **1989,** *140,* 491–497.

54. Mizuno, K.; Higuchi, O.; Tajima, H.; Yonemasu, T.; Nakamura, T. *J. Biochem.* **1993,** *114,* 96–102.

55. Takehara, T.; Matsumoto, K.; Nakamura T. *J. Biochem.* **1992,** *112,* 330–334.

56. Yanai, S.; Sugiyama, Y.; Iga, T.; Fuwa, T.; Hanano, M. *Pharm. Res.* **1991,** *8,* 557–562.

57. McClain, D. A.; Olefsky, J. M. *Diabetes* **1988,** *37,* 806–815.

58. Weigel, P. H.; Oka, J. A. *J. Biol. Chem.* **1983,** *258,* 5095–5102.

59. Moriarity, D. M.; Savage, C. R., Jr. *Arch. Biochem. Biophys.* **1980,** *203,* 506–518.

60. Weigel, P. H.; Oka, J. A. *J. Biol. Chem.* **1984,** *259,* 1150–1154.

61. Gladhaug, I. P.; Christoffersen, T. *J. Biol. Chem.* **1988,** *263,* 12199–12203.

62. Miyauchi, S.; Sawada, Y.; Iga, T.; Hanano, M.; Sugiyama, Y. *Pharm. Res.* **1993,** *10,* 434–440.

63. Ichikawa, M.; Tsao, S. C.; Lin, T.; Miyauchi, S.; Sawada, Y.; Iga, T.; Hanano, M.; Sugiyama, Y. *J. Hepatol.* **1992,** *16,* 38–49.

64. Clarke, B. L.; Weigel, P. H. *J. Biol. Chem.* **1985,** *260,* 128–133.

65. Hertel, C.; Coulter, S. J.; Perkins, J. P. *J. Biol. Chem.* **1985,** *260,* 12547–12553.

66. Gibson, A. E.; Noel, R. J.; Herlihy, J. T.; Ward, W. F. *Am. J. Physiol.* **1989,** *257,* C182–C184.

67. Griendling, K. K.; Delafontaine, P.; Rittenhouse, S. E.; Gimbrone, M. A., Jr.; Alexander, R. W. *J. Biol. Chem.* **1987,** *262,* 14555–14562.

68. Knauer, D. J.; Wiley, H. S.; Cunningham, D. D. *J. Biol. Chem.* **1984,** *259,* 5623–5631.

RECEIVED for review July 12, 1993. ACCEPTED revised manuscript January 31, 1994.

Index

Index

J

K

L

X

Y

Z

Production: Margaret J. Brown
Acquisition: Cheryl Shanks
Indexing: Colleen P. Stamm
Cover design: Cesar Caminero

Typeset by Maryland Composition Company, Inc., Glen Burnie, MD
Printed by United Book Press, Inc., Baltimore, MD
Bound by American Trade Bindery, Baltimore, MD

Bestsellers from ACS Books

The ACS Style Guide: A Manual for Authors and Editors
Edited by Janet S. Dodd
264 pp; clothbound ISBN 0-8412-0917-0; paperback ISBN 0-8412-0943-X

The Basics of Technical Communicating
By B. Edward Cain
ACS Professional Reference Book; 198 pp;
clothbound ISBN 0-8412-1451-4; paperback ISBN 0-8412-1452-2

Chemical Activities (student and teacher editions)
By Christie L. Borgford and Lee R. Summerlin
[illegible]

Chemical Demonstrations: A Sourcebook for Teachers
Volume 1, Second Edition [illegible]
Volume 2, [illegible]
[illegible]

[illegible]

[illegible]

[illegible]

[illegible]

[illegible]

The Basics of Technical Communicating
By B. Edward Cain
ACS Professional Reference Book; 198 pp;
clothbound ISBN 0-8412-1451-4; paperback ISBN 0-8412-1452-2

[illegible]

For further information and a free catalog of ACS books, contact:
American Chemical Society
Distribution Office, Department 225
1155 16th Street, NW, Washington, DC 20036
[illegible]

Other ACS Books

Biotechnology and Materials Science: Chemistry for the Future
Edited by Mary L. Good
160 pp; clothbound, ISBN 0–8412–1472–7, paperback, ISBN 0–8412–1473–5

Chemical Demonstrations: A Sourcebook for Teachers
Volume 1, Second Edition by Lee R. Summerlin and James L. Ealy, Jr.
192 pp; spiral bound; ISBN 0–8412–1481–6
Volume 2, Second Edition by Lee R. Summerlin, Christie L. Borgford, and Julie B. Ealy
229 pp; spiral bound; ISBN 0–8412–1535–9

The Language of Biotechnology: A Dictionary of Terms
By John M. Walker and Michael Cox
ACS Professional Reference Book; 256 pp;
clothbound, ISBN 0–8412–1489–1; paperback, ISBN 0–8412–1490–5

Cancer: The Outlaw Cell, Second Edition
Edited by Richard E. LaFond
274 pp; clothbound, ISBN 0–8412–1419–0; paperback, ISBN 0–8412–1420–4

Chemical Structure Software for Personal Computers
Edited by Daniel E. Meyer, Wendy A. Warr, and Richard A. Love
ACS Professional Reference Book; 107 pp;
clothbound, ISBN 0–8412–1538–3; paperback, ISBN 0–8412–1539–1

Practical Statistics for the Physical Sciences
By Larry L. Havlicek
ACS Professional Reference Book; 198 pp; clothbound; ISBN 0–8412–1453–0

The Basics of Technical Communicating
By B. Edward Cain
ACS Professional Reference Book; 198 pp;
clothbound, ISBN 0–8412–1451–4; paperback, ISBN 0–8412–1452–2

The ACS Style Guide: A Manual for Authors and Editors
Edited by Janet S. Dodd
264 pp; clothbound, ISBN 0–8412–0917–0; paperback, ISBN 0–8412–0943–X

Personal Computers for Scientists: A Byte at a Time
By Glenn I. Ouchi
276 pp; clothbound, ISBN 0–8412–1000–4; paperback, ISBN 0–8412–1001–2

Chemistry and Crime: From Sherlock Holmes to Today's Courtroom
Edited by Samuel M. Gerber
135 pp; clothbound, ISBN 0–8412–0784–4; paperback, ISBN 0–8412–0785–2
